W9-BKJ-707

A Dictionary of

TWENTIETH-CENTURY HISTORY 1914–1990

The Author

Peter Teed was educated at King's College School, Wimbledon, and Corpus Christi College, Cambridge. He taught at schools in both the UK and Australia, and was headmaster of Goole Grammar School from 1964 to 1985. He became 'A' - Level examiner in 1959 and was chairman of the Southern Group GCSE committee from 1984 until his retirement in 1985.

His other publications include *Britain 1906-1960: A Welfare State, The Move to Europe: Britain 1880-1972, Portraits and Documents 1868-1919* (with Michael Clarke), and *Portraits and Documents 1906-1960* (with Michael Clarke). He has also contributed to the *Oxford Illustrated Encyclopedia Volume 4* and was history consultant for the *Oxford Children's Encyclopedia.*

Archaeology is his main recreational interest, and he has travelled extensively in Europe.

The Advisers

On the UK and Europe: Professor Kenneth O. Morgan, Principal of University College of Wales, Aberystwyth

On North and South America: Professor Robert Burchell, Head of American Studies, University of Manchester

On Africa: Dr Anthony Kirk-Greene, Lecturer in the Modern History of Africa, St Antony's College, Oxford

On Australasia: Dr Colin Newbury, Fellow of Linacre College, Oxford

On Asia: Dr Stephen Tsang, Research Fellow at the Centre for Modern Chinese Studies, St Antony's College, Oxford

On the Middle East: Sir Gawain Bell, Specialist in Middle Eastern Affairs

A Dictionary of TWENTIETH CENTURY HISTORY

1914–1990

PETER TEED

Oxford New York
OXFORD UNIVERSITY PRESS
1992

Wingate College Library

Oxford University Press, Walton Street, Oxford OX2 6DP
Oxford New York Toronto
Delhi Bombay Calcutta Madras Karachi
Petaling Jaya Singapore Hong Kong Tokyo
Nairobi Dar es Salaam Cape Town
Melbourne Auckland

and associated companies in
Berlin Ibadan

Oxford is a trade mark of Oxford University Press

© Oxford University Press 1992

All rights reserved. No part of this publication may be reproduced, stored in a retrieval system, or transmitted, in any form or by any means, electronic, mechanical, photocopying, recording, or otherwise, without the prior permission of Oxford University Press

This book is sold subject to the condition that it shall not, by way of trade or otherwise, be lent, re-sold, hired out or otherwise circulated without the publisher's prior consent in any form of binding or cover other than that in which it is published and without a similar condition including this condition being imposed on the subsequent purchaser

British Library Cataloguing in Publication Data
Data available

Library of Congress Cataloging in Publication Data
Data available

ISBN 0–19–211676–2
ISBN 0–19–285207–8 pbk

Typeset by Alliance Phototypesetters
Printed in Great Britain by
Biddles Ltd.
Guildford and King's Lynn

Preface

This dictionary attempts to present to the reader, in concise form, the most historically significant features of the twentieth century, from the outbreak of the First World War to the end of 1990. That is to say that the inclusion of entries has been decided by judging the subjects to be significant within the framework of their particular nation, world region, or occasionally, in the global context. No such dictionary can be comprehensive of course, and this is no exception. Inevitably political and military entries dominate, but attempts have been made to include other areas of importance, and entries can be found on ideological, socio-cultural, economic and technological topics. That social change is subtle, often taking place undetected below the surface, and not lending itself to itemization, goes without saying: a dictionary such as this can only attempt to provide concise definitions (within historical contexts) of complex issues.

Every effort has been made to make this dictionary international, in line with the awareness, growing throughout the twentieth century, of global perspective.

PETER TEED

Almondsbury, Avon
July 1991

Abboud, Ibrahim (1900–83), Sudanese general and politician. Having served with the British army in World War II in Eritrea, Libya, and Ethiopia, he became Commander-in-Chief of the army in 1956, following independence. In November 1958 he carried out a bloodless coup. Proclaimed President, he banned all political parties and suspended newspapers. Under his government Islamic north Sudan imposed itself on the south, Arabic replacing English in schools and missionaries being expelled. This led to protests and riots and he resigned in 1964, retiring from public life.

Abd al-Aziz ibn Saud (*c.*1880–1953), King of Saudi Arabia. The Saud family had long been powerful in Arabia, but when he was a boy they had been deposed by a rival family, the Rashids. In 1901 he gathered forty camelmen, killed the governor of the castle of Riyadh, and within two years was ruler over half Arabia. A strong Muslim, he recruited Wahabist* extremists into his army and for ten years was a successful guerrilla fighter against the Turks. He accepted British arms and subsidies, but failed to support his rival Hussein when the Arab Revolt* began. In the years 1920–2 he extinguished Rashidi rule, having by now been recognized by the British as Emir of Nejd and al-Hasa. In 1924 he occupied the Hejaz, driving Hussein into exile. In January 1926 in the city of Mecca he proclaimed himself King of the Hejaz and Nejd. Many of his fanatical Wahabi followers wanted him to invade Iraq, where Hussein's son was installed, but he refused and with British help crushed his own supporters in 1929. In 1932 he renamed his country the Kingdom of Saudi Arabia and a year later signed his first agreement with an American oil company. Oil did not, however, flow for another twelve years, after which he became one of the richest men in the world; but as an austere idealist he deeply regretted the disappearance of the Arab way of life which resulted from the new wealth.

Abd al-Hadi, Awni (1889–1970), Palestinian politician. He was active as an Arab nationalist against the Turks and helped to organize the first Arab Congress in Paris in 1913. After the liberation of Palestine 1917–18 he served King Abdullah of Transjordan. He became one of the chief spokesmen of the Pan-Arab movement, joining the Arab Higher Committee in 1936 and helping foment the Arab rebellion 1936–9 (see PALESTINE). He was exiled by the British 1937–41 and helped to form the Arab League* 1944. He was Jordan's ambassador to Cairo 1951–5 and held various government posts in Jordan, but spent the later years of his life in Egypt.

Abd al-Ilah ibn Ali ibn Hussein (1913–58), Regent of Iraq 1939–53. On the death of King Ghazi of Iraq in 1939 he became regent for his 4-year-old cousin Faisal II*, ruling Iraq as a loyal ally of Britain. In 1941 he was expelled by a group of pro-German officers, but reinstated by the British. A supporter of moderate Iraqi nationalism, he relinquished office in 1953 but continued as chief adviser to King Faisal until both were killed in the Iraq revolution of 1958. He helped to negotiate the Baghdad Pact* in 1956, being a staunch opponent of the Soviet Union.

Abd al-Krim (Muhammad ibn Abd al-Karim al-Khattabi) (1881–1963), Moroccan nationalist. Accepted by the Berber Arabs of the Riff mountains as their resistance leader against both France and Spain, his troops heavily defeated a Spanish army in July 1921, inflicting 12,000 casualties. After three years of fighting he surrendered to a massive Franco-Spanish army led by Marshal Pétain in 1925. He was placed in detention on the island of Réunion until 1947, when he was allowed to return to France. On the way he escaped to Cairo, where he set up the Maghreb Bureau or Liberation Committee of the Arab West. After Moroccan independence (1956) he refused to return as long as French troops remained in Africa.

Abdication Crisis (UK). On 16 November 1936 King Edward VIII*, not yet crowned, let it be known that he wished to marry an American, Mrs Wallis Simpson, who had been divorced. This would require legislative approval from the British Parliament and from the dominions. The Prime Minister Baldwin*, the British cabinet, the Archbishop of Canterbury, and representatives of the dominions all strongly opposed, partly on the grounds that it would be inconsistent with his role as 'Supreme Governor' of the Church of England. The British Press only gave the news on 3 December, when public opinion supported the King. But he chose to abdicate, making a farewell broadcast to the nation, and commending his brother, the Duke of York, who succeeded him as George VI. Edward married Mrs Simpson in France on 7 June 1937.

Abdullah ibn Hussein (1882–1951), King of Transjordan. Son of Hussein ibn Ali*, Sherif of Mecca, he and his brother Faisal helped to lead the Arab Revolt* of 1916. In 1921 he was made Emir of the province of Transjordan, a region of Arabia created by Britain as a link between Egypt and Iraq and made a British protectorate in 1923. He employed Glubb Pasha* to build up an army, the Arab Legion*, and during World War II remained strongly pro-British, as a reward for which Britain ended its mandate in 1948, creating Abdullah King. In December 1948 he announced himself King of a united Palestine and Transjordan, but by then the Jewish part of Palestine had proclaimed the state of Israel. His troops occupied the West Bank area up to Jerusalem, and in 1949 he renamed his country the Hashemite Kingdom of Jordan. His aim was to be leader of a solid Arab bloc of Transjordan, Syria, Palestine, and Iraq, but he met strong opposition from other Arab leaders. He was assassinated by an Arab nationalist in the presence of his grandson Hussein ibn Talal*, who succeeded him.

Abdullah, Sheikh Muhammad (1905–82), Chief Minister of Jammu and Kashmir 1975–82. Educated at the Prince of Wales College and Aligarh University in physics, he became involved in politics and was founder of the Kashmir Muslim Conference in the 1930s, agitating against the arbitrary rule of the Hindu Dogra, Hari Singh, Maharaja of Kashmir*. Later, as leader of the Muslim National Conference, he established close links with the Indian National Congress* and with Jawaharlal Nehru*. In 1947 the Maharaja could not decide whether to join India or Pakistan. Tribal groups from the North-West Frontier, with Pakistani military support, invaded Kashmir, and Abdullah pressed for a decision to accede to India. As Chief Minister of the New Indian state of Kashmir he worked closely with Nehru, until he was deposed and imprisoned in 1953 on suspicion that he was planning a separatist movement. Briefly released in 1964, he was soon again detained and not finally released until 1968, when he formed a Plebiscite Front Party. In 1975 he was appointed Chief Minister of Jammu and Kashmir under Mrs Gandhi*, who granted a degree of autonomy to the state. His government was accused of corruption.

Aborigines (Australia). By 1914 the Aboriginal population, estimated in 1788 at 300,000, had declined to about 50,000, and it was widely assumed that they would disappear altogether. This decline had been caused by loss of land, adoption of European habits such as drinking alcohol, European diseases, declining birth rate, and violence between Europeans and Aborigines. During the 1930s, however, campaigns developed for aid to Aborigines in terms of health, education, and employment. At the same time reserves were created in central and northern Australia. Since World War II Aborigines have shown increasing determination to preserve their cultural heritage, and there has been a dramatic reversal in population trends. In 1967 Australians voted by a 90 per cent referendum for federal action on behalf of Aborigines, resulting in the establishment of the Office of Aboriginal Affairs, set up in September 1967, all Aborigines having been granted citizen rights in 1948. Alongside a cultural resurgence there have also been economic and social pressures—on the northern reserves demands for a share from companies engaged in mineral exploitation, growth of membership of trade unions on the sheep stations, and demands by the Pitjantjara in South Australia for the restoration of land freehold (granted in 1981). In the 1988 bicentenary celebrations there were widespread Aborigine protests.

Abortion controversy, a moral/political debate of the 20th century. The practice of abortion, the premature termination of pregnancy by expulsion of the foetus from the womb, goes back to prehistoric times. It was strongly condemned by Christian theology during the Middle Ages and efforts were made during the 19th century to deter the practice by severe criminal sanctions. During the 20th century controls have been steadily relaxed. In Britain, legislation passed in Parliament in 1967 allowed for a termination for up to twenty-eight weeks (later reduced to twenty four), when social or medical arguments justified it. In the USA a Supreme Court judgement *Roe* v. *Wade* of 1973 ruled in favour of a 'right to choose' and 'against unduly restrictive state regulations'. The debate has centred around when the foetus should be given the status of a living person. The Roman Catholic Church and other 'pro-life' humanitarian groups have argued strongly against the practice, which is seen as murder even within days of conception. The controversy remained a highly explosive social/political issue in the USA throughout the 1980s. In Britain efforts of the pro-life movement have so far failed to win significant support.

Achebe, Chinua (1930–), perhaps Africa's best-known novelist. Professor of English in the University of Nigeria 1976–81, he has also held a variety of academic posts in the USA and Nigeria, as well as posts in broadcasting. His first novel *Things Fall Apart* (1958), for which he was awarded a Rockefeller* Fellowship, sold 2 million copies in thirty languages. It is concerned with the collapse of the traditional Ibo life-style. His later works have increasingly attacked corrupt practices in government.

Acheson, Dean Gooderham (1893–1971), US Secretary of State 1949–53. Educated at Yale and the Harvard Law School, he built a successful New York law practice before becoming Assistant Secretary of State for President Roosevelt in 1941, and Under-Secretary for President Truman in 1945. He urged international control of atomic power in the Acheson–Lilienthal Report of 1946, outlined the Truman Doctrine* of US support for nations threatened by Communism, and helped to formulate the Marshall Plan*. He became Secretary of State in 1949, resigning 1953. He helped in the creation of NATO, but was attacked in Congress for his alleged liberalism and failure to support Syngman Rhee* in South Korea. He was a strong supporter of the French in Indo-China and of the Republic of China in Taiwan.

Action Française, an extreme right-wing group in France during the first half of the 20th century, and also the name of the newspaper published to promote its views. Founded by the poet and political journalist Charles Maurras, it aimed to overthrow the parliamentary republic and restore the monarchy. Although strongly nationalist, it was also anti-Semitic, and its relationship with royalist pretenders and the papacy was not always good. It became discredited for its overt Fascism and association with the Vichy regime 1940–4. Maurras was condemned as a collaborator and the movement broken up, but its nationalist paternalism continued to attract many on the extreme right of French politics.

Addams, Jane (1860–1935), US social reformer. With her friend Ellen Grates Starr, she opened Hull House in Chicago in 1889, a settlement-house for immigrants and workers, on the model of Toynbee Hall* in London. As a pioneer in the new discipline of sociology, she later had considerable influence over the planning of neighbourhood welfare institutions throughout the USA. She was a leader of the women's suffrage movement in the USA, a founder of the NAACP* in 1909, and, as a pacifist, of the Women's International League for Peace and Freedom, for which she received a Nobel Prize* in 1931.

Aden, a busy port and rocky peninsula in south-east Arabia. It was annexed by the British in 1839, growing rapidly as a coaling station. By 1914 it was still administered by the presidency of Bombay, but in 1932 it was placed directly under the government of India. In 1937 it became a Crown Colony. By that time Britain had established the Aden protectorate over the Arabian Emirates along the 600-mile coastline to the east. In 1959 this protectorate became the South Arabian Federation of Arab Emirates, which the colony of Aden joined in 1963. By now a republican movement had broken out under General al-Sallal in neighbouring Yemen, backed by Nasser's Egypt. The National Liberation Front (NLF) now developed, and there was civil war between republicans and the pro-British Emirates 1965–7. British forces were unable to keep the peace and the Federation collapsed. In October 1967 all seventeen emirs were deposed. There was a brief struggle between the NLF and a rival organization FLOSY (Front for the Liberation of Occupied South Yemen). In November the British left Aden and the People's Republic of South Yemen was proclaimed, which in 1970 changed its name to the People's Democratic Republic of Yemen. Backed by aid from the Soviet Union and China, a five-year plan (1975–80) brought considerable modernization to the hinterland. In 1979 there was a thirty-day war with the neighbouring Yemen Arab Republic, followed by intermittent discussions for union. This was finally achieved in May 1990 as the Republic of Yemen*.

Adenauer, Konrad (1876–1967), Chancellor of the German Federal Republic 1949–63. Born in Cologne and educated in Bonn and Munich, he practised law in Cologne, where he entered politics. He became a prominent member of the Centre Party in the Weimar Republic* and was Mayor of Cologne from 1917 until 1933, when he was removed by the Nazis for his opposition to their policies. He was subsequently twice arrested. In 1945 he again became Mayor, but was removed by the British for alleged inefficiency. He was a founding member of the Christian Democrat* Party, and in 1949, when the Federal Republic was created, he became its first Chancellor. During his period in office a sound democratic system of government was created; friendship with the USA and with France was established; and the West German people began to enjoy the fruits of the so-called 'economic miracle' of the Erhard* years. Under his guidance the Republic became a member of NATO and was accepted by the Soviet Union, which he visited in 1955. His critics accused him of being too autocratic in manner and too little concerned to achieve German reunification.

Affirmative action (USA). In 1943 President Roosevelt had established a fair Employment Practices Commission which had assisted the Black work-force to gain employment in industry. The 1964 Civil Rights Act established the Office of Federal Contract Compliance, and the Equal Opportunity Commission was set up in 1972. From now on institutions and companies which received federal funds or contracts were required to produce 'affirmative action' programmes, whereby management would regularly review the extent to which age, race, sex, and country of origin had been factors in decision-making. In 1978 the Supreme Court under Chief Justice Burger* supported 'positive discrimination' in education by its ruling on the *Bakke* case. In 1979 a court ruling, in the case of *United Steel Workers of America* v. *Weber*, established that both unions and employers should establish training programmes for 'minority employees'. During the 1980s the administration of President Reagan aimed to dismantle some 'affirmative action' procedures as being economically restrictive. On a number of occasions, however, the Justice Department was defeated by Supreme Court rulings. Nevertheless by the end of the decade statistics indicated a worsening gap between the Black and White communities in the USA.

Afghanistan. There had been British–Afghan wars 1838–42 and 1878–80, with Britain determined to maintain a friendly buffer-state against Russia. In 1914 the Emir Habi Habibullah (1901–19) had the backing of the British, but he was assassinated in 1919. His son Amanullah seized power with the help of the army, challenging British influence. After another brief war Britain confirmed the country's independence. Following the Bolshevik revolution in Russia a Russo-Afghan Treaty of 1921 confirmed frontiers. This developed in 1926 into a non-aggression pact. In 1927 Amanullah took the title of shah (king) and embarked on a programme of Westernization. This shocked the Muslim imams and he was forced to flee in 1929. After a chaotic interlude the British established General Nadir Khan as Shah, arming him with 10,000 rifles. He too aimed to Westernize, and alienated the Muslim clergy. He was assassinated in 1933, but his son Mohammed Zahir remained king until 1973. In 1947 a tense situation arose with Pakistan, when Afghanistan staked out a claim for a Pathan* state along the North-West Frontier, a claim rejected by Pakistan. The king's cousin General Mohammad Daoud was Prime Minister 1953–63, during which time he received aid from the Soviet Union. In 1964 Afghanistan became a constitutional monarchy and the first ever election was held in October 1965. On 19 July 1973, when the Shah was abroad, Daoud reasserted control, declaring the monarchy abolished, Afghanistan a republic, and himself President. On 27 April 1978 a left-wing 'Armed Forces Revolutionary Council' (Khalq) overthrew his regime. Daoud was assassinated and the Democratic Republic of Afghanistan proclaimed, with Nur Mohammad Taraki as President. The new regime embarked on reforms, but there was tension and rural unrest. In February the US ambassador was killed, and one month later Taraki was assassinated by supporters of the deputy Prime Minister Hafijullah Amin, who then sought US support. In December 1979 Soviet troops entered the country. Amin was killed and replaced by Babrak Karmal, who established a socialist regime. For the next ten years anti-government Mujaheddin* guerrillas waged war against Afghan troops armed and supported by Soviet tanks, aircraft, and equipment. Some 3 million refugees fled to Pakistan and Iran. In 1982 there was a massive Soviet attack against the insurgents established in the Panjahir valley, but with little success. In 1987 the Soviet Union began to seek ways to disengage from the conflict, which was finally done in February 1989. Although still receiving massive military aid from the USA through CIA channels, the Mujaheddin guerrillas failed to establish control or to oust Karmal's regime. In 1987 General Najibullah had become President and Commander-in-Chief of the armed forces. During 1990 he survived a coup by a fellow officer, ended the state of emergency, and in November began peace talks in Geneva with the Mujaheddin. US aid to the latter began to be reduced.

AFL *see* AMERICAN FEDERATION OF LABOR.

Aflaq, Michel (1910–), Syrian politician and founder of the Ba'ath Party*. After an early career as a schoolteacher in Damascus he entered journalism. During World War II he was a powerful influence on Arab nationalism, founding the Ba'ath Arab Renaissance Party with Salah-ul-Din al-Bitar. In 1953 this became the Arab Socialist Renaissance Party. Aflaq's ideology was for Arab unity within a socialist society, anti-Zionist and free from outside superpower pressure. In 1958 he helped to form the short-lived United Arab Republic*. In 1966 he was ousted from Syria by a more extreme Ba'ath faction, after which his political activity declined.

African National Congress (ANC). It was established in Bloemfontein in 1912 as the South African Native National Congress by Zulu Methodist minister J. W. Dube. Its original aim was to 'protect the interests of all coloured peoples in South Africa', having developed from the earlier Native Education Association of Cape Colony. In 1914 Dube led a deputation to Britain to protest against the Native Land Act (1913), which restricted the purchase of land by Black Africans. It was inspired by Mohandas Gandhi*, who was at that time campaigning for equal rights for Indians in South Africa. In 1926 the ANC established a united front with the Indian community, seeking a racially integrated, democratic southern Africa. From 1952 until 1967 it was led by the Natal chieftain Albert Luthuli*. Together with the more militant breakaway movement the Pan-Africanist Congress*, it was declared illegal by the South African government in 1960. Confronted by Afrikaner intransigence on racial issues, the ANC saw itself forced into a campaign of violence. A 'liberation army' Umkhonto Wesizwe ('Spear of the Nation') was formed, dedicated to the abolition of apartheid*. In 1962 its leader Nelson Mandela* was arrested, as were a number of its executive members. Mandela and many of his colleagues were convicted of sabotage and jailed for life. His friend and colleague Oliver Tambo escaped into exile in Zambia, from where he continued to lead the movement. In the early 1980s there were a number of acts of sabotage in South Africa, which prompted South African raids into neighbouring countries. Unofficial talks with the South African government began in 1988 and its leaders, including Mandela, were released from prison in 1989 and early 1990. Negotiations with the South African President de Klerk* began in May 1990.

Afrikaner (or Boer), the name generally given to the White Afrikaans-speaking population of South Africa. It is used particularly by descendants of the families which emigrated from The Netherlands, Germany, and France before 1806, when Britain seized Cape Colony. The Afrikaans language and membership of the Dutch Reformed (Calvinist) Church are the unifying factors, from which developed the concept of apartheid*. Afrikaans is the first official language in much of South Africa.

Aga Khan, a hereditary title first bestowed in Iran in the 19th century on the leader of the Ismaili sect of Shiite Islam. Many of the sect live in Bombay, the Khoja community, and in 1866 Aga Khan I settled there as world leader of the Ismailis. His grandson Aga Khan III, Aga Sultan Sir Mohammed Shah (1877–1957), played an active part in Indian politics, being founder in 1906 and first president of the All India Muslim League*. A strong supporter of Britain and also of the League of Nations, at the same time he was a moderate force working for independence in India. He was leader of the Muslim delegation to the Round Table Conference* in 1930–1 and welcomed the creation of Pakistan in 1947. He designated his grandson Prince Karim al Hussaini Shah as his successor, arranging for him to study at Harvard University. Karim's aim was to achieve a blending between the culture of Europe, where he mostly lived, and that of the Ismaili people, whose spiritual head he was. As Chancellor of the University of Pakistan, founder of the Aga Khan Foundation and of the Institute of Ismaili Studies, as well as 57th imam of the Ismailis, Aga Khan IV has continued his grandfather's support for education as well as his interest in horse-racing.

Aguinaldo, Emilio (1869–1964), nationalist leader from a wealthy Philippines family who had led a revolt against Spanish rule 1896–7. Although he supported the US invasion of the Philippines in 1898, he later resisted US occupation. Claiming to be the President of a Filipino Republic, he led his nationalist forces in guerrilla fighting against US forces from 1899 until 1901, when he was captured at Palawan. He then swore allegiance to the USA and retired into private life. When the USA granted autonomy to the Philippines in 1934 he unsuccessfully stood for President. Accused in 1945 of collaboration with the Japanese in World War II, he was arrested but never came to trial. On his release he was appointed a member of the Philippines Council of State and devoted the rest of his days to improving US–Philippines relations.

Ahidjo, Ahmadou (1924–89), President of the United Republic of Cameroon 1960–82. A Muslim, he was born and educated in Yaoundé, where he became a Post Office radio operator. He was early involved in politics and served in Paris (1953–6) in the Assemblée. At the age of 34 he became Premier of French Cameroons, taking a strong line against terrorist extremists. In 1960 he persuaded the United Nations to end mandate status and to grant independence, linking with a part of British Cameroons in 1961. He was elected President of Cameroon and held this office until 1982, when he failed to be re-elected. Accused of plotting, he retired to France. His main achievement was to create a United Republic in which racial intolerance was reduced.

AIDS (Acquired Immune Deficiency Syndrome). From the late 1970s a virus, given the name the Human Immunodeficiency Virus (HIV), began to spread infection through both Africa and particularly the west coast of the USA. A disease of the blood, it was at first spread unwittingly via blood transfusions. It can also be transmitted via anal or, less likely, vaginal sexual intercourse, or by infected needles. It is a wasting disease resulting in death. Its origins have been extensively debated, one theory being that it is the result of misjudged biological warfare research. Its spread into Australia and Europe developed during the 1980s, but the highest death toll has been in the USA and Africa. Extensive world-wide research for a cure has cost many millions of dollars.

Aircraft industry. It was the two world wars which most stimulated the development of the industry, which had only begun in 1910. During World War I Britain, France, and Germany competed in its development, with Britain producing 55,000 aircraft, at 3,000 per month by 1918, fighters and bombers. Production in the USA was stimulated by entry into the war in 1917 and some 11,000 aircraft were produced, mostly to British (de Havilland) design. Britain and France retained the lead in the early years after the war, with the RAF in Iraq maintaining a demand for military aircraft; but research was expensive and progress relatively slow, although by the end of the 1920s large comfortable seaplanes were in production, as well as small single-engine Moths. Following the world depression of 1929–33 it was the USA which began to take the lead. German rearmament stimulated British concentration on military aircraft design (Spitfire and Hurricane), but the Empire Flying Boat was also delivered, carrying twenty passengers at 175 m.p.h. in spacious accommodation. After war broke out in 1939 the USA was able to continue its development of air transports and civil airliners, while also mass-producing combat aircraft for the Lend-Lease* programme. Once again war stimulated development in terms of size, speed, and manœuvrability. The jet gas-turbine engine had been invented by F. Whittle in 1937, and the technology was made available to US engineers during the war. Britain's early post-war lead was lost in 1952, when metal fatigue resulted in disaster. Aircraft costs continued to mount and individual companies, for example Vickers, Shorts, Hawker, de Havilland, continued to amalgamate until the British Aircraft Corporation was formed in 1960, which itself became British Aerospace in 1977. But US competition, with its huge internal market, steadily grew, Lockheed, Douglas, and above all Boeing, with a total output of thousands of aircraft per year against a few hundred in Britain. The French industry was small but technically strong and the Concorde project of 1962, with joint research and construction in Bristol and Toulouse, was, at vast public expense, a powerful psychological boost for a combined European industry. In the Soviet Union aircraft production during World War II was on a par with American, but post-war developments of civil aviation were not as massive as in the USA or Europe.

Akihito (1933–), Emperor of Japan 1989– . Eldest son of Emperor Hirohito* and Empress Nagako, he attended Japanese High School before going to Gakushuin University, where he studied politics and economics, while also sharing his father's interest in marine biology. After his marriage in 1959 he and Crown Princess Michiko made official visits to thirty-seven countries, starting with the USA. From 1987 onwards he acted on behalf of his aged father in matters of state.

Alanbrooke, Alan Francis Brooke, 1st Viscount (1883–1963), British field marshal. Born of an Irish family, he was educated in France and at the Royal Military Academy, Woolwich, passing out in 1902. He served with distinction during World War I, and in the 1930s was noted as an artillery expert. In World War II he was commander of the 2nd Army Corps in the retreat from Dunkirk, then Commander-in-Chief Home Forces. Later, as Chief of the Imperial General Staff and chairman of the Chiefs of Staff Committee (1941–6) he represented the service chiefs in discussions with Churchill, who had an extremely high regard for his abilities. He accompanied Churchill to all his conferences with Roosevelt and Stalin. He served as Constable of the Tower of London 1950–5.

Åland Islands, a province of Finland. It consists of some 6,000 islands in the Gulf of Bothnia between Finland and Sweden, which had held them until 1809, when they were occupied by Russia. After the collapse of the Russian Empire in 1917, Finland declared the independence of the islands, and Swedish troops who were occupying the archipelago were ejected. The future of the islands was referred to the League of Nations (1921), which upheld their autonomy as part of Finland. Swedish is their first language and in 1945 there was some pressure for their restoration to Sweden.

Alaska. After its purchase from Russia in 1867 it remained an unorganized territory until 1884. The Klondike Gold Rush of 1896 and later extensive copper discoveries resulted in little population growth until World War II. In 1942 the 1,500-mile Alaska Highway was built and, as the 'Cold War' between the USA and the USSR developed, a strong military presence was established. It became the forty-ninth US state in 1959. The discovery of oil in 1969 resulted in the area having an important economic role as well as a strategic one, with the ice-free port of Valdez being developed.

Albania, South Balkan country on the Adriatic Sea. Having been part of the Ottoman Empire from the 15th century, it became an independent state as a result of the Balkan War of 1912. After a period of chaos it was proclaimed a republic in 1925, under the presidency of a large landowner, Ahmed Bey Zogu, who in 1928 announced himself King Zog. It was invaded by Italian troops in 1939, when Zog fled and a puppet government was installed, and it became an Italian protectorate. In 1941 German troops invaded, in the campaign to conquer Greece, and it was proclaimed a part of Greater Italy. However, partisan forces, whose leaders included Enver Hoxha*, were so strongly organized in their mountain bases that in November 1944 they captured Durazzo and proclaimed Albania's liberation. In November 1945 the provisional government of Hoxha was recognized by the Allies. Elections in December established the Communist People's Republic of Albania, which was at first under the strong influence of the Soviet Union, until a rift in 1958. It was expelled from Comecon* in 1961 and from the Warsaw Pact in 1968. Albania now became increasingly linked with China, until the death of Mao Zedong* in 1976. For the next decade it remained staunchly isolated and Stalinist in its policies, having cut its diplomatic links with most nations of the world. Enver Hoxha died in April 1985, although his wife Nexhmije remained a strong influence. The country's extreme isolationism soon began to ease. Under party leader Ramiz Alia it established diplomatic relations with neighbouring states and in 1988 took part in a conference of Balkan countries. In April 1990 Ramiz declared his aim to establish diplomatic relations with all European Community countries, to take part in the CSCE*, and to re-establish diplomatic relations with the USA and the USSR. In May 1990 the People's Assembly passed a wide range of social, economic, and judicial reforms, propaganda against the state no longer being a capital offence, religious services no longer being illegal, and foreign travel being made easier. Nevertheless there were widespread demonstrations demanding further easing of restrictions, with many seeking asylum in foreign embassies. There were demonstrations also of Albanians living in Kosovo in Yugoslavia. Diplomatic relations with the Soviet Union were in fact resumed in July 1990.

Alcalá Zamora, Niceto (1877–1949), President of the Spanish Republic 1931–6. He served as a monarchist in the governments of Alfonso XIII, being Minister for War in 1922, when Spain was defeated in Morocco. When Primo de Rivera seized power in 1924, he joined the Catalan socialist republicans and was imprisoned. In 1931 he was briefly Prime Minister and then first President of the unstable Spanish Republic. In spite of increasing disagreements with the extremists on the left, he survived until 1936, when he was obliged by the Cortes to resign. He went into exile, dying in South America.

Alemán, Miguel (1902–83), President of Mexico 1946–52. Son of a shopkeeper, he became a 'general' in the Mexican Revolution* and qualified as a lawyer, specializing in labour cases. In 1936 he became Governor of Vera Cruz and in 1940 helped President Ávila Camacho* to power, himself becoming Minister of the Interior. When elected President in 1946 he continued Camacho's conciliatory policies of Mexican reform and development, the process of industrialization being much accelerated. He remained a strong force in Mexican life, encouraging the growth of the steel, television, and radio industries.

Alessandri (Palma), Arturo (1868–1950), President of Chile 1920–4, 1925, 1932–8. Son of Italian immigrants, he graduated in law at the University of Chile in 1893 and entered politics in 1897, acting on behalf of the nitrate miners. He was elected President on a liberal policy, but, finding his attempts at reform blocked, he went into voluntary exile in 1924. The following year he was brought back by the army when a new constitution was adopted. He extended the suffrage, separated Church and State, while guaranteeing religious liberty, and made primary education compulsory. He resigned again in October 1925 and went to Italy. On his return he was re-elected President. By now the country was experiencing the Great Depression*. He reorganized the nitrate industry, developed schools, and improved conditions in agriculture and industry. Yet he lost much of his labour support, which was moving towards Marxism. He was elected to the Senate in 1946.

Alessandri (Rodríguez), Jorge (1896–1987), President of Chile 1958–64. Son of Arturo Alessandri*, he graduated in engineering at the University of Chile and entered politics when his father was President. He worked as an engineer during the 1930s and was Minister of Finance 1948–50. In 1957 he defeated Allende* in the presidential election. Although a right-wing politician backed by Conservatives and Liberals, he has been described as 'perhaps personally the most popular of recent Chilean leaders'. His unorthodox policies of vast public works and reduced taxation won some support from radicals as well as industrialists. But high inflation and an attempt to control salaries alienated his supporters, and he lost the election of 1964 to Eduardo Frei*. In the 1970 election he was narrowly defeated by Salvador Allende, whose overthrow by Pinochet* he supported.

Alexander I (1888–1934), King of Yugoslavia 1921–34. Educated in Geneva and St Petersburg, he returned to Serbia in 1909 recognized as heir to the Karageorgevic throne. He distinguished himself in the Balkan Wars of 1912 and 1913 and was Commander-in-Chief of the Serbian army in 1914, retreating with it to Corfu. Here his adviser Nikola Pašic* negotiated the Corfu Pact*. He returned to Serbia in 1921 when his father Peter died, and was proclaimed king of the Serbs, Croats, and Slovenes. With Pašic his Prime Minister, the next five years were spent in trying to overcome the ethnic, religious, and regional rivalries within the country of 'the southern Slavs'. After Pašic's death a period of instability followed until, following the assassination of the Croat leader Stjepan Radic, he established a personal dictatorship supported by the army. In the interest of greater unity he changed the name of the Kingdom of Serbs, Croats, and Slovenes to that of Yugoslavia. In 1931 some civil rights were restored, but they proved insufficient to quell rising political and separatist dissent, aggravated by economic depression. He was a keen supporter of the Little *Entente**, and was allegedly planning to restore parliamentary government when he was assassinated in Marseilles by a Croatian terrorist.

Alexander, Harold Rupert Leofric George, 1st Earl Alexander of Tunis (1891–1969), British Field Marshal. Son of the Irish peer the Earl of Caledon, he was educated at Harrow and Sandhurst. He served in the Irish Guards in World War I and later in Latvia and India. In World War II he commanded the rearguard at Dunkirk* and then the final retreat from Burma. As Commander-in-Chief, Middle East, in 1942, he turned the tide against Rommel*. In 1943 he was deputy to Eisenhower*, clearing Tunisia for the Allies. He then led the invasion of Sicily and commanded in Italy until the end of the war. He was Governor-General of Canada 1946–52 and briefly Minister of Defence under Churchill. Militarily he was a tenacious commander, with the ability to reconcile points of view and to gain the loyalty of all his subordinates.

Alfonso XIII (1886–1941) King of Spain 1902–31. A posthumous son of Alfonso XII, his mother was Regent until 1918 and remained a strong influence after that. Anti-clericalism* and socialist agitation were prominent in Spain before World War I, especially in Catalonia, and he survived five assassination attempts. In 1923 he accepted Primo de Rivera* as dictator and could not survive the resignation of the latter. He abdicated and his grandson Juan Carlos returned to Spain in 1975.

Algeria, a North African country. Colonized by the French 1830–60, it was made part of Metropolitan France in 1862. By 1914 the Arab population was still excluded from full political activity, although in 1919 educated Arabs could become eligible for French citizenship and full membership of the Algerian Assembly, provided they renounced Islam. A number of nationalist organizations began to organize themselves during the 1930s and were ultimately to coalesce into the FLN*. In November 1942 Allied troops landed at Algiers, where a Committee of National Liberation was set up by de Gaulle; Algerians were promised a full share in the social and political life of their country. This promise was never fulfilled, resulting in increasing instability and a rising tide of violence. In 1952 immense oil deposits were found, thus hardening the attitude of French settlers. On 1 November 1954 open insurrection broke out. The Algerian War* for independence lasted eight years. The FLN leader Ben Bella* became President of the new Republic, but he was deposed in 1985 by Hoari Boumédienne*. His more traditionalist neighbours Morocco and Mauritania were nervous of his socialist regime and there were border disputes with both these and Tunisia. His successor President Chadli eased relations with France and adopted a strong policy of 'non-alignment', acting as international mediator on a number of occasions, most notably over the Iran Hostage Crisis* of 1981. During 1986 there were riots against the higher cost of living, precipitated by a fall in oil revenue. In January 1987 six Islamic fundamentalists* were shot in a gun battle and 202 placed on trial, of whom four were condemned to death. A proposal in 1987 of President Qadhafi* for a federal union between Libya and Algeria did not at once develop, but in 1989 the Maghreb Union* was formed. In spite of reduction in state control over industry and agriculture there were serious riots in October 1988, which prompted President Chadli to move towards the recognition of opposition political parties, including the Islamic Salvation Front (FIS).

Algerian War (1954–62). Algeria being part of Metropolitan France, this was technically a civil war, but in another sense it was a classic struggle between the large well-equipped French army and a small nationalist group, the Front de Libération Nationale* (FLN), determined on what they regarded as decolonization. There had been intermittent acts of violence, especially in Oran and Constantine, since the late 1940s, but during the night of 1 November 1954 over sixty attacks against police and troops took place all over Algeria. This armed insurrection marked the beginning of a state of war with which the government of Mendès-France* had to grapple. Atrocities by the Armée de Libération Nationale (ALN) against civilians, particularly in Constantine in 1955, led to torture and extremist reactions from French troops under General Massu, who had full police powers against strikers and alleged terrorists. French settlers and the army were determined to keep Algeria French, but public opinion in France itself began to turn against the continued barbarity of the war. This helped to bring down the Fourth Republic*, and the new President de Gaulle* became convinced that the only solution lay in an open plebiscite for all adults in Algeria. He negotiated the Évian Agreements*, and the result gave 99 per cent in favour of independence, de Gaulle having survived the plotting of General Salan* and his OAS* supporters.

Allenby, Edmund Henry Hynman, 1st Viscount of Megiddo and Felixstowe (1861–1936), British field marshal. After graduating at Sandhurst in 1882, he served in the Boer War and commanded a cavalry unit in the British Expeditionary Force* in 1914. He remained in France until 1917 when he was appointed commander of the British forces in Egypt and Palestine. He defeated the Ottoman army in Palestine and Syria by skilful use of cavalry, capturing Jerusalem on 9 December 1917. Turkish resistance ended after the Battle of Megiddo (18–21 September 1918) and the fall of Damascus (1 October) and Aleppo. He was appointed Special High Commissioner for Egypt and the Sudan 1919–25, and in 1922 persuaded the British government to end its protectorate over Egypt.

Allende (Gossens), Salvador (1908–73), President of Chile 1970–3. Son of a wealthy Chilean family, he qualified as a doctor at the University of Chile in 1932, but then entered politics as a socialist. He ran unsuccessfully for the presidency in 1952, 1958, and 1964. In 1970 he won as leader of a Socialist/Communist/Christian Democrat Alliance, being the first avowed Marxist to win a Latin American presidency in a free election. During his tenure of office he set the country on a socialist path, seeking to nationalize mines and to break up the hacienda estates. In the process he incurred the antipathy of the wealthier middle class and of the military establishment. Under General Pinochet* a military coup (backed by the CIA*) overthrew him in September 1973. Allende died in the fighting, and was given a state funeral in September 1990.

Alliance for Progress, a US initiative under President J. F. Kennedy*, when twenty-two Latin American countries and the USA signed a Charter at Punta del Este in Uruguay. Its aim was 'the maintenance of democratic government', together with social and economic development. In 1963 an Inter-American Committee on the Alliance for Progress (CIAP) was formed. Although the programme, funded by loans from the international financial community, showed modest development of schools and hospitals, it lost impetus, with increasing tensions between the USA and many Latin American countries, together with CIA interference in Latin American politics. Its place was partially taken by the Economic Commission for Latin America and the Caribbean, a UN agency originally set up in 1948.

Al-Sabah Dynasty, the ruling family of Kuwait, descended from its 18th-century founder. There are currently some 1,200 members of the dynasty, headed by the Amir of Kuwait, Sheikh Jabir al-Ahmad Al-Sabah, head of state, 'immune and inviolable', and Supreme Commander of the armed forces. The Amir has ruled through his Prime Minister, usually the heir apparent, answerable to himself. Al-Sabah sheikhs have usually held the key government posts as Ministers of Defence, the Interior, Foreign Affairs, and Information. Sheikh Mubarak I (1896–1915) allied with Abd al-Aziz ibn Saud* in opposing Ottoman rule and in 1914 was accepted as 'an independent sheikh under British protection'. Kuwait remained a British protectorate until 1971. Oil revenues brought immense wealth to members of the dynasty, who fled into exile in August 1990 when Iraqi troops invaded Kuwait.

Alsace-Lorraine, a French region west of the Rhine. Alsace and the eastern part of Lorraine were ceded to Germany after the Franco-Prussian War in 1871 and held in common by all the German states of the Second Empire. Rich in both coal and iron ore, it enabled Germany to expand its naval and military power; but the policy of Germanization was resented by the French and the area was restored to France by the Treaty of Versailles*. In 1940 German troops occupied the region and it was restored as part of the Third Reich. In 1945 it was liberated for France by US and French troops, since when its leading city Strasbourg has become a symbol of European unity.

Ambedkar, Bhimrao Ramji (1893–1956), Indian statesman. Born into a harijan family (Untouchable, or low-caste Hindu) he won a series of scholarships to universities in the USA, Britain, and Germany. Humiliated as a civil servant, as a qualified lawyer he led the agitation for the constitutional rights of the harijans. In 1934 he accepted the Poona Pact, which would provide reserved seats for them in provincial legislatures, following a fast by Gandhi*. He was founder of the Scheduled Castes Federation, which as a party opposed the Indian National Congress*, but after independence he joined the Congress. As a leading constitutional lawyer and Law Minister 1947–51 he played a major role in formulating and drafting the Indian Constitution. In despair over the Hindu doctrine of untouchability, he became a Buddhist just before he died.

American Federation of Labor (AFL). Founded in 1886 after mass disorders culminating in the Haymarket Square Riot in Chicago in that year, from its formation until his retirement in 1924 it was decisively shaped by its president Samuel Gompers. Gompers wanted a radical organization of skilled workers committed to collective bargaining for better wages and conditions. The growing numbers of semi-skilled workers in mass-production industries at first found their champion in John L. Lewis*, leader of the more militant United Mine Workers. When he failed to convince the AFL of the need to promote industry-wide unions in steel, automobiles, and chemicals, Lewis formed (1936) the Committee (later Congress) of Industrial Organizations (CIO), its members seceding from the AFL. In 1955 these two rival organizations were reconciled as the AFL-CIO under George Meany and Walter Reuther*. With a total of over 15 million members, this body has remained the recognized voice of organized labour in the USA and Canada.

American Indian Movement (AIM), a militant organization founded in the USA in 1970. The publication of Helen Hunt Jackson's book *A Century of Dishonour* in 1881 had resulted in the formation of the Indian Rights Association and a feeling of national guilt over US treatment of Native Americans. Early policies accepted tribal ownership of reservation lands, but in 1889 the Dawes Severalty Act had attempted to introduce individual landholdings. In time this policy was conceived as undermining Indian communal culture, while failing to achieve assimilation. By the Snyder Act (1924) all Indians born in the USA were granted full citizenship. From the mid-1930s there was increasing pressure from Indians for a reversal of policies and for reform, the Indian Reorganization Act of 1934 recognizing the advantages of reinvesting land in tribal ownership. After World War II a Federal Indian Claims Commission was established, together with a Bureau of Indian Affairs. But militancy began to develop, resulting in AIM. Between 1969 and 1971 Alcatraz Island in San Francisco Bay was occupied, with AIM demanding it become an Indian Cultural Centre. They were unsuccessful, and it became the Golden Gate National Recreation Centre instead. The Washington offices of the Bureau of Indian Affairs were occupied in 1972, as was the village of Wounded Knee in 1973, scene of the last great US–Indian battle in 1890. In 1970 President Nixon formally repudiated the paternalistic policy of assimilation and adopted that of Indian self-determination. Since then AIM has achieved numerous grants of land to Indian tribes where areas have a religious significance, while Native American cultural awareness has steadily increased.

Amethyst Incident. In April 1949, in the closing months of the Chinese Civil War, a British frigate, HMS *Amethyst*, was fired on by Chinese Communists while sailing up the Chang Jiang (Yangtze Kiang) towards Nanjing, to help British nationals there. Attempts to rescue her failed and there was a total of some 100 casualties before the frigate successfully escaped. The Communists were seeking to establish sovereignty over what had been regarded as an international waterway.

Amin, Idi (1926–), President of Uganda 1971–80. With only rudimentary education, Amin joined the King's African Rifles at the end of World War II. He rose to the rank of major-general in the Ugandan army and in January 1971 the army and police overthrew President Obote* and proclaimed Amin head of state. His rule was characterized by the advancement of the interests of Muslim northern Uganda, the expulsion of non-Africans (most notably Ugandan Asians), and violence and murder on a massive scale, including the murder of the Archbishop in 1977. Although he was chairman of OAU* 1975–6, his regime provoked the hostility of both Kenya and Tanzania, and in 1979 Tanzanian forces invaded and overthrew his government. He fled to Libya and later to Saudi Arabia. It is estimated that some quarter of a million Ugandans were tortured and killed during his regime.

Amnesty International, an international pressure group. It was founded in 1961 by Peter Berenson, a defence lawyer, for political prisoners. Under its first chairman Sean MacBride (1961–75) it was built up into a major international agency, with a secretariat of 150 based in London. It has over 200,000 members with sections in over forty countries. It researches and then publicizes the cases of prisoners of conscience, where they have been deprived of basic freedoms of speech or religion, or where they have been subjected to torture. It also seeks to provide relief and support for families of political prisoners. It has achieved a number of successes in the continued battle for human rights, being prepared to indict governments of all political persuasions.

Amritsar Massacre (13 April 1919), a tragic incident in British India. Indian discontent had been mounting against the British as a result of the Rowlatt Act*. On 10 April rioting broke out in Amritsar, the capital of the Punjab, resulting in the death of five Englishmen and the beating of an Englishwoman. Gurkha troops under the command of Brigadier R. H. Dyer fired on peaceful, unarmed crowds gathered in the Jallianwala Bagh, an enclosed park, killing 379 and wounding over 1,200. Mounting agitation throughout India followed and Dyer was given an official, if belated, censure.

ANC *see* AFRICAN NATIONAL CONGRESS.

Andean Group ('Grupo Andino' or officially 'Acuerdo de Cartagena'), an agreement signed at Cartagena in 1969 between Bolivia, Colombia, Ecuador, Peru, and Venezuela. Chile signed, withdrew in 1977, but rejoined in 1990. Its aim was to enhance the competitive edge of its members in their economic relations with the more developed economies of Latin America. An Andean Council was established, a Parliament of five members from each country, and a Court of Justice. Early in the 1980s there were a number of political problems, Ecuador temporarily withdrawing over a border dispute. In 1984 a new Andean peso was established to reduce dependency on the US dollar. Since then greater progress has been made towards co-operation, programmes for agriculture, transport, social welfare, and education being instituted. A 1989 summit committed the group to nuclear arms control, peaceful coexistence, regional development, and political integration.

Anglican Communion, a world-wide family of Protestant Churches whose origins were in the 16th century. The Church of England was established under Elizabeth I, its 'Supreme Governor'. In 1784 an American episcopalian Church was founded, and other episcopal Churches followed in Australia, Canada, New Zealand, Africa, and elsewhere. During the 19th century attempts were made to restore continuity with the Roman Catholic Church and introduce its rituals into Anglican worship. This 'High Church' movement was countered by the 'Low Church' Evangelical movement, which was more fundamentalist in its views and very active in missionary work and social reform. Since 1868 bishops from all episcopal Churches in the world have gathered every ten years at Lambeth Conferences, under the presidency of the Archbishop of Canterbury. Continued controversies between its Anglo-Catholic and Evangelical wings, as well as over such issues as the ordination of women and homosexuality, have been deeply divisive, with the Church of England itself having a decreasing influence on the Anglican Communion as a whole. Since World War II the role of the General Synod of the Church of England has widened and the language of the Prayer Book has been modernized. The Church of England remains the established Church in England, with bishops members of the House of Lords.

Angola, a country in south-west Africa. The coastal strip was colonized by the Portuguese in the 16th century, but it was not until the 19th century that, following wars with the Ovimbundu, Ambo, Humbo, and Kuvale, Portugal began to exploit the mineral reserves of the hinterland. In 1951 Angola became an Overseas Territory and in 1955 an Overseas Province of Portugal. By then a nationalist movement had emerged and virtual civil war was to wage until November 1975, when independence was granted. Over 400,000 Portuguese were repatriated and almost total economic collapse followed. Internal fighting then developed between guerrilla factions, the Marxist Movement for the Liberation of Angola (MPLA), supported by Cuba and East Germany, and its opponent the National Union for the Total Independence of Angola (UNITA), led by Jonas Savimbi* and supported by South Africa and the USA. In February 1976 most countries recognized the MPLA as forming the official People's Republic, which took a strong line against the White mercenaries operating in the country. Punitive South African raids took place through the early 1980s, aimed at Namibian resistance forces operating from Angola. In August 1988 a Geneva Accord between the various parties aimed to end violence, but Savimbi for long refused to accept its terms.

Anguilla, the most northerly of the Lesser Antilles*, with a population of some 7,000, mostly descended from African slaves. A British colony since 1650, it became part of the colony of St Kitts–Nevis (St Christopher and Nevis*) in 1882 and as such formed part of the Federation of the West Indies* 1958–62. It was subsequently granted 'Associated State' status with St Kitts and Nevis, but this was highly unpopular and in 1967 Anguilla declared independence. Two years later it was occupied by British troops, who reduced the island to colonial status once again, it becoming a British dependency with an elected House of Assembly.

Animal rights movement. The first organization to protect animals was the Royal Society for the Prevention of Cruelty to Animals, founded in 1824. An American society was founded in 1866. By the late 20th century animal welfare societies and laws against cruelty to animals existed in nearly every country of the world, together with a number of international organizations such as the World Federation for the Protection of Animals, based in Zurich, and the World Wide Fund for Nature (WWF), founded in 1961 to protect threatened species such as the whale, the tiger, and the elephant. In Britain in the 1980s a militant animal rights movement resorted to terrorist methods such as letter bombs to highlight its case against the use of animals in laboratory experiments, particularly vivisection.

Anschluss (German; 'connection'), specifically applied to unification of Austria with Germany. The Second German Empire of 1871 did not include Austrian Germans, who remained in Austria-Hungary and in 1918 formed the new Austrian Republic, forbidden anschluss by the Treaty of Versailles. The French in 1931 prevented a projected customs union, and in 1934 a coup by Austrian Nazis failed. In February 1938 Hitler summoned Chancellor Kurt von Schuschnigg* to Berchtesgaden and demanded that he admit Nazis into his cabinet. Schuschnigg attempted to call a plebiscite on Austrian independence, failed, and resigned. German troops entered Vienna on 13 March 1938 and the Anschluss was proclaimed, being welcomed by a majority of Austrians. The ban on an anschluss was reiterated when the Allied powers recognized the second Austrian Republic in December 1945.

Antarctica. There were a number of scientific and geographical expeditions to the continent during the 19th century and three important overland expeditions early in the 20th century by Ernest Shackleton, Roald Amundsen, and Robert Falcon Scott. The main territorial claims, not recognized by the USA or the Soviet Union, have been those made by Britain (British Antarctic Territory) in 1908; New Zealand (Ross Dependency) in 1923; Australia (Australian Antarctic Territory) in 1933; France (Adélie Land) in 1938; Norway (Queen Maud Land) in 1939; Chile (Antarctic Peninsula) in 1940; and Argentina (Antarctic Peninsula) in 1942. The Antarctic Treaty signed in 1959 by twelve nations, including Britain, the USA, the Soviet Union, France, and Japan, guarantees the continuation of peaceful research, the banning of military or nuclear activity, and the establishment of a system of mutual inspection. In addition to the seven countries with territorial claims, by 1959 the following had also established scientific bases: the USA (including one at the Pole), the Soviet Union, Belgium, Japan, Poland, South Africa. Thirty-nine countries ratified the Treaty, which was for thirty years. During 1985 and 1986 a British Antarctic Survey calculated that the ozone layer at the Pole was being depleted at the rate of 6 per cent per year, a figure confirmed in March 1988 by the World Meteorological Association. In 1988 the United Nations General Assembly voted to restrain mineral exploitation and at the fifteenth biennial meeting of the Convention of the thirty-nine signatories of the 1959 Treaty, held in Paris, there was a Franco-Australian proposal to ban all mining and turn Antarctica into an international nature reserve. A special meeting of the Convention was held in 1990 to consider this, when Britain, the USA, and Japan forced a postponement of a decision until April 1991.

Anti-clericalism, opposition to the power and influence of the clergy, especially in politics. As a liberal movement aimed at the extreme political conservatism of the Catholic Church, it was a powerful force throughout the 19th century, especially in France, Spain, and also Bismarck's Germany. As a force of opposition to both Catholic and Orthodox clergy it contributed to the establishment of Communist regimes in Eastern Europe from 1917 onwards. It was particularly violent in Republican Spain 1931–9. In Latin America Liberation Theology* has in some ways been anti-clerical, in that it has been critical of the establishment of the Catholic Church. During the 1980s it was the Catholic Church in Poland which was to triumph over its anti-clerical and Communist foes. Increasing secularization of Western society has resulted in a steady decline in anti-clerical antipathy.

Anti-Comintern Pact (25 November 1936), an agreement between Germany and Japan, ostensibly to collaborate against international Communism (the Comintern*). Germany recognized Japan's regime in Manchuria, and Italy signed the Pact in 1937. In 1939 Hungary and Spain signed the Pact and in 1941 so did the Nazi satellites Slovakia, Romania, and Bulgaria. It was regarded in the West as a pact to overawe the nations of the free world.

Antigua and Barbuda, a small island state in the Lesser Antilles*, consisting of the islands of Antigua (pop. *c.*80,000), Barbuda (pop. *c.*1,500), and the uninhabited island of Redonda. First colonized by the British in 1632, it formed part of the Leeward Islands Federation 1871–1956 and then of the West Indies Federation* 1958–62. The majority of the population are descended from African slaves imported for the sugar industry; this steadily declined until it closed in 1982. Cotton and fruit partially replaced it, and since World War II the tourist industry has helped to boost the economy. In 1967 it became an Associated State of the United Kingdom, but there were demands for independence, which it achieved in 1981, as a member of the Commonwealth. In 1984 the Antigua Labour Party, whose leader is Vere C. Bird, won sixteen out of the seventeen seats in the House of Representatives. It was re-elected in March 1989.

Anti-nuclear organizations. Shocked by the potential of the hydrogen bomb*, Albert Einstein* and Bertrand Russell* drew up their Pugwash* manifesto signed by nine Nobel* Prize scientists in 1955. In 1958 Russell, Canon Collins, leading Labour politicians, and academics formed the CND* movement in Britain, with annual marches from Whitehall. From 1965 onwards a peace movement was developing in the USA; but this was initially directed against the Vietnam War. In 1969 a Union of Concerned Scientists was formed in the USA, opposed to nuclear armaments. The Nuclear Test Ban* (1963) and SALT I* (1972) treaties seemed to encourage a more optimistic view of the future, and in the 1970s the Green movement* was increasingly opposed both to nuclear energy* and to nuclear arms. By 1980 the Cold War seemed once again to be intensifying. At Strasbourg that year Lord Louis Mountbatten* spoke against the nuclear arms race, CND was revived in London, and the European Nuclear Disarmament (END) movement was started by the radical socialist historian E. P. Thompson, in response to the NATO decision to install medium-range missiles. In October there was a rally of 50,000 in London. END began a dialogue with Charter 77* in Czechoslovakia and with other dissidents in Eastern Europe, and in 1982 'Women for Peace' was formed in Holland, with a huge rally on 24 May, International Women's Day. It gained support in the USA and Australia as well as Europe. A hundred and fifty women chained themselves to the fence of the cruise missile base at Greenham Common, and in December 30,000 women surrounded the base. In the USA the Catholic bishops united in protest, supported by the Pontifical Academy of Science in Rome, and ANDEF (Action for Nuclear Disarmament Education Fund) was formed. In Germany both Catholic and Protestant Churches were united on the issue and West Germans became more and more reluctant for their country to be a nuclear battleground. It was against this background that the INF Treaty* was signed in 1988, to be followed by a steady reduction in tension and lessening of the threat of a nuclear confrontation in Europe. However, the threat of nuclear war remained, as nuclear capability was acquired by more and more countries.

Anti-Semitism, hostility towards Jews, perhaps the most notorious and long-lasting example of racism*. While its origins may be religious, in Christian abhorrence of the Jewish denial of Christ, it was already a social and political issue by the Middle Ages. It developed rapidly through the 19th century throughout Europe, where Jewish business acumen was deeply resented, resulting in many appalling pogroms* against Jews, especially in south-east Europe. The Nazi Holocaust* drew on a deep vein of hatred, especially in Eastern Europe, but the horror which this aroused failed to eradicate it. Anti-Semitism has survived in both Western and Eastern Europe and in the USA, and it contributes to the Arab–Israeli problem, even though both are Semitic peoples. There is no generally agreed explanation for such racial antagonism, although jealousy—of business skills, musical and artistic gifts, strong family ties, even of alleged sexual prowess—seems a common factor.

Antitrust laws (USA). After twenty-five years' agitation against monopolies, the US Congress passed the Sherman Anti-Trust Act in 1890, which declared illegal 'every contract, combination, or conspiracy in restraint of trade'. The Clayton Anti-Trust Act (1914), amended by the Robinson–Patman Act (1936), prohibited discrimination among customers, and mergers of firms that would lessen competition. At the same time it sought to protect 'Labor, agricultural and horticultural organizations'. After World War II there was a further growth in giant multinational corporations*, and the Celler–Kefauver Antimerger Act (1950) was intended to prevent tactics such as elimination of price competition, which it regarded as against the public interest. There have been numerous Supreme Court rulings through the 20th century which on balance have tended to weaken rather than strengthen legislation.

Antonescu, Ion (1882–1946), Romanian Dictator 1940–5. Born in Piteşti, he became a professional soldier and served as a colonel in World War I. After the war he continued to advance in rank and became a strong admirer of Mussolini, serving as a military attaché in Rome. In 1937 he became Chief of Staff in the Fascist regime of Carol II*. Although a strong supporter of the latter, he was briefly imprisoned in 1938 for criticizing his pro-France foreign policy. When Carol reversed policy and backed the Axis he became Defence Minister, and then in 1940 Prime Minister, with dictatorial powers, calling himself the Conducator. He failed to prevent the Vienna Award* of that year whereby Romania was pressurized into granting much of Transylvania to Hungary. In 1941 he sent thirty Romanian divisions against the Soviet Union; they captured Odessa, which was renamed Antonescu, but they suffered appalling losses before Stalingrad. In 1944 the Red Army crossed into Romania. He was arrested on orders of King Michael when Romania changed sides. He was tried and shot as a war criminal.

Anzac, an acronym derived from the initials of the Australian and New Zealand Army Corps, which fought during World War I. Originally it was applied to those members of the Corps who took part in the Gallipoli campaign*, and Anzac Day, 25 April, commemorates the landing at Gallipoli. Later the name came to be applied to all Australian and New Zealand servicemen. These fought in Palestine under Allenby* and on the Western Front* in World War I, and in North Africa, Greece, and Europe in World War II.

ANZUS, an acronym for the tripartite treaty between Australia, New Zealand, and the USA signed in San Francisco in 1951. Also known as the Pacific Security Treaty, it recognizes that an armed attack on any of the parties would endanger the peace and safety of the whole Pacific area, and declares that all parties would act to meet the common danger. Following New Zealand's anti-nuclear policy (1986), which included the banning of nuclear-armed ships from its ports, the USA suspended its security obligations to New Zealand, and all ministerial contacts. The latter, however, were restored in March 1990.

Aotearoa, the Maori name for New Zealand, 'the Long White Cloud'. Growing Maori nationalism in the 1970s and 1980s has urged that the term become the official name for the islands. It is increasingly used by pakeha (European New Zealanders) as well as by Maoris.

Apartheid (Afrikaans 'separateness'). The term was first used in South Africa politically in 1943, but as a concept it goes back to the time of the Dutch settlers of the 17th century, involving segregation of Black from White in land ownership, residence, marriage and other social intercourse, work, education, religion, and sport. From 1948 onwards it was expressed through statutes for race registration and control, in job reservation and trade union separation, and in the absence of parliamentary representation. Bantustans* were created, depriving the Bantu-speaking peoples of South African citizenship for an illusory 'independence'. From 1985 onwards restrictions began to be eased: subordinate chambers in Parliament for Indians and Coloureds (people of mixed descent) were created, but not for Blacks; rules for sport and leisure were relaxed; Pass Laws were abolished and the Group Areas Act (1950), for housing and business segregation, modified. International pressure continued for its total abolition, and during 1990 President de Klerk* accepted this in principle. Tentative negotiations began with the ANC*, the Separate Amenities Act of 1953 was repealed, and preparations made for the repeal of both the Group Areas Act and the Population Registration Act (1950). It was generally accepted, however, that full racial integration and democratic politics would take many more years to achieve.

APRA (Alianza Popular Revolucionaria Americana), a Latin American political party. It was founded in 1924 by Haya de la Torre*, then in exile in Mexico. As a left-wing party it developed in Peru under the title Aprista, having been much influenced by the ideas of the political scientist Carlos Mariátegui (1845–1930). Aprista was declared illegal in Peru 1931–45 and again 1948–56, after which it moved to the centre in policy and in the 1963 elections won fifty-eight out of the 108 seats in the Peru Chamber of Deputies. In 1981 there were party splits, but in the 1985 elections the party, now named People's Party, won a majority in both the Senate and Chamber of Deputies. The party did not have a significant influence in countries other than Peru.

April Fourth Incident (1976), an incident in China. Following the death of Zhou Enlai* (8 March), thousands of Beijing residents gathered in Tiananmen Square to mourn his death. Rioting developed when police removed wreaths and placards, the riots being officially described at the time as 'organized, premeditated and planned counter-revolution'.

Aquino, María Corazón (1933–), President of the Philippines 1986– . Of a wealthy landowning family, she married Benigno Aquino, a young politician. When the latter attempted to become President in 1972 he was imprisoned by President Marcos* for eight years. Allowed to go to the USA for medical treatment, he was shot by security guards on his return in 1983. Corazón became a symbol of opposition and in 1986 was recognized as President, when Marcos fled to the USA. Her position was confirmed under a new constitution. She successfully negotiated with a Muslim secessionist group and also with the Cordillera People's Liberation Army in north

Luzon, but failed to eradicate Communist insurgency, in spite of negotiations. She strengthened economic ties with the USA and Japan and strongly supported the Association of South-East Asian Nations*. Basic problems of land ownership remained, with over 70 per cent of the population below the poverty line. By December 1989 there had been no less than six attempted military coups against her. She only survived these with the strong support of US military forces on the islands, her negotiations over US bases continuing throughout 1990.

Arab Boycott. During the 1920s leaders of Palestine's Arabs tried to impose a boycott against both products and services (hospitals, lawyers, etc.) of the growing Jewish community. It was not very successful, but in 1945 the Council of the Arab League* proclaimed an official economic boycott against Jews in Palestine and from 1948 against the state of Israel. Although this adversely affected Israel, it failed to prevent its continued industrial and economic development.

Arab League. In September 1944 representatives from Egypt, Iraq, Lebanon, Saudi Arabia, Transjordan, and the Yemen, together with Palestinian Arabs, met in Alexandria to form a league to promote the interests of the Arab peoples. A pact was signed in March 1945 establishing a Council, which aimed to protect the independence and integrity of its members. Its main object was to organize opposition to Palestinian Jews, to Israel through the Arab Boycott*, and to France in Lebanon and North Africa. The League was later joined by Libya (1953); the Sudan (1956); Tunisia (1958); Morocco (1958); Kuwait (1961); Algeria (1962); South Yemen (1968); Bahrain, Qatar, Oman, and the Trucial States (1971); Mauritania (1973); Somalia (1974); the PLO (1976). In 1978 Egypt's membership was suspended following the Camp David Accord*, and the headquarters moved from Cairo to Tunis. At the League's summit in Amman in 1987 Egypt was readmitted. The League proved unable to resolve the Gulf crisis* in 1990.

Arab Legion. This began as a police force for the state of Transjordan, established by the British in 1921; a force of 1,000 volunteers formed by Colonel Frederick Peake to maintain internal law and order. In 1939 Major John Glubb* became its commander and he developed it into a highly disciplined military force which played a significant part in Middle East campaigns in World War II. In the 1948 war against Israel the Legion, then numbering over 7,000 volunteers with many British officers, won and held east Jerusalem and the West Bank territories, which Abdullah ibn Hussein* proclaimed part of the Kingdom of Jordan in 1949. Glubb resigned in 1956, by which time the Legion formed a part of the Jordanian army under Ali Abu Nawwar.

Arab Revolt (1916). In July 1915 Hussein ibn Ali*, Sherif of Mecca, offered help to the British in return for promises of independence for Arab peoples. He corresponded with the British High Commissioner in Egypt, Sir Henry McMahon, who wrote that Britain would 'stimulate national independence in the whole Arabic-speaking world' if he were to enter the war. The Revolt began in June 1916, when an Arab army of some 70,000 men, financed by Britain and led by Faisal, son of Hussein, moved against Turkish forces. They cut the Hejaz railway* and captured Aqabah. This enabled British troops under Allenby* to advance into Palestine and Syria. With the capture of Damascus (1 October 1918) Turkish hold on the Middle East ended. Independence was to be gained in some areas, but in others either France or Britain became the 'mandate' power under the League of Nations. This alienated many Arabs, while the Zionist Balfour Declaration* was seen as a direct contradiction to McMahon's assurances.

Arafat, Yasir (1929–), Palestinian politician. Born in Jerusalem, he studied in Cairo, where he joined the Muslim Brotherhood*. He served briefly in the Egyptian army before going to Kuwait as an engineer. Here he helped to form the al-Fatah* movement, emerging as its leader in 1968. The Palestine Liberation Organization* was founded in 1964; he became its chairman in 1969 and Commander-in-Chief of the Palestine Revolutionary Forces in 1971. As such he organized guerrilla raids on Israeli territory and assassination and anti-Israel terrorist attacks. Although gaining support from Eastern bloc countries, his ideology has always been right of centre. He attended periodic all-Arab 'summit' conferences and in 1974 was invited to address the UN General Assembly. In 1982 his organization was expelled by Israel from Lebanon. He was criticized by Marxist elements in the PLO as too conservative, failed to reach an accord with King Hussein ibn Talal* of Jordan, and quarrelled with President Asad* of Syria. In 1983 he set up a new base in Tunisia and in 1987 re-established contact with Asad. In December 1988 he was again invited to address the UN General Assembly, when he rejected military violence, recognized the existence of the state of Israel, and called for a political solution to the Palestine problem. Negotiations with the USA collapsed early in 1990, and he made tireless but fruitless efforts to prevent war, following the Iraqi invasion of Kuwait.

Arbenz Guzmán, Jacobo (1913–71), President of Guatemala 1951–4. Son of a Swiss immigrant, he was educated in the Guatemala Military Academy and took part in a military coup in 1944, becoming Minister of War. In 1950 he was elected President as the candidate of the Revolutionary Action Party. Comprehensive agrarian reform laws on behalf of landless Indians made possible the expropriation of large estates owned by the United Fruit Company* of the USA. When his administration was judged to be Communist by the Roman Catholic Church the Eisenhower administration in the USA began sending arms through the CIA to Guatemalan exiles in neighbouring countries. These arms enabled the exile Carlos Castillo Armas to lead a successful counter-revolution which deposed Arbenz in 1954. He fled to Mexico.

Arcos Raid (May 1927). Following the British General Strike*, a wave of anti-Red hysteria swept Britain over Soviet agents allegedly subverting the trade union movement. On 12 May 1927 200 police officers spent six hours ransacking the offices of a British company, Arcos Ltd., which was said to be acting as a front for Communist subversion, since it shared a building with the Russian trade delegation. They found nothing; but to justify the raid Prime Minister Baldwin* broke off diplomatic relations with the Soviet Union. They were restored by MacDonald's* Labour government in October 1929.

Ardennes campaign (also called the Battle of the Bulge) (December 1944–January 1945), the last important German counter-attack against Allied armies advancing towards Germany in World War II. Hitler ordered Field Marshal von Rundstedt* to make an attack through the hilly, wooded country of northern Luxembourg and through the Ardennes Forest on 16 December, taking advancing US forces by surprise. The latter fell back, but last-ditch resistance at several points, notably at Bastogne, held the Germans up long enough for the Allies to recover and prevent them from reaching their objective of Antwerp. A counter-offensive by Montgomery* and Bradley* cleared the Ardennes by 16 January with 120,000 German casualties.

Argentina. The republic consists of twenty-two provinces, the federal district of Buenos Aires, and the National Territories of Tierra del Fuego, the Antarctic Sector, and the South Atlantic Islands. The second half of the 19th century saw a demographic and agricultural revolution. The fertile plains (pampas) of the interior were transformed by means of foreign and domestic capital. An extensive railroad network was built, and the introduction of steamships and refrigeration vastly increased the export of cattle and grain. The influx of immigrants (mostly Spanish and Italian) contributed to an increase in population from 1.2 million to 8 million between 1870 and 1914. Argentina's export-orientated economy proved vulnerable to the fluctuations of the international market, and the Great Depression* resulted in a drop of 40 per cent in exports. A military coup saw the emergence of the armed forces as arbiter of Argentinian politics through the mid-20th century. For example the populist movement created with the support of the trade unions by Juan Domingo Perón* collapsed in 1955, with army officers acting in defence of 'Catholic Christianity'. Perón was re-elected in 1973 after an eighteen-year exile. His death in 1974 was followed by another period of military dictatorship (1976–83) in a particularly bitter and tragic period of authoritarian rule under a military junta. It is estimated that some 20,000 Argentinians lost their lives in the 'dirty war' of political terrorism of these years. In 1982 the armed forces of General Galtieri* suffered a humiliating defeat in the war with Britain over the Falkland (Malvinas) Islands*, and in 1983 a civilian administration was elected under President Raúl Alfonsín of the Radical Party. The process of redemocratization faced severe problems, most notably a virtually bankrupt economy, opposition from the Catholic Church, and political insensitivity in the armed forces. At the same time there was pressure from the civil courts to indict some 370 army officers for human rights offences. Alfonsín did not stand in the election of 1989, which was won by the Perónist Carlos Saul Menem, who inherited a debt of $US60,000m. and who restored diplomatic relations with Britain in February 1990.

Arias (Madrid), Arnulfo (1901–   ), President of Panama 1940–1, 1949–51, 1968. Younger brother of Panama President Harmodio Arias, he was educated in Chicago and at Harvard Medical School, qualifying as a surgeon. He served in his brother's government (1932–6) as Minister of Agriculture and Public Works. Elected President in 1940 on a far-right ticket, he disenfranchised Black West Indians in Panama, supported the Axis powers, and refused concessions to the USA. He was deposed with US assistance, but was re-elected in 1949 to lead a neo-Fascist, corrupt regime and was again deposed, this time by the police. Re-elected a third time in 1968, he was deposed by the army after eleven days.

Armenia, a region which comprised what is now north-eastern Turkey and the Armenian Soviet Socialist Republic. In 1828 north-east Armenia had been ceded to Russia by the Ottoman Turks. Agitation for independence developed in both Russian and Turkish Armenia, leading to massacres which culminated in the deportation, by the Young Turk* government of 1915, of Turkish Armenians to Syria and Palestine, over 1 million of them dying. In 1918 a short-lived independent Transcaucasian Federal Republic of Armenia came into being. This lasted until 1920 when, following the Battle of Kars, Turkish Armenia was renounced by the Russians, while the remainder was proclaimed a Soviet republic. Turkish Armenia still hoped for a separate existence, and this was guaranteed by the Treaty of Sèvres* in 1920. There were, however, further massacres and deportations, until, by the Treaty of Lausanne* (1923), it was absorbed into the new Republic of Turkey, with tensions continuing through the 20th century. In 1922 the Soviet Republic was absorbed into the Transcaucasian Soviet Federated Socialist Republic, one of the first four republics to join the USSR. In 1936 Armenia separated from Azerbaijan. Two separate republics were formed within the

Soviet Union, but with many Armenians remaining in the Karabakh area of Azerbaijan. This arrangement was to cause acute ethnic tension and violence 1989–90. During 1990 strong pressure developed for the Republic to become independent of the Soviet Union.

Armistices 1917–18, those armistices (agreements to cease fighting, pending a peace treaty) which ended World War I. War ended on the Eastern Front on 8 November 1917, followed by the Treaty of Brest-Litovsk* (March 1918). Fighting on the Bulgarian/Macedonian front ended with an armistice at Salonika on 29 September 1918; the Ottoman Empire made an armistice at Mudros* on 30 October; Austria-Hungary negotiated an armistice at Padua on 3 November. Germany accepted an armistice at Compiègne on 11 November. When France made peace in 1940 the armistice with Germany was signed in the same railway carriage preserved at Compiègne. The peace treaties confirming these armistices were negotiated at the Paris Peace Conference* 1919–20.

Arusha Declaration (1967), a policy statement in Tanzania. It proposed that TANU (Tanganyika African National Union) implement a socialist programme by which the major means of production would be under the collective ownership of the farmers and workers of the country, rather than, as in most of socialist Europe, under an urbanized, impersonal state. President Nyerere* was deeply committed to the concept of *ujamaa**, which sees all land and natural resources as belonging to the people within their village communities. The policy was moderated after 1977, to allow some private investment, and from 1987 was largely abandoned.

Asad, Hafiz al- (1928–), President of Syria 1971– . Of humble background, he joined the Ba'ath* Party when a student, in protest against social conditions in Syria. Graduating as an air force pilot in 1955, he was dismissed in 1961 for opposing Syria's secession from the UAR*. When the Ba'ath Party came to power in 1963 he became Commander-in-Chief of the Syrian armed forces and was leader of the radical Ba'ath faction which seized power in 1966. He became Prime Minister in 1970 and President March 1971. As such he remained hostile to a rival Ba'ath regime in Iraq, but improved relations with Egypt and gained the support of the Soviet Union. He became President of a Council for the Federation of Arab Republics and collaborated with Egypt in the unsuccessful war (Yom Kippur*) against Israel in 1973, losing the Golan Heights. His troops occupied much of Lebanon in 1976, but were largely driven out by Israel in 1982. In 1988 they occupied part of Beirut and in October 1990 enforced acceptance of the Taif Accord*. Asad has remained antagonistic to Iraq, and his regime appears to have sponsored various acts of international terrorism. It has repressed the Muslim Brotherhood* in Syria (Islamic Liberation Party), some 10,000 of whom have allegedly been killed. In August 1990 he supported UN policies against Iraq.

Asante (Ashanti), the largest and most prestigious Ghana chiefdom. After continued friction with the British during the 19th century, when their capital Kumasi was burned and their Golden Stool, symbol of the Asante people, confiscated, the Asantehene Prempeh I was exiled in 1896. In 1901 Britain annexed the region, linking it with the colony of the Gold Coast. In 1924 Prempeh was allowed to return and in 1935 an Asante Confederacy Council was established as an organ of local government, the Asantehene being head. In that year the Golden Stool was restored to Kumasi. After World War II the Asante region became a part of independent Ghana*.

ASEAN *see* ASSOCIATION OF SOUTH-EAST ASIAN NATIONS.

Ashes, the, a mythical trophy for which England and Australia have competed in test cricket matches over the past century. Begun in 1877, test matches between the two countries were at first annual, but were later played every two years. In 1882, after a disastrous series, the *Sporting Times* had written 'in affectionate memory of English cricket which died at the Oval 29 August 1882 . . . R.I.P.' Next year the English test team went to Australia 'determined to bring back the Ashes', and the term has remained. They in fact brought back the ashes of a cricket bail, which have remained at Lords' Cricket Ground ever since. Perhaps the two greatest batsmen in test cricket were the Australians Victor Trumper (1877–1915) and Sir Donald Bradman (1908–). During the 20th century test match cricket has developed, particularly in the West Indies, India, and Pakistan. South Africa was excluded from 1970 on grounds of its racial policy of apartheid*.

Asquith, Herbert Henry, 1st Earl of Oxford and Asquith (1852–1928), British Prime Minister 1908–16. Born in Yorkshire, he graduated from Balliol College, Oxford, and was called to the Bar in 1876. Ten years later he entered politics as a Liberal Member of Parliament. He served as Home Secretary 1892–5 and in 1905 joined the government of Campbell-Bannerman as Chancellor of the Exchequer. He introduced three skilful budgets, the third setting up old-age pensions, and he supported other important social legislation, such as the abolition of sweatshops and the establishment of labour exchanges. When Campbell-Bannerman fell ill (April 1908) he became Prime Minister, supporting Lloyd George* in his fight for the People's Budget in 1909 and the creation of a National Insurance scheme in 1911. Other important legislation included the Parliament Act (1911) to limit the power of the Lords and an Act to pay Members of Parliament. The later years of his ministry were beset with industrial unrest, violence in Ireland over his Home Rule Bill, and the dilemma over the issue of women's suffrage*. The Bill to disestablish the Anglican Church in Wales provoked much hostility before being passed. In 1915 he formed a coalition government with the Conservatives, but in the conduct of World War I he was considered too detached to provide dynamic leadership. His handling of the conscription issue and the Easter Rising* in Dublin provoked much criticism. Discontent grew and in December 1916 Lloyd George replaced him. The division in the Liberal Party at the 'Coupon Election'* (when Asquith lost his seat in Parliament) between his supporters (known as the Wee Frees) and those of Lloyd George lasted until 1926, when Asquith resigned the leadership of the party. The split marked the end of the Liberal Party's dominance in British politics.

Assam, constituent state of the Republic (Union) of India, to which it is linked by a narrow corridor through the lower Himalayas. Conquered by the British in 1842, by 1914 it had been incorporated into Bengal. This was unpopular, and in 1924 it became a separate province. Rice farming in the rich Brahmaputra valley expanded, as did tea plantations in the hills. During World War II the Japanese advance was halted here and Allied forces assembled for the reconquest of Burma. Since independence, Assam territory has shrunk through cessions to Pakistan (now Bangladesh), and the creation of the states of Nagaland, Meghalaya, Arunachal Pradesh, and Mizoram. In 1961–2 the frontier with China was disputed in the Indo-Chinese War*. Dominantly Hindu, Assam has absorbed many Muslim refugees, but not without periodic and serious rioting. It is subject to both earthquakes and floods.

Association of South-East Asian Nations (ASEAN). A regional organization formed in 1967 by Indonesia, Malaysia, the Philippines, Singapore, and Thailand, it replaced the earlier Association of South-East Asia (ASA). Brunei joined in 1984. Its summit meeting in 1976 resulted in a Treaty of Amity and Co-operation. Although aimed at accelerating economic growth, its main success has been in promoting diplomatic collaboration; for example it made a unified response to the 1979 Vietnamese occupation of Cambodia*, as it has done to policies of the superpowers. It has facilitated exchange of administrative and cultural resources, and co-operation in transport and communication. A permanent Secretariat was established in Jakarta. In November 1989 the ASEAN states joined with Australia, Canada, Japan, South Korea, and New Zealand to form a Council for Asia–Pacific Economic Co-operation.

Aswan High Dam. The proposal to build a High Dam some 4 miles south of the original Nile Aswan Dam (opened in 1920) precipitated the international Suez Crisis* (1956). The temple of Abu Simbel had first to be raised and the dam was opened in 1971 at a cost of $1,000m. It created the world's largest reservoir, Lake Nasser, which enables controlled irrigation over some 3 million acres of Egypt and the Sudan. It has resulted in an erosion of the Egyptian coastline and a diminution of sea-fish catches.

Atatürk ('Father of the Turks'), Mustafa Kemal (1881–1938), President of the Republic of Turkey 1923–38. An Ottoman army officer, he distinguished himself in wars against Italy (1911) and Bulgaria (1912), and at Gallipoli (1915). In May 1919 he was appointed Inspector-General of the 9th Army in Samsun, Anatolia, from where he organized resistance to the proposals of the Sèvres Treaty*. The defeat of the Greek army in 1922 was followed by the Chanak Crisis*; this paved the way for the Treaty of Lausanne* in 1923. Turkey was declared a republic and the sultanate abolished, together with the caliphate in 1924 (the temporal and spiritual leadership of Islam). As first President of the Republic Atatürk defined the principles of the state in the so-called six arrows of kemalism: republicanism, nationalism, populism, statism, secularism, and revolution. His policies involved a rejection of the Islamic past and the creation of a secular Turkish state, over which he ruled until his death in a highly autocratic fashion. He encouraged industrialization, the emancipation of women, and the use of the Latin alphabet.

Atlantic, Battle of (1939–45), the name given by Winston Churchill to the maritime struggle for supply routes in the Atlantic, the Caribbean, and northern European waters during World War II. German U-boats, sometimes assisted by Italian submarines, were the main weapon of attack, but aircraft, surface raiders, and mine-laying craft also participated. During the autumn of 1940 'wolf-packs' of U-boats caused havoc, for example sinking thirty-two British ships in a week of October. During 1942 long-range U-boats operated along the US east coast and in the Caribbean, with losses averaging ninety-six ships a month. A total of some 2,800 British and US ships were lost, placing the Allies in a critical situation. After summer 1943, with the introduction of better radar, the provision of longer-range aircraft and escort vessels, and the breaking of German codes, the situation eased, forty-one U-boats being destroyed during May 1943. Subsequent German technical innovations again increased U-boat effectiveness, and only the capture of their bases by Allied land forces in 1944 finally put an end to their threat.

Atlantic Charter (1941). This declaration of principles resulted from a meeting at sea, off Newfoundland, between Winston Churchill and President Roosevelt on 14 August 1941, before the USA had entered World War II. It stipulated freely chosen governments, free trade, freedom of the seas, and disarmament of current aggressor states; it condemned territorial changes made against the wishes of local populations. Both Britain and the USA renounced all territorial ambitions. Fourteen other nations fighting the Axis powers, including the Soviet Union, declared their support for these principles. The Charter provided the ideological base for the UN Charter of June 1945.

Atomic Energy Agency, a specialized agency of the UN with headquarters in Vienna and laboratories in Vienna, Monaco, and Trieste. It was established in 1957 following a conference at the UN in New York in October 1956. It gives advice and technical assistance on nuclear power (development and safety), radioactive waste disposal, prospecting for raw materials, use of radiation and isotopes in medicine, agriculture, and industry. By December 1987 a total of 166 nuclear safeguard agreements were in force with ninety-seven states.

Attlee, Clement Richard, 1st Earl Attlee (1883–1967), British Prime Minister 1945–51. Son of a successful lawyer, he was educated at Oxford and was successively a barrister, a social worker at Toynbee Hall*, and a lecturer at the London School of Economics, before entering politics. He served in the army during World War I and entered Parliament in 1922. He held office as a junior minister in the government of Ramsay MacDonald* 1930–1, when in a quiet way he distinguished himself in debate. In 1935, on the retirement of George Lansbury*, he was elected leader of the Labour Party on the second ballot. In 1937 he went to Spain to support left-wing volunteers fighting there. From May 1940 he held high office in the coalition government of Winston Churchill as Lord Privy Seal, deputy Prime Minister, and Dominions Secretary, chairing many vital Cabinet Committees. To the surprise of many, the Labour Party won the general election of 1945, on a programme of social reform and against a Conservative Party led by the victor of the war, Winston Churchill. Despite a war debt of $20,000m. and severe fiscal difficulties, the government embarked on an economic and social reform programme as advocated by J. M. Keynes*. It implemented the Beveridge Report* of 1942 through the creation of a Welfare State, supported by a policy of full employment. The National Insurance Act of 1946 was passed and the government also introduced a National Health Service*, a free medical and hospital service financed from general taxation. Public ownership was extended, the Bank of England being nationalized, as were key industries such as gas, coal, electricity, and railways, some of which had been successfully government-run during the war. The policy of full employment was rigorously pursued through relocation of industry and the creation of new towns. The wartime policy of subsidizing agriculture was extended. The economic stability of the country was underpinned by the international agreements reached at the Bretton Woods Conference*. The powers of the House of Lords were further reduced by the Parliament Act of 1949. Abroad the process of decolonization was begun with the granting of independence to India, Pakistan, and Burma in 1947 (in which Attlee played a key role), while British withdrawal from Palestine allowed the creation of Israel in 1948. In 1949, with the beginning of the Cold War, Britain helped to form NATO, Attlee's attempts to maintain friendly relations with the Soviet Union being increasingly unsuccessful. The second ministry, following an election in February 1950, gave Attlee a majority of only five. At home it faced fierce opposition over its austerity programme, while entry into the Korean War* necessitated increased rearmament. The Festival of Britain in the summer of 1951 encouraged a sense of national optimism in the future of the nation, but it did not prevent Labour from losing the election to the Conservatives in October. Attlee's health was poor at the time, but he remained leader of the opposition until succeeded by Hugh Gaitskell* in 1955, when he went to the House of Lords.

Auchinleck, Sir Claude John Eyre (1884–1981), British Field Marshal. Educated at Wellington College, he joined the Indian army, with whom he served in World War I. He commanded a British force in the unsuccessful operation at Narvik, in Norway (May–June 1940), and then became Commander-in-Chief in India. In July 1941 he exchanged commands with Wavell*. He mounted a major offensive in North Africa in November 1942, but a counter-offensive by Rommel* drove his forces back. He successfully halted the latter, however, in the first battle of El Alamein* (June–July 1942) before being replaced by General Alexander*. He returned to India as Commander-in-Chief, where he remained until 1947. His great abilities as a field commander were probably underrated by Churchill.

Aung San (1914–47), first Burmese Prime Minister 1947. When a student in Rangoon he achieved prominence in 1936 by organizing a student strike. He then joined the nationalist movement Dobama Asi-ayone, becoming secretary in 1938. After a period of secret military training under the Japanese, he returned to Burma in 1942 and became leader of the Japanese-sponsored Burma National Army, which actively assisted the Japanese advance through the country. He served as a minister in the Ba Maw puppet government, but became increasingly disillusioned, and defected to the Allies in the closing weeks of the Pacific War. He formed an Anti-Fascist People's Freedom League and led a post-war Council of Ministers. In January 1947 he negotiated in London a promise of full self-government. Denounced by the Communists, his AFPFL won the election of April 1947; but on 19 July 1947 he and six colleagues were assassinated at the behest of a political rival U Saw, who was later executed.

Auriol, Vincent (1884–1966), President of France 1947–53. Educated in Paris, he first entered politics in 1914 as a Socialist Deputy. He served as Minister of Finance (1936) and Minister of Justice (1938). During World War II he spent two years in internment before escaping to join General de Gaulle* and the Free French* in Britain. After liberation he served in de Gaulle's brief government 1945–6 and then played an active part in the formation of the Fourth Republic*. He was elected its first President, serving as a non-partisan statesman at a time of great political volatility in France.

Australia, Commonwealth of, a federation of six states (New South Wales, Victoria, Queensland, Tasmania, South Australia, Western Australia), together with the Northern Territory* and the Capital Territory. It had been inaugurated on 1 January 1901, with powers distributed between the Commonwealth and state governments, with the Crown, through its representative the Governor-General, retaining (until 1931) responsibility for defence and foreign affairs. The federal Parliament was to consist of a House of Representatives, with seats allocated to states in proportion to population, and a Senate with ten senators to each state and two to each Territory. The state Parliaments were to be bicameral except for Queensland, which opted for a single house. The Northern Territory was transferred from South Australia to the Commonwealth government in 1911, in which year land was transferred from New South Wales for the creation of the Australian Capital Territory of Canberra; this included Jervis Bay on the east coast, although the projected port and railway were never built. The Commonwealth Parliament continued to sit in Melbourne until 1927, when it was transferred to the new city of Canberra. In 1914 Andrew Fisher* was Prime Minister. As part of the British Empire, Australia found itself at war with Germany in August of that year. Patriotic response to Britain's needs was immense and more than 300,000 volunteers served in the Middle East (notably Gallipoli*) and on the Western Front. Fisher's successor Prime Minister Hughes* argued the case for conscription, but this was twice rejected by referendum (October 1916 and December 1917). As with Canada, Australia emerged from the war as virtually an independent nation, signing the peace treaties, becoming a member of the League of Nations, and being granted the 'mandate'* over German New Guinea and the Solomon Islands. The 1920s saw increased immigration from Britain, but in the early 1930s resentment at steadily increasing powers of the Commonwealth government, together with a drop in wheat and wool prices due to the Great Depression*, led to talk of secession, Western Australia petitioning the King in 1933. It was, however, ruled that this would only be allowed if supported by a referendum of all Australian people, and there was no referendum. During this decade reserves were established for the Aborigine* peoples, whose plight now began to be recognized. Although prices for gold, wool, wheat, and meat plummeted during the Depression, steel production boomed and manufacturing industries expanded. Australia declared war against Germany in September 1939 on the same day as Britain and France. Contingents were again sent to fight in the Mediterranean in the North African campaigns, and to Malaya; but as the Japanese threat extended Australia concentrated on its own defence, fought mostly in the jungles of New Guinea. The war greatly stimulated an expansion of industry, for which assisted immigration of 'New Australians' was needed in the decade following. Australia sent troops to fight in Korea, and, more unpopularly, in Vietnam, and strategic links with the USA became stronger, both the ANZUS* and SEATO* pacts being supported, as well as the Colombo Plan*. As trade with Europe declined, particularly after Britain's accession to the EEC (1972), so that with Asia expanded, particularly with Japan. Since World War II, in spite of massive development and diversification—in manufacturing industry, mining, hydroelectric installations, wheat cultivation—the economy has continued to suffer from fluctuations in world commodity prices. Until 1949 Labor* alternated in federal politics with various coalitions formed by Liberals, Nationalists, the Country Party*, and the United Australia Party. Since then it has alternated with the Liberal Party created by Robert Menzies*.

Austria. During World War I the Austrian Imperial Army came increasingly under German control. Defeat and revolution destroyed the Austro-Hungarian monarchy in 1918, and the first Austrian Republic was proclaimed on 12 November, a rump of the former state. A Constituent Assembly at once voted to join Germany, but this was forbidden by the Treaty of Versailles*. A vigorous socialist party developed in Vienna, where there were riots in 1927. In 1931 Austria's chief bank Kredit-Anstalt collapsed, and virtual civil war, socialists versus Fascists, followed. In 1934 German-backed Nazis murdered Chancellor Dollfuss* and attempted a coup. They were more successful in 1938, when Hitler's troops invaded and Anschluss* was declared. In 1945 Austria was invaded by Soviet troops and then divided into separate occupation zones, each controlled by an Allied power. In December 1945 the second Republic was formed, which gained full sovereignty in 1955 with the signing of the Austrian State Treaty*. It has remained neutral, democratic, and increasingly prosperous under a series of moderate socialist regimes. In 1986 a controversy over the election of President Kurt Waldheim* centred around anti-Semitism*, past and present. In 1989 Austria applied to join the European Community.

Austrian State Treaty (15 May 1955). The treaty, signed by the Soviet Union, the USA, Britain, and France, formerly recognized the second Austrian Republic and agreed that occupation forces would withdraw within five months. Western hopes that it signalled a softening of Soviet policy of an 'iron curtain' across Central Europe proved unfounded. The treaty prohibited the possession of major offensive weapons and required Austria to pay reparations to the Soviet Union, as well as to give assurances that it would ally itself with neither East nor West Germany, nor restore the Habsburgs.

Austro-Hungarian Empire (Dual Monarchy). It was established in 1867 by the *Augsleich* (Compromise), by which both Austria and Hungary became autonomous states but under a common sovereign, the ruling Habsburg. Each had its own Parliament for internal affairs, but foreign policy, war, and finance were to be decided by common ministers. As the Vienna bureaucracy grew, so did the dominance of Austria. The system came under increasing pressure from the subject nations, Croatians, Serbs, Slovaks, Romanians, Czechs, Poles. The Czechs from Bohemia-Moravia found a potent advocate in Tómaš Masaryk*. Bosnia and Hercegovina were formally annexed in 1908. It was a strong pro-Serbian nationalism within Bosnia that directly precipitated the outbreak of war in 1914. After the death of Francis Joseph* in 1916, his successor Charles I* promised constitutional reforms. The Allies gave their support to the emergent nations in the Fourteen Points* of January 1918, and the Empire was finally dissolved by the Treaties of St-Germain* and Trianon* at the Paris Peace Conference, the Emperor having been deposed in April 1919.

Automation, automatic control and performance of mechanical and manufacturing processes. During the 19th century a number of machines, such as looms and lathes, became increasingly self-regulating. At the same time transfer-machines were developed, whereby a series of machine-tools, each doing one job automatically, was linked into a continuous production line. Early control devices were pneumatic or hydraulic, but in the early 20th century, with the development of electrical devices and time-switches, more processes became automatically controlled, and a number of basic industries, such as oil-refining, chemicals, and food-processing, became increasingly automated. The development of computers after World War II resulted in more sophisticated automation in manufacturing industries, for example motor-car production, and resulted in work-force reductions. From the 1960s onwards the computerization of banking and office procedures greatly increased the speed of information processing.

Automobile industry. Although the motor car was invented in Europe, it was in the USA that its production grew fastest. The Olds Motor Vehicle company, formed in 1897, was producing 5,000 cars a year by 1904, while Henry Ford*, who formed his company in 1903, developed mass-production techniques from 1910 for the Model T, 15 million of which were built by 1927. In Britain Herbert Austin began car production in 1905 and William Morris (later Lord Nuffield*) in 1912. In Germany, where Daimler and Benz made quality vehicles (merging in 1926), the US General Motors* company took over the biggest company, Opel, in 1929, but it was challenged in the 1930s by Volkswagen's production of a 'people's car'. In France three companies emerged, which were to survive foreign competition into the 1980s, Renault, Peugeot, and Citroën. In Italy Fiat was founded in 1898 and retains 90 per cent of the Italian market, while the racing car Ferrari has dominated race-tracks through the century. Both world wars stimulated vehicle production, but since World War II the industry has become concentrated in the hands of the US giants Ford and General Motors, German, French, and Italian companies, and the hugely successful Japanese companies Toyota and Nissan. After a series of mergers and collapses, the British industry steadily shrank from 1960. By 1990 no independent British manufacturer survived, after Austin/Rover was absorbed by British Aerospace, but multinationals Ford, Peugeot, General Motors, Toyota all still assembled vehicles in Britain. The basic concept and design of this 'horseless carriage' altered remarkably little through the century. When it began, petroleum was a cheap fuel, so that alternative power systems, for example the electric battery, were never exploited. Automobile assembly-lines became increasingly automated from 1945 onwards, and this very factor made fundamental design changes unprofitable. As an industry it has relied on a wide range of components, manufacture and assembly often taking place thousands of miles apart. By 1990 in the USA one in every six businesses was dependent in one way or another on the industry, it being the largest consumer of steel, plate-glass, rubber, plastic, and paint. Its product was consuming the largest quantities of petroleum and its emissions polluting the atmosphere to a crisis level.

Ávila Camacho, Manuel (1897–1955), President of Mexico 1940–6. As a young member of the Partido de la Revolución Mexicana, he joined the army of Carranza* in 1914 and rose rapidly through the ranks. In the 1930s he was Minister of Defence and was elected President in 1940. A moderate statesman, he aimed to consolidate the achievements of the Mexican Revolution*, to rebuild relationships with the USA, and to achieve reconciliation with the Catholic Church. Succeeding in all these aims, his regime saw advances in social welfare, health care, and school construction, together with some land reform. In 1942 his government declared war against the Axis powers*. After his retirement he continued as an important Mexican political figure.

Awami League, a political party in East Pakistan (Bangladesh). It was founded by H. S. Suhrawardy and Mujibur Rahman* in 1952 as the Jinnah Awami Muslim League, although the movement behind it had existed informally before that date. It was renamed the Awami League under pressure from the East Bengali leader Maulana Abdul Hamid Bhashani, who left the party in 1957 to form the National Awami Party. During the 1960s the Awami League grew rapidly under Sheikh Mujibur Rahman, who had succeeded Suhrawardy as leader. In 1970 it completely dominated the election results in East Pakistan, which became Bangladesh in December 1971. In August 1975 it was disbanded, together with other political parties. It was later re-formed and became the largest opposition party in Bangladesh.

Awolowo, Obafemi (1909–87), Nigerian statesman. Son of a Yoruba* peasant, he was a newspaper reporter and trade union leader until getting to London in 1944, where he studied law. He was called to the Bar, and then returned to Nigeria to found the Action Group to preserve Yoruba culture. He became Premier of Western Nigeria in 1954. In 1963 he was imprisoned for allegedly conspiring to overthrow the federal government, but released in 1966. He supported General Gowon's* government in the Biafran War* against the Ibo*. In 1978 he founded the Nigerian Unity Party and stood for President. He was defeated and his party banned in 1983.

Axis powers, the term used in an agreement (October 1936) between Hitler and Mussolini proclaiming the creation of a Berlin–Rome Axis 'round which all European states can also assemble'. Japan joined on signing the Anti-Comintern Pact*

(1936). A full military and political alliance between Italy and Germany (the Pact of Steel) followed in 1939, and this became a Tripartite pact in 1940. In 1941 Hungary, Romania, and Bulgaria joined, as well as the Nazi-created states of Croatia and Slovakia.

Ayub Khan, Muhammad (1907–74), President of Pakistan 1960–9. Born on the North-West Frontier, son of a British NCO, he was educated at Aligarh University and the Royal Military College Sandhurst, being commissioned into the British Indian army in 1928. After distinguished service in World War II, he was given command of the military forces of East Pakistan (Bangladesh) in 1947 and in 1951 became the first Commander-in-Chief of the Pakistan army. As Minister of Defence 1954–6 he imposed severe repressive measures from October 1954 to August 1955. In October 1958 martial law was declared in Pakistan, and he was Chief Martial Law Administrator until 1960, when he was elected President, being re-elected in 1965. He pursued a policy of rapid economic growth and modest land reform. While allowing little opposition at national level, he encouraged what he called 'basic democracies' at the local level, introducing Pakistan's second constitution in 1962. In March 1968 he suffered a serious illness, and in October of that year there were student riots, provoked by imprisonment of opposition leaders. In March 1969 he was persuaded to resign, to be replaced by General Yahya Khan*.

Azaña, Manuel (1881–1940), President of the Spanish Republic 1936–9. A Spanish journalist and civil servant, in 1924 he founded a republican party and was imprisoned. In 1931, when Alfonso XIII* fled, he was released and was elected Prime Minister of the new Spanish Republic, succeeding Alcalá Zamora*. He was sympathetic to Catalan demands for autonomy, but his reforms antagonized too many; he was forced from office in 1934 and briefly imprisoned at a time of riots in Barcelona. In 1936 he was elected President, with a Popular Front government which soon found itself at war. He continued to hold office until 1939, when he fled to France.

Azerbaijan, an ancient Mongol land whose khanates were steadily conquered by both Russian and Persian armies in the 18th and 19th centuries. By 1914 it was the largest oil-producing area in the world, centred on Baku, which was conquered by the Red Army in 1920. A Soviet Republic was proclaimed, and it formed part of the Soviet Union from 1936. There is a large Muslim population in the republic, which feels akin to the Azerbaijani peoples of Iran. Its Armenian minority, belonging to the Christian Armenian Church, feels akin to the neighbouring Republic of Armenia*. In January 1990 there were serious disturbances in Baku, with an anti-Armenian pogrom* killing sixty Armenians. Pressure for more autonomy from Moscow developed and a state of emergency was declared. Azerbaijan then demanded the right to secede from the Soviet Union. A strike in Baku was broken by Soviet troops, but a tense situation remained into 1990.

Azhari, Ismail al- (1900–69), Sudanese statesman. He was one of a group of graduates of Gordon College, Khartoum, who in 1943 formed the first Sudanese political party, the Ashiqqa (brothers), demanding independence. His party won the elections of 1953 and he became the first Sudanese Prime Minister (1955). He lost his parliamentary majority in 1958 and was detained by General Abboud*. When the latter was deposed in 1964 he became President of the Supreme Council of the Sudan, but, following a new military coup in 1969, he was again placed under house arrest until he died.

Azikiwe, Benjamin Nnamdi (1904–), President of Nigeria 1963–6. An Ibo* (Igbo) born in Zunger in Northern Nigeria, he was educated at Lincoln College, Oxford, and then did postgraduate work at the University of Pennsylvania. He became a successful newspaper editor/owner and banker in both Nigeria and the Gold Coast. In 1944 he founded the Nigerian Youth Movement and in 1946 the National Council of Nigeria and the Cameroons (NCNC). His newspaper articles in the *West African Pilot*, signed Zik, exerted a powerful influence throughout the 1940s and 1950s, on emerging Nigerian nationalism. He held a number of top political posts before becoming the first Governor-General of independent Nigeria in 1960 and then President of the Republic in 1963. He was deposed in 1966 by a military coup, which ousted civilian government. Sympathetic to the Ibo cause, he at first supported the attempt to create an independent Biafra, but he deplored the war that developed and sought means to end it. He became leader of the Nigerian People's Party and was unsuccessful in a bid for presidency in 1979.

Baader–Meinhof Gang, byname for the West German anarchist terrorist group the Red Army Faction. Its leaders were Andreas Baader (1943–77) and Ulrike Meinhof (1934–76), and the gang set itself to oppose the capitalist organization of West German society and the presence of US forces. It engaged in murders, bombings, and kidnappings, two members being linked with Palestinian guerrillas who in 1976 hijacked an Air France plane, which was later rescued at Entebbe. Meinhof was arrested in 1972 and Baader in 1976. Their trials and deaths by suicide received considerable publicity. The gang continued its activities during the 1980s, forming a number of splinter cells. Claims that it received Soviet backing have not been substantiated.

Ba'athism, an Islamic ideology (resurrection) which emerged in the 1930s. It sought to synthesize a Marxist analysis of society with the concept of Arab unity—social, political, and cultural. It accepted Islam as the greatest achievement of Arab genius, but, unlike the Muslim Brotherhood* or later Islamic fundamentalism*, did not see it as the sole source of law. In 1943 the Ba'ath Arab Socialist (Renaissance) Party was founded by Michel Aflaq* and Salah al-Din al-Bitar. They regarded groupings of various Arab states as no more than branches of Pan-Arabism resulting from Western imperialism. From 1963 it formed the ruling party in Syria, and in Iraq from 1968. But in both countries it became no more than a cloak for military dominance, with intense hatred and rivalry developing between them.

Baden-Powell, Robert Stephenson Smyth, 1st Baron Baden-Powell of Gilwell (1857–1941), British soldier and founder of the Boy Scout movement. Son of an Oxford professor and godson of Robert Stephenson the railway engineer, he joined the army in 1876, serving in India, Afghanistan, and Africa. He was the successful defender of Mafeking in the Boer War, until relieved on 18 May 1900. He retired from the army in 1910, having in 1907 founded the Boy Scouts, following a successful boys' camp in Poole, Dorset. His knowledge of the skill of gaining information about hostile territory was fundamental to the teachings of the Scout movement, knowledge he had himself gained from his African scouts defending Mafeking. At the same time, training in self-reliance and a high code of moral conduct were to be the hallmarks of the movement. By the end of World War I the movement was developing on an international scale. With his sister Agnes he had founded the Girl Guide movement in 1910. As Chief Scout he travelled widely throughout the world and wrote numerous manuals. He died at his home in Kenya.

Badoglio, Pietro (1871–1956), Italian general and Prime Minister. A professional soldier, he fought at Adowa (1896), in Libya (1911), and distinguished himself at Caporetto (1917). By 1925 he was Chief of Staff. Governor of Libya 1929–33, he was sent by Mussolini in 1935 to rescue the faltering campaign in Ethiopia. He captured Addis Ababa and became Governor. He opposed Italian entry into the war in 1940 and resigned. In 1943, when leading Italians decided to depose Mussolini, he was chosen to lead a non-Fascist government. He made peace with the advancing Allies and declared war against Germany. Italian Resistance leaders refused to serve under him and in 1944 he retired.

Baghdad Pact, a treaty between Turkey and Iraq (February 1955), largely the achievement of Abd al-Ilāh* and at first aimed at mutual co-operation against left-wing militants. Later in the year Britain, Pakistan, and Iran signed the Pact, but there were riots in Amman when King Hussein proposed to join. At first the USA was reluctant to sign, and after the Suez Crisis* Britain's role was temporarily eclipsed. After Iraq withdrew in 1958 it became the Central Treaty Organization*.

Bahamas, the Commonwealth of the, an archipelago of some 700 islands (twenty-nine inhabited) running south-east from the coast of Florida, with 60 per cent of the population living on New Providence. It was a former British colonial territory, whose Parliament dates back to 1729. In the 1920s a degree of prosperity came to the islands during the US Prohibition*, and since World War II the tourist industry has greatly boosted the economy. Considerable racial tension built up in the 1930s, and the US civil rights movement* stimulated a demand by the Progressive Liberal Party (PLP) for an end to racial segregation. In 1964 internal self-government was granted, with the first elections under universal suffrage held in 1967. The Black-dominated PLP under Lynden Pindling won eighteen seats, as did the White-dominated UBP (United Bahamian Party), but Pindling formed a government, and in the next election in 1968 he won twenty-nine of the thirty-eight seats. Independence within the Commonwealth was granted in 1973, with the PLP maintaining its majority. During 1983–5 a Commission of Inquiry found widespread evidence of the illicit drug trade, with fifty-one suspects being indicted. The 1987 election was again won, however, by the PLP, with Sir Lynden Pindling remaining Prime Minister.

Bahrain, an independent state, consisting of a group of islands in the Persian Gulf traditionally under the Al Khalifa family. In 1914 it was under British protection and oil production was begun in 1934. In 1970 a UN Commission rejected claims on the islands from Iran and in August 1971 Sheikh Isa ibn Sulman al Khalifa declared independence, ending federal links with neighbouring emirates, joining the Arab League*, but maintaining a treaty of friendship with Britain. In 1975 tension between Shiite and Sunni communities resulted in the suspension of the National Assembly. During the Iran–Iraq War 1980–8 Bahrain resisted US pressure for the use of its bases, seeking to maintain strict neutrality. With oil reserves dwindling, petroleum-refining, aluminium smelting, banking, and other industries have been developed. In 1986 a causeway with Saudi Arabia was opened.

Balaguer, Joaquín (1907–   ), President of the Dominican Republic 1960–2, 1966–78, 1986–   . After studying law in Santo Domingo and Paris, he developed a successful practice and in 1932 entered politics, holding a number of government and diplomatic posts under the Trujillo* regime (1930–61). As Secretary of Education he greatly expanded education and library facilities. In 1960 he became titular President under the dictator Raphael Trujillo. When the latter was assassinated he tried to confiscate the vast Trujillo fortunes and to liberalize the regime, but after a year he felt obliged to go into exile. He was brought back on US intervention in the Republic in 1965 and was elected President. Having close links with the USA and the business community, he achieved steady economic growth and modest gains in social welfare. Re-elected in 1970 and 1974, his later governments were marred by political terrorism, fraud, and assassinations. He was, however, re-elected a fourth time in 1986, although now nearly totally blind, and yet again in May 1990.

Baldwin, Stanley, 1st Earl Baldwin of Bewdley (1867–1947), British Prime Minister 1923–4, 1924–9, 1935–7. Son of a wealthy Midlands ironmaster, after Cambridge he worked in his father's firm for twenty years before succeeding him as a Conservative Member of Parliament 1908–37. He was a junior member of Lloyd George's* wartime government and served as Financial Secretary to the Treasury and then President of the Board of Trade in the government of 1918–22. In October 1922 he was one of the leaders of the rebellion against Lloyd George and became Chancellor of the Exchequer under Bonar Law*. He was chosen as Prime Minister, in preference to Curzon*, when Law resigned from ill health. He lost the 1923 election over an attempt to introduce tariffs, but returned to office in October 1924. His premiership was marked by the return to the gold standard*, the General Strike* followed by the Trade Disputes Act of 1927, and a programme of social legislation introduced by Neville Chamberlain*. He lost the election of 1929, in spite of having given the 'flappers' (women aged 21–30) the vote the year before, but he served under Ramsay MacDonald* as Lord President of the Council in the National government* formed in 1931, succeeding him as Prime Minister in 1935. His last ministry witnessed the Abdication Crisis*, which he handled skilfully. In 1935 he at first approved of the Hoare*–Laval* proposal to allow Italy to annex part of Ethiopia, but later withdrew his support, while Italy went on to annex the whole country. As international relations deteriorated, with the German reoccupation of the Rhineland* and the outbreak of the Spanish Civil War*, he quietly initiated plans for rearmament, while accepting that public opinion would be reluctant as yet to accept these. He resigned the premiership in 1937 and moved to the House of Lords. With his experience in industry he understood and sympathized with the trade union movement more than many of his colleagues. Although many criticized his 'indolence' and failure to act decisively in foreign affairs, he has been described as the 'ablest politician of his day', fully 'master of events' at a time of crisis.

Balewa, Alhaji Sir Abubakar Tafawa (1912–66), Prime Minister of Nigeria 1957–66. From the Hausa* people, he became a teacher by profession, attending the London Institute of Education 1945–6. After this he was elected to the Northern Region House of Assembly. In 1947 he became a member of the Central Legislative Council and in 1952 joined the federal government. The first Prime Minister of the Federation of Nigeria, he retained his office when the country became independent in 1960. He was a founder and deputy president-general of the country's largest political party, the Northern People's Congress (NPC). He was killed in the army coup of 1966.

Balfour, Arthur James, 1st Earl Balfour (1848–1930), British Prime Minister 1902–5. Born in East Lothian, a nephew of Lord Salisbury, after Eton and Trinity College, Cambridge, he entered the House of Commons as a Conservative at the age of 26, becoming private secretary to his uncle, then Foreign Secretary. In 1887 he became Chief Secretary of Ireland, when his opposition to Home Rule* earned him the nickname 'Bloody Balfour'. He succeeded his uncle as Prime Minister in 1902. His government passed an Education Act establishing a national system of secondary education, but became increasingly split over Joseph Chamberlain's arguments for tariff reform*. He created a Committee of Imperial Defence and helped to establish the *Entente Cordiale* with France in 1904. The Conservatives were crushingly defeated in the election of January 1906 and Balfour then used the House of Lords, described by Lloyd George as 'Mr Balfour's Poodle', to attempt to block contentious Liberal legislation. He resigned the leadership of the Conservative Party to Bonar Law* in 1911, concentrating on writing books on philosophy. He returned to office in 1916 as Foreign Secretary in Lloyd George's wartime government, and is associated with the Balfour Declaration* of 1917, promising the Jews a national home in Palestine. Under Lloyd George* he was a leading British representative at the Paris Peace Conference* in 1919, a strong supporter of the League of Nations, and a British representative at the Washington Conference* 1921–2. As Lord President of the Council 1925–9 he was a strong supporter of the concept of dominion status, and the Statute of Westminster* of 1931 was to owe much to his inspiration.

Balfour Declaration, a letter from the British Foreign Secretary Arthur Balfour* to Lord Rothschild, a leading British Zionist*, dated 2 November 1917. In it he pledged Britain's support for the establishment of a Jewish national home in Palestine, 'without prejudice to the civil and religious rights of the non-Jewish people' (Arab) in Palestine, or to the status of Jews in any other country. The Declaration was confirmed by the Allies and formed the basis of the mandate for Palestine assigned to Britain by the League of Nations in 1920. At the same time Britain was making promises to the leaders of the Arab Revolt* that they would be recognized as rulers of Palestine!

Balkan Pact (1934). After World War I frontiers were redrawn throughout the Balkan peninsula and democratic regimes encouraged. These failed and authoritarian regimes emerged in all Balkan countries. In 1933 King Alexander* of Yugoslavia tried to link Romania, Greece, Turkey, Bulgaria, and Yugoslavia into a defensive pact. Bulgaria refused to take part because of a quarrel with Greece over Macedonia. The Pact established a permanent council of the 'Balkan *Entente*' with regular conferences of member states, but it fell apart in 1940 when members reacted in different ways to Hitler's Germany. There was a second short-lived pact in 1954, which quickly collapsed over the Cyprus* issue.

Banda, Hastings Kanuzu (1906–), President of Malawi 1966– . Son of a peasant in Nyasaland, he attended a Church of Scotland mission school until aged 12, when he walked to the Rand goldfield, where he became a clerk. With his savings he went to the USA to study political science and medicine. He continued his studies in Edinburgh and then worked as a GP in Liverpool, North Shields, and London; here his home became a centre for African exiles. In 1953 he went to Kumasi (Ghana) to work for five years, before returning to his homeland. Here he was arrested (1958) for an alleged plot and imprisoned. Although cleared of any plot by Justice Devlin, he was kept in prison for a year; when he was released he travelled to London for a constitutional conference. With self-government granted (1963) he became Prime Minister. In 1964, with independence, Nyasaland changed its name to Malawi. In 1966 it became a republic with Banda elected President, becoming President for life in 1971. His cautious policies towards South Africa caused him to be a somewhat isolated figure in the Commonwealth, and to be resented by some younger Malawian politicians. While immensely popular among his people, who call him 'the Little Messiah', he has not hesitated to use authoritarian methods in government.

Bandaranaike, Sirimavo Ratwatte Dias (1916–), Prime Minister of Sri Lanka 1960–5, 1970–7. Wife of S. W. R. D. Bandaranaike*, she succeeded him as Prime Minister when he was murdered, being the world's first woman Prime Minister. A strong supporter of Sinhalese culture and of Buddhism, she retained her husband's policies of neutralism and belief in socialism. In 1977 her party won only eight seats in the election for the National Assembly, when ethnic tension between Sinhalese and Tamil was worsening and economic distress increasing. She failed in a bid to be elected as President of Sri Lanka in December 1988 against Ranasinghe Premadasa.

Bandaranaike, Solomon West Ridgeway Dias (1899–1959), Prime Minister of Sri Lanka 1956–9. Educated in Colombo and Oxford, he qualified as a lawyer before entering politics. He presided over the Ceylon National Congress Movement during the 1930s, having founded the Maha Sinhala Party. In 1931 he was elected to the State Council of Ceylon and after independence became Minister of Health 1948–51, organizing an effective campaign against malaria. In 1952 he founded the Sinhalese Socialist Sri Lanka Freedom Party, which was the leading partner in the coalition which won the 1956 election, attracting both left-wing and Buddhist support. As Prime Minister he pursued a policy of promoting the Sinhalese language, Buddhism, nationalization of major assets, and neutrality, closing all British naval and military bases. His policy alienated Tamils and he was assassinated by a dissident Buddhist monk. His widow Mrs Sirimavo Bandaranaike* succeeded him as Prime Minister.

Bandung Conference (1955), an Afro-Asian conference. On 17 April 1955 President Sukarno* of Indonesia hosted a conference of representatives from twenty-nine Asian and African countries, together with observers representing Greek Cypriots, Black Americans, and the African National Congress. Its purpose was to form a non-aligned bloc opposed to both 'colonialism' and the 'imperialism' of the superpowers. The five principles of non-aggression, respect for sovereignty, non-interference in internal affairs, equality, and peaceful coexistence were adopted. Although both India and China were keen participants, the subsequent deterioration of relations between them weakened the effectiveness of the Conference. Yugoslavia later became a champion of the Non-aligned movement*, which subsequently broadened beyond its Afro-Asian origins. A series of later conferences followed, beginning in Belgrade in 1961.

Bangladesh. Formerly East Pakistan, on the Bay of Bengal, it was established in 1971, following nearly twenty years of agitation spearheaded by the Awami League*. There had been resentment among Bengali Pakistanis as to the status of Bengali languages, as well as a feeling that they were being economically exploited by Karachi. Following an electoral victory by the Awami League in December 1970, demands for greater autonomy were rejected by the Pakistani military government in March 1971. A Bangladesh government-in-exile was formed in Calcutta and civil war followed, with India giving support to the East Pakistani guerrillas, the Mukti Bahini. Nearly 10 million refugees poured into India, famine and cholera spread, and there was a brief war 3–16 December between Indian and Pakistani forces, before the latter surrendered. Bangladesh was formally recognized, with Mujibur Rahman* first Prime Minister. He struggled to establish a parliamentary democracy, but in January 1975 took dictatorial powers. This precipitated a military coup, and he was murdered. After a period of political chaos Major-General Zia Rahman proclaimed himself President, being later confirmed in an election. Zia was murdered in 1981 and during the 1980s the country suffered from floods and famine, but also from ethnic riots between Islamic fundamentalists* and Hindus. Martial law was imposed in 1982 by General Hussain Mohammad Ershad. It lasted until November 1986, when civilian rule was restored and elections held, while General Ershad retained the office of President. By 1987

Bangladesh was estimated to be the world's poorest country following terrible floods. In December President Ershad declared a new state of emergency, following extensive unrest. This lasted until April 1988. In March 1989 a fourteen-year conflict in the Chittagong Hills was ended by granting greater autonomy to the hill tribes, but in November 1990 Ershad declared a third state of emergency. This was to be followed by his deposition on 4 December. Chief Justice Shehabuddin Ahmed became acting President until elections could be held in February 1991.

Bantustan, a term used to describe a homeland reserved for Black Africans in South Africa. Following the Bantu Self-Government Act of 1959 ten homelands were created: Bophuthatswana, Ciskei, Gazankulu, KwaNdebele, KwaZulu, Lebowa, Qwa Qwa, Swazi, Transkei, and Venda. The Bantu Homelands Constitution Act of 1971 envisaged eventual independence, and this was granted as follows: Transkei in 1976 (Xhosa* people); Bophuthatswana in 1977 (Tswana* people); Venda in 1979 (Vhavenda people); Ciskei in 1981 (Xhosa people). The homelands are not unitary, and because their citizens live in the Republic they cannot be recognized in international law. Ciskei and Transkei are deep rivals and do not recognize each other. KwaZulu is the power-base of the Zulu* Inkatha movement. During the later 1980s there were military coups and counter-coups in all four 'independent' homelands. They all remain chronically underdeveloped and deeply resented by Black opinion.

Bao Dai (1913–), Emperor of Vietnam 1926–45. His initial aims to reform Vietnam did not receive support from the French authorities, and during the 1930s he steadily lost influence. During World War II he was accepted by the Japanese as an essentially powerless ruler. On 20 August 1945 he was obliged to abdicate by the VietMinh*, becoming Citizen Vinh Thuy and ending a thousand-year monarchy. In 1946 he fled to Hong Kong, where he gained a reputation as a playboy. In 1949 he was invited to return to Saigon as head of state within the French Union; but Vietnam was by then split by the French Indo-China War*. At its conclusion, following the partition at the Geneva Conference, he was for a short while President of South Vietnam before being deposed (1955) by Ngo Dinh Diem*. He retired to live in France.

Baptist Church, an evangelical Protestant Church, stressing adult baptism after a personal confession of faith, and the autonomy of local congregations. It originated in the 16th-century Reformation in Germany, where it suffered persecution. It spread rapidly in both Britain and the USA during the 19th century, particularly in the towns. In 1845 the Southern Baptist Convention split from the National Convention in the USA. Black Baptist congregations grew after the Civil War and contributed significantly to Black culture and to the civil rights movement*. Many of their ministers, such as Martin Luther King*, were at the forefront of that struggle during the 1950s and 1960s. At the same time the US Southern Baptists remain highly conservative, being anti-abortion and anti-evolution as well as a strong influence on education. As a result of missionary work there are strong Baptist Churches in Africa, Asia, and in the Soviet Union, where it is the largest Protestant group and where it was persecuted through much of the 20th century.

Barbados, the most easterly island in the Caribbean Sea. A British colony from 1627, with a Parliament dating from 1639, its economy for long depended on the sugar industry, for which African slaves had been imported. Partly as a result of the decline of the industry there were widespread riots and disorder in 1937–8, with bitter racial antagonism emerging. The British Colonial Welfare and Development Organization tried to diversify its economy, and after World War II universal adult suffrage was introduced, the Barbados Labour Party (BLP) winning the election of 1951. A member of the West Indian Federation*, it was granted full self-government in 1961, when an opposition party, the Democratic Labour Party (DLP), came to power. After the break-up of the Federation full independence within the Commonwealth was granted in 1966. The tourist industry and emigration to the United Kingdom both helped the economy. Since independence the BLP and DLP have alternated in power, both mildly socialist. The 1987 election was won by the DLP, with Erskine Sandiford Prime Minister. Barbados has been an enthusiastic member of CARICOM*.

Barbarossa, code-name for the largest military campaign in history—the German invasion of the Soviet Union on 22 June 1941. Plans were drafted in November 1940. The attack was to be three-pronged, north against Leningrad under von Leeb; centre against Moscow under von Bock; and south against the Crimea and on to oil-rich Baku under von Rundstedt*, with overall command under Field Marshal von Brauchitsch*. Early blitzkrieg* tactics were successful; the Baltic states were captured, the Ukraine overrun, and the industrial basin of the Don won. Thousands of prisoners were taken. Only von Rundstedt, however, achieved his early aims of capturing the Black Sea coast. Seventy-nine German divisions, aided by Romanian and Finnish forces, were involved. Hitler's personal intervention in the campaign was disruptive, but Russian resistance in the sieges of Leningrad and Moscow was decisive in preventing victory before the winter of 1941. In December Hitler replaced von Brauchitsch as personal commander in the field and the campaign of 1942 ended in the disaster of Stalingrad*.

Barton, Sir Edmund (1849–1920), Prime Minister of Australia 1901–3. Having held high political offices in New South Wales, following a successful career in the law, he became Australia's first Commonwealth Prime Minister. With Alfred Deakin* and Sir Henry Parkes he had been an architect of the federation of 1901. His ministry aimed to protect Australian industry and to preserve a White Australia. He became a High Court judge in 1903, seeking in his judgements whenever possible to preserve state autonomy. He came to England in 1915 and was made a Privy Councillor.

Baruch, Bernard Mannes (1870–1965), US financier and political adviser. From a poor Jewish family, he made a personal fortune on the New York Stock Exchange by 1914. Becoming the respected adviser of presidents from Wilson to Eisenhower, he preferred to be an *éminence grise* than to run for elective office. In World War I he served under President Wilson on the Council of National Defense and was the successful chairman of the War Industries Board 1917–19, exercising enormous power over the US wartime economy. Between world wars he was a member of various presidential conferences on capital, agriculture, transportation, and labour. During World War II he was special adviser to President Roosevelt* on manpower mobilization and post-war planning. In 1946 he was appointed to the UN Atomic Energy Commission, which proposed a World Atomic Authority with full control over the manufacture of atomic bombs throughout the world; a proposal rejected by the Soviet Union. He was a close friend of Winston Churchill from 1929 onwards.

Basques, a race of some three-quarters of a million people speaking a pre-Indo-European language and divided between France and Spain on the west Pyrenees. The main Basque city is Bilbao, which grew rapidly in the early 20th century. Under José Aguirre y Lecube (1904–60) a separatist movement developed, culminating in the declaration of a Basque Republic (Euzkadi) in Guernica* in October 1936.

Franco launched an air and land attack in April 1937, destroying Guernica and capturing Bilbao. Since 1975 Basque demands for independence have been led by a terrorist organization (EETA). A Basque Parliament was created in 1980, but terrorist activity continued.

Batistá (y Zaldívar), **Fulgencio** (1901–73), Dictator of Cuba 1933–8; President 1940–4, 1952–8. Born in Cuba, he joined the army and first came to prominence in 1933 when, as a sergeant, he led a successful revolt against the US-backed President Carlos Manuel de Céspedes. He promoted himself to colonel (later to general) and proceeded to rule through a sequence of puppet presidents. Although a neo-Fascist, he did allow a new constitution to be devised in 1938, and in 1940 was himself elected President. He established a strong and reasonably efficient, if corrupt, government, but retired in 1944. When he returned to power in 1952 he established a one-party dictatorship, increasingly using terrorist methods against his critics. He amassed vast fortunes for himself and his associates. The excesses of his second term abetted Castro's* revolution, which drove him from power on 31 December 1958. He fled to the Dominican Republic.

Batlle y Ordónez, José (1856–1929), President of Uruguay 1903–7, 1911–15. Son of an earlier Uruguayan President, during his terms of office he initiated legislation to increase public welfare. He believed that the Swiss Constitution, with its Bundesrat or federal council, was well suited to his country's needs, and during his second term he tried to have the office of President eliminated altogether. His political opponents compromised, agreeing to an executive branch in which power was shared between a President and a nine-man council. This decentralization of power placed Uruguay on a unique path in the 20th century. He himself served as President of the new National Council of Administration in 1926.

Bauhaus, the, a School of Art and Design founded at Weimar in 1919. It moved to Dessau in 1925, where it survived until closed by the Nazis in 1933. Headed by Walter Gropius 1919–28, its founding manifesto called for the unity of the creative arts, under the umbrella of architecture, based on a full awareness of the nature of materials and of perceptual relationships. The simple geometric forms of the Bauhaus had an immense influence on all 20th-century industrial design, as well as on its architecture.

Bavaria, formerly a kingdom within the Second German Empire. Ludwig III came to the throne in 1913, to be deposed at the end of World War I, when Bavaria became a republic. A Communist government was set up by the Spartakist* Kurt Eisner, and after the latter's assassination, a short-lived communist Soviet republic was proclaimed (1920–1). On its collapse Bavaria became the centre of right-wing politics, with the first attempted Nazi revolution in the Munich 'beer-hall' *putsch** (1923). In 1948 Bavaria became a state within the Federal Republic of Germany, minus the Rhenish Palatinate. It was for long the political power-base of the right-wing post-war politician Franz Josef Strauss*, leader of the Christian Social Union.

Bayar, Mahmud Celâl (1884–1986), third President of the Turkish Republic 1950–60. Following an early career in banking he joined the Young Turk* movement in 1908 and in 1919 the resistance movement led by Mustafa Kemal*. In 1922 he became a member of the latter's first government. During the 1930s, as Minister of the Economy and Prime Minister he organized successful development of many state industries. In 1945 he formed a new Democratic Party, which won the elections of 1950, when he became President. He was re-elected 1954 and 1957, now advocating development of private enterprise. Arrested in 1960, following a military coup, he was condemned to life imprisonment for alleged 'crimes against the state', but released in 1964.

Bay of Pigs incident. On 17 April 1961 a small force of CIA*-trained Cuban exiles from Miami landed from US ships, in Cochinos Bay, Cuba in an attempt to overthrow the revolutionary regime of Fidel Castro*, convinced that the people would rise to greet them. In fact the invaders were swiftly rounded up by Cuban troops, and the incident was a grave blow to the prestige of the USA and of President Kennedy*. It strengthened the Castro regime and tightened Cuba's links with the Soviet Union. Another result was that President Kennedy tried to tighten presidential control over the activities of the CIA.

Beatles, The, perhaps the most famous and enduring of all pop groups. The quartet of John Lennon, Paul McCartney, Ringo Starr, and George Harrison, working-class boys from Liverpool, dominated pop music from 1962 until 1970 on both sides of the Atlantic. Some of their songs, both lyrics and music, have been recognized as musical classics, for example 'Sergeant Pepper's Lonely Hearts Club Band' (1967) and 'Let it Be' (1970). The hysterical adulation of their teenage fans (Beatlemania) set a pattern of behaviour towards later pop groups.

Beatty, David, 1st Earl (1871–1936), British admiral. From an Irish family, he joined the Royal Navy as a cadet in 1884, and earned rapid promotion for his daring exploits in campaigns in Egypt and in the Boxer Rebellion in China in 1900. In 1913 Winston Churchill, First Lord of the Admiralty, secured for him command of the navy's battle-cruiser squadron. Beatty gained minor victories over German cruisers off Heligoland in 1914 and the Dogger Bank in 1915. He played a major role in the Battle of Jutland* in 1916, after which he was appointed Commander-in-Chief of the British Grand Fleet, when he received the surrender of the German Grand Fleet at Scapa Flow in November 1918. He was First Sea Lord 1919–27, with responsibility for reducing the navy to peacetime strength. He was the subject of considerable controversy over his conduct at the Battle of Jutland, being accused of rashness, incompetence, and failure to train his command adequately; but he had supporters who took exactly the opposite view.

Beck, Josef (1894–1944), Polish politician. Born in Warsaw, he joined the Polish Legion which was formed in 1914 by Josef Pilsudski* against the Germans, but which in 1919–20 then fought against the Bolsheviks in Russia. In 1932 he became Foreign Minister, when Pilsudski came to power, and concluded the non-aggression pact with Hitler in 1934. He remained in office under Smigly-Ridz* and on 25 August 1939 signed a military alliance with Britain, which brought the latter into war on 3 September. He fled to Romania, where he was interned until his death.

Begin, Menachem (1913–), Prime Minister of Israel 1977–83. Born and educated in Poland, he spent the years 1939–41 in a Russian labour camp in Siberia, before managing to emigrate to Palestine. Here he joined the Polish army in exile and became leader (1944) of the militant movement Irgun*, being involved in the destruction of the King David Hotel. On the creation of Israel he was elected leader of the Herut (Freedom) Party and a member of the Knesset. In 1967 he joined a National Unity government and in 1970 helped to form the right-wing coalition the Likud*, becoming Prime Minister in 1977. He negotiated the Camp David* Accord with President Sadat in 1978 for which, although a one-time terrorist, he was awarded the Nobel* Peace Prize. He retired into private life in 1983.

Beirut, capital of Lebanon. By 1914 it was one of the wealthiest cities of the Middle East, under what was effectively a French protectorate. It contained two missionary universities and was the intellectual centre of rising Arab nationalism. Occupied by the Allies in 1918, it was established by the French as the capital of the new state of Greater Lebanon. Under French mandate 1920–46 it grew rapidly as a centre for Arab banking and economic activity, but Christian–Muslim tensions steadily built up. In the 1975–6 Lebanese civil war large parts of the city were destroyed, leaving a vacant area between Christian East Beirut and Muslim West Beirut. There was further destruction in 1982 and again in 1984, by which time West Beirut had become a self-governing enclave separated from East Beirut by a line of destruction, the 'Green Line'. In March 1989 General Aoun from Christian East Beirut launched an all-out offensive, and intense shelling lasted through the summer. Aoun rejected the Taif Accord*, and fighting continued through early 1990 between his supporters and the Phalangist Lebanese Front. In October 1990 Syrian forces imposed the Accord and Aoun fled to France. The Green Line was demolished, but from being a rich cosmopolitan city Beirut had become little more than a dangerous ruin.

Belaúnde, Terry Fernando (1912–), President of Peru 1963–8, 1980–5. From an aristocratic Peruvian family, he trained as an architect in Mexico until 1936, when he returned to Peru to build a distinguished practice before going into politics. He was elected to the Chamber of Deputies 1945–8, when his father was Prime Minister. In 1956 he helped to found the moderate Acción Popular Party and in 1963 was elected President, with the support of the Christian Democrats and the backing of the army. His opponent was Haya de la Torre*, candidate for APRA*. His first term of office is remembered for its social, educational, and land reforms, as well as for industrial development and the construction of a vast highway system across the Andes. He was a strong supporter of the US Alliance for Progress*, but he lost popularity for granting oil rights to US companies and for continued high inflation. He was deposed by the army and fled to the USA. He returned briefly to Peru in 1970 but was deported. He returned again in 1976 and was again President 1980–5. During his second term inflation grew steadily worse, and he was unable to counter the terrorist activities of Sendero Luminoso*.

Belgium. In the early 20th century, when the River Congo (Zaïre) was opened for trade, appalling atrocities against Africans were revealed, leading to the transfer of Leopold II's (1865–1909) Congo Free State to the Belgian Parliament. In August 1914, when German troops invaded Belgium, King Albert I (1909–34) invoked the 1839 international Treaty of Neutrality and called on Britain and France for aid. German troops occupied the country and Albert led the Belgian army on the Western Front. When Germany invaded again in 1940 Leopold III* (1934–51) at once surrendered, himself becoming a prisoner. However, a government-in-exile in London continued the war, organizing a strong Resistance movement. After the war Leopold was obliged to abdicate (1951) in favour of his son Baudouin (b. 1930). Independence was granted to the Congo in 1960, to be followed by the Congo Crisis*. The main challenge since the war has been to unite the Flemish-speaking northerners with the French-speaking Walloons of the south. In 1977 the Pact of Egmont, introduced by Prime Minister Leo Tindemans, recognized three semi-autonomous regions: Flemings in the north, Walloons in the south, and Brussels. The latter has continued to expand as the headquarters of the European Commission. There is also a small German-speaking enclave Eupen-Malmédy*, ceded to Belgium after World War I. Belgium joined the Benelux Customs Union* in 1948, from which was to develop the European Community*, and was a founding member of NATO in 1949.

Belize, a Central American country. The British settled there in the 17th century and it became the Crown Colony of British Honduras in 1882, subject to the jurisdiction of the Governor of Jamaica. Grudging acceptance by its neighbours led to treaties recognizing its boundaries, although Guatemala never fully gave up its claims. In 1964 the colony was granted complete internal self-government. It adopted the name Belize in 1973, and in 1981 became an independent state within the Commonwealth. Guatemala, which bounds it on the west and south, had always claimed the territory on the basis of old Spanish treaties, and in 1981 it revived these claims, breaking off diplomatic relations with Britain, which then agreed to retain some 2,000 troops with air and sea support, to protect Belize's independence. The People's United Party (PUP), founded in 1950, had spearheaded the anti-British demands for independence under its leader George Price. After thirty years in office he and his party lost the election of 1984 to their rivals the United Democratic Party (UDP) under Manuel Esquivel. The latter made efforts to reduce the growth of marijuana for the illicit drug trade and to end the diplomatic impasse with Guatemala, which restored relations with Britain in 1986. In September 1989 the PUP won a general election, George Price returning as Prime Minister.

Bello, Alhaji Sir Ahmadu (1906–66), Nigerian statesman. As Sardauna of Sokoto (a traditional military title), he became leader of the Northern People's Congress and, in 1952, the first elected minister in Northern Nigeria. In 1954, when the Federation of Nigeria was formed, he became Premier of the Northern Region. When Nigeria became independent in 1960, his party combined with Azikiwe's* National Council of Nigeria and the Cameroons (NCNC) to control the new federal Parliament. Bello's deputy in the NPC, Abubakar Tafawa Balewa* was federal Prime Minister, while Bello himself continued to lead the party in the north. In 1966, when the army seized power, Bello was among the political leaders who were assassinated.

Ben Bella, Mohammed Ahmed (1916–), Algerian revolutionary leader and President of Algeria 1963–5. He served in a Moroccan regiment of the French army in World War II, receiving the Military Medal. In 1947 he became leader of a secret military wing of the Algerian nationalist movement which later became the Armée de Libération Nationale. In 1949 he took part in an attack in Oran and was imprisoned by the French 1949–52, but escaped to Cairo and then to Libya; from here he directed Algerian resistance 1952–4, which culminated in the insurrection of 1 November 1954. In that year he formed the Front de Libération Nationale*. In 1956, when on a Moroccan airliner, he was seized and interned in France. He was freed in 1962 and helped to negotiate the Évian Agreements*. In September he became Prime Minister of a provisional government and in 1963 was elected first President of the Algerian Republic. He was overthrown in a military coup by Colonel Houari Boumédienne* in June 1965 and kept in prison until July 1979. Under house arrest until October 1980, he was then freed and allowed to go to live in Paris. He returned to Algeria in September 1990.

Benelux Customs Union. It was formed on 1 January 1948 by Belgium, The Netherlands, and Luxembourg. Full economic union followed in 1960 within the European Economic Community*, providing free movement of capital, goods, and people within the three states, which would have a common commercial policy with other countries.

Beneš, Edvard (1884–1948), President of Czechoslovakia 1935–8, 1945–8. Of peasant stock, he was educated in Prague and at the Sorbonne, becoming a lecturer in economics at Prague University before World War I. In 1914 he fled from Prague to Paris, where he helped Masaryk* to form the Czech National Council and formed a Czech national army from exiles. He became the leader of the Czech National Socialist Party and was Czech delegate at the Paris Peace Conference*. He was appointed Foreign Minister of the new Republic of Czechoslovakia, which he remained until succeeding Masaryk as President. As Foreign Minister he formed a Little *Entente** with Yugoslavia and Romania in 1921, in an attempt to prevent the restoration of the Habsburg King Charles in Hungary. He strongly supported the League of Nations, working for the admission of the Soviet Union in 1934. In the same year he and the Greek jurist Nikolaos Politis drafted the abortive Geneva Protocol for the pacific settlement of international disputes. He resigned in 1938 in protest against the Munich Agreement* and taught in the USA until the outbreak of war, spending the war years in London, where he formed a government-in-exile. He returned to Czechoslovakia in 1945, but refused to sign Klement Gottwald's* Communist constitution and resigned in 1948, the year of his death.

Bengal, that part of the Indian subcontinent around the Ganges–Brahmaputra deltas with a Bengali-speaking population; its ancient name was Vanga. By 1764 the British had conquered the nawab dynasties of Bengal, Orissa, and Bihar and united them into the single province of Bengal, ruled from Calcutta. In 1905 the Viceroy Lord Curzon* proposed a division into East Bengal (majority Muslim and to be united with Assam) and the richer West Bengal (to include Orissa and Bihar). His action was violently opposed and helped to precipitate anti-British sentiment as well as Hindu–Muslim tensions, which in 1947 were to result in East Bengal becoming East Pakistan. In 1911 Britain had accepted Indian criticism and reunited East and West Bengal, separating Assam, and forming a new province of Bihar and Orissa. In 1947 these both became states within the Union of India, as did the state of West Bengal.

Ben-Gurion, David (1886–1973), Prime Minister of Israel 1948–53, 1955–63. Born in Russian Poland, he emigrated to Palestine in 1906, when he worked on a kibbutz and supported the move for the adoption of Hebrew as the Jewish national language. He entered Zionist* journalism and enrolled at the University of Constantinople to study law, but was expelled for Zionist activity. In 1917 he joined the Jewish Legion* in Palestine, and after the war became one of the organizers of the Jewish Labour Party (Mapai) and of the Jewish Federation of Labour (Histadrut), which he served as general secretary 1921–35. As chairman of the Jewish Agency* 1935–48, he was the leading figure in the Jewish community in Palestine. As Israel's first Prime Minister and Minister of Defence, he played the largest part in shaping Israel during its formative years. In 1965 he was expelled from the Labour Party and formed a new party known as Rafi.

Benin, a West African republic. Formerly ruled by Yoruba* kings, in 1904 it was constituted a Territory of French West Africa* under the name Dahomey. It became an autonomous republic within the French Community* in 1956, and independent in 1960. During the next decade civilian government alternated with military rule, but in 1972 a Marxist-Leninist republic was established under President Mathieu Kérékou; in 1975 he altered the name to Benin. Under his leadership it achieved increased domestic stability and international standing. During the 1980s offshore oil began to be exported, bringing greater prosperity. In 1989 political protest developed against the regime. There were widespread strikes and protests against an austerity IMF* programme, and, in December, a joint session of the People's

Revolutionary Party (Assembly and Council) agreed to abandon 'Marxist-Leninism' as the official Benin ideology. President Kérékou convened a constitutional conference early in 1990.

Benn, Anthony Wedgwood (1925–), British politician. From a family with a long history in Liberal politics, he was educated at New College, Oxford, immediately after World War II, in which he had served in the Royal Navy. He was first elected to Parliament in 1950 for the Labour Party. In 1960 he inherited the title Viscount Stansgate on the death of his father, and spent the next three years struggling to obtain a Peerage Act by which he could renounce the viscountcy and return to the House of Commons. At this time he renounced also the name Wedgwood, being later known simply as Tony Benn. He was Postmaster-General 1964–6 and then successively Minister of Technology 1966–70, Secretary of State for Trade and Industry 1974–5, and Secretary of State for Energy 1975–9. Although he was becoming increasingly left wing and opposed to nuclear weapons, his ministerial responsibilities had ironically involved considerable research and development in defence and nuclear power. He was also strongly opposed to the European Common Market. In 1979 his book *Arguments for Socialism* was a call for more open government and for 'socialism of the work-place'. Much influenced by the thinking of the 17th-century Levellers, his commitment to egalitarian democracy made him the target of considerable vilification. The publication of his *Diaries 1963–67* (1987) revealed much of the workings of British government.

Bennett, Richard Bedford, 1st Viscount Bennett (1870–1947), Prime Minister of Canada 1930–5. Born in New Brunswick, he was a successful lawyer and businessman before being elected to the Alberta Parliament in 1909 and to the Canadian House of Commons in 1911. In 1927 he became the leader of the Canadian Conservative Party, which won the general election of 1930. He was highly influential in discussions leading to the Statute of Westminster* (1931), and he presided over the Ottawa Conference (1932) which gave Canadian trade a preferential market in Britain. His government was regarded as reactionary and anti-American and his attempts to strengthen the federal Civil Service were unpopular, as was the so-called Bennett New Deal—proposals for social and economic legislation to meet the problems of the Great Depression*. He lost the election of 1935 and in 1938 retired from politics to live in England.

Ben-Zvi, Yitzhak (1884–1963), Hebrew scholar and President of Israel 1952–63. Born in Russia, he escaped a pogrom* and came to Palestine in 1907, where he founded a Hebrew high school. Exiled by the Turks in 1915, he joined Ben Gurion* in the USA and returned to Palestine in 1918 a member of the Jewish Legion*, fighting alongside the British. A founder of the Histadrut or General Federation of Jewish Labour he was also, with Ben Gurion, a founder of the General Council for Palestine Jews during the British mandate, being its chairman 1931–44 and president 1944–9. Consistently opposed to violence, he signed the Declaration of Independence 14 May 1948 and became President of Israel on the death of Chaim Weizmann* until his own death.

Beria, Lavrenti Pavlovich (1899–1953), Soviet commissar. Born in Georgia, he joined the Communist Party when a student in 1917 and quickly became head of the secret police (Cheka*) in that province (1921). He came to Moscow in 1938, appointed by his fellow Georgian Stalin to take charge of the NKVD* secret police as head of Internal Affairs, and to organize Soviet prison camps. During World War II he was a major figure in armaments production. After Stalin's death he became a victim of the ensuing power-struggle. In July 1953 he was arrested on charges of conspiracy. He was tried in secret and executed.

Berlin Airlift (1948–9). During 1948 the fragile co-operation between the occupying powers in Germany collapsed. In March the Russians imposed rail restrictions on traffic from the Western-occupied zones. In June the USA, Britain, and France established economic co-operation and currency reform within their zones, together with plans for a constituent assembly, which would enable a new federal state to emerge. The Russians retaliated by closing all land and water communication routes from the Western zones. The latter responded by supplying all necessities for their Berlin sectors by cargo airlift. By February 1949 the Soviet Union accepted that it could not unilaterally take over Berlin, and secret talks were begun. In May the Russians reopened all routes. The blockade confirmed that Berlin, and indeed all Germany, would be divided into two administrative blocks for the next forty years.

Berlin–Baghdad railway. A railway of which the planning and promotion by a German construction company was at one time seen as a significant factor precipitating World War I. In fact only a small part of the railway had been built, from the Bosporus through Anatolia, by 1914, when both British and Russian governments were satisfied that it did not represent a significant threat to their interests. It was completed to Baghdad and Basra during the 1920s by British companies. The other important Middle East rail project at this time was the Hejaz railway*.

Berlin Conference (1884–5). This conference of European states fundamentally affected the future of Africa. Chaired by Otto von Bismarck, it had been attended by representatives of all the European powers, together with the USA. Called on the initiative of Portugal to resolve the future of the Congo Basin, it had recognized the Congo Free State (later the Belgian Congo); in doing so it also confirmed, together with a later conference in Brussels in 1890, the arbitrary borders of the colonies of Germany, France, Belgium, Britain, Spain, and Portugal in West and Central Africa. These borders later became national frontiers.

Berlin Wall, built in August 1961 by the German Democratic Republic in order to stem the flow of refugees from East Germany into the West. Over 3 million had emigrated between 1945 and 1961. The Wall was heavily guarded and many people, especially in the 1960s, were killed or wounded while attempting to cross. In early November 1989 large numbers of East Berliners were moving to West Germany via Austria, and on 9 November a sudden decision was made that exit visas would no longer be necessary. East and West Berliners danced through the night on top of the Wall in celebration. Next day five crossing points were opened and some 40,000 crossed to gaze at the overloaded supermarkets, but only 1,500 elected to stay. Fresh openings were made and pieces of the Wall were chipped off and auctioned as souvenirs. On 22 December the Brandenburg Gate was ceremoniously reopened and plans made for the final demolition of all the Wall.

Bermuda, a British colony of some 150 islands in the West Atlantic. First settled by the Virginia Company, it has the oldest Parliament in the New World, dating from 1620. By the 20th century two-thirds of the population was of African or Indian descent, but culturally they feel little in common with the West Indian islands. Strategically important to the British navy, Bermuda grew rich on trade and tourism, with close ties with the USA. Britain granted the USA naval and air bases during World

War II in 1940. Universal adult suffrage was introduced in 1944 and the present Constitution in 1967, granting self-government. Political activity developed in the 1960s and there were sporadic race-riots, those of 1972–3 and 1977 being very bitter, when British troops were sent in. The dominant multiracial United Bermuda Party (UBP) supports the continued colonial status, while its rival Black Progressive Labour Party agitates for independence.

Bernadotte, Folke, Count (1895–1948), Swedish international mediator. Nephew of King Gustav V of Sweden, he was commissioned into the Swedish army as a young man. During World War I he worked for the Swedish Red Cross, which he continued to serve through the inter-war period, when he also became involved in the Boy Scout movement. During World War II he was again involved in Red Cross work, helping to exchange prisoners and settling refugees. In 1945 Himmler approached him to arrange peace talks with the Allies. As president of the Swedish Red Cross he was asked by Trygve Lie* to act as a mediator in Palestine in 1948 for the UN. He was murdered by Israeli terrorists determined to prevent the partition of Palestine.

Bernstein, Eduard (1850–1932), German political theorist. Son of an engine-driver, he became a political journalist and lived in London 1888–1901, when he was influenced by the Fabians*. An anti-military, international socialist, he rejected the Marxist idea of revolution, advocating co-operation between left-wing parties. He served in the Reichstag 1902–6, 1912–18, and 1920–8, and strongly influenced the thinking of the German Social Democratic Party.

Besant, Annie (1847–1933), British social reformer and theosophist. Married to a clergyman in 1867, she separated from him and joined the National Secular Society. She became a Fabian*, a trade union organizer (including the match-girls' strike of 1888), and a propagandist for birth control. She became a leading exponent of the religious movement of theosophy, and went to live in India, where in 1898 she founded what became the Hindu University of India. In 1916 she helped to form the All India Home Rule League. She was president of the Indian National Congress* 1918–19, one of only three Britons to hold this office. She published numerous books on theosophy.

Betancourt, Rómulo (1908–81), 'provisional' President of Venezuela 1945–8, President 1959–64. While a student at the University of Caracas he was imprisoned (1928) for demonstrating against the dictator Juan Gómez. He was later exiled until 1936 and then again in 1939 for alleged left-wing activities. He returned to Venezuela in 1941 and became a dominant figure in politics, founding the Acción Democrática (AD) Party, an anti-Communist left-of-centre party. In 1945 he was appointed 'provisional' President by the military and held office for three years, inaugurating a programme of social reform, including land redistribution, and imposing greater control over oil companies. Following a right-wing coup he was again exiled in 1948. He returned ten years later to be elected President, embarking on a programme of redemocratization, following the period of right-wing military juntas. He resumed his policies of land reform and higher taxation of oil companies, and secured a series of benefits for organized labour. Attacked both by right-wing supporters of his predecessors and by radical socialists, he nevertheless turned the presidential office over to a freely elected successor, Raúl Leoni, in 1964. He campaigned unsuccessfully for the presidency in 1973.

Bethlen, István, Count (1874–1944), Prime Minister of Hungary 1921–31. Born into an aristocratic family in Transylvania, he was first elected to the Hungarian

Parliament in 1901. A strong counter-revolutionary, he helped to support Admiral Horthy's* activities in deposing Béla Kun* and became Prime Minister in April 1921. With his belief in the need to preserve feudal aristocratic privileges, he ended land redistribution and the secret ballot in the country districts. An admirer of Mussolini, he signed a treaty of friendship with him in 1927. He had negotiated entry into the League of Nations, from which he obtained a reconstruction loan, but the worsening economic crisis after 1929 produced a situation with which he could not cope and he resigned in 1931. In 1944 German troops invaded Hungary and he went into hiding, but was caught by the Soviet forces who followed, and was taken to Moscow, where he died.

Bethmann Hollweg, Theobald von (1856–1921), Chancellor of the Second German Empire 1909–17. Born in Prussia, he served as a civil servant before becoming a Prussian Minister. As Chancellor he instituted a number of electoral reforms and gave greater autonomy to Alsace-Lorraine*. He greatly increased the German army, believing war with one or another neighbour inevitable. He hoped to retain Britain's neutrality and in 1912 worked successfully to restore peace in the Balkans. In July 1914 he and Kaiser William II, convinced of the need for a short, preventive war, promised support to Austria-Hungary. In 1917 he opposed unrestricted U-boat warfare, rightly foreseeing the entry of the USA into the war. He retired in July, when Hindenburg* and Ludendorff* established a virtual military dictatorship.

Bevan, Aneurin (1897–1960), British statesman and founder of the National Health Service*. Son of a Welsh miner, he started work in a coal-mine at 13, attended the Central Labour College, and became active in the National Union of Mineworkers. He was active among the Welsh miners in the General Strike* in 1926 and was elected to Parliament for Ebbw Vale in 1929 as a Labour Member; but his left-wing views and fiery personality made him a rebellious figure on domestic and foreign issues. He was briefly expelled from the Parliamentary Labour Party in 1939. A founder and editor of *Tribune* magazine, he was one of Winston Churchill's most vehement critics during World War II. As Minister of Health 1945–51, he was responsible for a considerable programme of house-building and for the creation, amid bitter controversy with the medical profession, of the National Health Service. In 1951 he became Minister of Labour, but resigned when charges were imposed by Gaitskell* for some medical services and prescriptions. The Bevanite group, opposed to high defence spending, was then formed within the party. In 1955 he failed against Gaitskell to win leadership of the party but in 1956 became Shadow Foreign Secretary after he had criticized Eden's activities in the Suez Crisis*. In 1957 he restated his position on defence and opposed the 'Bevanite' demand for unilateral renunciation of the hydrogen bomb. Brilliant in debate, fertile with ideas, but often in conflict with his own party, he was elected deputy leader in 1959. His wife Jennie Lee was also a notable Labour MP and minister.

Beveridge, William Henry, 1st Baron Beveridge (1879–1963), British economist and social reformer. Born in India, he was educated at Oxford, where he became a Fellow of University College. At the invitation of Winston Churchill he joined the Board of Trade in 1908, publishing a notable report *Unemployment* in 1909. In it he argued that the regulation of society by an interventionist state would strengthen rather than weaken the free market economy. He was instrumental in drafting the Labour Exchanges Act (1909), becoming the first Director of Labour Exchanges 1909–16. He also helped to draft the National Insurance Act of 1911. From 1919 to 1937 he was director of the London School of Economics and then master of University College, Oxford. During those years he chaired numerous committees related to the social services. In 1941 he was commissioned by the government to chair an inquiry into the social services and produced a report *Social Insurance and Allied Services* in 1942. This 'Beveridge Report' was to become the foundation of the British Welfare State and the blueprint for much social legislation 1944–8. Its main feature was to be a far-reaching, universal social insurance scheme 'against interruption and destruction of earning power and for special expenditure arising at birth, marriage and death'. Such provision of a whole range of benefits would check 'the twin evils of capitalist society, unemployment and poverty'. The Report was debated in the House of Commons in February 1943, when James Griffiths led a Labour revolt against what was alleged as government reluctance to implement it. It had widespread publicity and in retrospect it can be seen as a factor in the Labour election victory of 1945.

Bevin, Ernest (1881–1951), British trade union leader and politician. He began work in Bristol as a van-boy, but quickly emerged as a leader of unskilled workers in the city and docks. He became assistant secretary of the Dockers' Union in 1911, and then went on to found the Transport and General Workers' Union in 1921, being its secretary until 1940. His union supported the miners in the General Strike* of 1926, Bevin being active in negotiations for a settlement. Essentially a conciliator, as chairman of the TUC General Council from 1937 he helped the trade unions to be accepted as part of the British body politic during the 1930s. He ensured their full support for the war effort 1939–45, being appointed Minister of Labour and National Service by Churchill in 1940. As Foreign Secretary 1945–51 he reluctantly concluded that the wartime alliance with the Soviet Union had collapsed and supported the USA in early confrontations of the Cold War. For example he gave prompt support for the Marshall Plan*, and helped to negotiate the Treaty of Brussels in 1948 and to create NATO* in April 1949. In 1950 he convened a conference of Commonwealth foreign ministers which resulted in the Colombo Plan* for economic development of poorer Pacific nations. He had a strong commitment towards the Commonwealth, but was no enthusiast for the new Council of Europe and the idea of a United States of Europe. In 1945 he was strongly criticized for returning a shipload of Jewish emigrants to Europe, while his decision in 1947 to abandon the British mandate in Palestine resulted in the proclamation of Israel the following year.

Bhave, Vinoba (1896–1982), Indian leader. Born of a high caste brahmin family, in 1916 he became an ardent disciple of Gandhi*. Interned regularly in the 1920s and 1930s for his non-violent opposition to British rule, he was imprisoned 1940–4 for defying wartime regulations. After Gandhi's assassination in 1948 he was regarded as the leading exponent of Gandhism. In 1948 he founded the Sarvodaya Samaj to work among Indian refugees and in 1958 he began the Bhoodan or land-gift movement, with the object of acquiring land for redistribution among landless low caste villagers. At first his object was to acquire individual plots, but later he sought to transfer ownership of entire villages to village councils, walking some 40,000 miles in his campaign and being particularly successful in Bihar. He also led the Shantri Sena volunteer movement for conflict resolution and economic and social reform.

Bhopal, capital city of the state of Madhya Pradesh in India. On 3 December 1984 it was the scene of the worst industrial accident in history, when 45 tons of the toxic gas methyl isocyanate escaped from a plant owned by the US corporation Union Carbide. Probably 2,500 died within twenty-four hours and over 50,000 suffered greater or lesser lung disease and other complications as a result. Efforts to obtain adequate compensation for an accident caused by substandard levels of safety procedure were only partially successful.

Bhumibol, Adulyadej (1927–), King of Thailand 1946– . Absolute monarchy had ended in 1932, following a revolution, after which the main function of the King was to take part in ceremonies as head of state and Commander-in-Chief. The long experience of Bhumibol in office has, however, enabled him to act as a moderating force between extremist parties in Thai politics, successfully surviving two periods of military power 1957–73 and 1977–9.

Bhutan, a small state in the eastern Himalayas, governed as a Buddhist constitutional monarchy and with the world's lowest per capita income. In 1907 the hereditary Maharaja of Bhutan Ugyen Wangchuck took the title of Druk Gyalpo or king, under British patronage and subsidies. In 1949 Bhutan concluded a treaty with India, which took over the British role as patron and protector. In 1953 a National Assembly was restored and in 1969 King Jigme Dorji Wangchuck widened the constitution into a 'democratic monarchy' with a Council of Ministers and National Assembly of 140 members. His liberal policies have been followed by his son Jigme Singye Wangchuck, who has also gradually made diplomatic contacts separate from those of India. Since 1949 there has been a steady stream of Tibetan refugees into the country, whose presence has tended to destabilize the economy.

Bhutto, Benazir (1953–), Prime Minister of Pakistan 1988–90. Daughter of Zulfikar Bhutto* and educated at Oxford, she was placed under house arrest in 1977 when her father was deposed. She remained there until 1984, when she was released and allowed into exile in Britain. There she founded the Movement for the Restoration of Democracy in Pakistan. She returned to Pakistan in 1986 and, as leader of the Pakistan People's Party, became Prime Minister in December 1988, when she had to face the problem of over 3 million refugees from Afghanistan. In foreign affairs she aimed to improve relations with India, re-enter the Commonwealth, and, without losing the support of the army, reduce Pakistan's involvement in Afghanistan. After a year in office she was accused of failing to eradicate any of the country's domestic problems. In August 1990 she was dismissed by President Ghulam Ishaq Khan for 'abuse of power'. She lost the election of October and her husband was arrested on charges of corruption.

Bhutto, Zulfikar Ali (1928–79), President of Pakistan 1971–3; Prime Minister 1973–7. From an aristocratic family, he was educated in Oxford and America before becoming a barrister. In 1958 he joined the government of Ayub Khan* as Minister of Fuel and Power and then as Foreign Minister, when he improved relations with China. He resigned in 1966 over the Kashmir dispute with India, and formed the Pakistan People's Party with a policy of 'Islam, democracy, socialism and populism'. The PPP secured a majority in West Pakistan in the election of December 1970, and when the government was discredited over the loss of Bangladesh he became President, the first civilian to do so. When Britain, Australia, and New Zealand recognized Bangladesh he took Pakistan out of the Commonwealth (January 1972), but he improved relations with India by the Simla Agreement. In 1973 he introduced a new constitution making the Prime Minister chief executive, stepping down from the presidency and becoming Prime Minister and Minister for Foreign Affairs and for Atomic Energy. He introduced land reforms and the nationalization of major industries, but the failure of his economic programme precipitated a military coup in July 1977. He was subsequently hanged on the charge of complicity to murder (April 1979).

Biafran War (1967–70), a Nigerian civil war. In January 1966 an army mutiny resulted in the assassination of the Prime Minister Alhaji Sir Abubakar Tafawa Balewa* and the Northern and Western premiers. In July there was a counter-coup led by General Yakubu Gowon*. In September many Ibo people living in northern Nigeria were massacred, in revenge for the January killings, at a time when Gowon was trying to establish a new constitution in Lagos. In May 1987 Colonel Odumegwu Ojukwu* declared the eastern (Ibo) region to be the independent Republic of Biafra. Civil war followed. Britain and the Soviet Union supplied Gowon's federal government with arms, while France sent aid to Biafra. Enugu, the Biafran capital, was taken by federal troops and in May 1968 Port Harcourt, making the country land-locked. Gabon, Ivory Coast, Tanzania, Zambia, and Haiti all recognized Biafra, but it was doomed. In January 1970 resistance collapsed, Ojukwu fled, and an armistice was signed in Lagos by General Effiong. The war caused starvation and death to perhaps a million Ibo people.

Bidault, Georges (1899–1982), Prime Minister of France 1946, 1949–50, 1958. After serving in World War I he became a journalist and professor of history in Paris. During World War II he was a distinguished leader of the French Resistance movement. In 1944 he became active in the politics of the new Fourth Republic*, serving as Foreign Minister and also as Prime Minister on a number of occasions. In 1949 he founded a new party, the Mouvement Républicain Populaire. He bitterly opposed de Gaulle's policy for Algeria, supporting the OAS*. In 1962 he became president of the National Resistance Council. He was charged with plotting against the State and in 1965 went into exile in Brazil. He returned to France in 1968.

Bierut, Boleslaw (1892–1956), President of the Republic of Poland 1945–52, Premier 1952–4. Born near Lublin, when quite young he became interested in socialist ideas and joined the Polish Communist Party in 1920. He spent the next twenty years working for the party in Poland, Bulgaria, Czechoslovakia, and Austria, in between spells in prison. In 1938 he went to Moscow and returned to Poland in 1943. He organized the take-over of Poland by the party in 1944 following the Warsaw Rising* and became President of the Polish Republic in 1945, a loyal follower of Stalin. In 1948 he was responsible for the deposition of Wladyslaw Gomulka* for his attempts to deviate from Stalin's directives. He reorganized the party and the Constitution in 1952 and became Premier of Poland for two years. He was in Moscow in 1956 attending the Twentieth Congress* when he died.

Biko, Steve (1956–77), student leader in South Africa. A medical student at the University of Natal, he was co-founder and president of the all-Black South African Students' Association, whose aim was to raise Black consciousness. Active in the Black People's Convention, he was arrested on numerous occasions (1973–6). The manner of his death in prison at the age of 21 made him a symbol of heroism in Black South African townships and beyond. Following disclosures about his maltreatment, the South African government prohibited numerous Black organizations and detained newspaper editors, provoking international anger.

Binh, Nguyen Thi, Madame (1927–), Vietnamese guerrilla fighter and politician. As a schoolgirl in Saigon she joined an anti-colonial student movement, and was imprisoned 1951–4 for Communist militancy. In 1961 she joined the newly formed National Liberation Front* and was a member of its Central Command from 1962, taking part in the Paris Peace Talks 1968–73. She was Foreign Minister of the Provisional Revolutionary Government of South Vietnam, and served as Minister of Education 1976–87 and Councillor of State of the Socialist Republic of Vietnam.

Birkenhead, Frederick Edwin Smith, 1st Earl of (1872–1930), British Lord Chancellor 1919–24. Educated at Birkenhead School and Oxford, where he had a brilliant career, he was called to the Bar in 1899; he entered Parliament in 1906 as a Conservative who attacked the Liberal policy of free trade. He joined the wartime coalition in 1915 as Attorney-General and in 1919 became Lord Chancellor at the age of 46, one of his achievements being the Law of Property Act of 1922. He was a close associate of Lloyd George* at this time. A strong supporter of a united Ireland and of imperial India, and an opponent of Labour, which he regarded as 'Bolshevist', he nevertheless showed himself conciliatory over the General Strike* in 1926. He resigned from office in 1928.

Birla family, Indian commercial and industrial family of the Marwari or merchant caste. It is one of the two greatest Indian industrial families, second only to the Tata* family. Its best-known member was Ghanshyam Das Birla (1890–1975), who became Gandhi's* principal financial backer, paying most of the cost of the ashram (retreat), the harijan (Untouchables) organizations, the peasant uplift campaign, the national language movement, and many other Gandhian welfare projects. It was at Birla House, New Delhi, that Gandhi was killed. Like most Marwaris the Birlas are devout Hindus.

Bishop, Maurice (*c.*1945–83), Prime Minister of Grenada 1979–83. After graduating in law in the USA at the time of the Black Power movement*, he returned to Grenada in 1970 and was a founder of the New Jewel Movement (Joint Endeavour in Welfare, Education, and Liberty). In November 1973 he was arrested and severely beaten in police cells, for criticism of Sir Eric Gairy, accusing him of corruption and autocracy as Prime Minister. He was imprisoned for over a year. By 1979 Gairy's government was so widely discredited that an almost bloodless coup quickly brought the NJM to power. A highly popular figure, he suspended the Constitution and established the People's Revolutionary Government. Committed to a policy of non-alignment but of improved relations with Cuba, and of nationalizing the country's assets to solve economic problems, his left-wing policies alarmed the USA, which withdrew aid support in 1982. In October 1983, following a split within the PRG, he was forcibly removed from office and murdered by the military. Soon afterwards the US marines landed 'to prevent a Marxist revolution' and to 'restore order'.

Black and Tans, an auxiliary force of the Royal Irish Constabulary. The demands of the Irish republicans in 1919 for a free Ireland led to violence against the Royal Irish Constabulary, an armed British police force. Many of the policemen resigned, so the British government in 1920 reinforced the RIC with British ex-soldiers. Their distinctive temporary uniforms gave them their nickname of Black and Tans. They adopted a policy of harsh reprisals against the republicans, many people being killed in raids, and property destroyed. In December 1920 they burned down County Hall in Cork. Public opinion in Britain and the USA was shocked at their ill discipline and they were withdrawn after the Anglo-Irish truce of December 1921.

Black consciousness, an awareness of the identity, aptitudes, and aspirations of the Black peoples of the world. As a movement its origins go back to before 1914, but in the 1920s it developed rapidly in the USA, where Marcus Garvey* became a cult figure, while the Harlem Renaissance* brought Blacks into positions of considerable artistic and musical influence. During the 1930s in Africa and the Caribbean the writings of Aimé Césaire and Léopold-Sédar Senghor* helped to cultivate 'négritude'—the belief that the Negro has a distinct cultural heritage which had to be conserved against colonial pressures towards Europeanization. On both sides of the Atlantic there has been a nostalgia for the 'beauty and harmony of traditional African society', based on emotion, intuition, and spontaneous social interaction, as against the rationalistic tradition of European Hellenism. In the 1950s and 1960s it spawned a variety of movements in the USA such as the Black Power movement*.

Black Power movement, a movement among Black people in the USA in the 1960s. During the 1950s the civil rights movement* had gathered momentum under the inspiration of Martin Luther King*. There had also been an expansion of the Black Muslim movement (founded in 1930 by Elijah Muhammad), which emphasized Black separatism under Islam. By the mid-1960s many Blacks felt that the civil rights movement, in spite of its successes, was doing little to alter their lives. Under such leaders as Stokeley Carmichael they proposed that Black Americans in their own communities should concentrate on establishing their own political and economic power. In 1966 a Student Non-violent Co-ordinating Committee (SNCC) was formed by Carmichael to activate Black college students, while at the same time the Black Muslim movement was advocating Islam for Black salvation. More militant were the Black Panthers, founded in 1966, who advocated violence. All were concerned to stress the value of Black culture and all things Black. Riots in major US cities (1965–8), in Los Angeles, Chicago, Newark, and Detroit, seemed to herald a new wave of Black militancy. But the intensity of the Black Power movement tended to decline in the 1970s, when many Blacks began co-operating with White organizations against the Vietnam War.

Blackshirts (Italian: 'camicia nera'), the colloquial name given to the Squadre d'Azione ('Action Squad') founded in Italy in 1919 to maintain internal law and order. Organized along paramilitary lines, they wore black shirts and patrolled the cities to fight socialism and Communism, which they would do by violent means. In 1921 they were incorporated into the Fascist Party as a national militia. The term was also applied to the SS* in Nazi Germany.

Blitzkrieg (German; 'lightning war'). An Anglicized version, 'the Blitz', was coined by the British public to describe the German air assault on British cities 1940–1. As a military concept it was especially successful in the German campaigns in Poland, France, and Greece during World War II. It employed fast-moving tanks and motorized infantry, supported by dive-bombers. Superior but slower enemy forces were thrown off balance and rapid victory gained with small expenditure of men and materials.

Bloomsbury Group, a term given to a group of English artists and writers of the early 20th century. It began as friends of the children (Thoby, Vanessa, and Virginia) of Leslie Stephen, founder of the *Dictionary of National Biography*, who lived in Bloomsbury Square in London. Thoby died young, Vanessa married Clive Bell, and Virginia married Leonard Woolf. In art its members were strong supporters of Post-Impressionism* and in literature critics of what they regarded as Victorian philistinism and moral hypocrisy, demanding, in particular, more open discussion and tolerance of sexual mores. Virginia Woolf herself, Clive Bell, Lytton Strachey, J. M. Keynes*, E. M. Forster, Bertrand Russell*, and D. H. Lawrence were among members to have a lasting significance. Lady Ottoline Morrell was a generous patron of many of the group's members, while Cambridge University was for long closely associated with them.

Blue Shirts, a Chinese neo-Fascist organization. They were formed in 1932 by a group of Guomindang* officers, mostly graduates of the Whampoa Military Academy. Jiang Jiehi* (Chiang Kai-shek) was supreme leader of the organization and General Tai Li, his intelligence chief, played a leading role. Its origins can be traced to the advice which Jiang's men received from Nazi German military officers such as Colonels Max Bauer and Herman Kriebel. They were responsible for many acts of atrocity against Communist rivals, especially in the years preceding the Xi'an Incident*.

Blum, Léon (1872–1950), Prime Minister of France 1936, 1946–7. Of an Alsatian Jewish family, he became an established journalist and critic and was then drawn into politics by the Dreyfus Affair. In 1919 he was elected a Deputy and in 1936 he brought about a coalition of Radicals, Socialists, and Communists to form a 'Popular Front'* government. This granted workers a forty-hour week, paid holidays, and collective bargaining; but it provoked considerable hostility from industrialists and caused financial difficulties. Radicals refused to support intervention in the Spanish Civil War, while Communists withdrew support for his failure to intervene. His government fell. In 1940 he was arrested by the Vichy government and charged with causing France's defeat, but his skilful defence obliged the government to call off his trial (1942). When Germany occupied Vichy France he was interned in German concentration camps as a Jew (1943–5). He survived and returned briefly to office as the Prime Minister of a caretaker government 1946–7, when he helped to write the Constitution of the Fourth Republic*.

Boat people, term given to some of the half-million refugees who have fled from Vietnam since 1976. In the years 1979–82 thousands were found floating in hopelessly overcrowded boats; even in 1985 there were still some 1,000 leaving Vietnam each month. In that year the UN High Commission for Refugees negotiated a Rescue at Sea Resettlement Scheme, when thirteen member nations, including the USA, Japan, Britain, and Canada, agreed to take a quota of refugees rescued in the South China Sea. In addition $2m. were voted to the Thai navy to prevent pirates robbing and sinking boats. Wealthier refugees have been successfully resettled, but many poorer ones ended in camps, as for example in Hong Kong. Many earlier boat people were of Chinese origin, but since 1985 most have been of Vietnamese origin, a significant number from the north. Attempts in Hong Kong to repatriate these from camps have been deeply unpopular.

Bokassa, Jean Bedel (1921–   ), President of the Central African Republic* 1966–79. Educated in mission schools, he served in the French colonial army 1939–60. In 1966, as a colonel, he overthrew President David Dacko, first President of the Republic, and for the next thirteen years ruled autocratically. In 1976 he declared himself Emperor, crowning himself as had Napoleon and building himself a vast palace. In 1979 the French President Giscard d'Estaing, perhaps embarrassed by gifts of diamonds, ordered French paratroops to occupy the country and restore Dacko. Bokassa retired to the Ivory Coast and then to France. In 1986 he returned to the CAR, where he was tried and committed on charges of 'murder, cannibalism and embezzlement', being accused among other things of responsibility for the massacre of 100 schoolchildren. He was sentenced (1988) to forced labour for life.

Bolivia, a South American republic. This land-locked country had suffered the loss of its rich nitrate deposits to Chile in the War of the Pacific (1879–88), when access to the Pacific coast was also forfeited. Between 1884 and 1920 political stability under a two-party system, Conservative and Liberal, was established. Silver- and tin-mining were at a peak and the rural landowners and mining interests formed a governing élite. Between 1920 and 1932 however this consensus began to crumble. Trade unionism in the mines and railways and peasant resentment against landowners began to develop. In 1930 a popular revolution elected a reforming President Daniel Salamanca, but the disastrous Chaco War* (1932–5) against Paraguay precipitated an extended period of arbitrary rule. During this the Movimiento Nacionalista Revolucionaria (MNR) was founded by Dr Victor Paz Estenssoro*. In 1952 a National Revolution overthrew the dictatorship of the military junta; Paz Estenssoro returned from exile and was installed as President. Tin-mines were nationalized, adult suffrage introduced, and a bold programme of social reforms begun. Paz was re-elected in 1960, but overthrown in 1964 by another military coup. In 1967 a Communist revolutionary movement led by Che Guevara* was defeated. Military regimes followed each other quickly. Not all were right wing, and that of General Juan José Torres González (1970–1) aimed to replace Congress with workers' soviets. Democratic elections were restored in 1978, but there was a new military coup in 1980, and a state of political tension continued until 1982, when civilian rule was again restored. President Siles Zuazo was installed, but he could not secure national unity, going on hunger-strike in October 1984 in a bid to do so. In 1985 the MNR under Paz regained power, faced by massive debt and the highest inflation rate in the world. A collapse of world tin prices and Draconian domestic policies caused dismay and a drop in already dismally low standards of living. But Paz survived and by 1987 Bolivia was again receiving IMF* credit, with natural gas replacing tin as its main export. Following an inconclusive election in 1989, a coalition government under Jaime Paz Zamora of the MIR (Movement of the Revolutionary Left) was formed, committed to the elimination of the drug traffic.

Bolshevism (majority), a Russian political term. Its origins were as a grouping within the Russian Social Democratic Party. At a meeting in London in 1903 there was a majority for the view that violent revolution, rather than gradual change (advocated by those in the minority, the **Mensheviks**), was necessary in Russia and that party membership should be restricted to professional revolutionaries. The **Bolshevik** Party, an illegal political party, was formed in 1912. It was led by Lenin* and achieved power in October 1917 adding to its title the word Communist in 1919. In 1925, after the formation of the Soviet Union, the Soviet Union Communist Party of Bolsheviks was formed, which ruled without challenge until 1990. In 1952 the party abandoned the term Bolshevik in its title.

Bonhoeffer, Dietrich (1906–45), German theologian. He studied theology at the University of Berlin and briefly taught in New York before returning to a teaching post in Berlin in 1931. He became a leading spokesman for the Confessing Church, which later became a centre of German Protestant Resistance to the Nazi regime. In 1934 he signed the Barmen Declaration against attempts being made to reconcile Nazism with Christianity. In 1935 he became principal of the Finkenwalde Seminary, which was closed two years later, when he was forbidden to teach. Through contacts with double agents in the Military Intelligence Department, itself a centre of the Resistance movement in Germany, he established a dialogue with Bishop G. K. A. Bell of Chichester, whom he met in Sweden in 1942 with proposals for peace. These were rejected in London and in 1943 he was arrested. After the July 1944 bomb plot* he was detected as having been involved and he was executed. His book *Letters and Papers from Prison* is a moving and profound affirmation of faith.

Borden, Sir Robert Laird (1854–1937), Prime Minister of Canada 1911–20. Born in Nova Scotia, he practised as a successful lawyer before being elected to the Canadian House of Commons in 1896. In 1901 he became leader of the Canadian Conservative Party, then in opposition, and won the general election of 1911. His government raised an army of some half-million volunteers to serve with great distinction in World War I. A member of the Imperial War Cabinet, he returned from London in 1917 convinced of the need for conscription; the Military Service Act which followed was deeply unpopular, as was the introduction of income tax. In October 1917 he had created a coalition or union government with those Liberals willing to support conscription, and remained in office until 1920, introducing an act to give women over 21 the vote in March 1918. He attended the Paris Peace Conference and won the right of British dominions both to sign the peace treaties and to be separately represented in the League of Nations. He represented Canada at the Washington Conference* (1921) and on the Council of the League of Nations.

Borges, Jorge Luis (1899–1986), Argentine poet and novelist. Son of a schoolmaster in Buenos Aires, he was educated in Geneva and then spent some years in Spain, where he began writing at the end of World War I. He returned to Buenos Aires to become a librarian. After a severe accident in 1938 he began a series of fantastic tales based in a dreamworld, with its own language and symbols. Although he became blind he continued to dictate, his book *El libro de arena* (1955) revealing labyrinthine complexities within the psyche. From 1960 onwards his work became increasingly known in Europe, his 'nightmare fictions' being compared with those of Kafka*, and he was recognized as one of the greatest of Latin American writers.

Boris III (1894–1943), King of Bulgaria 1918–43. As Crown Prince he commanded Bulgarian troops on the Macedonian front in World War I, being defeated by Allied troops commanded by Franchet d'Esperey*. On succeeding his father Ferdinand I, he for long skilfully kept a low profile, surviving radical and Communist agitation. By 1935 he was virtually dictator of Bulgaria. Although he allied with Hitler in World War II, as part of his New Order for Europe, and dutifully declared war on Britain, the Soviet Union, and the USA, he failed to collaborate actively with Germany. An exasperated Hitler summoned him to his headquarters in East Prussia in August 1943. Three days later he died, being succeeded by his infant son Simeon.

Bormann, Martin (1900–*c.*1945), German Nazi leader. Born in Halderstadt, he joined the paramilitary Freikorps which roamed Berlin after World War I. He was briefly imprisoned for his part in a political murder in 1924 and then joined the National Socialist Party. In 1928 he became a member of Hitler's personal staff. He held offices in the SA (Brownshirts*) and in 1933 became a member of Hess's staff. After the departure of the latter in 1941 he headed the party's Chancery. His intimacy with Hitler enabled him to wield great power unobtrusively. He was an extremist on racial questions, being a major advocate of the extermination of both Jews and Slavs. He was also behind the offensive against the churches. He was sentenced to death *in absentia* at the Nuremberg Trials*. In 1973, after identification of a skeleton exhumed in Berlin, the West German government declared that he had committed suicide after Hitler's death. There had been persistent rumours that he had escaped to South America.

Borneo, a large island in the South China Sea. By 1914 it was divided between Britain and The Netherlands. In Dutch Borneo there had been a nationalist uprising 1903–8. Britain had three protectorates, Sarawak*, Brunei* (off which lay the Straits

Settlement island of Labuan), and British North Borneo, which had in fact been bought by the British North Borneo Company from the Sultan of Brunei in 1877. Railway development and deforestation were economic features before World War II, when the island was occupied by Japan (January 1942–September 1945). Afterwards both Sarawak and North Borneo became British Crown Colonies (1946). Dutch Borneo, as Kalimantan, associated itself with the struggle for Indonesian independence, which was finally won in 1949. In 1963 British North Borneo joined the Federation of Malaysia as Sabah*, as did Sarawak, a decision challenged by Indonesia in the Konfrontasi*. The Sultanate of Brunei eventually became an independent state.

Bosch, Juan (1909–), President of the Dominican Republic 1963. Active as a writer and journalist, he went into exile during the regime of the dictator Rafael Trujillo*, founding a left-wing party, the Partido Revolucionario Dominicano (PRD), while abroad. He returned to the Republic in 1961 after Trujillo's assassination and, appealing to the peasantry and the more liberal middle class, was elected President after a dazzling campaign. He took office in February 1963 after the first free elections for nearly forty years. He introduced sweeping liberal and democratic proposals for reform. In doing so he alarmed the USA and alienated the Catholic Church, the great landowners and industrialists, and the military. With probable CIA connivance he was overthrown in September by the army. His supporters launched a protest revolt in 1965, a movement which prompted a military intervention by the USA. In 1966 he was defeated for the presidency by Joaquín Balaguer*, who had heavy US backing. He remained active in politics and as a writer, and was again defeated by Balaguer in the presidential election of May 1990.

Bose, Subhas Chandra (1897–1945), Indian politician. Educated at Cambridge, he returned to India to become chief executive of Calcutta city. He became a supporter of Gandhi* and a member of the Swaraj Party*, and was deported to Burma 1924–7. On his return he was a member of the Indian National Congress*, where he pressed for immediate independence and became president of the All-India Trades Union Congress 1929–31. Rearrested, he was allowed to go into exile in Britain 1931–6. On his return he was again imprisoned, but then served as president of the Indian National Congress 1938–9, disagreeing with moderates over the pace of the independence movement. Placed under house arrest early in World War II, he escaped and went to Germany. Having failed to gain support there he travelled to Japan, where he announced the formation of an Indian National Army to drive the British from India. He recruited Indian prisoners of war and formed a provisional government-in-exile. His army suffered defeat before Imphal early in 1944. Deeply patriotic to the cause of India, he died in an air crash in Japan.

Bosnia and Hercegovina, one of the six constituent republics of Yugoslavia. Pan-Slavonic nationalism had provoked unrest through much of the 19th century. The area was occupied by Austro-Hungarian troops in 1875 and this was confirmed by the Congress of Berlin in 1878; in 1908 it was formally annexed to the Austro-Hungarian Empire. This provoked protests from both Russia and Serbia, and an international crisis was only averted when Germany threatened to intervene. Serbian terrorist activity continued, however, culminating in the assassination at Sarajevo*, which sparked off world war. In 1918 it was integrated into the new Kingdom of Serbs, Croats, and Slovenes, later renamed (1929) Yugoslavia. During World War II the two provinces were incorporated into the German puppet state of Croatia, and were the scene of bitter fighting by Yugoslav partisans. Along with the rest of Yugoslavia*, separatist pressures began to emerge in the late 1980s. In elections in December 1990 the Communists (LCY) were rejected and three nationalist parties formed a coalition.

Botha, Louis (1862–1919), 1st Prime Minister of the Union of South Africa 1910–19. An Afrikaner farmer, he fought in the Boer War, gaining promotion to the rank of general. After the Peace of Vereeniging (1902) he worked tirelessly for reconciliation with Britain. In 1910 he became Prime Minister and in 1911 founded the moderate South African Party. In 1915 some of his former supporters led a rebellion against him which he suppressed. He then led a successful campaign against German South-West Africa, which capitulated in July 1915. He attended the Paris Peace Conference in 1919, where he advocated, without success, leniency towards a defeated enemy.

Botha, Pieter Willem (1916–), Prime Minister of South Africa 1978–84; President 1984–9. Although from the Orange Free State, he moved to Cape Province in 1936, where he became an officer of the National Party, was elected to Parliament in 1948, and, a staunch supporter of apartheid*, held various government offices. In 1978 he succeeded B. J. Vorster* as Prime Minister. He continued the policy of granting 'independence' to Black homelands and made limited concessions to Asians and Coloureds; but he remained firmly opposed to enfranchisement of Blacks. As Commander-in-Chief of the armed forces he embarked on military involvement in South West Africa* (Namibia), Angola*, and Mozambique*, and encouraged destabilizing ventures in Zimbabwe and other southern African states. His policies precipitated right-wing opposition from White extremists, who formed the Conservative Party in 1982. Botha responded with the Constitution of 1983, which made him President. Meanwhile all Black political activity was banned under the state of emergency edict of July 1985. Following a stroke in December 1988, he lost his grip on his party and in September 1989 retired, a bitterly resentful man.

Botswana, a country in southern Africa, formerly known as Bechuanaland. British missionaries first arrived in 1801. In 1885 Britain declared it a protectorate to be administered from Mafeking. The success of the cattle industry in the early 20th century led the Union of South Africa to seek to incorporate Bechuanaland, along with Basutoland and Swaziland, but this was rejected by the British government in 1935; no transfer would be tolerated until the inhabitants had been consulted and agreement reached. The Tswana* people are dominant. Seretse Khama*, chief of one of its tribes, the Ngwato, was banned from the country from 1948 until 1956, allegedly for marrying an Englishwoman. By now a nationalist movement was developing, which culminated in a democratic constitution in 1965, followed by independence on 30 September 1966, as the Republic of Botswana, with Seretse Khama as President. He was succeeded on his death in 1980 by the Vice-President Dr Quett Masire (re-elected 1984). The country retains economic links with South Africa, although it is a member of SADCC* and since 1980 has moved closer to Zimbabwe.

Boumédienne, Houari (1927–78), President of Algeria 1965–78. Born in Algeria and educated in Paris and Cairo, he served in the French army in World War II before attending university in Cairo and becoming a schoolmaster. In the early 1950s he joined a group of expatriate Algerian nationalists in Cairo, which included Ben Bella*. In 1955 he joined the FLN* and was assigned to the district of Oran. In 1960 he became the FLN chief-of-staff, having been expelled to Tunisia. In March 1962 his forces occupied Algiers for Ben Bella, then still in France. He served the latter as his Minister of Defence, but, opposed to what he regarded as Westernizing policies, overthrew him in June 1965. He governed Algeria first as President of the Council of Revolution, but from December 1976 as the elected President of the Republic. His four-year plan 1969–73, which included nationalizing the oil industry, was backed by the Soviet Union, but he also managed to maintain economic links with Western countries. Essentially a Pan-Arab idealist, he was highly influential within the Arab League* and broke with Sadat* over the Camp David Accord* of 1978. His dream of a North African socialist federation of Arab states remained unrealized when he died. He was succeeded by President Chadli.

Bourguiba, Habib ibn Ali (1903–   ), President of Tunis 1957–87. Born in Tunis, he studied in Paris and on his return in 1921 joined the moderate nationalist party the Tunisian Constitutional Party (Destour). In 1934 he formed a more extreme group, the Neo-Destour Party, and was imprisoned by the French 1934–6 and 1938–42. Tunisia was occupied by Italian and then German forces, when he was again imprisoned for non-co-operation 1942–3. On liberation, French colonists again refused to recognize him and he moved to Cairo 1945, by now convinced of the need for total Tunisian independence. In 1949 he returned to Tunis only to be again imprisoned. The government of Mendès-France* (1954–5) recognized him as a moderate with whom it could negotiate. There followed an agreement leading to total independence (1956), when he was elected Prime Minister by a Constituent Assembly. In 1957 the Tunisian National Assembly proclaimed Tunis a republic, electing him first President. Re-elected four times, he was proclaimed President for life in 1975. His policy was basically that of moderate and gradualist socialism, for which he was criticized by his neighbours Boumédienne* and Qadhafi*, and by socialist extremists at home, who organized riots in 1975 and 1980. From 1981 he democratized the National Assembly, recognizing the right of opposition parties to exist. In failing health, he was deposed in 1987.

Bradley, Omar Nelson (1893–1981), US general. A graduate of the US Military Academy in 1915, he fought in France during World War I and rose steadily in rank to become commander of US forces in Tunisia and Sicily in World War II. He went on to command the US land forces in the Normandy campaign (1944), and later, following the Ardennes campaign*, as commander of the 12th Army Group of some 1¼ million men, swept through Europe to link up with Soviet forces on the Elbe in 1945. Chairman of the US Joint Chiefs of Staff 1948–53, he was responsible for the build-up of NATO and for formulating US global strategy. He supported the US involvement in the Korean War, but deeply distrusted General Douglas MacArthur*. Shortly after the latter had been dismissed by President Truman for wanting to carry the war into China, he told a Senate Committee that this would have been 'to fight the wrong war, at the wrong place, at the wrong time, with the wrong enemy'.

Brandt, Willy (1913–   ), West German Chancellor 1969–74. Born in Lübeck Herbert Ernst Karl Frahm, he worked there as a social worker and a keen Social Democrat, until in 1932 he had to flee from the Gestapo. He assumed the name of Willy Brandt and lived in Norway. In 1940 he moved to Sweden to work with the German Resistance movement. He returned to Berlin in 1945 and, as Mayor of West Berlin 1957–66, he resisted Soviet demands that Berlin become a demilitarized free city in 1958. He successfully survived the crisis arising from the building of the Berlin Wall* in 1961. In 1964 he became chairman of the Social Democratic Party, an office he held until 1987. He was elected Federal Chancellor in 1969, his main achievement being that of *Ostpolitik** towards Eastern Europe. In 1970 he negotiated an agreement with the Soviet Union accepting the *de facto* frontiers of Europe, making a second agreement in 1971 on the status of Berlin. In that year he also signed non-aggression pacts with Poland and the Soviet Union, accepting the Oder–Neisse boundary with Poland. In 1972 he negotiated an agreement with the German Democratic Republic recognizing the latter's existence and establishing diplomatic relations between the two nations. He resigned in 1974 over a spy scandal in his office, but accepted an invitation to chair the Independent Commission on International Development Issues, whose findings were published in 1980 as the Brandt Report*. He remained a highly popular figure in his adopted city of Berlin.

Brandt Report. Convened by the UN, an International Commission on the state of the world economy met 1977–9 under the chairmanship of Willy Brandt*. In its report *North–South: A Programme for Survival 1980*, it recommended urgent improvement in trade relations between the rich northern and poor southern hemispheres. Governments in the north were reluctant to accept its recommendations and members of the Commission therefore reconvened to produce a second report, *Common Crisis North–South: Co-operation for World Recovery, 1983*. This perceived 'far greater dangers than three years ago', forecasting 'conflict and catastrophe', unless the imbalances in international finance could be redressed.

Brătianu, Ionel (1864–1927), six times Prime Minister of Romania 1909–27. Son of a leading Romanian statesman Ion Brătianu, he became leader of the Romanian Liberal Party, whose aim was that of Greater Romania, that is, to regain Moldavia* and Transylvania*; it was also, however, an advocate of land reform on behalf of the peasantry. Prime Minister through World War I he tried to maintain neutrality, but in 1916 was persuaded to enter the war on the side of the Allies in the hope of territorial aggrandizement. This he fought for successfully at the Paris Peace Conference*, winning much of what he wanted: Transylvania from Hungary, Moldavia from the new Soviet Union, and Dobrudja* from Bulgaria. He continued in office after the war, encouraging Ferdinand I to exile the Crown Prince Carol* in 1925.

Brauchitsch, Walter von (1881–1948), German field marshal. A professional soldier serving in World War I and between the wars, by 1938 he was Commander-in-Chief of the German army. He directed operations for the Austrian Anschluss* and the occupation of Sudetenland. Using blitzkrieg* techniques, his armies successfully conquered Poland, Norway, Denmark, The Netherlands, and France within nine months of war being declared in September 1939. However, he was relieved of his command by Hitler, as a scapegoat for the German failure to capture Moscow in 1941.

Brazil, a South American republic comprising 48 per cent of the continent. Under its 1988 Constitution it is a federal republic of twenty-three states: Acre, Alagoas, Amazonas, Bahia, Ceará, Espírito Santo, Goias, Maranhão, Mato Grosso, Mato Grosso do Sul, Minas Gerais, Pará, Paraîba, Paraná, Pernambuco, Piauí, Rio de Janeiro, Rio Grande do Norte, Rio Grande do Sul, Rondônia, Santa Catarina, São Paulo, Sergipe. In addition there is a Federal Territory of Amapá Roraima. During the first two decades of the Republic (established 1889), with its main exports of coffee relatively steady, there was reasonable political and social stability. During the 1920s, however, there was mounting social unrest, culminating in a major revolt in 1930 and the establishment by the military of Getúlio Vargas* as provisional President. With continued military support he introduced his *Estado Novo* in 1937, being virtual dictator until 1945 and bringing Brazil into World War II in 1942. Vargas resigned in 1945 but was re-elected in 1951, his second presidency being marred by extensive corruption. His successor Juscelino Kubitschek* embarked on an ambitious programme of economic expansion, including the construction of the futuristic capital Brasilia, intended to encourage development of the interior. His successors Jânio Quadros (1960–1) and João Goulart (1961–4) had to face the consequent inflation and a severe balance-of-payments deficit. In rural areas peasant leagues mobilized behind the cause of radical land reform. Faced with this threat, Brazil's landowners and industrialists backed a military coup in 1964. Over the next twenty-one years a series of authoritarian military regimes sought to attract foreign investment, promoting industrialization, a policy which increased the inequalities of wealth. Under President General Figueiredo* (1978–84) steady moves were made towards re-establishing democracy. In January 1985 President Tancredo Neves of the Democratic Alliance party was elected, the first civilian President since Goulart. He died without taking office, but his successor Vice-President José Sarney (1985–9) sought to grapple with the problems of this vast country: a mounting external debt ($120,000m. in 1987); urban and industrial unrest; Indian displacement; and large-scale deforestation (200,000 sq. km. of virgin tropical forest in 1987). With inflation running at 85 per cent p.a. Fernando Collor de Mello won the 1989 elections as President. A wealthy landowner from Alagoas, his conservative National Reconstruction Party (PRN) introduced stringent economic measures, including prices and incomes freeze and increased taxes.

Brazzaville Bloc. It was a declared policy of the UN that as former colonies became independent there should emerge regional groupings to aid economic and political development. One of the first of these was the Brazzaville Bloc, which resulted from a conference in Brazzaville in 1960. The following states joined: Madagascar, Congo, Gabon, the Central African Republic, Chad, Niger, Dahomey (Benin), Upper Volta (Burkina Faso), Togo, Ivory Coast (Côte d'Ivoire), Senegal, Mauritania. A further Conference of Heads of African States in May 1961 in Monrovia established an Organization of Co-operation of the African and Malagasy States. This included all the Brazzaville states but also Ethiopia, Liberia, Libya, Niger, Nigeria, Sierra Leone, Somalia, and Tunisia. It was this organization which developed into the OAU* in 1963, which then superseded the Brazzaville Bloc.

Brazzaville Declaration. Leaders of French West Africa and Equatorial Africa met General de Gaulle in Brazzaville in July 1944. They called for reform of French colonial rule. The Declaration at the end of the meeting confirmed the continuation of the French colonial Empire, rejecting the foreseeable possibility of independence. It did, however, recommend the establishment of territorial assemblies, participation in elections for the Assemblée Nationale in Paris, and the guarantee of equal rights for all citizens. The French Union* of 1946 confirmed the Declaration. This failure to accept the pressures for decolonization paved the way for colonial wars in Indo-China, Algeria, and elsewhere.

Brecht, Bertolt (1898–1956), German poet and playwright. Born in Augsburg, when a student he was strongly influenced by Marxism* and by Expressionism*. He adopted the aesthetic of socialist realism* and aimed to develop an 'epic theatre' which would arouse a direct political judgement in the audience faced with moral and political dilemmas. To achieve this he rejected the traditional Aristotelian theory of drama by seeking an 'alienation effect' (*Verfremdungseffekt*): distancing that audience from the action and characters, so that they could see more clearly the social and political contexts in which the play is set. His close collaboration with Kurt Weill resulted in such masterpieces as *The Threepenny Opera* (1928) and *Mahagonny* (1930). Persecuted by the Nazis, he fled to the Soviet Union in 1933, and in 1941 he went to the USA. As artistic director of the Berlin Ensemble in East Berlin 1949–56 his productions of such plays as *The Life of Galileo*, *Mother Courage and Her Children*, and *The Caucasian Chalk Circle* were seminal events for 20th-century theatre. He refused to condemn the repressive regime of the German Democratic Republic.

Brest-Litovsk, Treaty of (1918). In December 1917 a conference was convened in the Polish city attended by Germany, Austria-Hungary, and Bolshevik Russians, in order to end the war on the Russian front. Trotsky* skilfully prolonged discussions in the hope of Allied help, or of a socialist uprising by German and Austrian industrial workers. Neither happened and the German army resumed its advance. Lenin* then ordered his delegates to accept the German terms. Russia was to surrender nearly half of its European territory: Finland, the Baltic states, west Belorussia, Poland, the Ukraine, and parts of the Caucasus. The treaty was signed on 3 March 1918 and was annulled by the German armistice in November, and the Treaty of Versailles. War between the Allies and the Bolsheviks (War of Intervention) followed 1919–21, but only the Ukraine was regained, the Soviet Union failing to defeat Poland 1920–1.

Bretton Woods Conference, an international finance conference. In July 1944 representatives from forty-four nations met at Bretton Woods, NH, to consider stabilization of world currencies and the establishment of credit for international trade in the post-war world. The Bretton Woods Final Act proposed an International Bank for Reconstruction and Development (the World Bank*) and a plan for an International Monetary Fund* to finance short-term imbalances in international trade. The conference hoped to see exchange controls and discriminatory tariffs ended as soon possible. The Bank and the IMF became specialized agencies of the UN.

Brezhnev, Leonid Ilyich (1906–82), Soviet statesman. Born in Dneprodzerzhinsk in the Ukraine, he graduated as an engineer, after which he specialized in agricultural engineering, becoming a Communist Party official in 1938. He was elected to the Supreme Soviet of the Union in 1950. In 1960 he became President of the Praesidium and in 1964 succeeded Khrushchev* as first secretary of the Communist Party. Through these two offices he came to exercise effective control over Soviet policy, although initially he shared power with Kosygin*. He was largely responsible for the decision to invade Czechoslovakia in 1968, maintaining that one socialist state may interfere in the affairs of another if the continuance of socialism is at risk. In foreign affairs he initiated a policy of *détente*, culminating with SALT I* in 1972. But his decision to send troops into Afghanistan (1979) intensified the Cold War* throughout the last years of his life.

Briand, Aristide (1862–1932), eleven times Prime Minister of France. The son of a Breton farmer, he first became a Deputy in 1903, a strong socialist and impressive orator. In 1905 he took a leading part in achieving the separation of Church and State in France. He was Prime Minister 1909–10, when he lost popularity for using the military against industrial strikes. His wartime ministry 1915–17 was accused of a lack of vigour. In the 1920s he was a powerful advocate of peace and international co-operation through the League of Nations. The cabinet he headed in 1921–2 fell because of his criticism of French demands for retribution against the Germans, following the Treaty of Versailles. Working closely with Austen Chamberlain* and Gustav Stresemann*, the British and German foreign ministers, his greatest achievements were the Locarno Pact* of 1925 and the Kellogg–Briand Pact* of 1929, which represented a return of the USA to international commitments, after its period of isolationism. He had dominated French foreign policy throughout most of the 1920s. In 1930 he put forward a plan for a

Union of European States, but the onset of the Depression defeated him and he retired to his Normandy farm.

Britain, Battle of (August–October 1940), a series of air battles between the Royal Air Force and the German *Luftwaffe*. After the fall of France in June 1940 German aircraft launched an offensive against British coastal shipping, as a prelude to launching an invasion. These attacks then moved against southern England, the main air defence being a force of some 700 Hurricane and Spitfire fighters, under the command of Air Chief Marshal Hugh Dowding*. German dive-bomber losses were heavy, seventy-five being shot down on 15 August. In late August and early September mass bomber attacks were launched against fighter airfields, installations, and factories. Although there were heavy British losses, Fighter Command on 15 September shot down fifty-six bombers (claiming at the time a score of 185). Hitler now ordered the offensive to be diverted to British cities, the day-bombing of London again causing heavy German casualties. On 1 October day-bombing of cities was replaced by night-bombing, but by this time it was clear that German losses were so high that the attempt to destroy British air power had failed. Consequently on 12 October plans for the invasion of Britain were indefinitely postponed. Though heavily outnumbered by the Germans, the British lost some 900 aircraft against some 1,700 German planes. Radar, used by the British for the first time in battle, made a significant contribution. Four hundred and forty-nine British fighter pilots were killed.

British Empire. In 1884 an Imperial Federation League put forward a proposal for political federation between Britain and its colonies. The scheme was rejected by the colonial premiers when they gathered in London for Queen Victoria's jubilees. Strategically the key area of the late 19th-century British Empire was seen to be southern Africa, and it was the dream of Cecil Rhodes and Alfred Milner* to create a single Cape-to-Cairo dominion, linked by railway and acting as a pivot for the whole Empire. The dream faded with the Boer War 1899–1902. Another result of this war was the creation of a permanent Committee of Imperial Defence, which was to continue until 1938. The Empire reached its zenith *c.*1920, when German and Turkish mandates* were acquired and over 600 million people were ruled from London. In the later 19th century movements for home rule had begun in all the White colonies, starting with Canada, but spreading to Australasia and South Africa, such moves resulting in 1931 in independent dominion status for these lands. Although the Indian National Congress* had been formed in 1885, achievement of similar self-government for the non-White peoples of the Empire proved much more difficult. It was only after 1945 that the process of decolonization* began; by 1980 this was largely complete.

British Expeditionary Force (BEF). Following army reforms of Richard Haldane* in 1906–7, a territorial reserve army had been created in Britain. This was immediately mobilized when war was declared on 4 August 1914 and, together with regular troops, sent to France under Sir John French. As the Germans advanced into France, this BEF of six divisions moved up the German flank towards Belgium before being halted and defeated at the Battle of Mons (23–4 August). From here they steadily retreated to Ypres*, where they took part in the first Battle of Ypres (20 October–17 November). It is estimated that by the end of November survivors from the original force averaged no more than one officer and thirty men per battalion of approximately 600 men. An expeditionary force was again mobilized and sent to France in September 1939.

British Raj, a term for the British Empire of India, which perhaps reached its apogee when Lord Curzon* was Viceroy (1899–1905), or at the durbar of George V in

1911. The Indian National Congress*, which initiated nationalist criticism of the Raj and eventually succeeded it, had been founded in 1885. British control over the directly ruled territories (about three-fifths of the total area) was exercised by a Secretary of State in the British cabinet and a Viceroy in India. The administration was staffed by the Indian Civil Service, which in later years was gradually opened to Indians, although Europeans and Indians remained socially separate. In addition there were in India some 700 princely states, bound by treaty to the Crown, where the princes preserved control over domestic affairs. There was a large British army garrison, together with the Indian army*, with British officers in controlling positions until the 1920s and recruited from British and Indian ranks. From 1920 onwards the position of the Raj was steadily weakened, and when Britain transferred power in 1947 to the new states of India and Pakistan, all British personnel were withdrawn. Yet Western modes of thought and practice, channelled through the educational system, made a lasting contribution to the subcontinent's character.

Brownshirt, member of the Nazi paramilitary organization the *Sturmabteilung*, or SA. Fitted out in brown uniforms reminiscent of Mussolini's Blackshirts*, the SA was recruited from various rough elements of society, being founded by Adolf Hitler in Munich in 1921. They figured prominently in early Nazi marches and rallies, and their methods of violent intimidation and Jewish persecution played a key role in Hitler's rise to power. The SA was led from 1931 by a radical anti-capitalist, Ernst Röhm, whose ambition was that the SA should achieve parity with the army and the Nazi Party, and serve as the vehicle for a Nazi revolution in state and society. For Hitler the main consideration was to ensure the support and loyalty of the German officer corps and of leading capitalist industrialists. In June 1934 therefore he had Röhm and more than seventy leading members of the SA summarily executed by the SS on the 'Night of the Long Knives'*. The anti-capitalist revolutionary phase of Nazism was ended.

***Brown* v. *Board of Education of Topeka,* Kansas** (1954), a US Supreme Court* case. The Board of Education in Topeka, eastern Kansas, had established separate White and Black schools, in accordance with the Supreme Court decision of *Plessy* v. *Fergusson* of 1896. The Board's policy was challenged by the National Association for the Advancement of Colored People*. The case was brought before the Supreme Court, where it was argued by a Black lawyer, Thurgood Marshall, who was himself to be the first Black justice appointed to the Court in 1967. The Court under Chief Justice Earl Warren* found unanimously that racial segregation in schools violated the Fourteenth Amendment of the Constitution, thus reversing the decision of 1896 and opening the way for desegregation not only in schools but in other public facilities. A second case before the Court in 1955 produced the phrase 'desegregation with all due speed'.

Bruce, Stanley, 1st Viscount (1883–1967), Prime Minister of Australia 1923–9. From a wealthy Victoria family, he was educated at Cambridge and called to the English Bar before serving in World War I, being wounded at Gallipoli. On returning to Melbourne in 1918 he joined the Nationalist Party of William Hughes*, but quarrelled with him in 1923, believing more strongly in free enterprise. He formed a coalition government with the Country Party* led by Earle Page*, encouraging immigration and economic expansion. He was defeated by Labor in the election of October 1929 and in 1931 handed the leadership of the Nationalists to J. A. Lyons*. He remained a much-respected Australian figure being President of the Council of the

League of Nations 1936. He served as High Commissioner in London 1933–45, attending meetings of the war cabinet and the Pacific War Council (1942–5). He lived in England for the rest of his life, becoming chairman of the World Food Council 1947–51 and of the Finance for Industry Corporation 1947–57.

Brundtland, Gro Harlem (1939–), Prime Minister of Norway 1981, 1986–9, 1990– . Born in Oslo, daughter of a Labour Party politician, she attended medical school, where she married, and went on to become Medical Officer of Health in Oslo. In 1974 she entered politics, being invited by the Labour government then in power to become Minister of the Environment (1974–9) and being elected to the Storting. Her main interests have been 'women's values', preventive medicine, and the environment. In 1985 she was appointed chair of the UN World Commission on Environment and Development, whose report was published in 1988. It argued that the protection of traditional, life-sustaining rights of land and resources of such areas as the Arctic, Australia, tropical Africa, and the Amazon Basin was critical for the preservation of these environments and for global survival. After a brief period as Prime Minister in 1981, she came into office in 1986 to lead a minority government. She survived two votes of no confidence, but lost the election of September 1989 to a conservative coalition. In November 1990 she returned to office again to lead another minority government.

Brunei, an independent sultanate in north Borneo. In 1888 the Sultan had accepted British protection, which in 1906 was extended through the appointment of a British Resident. The protectorate's economy was revolutionized by the discovery of substantial onshore oil deposits in 1929. Occupied by the Japanese in 1942, it was liberated by the Australians in 1945. In 1965 offshore oil and gas fields were found. Partly because of these natural resources and the enormous wealth they have provided, the sultanate resisted pressure to join the newly formed Federation of Malaysia* in 1963. Following a treaty in 1979, full independence from Britain was achieved 31 December 1983. A state of emergency has been in force since December 1962.

Brüning, Heinrich (1885–1970), Chancellor of the Weimar Republic 1930–2. He graduated in economics at Bonn before serving in World War I. He entered the Reichstag in 1924, a Catholic member of the Centre Party*. Appointed Chancellor to stabilize the economy, he tried to solve Germany's economic problems by unpopular deflationary measures such as higher taxation, cuts in government expenditure, and attempts to reduce reparations* payments. After the elections of 1930 he lost majority support in the Reichstag and ruled by emergency decrees. He was forced to resign in 1932 by President Hindenburg*, whose confidence he had lost. He escaped the 1934 purges by fleeing to Holland and then to the USA, where he became a lecturer at Harvard (1939–52).

Brusilov, Aleksey (1853–1926), Russian general. Commissioned a cavalry officer into the Russian Imperial Army, he fought against Turkey and Japan before World War I. In June 1916 he launched an offensive against Austria between the Pripet Marshes and the Carpathians, capturing some 250,000 prisoners at Lutsk. The offensive cost even more Russian lives and came to a halt because of a failure of Russian supply lines. Yet it forced the Germans to withdraw men from the Somme and prompted Romania to enter the war on the Allied side. After the 'February' Revolution of 1917 he held supreme command under the provisional government* and stayed loyal to the Bolsheviks after they took power. He was adviser to the Red Army in its war against Poland 1920–1 and then inspector of cavalry 1923–6.

Bryan, William Jennings (1860–1925), US Secretary of State 1913–15. Born in Illinois, he built up a successful law practice in Nebraska before being elected to the US Congress. Here he attacked the policy of maintaining a gold standard* ('you shall not crucify mankind upon a cross of gold'), and winning the presidential nomination for the Democrats in 1896. He failed to get elected then or in 1900 and 1908, but supported Woodrow Wilson for the 1912 election, becoming his Secretary of State. He wished to keep the USA out of World War I and resigned in June 1915 over Wilson's letter to Germany protesting at the sinking of the *Lusitania**. In 1925 he appeared for the prosecution in the celebrated Scopes Trial*. He won the case but died five days after the trial ended.

Buck, Sir Peter (1880–1951), New Zealand politician and anthropologist. Illegitimate son of a William Buck, he was born Te Rangi Hiroa. He was brought up by a Maori chieftain who sent him to Te Aute College. He trained as a medical officer of health, joined the Young Maori movement*, and with Maui Pomare, a fellow health officer, launched a successful campaign 1905–14 to improve Maori health. In 1909 he entered Parliament for the North Maori electorate and was Minister for the Maori Race 1912–14. He resumed his public health work after World War I as Government Director of Hygiene, before becoming a major scholar on Maori and Polynesian culture. From 1927 he spent the rest of his life working at the Bernice Bishop Museum of Ethnology in Honolulu, becoming its director and visiting professor of anthropology at Yale University.

Buddhism, a major world religion. Like other eastern religions this was regarded by many in 1914 as of decreasing influence in the world. In fact it was to see a great revival in the 20th century in south and south-east Asia and in Japan, as well as growth in Europe and the USA. An eightfold path to morality, described in the *Samyutta-nikaya* (right views, intention, speech, conduct, occupation, effort, mindedness, and concentration) leads ultimately to nirvana, when the soul is released from the cycle of birth and death. The monastic vocation has grown in popularity, *bhikkus* or monks having to observe over 200 rules, which include sexual abstinence, no taking of life, possessing nothing that has not been freely given, and making no claims regarding spiritual attainments. They may not eat between noon and the following morning; food must be freely donated and must be vegetarian. The sixth Buddhist Council was held in Rangoon, May 1954–May 1956, to commemorate the 2,500th anniversary of the death of Gautama Buddha. It was attended by monks from Burma, India, Pakistan, Sri Lanka, Nepal, Cambodia, Laos, and Thailand. Theological differences exist between Buddhists from Sri Lanka, Burma, and Thailand (the Theravada or Little Vehicle) and those from Japan, Korea, China, Tibet, and Nepal (the Mahayana or Great Vehicle). Zen Buddhism developed in Japan, with a strong emphasis on mysticism and the need to open the 'doors of perception' into a world of wonder. It was particularly influential in the USA during the 1960s. Buddhism has continued to prove an attractive alternative to many who are unable to accept the more theologically based morality of Christianity or Islam.

Bukharin, Nikolai Ivanovich (1888–1938), Soviet leader and theoretician. Born in Moscow, he trained as an artist, and then studied economics when he became a member of the Bolshevik faction of the Social Democratic (Workers') Party in 1906. He first met Lenin* in Cracow in 1912, where he helped to edit the party newspaper *Pravda*, and then went to New York, where in 1916 he edited *Novy mir* (New World). He returned to Russia in 1917 and played an active part in events leading to the 'October' Revolution, being a close friend and aide of Lenin. He became editor of *Pravda* and argued with Lenin over Russia's withdrawal from World War I, pressing for a mass European revolution. During the 1920s he published a number of important theoretical books, including *The ABC of Communism*, and supported the New Economic Policy, arguing that industrialization required a healthy agricultural base. He opposed those who were pressing for collectivization of agriculture. At that time Stalin* agreed with him, against arguments being put by Trotsky*, Zinoviev*, and Kamenev*. In 1926, as a member of the Politburo, he became chairman of the Comintern's executive committee. In 1928 Stalin reversed his views on collectivization and Bukharin was expelled from the Politburo for opposing him. He became editor of *Izvestia* and retained some influence, helping to draft the 1936 Constitution of the USSR. In January 1937, however, he was arrested, expelled from the party, and accused of wishing to restore bourgeois capitalism and of espionage. He was convicted and executed following his trial in March 1938. In 1989 he was rehabilitated as a major socialist theoretician and accepted in the Soviet Union as a martyr to Stalin's craze for power.

Bukovina (beech-tree country), a disputed area in the Carpathian foothills. It was part of the Austro-Hungarian Empire, inhabited mostly by Ukrainians. In 1919 it was ceded to Romania, but in June 1940 Hitler put pressure on Romania to cede it to the Soviet Union, as part of the Nazi–Soviet Pact. Romania agreed to cede north Bukovina but reclaimed it in 1941. In 1944 Soviet troops occupied it and it became part of the Ukrainian Soviet Socialist Republic. South Bukovina remains part of Romania.

Bulganin, Nikolai Alekandrovich (1895–1975), Soviet statesman. Born in Nijni Novgorod, he fought in World War I and joined the Communist Party at the time of the Revolution in 1917, after which he joined the secret police Cheka*. In 1931 he became Mayor of Moscow and held various offices in the city during the next twelve years; during the siege of the city 1941–3 he organized its defence. After the war he was appointed a marshal of the Soviet Union for his wartime achievement and in 1946 he succeeded Stalin as Minister of Defence. He was Premier and Chairman of the Council of Ministers 1955–8, during which time he shared power with Khrushchev*. The latter replaced him as Premier in March 1958; he lost his membership of the party Praesidium and became chairman of the Soviet State Bank.

Bulgaria. In 1908 the German Prince Ferdinand of Saxe-Coburg, taking advantage of the Young Turk* movement, formally proclaimed Bulgarian independence from Ottoman rule and was crowned King Ferdinand I (1908–18). Participation in World War I on the side of the Central powers led to invasion by the Allies in 1916 and the loss of territory, mostly to Romania. Between 1919 and 1923 Bulgaria was virtually a peasant dictatorship under Alexander Stamboliisky*, leader of the Agrarian Union. He was murdered and an attempt by Communists under Dimitrov* to seize power followed. Military and political instability persisted until 1935, when an authoritarian government was established by King Boris III*. His half-hearted co-operation with Germany was followed in 1944 by invasion by the Soviet Union. In 1946 the monarchy was abolished and a Communist state proclaimed. Bulgaria became a staunch member of the Warsaw Pact*, which it remained until disturbances in 1989. On 3 November some 4,000 people demonstrated in Sofia, the largest such event since 1947. On 9 November Todor Zhivkov*, who had been general secretary of the Bulgarian Communist Party since 1954, was ousted from power. His successor was Petur Mladenov, who promised democratization and reform. Hard-liners were dismissed and an amnesty declared for political prisoners. In January 1990 there was a demonstration by some 40,000 in Sofia, following which the National Assembly revoked the

Communist Party's monopoly of power. Zhivkov was indicted and the secret police disbanded. In April Mladenov was elected President by the Assembly, which then dissolved itself, the Communist Party changing its name to the Bulgarian Socialist Party (BSP). Seven parties took part in the June elections, which the Socialists narrowly won in what were considered fair proceedings. Mladenov was later obliged to resign, to be replaced by Zhelyn Zhelev of the Union of Democratic Forces, with the BSP in coalition. A massive general strike in November 1990 resulted in both a political and economic crisis, the BSP losing all credibility.

Bunche, Ralph (1904–71), UN diplomat. Born in Detroit, he won a scholarship to Harvard University and had a distinguished career there before joining the staff of Howard University, where he taught political science. He travelled widely in Africa studying French colonial policy and published a major book on race relations, *An American Dilemma* (1944). During World War II he served with the US Joint Chiefs of Staff and the State Department. In 1946 he joined the Secretariat of the UN and served on the Palestine Peace Commission in Palestine in 1947. After Count Bernadotte* was assassinated in 1948 he carried on negotiations with such skill that he was able to arrange an armistice between the warring Arabs and Jews. For this he was awarded a Nobel* Peace Prize, the first awarded to a Black American. He served as Director of the Trusteeship Division of the UN 1948–54 and then became Under-Secretary for Political Affairs. As such he was responsible for UN peace-keeping ventures in Suez (1956), the Congo (1960), and Cyprus (1964). He was a staunch supporter of NAACP*.

Burger, Warren E. (1907–), Chief Justice of US Supreme Court 1969–86. After a distinguished legal career he was called to the Minnesota Bar in 1931 and in 1955 was appointed judge in the Court of Appeal in the District of Columbia. In May 1969 President Nixon appointed him Chief Justice to succeed Chief Justice Warren*. In 1971 the Court upheld a policy of busing*, although to little avail, as it recognized in 1974 in its judgement on *Milliken* v. *Bradley*, which accepted the reality of racial segregation in housing. To the annoyance of the law-enforcement lobby, it upheld the ruling of *Miranda* v. *Arizona**. In the 1978 *Bakke* case it supported 'positive discrimination' in favour of disadvantaged candidates for college admission, that is, Blacks or Hispanics, even though it also ruled that in this particular case a rejected White candidate Allan Bakke be admitted. In 1974 Burger wrote a judgement for the case of *United States* v. *Richard M. Nixon*, in which he confirmed that the Supreme Court and not the President was the final arbiter of the US Constitution.

Burkina Faso, an inland country of western Africa. Known as Haute-Volta (Upper Volta) until 1984, it was a French protectorate from 1898, originally attached to Soudan (now Mali*). In 1958 it became an autonomous republic within the French Community* and it became independent in 1960. In 1966 a military coup overthrew President Maurice Yaméogo, and a new constitution was adopted in 1977. Since then there have been a series of military governments, 1980, 1982, 1983, and 1987, when President Thomas Sankara was assassinated. He was succeeded by Captain Blaise Compaoré, leader of a left-wing Popular Front. He has sought to follow moderate policies, with some restoration of both the authority of local chiefs and the activities of trade unions.

Burma (Myanmar), a country in South-East Asia. Britain had annexed Burma in three stages, 1824–6, 1852, and 1885, when a war lasting a week had given it all the

Kingdom of Upper Burma, whose King Thibaw was exiled. Guerrilla resistance to this annexation had continued, while British exploitation of the country was rapidly advancing—timber, rubies, oil, rice. Graduates of the first generation of students from Rangoon Liberal Arts College were resentful of British colonialism and during the 1920s student strikes increased as part of the so-called Thakin movement. In 1931 there was an uprising of Burmese peasants deprived of land by European companies; armed with sticks and stones they resisted British and Indian troops for two years. In 1936 Aung San* and U Nu* emerged as student leaders. Britain now moved to some form of home rule, recognizing Burma as a Crown Colony separate from India. Under a new constitution Ba Maw was elected Prime Minister. By now, however, there were demands from the Dobama Asi-ayone (Thakin) Party for full independence. A warrant was issued for the arrest of Aung San, but he escaped to China and then to Japan, which promised to sponsor independence. Aung returned in secret to Burma to recruit twenty-nine fellow students; these 'thirty comrades' were to form a cadre for a new government. When Japanese troops moved into Malaya Aung announced the formation of the Burmese Independence Army, while Ba Maw was proclaimed head of state. By 1943, however, disillusion with Japan had set in and Aung San's forces joined the British side. The Anti-Fascist People's Freedom League was formed, which successfully negotiated independence (1948) following the assassination of Aung San. Burma elected to remain outside the Commonwealth. The government of U Nu faced many challenges, particularly from the Karens in the Irrawaddy delta and the Chin, Kayah, and Kachin hill tribes. It succumbed in 1962 to an army coup led by Ne Win*, who had been fighting the guerrillas. He established an authoritarian state based on quasi-socialist and Buddhist principles, and maintained a policy of strict neutrality and limited foreign contact. When he retired in 1986 U San Yu became chairman of the governing Burma Socialist Program Party. The Burmese Communist Party and various ethnic insurgent groups which had been in revolt since independence remained a major challenge to his government. From being one of the richest countries of South-East Asia Burma had become one of the poorest and most isolated during the twenty-five years of Ne Win's power. In September 1988 General Saw Maung seized power, imposing martial law, with military tribunals executing hundreds of civilians. The country adopted the name Myanmar, but its social, economic, and political problems only worsened. During 1989 Aung San's daughter Daw Aung San Suu Kyi emerged as a popular figure. Although she was under house arrest throughout 1990, her party, the National League for Democracy (NLD), won a two-thirds majority for a Constituent Assembly in elections in May. Saw Maung's State Law and Order Restoration Council (SLORC) refused, however, to allow the Assembly to meet, and arrested more NLD leaders.

Burma campaigns, World War II (January 1942–May 1945). On 20 January 1942 two Japanese divisions advanced into Burma, accompanied by the Burma National Army of Aung San*. They captured Rangoon and forced the British garrison to begin the long evacuation. They quickly reached Lashio at the southern end of the 'Burma Road', thus cutting off the supply link from India to China. In May they took Mandalay and the British forces under General Alexander* withdrew to the Indian frontier. During 1943 there were attempts to reassert control over the Arakan, but these failed, although Wingate*, with his Chindit units, organized effective guerrilla activity behind Japanese lines, where an originally pro-Japanese population was becoming increasingly disillusioned. Early in the spring of 1944 heavy fighting took place in defence of Imphal, when an attempted Japanese invasion of Assam and north India was deflected in a series of bloody battles, of which Kohima was the most important. In

October an offensive was launched by British and Commonwealth troops, and US-supported Chinese Nationalists under General Stilwell*. The Burma Road was reopened in January 1945. By now a discontented Aung San had contacted Mountbatten* and in March his troops joined the Allies. As General Slim's 14th Army advanced down the Irrawaddy, a force of Indian, Gurkha, and West African forces moved through the jungle of the Arakan, supported by air-drop and amphibious operations. Here the Japanese headquarters Akyab fell in January, while inland Mandalay fell to Indian and British troops, after fierce fighting, on 20 March. Rangoon was attacked by land and sea and fell on 2 May, and by 17 May Burma had been recaptured.

Burnham, Linden Forbes Sampson (1923–85), Prime Minister of Guyana 1964–80; President 1980–5. Born in Kitty, British Guiana, he graduated in law at the University of London (1947) and then returned home to practise law and enter politics. In 1950 he was one of the founders of the People's Progressive Party (PPP). The party split in 1955, a more radical wing under the Marxist Cheddi Jagan, which attracted the Indian vote, going on to win elections to the House of Assembly. A constitutional change by Britain, introducing proportional representation, enabled Burnham to become Premier and, with independence in 1966, Prime Minister. From then onwards his policies became increasingly left wing. He established relations with Cuba and nationalized US bauxite mines and British sugar plantations and refineries. In 1980 under a new constitution he was declared President. Economic stagnation, resulting partly from his policies of 'co-operative socialism', led to considerable unrest and fears of a military coup at the time of his death.

Burns, John Elliott (1858–1943), British trade union leader and politician. Born in London of Scottish descent, he was largely self-educated, having worked in a factory as a child and then been apprenticed as an engineer. He was a radical socialist as a young man, and the London Dockers' Strike of 1889 owed much of its success to his leadership. Burns was one of the first Labour representatives to be elected to Parliament in 1892, but he fell out with Keir Hardie and turned his back on socialism. As a supporter of the Liberal Party he became president of the Local Government Board, introducing the first Town Planning Act in 1909. In 1914 he was President of the Board of Trade, but resigned in protest against Britain's entry into World War I. He remained in Parliament until 1918 and in later years lived in retirement, never joining the Labour Party.

Burundi, a country in East/Central Africa. It was ruled as a monarchy in the 19th century by *mwamis* (kings) of the Tutsi* people, who dominated a peasant population of Bahutu (Hutu*). Germany annexed it in the 1890s and in 1919 Belgium obtained a League of Nations mandate* and ruled it as part of Ruanda-Urundi, which in 1962 became independent. In 1964 its union with Rwanda* was dissolved. Burundi became a republic in 1966, when the traditional role of the *mwamis* was ended. Tribal rivalries and violence continued. In 1972 there was an alleged plot against President Micombero to restore the ex-Mwami Ntare V. This sparked off terrible violence. Some 5,000 Tutsi died and over 100,000 Hutu peasants were killed, but Tutsi dominance survived. In 1976 and 1987 there were military coups, the latter led by Major Pierre Buyoya, who became President. In 1988 there was renewed violence, with some 5,000 Hutu dead.

Bush, George (1924–), President of the USA 1989– . Son of a successful banker and Senator for Connecticut, he served in the US navy as a pilot in World War II.

After graduating at Yale University he moved to Texas, where his success in the oil industry took him into politics. In 1966 he was elected as a Republican to the House of Representatives, where he served until 1971, when President Nixon appointed him US ambassador to the UN. In 1974 he headed the Republican Party Committee calling on Nixon to resign. He headed the US Liaison Office in China 1974–5, and was then placed in charge of the CIA 1976–7. In 1980 he was elected Vice-President under Ronald Reagan, an office he held until his election as President in 1988. Since taking office he has grappled with the problem of the USA's balance-of-payments deficit, while supporting President Gorbachev's attempts to reduce global tensions. He gained considerable popularity in the USA for his intervention in Panama* 1989–90, and in August 1990 faced the tremendous challenge of the Gulf Crisis*.

Busing, an educational policy in the USA during the later 1960s. Children were taken by bus from Black or Hispanic neighbourhoods usually to what had been White suburban schools and vice versa, in order to secure racially integrated schooling. The desegregation movement had mainly affected the Southern states, where busing was first introduced against strong opposition from White families. But it was also used in many northern cities where *de facto* segregation also existed, the central areas being inhabited almost entirely by Blacks. In 1971 the Supreme Court approved the principle of busing, but in 1972 Congress ordered that further schemes should be delayed. Busing remained a controversial issue, but its use steadily declined as a means of racial integration.

Bustamante, Sir William Alexander (1884–1977), Prime Minister of Jamaica 1962–5. Born in Jamaica of Irish parents, he had an adventurous career in Latin America before returning to Jamaica to become a trade union organizer. In 1938 he founded the Jamaica Labour Party, but was interned in 1940 as a 'militant'. By 1949, however, he was recognized by the British as the leading political spokesman of Jamaica. As such he opposed the West Indian Federation*, and when this broke up he became his country's first Prime Minister on independence from Britain in 1962. During his years in office he initiated an ambitious five-year plan which included major public works, agrarian reform, and social welfare. However, he was obliged to retire in 1965 through ill health.

Buthelezi, Gatsha Mangosuthu (1928–), Zulu tribal chief (1952) and Chief Minister of KwaZulu (1972). He is the grandson of King Dinizulu who had founded Inkatha as a Zulu* cultural movement. While a university student he joined the ANC* youth wing, but was expelled in 1950. After this he quarrelled with most Black African groups. In 1974 he revived Inkatha as a 'liberation movement', opposed both to any policy of 'armed struggle' and to economic sanctions against South Africa. There were serious inter-Black riots in Natal in the late 1980s and in 1990, mostly between members of ANC and of Inkatha, which now proclaimed itself a 'political party'. With some reluctance he agreed to meet Nelson Mandela* in an effort to reduce the inter-Black violence.

Butler, Richard Austen, 1st Baron Butler of Saffron Walden (1902–82), British statesman. Born in India and educated at Cambridge, he was first elected to Parliament in 1929 as a Conservative. He held minor office during the 1930s and as a Foreign Office minister supported the Munich Agreement* of September 1938. As president of the Board of Education 1941–5, he was responsible for the Education Act of 1944, which laid down the framework for the post-war English free secondary education system, introducing the '11-plus' examination for the selection of grammar school children. He was an important influence in persuading the Conservative Party to accept the principles of the Welfare State. He held a number of ministerial posts between 1951 and 1964, being Chancellor of the Exchequer 1951–5 and Home Secretary 1957–62. He wound up the Central African Federation* 1962–3 and was Foreign Secretary 1963–4. He was twice defeated in contests for the leadership of the Conservative Party, first by Harold Macmillan in 1957 and then by Sir Alec Douglas-Home in 1963. He became master of Trinity College, Cambridge, and a life-peer in 1965.

Byrnes, James Francis (1879–1972), US Secretary of State 1945–7. From a modest background in South Carolina, he trained as a lawyer and built up a practice to become a public prosecutor in 1908. He sat in the House of Representatives as a 'Southern Democrat' from 1911 to 1925 and in the Senate 1931–41. Although on the right of his party he strongly supported President Roosevelt's New Deal*, piloting a number of projects through Congress and master-minding Roosevelt's campaign for a third term in 1940. Briefly a Supreme Court judge, he became director of the Office of War Mobilization 1943–5 and was then appointed Secretary of State by President Truman. A strong believer in the UN, he helped to ensure economic recovery in Germany and was reluctant to accept the division of that country. No enthusiast for the Truman Doctrine*, he was succeeded by George Marshall* and was later elected Governor of South Carolina, where he supported racial segregation.

Cabral, Amilcar (1924–73), Guinean revolutionary. He was educated in Lisbon, where in 1948 he founded a Centre for African Studies. In 1956 he founded the African Party for the Independence of Guinea and Cape Verde, and from 1962 until his death he led a successful guerrilla campaign. This had gained control of much of the interior before he was assassinated, supposedly by a Portuguese agent. In the following year Portuguese Guinea became the independent Guinea-Bissau*.

Caetano, Marcello José das Neves Alves (1904–81), Prime Minister of Portugal 1968–74. Born in Lisbon, he trained as a lawyer and worked closely with Salazar*. He helped to draft the 1933 Constitution and in 1944 drafted the law integrating Portuguese overseas territories with metropolitan Portugal. He became deputy Prime Minister in 1955, but in 1959 left politics to become rector of Lisbon University. When Salazar had a stroke in 1968 he returned and became Prime Minister. He sought to liberalize Portuguese life, redrafting the Constitution, but continuing a repressive policy in the colonies. In 1974 he was ousted in a military coup by General Spinola and went into exile, becoming principal of a Brazilian Institute of Comparative Law.

Caillaux, Joseph (1863–1944), Prime Minister of France 1911–12. Having been a tax inspector, he quickly became involved in government finance on being elected a Deputy in 1898. A member of the Radical Party, he served as Finance Minister on six occasions, as well as being briefly Prime Minister. During World War I he was suspected of pro-German sympathies and was briefly imprisoned in 1918. He entered the Senate for his home *département* in 1925, serving there with distinction until World War II.

Cairo Conference, World War II (22–6 November 1943). This was attended by President Roosevelt, Winston Churchill, and Jiang Jiehi (Chiang Kai-shek) to decide on policy for the Far East. Unconditional surrender by Japan was to be a prerequisite; Manchuria would be returned to China; Korea, occupied by Japan since 1905, was to be returned to its own people. Formosa (Taiwan), occupied by Japan since 1895, would return to China, and Japan would be 'stripped of all the islands in the Pacific' which had been mandated to it after World War I, that is, most of Micronesia*. These resolutions were incorporated into the Cairo Declaration of 1 December 1943. The Tehran Conference* immediately followed with Stalin.

Cajuns, descendants of French Canadians whom the British exiled from Acadia (Nova Scotia) in the 18th century. They were settled in southern Louisiana, where they have created a small self-sufficient agricultural community raising cattle, cultivating yams, corn, and cotton, and developing home crafts. Cajuns speak a patois based on 18th-century French and remain intensely separate from other communities in Louisiana. They have their own style of country music and cuisine.

Callaghan, Leonard James, 1st Baron Callaghan of Cardiff (1912–), British Prime Minister 1976–9. Son of a naval petty officer, he grew up in Portsmouth and worked as a clerk in the Inland Revenue before joining the Royal Navy in 1939. He was elected to Parliament in 1945 as Labour MP for Cardiff South. He held minor offices 1947–51 and was Chancellor of the Exchequer 1964–7, Home Secretary 1967–70, and Foreign Secretary 1974–6. At the Home Office he took major initiatives in reforming the political structure of Northern Ireland. When Harold Wilson announced his retirement in March 1976 he was elected leader of the Labour Party and formed a government as Prime Minister. His government was cool towards the rest of the European Economic Community and divided over defence policies, Callaghan himself being a staunch opponent of unilateral nuclear disarmament. Domestically it did not command a majority in the House of Commons and therefore entered into a 'Lib.–Lab.' pact with the Liberal Party 1977–8. Partly to meet Liberal interests, devolution bills were introduced for Scotland and Wales; that for Wales was defeated in a referendum and that for Scotland in the House of Commons. His government's position was weakened by widespread strikes, during the so-called 'winter of discontent' 1978–9, in protest against wage restraint, and it was defeated in the House of Commons on the devolution issue. The Conservatives won the general election with a large majority for Margaret Thatcher*. He resigned the Labour leadership in October 1980 and moved to the House of Lords.

Calles, Plutarcho Eliás (1877–1945), President of Mexico 1924–8. As an elementary schoolmaster, in 1910 he took part in the struggle against the dictator Díaz. He rose to the rank of 'general' in the Mexican Revolution* and at first supported Carranza* but later helped to overthrow him. As President he sought to implement the ideals of the Revolution by supporting agrarian reform, organized labour, economic nationalism (particularly in the oil industry), increased education, and secularization. During 1928–34, although not in office himself, he continued to exert a powerful influence, being a joint founder in 1929, with Cárdenas*, of the Partido Nacional Revolucionario (PNR). His policies had, however, alienated the Catholic Church and his reputation declined. He lived in exile 1934–41.

Calwell, Arthur (1896–1973), Australian politician. Born and educated in Melbourne, he became a senior civil servant before entering politics in 1940, as a member of the Labor Party*. He was Minister of Information during World War II, after which he served as Minister of Immigration. Australian industry was suffering from a labour shortage and he initiated a system of subsidized immigration* from Europe (£10 from London to Melbourne), thus bringing to Australia many thousands of new inhabitants from the Mediterranean countries as well as refugees from Poland, Germany, and Austria. At the same time he firmly adhered to the policy of 'White Australia'. He was leader of the Labor Party in opposition 1960–7.

Cambodia (Kampuchea), a South-East Asian country. In the 12th century it had been the centre of a great empire, which had steadily been eroded by its neighbours Vietnam and Thailand. By 1914 it was a French colony (1884) and administered as part of French Indo-China*, having survived a major revolt in 1885. The retention of the Cambodian monarchy provided a national focus and the years between the wars were relatively calm. In 1941 the French Governor-General elevated the young Prince Sihanouk* to the throne, in the hope that he would prove amenable to the Japanese occupation. In March 1945 Sihanouk declared independence from French control, which was overthrown throughout Indo-China. Although efforts were made to re-establish this in September 1945, in 1949 Cambodia was declared an 'autonomous state within the French Union*' and in 1953 it obtained full independence under Sihanouk. Following the Geneva Agreements* all French troops were withdrawn in 1955. In 1960 Sihanouk, who had resigned as king in 1955, was named head of state in the Republic of Cambodia and leader of the Popular Socialist Community. His attempts to maintain neutrality were ended in 1970. The USA felt his policy was favouring VietMinh forces, who were allegedly finding refuge in the country, and he was overthrown by Lon Nol*, who proclaimed the Khmer Republic. His regime was steadily built up by the USA, which at the same time dropped 539,129 tons of bombs on Khmer Rouge* targets in Cambodia between 18 March 1969 and 15 August 1973. By then the US Congress was challenging the legality of these raids, since the USA was not in a state of war with Cambodia, and they ceased, following the success of the Paris Peace Accords*. By then Khmer Rouge had 60,000 men organized into 175 battalions; in spite of bombing they held most of the countryside, but had failed to capture Phnom Penh, the capital. In April 1975, however, the USA decided to evacuate. Within weeks the city fell. Hospitals were emptied and the whole population forced to leave the city, which Khmer Rouge regarded as decadent. A new communist society was to be created, based on the land. The Democratic Republic of Kampuchea was proclaimed, with Pol Pot* Prime Minister and the recalled Prince Sihanouk as head of state. Deaths from the reign of terror which followed are still disputed, as the regime tried to create a new communist society, deliberately cut off from international contact. In the first two years some progress was made in repairing the ravages of war, with massive use of peasant labour. But the regime began to lose the support of its earlier ally Vietnam over a frontier dispute along the Mekong River and by December 1977 there was effectively open war along the border. During 1978 Pol Pot sought to gain some credibility by re-establishing Sihanouk, who spoke out in his favour, but in December the country was invaded by Vietnam troops supporting Heng Samrin, leader of the Kampuchean National United Front for National Salvation. Phnom Penh fell on 7 January 1979. Pol Pot took refuge in Thailand, while a new People's Republic of Kampuchea was proclaimed under Heng Samrin. International relief organizations were active from 1980 onwards; perhaps as many as 3½ million people had died of starvation, disease, political killing, and war over the decade. In 1982 a Coalition Government of Democratic Kampuchea (CGDK) in exile was formed, by the Armée Nationale Sihanoukist (ANS), the Kampuchean People's National Liberation Front led by Son Sann, and the Khmer Rouge. This government was recognized by the United Nations in October 1983 and strongly supported by ASEAN*, as against the People's Republic of Kampuchea. Border fighting continued while efforts to achieve a peaceful solution to the country's problems were made by both the Soviet Union and China. In December 1987 inconclusive discussions took place near Paris and later in Indonesia between Prince Sihanouk and Hun Sen, Prime Minister of the People's Republic. A conference in Paris and a series of further meetings followed. In January 1990 an Australian peace plan was accepted by the UN Security Council. Sihanouk resigned from the CGDK, which also in July lost the support of the USA. The latter now gave its support to Vietnam- backed Hun Sen in Phnom Penh, while recommending a UN peace- keeping force.

Cambrai, Battle of, World War I (November 1915), the first successful tank battle in history. On 20 November 500 British tanks of the new Tank Corps broke the German line and captured 10,000 prisoners. Unfortunately a German counter-attack early in December regained their line and took as many British prisoners, the British having failed to exploit their advantage.

Cameroon, Republic of, a west-coast country of Central Africa. The Germans started trading there *c.* 1860 and the German protectorate of Kamerun was confirmed by a Franco-German Treaty of 1911. In 1916 an Anglo-French force occupied it and from 1919 it was administered under League of Nations mandate, divided between the British and the French. In 1960 the French Caméroun became independent. Following a plebiscite, it was joined by part of the British Cameroons, the remainder becoming part of Nigeria. The merged French and British territories were renamed in 1984 the Republic of Cameroon. In 1982 the first President Ahmadou Ahidjo* was replaced by President Paul Biya.

Campaign for Nuclear Disarmament (CND). It was created in 1958, with the philosopher Bertrand Russell* as president and the churchman Canon Collins its chairman, pledged to the abandonment of British nuclear weapons and reduction in defence spending. At Easter each year 1958–64 there was a mass march from Trafalgar Square in London to Aldermaston, Berks., the Atomic Weapons Research Establishment. Frustration at lack of progress led to the creation of a splinter group in 1962, the Committee of 100, led by Russell and pledged to civil disobedience. From 1963 to 1980 it was somewhat in eclipse, but it revived, with 50,000 attending a rally in London 1 October 1980, as a protest against the deployment of US cruise missiles at Greenham Common. Similar protest peace movements developed in the USA, Australasia, and many European countries, including some communist ones.

Camp David Accord, a Middle East peace agreement. In September 1978 negotiations took place in the official country home of the US President in Maryland between President Carter*, President Sadat* of Egypt, and the Prime Minister of Israel Menachem Begin*. A peace treaty was finally signed in Washington 26 March 1979, ending thirty years of conflict. Israel restored Sinai to Egypt, but opposition from Israeli zealots prevented the promised development of Arab autonomy on the West Bank. The agreement was opposed by the PLO*, and Egypt was expelled from the Arab League*.

Canada, a federation of ten North American provinces (Quebec, Ontario, Nova Scotia, New Brunswick, British Columbia, Prince Edward Island, Manitoba, Alberta, Saskatchewan, and Newfoundland*/Labrador), together with the Yukon Territory* and the North West Territories*. The first four provinces had formed the original 'dominion' in 1867 and by 1914 only Newfoundland remained a separate British colony, joining in 1949. Each province has a Legislative Assembly and Executive Council led by a Premier (women first gaining the vote in Manitoba in 1916 and finally in Quebec in 1940). The franchise for the House of Commons has steadily widened (women over 21 in 1918), seats being allocated to provinces on a population basis, with the Territories each having a single seat. The Upper House or Senate of 102 members is recruited by royal appointment, on advice of the Prime Minister.

Until the mid-20th century Canadian Indians* had few rights, their lands being taken over whenever White settlers coveted them. Immigration of Europeans, predominantly Scottish, continued throughout the 19th century, but by 1901 only 12 per cent of the population was of non Franco-British origin. The peak year for immigration was 1913, when half a million arrived. Since 1945 over 3 million immigrants have been absorbed, with the proportion of non Franco-British rising to 30 per cent—German, Ukrainian, Italian, Irish, etc. Rivalry between French- and English-speaking Canada goes back to

1763, when Britain seized the French colony of Quebec. It remained a strong element in Canadian politics throughout the 20th century.

In August 1914 Britain declared war against Germany in the name of the British Empire, so that Canada was automatically at war. One in eight of those Canadians who served in France was killed. Canada suffered severely from the Great Depression*, but under the Liberal governments of Mackenzie King* (1935–48) and Louis St Laurent* (1948–57) it emerged as a major industrial and trading nation. The Statute of Westminster* (1931) granted virtual independence, so that in 1939 it was a Canadian decision to enter the war against Germany. Once again large numbers of Canadians served all over the world, particularly in Europe. After the war Canada accepted, with some reluctance, the need for NORAD*; it gave strong commitment to NATO and to the UN, sending troops on more than fifteen UN peace-keeping missions. Since 1957 the two major federal parties, Liberal* and Progressive Conservative*, have alternated in office, with the Liberal Pierre Trudeau* in office during the years when the Parti Québécois* was particularly active. In 1982 the British Parliament accepted the 'patriation' to Canada of the British North America Act of 1867, and Queen Elizabeth II was installed as Queen of Canada in Ottawa in April 1982. Canada has remained a staunch member of the Commonwealth but often at odds with Britain, as for example over policies towards South Africa. An increasing proportion of its economy is now controlled by US-centred multinational companies, particularly for the exploitation of its rich mineral resources. In June 1990 a deep constitutional crisis developed. The so-called Meech Lake Accord of Brian Mulroney* collapsed when Manitoba and Newfoundland failed to ratify it. At the same time the Parti Québécois* revived its separatist demands.

Canadian Constitution, patriation of, the 'bringing home' of the Constitution from Britain to Canada. The Constitution of Canada was first defined by the British North American Act of 1867. During the 1970s the government of Pierre Trudeau* pledged that it would 'repatriate' the Act, so that the Constitution would be directly controlled by the Canadian Parliament rather than in Westminster. A Bill was passed by the British Parliament and a new Canadian Constitution enshrining a Canadian Charter of Rights and Freedoms (ethnic, sexual, language rights, etc.) was proclaimed on 17 April 1982. There were repeated disputes between the federal and provincial governments over ways and means, but the Meech Lake Accord (see MULRONEY) of 1987 seemed to resolve these, granting Quebec 'distinct society' status. The Accord collapsed in June 1990 when Newfoundland and Manitoba failed to ratify it.

Canadian Indians. The native peoples of Canada fall into four main groups: some 25,000 Innuit or Eskimo in the far north; the Athabascan-speaking peoples of the north-west who spread through the Yukon Territory into Alaska; the Algonquian-speaking peoples of eastern and woodland Canada, including the Naskapi of northern Quebec and spreading into the eastern states of the USA; fourthly the Indians of the Pacific coast of the north-west, who retain some highly sophisticated cultural patterns. Until the mid-20th century it was widely assumed that these native peoples, like those of the rest of North America, would be steadily 'acculturated', that is, assimilated into White society, many tribes and languages disappearing and interracial marriage becoming common. In more recent decades there has been a reversal, with both a rapid population increase (413,380 in 1989) and a revival of Indian cultures, particularly in music and dance. Although treaties had been made between tribes and the Crown reserving areas for Indian use, these were never adhered to, White settlers moving in whenever mining or other prospects loomed. Nevertheless there remain some 2,000 reserves and settlements to which some 500 different Indian bands have access. The Canadian government resisted separatist demands for an Indian Claims Commission, although in 1966 it created a Department of Indian Affairs and Northern Development. The franchise was granted to all native peoples in 1960.

Canaris, Wilhelm (1887–1945), German admiral and Resistance leader. Born in Westphalia, he was commissioned into the German Imperial Navy before World War I and quickly rose to the rank of admiral. An early supporter of the National Socialist Party, in 1935 he was appointed head of German military intelligence, *Abwehr*, and as such organized aid and support for Franco in the Spanish Civil War. He came, however, to regard the Nazi regime as evil and dangerous and began a policy of recruiting anti-Nazi agents into the *Abwehr*, which became the centre of the German Resistance. In 1942 he dispatched one of his agents, Dietrich Bonhoeffer*, to sound out British opinion for a negotiated peace. By February 1944 the SS had developed suspicions of the *Abwehr* and he was transferred. He helped to organize support for the July 1944 Plot*, and was arrested and executed in Flosenburg concentration camp.

Cape Verde, an island state in the Atlantic. Long occupied by the Portuguese, it became an Overseas Province in 1951 and was granted independence in 1975. In 1981 its inhabitants rejected a proposal that it be united with Guinea-Bissau. Its population of some 350,000, who still see Amilcar Cabral* as their hero, suffered from severe droughts during the early 20th century. Traditionally its people have emigrated to seek employment in Brazil and Venezuela.

Capitalism, an economic system. Workers in an enterprise exchange labour hours for a wage provided from capital accumulated by the owner(s) of the enterprise, who then sells the goods produced in a free market against competition. The price obtained must both provide for the needs of the owner and leave a balance (profit) to be added to the accumulation of capital, enabling the enterprise to expand, further expansion being achieved through interest-paying loans from bank and/or stock market. The 'small business' of the 20th century retains these early characteristics, whether retail or manufacturing, but already by 1914 the simple relationships of early capitalism, exemplified by the mill-owner and his hands, were rapidly being overtaken by the growth of corporations and joint-stock companies. In the USA the 'gilded age' of the 1890s had seen vast wealth being accumulated by such giants as Carnegie, and attempts were being made, through antitrust* legislation, to reduce the growth of inter-state monopolies. On the whole these failed, and a characteristic of 20th-century economies has been the further growth of these enterprises into global giants or multinational corporations* over whose operations national governments have little control. Ideologically capitalism has sought to restrict the role of the State in economic policy to a minimum. In practice the 20th century has seen the emergence of 'mixed' economies in many Western European states, whereby certain goods and services are provided directly by government agencies or corporations, for example the Tennessee Valley Authority in the USA. During the 1960s Marxists talked of a 'crisis of capitalism', while others began to popularize the term 'post-capitalist society'. In fact capitalism showed remarkable resilience, particularly in its successful challenge to the East European economies 1989–90.

Capone, Al(fonso) (1899–1947), US gangster. His parents came from Naples in 1893 and he grew up in Brooklyn, where he acquired a gash which earned him the nickname 'Scarface'. He moved from the New York underworld in 1919 to Chicago, and became the most flamboyant and widely publicized criminal of the Prohibition era*. In

1925 he took over Chicago's South Side gang from Johnny Torrio, and dominated that city's underworld, dealing in bootleg liquor, extortion, white slavery, and other rackets, and controlling the corrupt administration of Mayor Bill Thompson. His war on other syndicates, culminating in the St Valentine's Day Massacre of 1929 against the North Side, went unchecked until his indictment in 1931 for income tax evasion, which led to a prison sentence. Physically and mentally broken by syphilis, he was released in 1939.

Caporetto, Battle of, World War I (24 October–18 November 1917). Caporetto (now in Yugoslavia: Kobarid) is a small town on the River Isonzo north of Trieste. In the autumn of 1917 three Italian armies manned a front against Austria-Hungary stretching from the Isonzo to the Swiss Alps. With forces freed from the Russian front following the Revolution there Ludendorff* bombarded the town on 24 October and crossed the river. A rapid advance threatened to outflank the entire front and General Cadorna ordered retreat. Fighting continued until mid-November, when river flooding brought it to an end and the Italians regrouped on a line running north of Venice, which was later strengthened by British, French, and US reinforcements. Some 300,000 Italians were taken prisoner. The German offensive for March 1918 on the Western Front could now be planned.

Cárdenas, Lázaro (1895–1970), President of Mexico 1934–40. Of Indian descent, he joined the army of the Mexican Revolution* led by Carranza in 1913 and rose to the rank of general. In 1929, with Plutarcho Calles*, he founded the Partido Nacional Revolucionario (PNR), which as the PRI has dominated Mexican politics ever since. He was elected to the presidency with the support of Calles, although once in office he broke with his old comrade, who was forced into exile. Cárdenas proceeded to carry the Mexican Revolution to the left during his six-year administration. He distributed more land to the peasantry than all of his predecessors combined, encouraged organized labour through support for the Confederación de Trabajadores de México, and nationalized the property of foreign-owned railway and oil companies (1938). As a mestizo (mixed Indian and Spanish descent), he won the support of the Indian and Mexican working classes. In 1942 he returned to politics as Minister of National Defence and remained a major political figure throughout his life. A strong supporter of co-operative reform, he was often a critic of the USA and a supporter of the Cuban Revolution. In his later years he won hydroelectric schemes and a steel plant for his home state of Michaocán.

Caribbean Community and Common Market (CARICOM). This was formed in 1973, replacing the former Caribbean Free Trade Association (CARIFTA). By 1990 the following states had become members: Antigua and Barbuda, the Bahamas, Barbados, Belize, Dominica, Grenada, Guyana, Jamaica, Montserrat, St Christopher and Nevis, St Lucia, St Vincent and the Grenadines, Trinidad and Tobago; with the Dominican Republic, Haiti, and Surinam as observers. A permanent secretariat was established in Guyana with a Conference of Heads of Government its authority. CARICOM aims (1) to co-ordinate foreign policy; (2) to co-operate in such areas as health, education, sport, tax administration; (3) to operate a Caribbean Common Market to co-ordinate trade, fiscal, and industrial policies. The Bahamas does not participate in the Common Market.

CARICOM *see* CARIBBEAN COMMUNITY AND COMMON MARKET.

Carmona, Antonio Oscar de Fragoso (1869–1951), President of Portugal 1927–51. He graduated from the Royal Military College of Portugal in 1888 and by

1922 held the rank of general. He took part in a military coup in 1926 and became Prime Minister, with Salazar* his Finance Minister. In 1927 he was elected President, adopting a new constitution in 1933 for the 'New State'. He remained President, with Salazar wielding dictatorial powers as Prime Minister.

Carol II (1893–1953), King of Romania. Great-nephew of Carol I (1839–1914), as Crown Prince he was exiled by his father Ferdinand I (1914–27) in 1925 for his scandalous domestic life, living with his mistress in France. His 6-year-old son Michael became King in 1927, but in 1930 Carol flew back to Bucharest and was accepted as King. In 1937, inspired by his admiration for Mussolini, he openly assumed dictatorial powers along neo-Fascist lines. In spite of this admiration, however, he found in 1940 that Hitler had agreed in the Nazi–Soviet Pact to force him to make large concessions to the Soviet Union. He and his dictator minister Antonescu* were forced to concede Bessarabia (Moldavia*) and north Bukovina* to the Soviet Union, and half of Transylvania to Hungary. Humiliated, he abdicated and went again into exile.

Carranza, Venustiano (1859–1920), President of Mexico 1917–20. Son of a landowner, he became active in politics from 1880, becoming Governor of the province of Coahuila in 1910. He supported Francisco Madero and played a minor role in the revolution against Porfirio Díaz. When Madero was assassinated by Huerta's* forces, he recruited a constitutional army, and when Huerta fled he became provisional President in 1914. A voice for moderation in a violent decade, he favoured political but not social reform. He was thus opposed to both Pancho Villa* and Emiliano Zapata*, who championed radical policies of land redistribution. He defeated Villa in 1915 at Celaya and again in 1917, and in 1919 Zapata was assassinated. In 1917 he reluctantly accepted the leftist Constitution under which he became President. He alienated the USA by refusing to bring Mexico into World War I and by seeking control over US oil interests in the country. His revolutionary colleagues accused him of procrastination and in April 1920 his erstwhile colleague General Obregón led an armed rebellion against him. Carranza was assassinated in the village of Tlaxcalantongo on his way into exile.

Carson, Edward Henry, 1st Baron Carson (1854–1935), Anglo-Irish statesman. Born and educated in Dublin, he was called to the Irish Bar and became an Irish QC in 1889 and an English QC in 1894, described as 'one of the most formidable orators of his day'. He represented Trinity College, Dublin, in the House of Commons 1892–1918 and held the office of Solicitor-General 1900–5. He was determined to preserve Ireland's constitutional relationship with Britain and in 1912 strongly opposed the Third Home Rule Bill. In the years just before the outbreak of war he organized in Belfast a covenant, with thousands pledging themselves not to recognize Home Rule*. He helped to establish the Ulster Volunteers*, a private army of some 80,000 pledged to resist Home Rule, and, as president of the Ulster Union Council, he threatened to establish a separate provisional government if the Bill proceeded. He had brought Ireland to the brink of civil war, but when Britain went to war against Germany in August 1914, he gave his full support and joined the coalition government of May 1915. He later served at the Admiralty in the cabinet of Lloyd George 1917–18. By the end of the war he had reluctantly agreed to Home Rule for southern Ireland, but insisted that Ulster, including the predominantly Catholic counties of Tyrone and Fermanagh, should remain under the British Crown. By 1920 he was seeking a compromise to the Irish problem but he was most reluctant to accept the Anglo-Irish Treaty of December 1921. He served as a Lord of Appeal 1921–9, but never ceased to speak for the interests of Ulster.

Carter, James Earl ('Jimmy') (1924–), President of the USA 1977–81. Born and educated in Georgia, he attended the US Naval Academy and served in the US navy 1947–53. He then became a peanut farmer and a state Senator. In 1971 he was elected Governor of Georgia, going on to be nominated by the Democrats for President with Walter Mondale Vice-President. He defeated Gerald Ford* in November 1976 and took office in January 1977. His southern Baptist Christian background and dissociation from the US political establishment tainted by Watergate* had helped him to win. However, inexperience of Washington politics partly explains his failure to establish good relations with Congress. Initially his policy was to deprive of US aid those regimes, such as many in Latin America, which failed to respect basic human rights as agreed by the Helsinki Conference (1975); but this was a policy which he could not maintain. His measures to pardon draft-dodgers (young men imprisoned for evading conscription in the Vietnam War) and to introduce administrative and economic reforms were popular. He failed, however, to obtain congressional approval for his energy policy, which sought to reduce oil consumption, while the Senate in 1979 refused to ratify the agreement on Strategic Arms Limitation (SALT II*). In foreign affairs his administration achieved the Camp David Accord* between Israel and Egypt, and also a treaty promising the transference of the Panama Canal to Panama*. The failure of a helicopter rescue mission in April 1980 to resolve the Iran Hostage Crisis* resulted in a disastrous loss of public regard for his abilities. Although renominated by the Democrats in 1980, he was defeated by the Republican Ronald Reagan*. Since leaving office he has engaged in numerous peace missions and regained much of his earlier reputation.

Casablanca Conference, World War II (14–24 January 1943). This Conference between Winston Churchill and President Roosevelt was convened in order to plan strategy for the continuation of World War II. There was to be an increase of US bombing of Germany; plans were agreed for the forthcoming invasion of Sicily and for the transfer of British forces to the Far East on the defeat of Germany. Roosevelt issued a statement, with which Churchill concurred, that there would be no negotiated settlement with Germany, only 'unconditional surrender'. General de Gaulle* was present, but he resented an Allied conference being held on French territory without his prior knowledge.

Casement, Roger David (1864–1916), Irish patriot. Born in County Antrim, he joined the British consular service in 1892, travelling extensively in the British Niger Coast protectorate and in the Belgian Congo. He won respect for his reports exposing cases of ill treatment of native labour in Africa (1903), particularly in the Upper Congo, and of rubber plantation workers in Peru (1912). In 1911 he was knighted by the British government and he retired to Ireland from the consular service in 1913. An Ulster Protestant, he supported the demand for Irish independence and in 1914 went to Germany and the USA to seek help for an Irish uprising. His attempt to recruit Irish prisoners of war in Germany to fight against the British in Ireland was not a success, nor would the Germans provide him with troops. He was landed on the Irish coast in County Kerry from a German submarine in 1916, with the aim of trying to persuade the organizers of the planned Easter Rising* to postpone the attempt. He was arrested, tried, and executed for treason. Alleged passages from his diary describing homosexual acts were circulated by British agents in the USA to discredit him among Irish Americans. The request that he be buried in Ireland, rejected at the time, was fulfilled in 1965.

Casey, Richard Gardiner, 1st Baron (1890–1976), Australian statesman and Governor-General of Australia 1960–5. Son of a Queensland pastoralist, he was educated in Melbourne and at Oxford. He worked in London as an Australian 'liaison officer' until entering politics in 1931 as a member of the new United Australia Party, formed by the Liberal J. A. Lyons* at the time of the Great Depression*. He held various portfolios under Lyons and in 1940 was sent to the USA as Australia's minister, until invited by Winston Churchill to join his war cabinet (1942–3) and then to become Governor of Bengal (1944–6) in the difficult years preceding independence. In 1949 Robert Menzies won the Australian election with an anti-Labor coalition, for which he revived the label 'Liberal'. Casey returned to the Australian House of Representatives, serving as Minister of External Affairs, before being appointed Governor-General.

Castro (Ruz), Fidel (1927–), Prime Minister of Cuba 1959–76; President 1976– . Son of an immigrant Spanish sugar planter, he graduated in law at the University of Havana. In 1947 he joined the Cuban People's Party and in 1953 led an attempted revolution in Santiago, Cuba, for which he was imprisoned. His self-defence at his trial, known by its concluding words, 'History will absolve me', became his major policy statement. Exiled in 1955, he went to Mexico and in 1956 landed on the Cuban coast with eighty-two men, including 'Che' Guevara*. Only twelve survived the landing. He conducted successful guerrilla operations from the Sierra Maestra mountains, and in December 1958 led a march on Havana. The dictator General Batistá* fled and on 1 January 1959 Castro proclaimed the Cuban Revolution, ordering the arrest and execution of many of Batistá's supporters. He declared himself Prime Minister, and, unable to establish diplomatic or commercial agreements with the USA, negotiated credit, arms, and food supplies with the Soviet Union. He expropriated foreign industry and collectivized agriculture. The USA having cancelled all trade agreements from 1961, Castro was openly aligned with the Soviet Union, emerging more and more strongly as a Marxist. The abortive US–Cuban invasion (April 1961) of the Bay of Pigs* boosted his popularity, as did his successful survival of the Cuban Missile Crisis* (October 1962) and of several assassination attempts. A keen supporter of revolution in other Latin American countries, and of liberation movements in Africa, particularly Angola and Ethiopia, he achieved considerable status in the Third World as a leader of the Non-aligned movement. However, his authoritarian regime of one-party government seemed increasingly out of fashion in the later 1980s following events in Eastern Europe.

Cavell, Edith (1865–1915), English nurse. Daughter of a Norfolk vicar, in 1906 she helped to establish a training school for nurses in Brussels. Left in charge of her Institute, after it became a Red Cross hospital in World War I, she nursed German and Allied soldiers alike. She believed it was her duty to help British and French soldiers to escape to neutral Holland. Unable to conceal these activities, she was arrested, court-martialled, and executed on 12 October 1915. The German Kaiser considered such action a serious political blunder. Her last words were, 'I realize that patriotism is not enough. I must have no hatred or bitterness towards anyone.'

Ceauşescu, Nicolae (1918–89), President of Romanian Socialist Republic 1967–89. Born in rural Romania, he joined the Communist Party when 18 and became responsible for building up a Union of Communist Youth. He worked underground until the entry of Soviet troops in 1944 and the establishment of the Romanian Republic in 1948. He became general secretary of the party, working closely with President Gheorghiu-Dej*, whom he succeeded in 1965, being elected President in 1967. He denounced the Soviet occupation of Prague in 1968 and was for long admired in the

West for his lack of enthusiasm for the Soviet Union. When revolution broke out in Romania in December 1989 he and his wife were shot on Christmas Day. The full extent of his personal aggrandizement and callous and brutal regime then emerged.

CENTO *see* CENTRAL TREATY ORGANIZATION.

Central African Federation (1953–63). It consisted of the self-governing colony of Southern Rhodesia (Zimbabwe*) and the British protectorates of Northern Rhodesia (Zambia*) and Nyasaland (Malawi*). In the 1920s and 1930s European settlers had pressed for closer union, but Britain had rejected them because of responsibilities for Africans in Nyasaland and Northern Rhodesia. In 1953 the Conservative government in Britain allowed economic arguments to prevail, and a federal constitution was introduced whereby the federal government handled external affairs, defence, currency, intercolonial relations, and federal taxes. Riots and demonstrations by African nationalists followed (1960–1). Macmillan had made his 'Wind of Change' speech* in February 1960, and after the Monckton Report in 1962 Britain accepted in principle the right of secession. A meeting of the four governments concerned, at the Victoria Falls Conference, agreed to dissolve the Federation, which officially came to an end in 1963. Nyasaland and Northern Rhodesia became independent, but Southern Rhodesia made a unilateral declaration of independence (UDI) from Britain in 1965, calling itself Rhodesia. The ensuing political impasse did not end until the creation of Zimbabwe in 1980.

Central African Republic, formerly the French colony of Ubangi Shari. This land-locked country formed part of French Equatorial Africa*, and in 1958 it became a republic within the French Community and fully independent in 1960. In 1976 its President Jean Bedel Bokassa* declared it an empire and himself the emperor. He was deposed in 1979 and the country reverted to being a republic. Political instability persisted, and in 1981 General André Kolingba seized power from the civilian government. This was restored in 1986 with Kolingba, as President, rejecting demands in 1990 for multi-party politics. Diamonds and uranium are sources of revenue in the economy.

Central American Common Market (CACM). In 1960 a treaty between Guatemala, Honduras, El Salvador, Nicaragua, and Costa Rica sought to reduce trade barriers, stimulate exports, and encourage industrialization by means of regional co-operation. With a permanent Secretariat in Guatemala City, it also aimed at co-operation with the Latin American Free Trade Association (see LATIN AMERICAN INTEGRATION ASSOCIATION). During the 1970s, however, it lost impetus, with ideological differences between members, a draft treaty for monetary union and common social and economic programmes failing to gain ratification. During the early 1980s there was a sharp rise in balance-of-payments deficits in the area, but a new tariff and customs agreement came into effect in 1986, when regional trade improved. Nevertheless CACM's achievements remained very limited.

Central Intelligence Agency (CIA). Developing from a US wartime agency, the Office of Strategic Services (OSS), it was established by Congress in 1947 'to collect, co-ordinate and analyse foreign intelligence' and to be solely responsible to the President. Under an early director Allen Dulles (1953–61) its role was widened to include 'operational planning', whereby 'covert action' projects against non-friendly states would be planned and financed, as for example against Prime Minister Musaddiq* of Iran in 1953 and President Arbenz* of Guatemala in 1954, and deep involvement in the Congo Crisis* in 1960. Following the fiasco of the Bay of Pigs* operation in Cuba in 1961 President Kennedy sought to curb 'covert action' projects, but to improve intelligence gathering. He was not very successful; while backing the plan in Chile to depose President Allende*, the CIA failed to provide intelligence for the Yom Kippur War* in 1973 or the Cyprus War* of 1974. A report of a Senate Committee in April 1976, under the chairmanship of Senator Frank Church, was highly critical and called for stricter control, the prohibition of assassination attempts, and greater accountability through Congress. During the 1950s the agency had developed four subdirectorates: Intelligence, Operations, Science and Technology, and Administration. Following the Church Report it has been obliged to reveal more of the activities of all these areas. During the 1980s it commanded a vast budget and had immense influence, particularly in Pakistan and Afghanistan, Nicaragua, and many Central and South American countries.

Central Treaty Organization (CENTO), successor to the Baghdad Pact* organization. The Baghdad Pact had at first been for mutual support between members. Increasingly, however, it sought to provide an integrated defence policy against the Soviet Union. The 1958 Iraq revolution resulted in the latter's withdrawal. The Pact's organization moved its headquarters to Ankara and became known as the Central Treaty Organization, with US participation paramount (as an associate member). It was dissolved after the Iranian revolution of 1979.

Centre Party, German political party 1871–1933. It had been formed to protect Catholic interests against Bismarck's aggressive Protestantism, but had gradually become a moderate party of the centre in German politics. In 1917 it sought to muster opinion for a negotiated peace to end World War I. During the Weimar Republic, it was right of centre but strongly committed to democracy. It was dissolved by the Nazis in 1933. In 1945 the Christian Democrat Party of Konrad Adenauer* revived much of the ideology of the Centre Party.

Chaco War (1932–5), fought between Bolivia and Paraguay. The Gran Chaco, or Chaco Boreal, is an extensive lowland plain at the foot of the Andes. It had been an object of dispute between the two countries since the early 19th century. Bolivia's final loss of its Pacific coast in 1929, when the Tacna–Arica dispute (see FRONTIER DISPUTES) was settled, prompted it to push its claims to the Chaco, where oil deposits were rumoured. Border clashes in the late 1920s led to outright war in 1932. Bolivia had the larger army and superior military equipment, but the Aymará and Quechua Indian conscripts from the Andean highlands did not fare well in the low, humid Chaco. The Paraguayans, although poorly trained and equipped, were fighting closer to home, and their Colonel José Félix Estigarribia drove the Bolivians across the Chaco and forced them to sue for peace. Paraguay gained most of the disputed territory, although Bolivia gained use of the River Paraguay to the south Atlantic coast. The price was immense for both countries. More than 50,000 Bolivians and 35,000 Paraguayans lost their lives. Economic stagnation was to plague both countries for years to come.

Chad. In 1908 Chad was proclaimed part of French Equatorial Africa*. It became a colony under French administration in 1920, its rich mineral deposits being rapidly exploited. In 1940 it was the first French colony to declare for the Free French*. It became autonomous within the French Community* in 1958 and fully independent in 1960, with François Tombalbaye its first President. Since then the country has struggled to maintain unity between the Arabic-speaking Muslim north and the economically more sophisticated Black Africans of the south and west, mostly Christian. Tombalbaye was assassinated in 1975 and in 1980 Libyan forces invaded, giving backing to Goukouni Oueddei as President. In 1982 the rebel Armed Forces of the North installed the nationalist Hissène Habré as President in place of Goukouni. French and US help led to a Libyan withdrawal in 1987 and an OAU* peace-keeping force in 1988. After this oil and mineral exploitation was resumed, with Habré accepted as President. In August 1990 he agreed to ask the International Court of Justice to adjudicate on the border with Libya, but five months later he had been deposed.

Chamberlain, Arthur Neville (1869–1940), British Prime Minister 1937–40. Son of Joseph Chamberlain and half-brother of Austen Chamberlain*, he spent seven years as a young man working in the Bahamas, before returning to his native Birmingham, where he became Lord Mayor in 1915. He was Director-General of National Service 1916–17, but was dismissed by Lloyd George*, with whom he could never agree. He first entered Parliament in 1918 as a Conservative. As Minister of Health 1923 and 1924–9, he was responsible for the reform of the Poor Law, the promotion of council-house building, and the systematizing of local government. A skilful Chancellor of the Exchequer (1931–7), he steered the British economy back towards prosperity with a policy of low interest rates and easy credit. In May 1937 he succeeded Baldwin* as Prime Minister. His hope for a large programme of social reform was ended, however, by the necessity for rearmament, which had already begun. His policy, largely popular at the time but later denigrated as 'appeasement', was to accommodate the European dictators in order to avoid war. At his three meetings with Hitler, at Berchtesgaden, at Godesberg, and at Munich*, he made increasing concessions. In spite of these he did not save Czechoslovakia from German invasion in March 1939. He was reluctant to negotiate seriously with the Soviet Union, but did pledge military support to Poland. When Germany invaded Poland later in the year, Chamberlain had little choice but to declare war. In May 1940, following the routing of British forces in Norway, his own party rebelled against him and he was forced to resign in favour of Winston Churchill*, whom he whole-heartedly supported until his death later in the year.

Chamberlain, Sir Joseph Austen (1863–1937), British statesman. Eldest son of Joseph Chamberlain and half-brother of Neville Chamberlain*, he was educated at Rugby School and Cambridge before entering Parliament in 1892 as a Liberal Unionist (later as a Conservative). He was Chancellor of the Exchequer 1903–5 and Secretary of State for India 1915–17, resigning over alleged blunders in the Mesopotamia campaign*. He was again Chancellor of the Exchequer in 1919 and leader of the Conservative Party in 1921, but loyalty to Lloyd George* led to his resignation in 1922. Returned to favour, he was Foreign Secretary 1924–9, playing a major part in securing the Locarno Treaties*, for which he received the Nobel* Peace Prize.

Chamorro Vargas, Emiliano (1871–1966), President of Nicaragua 1917–21. Son of a wealthy and distinguished Nicaraguan family, as a young man he was active in organizing revolt against the corrupt regime of José Zelaya. In 1917 he himself became President, having earlier served in Washington as a diplomat, where he had signed a treaty giving the USA the right to build a canal across Nicaragua in exchange for $3m. aid. As president of the Nicaragua Conservative Party he attempted a coup in 1926 and was forced into exile. Returning to Nicaragua he gave reluctant support to the Somoza regime while seeking to moderate its methods. In 1954 he was again exiled for alleged involvement in an assassination plot against the President. He lived in retirement after his return.

Chamoun, Camille (1900–87), Lebanese statesman; President 1952–8. Of wealthy Maronite* Christian background, he served in Lebanese administrations before World War II and was ambassador to Britain 1944–7. Elected President in 1952, his policies were for peaceful coexistence between Muslim and Christian. At the same time he attempted to reorganize government and to strengthen democratic procedures. In 1956 he faced a crisis. His support for France and Britain in the Suez Crisis* alienated Muslim opinion and weakened his support from the army. His regime was saved by the intervention of the US marines, 1958, and reluctantly he did not seek re-election. He strongly opposed PLO* activities in Lebanon, and when civil war broke out in 1975 he joined various governments anxious to prevent Syrian domination, while also recruiting guerrilla Maronite fighters, the Chamounists. In 1980 his more moderate National Liberal Party split from the main Phalangist Party and there was fierce fighting for control of East Beirut. In 1983 and 1984 he attended two unsuccessful reconciliation conferences in Switzerland. He narrowly escaped assassination in January 1987, by which time, as leader of the National Liberal Party, he was coming to support the need for Syrian intervention.

Chanak Crisis, an international crisis September–October 1922. The Treaty of Sèvres in 1920 had allocated European Turkey, as well as the city of Smyrna (İzmir), to Greece. Constantinople and the Dardanelles Straits were to be neutralized and occupied by British and Allied troops. In 1922 Mustafa Kemal* defeated the Greeks, occupied Smyrna, and threatened to cross to Constantinople. British troops at Chanak on the Dardanelles were reinforced by Lloyd George*. There was a tense military confrontation before Kemal accepted a compromise, whereby a new treaty would be negotiated, with the Dardanelles and the Bosporus neutralized. In Britain Conservative rebels led by Stanley Baldwin* used the crisis as a pretext for unseating Lloyd George as Prime Minister, alleging he had been irresponsible in bringing Britain to the verge of war. They had been increasingly critical of his style and of his cynical use of the honours system. The Treaty of Lausanne* in July 1923 replaced that of Sèvres, granting Turkey eastern Thrace and Constantinople (Istanbul).

Chang Hsueh-liang (1898–   ), Chinese warlord* known as the Young Marshal. Son of warlord Chang Tso-lin, he inherited control of Manchuria in 1928 when his father was murdered by officers of the Japanese Guandong Army*. He pledged support to the Nationalist cause, but following the Mukden Incident (see MANCHUKUO) he was driven from Manchuria in 1931, withdrawing into the Shanxi province. Here (1936) Jiang Jiehi* (Chiang Kai-shek) met him at his headquarters at Xi'an to persuade him to attack the Communists in Yan'an. Instead Jiang was arrested and only released on a promise to recreate a 'United Front'* against Japan. Jiang never forgave him. He was arrested in Nanjing and held a prisoner throughout the Sino-Japanese and civil wars. Transferred to Taiwan in 1949, he was kept under house arrest until the death of Jiang.

Charles I (1887–1922), last Habsburg Emperor of Austria-Hungary 1916–18. Heir on the death of his uncle at Sarajevo*, he succeeded his great uncle Francis Joseph*. He at once began negotiations to take the Empire out of the war, dismissing his more militant advisers and establishing contact with Britain and France. This peace initiative failed, as did last minute attempts to introduce constitutional reforms. He withdrew to Switzerland in November 1918 but refused to abdicate. Technically his heirs have a claim to an empire which was broken up by the Paris Peace Conference. He was twice involved in attempts to regain power in Hungary.

Charter 77, a Czechoslovak human rights movement. Named after a document delivered in 1977, initially signed by 242 academics, intellectuals, and churchmen, the charter appealed to the Czech government to adjust the country's laws in conformity with the Universal Declaration of Human Rights* and to respect in practice the agreement of the Helsinki Conference*. More signatories followed, even though it cost jobs, freedom, and in the case of Professor Patocka his life. It was Charter 77 which inspired the successful Czech revolution of 1989.

Chavez, César Estrada (1927–), US labour union organizer. Born in Arizona of Mexican descent, he spent his childhood living and working in Chicano* migrant labour camps in Arizona and California. After serving in the US navy he returned to California and began trying to organize migrant farm labour. In 1962 he created the National Farm Workers' Association (NFWA), which in 1971 became the United Farm Workers of America (UFW). In 1965 he had led a five-year grape-pickers' strike in California, which won much public support. During the early 1970s the UFW was competing with the Teamsters* for members, but in 1977 it won an agreement as having sole right to organize field labour. During the 1980s Chavez lost much personal support owing to his preoccupation with faith-healing and holistic religion, and membership of the UFW declined. His work, however, had greatly boosted the self-esteem of American Chicanos.

Cheka, an acronym for the All-Russian Extraordinary Commission for the Suppression of Counter-Revolution and Sabotage. It was instituted by Lenin* in December 1917 and run by Dzerzhinsky*, a Pole. Lenin had always accepted that his revolution would need protection through terror, and this was its purpose. Its headquarters, the Lubyanka prison in Moscow, contained offices and places for interrogation under torture and for execution. In 1922 Cheka became GPU and later OGPU or United State Political Administration. Its successor was NKVD*.

Chemical warfare. Chlorine gas was first used by German troops in January 1915 against the Russians, but with little effect. In April 1915 it was used against British and French troops with some tactical success. To counter its effect gas masks were devised, further encumbering troops in trench warfare. In 1917 both sides used mustard gas, which severely burned and blistered the skin and caused blindness, though seldom death. The League of Nations outlawed the use of gas, and public outcry against it led to this ban holding throughout World War II. The German discoveries of a gas which attacked the nervous system and of another which attacked the blood were never used; nor were British developments in biological warfare. Tear gas has continued to be used in dispersing rioters by police forces all over the world. There was extensive use of 'defoliation agents' such as agent orange during the Vietnam War by the US air force, in futile attempts to reveal Vietcong military centres. Iraq used chemical weapons in the Iran–Iraq War* and also against Kurdish rebels. Stockpiling by NATO and the Warsaw Pact of potentially ever more effective weapons, including nerve gases and virus diseases, continued throughout the years of the Cold War, until a mutual agreement in June 1990 to destroy them. A technological problem then was how safely to do this. The possibility of chemical and biological warfare in the Gulf Crisis* in 1990 was a major threat.

Chen Duxiu (Chen Tu-hsiu) (1879–1942), leading Chinese Communist intellectual. Born of wealthy parents, he sat the Civil Service examination before going to Japan to further his studies. In 1915 he established a youth magazine, *Hsin ch' ingnien*, which was to be highly influential. In 1917 he was appointed Dean of the

College of Letters at Beijing University. As such he was a leader of the cultural revolution which resulted in the May Fourth Movement* in 1919. He became a Communist and was the first head of the CCP from 1921 until 1927, when the party's alliance with the Guomindang* collapsed. He was expelled from the party in 1930 as a 'Trotskyite', and in 1932 arrested and imprisoned by the Nationalists. Released in 1937, he gave his support to the United Front* against Japan, moving to Chongqing (Chungking) in 1938, where he became a schoolteacher and continued to publish articles. He became increasingly critical of Soviet 'democratic dictatorship' and an advocate of parliamentary democracy.

Chen Li (Ch'en I) (1901–72), Chinese general. From the Sichuan province, he studied in France 1919–21 and on his return joined the newly formed Chinese Communist Party. In 1928 he joined Mao Zedong* and Zhu De* in the mountains and soon became a leading figure in the Communist army. He did not take part in the Long March*, but went underground to organize guerrilla groups in central and south China. In the Sino-Japanese War he became commander of the New 4th Army on the Lower Chang Jiang (Yangtze), and in the civil war which followed he played a major role in the victory of Huai-Hai*. On the establishment of the People's Republic he became Mayor of re-occupied Shanghai and was then Foreign Minister 1958–66. He was fiercely attacked, however, in the Cultural Revolution* and stripped of all his offices. He was an outstanding military commander.

Chernobyl, scene of the worst nuclear disaster to date on 25–6 April 1986. It began when a steam turbine test got out of control and the nuclear reactor became overheated. An explosion blew the roof off the generator building, there was a further reaction generating hydrogen, which then ignited, causing a fire which lasted four hours. The rogue generator was buried in concrete, where it will remain radioactive for centuries. There were two immediate deaths, twenty-nine died from radiation sickness, and 200 more were treated for the same condition. An immediate evacuation on a radius of 30 km. took place, with the land permanently contaminated. The Soviet Republics of the Ukraine and Belorussia were worst hit and abnormal births and inexplicable illnesses followed. By February 1989 it was clear that some 25 per cent of the land of Belorussia was unfit for agriculture. Beyond this, radioactive fall-out affected all Europe and reached into Asia Minor and Siberia. North Wales sheep farmers suffered particularly, as did all of Scandinavia, where thousands of reindeer had to be slaughtered.

Chetniks, Serbian nationalist guerrillas. They began to operate during the 19th century against Ottoman Turks, and in World War I chetnik bands attacked Austrian and German forces which had occupied Serbia. In World War II chetnik guerrillas were backed by the British 1941–4, operating against the puppet state of Croatia. Fiercely pro-Serbian and anti-Communist, and led by General Mihailovich*, they opposed the Communist partisans of Tito*. When the Soviet Red Army arrived to liberate the Balkans many were shot, accused of collaborating with the Germans. Serbian nationalism re-emerged in the late 1980s.

Chiang Ching-kuo (1909–88), President of the Republic of China (Taiwan) 1978–88. A student Communist at the University of Shanghai, he later studied and worked in the Soviet Union 1927–37, having been made to denounce his father Jiang Jiehi* (Chiang Kai-shek) for breaking with the Communists in 1927. Reconciled with his father in 1937, he returned to China to hold office in the National government. In 1949 he fled to Taiwan, where he became head of secret police, Minister of Defence

(1965–72), Premier (1972–8), and then President. His ten years in office saw suppression of political opposition at first, but later he lifted martial law, and he allowed opposition parties to emerge before his death. At the same time Taiwan's economy continued to expand, based on US capital investment.

Chicanos, descendants of Mexicans living in the area of the USA occupied in the Mexican–American War of 1846–8. They were long an underprivileged group, but were encouraged by the civil rights movement* from the 1950s to launch Chicano organizations to secure improvements. In the 1960s César Chavez* wrung concessions for the California grape-pickers by a series of strikes and boycotts, while in the 1970s there were advances in education, with bilingual programmes being established.

Chifley, Joseph Benedict (1885–1951), Prime Minister of Australia 1945–9. Born in Bathurst, NSW, son of Irish Catholic immigrants, he was dismissed as an engine-driver for taking part in a rail strike in 1917. He became a rail union official and Labor Party activist. In 1928 he was elected to the federal Parliament and served as Minister of Defence 1929–31. He returned to Parliament in 1940 and as Finance Minister (1941–5) introduced a uniform tax scheme. When Prime Minister J. J. Curtin* died suddenly in 1945, Chifley succeeded, a highly popular figure among Labor supporters. His attempt to nationalize the banks was defeated, but his comprehensive social insurance scheme for sickness and unemployment was adopted. In 1949 he lost popularity over a Communist-inspired coal strike when he sent in troops, and he lost the election in December.

Chile. During the later 19th century Chile was politically stable, with a liberal and democratic system of government. British capital was extensively invested, building railroads and opening copper and nitrate mines. Following war with Bolivia and Peru (1879–83), rich natural nitrate (saltpetre) deposits had been annexed and the country experienced a fifty-year economic boom, particularly during World War I with its insatiable demand for armaments. The 1920s saw economic problems, synthetic nitrates replacing Chile's natural saltpetre and world copper prices fluctuating. Politically there emerged both Communist and Fascist parties. Arturo Alessandri* sought to introduce moderate social reforms, but he met opposition from the military and from Communists. Left-wing Popular Front* governments dominated from 1938 until 1952, although the Communist Party was outlawed in 1938. The right-wing Jorge Alessandri* came to power in 1957 with a massive programme of public works, which themselves helped to generate heightened inflation. In 1964 the Christian Democrats led by Eduardo Frei* gained office. Their six years in government saw considerable reform in health, housing, and education. They passed an Act for Agrarian Reform, and the government took a 50 per cent stake in US copper mines. By 1970, however, political polarization was again taking place, a right-wing National Party attracting Fascist support and a left-wing Frente de Accíon Popular backed by many Communists. It chose Salvador Allende* for President and he was narrowly elected. As head of a Popular Front (Unidad Popular), he faced a majority opposition in the Chile Congress and the hostility of the USA. He was increasingly frustrated in his attempts to implement his radical programme of nationalization and agrarian reform. Inflation, capital flight, and a rapidly rising balance-of-payments deficit contributed to an economic crisis in 1973. In September the army Commander-in-Chief Augusto Pinochet* led the military coup, which cost 15,000 lives and prompted one-tenth of the population to emigrate. The authoritarian regime brutally suppressed all labour unions and opposition groups, while pursuing the goal of a free market economy. Although inflation was dramatically reduced, so were demand, output, and employment. The economy began to spiral downwards, producing the world's highest per capita level of external debt. By 1983, in spite of ruthless suppression of criticism, the situation was such as to oblige Pinochet to open a 'dialogue' with opposition groups. Public protests and bomb outrages began, and a 'state of siege' was proclaimed in November 1984. In May 1986 15,000 people were arrested for demonstrating against the government. Strikes (illegal) and an attempted assassination of Pinochet followed. Although retaining the support of the army and police, Pinochet reluctantly accepted a plebiscite decision of October 1988 for the 're-establishment' of 'workable democracy'. He agreed not to stand for re-election as President while remaining Commander-in-Chief. In December 1989 the Christian Democrat candidate Patricio Aylwin was elected President.

Chilembwe, John (*c.*1871–1915), an early Malawian nationalist. A servant of the Baptist missionary Joseph Booth, who sent him to a Negro theological college in the USA, he became a church minister and in 1900 established the Providence Industrial Mission in Nyasaland. He began increasingly to protest in churches and schools against the injustices of colonial rule and in 1915 started a rebellion. It failed to win enough African support, and he was shot by the police.

China, People's Republic of. Founded on 1 October 1949, it is one of the largest countries in the world, with a population of over 1,000 million. It is divided into twenty-two provinces: Anhui, Fujian, Gansu, Guangdong, Guizhou, Hainan Island, Hebei, Heilongjiang, Henan, Hubei, Hunan, Jiangsu, Jiangxi, Jilin, Liaoning, Qinghai, Shaanxi, Shandong, Shanxi, Sichuan, Yunnan, Zhejiang. There are also five autonomous regions: Inner Mongolia*, Guangxi, Xizang (Tibet*), Ningxia, Xinjiang. A National People's Congress* with 2,700 members elects the highest state dignitaries, including the state President, and a Standing Committee which supervises the State Council headed by the Premier (Li Peng 1988–   ), with various vice-premiers and commissioners. The Chinese Communist Party*, with 44 million members, is the sole effective political party, whose chairman, general secretary, and Politburo wield ultimate political power. Since its establishment the Republic has been actively engaged in war in Korea, and supported the VietMinh* and North Vietnam* in the Vietnam struggle. During the 1960s there were border disputes with India (1961–2) and with the Soviet Union (1968–9). There was also a dispute with Vietnam at the close of the Vietnam War, over Vietnamese policy towards Cambodia*. After the early land reforms, the government of Chairman Mao Zedong* tried to speed up economic development through the creation of large self-sufficient rural communes* in the 'Great Leap Forward'*. When this largely failed, ideological dispute developed between Mao, who retained his egalitarian idealism, and pragmatists in both army and industry, who wanted incentives and the development of managerial élites. The 'Cultural Revolution'* was in part a protest against policies of élitism. The deaths of the three giants Mao Zedong, Zhu Enlai*, and Zhu De* in 1976 left something of a power vacuum, with an intense struggle between the surviving party veterans, as well as aspiring younger politicians. One of the last of the veterans, Deng Xiaoping*, was committed in the early 1980s to economic reforms and to improving relations with the Soviet Union. But his decisions over the Tiananmen Square Massacre* left a confused and complex situation which would need to await the passing of the veterans and the emergence of new leadership. In early 1990 there were renewed CCP efforts to purge corruption, as well as bitter criticism of deposed general secretary Zhao Ziyang. International sanctions, imposed after June 1989, were gradually eased.

China–Japan Peace and Friendship Treaty (1978). A feature of post-war Japanese foreign policy was the tension between dependence upon the USA and popular pressure for closer relations with the People's Republic of China. However, the growing Western inclinations of Chinese policy, the thaw in US–Chinese relations following the Nixon visit of 1972, and increasing Japanese dependence on Asia for its foreign trade all contributed to improved Sino-Japanese relations. One of the major aims of this Treaty of 1978 was the establishment of closer trading links through economic co-operation.

Chinese Civil Wars, a series of conflicts in China between Communist and Nationalist forces. Hostilities first broke out in 1927 during the Northern Expedition*, with purges of the Guomindang* and a series of abortive Communist urban uprisings, ruthlessly suppressed in Shanghai, Nanchang, Guangzhou (Canton), and other cities. Having moved into rural areas, the Communists utilized guerrilla tactics to neutralize the superior Nationalist strength. After a three-year campaign Jiang Jiehi* (Chiang Kai-shek) finally managed to destroy the Jiangxi Soviet* which had been established by Mao Zedong*; but after the Long March* the Communists re-established themselves in Yan'an*, in the north-west of the country. Hostilities between the two sides largely ended following the Xi'an Incident* and the Japanese invasion of 1937, an uneasy truce being maintained while largely separate campaigns were fought against the common enemy. Violence broke out briefly immediately the war ended in August 1945, resuming on a wide scale in April 1946 particularly after the US general George Marshall* had failed to arrange a lasting compromise settlement by January 1947. During the first year of the renewed conflict, numerically superior nationalist troops made large territorial gains, including the Communist capital in Yan'an. Thereafter Guomindang organization and morale crumbled in the face of successful guerrilla and conventional military operations by the Communists. Hyperinflation developed and there was a growing loss of confidence in a manifestly corrupt administration. By the end of 1947 a successful Communist counter-offensive was well under way. In November 1948 Lin Baio* completed his conquest of Manchuria, where the Nationalists lost half a million men, many of whom defected to the Communists. In central China the Nationalists lost Shandong province, and in January 1949 they were defeated in the campaign of Huai-Hai*, near Xuzhou. Beijing fell in January and Nanjing and Shanghai in April. The People's Republic of China was proclaimed on 1 October 1949, and the Communist victory was complete when the Nationalist government fled to Taiwan in December.

Chinese Communist Party (CCP). Interest in Communism was stimulated in China by the Russian renunciation of privileges previously held by tsarist Russia, following its Revolution (1917) and by the May Fourth* Movement (1919). It was promoted by Li Dazhao*, librarian of Beijing University, and Chen Duxiu*. They were co-founders of the party, which held its first Congress in Shanghai in July 1921. Under Comintern* direction, CCP members at first joined the Guomindang* and worked in it for national liberation. Early activities concentrated on trade union organization in Shanghai and other large cities but a peasant movement was already being developed by Peng Pai*. Purged by the Guomindang with massive casualties in 1927 and forced out of the cities, the CCP had to rely on China's vast peasant population for its revolutionary base. It set up the Jiangxi Soviet* in southern China in 1931, but was forced to move north under Mao Zedong* in the Long March* (1934–5). Temporarily at peace with the Guomindang after the Xi'an Incident* of 1936, the CCP provided an effective resistance force when the Japanese invaded in 1937. After the end of World War II, the party's military strength and rural organization enabled it to triumph over the Nationalists in renewed civil war and to proclaim the People's Republic in 1949. The party has ruled

China ever since. Internal arguments over political doctrine and economic policies led to the chaos of the Cultural Revolution*, during which the CCP appeared to turn in on itself. After the death of Mao Zedong (1976) and the purge of the Gang of Four*, the CCP seemed to be pursuing a more stable political direction, having adopted the Four Modernizations* under the leadership of Deng Xiaoping*. But allegations of corruption and demands for more open government led to a prolonged crisis 1987–9.

Chinese People's Political Consultative Conference (CPPCC). On 21 September 1949 Mao Zedong* opened a Conference of people's delegates in Beijing which was charged with drawing up a constitution for the People's Republic and a Common Programme for the Future. Apart from the CCP many smaller parties had been invited to send delegates, and eleven out of the twenty-four ministers of the first government of the Republic came from these delegates. Its draft constitution was adopted in 1954, to be replaced in 1975, then in 1978, and again in 1982. The Conference has a number of standing committees and has met regularly over the years.

Chinese Revolution (1911). In the early years of the 20th century, after a half-century of anti-Manchu agitation, the Manchu government under the Qing Empress Dowager Cixi (1834–1908) granted limited authority to provincial assemblies, and these became power-bases for constitutional reformers and republicans. In October there was a republican uprising in the city of Wuchang which the Qing government failed to suppress. By the end of November fifteen provinces had seceded, and on 29 December 1911 provincial delegates proclaimed a republic with Sun Yixian* as provisional President. In February 1912 the last Qing Emperor Pu Yi* was obliged to abdicate, and Sun stepped down to allow Yuan Shikai* to become President. The Provisional Constitution of March 1912 allowed for the creation of a democratically elected Parliament, but this was ignored by Yuan Shikai after an abortive second revolution in 1913 which challenged his authority. Yuan had himself proclaimed Emperor in 1916, but by then central government was ineffective and China had become controlled by warlords*. Not until 1949 and the triumph of the Communist Party was China to be effectively united again under a single administration.

Christian Democrats, moderate Roman Catholic political parties of the 20th century. Early parties, such as in Uruguay in 1910 and Italy in 1919, were left of centre, but since 1945 they have tended to be to the right. They are hostile to socialism while anxious for social reform, together with moral and political stability. They have been in power in the German Federal Republic, Belgium, France, Italy, and The Netherlands, being especially strong in the first decade after World War II. In the 1990 elections the German Christian Democrat Union Party made strong gains in the five reconstituted eastern states, and won a majority in the federal Bundestag.

Christian Science, a religious movement founded in 1879 by Mrs Mary Baker Eddy. Her book, published in 1875, *Science and Health, with Key to the Scriptures*, teaches that God is divine mind. Only mind is real; matter, evil, sin, disease, and death are all unreal illusions. Membership of the Church has declined in the USA and Europe since 1950, but there has been considerable growth in Africa and South America. Mrs Eddy, a frail lady deeply committed to religious healing, founded the international daily newspaper the *Christian Science Monitor* in 1908.

Churchill, Sir Winston Leonard Spencer (1874–1965), British Prime Minister 1940–5, 1951–5. Born in Blenheim Palace, Oxon., the son of Lord Randolph Churchill and Jenny Jerome of New York, he was commissioned from Sandhurst into the 4th

Hussars in 1895 and fought at the Battle of Omdurman in 1898. He covered the Boer War (1899–1902) as a journalist, being captured and managing to escape. Elected as Conservative Member of Parliament in 1900, he switched to the Liberals in 1904 as a supporter of free trade. He served as Under-Secretary of State for the Colonies 1906–8 and then in Asquith's great reforming government, first as President of the Board of Trade 1908–10, when he introduced measures to improve working conditions, and established labour exchanges. As Home Secretary 1910–11 he supported Lloyd George's Insurance Bill, but was criticized at this time for over-reacting to events by calling in troops to combat industrial disputes. At the Admiralty from 1911 until 1915, it was largely due to him that the Royal Navy was modernized in time to meet Germany in World War I. Resigning after the Gallipoli* evacuation, for which he was blamed, he served briefly on the Western Front. In 1917 he became Minister of Munitions, in 1918 Minister for War and Air, and Colonial Secretary 1921–2, a strong supporter of Lloyd George*. Although he did not support the way the latter was deposed by Conservative rebels in October 1922, he did in fact rejoin the Conservative Party and served as Chancellor of the Exchequer 1924–9. His return to the gold standard* brought serious economic consequences including, indirectly, the General Strike* of 1926, in which his attitude was unhelpful. During the 1930s he was out of office largely because of his bellicose attitude towards Indian demands for independence, but his support for rearmament against Nazi Germany ensured his inclusion, as First Lord of the Admiralty, in Neville Chamberlain's* wartime government of September 1939. In May 1940 he became Prime Minister and Defence Minister of a coalition government of Conservative, Liberal, and Labour members. As a war leader, Churchill was superb in maintaining popular morale, together with close relations with the USA and Commonwealth countries. In August 1941, with President Franklin D. Roosevelt, he was instrumental in drawing up the Atlantic Charter* as a 'buttress of the free world'. Wary of Soviet expansionism, he was concerned that the USA should not concede too many of Stalin's demands as the war drew to a close. He lost office in the 1945 election, but returned as Prime Minister in 1951. In failing health, he was preoccupied with the need for Western unity in the Cold War*, and for the maintenance of a 'special relationship' between Britain and the USA, which granted him honorary citizenship. Under his premiership the economy made a significant recovery and he seemed to revert to the 'social Liberalism' of his youth. He suffered a stroke in 1953 and resigned two years later.

Church of Scotland, the Presbyterian Church in Scotland. During the 19th century it had been split between the official, conservative, Moderators' Church of Scotland and a more evangelical Free Church of Scotland. In 1921 the former was disestablished from the Crown and in 1929 the two rival churches were reunited as the Church of Scotland. It derives its theology from the 16th-century Calvinism of John Knox and over the centuries contributed to the high levels of Scottish education and mission work. Attempts to unite it with the Church of England were defeated in 1959 and 1971. In England and Wales the Presbyterian United Reformed Church (formed in 1972) is allied to it. There is a strong Presbyterian Church in the USA.

CIA *see* CENTRAL INTELLIGENCE AGENCY.

Ciano, Count Galeazzo (1903–44), Italian Foreign Minister 1936–43. Wealthy son of an Italian admiral, he served in the diplomatic corps during the 1920s and in 1930 married Mussolini's* daughter. His diaries reveal that although as Foreign Minister he negotiated pre-war agreements with Nazi Germany, he soon came to resent the German connection. He was among those leaders who voted for the deposition of Mussolini in 1943. He was arrested by the Gestapo while visiting Munich and sent back to Italy, where he was tried for treason by a Fascist court of the puppet government of Mussolini at Salo. He was shot outside Verona.

Civil rights movement (USA). This developed soon after World War II, with racial equality established within the armed services in 1948. The Supreme Court decision in *Brown* v. *The Board of Education of Topeka, Kansas** (1954) was crucial, provoking a mass movement through the Southern states. In December 1955 Blacks boycotted the bus system in Montgomery and won. Baptist minister Martin Luther King* founded the Southern Christian Leadership Conference and personally took part in the Montgomery boycott. In September 1957 federal troops were sent to Little Rock*, Ark., to enforce school desegregation and Congress passed a Civil Rights Act establishing a Federal Agency of six Commissioners to investigate complaints. A further act was passed in 1960. During 1960–1 there were student sit-ins forcing desegregation of lunch-counters, cinemas, supermarkets, libraries, etc. Freedom Riders* came south and there were some 3,600 arrests. In August 1963 some quarter-million took part in the peaceful 'March on Washington', when King made his 'I have a dream' speech. Soon afterwards Kennedy was murdered, but the Civil Rights Act of July 1964 was the most far-reaching Bill of its kind ever, forbidding racial discrimination in employment, education, or accommodation. The Voting Rights Act of 1965 enabled Blacks to register without harassment and to participate in the democratic process. The movement fragmented, however, in the mid-1960s, with disagreement on how to define and acquire non-political civil rights, such as those to housing and employment.

Clark, Charles Joseph (1939–), Prime Minister of Canada 1979. After being a student politician at the University of Alberta he became a journalist and political adviser, before being elected to the Canadian House of Commons in 1972. In 1976 he was elected leader of the Progressive Conservative Party and was Prime Minister in 1979, as leader of a minority government. His party lost the general election in 1980 and he lost the leadership to Brian Mulroney* in 1983.

Clemenceau, Georges (1849–1929), Prime Minister of France 1906–9, 1917–20. Son of a doctor, he graduated in Paris and practised in Montmartre until entering politics at the time of the Franco-German War. He was elected to the National Assembly in 1871 as a Radical, being strongly anti-clerical. He fought for justice in the Dreyfus Affair (1897), but as Prime Minister and Minister of the Interior 1906–9 he ruthlessly suppressed popular strikes and demonstrations. In 1917, with French defeatism at its peak, he formed his 'victory cabinet' with himself as Minister of War, persuading the Allies to accept Foch* as Commander-in-Chief. Nicknamed 'The Tiger', he became chairman of the Peace Conference at Versailles in 1919, where in addition to the restoration of Alsace-Lorraine to France, he demanded the Saar Basin and the permanent separation of the Rhine left bank from Germany, which should also pay the total cost of the war. Failing to get all that he demanded he lost popularity and was defeated in elections in 1920. His autocratic style had been increasingly resented by political colleagues.

CND *see* CAMPAIGN FOR NUCLEAR DISARMAMENT.

Coates, Joseph Gordon (1878–1943), Prime Minister of New Zealand 1925–8. Son of a farmer near Auckland, he was first elected to Parliament in 1911 before serving in World War I. After the war he returned to Parliament in the Reform Party*, serving under W. F. Massey* as Minister of Native Affairs. On Massey's death he became Prime Minister. Although committed to economic expansion and free enterprise, his government also developed public services, for example roads, railways, and hydroelectricity. He was deeply committed to the plight of the Maori and with advice from Apirana Ngata* assisted Maori farming schemes. In 1931 he joined G. W. Forbes of the United Party in a coalition government facing the Great Depression*. This established a Reserve Bank and laid the foundations for economic recovery, but could not prevent a Labour landslide in 1935. In 1936 he and Forbes together formed the National Party* as an anti-Labour party. In 1940 he joined Fraser's wartime government as Minister of Armed Forces and War Co-ordination and died in office.

Cod War (1972–6), the popular name given to the period of antagonism between Britain and Iceland over fishing rights. The cause was Iceland's unilateral extension of its fishing limits, to protect against over-fishing, from 12 to 50 miles. Icelandic gunboats harassed British trawlers, particularly in the years 1975–6, when Royal Navy frigates were deployed as escort. A compromise agreement was reached in 1976, whereby twenty-four British trawlers would be allowed to fish within an agreed 200-mile zone. It hastened the decline of the Hull and Grimsby fishing industries.

Cold War, a period of international tension 1946–89. The Soviet Union, the USA, and Britain had been wartime allies 1941–5, but already, before Germany was defeated, they began to differ about its future and that of Eastern Europe. The summit meetings at Yalta* and Potsdam* had established certain agreements, but as Communist governments seized power in Eastern Europe and both Greece and Turkey were threatened with similar take-overs, the Western powers became increasingly alarmed. The term Cold War, as opposed to a Hot Atomic War, was in use in the Press from 1946 onwards. The USA had at first assumed that its commitment to Europe had ended with victory in 1945. But by 1947 it accepted a global role, putting forward both the Truman Doctrine* and the Marshall Plan* to bolster economies and to provide military aid where necessary 'in support of free peoples'. In 1949 NATO* was formed as a defence against possible attack from the Soviet Union. The Communist bloc countered with the establishment of Comecon* in 1949, and, when West Germany joined NATO in 1955, with the Warsaw Pact*. Over the following decades the Cold War spread to every part of the world, although the extent of Moscow control over other Communist parties remains debated. The USA developed a strategy of 'containment' of the Soviet Union by a series of alliances in the Pacific, South-East Asia, and the Near East. There was a succession of world crises, all of which were seen as a part of the Cold War: Korea, Indo-China, Hungary, the Cuban Missile Crisis*, the Vietnam War*. There were, however, also times when *détente* reduced the tension, as for example the Rapacki Plan* of 1955 followed by Khrushchev's speech to the Twentieth Congress*. The development of the nuclear arms race from the 1950s, only slightly modified by the SALT* talks 1969–79, maintained tension through the decades. The Helsinki Final Act* of 1975 raised hopes for peace, but tension was again heightened by the Soviet invasion of Afghanistan in 1979, followed by increases in nuclear missiles in Europe. The installation of US cruise missiles in the early 1980s and the announcement of the Strategic Defense Initiative* further heightened tension, but from 1985 onwards the threat of European war began to recede, with START talks resumed and the 1988 INF* treaty signed. During 1989 the pace of change quickened, as Western powers began to feel convinced that the Soviet Union was genuine in its offers to reduce tension. In February Soviet forces withdrew from Afghanistan and pacification in such troubled areas as Angola and Nicaragua seemed more possible. In December Presidents

Bush and Gorbachev met at Malta and pronounced the Cold War officially over. By that time the Communist regimes of the Warsaw Pact* countries were collapsing and the Berlin Wall* had fallen. During 1990 NATO began to address itself to the prospect of peace in Europe, while the CSCE* was seen as perhaps the European security organization of the future.

Collectivism, a loose term to describe a variety of 20th-century experiments to implement the 19th-century politico-economic theory that the means of production/distribution within a society should be under the control of 'the People', and not left to individuals pursuing self-interest. The Co-operative movement*, in both Communist and capitalist societies, has been a form of collectivism, as are the Israeli kibbutz*, a Soviet *Kolkhoz* (collective farm), and a Chinese commune*. The huge apparatus of state collectivism, which became a characteristic of 20th-century Communist societies, and to a lesser extent of Western mixed economies, was by the 1980s seen as bankrupt, in that 'the People' remained yet further from any control than ever. Guild Socialism* had tried to address this problem in the early 20th century, but had failed. The concept of 'worker participation' appealed to many European social democrats, but was rejected by more Thatcherite* capitalism. Market Socialism* was yet another attempt to reconcile the desire for social justice with economic realism.

Collectivization, a term used to describe the process by which private landowners or corporations have been expropriated in order to establish collective, or co-operative, farms, which are then the joint property of all the members, who share any profits. Collectivization in the Soviet Union in the 1930s led to great hardship among wealthy kulaks*, millions of whom were to die of destitution and brutality. A modern collective farm (*Kolkhoz*) in the Soviet Union is about 6,000 hectares (15,000 acres) in extent, nine-tenths cultivated collectively, but each family owning a small plot for its own use. Since World War II the process of collectivization has gone on world wide: in China, North Korea, Vietnam, Cuba, and many African and Latin American countries. Attempts to impose it in East Europe were never popular and it was abandoned altogether in Poland and Yugoslavia. In China the Soviet example was followed by Mao Zedong* in his first five-year plan of 1953, but without the ruthless subordination of agriculture to industry which Stalin had imposed. In the Soviet Union many collective farms have proved extremely inefficient and widespread efforts have been made to devise greater incentives; there are also huge state farms, on which peasants are employees of the state. These have particularly developed in virgin areas of Siberia and in the Asian republics.

Collins, Michael (1890–1922), Prime Minister of the Irish Free State 1922. Born in Cork, son of a farmer, he worked in London as a postal clerk and in a bank before returning to Ireland as a member of the Irish Republican Brotherhood. He assisted in organizing the Easter Rising* of 1916 and was imprisoned in 1916 and 1918. Elected to the Westminster Parliament in 1918 for Sinn Fein*, he helped the latter party to establish the unofficial Dáil Éireann in 1919. Next year he acted as Finance Minister in Arthur Griffith's* unofficial government against which the Black and Tans* operated, the British government at this time offering a reward of £10,000 for him dead or alive. In October of 1921, however, he arrived alive in London to attend the Conference which was to result in the offer of dominion status, playing a large part in the negotiations leading to the Anglo-Irish Truce of December 1921. In January he became Prime Minister of the Free State, winning the approval of the Treaty by the now official Dáil Éireann. But he was challenged by the extremist Irish Republican Society, who killed him in an ambush at Beal-na-Blath, County Cork, in August 1922. A civil war followed in Ireland.

Colombia. Named New Granada by the Spanish, it was renamed Colombia in 1886, when a presidential republic was established. The War of the Thousand Days (1899–1902), precipitated by the USA, led to the separation of Panama* in 1903. During the first half of the 20th century two political parties, Liberal and Conservative, operated the political system on a limited franchise, with occasional bouts of urban violence. In 1948 violence moved from urban to rural areas, with protests against the grip of the hacienda landowners. This precipitated a military government 1953–7 under General Gustavo Rojas. In December 1957 universal adult suffrage was introduced and the two parties agreed to alternate power, the so-called National Front. This they did until the early 1970s, when Rojas returned to politics to lead a neo-Fascist Alianza Nacional Popular (ANAPO). This party, beaten at the polls, formed a guerrilla group, the FARC, which was active through the 1970s until 1982, when it resumed a political role. Other guerrilla groups, however, increasingly emerged, some as agents of the illegal drug industry. Colombia's economy recovered from the set-backs of the early 1970s as diversification of production and foreign investment increased, but with agriculture still its chief source of income it is estimated that by 1980 the country's illegal drug trade was supplying 80 per cent of the world's cocaine market. During the mid-1980s Colombia achieved some economic growth and a successful record of external debt management, but the drug trade increasingly dominated both internal affairs and its relations with the USA. In 1986 the Liberal Virgilio Barco Vargas was elected President. In July of that year there were 130 violent deaths through the action of guerrilla groups, as the new President grappled with ruthless and immensely rich drug dealers. One presidential candidate for the 1990 election was shot dead; it was won by César Gaviria Trujillo of the Liberal Party. At the same time a referendum voted for the establishment of a Constitutional Assembly, while a peace process with left-wing guerrillas was begun.

Colombo Plan. Based on an Australian initiative at a meeting of Commonwealth ministers in Colombo in January 1950, the Plan was originally intended as a means by which richer Commonwealth countries, the UK, Canada, Australia, and New Zealand, could assist the poorer members in South-East Asia. As such it was originally a 'Plan for Economic development in South and South-East Asia' (1951–77). The USA and Japan joined and became major donors for both capital development and training programmes, arranged at annual meetings. By 1986 there were twenty-six members, who had voted to give the Plan an 'indefinite span of life'. Many later members are non-Commonwealth and the title was changed in 1977 to a 'Plan for Co-operative Economic and Social Development in Asia and the Pacific'.

Comecon. In January 1949 Stalin formed this Council for Mutual Economic Assistance to improve trade between countries of the Soviet bloc. Its original members were Albania, Bulgaria, Czechoslovakia, Hungary, Poland, Romania, and the Soviet Union. East Germany joined in 1950, Mongolia in 1962, Cuba in 1972, and Vietnam in 1978. Albania was expelled in 1961, while Yugoslavia had associate status. During the years 1955–60 Khrushchev* tried to use it to encourage interdependence in trade and production, with proposals for supranational economic planning put forward in 1960. There was considerable opposition by Bulgaria and Romania, which had been declared primarily agricultural lands, but in 1962 some agreement was reached as to the role of the economies of the satellite countries of Eastern Europe. In 1987 Comecon agreed to discuss co-operation with the European Community. After the collapse of the East

Europe Communist regimes in 1989, its activities began to disintegrate, and by the end of 1990 it had ceased to function.

Cominform (Communist Information Bureau). It was created by Stalin in 1947 as an organization to co-ordinate Communist Party activities throughout Europe. It assumed some of the functions of the Internationals*, which had lapsed with the dissolution of the Comintern* in 1943. After Tito quarrelled with Stalin in 1948 Yugoslavia was expelled. The Cominform was abolished in 1956, partly as a gesture of renewed friendship with Yugoslavia and partly to improve relations with the West.

Comintern (Communist International). It was established by Lenin* in Moscow in 1919 at the Congress of the Third International under Zinoviev* as chairman. It was to be an organization of all national Communist parties with the ultimate purpose of bringing about world revolution. Twenty-seven countries attended its second meeting in Moscow in 1920. Here Lenin put forward his Twenty-One Points, which required all national parties to model their structure on disciplined lines, in conformity with the Soviet pattern, and to expel all moderate ideology. In 1943 Stalin dissolved the Comintern, but it was revived in a modified form in 1947 as the Cominform*.

Commonwealth of Nations, an international group of nations with its secretariat in London. It consists of the UK and former members of the British Empire, all of whom are totally independent, but who, for historical reasons, accept the British monarch as head of the Commonwealth and symbol of the free association of its members. The term British Commonwealth began to be used during the 1920s to refer to Britain and its dominions, Canada, Australia, South Africa, New Zealand, Newfoundland, and Eire. Their independence was given legal authority by the Statute of Westminster* in 1931. The right to make independent decisions was made clear over the abdication of Edward VIII in 1936 and the decision whether or not to go to war in 1939. Beginning with the granting of independence to India, Pakistan, and Burma in 1947, the title Commonwealth of Nations was adopted; since then some members have opted to be republics, while others are constitutional monarchies under the British Crown. A minority of countries from the old Empire have opted to withdraw from the Commonwealth, for example, Burma in 1947, the Republic of Ireland in 1949, South Africa in 1961, Pakistan in 1972, and Fiji in 1987. During the later 1950s pressure built up in Britain for an end to free immigration of Commonwealth citizens, which in 1959 was running at some 115,000, mostly from the West Indies, India, and Pakistan. From 1962 onwards increasing immigration restrictions have been reciprocally imposed, and this has greatly weakened the popular appeal of the Commonwealth. Nevertheless regular conferences and financial and cultural links have helped to maintain some degree of unity among its members, whose population comprises one quarter of mankind.

Communes, Chinese. In the early days of the People's Republic 'mutual-aid' teams emerged to implement land reform. In the years 1955–6, in the 'high tide of socialism', these were merged into co-operatives. During the 'Great Leap Forward'* of 1958 these were themselves combined into larger units known as communes, which were responsible for planning local farming and running public services. Commune power was gradually devolved to 'production brigades' after the disastrous harvests of 1959–61.

Communism, a 19th-century ideology which advocated that all property and authority be vested in the community, where all are regarded as equal and work for the common benefit. It argued that 'the state' was no more than a bourgeois device to retain power and that, with the establishment of a classless society where all are equal, it would 'wither away'. Paradoxically, however, when the means of production were taken from the capitalists and given to 'the people', it was the state, acting in the name of the people, which hugely gained in power. Furthermore, once Communist regimes were established, the state further increased its power through the establishment of paramilitary police pledged to 'preserve the revolution'. At the same time the concept of *égalité*, once seen as the basis of all Communist belief, steadily lost its appeal.

Communist parties. In 1914 there was no officially termed Communist Party, although in most European countries there was a Social Democratic* Party which subscribed in part at least to Marxist ideas. Communism had been a popular political term since at least 1848. Following the Russian Revolution Lenin suggested in 1919 the adoption of the term 'Communist' along with that of 'Bolshevik' to describe his triumphant party, and Communist parties were quickly founded through much of Europe: in Germany, France, Poland, Britain, Italy, and in most East European states, one holding brief office under Béla Kun* in Hungary in 1919. By 1939 almost every country in the world had a Communist Party, either legal or illegal, but it is far from clear how far these were controlled and manipulated from Moscow. In Britain the Communist Party became intellectually fashionable in the 1930s, but failed ever to win a seat in Parliament. Between 1941 and 1945 the Communist parties of countries occupied by the Germans or Japanese were often the main organizers of Resistance movements—sometimes in alliance with nationalists, as for example the VietMinh* in Vietnam or the Hukbalahap* in the Philippines, but often acting, as in France, in rivalry with them. Many such Resistance movements received active aid and support from the USA and Allies, as for example in Malaya, but as soon as war ended, fundamental differences arose, usually over land, which the Communist parties were pledged to expropriate. In Western Europe there was considerable tension in both France and Italy, and in Greece civil war for two years. In Asia the Chinese party had won power by 1949, but the Malayan Emergency* lasted ten years, the Huks were ruthlessly suppressed in the Philippines, the Korean War* lasted three years, while French Indo-China was to see war drag on until 1975 and beyond. In all countries where the Communist Party triumphed it immediately outlawed all other parties, becoming the sole vehicle of government, but seldom advancing towards the Communist ideal whereby all property is held in common. In Eastern and Central Europe the parties seized and held power under close Soviet surveillance, except in Yugoslavia and Albania, both of which remained sturdily independent. In Western Europe, where the parties remained a minority, the concept of Eurocommunism emerged, resulting in November 1975 in a Eurocommunist Manifesto, issued jointly by the French and Italian parties, which abandoned the postulate of the dictatorship of the proletariat, advocating the parliamentary road to socialism and Communism. During the later 1980s Communist parties throughout Eastern Europe were challenged. That of East Germany virtually disappeared in 1989, while in most other countries they became an opposition minority party, usually with a change of name, in the new era of democratic freedom.

Communist parties of India. A small Communist Party was formed in India in the early 1920s, attracting support from anti-British intellectuals and in the urban ghettos of Bombay and Calcutta. It affiliated to the Indian National Congress*, whose efforts to win independence it supported, although in many ways it opposed the philosophy of the Mahatma Gandhi*. It failed to win mass support from the Indian peasantry in the way that neighbouring parties in China and Indo-China did. In 1941 it declared support for Britain, now allied to the Soviet Union, and was expelled from Congress, who opposed the war. At independence this Communist Party of India (CPI) sponsored a programme of violence and there was an unsuccessful agrarian revolt in Telengana in Andhra Pradesh, suppressed by the Indian army. In 1956 the party first won a majority in the Legislative Assembly of the densely populated state of Kerala, governing until 1959; but elsewhere it still failed to attract mass support. In 1964 it split over the question of its relation with the dominant Congress Party, and the Communist Party of India (Marxist), or CPI(M), emerged as the more powerful and militant wing, being strong in both West Bengal and Kerala. Since 1977 it has been the dominant partner of a coalition in West Bengal, while in Kerala an alliance, CPI and CPI(M), has alternated in power with the Congress Party. More extreme splinter groups, for example the Naxalite* movement and the Communist Party of India (Marxist and Leninist), have advocated terrorism and have been consistently harassed. But India has accepted both the CPI and the CPI(M) as legitimate parties within the democratic process.

Comoros, a small Islamic island republic in the Indian Ocean. When it was granted independence by France in 1975 Ahmed Abdallah Abderrahman was elected President. He was assassinated in November 1989 by supporters of a French mercenary, 'Bob' Denard, who worked for the Mozambique* rebel movement the MNR. French paratroopers soon arrived, and after a brief period in power Denard was deported to South Africa. Said Mohammed Djohar was elected President in March 1990.

Computer industry. This was the fastest growing industry in the last decades of the 20th century. Automatic calculators with electromechanical relays had been developed in the USA and Germany during the 1930s. During World War II both Britain and Germany were developing analogue computers for decoding and other intelligence work. The essential component of a computer is its ability to store a programme in its own memory, and the first such was the digital computer UNIVAC 1 Eckert-Mauchly in 1951 in the USA. The invention of the transistor to replace vacuum tubes in the early 1950s enabled much smaller machines to begin to develop, for example the Philco 2000 in 1958, which began to be installed in banks and offices. Ten years later International Business Machines (IBM) had gained 70 per cent of the US market with such 'main-frame' machines, recording a profit of $871m. in 1968. By the early 1970s it had captured 80 per cent of the world market, operating in West Germany, Britain, France, Japan, and elsewhere. From the mid-1970s and through the 1980s, however, microelectronics transformed the industry, following the invention in 1971 of densely packed integrated circuits on 'silicon chips'—the microprocessor Intel4004. The micro-computer was quickly developed by Apple Computer Inc., and for a decade a host of competing firms crowded into 'Silicon Valley' in northern California, transforming US and European life-styles, as personal computers and word-processors entered almost every home and office. However, it was Japan's electronic giants Fujitsu and Toshiba which advanced R. & D. fastest in this new phase of the industry, with applications of the microprocessor to a whole range of household appliances and activities, capturing 50 per cent of the US electronic cash register market by the early 1980s. Not only had the industry transformed manufacturing processes, by means of robotic developments, but office practice, banking, public administration, telecommunications, journalism, publishing, warehouse operating, and many other activities had been equally changed out of recognition.

Concentration camps (Germany 1933–45). Described by Goebbels* in 1934 as 'camps to turn anti-social members of society into useful members by the most humane means possible', they came in fact to witness some of the worst acts of torture, horror, and mass murder in the 20th century. With administration in the hands of the SS*, early inmates included trade unionists, Protestant and Catholic dissidents, communists, gypsies, and Jews. By 1939 there were six camps through which some 200,000 had passed. With the outbreak of war these were increased in both number and size, especially after 1941, when a policy of extermination of Jews and unwanted elements was adopted. In Eastern Europe prisoners were used initially in labour battalions or in the tasks of genocide, until they too were exterminated. In camps such as Auschwitz in Poland, gas chambers could kill and incinerate some 12,000 people a day. In the West, Belsen, Dachau, and Buchenwald (a forced labour camp where doctors conducted medical research on prisoners) were notorious. An estimated 4 million Jews died in the camps, as well as some half-million gypsies. In addition millions of Poles, Soviet prisoners of war, and other civilians perished. Jews from occupied France were rounded up and shipped to the camps, often with the connivance of anti-Semitic elements in France. After the war many camp officials were tried and punished, but many also escaped. Maidanek was the first camp to be liberated (by the Red Army), in July 1944. After 1953 West Germany paid $37 billion in reparations to Jewish victims of Nazism.

Condor Legion, a unit sent by Hitler to aid Franco in the Spanish Civil War*. The Legion played an important role in the early days of the war in transporting troops from Morocco. Later its aircraft bombed rebel lines and civilian centres, notably Guernica* on 27 April 1937. The Legion remained under German command and was regarded by Hitler as a useful means of testing new weapons and aircraft.

Conference on Security and Co-operation in Europe (CSCE). When the thirty five countries, including all of Europe except Albania, and Canada and the USA, signed the Final Act of the Helsinki Conference*, it was agreed that there should be follow-up conferences to maintain a momentum. One was held in Belgrade in 1977, another in Madrid in 1980. A third met in Vienna 1986–9, holding seven sessions on Confidence-Building and Disarmament in Europe and agreeing on a Human Rights Conference for 1991. Nevertheless CSCE had no permanent secretariat and in some ways had little effective influence, apart from bringing the sixteen NATO and seven Warsaw Pact nations together. As the latter disintegrated, however, in 1989–90, there was a flurry of development, with Czechoslovakia proposing at the Bratislava Conference of Central European Powers that it take on an expanded role. Preparations for this would begin, to be presented to a Helsinki Conference in 1991. The European Community accepted that the CSCE could 'serve as a framework for reform and stability in our continent', and during 1990 a European Bank for Reconstruction and Development was established and the Charter of Paris for a New Europe signed by the CSCE states.

Congo, People's Republic of, a west coast country of central Africa. The French explorer de Brazza first brought it under French control in 1880 and it was absorbed as a territory of French Equatorial Africa* (1910–58). The Declaration of Brazzaville* was issued in its capital in 1944. It became a member of the French Community* in 1958 and fully independent in 1960. In the 1960s and 1970s it suffered from unstable governments, military and civilian, but greater stability has been achieved since 1979 under the regime of Colonel Denis Sassou-Nguesso. The Republic has maintained close links with France for economic assistance and was also one of the founding members of the OAU* with the establishment of the Brazzaville Bloc*.

Congo Crisis (1960–5). The sudden decision by Belgium to grant independence to the vast country along the River Congo was taken in January 1960. A single state, the Congo Republic, was to be created, governed from Leopoldville (Kinshasa) under President Kasavubu*, with the radical nationalist Patrice Lumumba* as Prime Minister. Fighting began between tribes during parliamentary elections in May and further fighting occurred at independence (30 June). The Congolese troops of the Force Publique (armed police) mutinied against their Belgian officers. Europeans and their property were attacked, and Belgian refugees fled. In the rich mining province of Katanga (Shaba) Moïse Tshombe*, supported by Belgian troops and White mercenaries, proclaimed an independent republic. The Congo government appealed to the UN for troops to restore order, and the UN secretary-general Hammarskjöld* dispatched a peace-keeping force. A military coup led by Colonel Mobutu* ousted Lumumba, who was later murdered with the connivance of Tshombe, and Hammarskjöld died in an air crash in September on a visit to the Congo. Fighting continued and independent regimes were established at different times in Katanga, Kisangani (Stanleyville), and Kasai. In November 1965 the Congolese army under Mobutu staged a second coup and Mobutu declared himself President in place of Kasavubu.

Congress, Indian National, the principal Indian political party. It had been founded in 1885 as an annual meeting of educated Indians wanting a greater share in the government of India. The decision of the Viceroy Curzon to split Bengal* in 1905 had resulted in the precipitation of more extremist policies under the leadership of Bal Gangadhar Tilak. In 1918 the Englishwoman Annie Besant* became one of three Britons to hold the office of President. Tilak died in 1920, and under the leadership of Mohandas Gandhi* Congress developed a powerful central organization, with elaborate branch organizations in provinces and districts, and acquired a mass membership. Throughout the 1920s it was conducting major political campaigns for self-rule and independence. In 1937 it easily won the elections, held under the Government of India Act* (1935), in a majority of provinces. In 1939 it withdrew from government in protest over India being declared 'at war' with Germany. Many of its leaders were imprisoned 1942–5 during the 'Quit India' campaign. Released from prison, its leaders collaborated with the last Viceroys Wavell* and Mountbatten* to negotiate independence. Under Jawaharlal Nehru* the party continued to dominate the politics of the Republic. After his death in 1964 a struggle ensued between the 'Old Guard' or Syndicate, and younger more radical elements, of whom Indira Gandhi* became the leader. In 1969 it split between these two factions, but was quickly rebuilt under Mrs Gandhi's leadership. From 1975, however, new splits emerged, resulting in a suppression of democracy, followed by the victory in 1977 of the Janata* Alliance Party, which remained in power for two years. In 1978 Mrs Gandhi formed a new party, the 'real' Indian National Congress, or Congress (I) (for Indira). In 1979 she led this faction to victory in elections and again became Prime Minister in 1980. After her assassination in October 1984 the splits between factions largely healed and the leadership of the Congress (I) Party passed to her son Rajiv Gandhi*, who became Prime Minister.

Congress of the United States. Provided for by Article I of the US Constitution, it consists of two houses. The House of Representatives, whose members serve a two-year term, originates all revenue Bills (i.e. federal taxes and duties), can initiate other Bills, and must give majority approval to all legislation; it is presided over by a Speaker elected from the majority party. The Senate, whose members serve for six years, must ratify all treaties and approve senior presidential appointments, and also give majority approval to all legislation; it is presided over by the US Vice-President. Congress alone can declare war, is responsible for raising and maintaining the armed services, for the regulation of commerce, patents, and copyrights, and for the establishment of post offices and federal courts. During the 20th century much of the effective work of Congress has been done through powerful standing committees, whose proceedings since 1951 have been nationally televised. While the presidency has moved between the Democrat and Republican parties, the Congress has tended to be Democrat controlled through most of the century. Yet tension between the powerful presidential executive and determined congressional legislature has been a continuous feature of US government whichever party had the majority. In 1973 the War Powers Act reclaimed some congressional control over US foreign commitments, since it restricts to sixty days the time a President can unilaterally commit US troops overseas.

Conrad von Hötzendorf, Franz, Baron (1852–1925), Chief of Staff of the Austro-Hungarian army 1906–17. A professional soldier, he became a close friend of the Archduke Francis Ferdinand. Appointed in 1906, he did much to reorganize the imperial army and was responsible for its mobilization after the Archduke's death at Sarajevo*. In 1915 he took the field and broke the Russians in Galicia. He was dismissed, however, by the new Emperor Charles I*. In June 1917 he commanded in the field and was defeated at Asiago in Italy. He then retired and wrote his memoirs.

Conscription, compulsory enlistment for military service. It began in modern times in France in 1793 with the *levée en masse*. Since then most European countries have had a system of conscription for eighteen months or two years. In 1916 casualties on the Western Front* were such that Britain introduced conscription in February against strong opposition from the Labour Party, many conscientious objectors being imprisoned. Canada, New Zealand, and Australia were all asked for more troops, in all of which countries it became a highly divisive issue—a future New Zealand Prime Minister, Peter Fraser, was imprisoned, while fierce rioting in Quebec resulted in four deaths. Germany was forbidden conscription by the Treaty of Versailles, but it was reintroduced by Hitler in 1935. In 1939 it was reintroduced in Britain. Here National Service was to last until 1960. In the USA a 'draft' system was introduced in May 1917 (until 1919) and again in 1940, its first peacetime use, until 1947. It was reintroduced by the Universal Military Training Act of 1948, implemented for both the Korean and Vietnam Wars. As military technology has advanced, huge conscript armies have had a decreasing purpose, their main role being frontier guarding, as in China and India. The Israeli conscript army is one of the world's most intensely trained armies. Western nations have adopted steadily more sympathetic attitudes to pacifists and conscientious objectors. In 1918 CO's in Britain were denied the right to vote for five years.

Conservative Party (UK). It was the first British political party to create a national organization, with the formation of the National Union in 1867 and a Conservative Central Office in 1870. Allied from 1886 with those Liberals who opposed Home Rule* in Ireland, its full title became the National Union of Conservative and Unionist Associations. The party was strongly imperialist throughout the first half of the 20th century, although splitting in 1903 over the issue of imperial preference or free trade. Until World War II it strongly resisted many demands from within the Empire for independence. It was in power throughout the period 1915–45, except for 1924 and 1929–31, either alone or in coalition, as in 1915–22, 1931–5, 1940–5. Since World War II it has alternated in government with the Labour Party 1951–64, 1970–4, 1979– . Before the 1970s the party's post-war policies tended to be pragmatic, accepting the inevitability of decolonization and the basic philosophy of the Welfare State, and being prepared to adjust to a consensus of public opinion. Under the leadership of Margaret Thatcher, however, it claimed to be reasserting the 19th-century liberal emphasis on individual free enterprise, challenging the need for state support and subsidy, while combining this with a strong assertion of state power against left-wing local authorities. With the growing crisis in Northern Ireland after 1968 the Ulster Unionists dissociated themselves from the party. In recent years it has tended to return to the early 19th-century use of the term Tory.

Conservative Party of Canada (since 1942 the Progressive Conservative Party). The party traces its origins back to the Liberal-Conservatives of 1854, whose leader John MacDonald achieved the Confederation of 1867. In 1914 the party was in office under Robert Borden*, who in 1917 made an alliance with those Liberals willing to support conscription. It was in office again under R. B. Bennett*, but was eclipsed by Mackenzie King's Liberals until 1957, having changed its name to the Progressive Conservatives in 1942, at the insistence of its new leader John Bracken from the Progressive Party*. Under Diefenbaker*, its next leader, who was Prime Minister 1957–63, it gained support from the western provinces at the expense of the urban vote of the east. It was again eclipsed by the Liberals under Trudeau* until 1984, when the Quebecois Brian Mulroney* took office, re-elected November 1988. The party has been criticized as being too closely associated with high finance and big business. A strong supporter of the Commonwealth, it has in fact achieved much progressive legislation and in recent years advocated North American free trade.

Constantine II (1940–), King of the Hellenes 1947–73. Son of King Paul I (1947–64), he spent his childhood in exile from Greece and was a highly popular heir to the throne: a keen sportsman who married a Danish princess. Unfortunately, he at first supported the military take-over of 'the Colonels'* in April 1967, only realizing his error in December, when he went into exile in Italy. Greece became a republic in 1973, and on his return from exile in 1974 a national referendum failed to give him a majority for restoration of the Greek monarchy.

Constantinople Agreement (April 1915), a secret agreement between Britain, France, and tsarist Russia that after the war Constantinople and the Dardanelles would be incorporated into the Russian Empire, a dream going back to Catherine the Great. In 1918 the Bolsheviks published the agreement, as they had done in November 1917 the equally embarrassing Sykes–Picot Agreement*. Knowledge of the agreement strengthened Mustafa Kemal's* determination to regain Constantinople for the new Turkish Republic, even at the risk of war with the imperial powers.

Consumer protection. Legislation to protect consumers goes back into the 19th century when, for example, there was a British Pure Foods and Drugs Act in 1872, a Sale of Goods Act in 1893, and the establishment of the British Standards Institute in 1901. In the USA the Federal Trade Commission was established in 1914 and the Food, Drug and Insecticide administration in 1927. In Britain the Consumer Association was formed in 1957 and a Consumer Protection and Advisory Committee ten years later. While government actions have for the most part controlled the worst abuses and provided basic safeguards, nevertheless high-powered advertising and skilful promotion have prompted the formation of a variety of voluntary organizations in countries throughout the world, many of these being linked by the International Organization of Consumer

Unions. These seek to persuade particular industries to adopt voluntary codes and to be more honest in product labelling. In the USA the lawyer Ralph Nader pioneered the movement for consumer protection, attacking particularly such giant corporations as General Motors*, which also controls Du Pont, the drugs company. His book *Unsafe at Any Speed* (1965) criticized standards in the US automobile industry and resulted directly in the establishment of a National Highway and Traffic Administration in 1970.

Contadora Group. In 1983 a group of Latin American countries, Mexico, Colombia, Panama, Venezuela, and Panama, with a 'support group' of Argentina, Brazil, Peru, and Uruguay, arranged a meeting on the island of Contadora, pledged to find a peaceful solution to the problems of Central America. The Group produced a draft agreement, but it failed to gain the support of the USA. The Group, however, gave full support to the August 1987 Peace Plan 'Esquipulas II' of President Arias of Costa Rica*.

Containment, the basic strategy of US policy during the years of the Cold War*. NATO* would provide a means of containment in Europe, armed with conventional and nuclear forces and stretching from the Arctic Circle to Turkey. Similar pacts in the Far East were ANZUS* (1951) and SEATO* (1954) and for a short while in the Middle East CENTO*. In the 1960s the policy was extended to include the need to prevent Soviet participation in the affairs of Latin America or Africa, the most dramatic episode perhaps being the Cuban Missile Crisis* of 1962.

Control commissions (Germany). After World War I an Allied Control Commission was established to supervise German demilitarization. At the end of World War II the USA, the Soviet Union, and Britain agreed that Germany should be occupied in four zones, to be administered by themselves, together with France. Berlin likewise was to be divided. Four military commanders would form a supreme Control Council or Kommandatura. The responsibility of these four commanders was in theory to deal with matters relating to the whole of Germany. In practice each occupying power administered its zone independently, although the British and US zones were merged early in 1947. However, the Control Commission did do significant work in removing members of the Nazi Party from important posts. In 1948 in Berlin the Russians withdrew from the Kommandatura and established a separate municipal government. The three Western zones were then merged into West Berlin. Increasing tension between the Soviet and Western zones led in 1949 to the creation of the Federal and the Democratic Republics.

Convoy system, merchant ships in wartime sailing in groups under armed naval escort. In 1917 Germany's policy of unrestricted submarine (U-boat) warfare nearly defeated Britain. One ship in four leaving British ports was sunk and new construction only replaced one-tenth of tonnage lost. The loss of Norwegian pit-props threatened the coal industry, while only six weeks' supply of wheat remained. Faced with such a crisis Lloyd George* overruled the Admiralty's refusal to organize convoys, and by November 1918, 80 per cent of shipping, including foreign vessels, travelled in convoy. In World War II transatlantic convoys were immediately instituted, in spite of a shortage of destroyers, using long-range aircraft for protection. During 1942 the system was extended to US shipping, as the Allies at that time were losing an average of ninety-six ships a month.

Cook Islands, a group of some sixteen fertile Polynesian islands lying between the Society and Samoan groups, discovered by Captain Cook in 1773. In 1901 Britain had declared them part of the Dominion of New Zealand. They continue to be the responsibility of the New Zealand Department of Island Territories, but since 1965 have had full internal self-government under a new constitution.

Co-operative Commonwealth Federation (CCF), a Canadian political party. Founded in 1932 in Calgary, Alberta, by a number of labour and farm groups, its aim was co-operative production 'for supplying of human needs instead of making profits'. Branded at the time as a dangerous Bolshevik party, its programmes for workers' compensation, unemployment insurance, and health and welfare insurance were all later to be adopted. The party was in office in Saskatchewan in 1944 and in opposition in British Columbia, Manitoba, and Ontario, but its appeal declined in the 1950s and in 1961 it joined the Canadian Labour Party to found the New Democratic Party* of Canada.

Co-operative movement, the development of business ventures owned and run jointly by members. It first appeared in new industrial towns in Britain at the end of the 18th century, as an alternative to competitive capitalism. The social reformer Robert Owen made many attempts to set up co-operative communities, but it was the founding of the Rochdale Pioneers in 1844 which first successfully established the movement in Britain. Here in 1864 a federation of co-operative societies, the Co-operative Wholesale Society (CWS), was formed, which developed as a manufacturer and wholesale trader, opening factories and developing its own farms. The Co-operative Party was established in Britain in 1917 to represent its members' interests in Parliament, subsequently contesting elections in alliance with the Labour Party. The movement spread rapidly through northern Europe, and in the USA the first co-operatives were established in the early 19th century to open up the prairies. During the 20th century the process of collectivization* and the break-up of private estates in both Communist and capitalist societies has resulted in a massive development of co-operative farming. At the same time agricultural, marketing, and purchasing co-operatives have developed throughout the world, together with credit co-operatives (over 22,000 credit unions in the USA) and housing co-operatives, which began in Scandinavia, but have rapidly developed in urban Britain and the USA.

Coral Sea, Battle of the, World War II (4–8 May 1942). Following their spectacular successes in South-East Asia and in the Pacific since Pearl Harbor* in December 1941, the Japanese were preparing to invade Australia from New Guinea. The US commander Admiral Nimitz* sent a strong naval force, under Admiral Fletcher, into the Coral Sea, consisting of two carriers and seven cruisers, with support ships. Here they faced three Japanese carriers and six cruisers. The battle was largely fought by aeroplane, all three Japanese carriers being immobilized and one sunk, together with three heavy cruisers, one light cruiser, and two destroyers, at the cost of the US carrier *Lexington* and one destroyer. Japanese expansion southwards was effectively checked.

Corfu Incident (August 1923). On 27 August an Italian general and four members of his staff, on an international mission to determine the Greek–Albanian frontier, were found murdered in Corfu. Mussolini retaliated by bombarding the island, killing sixteen people; he also demanded a large indemnity from the Greek government. The latter appealed to the League of Nations*, which referred the dispute to its Council of Ambassadors. The Council ordered Greece to pay 50m. lire. Under pressure from Britain and France, Italian troops were withdrawn on 27 September. The outcome of the dispute raised doubts as to the strength and efficiency of the newly created League of Nations.

Corfu Pact (20 July 1917). When Serbia was overrun in World War I the defeated Serbian army occupied the island of Corfu, which became the seat of a government-in-exile. Here a Pact was made between Nikola Pašić*, Serbian Prime Minister, and other south Slav refugees, that the Karageorgević dynasty of Serbia should establish a unified state of south Slavs after the war. This in 1929 was to adopt the name Yugoslavia.

Corporatism, an ideology which sees a political community as composed of various economic and functional groups, syndicates, or corporations, which then represent the interests of their members. It was developed in Fascist Italy where, with all officials within syndicates and corporations members of the party, the corporate state greatly increased its control.

Cosgrave, Liam (1920–), Prime Minister (Taoiseach) of the Irish Republic 1973–7. Eldest son of William T. Cosgrave*, he was born and educated in Dublin and became a lawyer, being called to the Irish Bar in 1943. He sat in the Dáil from that year until his retirement in 1981 as a member of Fine Gael*, the party led by his father until his retirement. He held minor office when Fine Gael formed a coalition government 1948–51 and again 1954–7, leading an Irish delegation to the UN in 1956. He became leader of his party in 1965 and held office as Taoiseach in a coalition government 1973–7, being defeated on economic issues.

Cosgrave, William Thomas (1880–1965), President of the Executive Council of the Irish Free State 1922–32. Son of a Dublin councillor, he worked in the grocery trade and also became a councillor. In 1913 he joined the Irish Volunteers, determined to gain independence from Britain. He took part in the Easter Rising* of 1916, and was captured and imprisoned for a short while. In 1917 he was elected to the Westminster Parliament for Sinn Fein in a by-election, and in 1919 became Minister for Local Government in Arthur Griffith's* unofficial government of the Dáil Éireann, being among those proscribed by the Black and Tans* in 1920. He reluctantly accepted the Anglo-Irish Treaty of December 1921, was elected to the first official Dáil Éireann in January 1922, and was subsequently President of the Free State 1922–32. During these years he did much to enhance the international standing of the new state, working actively with the League of Nations. He was leader of the opposition in the Dáil 1933–44, and was the father of Liam Cosgrave*.

Costa Rica, a Central American republic, independent since 1838, with a population of mostly Spanish descent. A policy of isolation and stability, together with agricultural fertility, brought considerable British and US investment in the 19th century. Apart from a brief dictatorship under Frederico Tinoco Granados (1917–19), Costa Rica was remarkable in the early 20th century for its stable liberal political system. After World War II, however, left-wing groups emerged, including a Communist Party. A new constitution, granting universal suffrage, was introduced in 1949. The socialist Presidents Otilio Ulate (1948–53) and José Figueres* (1953–8, 1970–4) tried to reduce the influence of the army, to nationalize banks and industry, and to reduce the power of US investment. Political tensions in the 1970s were aggravated by economic problems and the arrival of fugitives from neighbouring states. President Luis Alberto Monge (1982–6) had to impose severe economic restraint, even though economic prospects remained better than in most Latin American states. President

Oscar Arias of the Partido Liberación Nacional was elected President in 1986. In August 1987 he produced a peace plan for Central America for which he was awarded a Nobel* Peace Prize. All Central American countries accepted it, but it never won the full support of the USA, President Reagan remaining eager to support the Nicaraguan Contras (Freedom Fighters) and reducing the level of US aid to Costa Rica. In the 1990 presidential election Oscar Arias was succeeded by Rafael Angel Calderón Fournier of the Social Christian Unity Party.

Côte d'Ivoire, République de la, a West African country. France had established a colony in 1893. It became a Territory of French West Africa in1904, but was not fully pacified until 1912. In 1933 most of the area of the Upper Volta was added to the Ivory Coast, but in 1948 this was returned to the reconstituted Upper Volta, today Burkina Faso*. The colony became an autonomous republic within the French Community* in 1958 and independent as the République de la Côte d'Ivoire in 1960. It is governed by the moderate Parti Démocratique de la Côte d'Ivoire, with Félix Houphouët-Boigny* its President. The latter organized a loose confederation of neighbouring states, with Benin, Burkina Faso, Niger, and Togo, but its Conseil d'Entente failed to develop a unified market. The country has an expanding economy, with large oil deposits and a developing industrial sector, but falling cocoa and coffee prices caused considerable economic hardship in the 1980s.

Coudenhove-Kalergi, Count R. N. von (1894–1972), Austrian diplomat. Born in Poland, he was educated in Vienna, which was for long his home. His book *Pan-Europa* (1923), dedicated to the Youth of Europe, urged the formation of 'a European organization to fill the void left by European anarchy'. In 1924 he used his modest fortune to fund a secretariat in Vienna which attracted representatives from twenty-four countries to a Pan-European Congress in October 1926. His ideas appealed to both Stresemann* and Briand*, and the latter helped to organize a second Congress in Berlin in 1930, when Coudenhove-Kalergi urged the formation of a United States of Europe in order to prevent the 'formation of two international factions, one claiming revision of the Treaty of Versailles, the other standing for its provisions, with a danger of unemployment, poverty and social revolt'. He lived in the USA from 1938, where he continued to urge the idea of Pan-Europe. He returned to live in Switzerland in 1946 and helped to organize the Congress of The Hague in 1948 from which the Council of Europe* resulted in 1949. In 1950 he was awarded the Charlemagne Prize at Strasburg for his work towards achieving European unity.

Council of Europe, an association of European states independent of the European Community. It was founded in 1949, committed to the principles of freedom and the rule of law, and to safeguarding the political and cultural heritage of Europe. With a membership of twenty-one European democracies, it is served by a Committee of Ministers, a Parliamentary Assembly at Strasburg, the European Court of Human Rights, and the European Commission of Human Rights. Although without legislative powers, treaties have covered the suppression of terrorism, the legal status of migrant workers, and the protection of personal data. The Council was addressed by President Gorbachev in 1989 and in 1990 its role was expected to expand as East European democracies sought membership.

Country Party (since 1982 the National Party of Australia). It was founded in 1913 by a group of New South Wales sheep and wheat farmers and elected its first members to the federal Parliament in 1919. It survived as a small right-of-centre party in various coalitions in both state and federal politics throughout the century, consistently voicing the causes of the rural voter. Its first federal coalition was with Stanley Bruce* 1923–9, with fourteen members of the House of Representatives, led by Earle Page*. The second coalition was with J. A. Lyons*, joining his United Australia Party in 1934 and remaining in power with them until 1941, with Page at one time acting as Prime Minister. After the war the party supported Robert Menzies and his new Liberal Party, being in a ruling coalition with them 1949–72. In 1975 it joined Malcolm Fraser*, following the Whitlam* Crisis, in yet another coalition which lasted until 1983, having adopted the title of National Party of Australia in 1982.

Coupon Election (UK) (December 1918). After the armistice of 11 November 1918 Lloyd George* immediately called an election, in order to go to the Peace Conference with the authority of a new Parliament. There had been a parliamentary debate in May 1918 over an allegation by Major-General Sir Frederick Maurice ('the Maurice Debate') that Lloyd George's government had failed to support General Haig* during the German offensive of March of that year. The 106 Liberals who had voted for a select committee of inquiry were denied 'a coupon', or letter of commendation from the party whips, for this election. It was this more than anything else which fatally split the Liberal Party. The election campaign was a bitter one, with Lloyd George declaring that Germany would have to pay 'the whole cost of the war', and others campaigning to 'hang the Kaiser'. The demand for vengeance against Germany was fanned by the popular Press, thus preparing the way for the reparations demands of the Paris Peace Conference*. The election result was an overwhelming vote for Lloyd George, the coalition government, and for the Conservatives. The Liberals halved their parliamentary strength and never regained it.

Cox, Sir Percy Zachariah (1864–1937), British civil servant and diplomat. After attending Sandhurst he served in India in the army and then transferred to the Indian political service, working in the Persian Gulf. During World War I he was actively involved in liaison with Iraqis, as the Indian Expeditionary Force advanced into Mesopotamia. After the war he was appointed High Commissioner in Baghdad, where he supervised the election of Emir Faisal* as King (1921–33), created the Iraqi army, devised an Iraqi constitution, and negotiated the Anglo-Iraq Treaty of Alliance which would provide a British military presence. He retired in 1923.

Creole, a term loosely applied to persons of predominantly European stock born in a society with a large Black or indigenous population. It is used in countries such as Angola, Madagascar, Mauritius, Mozambique, Sierra Leone, and other African countries where European colonists were long established. It is also used in the West Indies, Brazil, Venezuela, and other South American countries.

Crete. It was declared a part of Greece by the Treaty of London in 1913, after being under Turkish rule for centuries. In 1941 it was attacked by German troops from the mainland. Despite resistance by Greek and Commonwealth forces the Germans successfully made the first airborne invasion in military history. A bloody twelve-day battle followed, when some 18,000 Allied troops were captured. Since 1945 it has become the centre for extensive tourism and also a NATO missile base. But much of it retains its unique cultural heritage.

Cripps, Sir Richard Stafford (1889–1952), British statesman. A nephew of Beatrice Webb, he was born in London and educated at Winchester and the University of London. He was called to the Bar in 1913 and during World War I, as a pacifist, he served in the British Red Cross. A brilliant barrister specializing in company law, he was

Solicitor-General 1930–1 and in January 1931 elected to Parliament as a Labour Member for Bristol. He refused to join the National government in 1931 and in 1939 was, for a short while, expelled from the Labour Party for advocating a popular front* with the Communists. In 1940 he was appointed ambassador to Moscow and then served as Minister for Aircraft Production 1942–5, now a much more centrist figure. He served in Attlee's* Labour government successively as President of the Board of Trade and Chancellor of the Exchequer, where he showed a unique mastery of economic detail. In these posts he was responsible for programmes of austerity: rationing, strict taxation, wage and other controls, necessitated in order to adjust Britain to its reduced status, following the withdrawal of US Lend-Lease* in September 1945. He also directed a notable expansion of exports, especially after devaluation of the Pound in 1949.

Croatia. As Croatia-Slovenia it was an autonomous region of the Austro-Hungarian Empire. In October 1918 an independent Croatia was proclaimed, but this somewhat reluctantly joined the Kingdom of Serbs, Croats, and Slovenes in 1921 under Alexander I*; the country was renamed Yugoslavia in 1929. Strong antipathies against Serbia survived, fanned by Fascist terrorists, the Uštaše*. In 1941 the leader of the latter Ante Pavelić* established a Nazi puppet 'kingdom' with an Italian duke as King of Croatia. Appalling anti-Serbian atrocities followed. In 1946 Croatia became a Socialist People's Republic as part of the Federal Republic of Yugoslavia. A movement for Croatian independence re-emerged in the late 1980s and a non-Communist government was formed in May 1990. By the end of the year anti-Serbian partisans were increasingly evident.

Crosland, Charles Anthony Raven (1918–77), British politician. Born in Sussex, son of a senior civil servant, he was a scholar at Trinity College, Oxford, at the outbreak of World War II, in which he served in the Parachute Regiment. Returning to Oxford, he became a Fellow of his college, and in 1950 entered Parliament as a Labour MP. His book *The Future of Socialism* (1956) was a highly influential revisionist document, with its optimistic forecast of continuing economic growth. As Secretary of State for Education 1964–7, his strongly held egalitarian principles led to the closure of a majority of grammar schools and the establishment of a comprehensive state school system, together with the establishment of the system of polytechnics. During 1967–70 and 1974–7 he held a number of cabinet posts, being Foreign Secretary 1976–7 before his sudden and early death. He was described as having one of the 'strongest intellects in post-war politics'.

Crossman, Richard Howard Stafford (1907–74), British politician. Son of a successful barrister, he was educated at Winchester and New College, Oxford, where he became a Fellow in 1930. As a supporter of Labour he championed rearmament in the later 1930s and in World War II was assistant chief of the Psychological Warfare Division. He entered Parliament as a Labour MP in 1945 and was an active 'Bevanite' in the early 1950s. During the administrations of Harold Wilson* he was successively Minister of Housing and Local Government, Leader of the House of Commons, and Secretary of State for Social Services. His posthumous *Diaries* (1975–7) provide revealing insights into the working of government.

CSCE *see* CONFERENCE ON SECURITY AND CO-OPERATION IN EUROPE.

Cuba, one of the Greater Antilles* islands. Heavy US investments in the late 19th century had made Cuba the largest sugar-producing country in the world. In 1898 a War of Independence against Spain (1895–1901) was joined by the USA after a well-orchestrated Press campaign, and the island was occupied by US troops in 1899. In 1902 the Republic of Cuba was proclaimed under US 'protection'. A series of corrupt and socially insensitive governments followed, with periodic interventions by US marines, 1906, 1912–13, 1917, culminating in the brutal regime of Gerardo Machado (1925–33). This prompted a student revolution and further US intervention. Sergeant Fulgencio Batista*, with US assistance, overthrew the student revolutionary government, promoted himself general, and was to dominate Cuban politics until 1958. Having crushed student paramilitary groups, he produced a new constitution under which he became President. In his first period in office considerable social reform was achieved, but a nascent Cuban democracy foundered under the corrupt regimes of his two successors. When Batista returned to power in 1952, again with US help, it was to lead an equally corrupt, cynical, and brutal government. In 1956 Fidel Castro* initiated a guerrilla war which led to the establishment of a socialist regime under his leadership. He repulsed the invasion by Cuban exiles at Cochinos Bay (the Bay of Pigs*) in April 1961 and survived the Missile Crisis* of 1962. The early achievements of his one-party state were considerable in public health, education, and housing, while industries were nationalized and vast sugar plantations brought under state control. After the turbulence of the 1960s the revolution stabilized, with the establishment of more broadly based representative assemblies at municipal, provincial, and national levels. In 1975 a further round of nationalization of commercial enterprises took place, but initial hopes of economic diversification and industrialization were not realized, and Cuba continued to rely on the export of sugar at prices subsidized by the Soviet Union. With rationing and food shortages continuing, there was a sense of frustration, leading to some 125,000 emigrants leaving the country in 1980. Relations with the USA continued to be bad throughout the 1980s, with Cuba giving support to the Contadora Group's* peace initiative for Central America, and numerous CIA attempts to assassinate Castro. During World War II a US military and naval base was built at Guantanamo. Perhaps surprisingly this continued to function throughout the period of ideological disagreement.

Cuban Missile Crisis (October 1962). It was precipitated when the US government learned that Soviet ballistic missiles with atomic warheads capable of hitting the USA were being installed in Cuba. President Kennedy* reinforced the US naval base at Guantanamo, ordered a naval blockade against Soviet military shipments to Cuba, and demanded that the Soviet Union remove its missiles and bases from the island. On 27 October there seemed a real danger of nuclear war as the rival forces were placed on full alert. The crisis sharpened as Soviet merchant vessels, thought to be carrying missiles, approached the US blockading forces. However, on 28 October the Soviet ships were ordered by First Secretary Khrushchev* to turn back. The Soviet Union agreed to US demands to dismantle the rocket bases in return for a US pledge not to attack Castro's Cuba. The naval blockade was lifted on 20 November. One outcome of the crisis was the establishment of a direct, exclusive line (the 'hot line') of communication, to be used in an emergency, between the President of the United States and the leader of the Soviet Union.

Cubism, a Post-Impressionist style of painting 1907–14. It was developed by Picasso* and Georges Braque in Paris, emphasizing the two-dimensional nature of the picture plane and presenting different facets of an object simultaneously. It was to influence modernist painters in the 1920s, as well as sculptors and architects, notably Le Corbusier*.

Cultural Revolution. In the early 1960s in China there was oblique criticism of Mao Zedong's* egalitarian ideology from younger, pragmatic bureaucrats. Mao retaliated by utilizing discontented students and young workers, who became his Red Guards*. On 3 September Marshal Lin Baio* made a speech urging students to return to basic revolutionary principles and for the next three years teachers, artists, and intellectuals were subjected to harassment, many schools and colleges being closed. State Chairman Lin Shaoqi was formally dismissed from all his posts in October 1968 and Lin Baio seems to have emerged as Mao's choice to succeed him. By the Ninth Party Congress of 1969 the Cultural Revolution was adversely affecting the economy. From then on its more fanatical aspects can be said to have ended, although many radical policies continued until Mao's death.

Cunningham, Andrew Browne, 1st Viscount Cunningham of Hyndhope (1883–1963), British First Sea Lord 1943–6. Born in Scotland he joined the Royal Navy in 1898 as a midshipman. He won distinction in World War I and by the beginning of World War II was commander of the 1st Cruiser Squadron in the Mediterranean. He moved to the North Sea before returning to the Mediterranean as naval Commander-in-Chief. Here he was faced with an Italian fleet that was numerically superior to his own. However, he asserted British domination by his air attack on the Italian base of Taranto in 1940, and at Cape Matapan in 1941, where his victory effectually neutralized the Italian fleet for the rest of the war. As First Sea Lord he was responsible for naval strategy and attended all the meetings of Allied heads of government.

Curragh Incident, a 'mutiny' at the British military headquarters on the Curragh plain near Dublin. In March 1914 the British military commander there, General Sir Arthur Paget, allegedly on instructions from the Secretary of State for War, Colonel John Seely, informed his officers that since military action might be necessary against private armies in Ulster, officers with Ulster connections were to be allowed to 'disappear' or resign. Whereupon General Hubert Gough of the 3rd Cavalry Brigade, together with fifty-eight out of his seventy-one officers, tendered their resignations. On its own initiative the War Office then agreed that the British government would not in fact use force to coerce Ulster into Home Rule*. All the officers were then reinstated. Prime Minister Asquith* rightly regarded the incident as a threat to army discipline and demanded the resignations of Seely, and of the Chief of Staff Sir John French and the Adjutant-General Sir Spencer Ewart, although the latter officers were soon reinstated.

Curtin, John Joseph (1885–1945), Prime Minister of Australia 1941–5. From an Irish Catholic family in Victoria, he became secretary of the Timber Workers' Union and then moved to Western Australia, where he became a Labor Party activist. In 1928 he was elected to the federal Parliament. In 1935 he became Labor leader and in 1941 Prime Minister. Having opposed conscription in World War I, he felt it necessary in World War II once the Japanese threat developed. He worked closely with the USA in its defence strategy for the Pacific, but was suspicious of the ever-mounting powers of General MacArthur*. At the conference of Commonwealth Prime Ministers in London in 1944 he urged the strengthening of links between the UK and the dominions, in some ways presaging the Colombo Plan*, which Australia was to put forward in 1950. He died suddenly in July 1945, being succeeded by Joseph Chifley*.

Curzon, George Nathaniel, 1st Marquess of Kedleston (1859–1925), Viceroy of British India 1899–1905, British Foreign Secretary 1919–23. Born in

Kedleston Hall, Derbyshire, and educated at Eton and Oxford, he was a Conservative MP 1886–98 and Under-Secretary for Foreign Affairs 1895–8, before becoming Viceroy. His seven years in India were controversial, his partition of Bengal* in 1901 causing long-standing resentment. More positive achievements included reforms in education and agriculture, the extension of railways and canals, and the strengthening of the North-West Frontier. With an autocratic manner, he clashed with the equally domineering Kitchener*, Commander-in-Chief in India, over control of the Indian army, and in 1905 returned to private life at Kedleston. In 1915 he joined Asquith's coalition government as Lord Privy Seal and was a member of Lloyd George's* war cabinet 1917–19. He then became Foreign Secretary 1919–23, during which time Britain became involved in the unsuccessful War of Intervention in Russia. As a strong imperialist, he supported Milner's* aims further to extend the Empire, creating a Middle East Department and aiming to weld Britain's mandates and protectorates into a single Middle East Dominion to link up with India. He deplored what he called 'the uppishness' of the White dominions now seeking independence. In July 1923 he was largely responsible for the Treaty of Lausanne*, which recognized the Republic of Turkey. Lloyd George's tendency to interfere in foreign affairs irritated him, and in October 1922 he supported the Conservative rebellion against him, continuing as Foreign Secretary under Bonar Law*. When the latter resigned he hoped to succeed him, but his haughty, aristocratic style led the Conservatives to prefer Baldwin*, under whom he continued as Foreign Secretary. He was Lord President of the Council from 1924 until his death. He was described as both the 'last British aristocratic proconsul' and 'the finest orator of his times'. His name was given in 1920 to a proposed frontier between Poland and Bolshevik Russia.

Curzon Line, frontier between Poland and the Soviet Union since 1945. It was first suggested by the British Foreign Secretary Lord Curzon* in July 1920 as a possible frontier for the new Republic of Poland. In fact it was rejected and Poland claimed and obtained more lands to the east. It became, however, the boundary between the German- and Russian-occupied spheres in 1939 and then the definitive frontier.

Cyprus, an island in the east Mediterranean. Having formed part of the Ottoman Empire until 1879, it was formally annexed by Britain in 1914 and in 1925 declared a Crown Colony. From the outset there was rivalry between the Greek-speaking majority, many of whom wanted union (*enosis*) with Greece, and the Turkish minority. There were serious pro-*enosis* riots in 1931, which the British firmly suppressed. After World War II a terrorist organization, EOKA*, was formed under the leadership of George Grivas. Between 1955 and 1959 this committed numerous acts of sabotage and terror against what it regarded as the occupying power, Great Britain. In 1959 Britain achieved a compromise, its offer of membership of the Commonwealth as an alternative to *enosis* being accepted by Archbishop Makarios*. The latter became President of the new state, but by 1964 the government was in chaos and a UN peace-keeping force was sent to the island to maintain peace between the communities. In 1974 a coup for *enosis* organized by a group of Greece-born Cypriot officers and backed by the 'Greek Colonels'* overthrew the President, who went into exile. Turkish forces invaded the island, gaining control over most of it, while in Greece the government of the Colonels collapsed. Talks in Geneva between Britain, Turkey, Greece, and the two Cypriot communities failed and, although Makarios resumed the presidency in 1975, the Turkish Federated State of Cyprus was formed on the north of the island, comprising some 35 per cent, with its own President. Britain retained an RAF base, Limassol, on the south (Greek) side; this has been a key intelligence centre.

Czechoslovakia. It was created from the Austro-Hungarian Empire in 1918, incorporating the Czechs of Bohemia-Moravia and the Slovaks in the east. Alliances with Yugoslavia and Romania (the Little *Entente**) and with France (1924) ensured a degree of stability, but national minorities of Germans and Hungarians created tensions. In 1938 the Munich Agreement* was made by Britain and France with Hitler, and by March 1939 German troops had occupied Bohemia-Moravia, and Slovakia had become a German puppet protectorate. A government-in-exile in London was led by Beneš*, who returned in 1945 to form a brief coalition government with Communist Resistance leaders. In 1948 Klement Gottwald, with Soviet backing, gained control and Czechoslovakia became a satellite of the Soviet Union. In the 'Prague Spring'* of 1968 an attempt by Alexander Dubček*, and other liberal Communist reformers, to gain a degree of independence was foiled by the intervention of Warsaw Pact* troops. The hard-liner Gustav Husák became President, but a continuing wish for independence continued, as was shown by the Charter 77 movement*. In January 1989 demonstrations took place on six consecutive days, to mark the anniversary of the suicide of a student, Jan Palach, who had burned himself alive in protest in 1969. The demonstrations were broken up by riot police and many were imprisoned, including Vaclav Havel*. A week of street protest began on 17 November, when 30,000 people took part in a march organized by the official Socialist Youth Union. This was brutally suppressed. Two days later twelve opposition groups joined to create Civic Forum, for whom Havel, who had been released in May, became the spokesman. Their immediate aim was the resignation of Milos Jakes, secretary of the Communist Party. This they achieved on 24 November, the day that Alexander Dubček returned to speak at a rally of 250,000 in Wenceslaus Square. On 28 November Prime Minister Ladislav Adamec announced the end of the monopoly of power of the Communist Party. On 9 December President Husák resigned. Havel was nominated for President and Dubček as chairman of the Federal Assembly, until elections could be held. On 17 December the wire border with Austria was ceremoniously cut and Czechoslovakia was a free country. In 1990 the secret police and the death penalty were abolished and Czechoslovakia left Comecon*. Vaclav Havel was elected President, on a popular vote, of what was henceforth to be known as the Czech and Slovak Federative Republic. May elections were won by Civic Forum in the Czech lands and by the Public Against Violence Party in Slovakia.

Czernin, Ottokar, Count von (1872–1932), Austro-Hungarian statesman. Born in Bohemia into a wealthy aristocratic family, Count Czernin entered the imperial diplomatic service as a young man of 23. He served in Paris and The Hague and he became a close friend of the Archduke Francis Ferdinand. He believed in the need to restore the authority of the aristocracy and was opposed to the introduction of universal suffrage in 1908. When the Emperor Charles I succeeded, he was appointed Foreign Minister, anxious to realize the new Emperor's aim of an armistice. In this he failed, through an indiscreet leak of information to Germany, and he resigned in April 1918. He served as a right-wing member of the Austrian National Assembly 1920–3.

Daladier, Édouard (1884–1970), Prime Minister of France 1933, 1934, 1938–40. Son of a baker in Provence, he became a teacher and served with distinction in World War I. In 1919 he was elected a Deputy as a Radical Socialist and held office for brief spells on a number of occasions, the first time as Minister for the Colonies in 1924. Strongly anti-Communist, he supported the policies of appeasement towards Nazi Germany and with Neville Chamberlain* signed the Munich Agreement* in 1938. In March 1940 he was replaced by Paul Reynaud* as Prime Minister. In June he tried to form a government-in-exile in North Africa, but he was arrested by the Vichy government and was tried at Riom, together with other democratic leaders, accused of responsibility for France's disasters. Although acquitted, he remained imprisoned in France and Germany. In 1946 he was elected to the Constituent Assembly and was a member of the new National Assembly of the Fourth Republic* 1947–58.

Dalai Lama, title of the abbot of the Dge-lugs-pa (Yellow Hat) order of Tibetan Buddhist monks who, since the 17th century, has claimed to be both temporal and spiritual head of Tibet. In 1914 the 13th Dalai Lama proclaimed Tibet as a sovereign state, following the collapse of the Manchu dynasty in China. The 14th Dalai Lama Bstan' dzin-rgy a-mtsho (Ten zin Gyat so) was enthroned in 1940, but fled into exile in India in 1959, following an unsuccessful revolt against the Chinese, who had reoccupied the country in 1950. He actively lobbied world opinion on behalf of Tibet and in 1989 was awarded the Nobel* Peace Prize.

D'Annunzio, Gabriele (1863–1938), Italian political adventurer and novelist. Born of a wealthy and aristocratic family, by 1914 he had won a reputation for his erotic novels, as well as for poetry. He urged Italy to enter World War I in the hope that it would gain pickings from the Austro-Hungarian Empire. He himself fought with spectacular daring in the Italian air force 1915–18. In 1919 he raised an irregular band of troops (the early Blackshirts*) and, in defiance of the Versailles Peace Settlement, seized the Adriatic port of Fiume (Rijeka). He imposed an authoritarian and Fascist-type government on it, until fifteen months later it was starved into surrender by forces of the liberal government of Giovanni Giolitti*. He welcomed Mussolini's rise to power, but took no further part in politics, retiring to his luxury villa on Lake Garda.

Danquah, Joseph Boakye (1895–1965), Ghanaian statesman. Educated in an elementary mission school, he became a law clerk at 17 and then secretary to his brother, a paramount chief. He gained entrance to University College London, where he read law and philosophy, and returned to the Gold Coast in 1927 to set up a law practice. In 1931 he founded the daily *Times of West Africa*, which was to have immense popular influence. Regarded by the British as a moderate and reliable nationalist, he was a member of various delegations and of the Gold Coast Legislative Council. In 1947 he founded the United Gold Coast Convention with Kwame Nkrumah*, who later left to found the more radical Convention People's Party. Imprisoned in 1948, next year he was a leading figure in constitutional negotiations and an advocate of the country's name of Ghana. He failed to be elected to the Legislative Assembly in 1954 and 1956, and became increasingly opposed to government policies. He was imprisoned 1961–2 and again in 1964 for criticism of Nkrumah's dictatorial methods. He died in prison.

Dardanelles, the 61-km. (38-mile) strait between the Aegean and the Sea of Marmara. By the 1841 Convention of London, it was closed to all warships in time of peace. In 1914, however, German warships anchored in Constantinople, and in November the Ottoman government closed the Strait to all commerce, thus cutting Russia off from her allies Britain and France. The latter declared war and decided to force a passage. Elaborate, but open, plans were made, a naval force being gathered in Alexandria. When this arrived in February 1915 (fifteen British ships and three French), Krupp guns had been installed by the Turks on land. These did disastrous damage, three ships being sunk and two more severely damaged. The forced landing at Gallipoli* followed, when three more British battleships were sunk. The Gallipoli withdrawal took place in January 1916 and in 1918, with the collapse of the Ottoman Empire, the Straits were placed under an international commission and opened to all vessels (including warships) at all times. This arrangement was modified at Lausanne* in 1923 to permit the passage of warships of under 10,000 tonnes only, and that in peacetime only, and to reduce the power of the Commission. This was abolished in 1936 by the Montreux Convention*, which restored control of the Straits to Turkey. The Convention has remained effective despite Soviet attempts to have it revised in its favour.

Darlan, Jean Louis Xavier François (1881–1942), French admiral. Born in Aquitaine, he was commissioned into the French navy in 1903 and served in China. During World War I he transferred to the army and fought at Verdun. Returning to the navy, he rose to the rank of vice-admiral between the wars, and was the virtual creator of the French navy which entered World War II. In 1939 he was promoted admiral and Commander-in-Chief of all naval forces. After he became Minister of Marine in the Vichy government in June 1940 he was regarded by the British as pro-German. His secret order to his commanders, to scuttle their ships should the Germans try to take them over, was not of course known to the British, who in July destroyed much of his fleet at Mers el-Kebir*. Darlan, together with Laval*, now actively collaborated with the Germans, and as Vice-Premier (1941–2) was the real power in France. When the British took over Madagascar in May 1942 he urged the garrison to resist to the utmost. When, however, the Allies landed at Casablanca and Algiers in November of that year he opened negotiations with General Eisenhower. He ordered the Vichy French forces to cease fire and had himself proclaimed head of state in French Africa. A month later he was assassinated.

Dawes Plan, US proposals to solve German economic crisis. In 1923 the German government applied to the Allied Reparations Commission for an investigation into the German economy, which could no longer pay reparations in kind. A committee of experts was appointed under the chairmanship of the US banker C. G. Dawes. In April 1924 it recommended fixed annual payments of 2½ million gold marks, to be raised from taxation. At the same time international loans would be made to stabilize the economy. The figure of 2½ million was achieved in 1928–9, but after the financial collapse of that year payments lapsed. In 1930 the Dawes Plan was superseded by the Young Plan*.

Dayan, Moshe (1915–81), Israeli general. Born in Palestine, he was educated at the Hebrew University of Jerusalem, where he joined Haganah*. He was imprisoned by the British 1939–41, but then released, joining an auxiliary unit of the British army and serving in Syria against Axis supporters. Here he lost an eye. When Israel declared independence in May 1948 he took part in a military counter-offensive against the Arab invasion, later serving on the armistice commission of 1949. As Chief of Staff he directed military operations during the Suez War 1956, reaching the Suez Canal within one week. He was Minister of Agriculture 1959–64 and Minister of Defence 1967–74. His campaigns in 1967 showed military brilliance, the West Bank, Golan Heights, and Sinai all being occupied. He resented the UN cease-fire for the Yom Kippur* (October) War of 1974 and moved into opposition, strongly resenting the Camp David Accord*.

Deakin, Alfred (1856–1919), Prime Minister of Australia 1903, 1905–8, 1909–10. A successful Melbourne lawyer, he had entered state politics and been an architect of Australian federation (January 1901). As Prime Minister he had introduced old age and sickness benefits. A strong supporter of the British Empire, he was an advocate of imperial preference*. In 1914 he was a much-respected Australian 'elder statesman'.

Debs, Eugene V(ictor), (1855–1926), US labour organizer. Starting at 14, he devoted his early life to working on the US railroad system and became an advocate for labour organization. He became president of the American Railroad Union and was imprisoned in 1894 for 'conspiracy' over the unsuccessful Pullman strike. While in prison he became a socialist and founded the Socialist Party of America. A highly effective public speaker, he was socialist candidate for President five times between 1900 and 1920. He was imprisoned again in 1917 for criticizing the Espionage Act and, as a pacifist, for discouraging recruitment to the US armed services. He was released in 1921 on orders of President Harding, but his citizenship was never restored.

Declaration of Human Rights. This was adopted by the General Assembly of the United Nations in Paris in 1948, with the Soviet Union, the Eastern bloc countries, Saudi Arabia, and South Africa abstaining. The notion of basic political and social rights dates from the late 18th century. This 'Universal Declaration' has thirty Articles and begins 'all human beings are born free and equal in dignity and rights'. It goes on to rehearse the basic political right to participate in the democratic process, without fear of arbitrary arrest and with total freedom of thought and religion. Then, however, it lists a series of social rights, which include the right to marriage, to education, to own property, to work, to receive social security, and to participate in the cultural life of the community. Such a comprehensive list, with egalitarian overtones, could only be an ideal to which a nation might aspire, and certainly none has ever fully realized that ideal. Nevertheless it became a standard demand of the West, during the Cold War years, that Soviet bloc countries should recognize more human rights.

Decolonization. The term became current in the later 1950s, although pressures to achieve it began much sooner. The colonial empires of Britain, France, The Netherlands, Belgium, Spain, Portugal in Africa, the West Indies, and in the Far East reached their maximum size when the League of Nations 'mandated' German colonies to the victors after World War I. The Middle East mandates of Iraq, Syria, and

Transjordan all gained independence between the wars, while pressures for Indian independence led to major concessions. World War II was the major factor for decolonization in Asia, even though the process was protracted and bloody in both Indonesia and Indo-China. Between 1956 and 1990 all African colonies became decolonized in the political sense, Namibia* being the last; in the Caribbean only a handful of territories retained colonial status, the largest being French Guyane; in the Pacific, however, both France and the USA retained control of significant island areas. Economically decolonization at times worsened rather than improved the quality of life for people of the Third World.

Defence of the Realm Acts (DORA), legislation by the British Parliament during World War I. Under a series of Acts 1914, 1915, and 1916 the British government took powers to commandeer factories and directly to control all aspects of war production, making it unlawful for war-workers to move elsewhere. Left-wing 'agitators' were deported to other parts of the country, especially from Clydeside. Strict Press censorship was imposed. All Germans had already been interned, but war hysteria led tribunals to harass anyone with a German name or connection (for example the writer D. H. Lawrence) and to imprison or fine pacifists (for example Bertrand Russell*). The Act of May 1915 gave wide powers over the supply and sale of intoxicating liquor, powers which were widely resented, but which long survived the war. An Emergency Powers Act of 1920 confirmed the government's power to issue regulations in times of emergency and in 1939 many such regulations were reintroduced.

Deforestation. The process began some 10,000 years ago with forest clearances for early farming. In Britain it had gone so far by the 16th century that Henry VIII ordered 'no further conversion of wood to pasture', but to no effect. The 19th century saw a massive growth in demand for tropical woods for building and furniture, especially teak and mahogany, but the pace of deforestation of rain forests in Latin America, Africa, Indonesia, and elsewhere accelerated so rapidly after World War II as to produce a global crisis. Not only are species of both flora and fauna being rapidly destroyed, but the whole ecological balance of the globe is threatened by an excess of carbon dioxide no longer being reprocessed. In 1988 in Brazil the president of the Union of Forest Peoples, Francisco Mendes, was openly murdered by agents of cattle rangers, who fell and burn thousands of acres a day. The issue of deforestation has become an increasing concern of the UN Environment Programme, and in June 1990 Operation Amazonia was launched in Brazil to try to reduce forest clearance.

De Gasperi, Alcide (1881–1954), Italian Prime Minister 1945–53. Born in Austrian Trentino, he was educated in Vienna and was elected to the Austro-Hungarian Parliament in 1911. From 1919 until 1925 he was a Deputy in the Italian Parliament and secretary of the Italian People's Party. Mussolini* then abolished all political parties and de Gasperi was imprisoned in 1926. On his release he took refuge in the Vatican, where he remained until 1943 and the fall of Mussolini. He played an important part in creating the Christian Democrat Party as a focus for moderate Catholic opinion after the Fascist era. As Prime Minister he adopted a strong stand against Communism and supported co-operation among the European powers. He took Italy into NATO*, the Council of Europe*, and the European Coal and Steel Community*.

De Gaulle, Charles André Joseph Marie (1890–1970), President of France 1958–69. Born in Lille, he graduated into the French army in 1909, serving under Colonel Pétain. He was wounded at the Battle of Verdun* and after the war gained a military reputation as a theorist, arguing the case for the greater mechanization of the

French army. When France surrendered in 1940 he fled to Britain, from where he led the Free* or Fighting French forces. Marching into Paris in June 1944, he formed a provisional government (1944–6), becoming provisional President (1945–6). However, he then retired into private life following disagreement over the Constitution adopted for the new Fourth Republic*. In 1947 he was back in politics, forming the Rassemblement du Peuple Français, a party advocating strong government. It only met with modest success, however, and he disbanded it in 1953 and again retired. In 1958 he re-entered public life at the height of the Algerian crisis. The Fourth Republic was dissolved and a new constitution was drawn up with greatly increased presidential powers. Thus the Fifth Republic* came into being with de Gaulle its first President. He conceded independence to France's African colonies in 1960, and following the Évian Agreements* ended the Algerian War. De Gaulle dominated the European Economic Community* in its early years, excluding Britain from membership. He developed an independent French nuclear deterrent and in 1966 withdrew French military support from NATO*. Between 1966 and 1968 he travelled extensively in Eastern Europe, urging a Europe 'from the Atlantic to the Urals', without apparent response. His position was shaken by a serious uprising in Paris May–June 1968, by students shocked by the Vietnam War and discontented by the contrast between high expenditure on defence and that on education and social services. They were supported by industrial workers in what became the most sustained strike in France's history. De Gaulle was forced to liberalize the higher education system and make economic concessions to the workers. In 1969, following an adverse national referendum, he resigned from office. In the 1990 celebrations of his birth all sections of French society regarded him as a national hero.

De Klerk, Fredrik Willem (1936–), President of South Africa 1989– . Son of a distinguished Afrikaner family, he was born in Johannesburg and graduated in law at Potchefstroom University. He practised law until entering politics in 1972. He served under Prime Minister Vorster and under P. W. Botha*, being Minister for Internal Affairs in 1982. At that time he became leader of the National Party of Transvaal, and pressed the concept of 'limited power-sharing' between the races. On becoming President in September 1989 he appeared to move steadily towards the position of universal suffrage, while being threatened from the right by Conservative and extremist groups. He opened discussions with Nelson Mandela* and the ANC* in 1990.

Democracy Wall movement (1978–9). In 1957, as part of the Hundred Flowers Campaign*, Beijing University students had used wall posters to criticize their courses. In 1978–9 students again used wall posters to demand greater democracy. In November 1978 a fourteen-page poster attacked the late Mao Zedong* and the Gang of Four*. Deng Xiaoping* at first defended this Democracy Wall movement as 'showing the stable situation of China'. Later, however, after the Gang of Four had been deposed in 1979, he put an end to the movement as student demands developed.

Democratic Party (USA). A major political party, known in the 1790s as the Democratic-Republican Party, it split in 1860 over slavery. It reunited after the Civil War, gaining support from the ever-expanding West and from the immigrant working classes of the industrialized north-east, while retaining the loyalty of the deep South. In the early 20th century it adopted many of the policies of the Progressive Movement and its candidate for President, Woodrow Wilson*, was elected in 1912. Although in eclipse during the 1920s, it re-emerged in the years of the Great Depression*, with policies to end unemployment and stimulate industry. It captured both Congress and the presidency, with Franklin D. Roosevelt* the only President to have been re-elected three times. Since then it has tended to dominate the House of Representatives and has generally held the Senate as well, supporting civil rights, social welfare, and Third World aid. Following the civil rights movement* and desegregation in the 1950s and 1960s, it lost much of its support from the Dixiecrat* Southern states, becoming less of a coalition party and more one which favours the working classes of the big cities and the small farmers, as against big business and the middle classes. The Democratic presidencies of John F. Kennedy* and Lyndon B. Johnson* saw fruitful partnerships between Congress and President. The Vietnam War, however, badly divided the party in 1968. Under the Republican President Nixon* it retained control of Congress and it won the election of President Carter*. It lost control of the Senate in 1980 but regained it in 1986, and retained majorities in both Houses of Congress in the 1988 election.

Demographic change, fluctuations in world population. Thomas Malthus in 1803 had propounded the theory that a rapidly growing population would soon increase beyond the capacity of the world to feed it, and that, if war and famine were not to reduce it, there would be mass starvation, unless other measures could be devised to prevent catastrophe. Western Europe saw massive population increases during the 19th century, but also vast population movements, some 36 million to the USA and 20 million Russians to beyond the Urals into Siberia. By 1914 artificial birth control was still regarded as immoral (as it remained with the Roman Catholic Church), its advocate Marie Stopes being vilified. It was not until after World War II that birth-control clinics became a norm, Japan instituting the first. Since then, however, a population explosion has taken place in Latin America, Africa, and Asia, where over half the world's population now lives. In Kenya, which has the highest birth rate in the world, more than 50 per cent of the population is now under 15. By the late 1980s the total world population was over 5,000 million and rising still at the rate of 1.7 per cent. By the year AD 2000 it was expected to reach 6,000 million, posing massive political and economic problems.

Deng Xiaoping (Teng Hsaio-p'ing) (1904–   ), Chinese statesman. He studied with Zhou Enlai* in France in the early 1920s and spent some time in the Soviet Union before returning to China and working for the Communists in Shanghai and Jiangxi. During the war years 1937–49 he rose to prominence as a political commissar, and afterwards he held the senior party position in south-west China. He moved to Beijing in 1952 and became general secretary of the Chinese Communist Party in 1956 and a member of the Politburo. With the 'Great Leap Forward'* he was identified with the pragmatic wing of the party, pressing for the need to create technical and managerial élites, thus conflicting with Mao Zedong's* stress on egalitarianism and revolutionary zeal. He was discredited during the Cultural Revolution*, but re-emerged in 1973 under the patronage of the ageing Zhou Enlai. On the death of the latter he was again discredited by the Gang of Four*, but rehabilitated in 1977 in the struggle for power within the administration of Hua Guofeng*; he became prominent in pressing the need for the Four Modernizations* first promulgated back in 1963, and also for improved relations with the West and the Soviet Union. Although he refrained from personally taking top party or government posts, he took control of the armed forces as chairman of the Central Military Commission. From 1981 his policies were to decentralize economic management, to stimulate agricultural production by giving individual farmers control over their land, and to introduce incentive schemes into industrial production. He also sought (unsuccessfully) to purge corruption and to retire ageing party veterans. With all this, however, he was not prepared to allow political modernization, whereby the CCP would lose its monopoly of power, and this determination led directly to the Tiananmen Square Massacre*, for which he seems to have been directly responsible. In March 1990 he resigned as chairman of the Central Military Commission.

Denikin, Anton Ivanovich (1872–1947), Russian general and counter-revolutionary. Son of a serf, he joined the Imperial Army when he was 15 and by 1917 had reached the rank of general. He served the provisional government* February–October 1917, but when the Bolsheviks seized power in the 'October' Revolution he assumed command of a 'White' army of counter-revolutionaries. During 1918 his Armed Forces of the South gained control of a large part of southern Russia and in May 1919 he launched an offensive against Moscow. He was defeated by the Red Army of Trotsky* at Orel. He retreated to the Caucasus, where in March 1920 his army disintegrated and he fled to France. He died in the USA.

Denmark. In 1863 this constitutional monarchy had incorporated Schleswig, which the Danish king ruled personally as its duke. This was opposed by Prussia and Austria, whose troops invaded in 1864. In 1871 Schleswig was absorbed into the Second German Empire. After World War I, North Schleswig voted to return to Denmark, which had remained neutral during the war. Between the wars under the Social Democrat Party (1929–40) a system of social welfare was developed, particularly under Prime Minister Theodore Staunding (1933–5). Despite another declaration of neutrality at the start of World War II, the Germans occupied the country 1940–5. A Danish Freedom Council was established in London and closely collaborated with a strong Resistance movement. King Christian X (1912–47) took a brave stand against Nazi pressures, especially in helping to protect Jews. In 1946 all Schleswig-Holstein, crowded with German refugees, voted to become permanently part of Germany. In 1949 Denmark entered NATO* and in 1960 joined EFTA*. Like Britain it then joined the European Community* in 1973. The Social Democrats have remained in power through most of the post-war period.

Depression, the Great, popular term for a world economic crisis 1929–33. It began in October 1929, when the New York Stock Market collapsed. As a result US banks began to call in international loans and were unwilling to continue loans to Germany for reparations* and industrial development. In 1931 discussions took place between Germany and Austria for a customs union. The French saw this as a first step towards Anschluss* and in May they withdrew funds from the large bank of Kredit-Anstalt, controlled by the Rothschilds. The bank announced its inability to fulfil obligations, and soon other Austrian and German banks were having to close. Although President Hoover in the USA negotiated a one-year moratorium on reparations*, it was too late. Because Germany had been the main recipient of loans from Britain and the USA, the German collapse was soon felt in other countries. Throughout the USA and Germany members of the public began a 'run on the banks', withdrawing their personal savings; more and more banks had to close. Farmers could not sell crops; factories and industrial concerns could not borrow and had to close; workers were thrown out of work; retail shops went bankrupt; and governments could not afford to continue unemployment benefits, even when these had been available. In the colonies of the European powers and in Latin America demand for basic commodities collapsed, increasing unemployment but also stimulating nationalist agitation. Unemployment in Germany rose to 6 million, in Britain to 3 million, and to 14 million in the USA, where by 1932 nearly every bank was closed. In Europe the effect was everywhere to foster political extremism. Renewed fears of a Bolshevik uprising produced extreme right-wing militarist regimes, inspired by Fascism*, throughout the Balkans and also in the Baltic states. In 1932 Franklin D. Roosevelt* was elected President of the USA and financial confidence was gradually restored, but not before the Third Reich* in Germany had established itself to seek to revitalize the German economy.

Desai, Morarji Ranchhodji (1896–), Prime Minister of India 1977–9. Son of a village teacher, he was a civil servant before becoming a disciple of Gandhi* in 1930, spending a total of ten years in British prisons for civil disobedience. He made his reputation as Finance Minister of Bombay 1946–52, before becoming Bombay's Chief Minister 1952–6. He joined the Indian government in 1956 as Minister of Commerce and Industry, and was then Minister of Finance 1958–63, overseeing a series of five-year plans which led to a doubling of industrial output in ten years. After the death of Jawaharlal Nehru* (1964) he was a contender for the succession, but his austere style failed to win him support in the Congress Party against Indira Gandhi*, whom he increasingly opposed, being interned 1975–7. In 1977 he was the obvious candidate to lead the Janata* opposition to Mrs Gandhi, and he won the election of that year. His government restored parliamentary democracy, allowing a free Press and releasing political prisoners; but as Prime Minister he failed to deal with the economic and factional problems confronting him and he resigned in 1979.

Desegregation, abolition of racial segregation. A US Supreme Court decision in 1896 had ruled as constitutional a Louisiana law requiring 'separate but equal' facilities for Blacks and Whites in trains. For the next fifty years many Southern states continued to use the 'separate but equal' rule as an excuse for requiring segregated facilities. With the founding of the National Association for the Advancement of Colored People* (NAACP) in 1909, Black and White Americans had begun efforts to end segregation, but they met with fierce resistance, especially in the South. During World War II over 1 million Blacks served in the US armed forces, and in 1948 President Truman ended segregation in the forces. The Supreme Court decision of 1954 *Brown* v. *Board of Education of Topeka**, ending school segregation, was a landmark. The efforts of Martin Luther King* and the civil rights movement* resulted in the Civil Rights Act of 1964 and the Voting Rights Act of 1965, which effectively outlawed segregation and ended literacy tests. There were still Black ghettos in cities, but the purely legal obstacles to the equality of the races was now removed.

Destroyer Bases Deal. In September 1940 Britain and the USA negotiated a Destroyer Transfer Agreement, whereby the USA would transfer to Britain fifty much-needed destroyers. These would be in exchange for leases of bases in the West Indies, Newfoundland, and British Guiana. Being of World War I design, the ships became surplus to British needs later in the war, but at this critical stage, with high U-boat losses, they were an invaluable supplement to available escort vessels.

Détente. The term was first applied to international relations in 1908. It is particularly associated with the 'thaw' in the Cold War in the later 1960s and with the policies of Richard Nixon and Henry Kissinger*. More relaxed relations were marked by the holding of the Helsinki Conference* 1972–5; by the signing of SALT I* in 1972; and by Chancellor Brandt's* policy of *Ostpolitik**. The Soviet invasion of Afghanistan in 1979 worsened the Cold War into the early 1980s.

De Valera, Eamon (1882–1975), President of the Republic of Ireland 1959–73. Born in New York of an Irish mother and Spanish father, he was educated in Ireland and became a mathematics tutor at Maynooth College, a seminary for intending

Catholic priests. In 1913 he joined the Irish Volunteers and distinguished himself as a battalion commander in the Easter Rising* in Dublin in 1916. He was captured and imprisoned, and would have been executed but for his American birth. He served a year in Lewes Prison. Released in June 1917, he was rearrested in May 1918, but escaped from Lincoln Prison in February 1919 and returned to the USA, where he raised £1m. in support of the IRA*. Although in prison, he had been elected as a Sinn Fein* Member of the Westminster Parliament in December 1918 and became President of the rebel independent government (the Dáil Éireann) set up by Sinn Fein in 1919, with Arthur Griffith* his Vice-President. He returned to Ireland in 1920 at the time of Black and Tan* activity against the IRA. He did not attend the negotiations leading to the Anglo-Irish Treaty of December 1921, and repudiated its concept of an Irish Free State from which six Ulster counties were to be excluded. He took part in the Irish Civil War 1922–3, being imprisoned by the government of William Cosgrave* from August 1923 to July 1924, when he took an oath of allegiance and abandoned the policy of violence. Having been President of Sinn Fein from 1917, in 1926 he formed his new party Fianna Fáil* and was the leader of the opposition to Cosgrave in the Dáil until 1932, when his party won the election and he became President of the Executive Council of the Irish Free State. In 1937 he repudiated the oath of allegiance to the British Crown, categorizing his country as the 'sovereign independent democratic state' of Ireland or Eire, yet still theoretically within the British Commonwealth. He stopped the payments of annuities to Britain and negotiated the return of naval bases held by Britain under the 1921 treaty. He was Prime Minister or Taoiseach throughout World War II, during which he kept Ireland neutral, but was defeated in 1948. Ireland left the Commonwealth in 1949, becoming a totally independent republic. De Valera was re-elected as Taoiseach 1951–4 and again 1957–9. He continued to have public support and was twice elected President of the Republic, 1959 and 1966. His last presidency, until 1973, took him into his ninetieth year.

Diagne, Blaise (1872–1932), Senegalese politician. He was the first Black Deputy of Senegal (1914) and in 1918 was appointed High Commissioner of French West Africa. As such he was responsible for the recruitment of African troops to fight in World War I. Regularly re-elected Deputy, he was appointed Under-Secretary of State for the French Colonies in 1931.

Diefenbaker, John George (1895–1979), Prime Minister of Canada 1957–63. Born in Ontario, his father moved to the North West Territories and he was educated in Saskatchewan. After serving in World War I he became a successful lawyer. He joined the Conservative Party, but for long failed to be elected either provincially or federally, only entering the Canadian House of Commons in 1940. In 1956 he became leader of his party and next year led a minority government. In 1958 his party won a majority of 158 seats and 'the Chief' proclaimed his vision of a 'new Canada'. Although there were some important measures of social reform, the vision soon faded and unemployment rose. As a strong supporter of the Commonwealth he opposed Britain's wish to enter the EEC*; and his efforts to reduce Canada's dependence on the USA for defence failed, in that US pressure obliged him to accept nuclear warheads at the time of the Cuban Missile Crisis* in 1962. He was defeated in Parliament in February 1963 and lost the general election that followed, becoming leader of the opposition until 1967. He remained MP for Prince Albert and became a much respected national figure.

Diem, Ngo Dinh (1901–63), President of South Vietnam 1955–63. Of a wealthy Vietnamese family and brought up a Roman Catholic in Hue, he graduated in the French University of Hanoi. He served as a French Indo-China provincial governor 1919–32, and was accused of ruthless repression of the 1930–1 uprising. For a short while he was Minister of the Interior for Emperor Bao Dai*, but felt the latter was too subservient to the French. He resigned and spent the years 1933–45 living in Hue. Captured by forces of Ho Chi Minh* in 1945 he refused the latter's invitation to join a government. Instead, in 1947 he founded the National Union Front, a non-violent, anti-Communist party, which was equally anti-French. He was exiled by the French authorities and lived in the USA and as a monk in Belgium. In 1954 he was invited to return to Vietnam to join an anti-Communist government, following the disaster of Dien Bien Phu*. In October 1955, following a probably rigged referendum, he replaced Bao Dai as President of the new Republic of (South) Vietnam. Personally an austere man, described as 'a mandarin Catholic nationalist', his harsh, repressive regime, with his brother head of the political police, was far from popular. From 1960 onwards it was virtually at war with the National Liberation Front*. Promises of land reform were not kept and the killing of hundreds of Buddhists for allegedly assisting Communists led to a military coup, backed by the CIA*. On 2 November he was assassinated along with his brother.

Dien Bien Phu, a village in North Vietnam and scene of a decisive battle in the French Indo-China War* (1946–54). The French Commander-in-Chief General Henri Navarre planned to send airborne troops under Colonel de Castriès to seize the village in November 1953, it being in a strategic position, 75 miles south of China and commanding the route from Hanoi to Laos. The village would be strongly fortified and, relying on superior fire-power, the French garrison would destroy the VietMinh army of General Giap*. The plan proved a disaster. Giap mobilized thousands of peasants to manhandle siege guns into the surrounding hills. With these he dominated the airstrips so that the French forces were cut off. After a siege lasting 13 March–7 May 1954 the garrison of 16,500 men surrendered, only 3,000 defenders surviving. Within two months the war had ended. The Geneva Agreements* which followed effectively ended French rule in Indo-China.

Dieppe Raid, World War II (18–19 August 1942), an amphibious raid on the French coast to destroy the airfield, port, and radar installations in Dieppe. Some 1,000 British commando and 5,000 Canadian infantry troops were involved. There was considerable confusion as landing-craft approached the two beaches, where they met heavy fire. The assault was a failure and the order was given to withdraw. Not only were over two-thirds of the troops lost, but German shore guns sank one destroyer and thirty-three landing-craft, and shot down 106 aircraft. Although a disaster, the raid taught many useful lessons for later landings in North Africa, Italy, and Normandy, not least the need for meticulous planning and total security.

Diet of Japan. The term Diet, originally meaning a representative assembly meeting for a day, was adopted by the Japanese Parliament under the Meiji Constitution of 1889. It was relatively impotent under the supreme powers of the Emperor, but the post-war Constitution of 1947 made both houses elective and it became the 'highest organ of state power'. Members of the Lower House of Representatives, at first with 466 seats, now 512, are elected for four years, or less if the Prime Minister calls an election, constituencies being defined on a population basis. The 252 elected members of the Upper House of Councillors sit for six years, half at a time being elected every three years, candidates coming from 'prefecturial constituencies'. They are usually well-known national figures.

Dimitrov, Georgi (1882–1949), Prime Minister of Bulgaria 1946–9. Born at Pernik, he became a printer and in 1905 was organizing trade union strikes. He became a secret Communist in 1917 and fled to Moscow in 1923. From 1929 he was head of the Bulgarian sector of the Comintern* in Berlin. When the German Reichstag was burned down in 1933 he was accused with other Communists of complicity. His powerful defence at his trial, however, forced the Nazis to release him, and he settled in Moscow. In 1945 he was appointed head of the provisional government in Bulgaria, which was followed in 1946 by the establishment of the Bulgarian People's Republic. His premiership was marked by ruthless policies of sovietization, but also by a plan for a federation of Balkan states, a plan which collapsed when Yugoslavia broke with the Soviet Union in 1948.

Dinh, Nguyen Thi, Madame (1920–), Vietnamese revolutionary. Born in the Mekong delta of peasant stock, she joined the revolutionaries when only 11, after her brother had been captured and tortured, following an uprising 1930–1. Arrested in 1940, at the time of a rising in the Mekong delta, she was imprisoned until 1943. After World War I she took an active part in the war against the French, resenting the Geneva Agreement and the establishment of the South Vietnamese regime of Ngo Dinh Diem*, against whom she actively worked. In 1960 she helped to stage an uprising at Ben Tre on 17 January, which in many ways marks the beginning of the Vietnam War*. In December 1960 she helped to form the National Liberation Front*, becoming a member of the Praesidium of the NLF and chairman of the South Vietnam Liberation Association, organizing arms runs and many times only narrowly escaping with her life. In 1976 she was elected to the Central Committee of the Communist Party in Hanoi, writing her memoirs.

Disarmament and arms control. Attempts to achieve disarmament by international agreement began before World War I, after which the Covenant of the League of Nations raised hopes for international disarmament. The Washington Conference* 1921–2 sought to limit naval arms. The League convened a conference of sixty nations 1932–4. Its failure was caused partly by French caution, but more especially by Germany's withdrawal in October 1933. In 1952 the United Nations established a permanent Disarmament Commission. Although the Soviet Union left it in 1957, a Geneva Disarmament Conference was successfully convened in 1966, and helped to bring about the nuclear Non-proliferation Treaty* of 1968. Since then most of the important negotiations on arms control have taken place in direct talks between the USA and the Soviet Union. Strategic Arms Limitation Talks* between the two led to some limitation of nuclear arms in SALT I (1972) and SALT II (1979), never ratified by the US Senate. Member states of NATO and the Warsaw Pact met in Vienna from 1973 in the Mutual and Balanced Force Reduction Talks (MBFR), concerned with limiting ground forces in Central Europe. The final act of the Helsinki Conference* of 1975 provided for 'confidence-building measures' (such as notification of military manœuvres and even the presence of opposite numbers to observe these). In 1982 Soviet and US negotiations on strategic nuclear weapons restarted under the title of START (Strategic Arms Reduction Talks). They were suspended in 1983 but resumed in 1987. In 1988 the INF Treaty* on eliminating intermediate-range weapons, was signed in Moscow, while talks on the elimination of short- and strategic-range weapons were resumed, as well as negotiations for the elimination of chemical weapons. The collapse of the Warsaw Pact in 1990 led to the hope of substantial disarmament and on 19 November a Conventional Arms in Europe Treaty (CFE) was signed in Paris by NATO and Warsaw Pact states, imposing ceilings on non-nuclear arms and allowing site inspection. At the same time arms build-up in Israel and Saudi Arabia accelerated.

Divine, Father (*c.*1882–1965), Black American evangelist. Born George Baker in Georgia, he became a popular preacher in Baltimore and in 1915 moved to New York. Here he founded a Peace Mission movement in 1919. He opened a free employment bureau and fed the destitute in his house in Sayville, Long Island, which became known as 'Heaven'. It was now that he began styling himself at first Major and then Father. By the 1940s his followers numbered thousands, both Black and White, and some 200 'Heavens' were established as centres for communal living in cities across the USA. He forbade his followers to drink, smoke, or use cosmetics, and many regarded him as a personification of God, revering his remarkable powers of healing. His cult did not long outlast his death.

Dixiecrat, popular name in the USA for a Democrat in a Southern state opposed to desegregation*. In 1948 the States Rights Democratic Party was founded by die-hard Southern Democrats opposed to President Truman's* renomination as Democrat candidate, on account of his stand on civil rights. After Truman's victory they abandoned their presidential efforts, but continued to resist civil rights programmes in Congress. Many Dixiecrats moved to support the Republican Party in the 1960s and 1970s. A short-lived Dixiecrat American Independent Party was formed for the 1968 elections, with George Wallace* candidate for President.

Djibouti *see* JIBOUTI.

Djilas, Milovan (1911–), Yugoslav politician. Born in Montenegro, he became a communist while studying at the University of Belgrade. He was imprisoned by the royalist regime 1933–6, but was then released, when he worked as a journalist. A supporter of Tito*, with whose partisans he operated 1941–4, he became Vice-President of the Federal Republic of Yugoslavia in 1946. In 1953, however, he openly criticized the party leadership for corruption and abuses. He was expelled from the party, lost all his positions, and, between 1956 and 1961, was in prison. Rearrested in 1962, he was placed under house arrest. In his book *The New Class* (1955) he condemns the abuse of power of which communists in East Europe had been guilty. Courted by the West, he nevertheless remained an idealistic socialist, last arrested in 1982 for giving an unauthorized lecture. In many ways he predicted the kind of collapse of communist regimes seen in 1989. In May 1989 he was rehabilitated by the Yugoslav Communist Party.

Dobrudja, the Black Sea littoral at the mouth of the Danube, scene of heavy fighting between Romanian and Austrian troops in World War I. South Dobrudja has been claimed by Bulgaria since 1878. It was annexed by Romania in 1913; recovered by Bulgaria in 1918; lost to Romania in 1919; recovered in 1940 by Bulgaria, whose possession was confirmed in 1947. Since then Romania has abandoned any claim.

Dodecanese, twelve islands in the east Aegean Sea. They were ruled by the Ottoman Turks from the 16th century, but were conquered by the Italians in 1912. After World War I Italy failed to honour its promise to hand them to Greece. Instead Mussolini developed both ports and airfields as jumping-off points for Italian imperial ambitions in the eastern Mediterranean. In 1943 the islands were taken by the Germans and in 1947 ceded to Greece, with Leros, only 30 miles from Turkey, being demilitarized. They have developed as popular holiday resorts.

Doenitz, Karl (1891–1980), German admiral. Born in Prussia, he was commissioned into the Imperial German Navy and specialized in submarine warfare. During the 1930s he supervised the U-boat construction programme and on the outbreak of war co-ordinated the strategy for the Battle of the Atlantic*. In 1943 he succeeded Admiral Raeder as Commander-in-Chief of the German navy. On Hitler's suicide on 30 April 1945 he took the office of head of state and on 7 May agreed to unconditional surrender. Sentenced at the Nuremberg Trials* to ten years' imprisonment, he served all these, being released in 1956.

Dollfuss, Engelbert (1892–1934), Chancellor of the Republic of Austria 1932–4. From a farmer's family, he studied theology and law at the University of Vienna before World War I, in which he won eight decorations for bravery. He was elected to the new Austrian Parliament for the right-wing Christian Social Party and in 1931 became Minister of Agriculture. His term of office as Chancellor was troubled by his hostility to both Communists and extreme nationalists seeking anschluss* with Germany. In addition he had to face the social unrest arising from the Great Depression*. Terrorist activities led him to suspend Parliament in March 1933, and in February 1934 he ordered the bombardment of a strongly socialist housing estate in Vienna, whose inhabitants were demonstrating. There was fierce fighting and deep resentment. In May he put forward a highly authoritarian constitution, but, by antagonizing the working classes, he had deprived himself of an effective political base. On 25 July 1934 he was assassinated by Austrian Nazis seeking anschluss with Germany.

Dominica, a small island republic in the Lesser Antilles*. A British colony from 1805, its main exports are of fruit, which has replaced sugar. After the break-up of the West Indian Federation* in 1962 it gained self-government in 1967, its first Premier being Edward Le Blanc, leader of the Dominican Labour Party (DLP). In 1978 it became an independent republic within the Commonwealth, now led by Prime Minister Patrick John of the DLP. When a rival party, the Dominica Freedom Party (DFP), came to power in 1980 there were two attempted coups for which Patrick John was imprisoned. The DFP remained in power through the 1980s under Mary Eugenia Charles as Prime Minister; she won a third term of office in May 1990.

Dominican Republic, the eastern half of the Caribbean island of Hispaniola. As Santo Domingo it gained independence from Spain in 1821, but after an anarchic century the country was bankrupt by 1905, when the USA assumed fiscal control. Disorder continued and it was occupied by US marines 1916–24. A constitutional government was then established, but this was overthrown by Rafael Trujillo*, whose military dictatorship lasted from 1930 until 1961. On his assassination President Juan Bosch* established a brief democratic government until deposed by a military junta. Civil war and fears of a Communist take-over brought renewed US intervention in 1965 and a new constitution in 1966. Backed by the USA, President Joaquín Balaguer* was installed, and he remained in office until 1978, during which time there was a process of redemocratization. Balaguer's successor President Guzman succeeded in reducing the power of the military and the extent of corruption. The mid-1980s, however, saw renewed violence, and in 1986 the aged Balaguer of the Partido Reformista returned to office to face severe economic problems.

Douglas-Home, Alexander Frederick, 14th Earl of Home, Baron Home of the Hirsel (1903–), British Prime Minister 1963–4. Educated at Eton and Oxford, he was elected to the House of Commons as a Conservative in 1931, becoming parliamentary private secretary to Neville Chamberlain* at the time of the Munich

Agreement*. In 1951 he succeeded to the earldom, but served in both the Eden and Macmillan governments, being Foreign Secretary 1960–3. Anxious to return to the House of Commons, he disclaimed his peerage and was created a Knight of the Thistle, being known as Sir Alec Douglas-Home. As such he succeeded Macmillan as Prime Minister. His short ministry is remembered for its risky policy of monetary expansion, under Reginald Maudling as Chancellor of the Exchequer, and for its acceptance of the Robbins Report on expansion of higher education. The Conservatives were narrowly defeated by Harold Wilson in the 1964 election. Douglas-Home served again as Foreign Secretary under Edward Heath 1970–4. He then returned to the House of Lords as Baron Home.

Dowding, Hugh Caswall Tremenheere, 1st Baron Dowding (1882–1970), British air chief marshal. Born in Scotland and educated at Winchester and the Royal Military College, Woolwich, he became an officer in the Royal Artillery, serving in India until 1912 when, while at Camberley Staff College, he taught himself to fly, quickly becoming proficient. On the outbreak of war in 1914 he was appointed commandant of the Dover camp of the newly formed Royal Flying Corps and then served as a squadron commander in France. By 1917 he held the rank of brigadier-general and he became a permanent officer in the independent Royal Air Force in 1918. In 1936 he was appointed Commander-in-Chief of Fighter Command. During the next three years he built up a force of Spitfire and Hurricane fighter aircraft, encouraged the key technological development of radar, and created an operations room which would be able to control his command. From here he fought the Battle of Britain*, September–October 1940. Mentally and physically exhausted, he was replaced in November 1940, when the German *Luftwaffe* had abandoned its daylight bombing offensive and switched to night raids. He held various senior commands until he retired in 1942, a legendary figure for Battle of Britain pilots.

Dresden Raid, World War II (February 1945). The main air raid on the city of Dresden, until then considered safe because of its beauty, was on the night of 13–14 February 1945. 805 planes from Britain's Bomber Command attacked the city. This was followed by three more raids in daylight by the US 8th Air Force. The Allied Commander-in-Chief General Eisenhower was anxious to link up with the advancing Red Army and Dresden was seen as strategically important as a communications centre, as well as being a centre for industry. It was known to be overcrowded with some 200,000 refugees, but it was felt that the inevitably high casualties might in the end help to shorten the war. Over 30,000 buildings were flattened. Numbers of dead and wounded are still in dispute, estimates varying from 55,000 to 400,000.

Druze, member of an Islamic sect founded in the 11th century, which survived unchanged into the 20th century as a close-knit community. The belief is that whenever a Druze dies another Druze is born, the soul migrating into the newborn infant. Much of the doctrine remains a secret. Druzes traditionally lived in the villages of Mount Lebanon, but the entry of France into the Lebanon in 1860, following a period of bloody conflict between the Druzes and Maronite* Christians, led to a large number of Lebanese Druzes migrating into the basalt hills south of Damascus. Following French detention of a number of Druze leaders, a full-scale rebellion under Sultan al-Atrash occurred in Syria 1925–7, when Druze areas of Damascus were twice bombed by the French. The rebellion failed when Lebanese and Syrian Druze could not agree. During the 20th century the Druzes have become more urbanized, adopting Western education and life-styles. In recent times Druze leaders have been active in Lebanese politics, for example Majd Arslan, Khaled Jumblat, and Kamal Jumblat*, who, as leader of the Progressive Socialist Party, was Minister for Home Affairs and played a key role in the civil war 1975–6. After his assassination his son Walid, who became recognized as Druze leader, supported Syrian intervention in Lebanon, but was reconciled to President Jumayyil* in 1984 and served in his government. Traditional enmities between Druze and Maronite Christians go back centuries and an aspect of the war 1975–6 was Druze fear that the Phalangist Maronites were trying to dispossess them of their land.

Dubček, Alexander (1921–), Czechoslovak statesman. He was born in Uhrovek in Slovakia; his family moved to the Soviet Union in 1925 and he was educated there. He returned to Slovakia in 1938 a member of the Communist Party, and he operated with the Slovak Resistance throughout World War II. He held several Communist Party posts between 1945 and 1968, when he became first secretary of the party. In what became known as the 'Prague Spring'* he and other liberal members of the government set about freeing the country from the rigid political and economic controls which had been strangling it since 1948. He promised a gradual democratization of politics and a foreign policy independent of the Soviet Union. This was unacceptable to Brezhnev* in Moscow and the Warsaw Pact was organized to invade Czechoslovakia. Together with other leaders he was summoned to Moscow and forced to agree to rescind key reforms. He was removed from office in 1969 and expelled from the party in 1970. He held minor administrative posts in Bratislava until 1989, when he re-emerged as a popular figure, being elected Speaker of the National Assembly in 1990.

Du Bois, William Edward Burghardt (1868–1963), sociologist and US civil rights leader. Born in Massachusetts, he graduated at Fisk University, Tennessee, in 1888 and in 1895 received a Harvard Ph.D. As a professor at Atlanta University 1897–1914 he became a recognized scholar in the field of race relations. His book *The Souls of Black Folk* (1902) established him as leader of a movement for civil rights agitation, and in 1909 he was a founder of the NAACP*. At the same time he was a pioneer for Pan-Africanism* and organizer of the Pan-African Conference in London in 1900. After active work with NAACP 1909–34 he returned to Atlanta, where he wrote his most important book, *An Essay towards an Autobiography of a Race Concept*. In 1945 he was an organizer of the Pan-African Conference in Manchester, where he worked closely with Kwame Nkrumah*. During the 1950s ideologically he moved left, and in 1951 he was indicted (but acquitted) as 'an unregistered agent of a foreign power'. In 1961 he joined the Communist Party and received a Lenin Peace Prize. He moved to Ghana, where he became a citizen and where he died.

Dulles, John Foster (1888–1959), US Secretary of State 1953–9. Born in Washington, DC, he was educated at Princeton and in Paris. He became a successful New York lawyer and was invited to attend the Paris Peace Conference* in 1919. He attended numerous international conferences between the wars and served as an adviser to the US delegation at the San Francisco Conference (1945), which set up the United Nations. He was author of the draft for the Japanese Peace Treaty of 1951. As Secretary of State under Eisenhower* he became a keen protagonist of the Cold War*, building up NATO* and being largely responsible for SEATO*. He advanced beyond the Truman Doctrine* of 'containment' by urging that the USA prepare a massive nuclear arms build-up as a deterrent to Soviet aggression. In 1957, having been an opponent of the Anglo-French Suez venture in 1956, he helped to formulate the Eisenhower Doctrine* of economic and military aid to halt aggression in the Middle East. He gave clear assurances to the Soviet Union that the USA would be prepared to defend West Berlin against any encroachment. His aggressive anti-Soviet impulses were, however, checked by President Eisenhower.

Duma. In 1906 Nicholas II* of Russia established an elective assembly, or Duma, following the first Russian Revolution in 1905. Its efforts to introduce reform were nullified by reactionary groups at court which persuaded the Tsar to dissolve three successive Dumas. The fourth Duma of 1912–17 was suspended by the Tsar for much of the war. It refused an imperial decree for its dissolution in February 1917. Instead it formed a provisional government*, under Prince Lvov, which then persuaded Nicholas to abdicate.

Dumbarton Oaks Conference. During August–October 1944 an international conference met at Dumbarton Oaks, Washington, DC. Representatives of the USA, Britain, the Soviet Union, and China drew up proposals which served as the basis for the Charter of the United Nations* at San Francisco the following year. Attention was focused particularly on measures to secure 'the maintenance of international peace and security'. One of its main achievements was the planning of the Security Council with its mechanism of the veto.

Dunkirk, Treaty of (4 March 1947). The Fourth Republic in France, in an effort to revive the *Entente* of 1904, made an alliance with Attlee's government in Britain whereby the two governments pledged 'constant consultation' in economic matters and joint military action against any aggressor. The Pact was enlarged in March 1948 by the Treaty of Brussels, when the Benelux* countries joined.

Dunkirk Evacuation, World War II (26 May–4 June 1940). In May 1940 German tank forces, advancing rapidly through Belgium and northern France, cut off large numbers of British and French troops. General Gort, commanding the British Expeditionary Force, organized a withdrawal to the port and beaches of Dunkirk. Here naval and merchant vessels, aided by hundreds of small private boats from English sailing ports, carried off some 330,000 men—most, but not all, of the troops. Vast quantities of arms and supplies had to be abandoned. Fighter patrols from England successfully held off German bombers.

Dutch East Indies. The 'East Indies' was the 17th-century name given to the Malay Archipelago of some 2,000 islands extending some 3,500 miles east–west and 1,300 miles north–south. By 1914 they had been colonized by Britain, Portugal, Germany, and The Netherlands, who controlled by far the largest area. The island of Java was both the largest and richest of their islands, with outlying islands being known as the 'Outer Provinces', in many of which there had been fierce rivalry with British traders. The movement for independence in the Dutch East Indies had three strands: European settlers, who established a *Volkstaad* and wanted 'to liberate Batavia (Jakarta) from The Hague', and among whom there developed a neo-Fascist group 'the Vaterlandsche', whose ideology was similar to that of the South African Boers; an Islamic Party, Sarikat Islam*, founded in 1911 and increasingly racial in attitude; and a nationalist party, Budi Utomo ('Glorious Endeavour'), which appealed to unemployed Western-educated intelligentsia with allies among European socialists. In 1926 the Partai Nasionalis Indonesia was formed, drawing supporters increasingly from Budi Utomo. This was partly in protest against the Constitution of 1925, which had established a *Volkstaad* with European control. The islands were adversely affected by the Great Depression* and this stimulated racial discord, particulary between Malays and Chinese, and nationalist demands for full independence. By 1939, however, policies of conciliation were emerging, which were cut short by war and the Japanese invasion of 1942. On Japan's surrender Achmad Sukarno* proclaimed the Indonesian Republic, and four years of struggle followed before The Hague agreed to its existence.

Dutra, Eurico Gaspar (1885–1974), President of Brazil 1946–51. He served in the Brazilian army for twenty-two years, being commissioned in 1910. An early supporter of Vargas* and the *Estado Novo*, he served as Minister of War until 1945, when he broke with the President and staged a coup. He was elected President with the support of the Partido Social Democrático in 1946. He had a more democratic constitution adopted during his first year in office and promoted economic nationalism and industrialization. Although a political conservative, he did not repudiate the idea of state participation in the economy. In 1951 Vargas returned briefly to office, when Dutra retired.

Duvalier, François (1907–71), President of Haiti 1957–71. Although a graduate of the Haiti School of Medicine he came to regard voodoo as a key source of Haitian culture. In 1946 he became Director of Public Health, and his election in 1957 as President was the first held under the rule of universal adult suffrage. Within a year after coming to office he suspended all constitutional guarantees and established a reign of terror based on the Tontons Macoutes, a notorious police and spy organization. In 1964 he was declared President for life. The economy of Haiti declined severely and 90 per cent of the people remained illiterate. He lost support and aid from the USA, was opposed by the Roman Catholic Church, and was internationally isolated. Yet he survived, and by the time of his death in 1971 'Papa Doc', as he was called, had assured the succession of his son Jean-Claude. The latter was deposed and fled to France in 1986.

Dyarchy, a system of government formally introduced into British India by the 1919 Government of India Act*. Government of the thirteen Indian provinces was to be on a dual basis: a reserved area of finance, police, and justice, under the control of the Governor; and a 'transferred' area, such as local government, education, and health, under the control of Indian ministers chosen from the elected members of a Legislative Council. The system was devised by Lionel Curtis, a founding member of the Royal Institute of International Affairs. It was criticized by Indian nationalists, but was quite successful in Madras and the Punjab before being ended in 1935.

Dzerzhinsky, Felix Edmundovich (1877–1926), Soviet politician. Born in Vilna, son of a Polish nobleman, he joined the Lithuanian Social Democratic Party in 1905. Arrested and exiled to Siberia five times by the Russian imperial police, he helped to organize the 1905 Revolution in St Petersburg and was arrested a sixth time in 1912. Released by the provisional government* in 1917, he was one of the Bolshevik organizers of the 'October' Revolution and in December became head of the new secret police Cheka*. He organized prison camps for counter-revolutionaries, whom he ruthlessly suppressed. In 1921 he also took on the task of reorganizing the railway system of Russia. A supporter of Stalin, he was chairman of the Soviet Supreme Economic Council 1924–6 and elected to the Politburo the year of his death.

Earhart, Amelia (1897–1937), pioneer aviator. Born in Kansas, USA, she worked as a nurse and social worker in World War I and the 1920s. Having married a wealthy businessman she became keenly interested in flying, becoming the first woman to fly the Atlantic solo in 1932. A critic of male domination within the aviation industry, she was also a strong advocate of its commercial development. She was the first person to fly California to Hawaii in 1935; her plane disappeared over the Pacific in 1937 while on an around-the-world flight.

East African Community (1967–77). It began with a declaration of intent (June 1963) between Kenya, Tanganyika, and Uganda to improve trade, communications, and economic development. This was the East African Common Services Organization. It was developed by the Treaty of Kampala (1967) into the East African Community with headquarters in Arusha. It made considerable economic headway before 1971, when Uganda came under Amin's* regime. The Community broke up in 1977.

Eastern Front campaigns, World War II (1939–45). The first campaign in September 1939 followed the Nazi–Soviet Pact*. On 1 September German troops moved into Poland, capturing Poznań, the Polish Corridor*, and Gdansk. At the same time Soviet forces invaded from the east, moving up to the disputed frontier of the so-called Curzon Line*. Poland collapsed. Finland was defeated in the Finnish–Russian War* of November 1939–March 1940, and by the summer of 1940 the three Baltic republics had all succumbed to the Soviet Union. In June 1941 Hitler launched Operation Barbarossa* against his one-time ally, and Italy, Hungary, Romania, Finland, and Slovakia all joined the invasion of the Soviet Union. By the end of 1941 German 'blitzkrieg'* tactics had won Belorussia and most of the Ukraine; Leningrad was under siege and German forces were advancing on Moscow. The Russian winter halted the advance and the attack on Moscow was foiled by a Soviet counter-offensive. Britain, now allied with the Soviet Union, launched a joint British–Soviet occupation of Iran (1941), thus providing a route for British and US supplies to the Red Army, as an alternative to ice-bound Murmansk. During 1942 Leningrad continued to be besieged, while a massive German offensive was launched against Stalingrad* and the oilfields of the Caucasus. Kursk, Kharkov, and Rostov all fell, as did the Crimea, and the oil centre of Maikop was reached. Here the Soviet line consolidated and forces were built up for a counter-offensive, which began in December 1942, the relief of Stalingrad following in February 1943. The loss there of some 300,000 German troops marked a turning-point in the war. A new German offensive regained Kharkov from the Russians, but German forces were heavily defeated at the Battle of Kursk* in July. The Red Army now resumed its advance and by the winter of 1943–4 it was back on the River Dnieper. In

November 1943 Hitler ordered forces to be recalled from the Eastern Front to defend the Atlantic. Soviet offensives from January to May 1944 relieved Leningrad, recaptured the Crimea and Odessa, and re-entered Poland. In the south they invaded Romania, Bulgaria, and Hungary. Through the rest of the year and into 1945 the Red Army continued its advance, finally entering Germany in January 1945. By April it was linking up with advance troops of the Allied armies from the west, and on 2 May Berlin surrendered to Soviet troops. Victory on the Eastern Front had cost at least 20 million Soviet lives.

Eastern Orthodox churches, the ancient Christian churches of the Middle East and Eastern Europe. Each church is independent, but a special honour is accorded to the Ecumenical Patriarch of Constantinople. There is a sizeable Orthodox population in the USA, but the majority of the 150 million Orthodox members in the world live in the Soviet Union, Bulgaria, Romania, parts of Yugoslavia, and in Greece, where it is the overwhelmingly dominant church. The Soviet and East European churches suffered considerable persecution under Communist regimes. In recent years there has been a *rapprochement* between the Orthodox churches and the Roman Catholic Church, following the lifting in 1965 of mutual excommunications imposed in 1054. Eastern Orthodoxy accepts the Nicene Creed; its worship is centred around the Eucharist, the various liturgies used being longer and more elaborate than those of the Catholic Church. There is a strong tradition of monasticism and of mysticism, especially in the Greek Church.

Easter Rising, an insurrection in Dublin 24–9 April 1916. The Irish Republican Brotherhood had planned the uprising, supported by Sinn Fein*, in which members of the Irish Volunteers and the Irish Citizen Army would take part. A German ship, the *Aud*, carrying a large consignment of arms, was intercepted on 20 April by the British navy. Roger Casement* of the IRB, acting as a link with Germany, was arrested soon after landing from a German U-boat, allegedly to advise postponement. The military leaders, Pádriac Pearse and James Connolly, decided nevertheless to continue with the plan for Easter Monday 24 April. Fifteen hundred volunteers turned out, the General Post Office in Dublin was seized, along with other strategic buildings including a workhouse and Boland's Mill, where de Valera* was in command. The Irish republic was proclaimed and a provisional government set up with Pearse as President. There was some fighting with British troops before Pearse announced unconditional surrender on 29 April. At first the rising had little public support, with many Irishmen loyally serving in British forces on the Western Front in France. Pearse, Connolly, and thirteen other leaders were executed in prison and over 2,000 men and women imprisoned. It was the executions which led to a fundamental change of feeling in Ireland, even though all prisoners were released by an amnesty in June 1917. In the 1918 general election the Sinn Fein (Republican) party won a massive majority vote.

Eban, Abba Solomon (1915–), Israeli politician. Educated in England, he served in the British army in Egypt during World War II and then joined the Jewish Agency*. A member of the UN Commission on Palestine in 1947, he played an important role in the creation of Israel, after which he served for ten years as joint ambassador to the UN and USA. As Minister for Foreign Affairs 1966–74, his many speeches to the UN General Assembly had a powerful effect on US Jewry. Since retiring from active politics he has lived much in the USA, holding various university academic posts.

Ebert, Friedrich (1871–1925), President of the Weimar Republic 1920–5. The son of a tailor, he was an itinerant saddler before becoming a journalist. Editor of a socialist paper in Bremen in 1894, he entered the Reichstag in 1916, where he quickly made his mark. In November 1918 he succeeded Prince Max as the last Chancellor of the Second German Empire and on 9 November in Berlin he proposed the new Republic. This was duly formed in Weimar, and he became its first President. He steered a difficult course between revolution and counter-revolution, in order to give Germany a liberal, parliamentary constitution. However, he lost support from the left for crushing the Communists, and from the right for signing the Treaty of Versailles*.

Éboué, Félix (1884–1944), French colonial administrator. Born in French Guiana, descended from African slaves, he was educated in Paris and in 1908 was sent to Ubangi Shari (now Central African Republic), where he served in the French colonial service. He was appointed acting Governor of Martinique in 1932, Governor of Guadaloupe in 1936, and then of Chad in 1938. In World War II he supported the Free French and was appointed by General de Gaulle Governor-General of French Equatorial Africa. He died suddenly in Cairo in 1944, and after the war was given a hero's funeral in the Pantheon in Paris.

Echeverría Álvarez, Luis (1922–), President of Mexico 1970–6. He studied and taught law at the University of Mexico before entering politics, joining the Partido Revolucionario Nacional, which in 1946 became the Partido Revolucionario Institucional (PRI). In 1964 he became Secretary of the Interior and was criticized for his harsh treatment of student riots in 1968. After being elected President, however, his politics moved left. He expanded social welfare, sponsored a public works programme, and resumed the policy of President López Mateos* of redistribution of land to peasants. He opened trade with Cuba and China and took a firm line against US involvement in Latin America. His policies, however, lost the support of the wealthier middle class. He was replaced by President José López Portillo in 1976, also of the PRI. He then worked with the United Nations and was Mexican ambassador to Australia and New Zealand.

Economic and Social Council of the United Nations. This major organization has fifty-four members elected by the General Assembly* for three-year terms and acts as a policy-making Council with a number of subsidiary Commissions and Committees; for example, the Commission on Human Rights, Commission on Narcotic Drugs, Economic Commission for Europe, Economic and Social Commission for Asia and the Pacific, Committee on Natural Resources. In 1965 the Council set up the UN Development Programme, the world's largest source of grants to assist developing countries. Another related body is Unicef, established in 1946 as a relief agency*.

Economic Community of West African States (ECOWAS). It was constituted as an economic grouping, largely on the initiative of General Gowon*, at Lagos in 1975, by fifteen West African states, joined by Cape Verde* in 1977. Its object was to achieve a liberalization of trade and an eventual customs union. There would be a common fund to promote development projects and to compensate states which suffered losses, with specialized commissions for trade, industry, transport, and social and cultural affairs. In 1987 at the tenth summit meeting, held in Abuja, Nigeria, a $926m. programme was launched with 136 projects. It was agreed that a single monetary zone would emerge on 'a gradual and pragmatic basis'.

ECOWAS *see* ECONOMIC COMMUNITY OF WEST AFRICAN STATES.

ECSC *see* EUROPEAN COAL AND STEEL COMMUNITY.

Ecuador. In 1880 it became a separate South American republic whose politics reflected the tension between the conservative landowners of the interior and the more liberal business community of the coastal plain, with the majority of the population illiterate American Indians. In the early years after World War I the economy was expanding, with cocoa exports and railroad construction. By the mid-1920s, however, a depression had begun and the increasing poverty of the masses led to political turbulence. Although US military bases in World War II brought some economic gain, a disastrous war with Peru in 1941 forced Ecuador to abandon claims on the Upper Amazon. Between 1944 and 1972 José Maria Velasco Ibarra alternated with military rule, being elected President five times. The discovery of oil in the 1970s might have brought new prosperity, but in fact the mass of the population remained poor, with the great hacienda estates staying intact. The election of the Social Democrat Jaime Roldós Aquilera as President in 1979 promised reform, but he died in a mysterious air crash in 1981. His successor Osvaldo Hurtado Larrea (1981–4) was accused of embezzlement, and President Febres Cordero (1984–8) faced military intervention, a crisis of external debt, trade union unrest, and a decline in oil prices. In 1988 he was succeeded by President Rodrigo Borja, whose Social Democrat government took over management of foreign oil companies. The growth of education in the 1970s and 1980s resulted in a greatly increased electorate.

Ecumenical movement, seeking world-wide Christian unity. The modern ecumenical movement began in 1910 with a World Missionary Conference in Edinburgh, which established an International Missionary Council. Between the wars there were a number of world church conferences on such themes as 'Life and Work', 'Faith and Order'. In 1948 the movement resulted in the formation in Amsterdam of a World Council of Churches, whose ultimate aim was the reunion of all Christian churches or denominations. The Roman Catholic Church and the Unitarians were the only two Western churches not to join the Council, although from 1961 the Catholic Church sent accredited observers and in 1965 it achieved a *rapprochement* with the Eastern Orthodox churches*. The Council has a strong advisory role. Although its efforts to achieve unity have seemed slow, a great deal of inter-communion intercourse and activity has developed, with Christians everywhere more than ready to share their beliefs and give service in the modern world. More formal attempts to unite the Anglican Church with the Church of Scotland and with the Methodist Church have however failed.

Eden, Robert Anthony, 1st Earl of Avon (1897–1977), British Prime Minister 1955–7. Educated at Eton and Oxford, he served with distinction on the Western Front 1915–18. Elected to Parliament in 1923 as a Conservative, his main interests were always foreign affairs. At Oxford he had studied oriental languages. Foreign Secretary 1935–8, 1940–5, and 1951–5, he was noted for his support for the League of Nations in the 1930s, resigning from Neville Chamberlain's* government in protest over the latter's policy towards Mussolini. He was deputy to Churchill 1951–5, and succeeded him as Prime Minister. His premiership was dominated by the Suez Crisis*. He had opposed appeasement of the dictators in the 1930s, and was determined to stand up to President Nasser* of Egypt, whom he perceived as a potential aggressor after the nationalization of the Suez Canal. Widespread opposition to Britain's role in the Crisis, together with his own failing health, led to his resignation and retirement.

Edward VIII (1894–1972), King of Great Britain and Northern Ireland and of Dependencies overseas, Emperor of India, 1936. Eldest son of George V* and Queen

Mary, he served as a staff officer in World War I. As Prince of Wales he made a series of Empire tours and became increasingly concerned over levels of poverty in Britain. The Abdication Crisis*, provoked by his determination to marry the divorced Mrs Simpson, led to his abandoning the crown. Created Duke of Windsor, he was Governor of the Bahamas during World War II. He lived in France, but was buried at Windsor in 1972, together with the Duchess after her death in 1986.

EEC *see* EUROPEAN ECONOMIC COMMUNITY.

EFTA *see* EUROPEAN FREE TRADE ASSOCIATION.

Egalitarianism, doctrine of human equality. The belief that all human beings are of equal value in the eyes of God, and that it is morally wrong for anyone to possess more of the world's goods than another, has had a long history, especially within the monastic tradition. Although such an ideal of *égalité*, that all are inherently equal, helped to inspire the French Revolution and much 19th-century thought, in fact it failed to be realized in any secularized society, even those of communist countries. Yet the concept of 'equal opportunity' was dominant throughout the 20th century, inspiring both educational practice and demands for sexual and racial equality of opportunity. At the same time, as a common general life-style emerged in Western society it became markedly more egalitarian in both fashion and behaviour; domestic service almost disappeared, while mass-marketing produced single fashions in clothes, with hats disappearing so that they need no longer be 'doffed', and cowboy jeans becoming widespread throughout the world.

Egypt. Occupied by the British in 1882, following a nationalist revolt, it was declared a British protectorate in December 1914. Martial law was imposed, but nationalist, anti-British riots occurred in 1919. In 1922 Britain established a constitutional monarchy under Sultan Ahmad Fuad*, while retaining control over defence and imperial communications, as well as the Sudan. Nationalist agitation led by the Wafd* Party resulted in a large Wafd majority in a general election in 1936, as well as the growth of the Muslim Brotherhood*, reflecting a strong revival of Islam. As Prime Minister, Nahhas Pasha* negotiated an agreement for the retirement of British forces to the Canal Zone and for their eventual withdrawal over twenty years. In 1938 the young King Farouk* dismissed Nahhas, but was obliged to reinstate him in 1942, when German troops threatened the country. As the centre of the Allied strategic defence of the Middle East during these years, vast wealth came into the country. In 1948 Egyptian forces invaded Palestine, but failed to prevent the establishment of Israel, for which Farouk's incompetent military leaders were blamed. In 1952 a group of young officers, headed by General Neguib*, overthrew the monarchy, and Egypt was declared a republic. In 1954 Colonel Nasser* overthrew Neguib and embarked on a programme of rearmament and economic expansion, which included the Aswan Dam* project. This helped to precipitate the Suez Crisis*, which itself greatly increased Egypt's prestige in the Arab world. In 1958 Nasser negotiated a link with Syria, creating the short-lived United Arab Republic*. On 5 June 1967 Israel attacked both Jordan and Egypt, whose air force was largely destroyed on the ground. Egypt was disastrously defeated. On Nasser's death in 1970 his successor Anwar Sadat* at first continued the policy of confrontation with Israel, but after the failure of the Yom Kippur War* (1973), he became more flexible. Following the Camp David Accord*, Egypt was expelled from the Arab League and President Sadat was assassinated in 1981. Although the Aswan Dam project greatly increased productivity of the land, poverty remained a major problem, together with a steady growth of Islamic fundamentalism*. Egypt was formally readmitted to the Arab League in 1989, and under President Mubarak* opposed Iraqi occupation of Kuwait.

Eichmann, Karl Adolf (1906–62), Nazi administrator. Born in Bavaria, he was educated in Linz in Austria, where he became a travelling salesman. In 1932 he joined the Austrian Nazi Party. After working at Dachau camp he joined the German secret service, going on a secret mission to Palestine. After the Anschluss* in 1938 he supervised anti-Jewish action in Vienna and later in Prague. In 1941 he became head of the Gestapo department responsible for Jewish affairs. In 1942, following a conference at Wannsee on the 'final solution to the Jewish problem', he was appointed to organize the logistic arrangements for the dispatch of Jews to death camps and to set up gas chambers for mass extermination. He escaped from custody in 1945, going to South America. In 1960 he was abducted from Argentina by Israelis and was executed in Israel after trial.

Einstein, Albert (1879–1955), mathematical physicist. Born in Ulm of Jewish parents, he was educated in Munich and then in Zurich, where he attended the technical school when he was 17. He then became a schoolmaster. In 1905 he formulated his first special theory of relativity and in 1909 became a Professor of Physics in Zurich. In 1912 he put forward a theory of photochemical equivalence, and he was director of the Kaiser Wilhelm Institute in Berlin 1914–33. In 1916 he published *Die Grundlagen der allgemeinen Relativitätstheorie*, in which his general theory of relativity revolutionized the previous Newtonian theory of the universe. Observations at a total eclipse in 1919 confirmed his theories. In 1933 he fled Hitler's anti-Semitism and became a US citizen, teaching at Princeton. In 1939 he warned President Roosevelt of German research into the possibilities of an atomic bomb and after 1945 became increasingly alarmed at the potential threat to mankind through atomic warfare. In 1955 he helped to establish Pugwash*.

Eisenhower, Dwight David (1890–1969), President of the USA 1953–61. Born in Texas, he grew up in Kansas and graduated from West Point in 1915. During World War I he commanded a tank training unit and had numerous assignments between the wars, including service with Douglas MacArthur* in the Philippines. In 1942 General George Marshall* selected him over 366 more senior officers to be commander of US troops in Europe. As a lieutenant-general he went on to command Operation Torch in November 1942, the Allied landing in North Africa. In December 1943 he was appointed Supreme Commander of the Allied Expeditionary Forces. As such he was responsible for the planning and execution of the Normandy landings and subsequent campaigns in Europe. In command of men of many nationalities and backgrounds, his qualities of friendliness and optimism enabled him to secure inter-Allied collaboration and to avoid confrontation. In 1951 he was persuaded to return to active service as Supreme Commander of NATO*, a command he held for fifteen months, finally retiring from the army in June 1952. He had been approached by both Democrat and Republican parties to stand for President but was elected for the Republicans in November 1952, with Richard Nixon* as Vice-President. He promised to seek an end to war in Korea if elected, which he did in July 1953. His 'modern republicanism' sought reduced taxes, balanced budgets, and a decrease in federal control of the economy. His administration was embarrassed by the right-wing 'witch-hunt' of Senator McCarthy*, with its anti-communist hysteria. In spite of tough talk there was a move towards reconciliation with China and a decision not to become engaged in Indo-China, following the defeat of France at Dien Bien Phu*. With John

Dulles as Secretary of State, the NATO and ANZUS* pacts were extended by the SEATO* Pact of 1954. After Dulles's death in 1959 Eisenhower's keen interest in foreign affairs became more noticeable as he sought to negotiate with the Soviet Union. In May 1960, however, a US high-altitude reconnaissance U-2* plane was shot down over Soviet territory, an incident which destroyed these hopes. The construction of the Berlin Wall* in August 1961 reflected a steady deterioration in Soviet–US relations at the close of his years in office.

Eisenhower Doctrine (1957). Following the Suez Crisis* President Eisenhower proposed to give US economic and military aid to Middle East governments who felt their independence threatened. In 1958 the USA sent 10,000 troops to Lebanon to support President Camille Chamoun*, who feared revolution. Britain had also sent troops in 1957 to protect the regime of King Hussein ibn Talal* of Jordan. The assumption that Arab nationalism was Soviet inspired came to be seen as fallacious and the Doctrine lapsed with the death of Secretary of State Dulles* in 1959.

El Alamein, Battle of, World War II (23 October–4 November 1942). In June 1942 the British under General Auchinleck* took up a defensive position on the Mediterranean at El Alamein, some 80 km. from Alexandria. During the first battle 30 June–25 July the German and Italian forces under Field Marshal Rommel* were effectively halted. A front was established between the sea and the salt marshes of the Qattara Depression. In August General Montgomery* was appointed to command the 8th Army. On 23 October he launched an offensive in which, after a heavy artillery preparation, about 1,200 tanks advanced, followed by infantry, against the German Afrika Korps. Rommel was handicapped by a fuel shortage and had only about 500 tanks in effective use, and he never regained the initiative. On 4 November he ordered retreat, withdrawing back into Libya. This battle marked the beginning of the end of the North African campaign for Germany.

ELAS. The initials stand for the Greek words meaning National People's Liberation Army. It was created during World War II by the Communist-controlled National Liberation Front (EAM) to fight against German occupation forces. By the time of the German defeat and withdrawal in 1945 EAM/ELAS controlled much of Greece and opposed the restoration of the monarchy. A bitter civil war was fought over three years. In 1946 a plebiscite voted for the restoration of King George II, whose government then received massive US aid in accordance with the Truman Doctrine*. Stalin's unwillingness to support the Greek Communists contributed to their defeat.

Electoral college (USA). The device of an electoral college was adopted by the framers of the US Constitution in 1787, each state to receive as many electors as it had members of Congress. These electors would then meet to elect the President. As states extended their franchise these electors came to be chosen by direct election rather than by state legislatures. However, with the emergence of organized political parties, the holding of national party conventions to select presidential candidates developed. Today the candidate who wins a majority of the popular vote in a state gains all that state's electoral votes (except in Maine since 1969). Members of the 'college' actually meet in each state capital on the second Tuesday of December following a presidential election. The 'winner takes all' effect of the present system, which allegedly favours small states, has often been criticized, and the 1960, 1968, and 1976 elections all produced demands for its abolition in preference for direct election. In 1970 a Bill for abolition of the college was defeated in the Senate.

Electrification. One of the most widespread technical achievements of the 20th century was that of electrification. Early in the century only the world's larger cities had electricity—produced by coal-fired generating installations. In such cities main streets, public buildings, hotels, and wealthier homes were lit by electricity, which was also beginning to power tramways and railways, as well as escalators, cooking ovens, and other devices; but in ordinary city homes cooking was by gas or solid fuel (coal or wood) and lighting by gas, oil lamps, or candle-light. The vast majority of mankind in small towns, villages, and farms subsisted on open fires and candle-light. By the end of the century almost the entire surface of the globe had been electrified. Coal, oil, nuclear-powered generators, together with hydroelectric and tidal schemes, sought to satisfy the ever-increasing and profligate demand for electrical power. Meanwhile the debate for future energy provision, when all fossil fuels will have been consumed, had begun.

Eliot, Thomas Stearns (1888–1965), poet and literary critic. Born in St Louis, USA, he was educated and taught at Harvard before going to live in Paris and then London. His publication of *The Waste Land* in 1922 marks a turning-point in English poetry, concerned as it is with the dehumanization and alienation of 20th-century urban life, against the background of the slaughter of World War I. His concept of 'the dissociation of sensibility' profoundly affected cultural and aesthetic perceptions, and valuations of recent centuries. *Four Quartets* (1943) is considered his greatest poem, Christian mysticism having become an increasing part of his interests and writing in his later years.

Elizabeth II (1926–), Queen of Great Britain and Northern Ireland and Dependencies overseas, head of the Commonwealth of Nations 1952– . Elder daughter of George VI*, she became heir to the throne on the abdication of her uncle in 1936. She was trained in motor transport driving and maintenance in the Auxiliary Territorial Service (ATS) late in World War II, and in 1947 married a distant cousin, Philip Mountbatten, formerly Prince Philip of Greece and Denmark. Their first child and heir to the throne, Prince Charles, was born in 1948. Her coronation in 1953 was the first major royal occasion to be televised. Since then she has devoted much of her reign to ceremonial functions and to tours of the Commonwealth and other countries. While adhering strictly to the conventions of the British constitution, she has always held a weekly audience with her Prime Minister and shown a strong personal commitment to the Commonwealth.

Ellis Island, an island in New York Bay off Manhattan Island. Long used as an arsenal and a fort, from 1892 until 1943 it served as the centre for immigration control. From 1943 until 1954 it was a detention centre for deportees. In 1956 it became part of the Statue of Liberty National Monument, and open to sightseers.

El Salvador, a Central American republic. After independence from Spain in 1821 internal struggles between liberals and conservatives and a series of border clashes with neighbours retarded its development. By the early 20th century the conservatives had gained ascendancy and the presidency remained within a handful of élite families as if it were their personal patrimony. A fall in coffee prices in 1931 sparked off brutal suppression by General Hermández Martínez, President 1931–44. While some post-war presidents, such as Oscar Osorio (1950–6) and José M. Lemus (1956–60), appeared mildly sympathetic to badly needed social reform, they were held in check by their more conservative military colleagues in concert with the civilian oligarchy, the army's stance moving steadily to the right. Repressive measures and violations of human rights by the army through the 1970s were documented by a number of international agencies, and posed a large refugee problem. In 1980 the Frente Farabundo Martí de Liberación Nacional (FMLN) was challenged by paramilitary death squads and internal civil war developed, following the open murder of Archbishop Oscar Romero in March 1980. In 1982 120 political murders were being committed by death squads each week. In 1984 the Christian Democrat José Napoleón Duarte was elected President with US support. Regarded by the USA as President of a 'Front-Line State' against Nicaragua, Duarte was discouraged from supporting the peace plan of President Arias of Costa Rica*; yet negotiations with the FMLN were opened in October 1987. In the presidential election of March 1989 the right-wing candidate of the Alianza Republicana Nacionalista (Arena), Alfredo Cristiani, was declared elected. FMLN boycotted the election, which was followed by renewed violence. The new President acted vigorously against both the death squads and FMLN guerrillas. In September 1989 peace talks began in Mexico City with FMLN representatives, meetings continuing through 1990.

Enver Bey (Pasha) (1881–1922), Turkish general and statesman. Born in Apana, he enlisted in the Ottoman army and was trained in Germany. He played a prominent part in the 1908 Young Turk* revolution and was then military attaché in Berlin. He fought in the Balkan War of 1913 and subsequently led a successful coup, becoming Minister of War. He arranged the appointment of the German General Liman von Sanders as Commander-in-Chief of the Turkish army and played a leading role in determining the entry of the Ottoman Empire into World War I in November 1914 on the side of the Central powers. He was virtual ruler of the Empire, but also personally took the field in the Turkish campaign against the Russians in the Caucasus. He resigned in October 1918. Condemned by the new Republic, in 1921 he fled to Turkistan, where he was killed while leading opposition to Soviet rule.

Environmental conservation. During the 19th century vast areas of the world suffered from deforestation*, pollution, and the slaughter of millions of animals, for example the American bison, while industrial waste polluted both air and water. By 1900 the movement to create national parks had begun, the first being Yellowstone in the USA (1872). The 20th century saw an increase in all the problems and since World War II a growing realization of both the effects of pollution (acid rain, poisoned algae-filled lakes, the 'greenhouse effect', etc.), and the overconsumption of fossil fuels, tropical woods, and minerals. There were modest attempts towards environment conservation, for example clean-air bills and encouragement of recycling processes. Numerous Green movement* pressure groups, backed by scientific evidence, lobbied the United Nations as well as individual governments for more radical global action. There have been numerous conferences. But while all accepted the scale of the problems facing mankind in the 1990s, few were willing to accept politically the social and economic consequences of necessary action, for example the banning of the private motor car.

EOKA (National Organization of Cypriot Fighters). As the militant wing of the *enosis* movement in Cyprus* demanding union with Greece, it organized five years of guerrilla warfare 1954–9. Its commander was Colonel Georgios Grivas, whose terrorist attacks were aimed mostly at the British army. In 1956 Makarios* was exiled on the charge of being linked with EOKA. After independence in 1960 it continued to operate underground, as EOKA-B, renewing demands for *enosis*. In 1974 it proclaimed union with the regime of the Greek Colonels*, but the coup collapsed with a Turkish invasion.

Equatorial Guinea, a west-coast country of Central Africa. Formerly the Spanish colonies of Fernando Póo (Bioko) and Guinea, the mainland was not effectively occupied by Spain until 1926. It was declared independent in 1968 following a campaign led by Macias Nguema, who became the first President. Racial tension with Spanish colonists led to riots and repressive action by Nguema, who relied heavily on Cuban and Chinese advisers. He was overthrown and executed (1979) by his nephew Colonel Obiang Nguema. His Supreme Military Council built close links with Spain and Morocco and pursued rather less repressive domestic policies.

Erhard, Ludwig (1897–1977), Chancellor of the German Federal Republic 1963–6. Born in Bavaria, he studied economics at Munich and became head of an economic research unit there before and during the war. In 1945 he became Professor of Economics at Munich University and chairman of the Economic Executive Council of the US–British zone of occupation. He entered the Bundestag as a Christian Democrat in 1949 and became Federal Minister of Economics, an office he held until 1963. During these years he assisted in his country's 'economic miracle' (*Wirtschaftswunder*), which trebled the gross national product. He succeeded Adenauer* as Chancellor, but lacked his predecessor's political skills. He resigned in November 1966 following unpopular tax increases.

Eritrea, a province of Ethiopia on the Red Sea. By 1889 Italy had occupied the region and declared it a colony. From here the Italians launched campaigns against Ethiopia, 1895–6 and 1935–6. Under British military administration 1941–52, a plan to join the Muslim west with the Sudan and the Christian centre with Ethiopia failed. Instead the UN voted to make it a federal region of the Ethiopian Empire. In 1962 Emperor Haile Selassie* declared it a province, and the Eritrean People's Liberation Front (EPLF) then emerged, seeking secession. From 1974 there was fierce fighting, in spite of drought and famine, between the EPLF and Ethiopian regimes, with a major EPLF success at Nakfa in 1979 followed by brief peace talks. Fighting resumed in 1981 and continued throughout the 1980s. A new EPLF victory on the Nakfa front in 1988 and the capture of Massawa in February 1990 were followed by further EPLF advances.

Erlander, Tage Fritiof (1901–85), Swedish Prime Minister 1946–68. After graduating at Lund University in 1928, he became an encyclopaedia editor and in 1933 entered politics, being elected to the Riksdag as a Social Democrat. He held various offices, including that of Minister of Education in 1945. In 1946, on the sudden death of Prime Minister Albin Hansson, he became Prime Minister, continuing with his educational reforms, which included extending compulsory education and expanding higher education provision. His government extended social welfare provision in Sweden, with higher pension benefits, child allowances, and rent subsidies. He was always a strong supporter of Swedish neutrality.

Eshkol, Levi (1895–1969), Prime Minister of Israel 1963–9. He came to Palestine from Russia in 1914, founding an early kibbutz*. When the British entered Palestine in 1917 he enlisted in the Jewish Legion*. With Ben Gurion* and Ben-Zvi* he was a founder of the Histadrut or General Federation of Jewish Labour. During World War II he organized immigration from Germany and later helped the arming of Haganah*. He served under Ben Gurion, whom he succeeded as Prime Minister in 1963. He united various Labour groups to form the Labour Party of Israel* and successfully led the country through the Six-Day War*.

Estonia, a Baltic republic. Annexed by Russia in 1709, it regained its independence in 1918, at the time of the Bolshevik Revolution. Its history in the 1920s was of an agrarian revolution whereby the great estates of the Baltic Barons were broken up, creating a prosperous peasantry. An attempted Communist uprising in 1924 was suppressed. From 1934 until 1939 the regime of Konstantin Paets was highly autocratic. Paets admired Hitler, but his attempt to make a pact was invalidated by the Nazi–Soviet Pact of August 1939. In September Soviet troops occupied key ports and in 1940 the whole country. It welcomed German troops in 1941 but its anti-Bolshevik Resistance forces could not prevent the Red Army from reoccupying it in 1944. It became a constituent republic of the Soviet Union. In February 1990 a mass rally in Tallinn called on the Supreme Soviet of the USSR to enter negotiations for the restoration of independence, the demand being endorsed by the Estonia Supreme Soviet, which also ended the monopoly of power of the Communist Party. In May the Supreme Soviet reinstated the 1920 Constitution of the Republic of Estonia and talks began for the establishment of a 'special relationship' with the Soviet Union, within a Confederation of Baltic States.

Ethiopia, a north-east African country, also called Abyssinia. Under the Emperor Menelik II (1844–1913) it defeated an Italian attempt at colonization 1895–6, thus becoming the only African kingdom to survive into the 20th century. Menelik began a policy of modernization, encouraging mission schools and bringing a railway from Jibouti. On his death his grandson Lij Yasu succeeded him. He favoured Germany in World War I and was deposed in 1916. A great nephew of Menelik, Ras Tafari, was appointed regent and in 1930 was crowned Emperor Haile Selassie*. He extended modernization, but was defeated by Italian forces invading from Eritrea and Somaliland 1935–6. Restored by a combined Allied force of British and African troops under Sir Alan Cunningham, the Emperor gained Eritrea* in 1952 as a federal region. A severe drought in 1973 caused great distress and Haile Selassie was deposed. A short-lived liberal administration by Endel Katchew Makonnen was swept aside in September 1974 by a group of radical officers who, under Brigadier Teferi Banti, executed some sixty former leaders. Banti was himself killed in a coup in 1977 led by Colonel Mengistu Haile Mariam, who became chairman of the Provisional Military Administrative Council. His government was faced by guerrilla fighting in the province of Eritrea and a fierce border war dispute in the Ogaden desert with the Somali Republic*. In spite of Cuban and Soviet aid, and international relief efforts, the Ethiopian Famine* of 1984 was an appalling disaster, while 1987 again saw millions of drought victims. Fighting continued into 1990 in the north in Eritrea* and the Tigre region, the Tigreans seeking greater autonomy.

Ethiopian Famine (1984). Following drought conditions from mid-1982 there was an almost complete crop failure in northern and eastern Ethiopia in 1983. Thousands of refugees from Eritrea and Tigre had already been victims of the civil war between the Ethiopian government and the Eritrean People's Liberation Front (EPLF) and the Tigre People's Liberation Front (TPLF). In addition pests had ravaged crops elsewhere in the country. In May 1984 the Ethiopian Commission for Relief and Rehabilitation appealed for international aid to save some 5 million drought victims from famine. By October, when some 1,000 victims were dying each week, massive aid began to arrive from the EEC, Australia, Canada, the USA, and elsewhere. The pop group Band Aid raised £6m., in West Germany some £29m. was raised in a single day (23 January 1985), and in August 1985 Bob Geldof's Live Aid concert raised over £40m. Some 407, 689 tonnes of food were delivered October–January, but war conditions prevented some of this from reaching victims. A resettlement plan by the Ethiopian government of 'villagization' of refugees was criticized by relief agencies. Large numbers fled to Jibouti, Somalia, and drought-torn Sudan. There was further drought in 1987 and a new famine appeal early in 1990.

Eupen-Malmédy, an area of German-speaking Belgium. It had been part of the Austrian Netherlands, but was ceded to Prussia in 1815. In 1919 it was awarded to Belgium, the city of Eupen having valuable industrial assets. In 1972 a Cultural Council was established, which seeks to protect German language rights.

European Atomic Energy Commission (Euratom). Established in March 1957 by the second of the Rome treaties, its aim is co-operation of member states in nuclear research for the production of nuclear energy. In 1967 it was incorporated into the European Community*.

European Coal and Steel Community (ECSC). Following proposals in 1950 in the Schuman* Plan, the Treaty of Paris of April 1951 established a single authority which was to merge the coal, iron, and steel industries of West Germany, France, Italy, and the Benelux countries, eliminating tariffs and creating a free labour market. Britain refused to take part. The Community began to function in July 1952. It was to merge with the EEC and Euratom in 1967 within the European Community*.

European Community (EC). It came into being in 1967 through the merger of the European Economic Community, the European Atomic Energy Commission, and the European Coal and Steel Community, committed to economic and political integration as envisaged by the Treaties of Rome. Its membership is identical with that of the EEC. It operates, within the framework of the European Commission (the executive body with powers of proposal), an Economic and Social Committee, a European Investment Bank, and the European Court of Justice. The ongoing decision-making body is the Council of Ministers (foreign and other ministers of the member states), together with a European Council of Heads of Government which meets three times a year, the presidency rotating every six months. The European Parliament, which held its first direct elections in 1979, has powers of supervision and consultation, and a measure of control over the Community's budget. Decisions of the Court of Justice are directly binding on member states and are superior over any national Act of Parliament inconsistent with it. Following the collapse of the East European Communist regimes strong pressures developed towards accelerating political integration. Mrs Thatcher and many within her Conservative Party fiercely opposed any such 'loss of sovereignty'.

European Defence Community. During 1950 Winston Churchill was pressing for a Common European Army and in October the French Foreign Minister René Pleven put forward a plan for a European Defence Community, whose membership would be identical with that of the newly proposed ECSC. Britain refused to take part, and in 1954 the French Assemblée rejected it. By then NATO* existed. The French reluctantly acceded to US pressure that West Germany be allowed to join NATO, which it did in May 1955, whereupon the Warsaw Pact* was created.

European Economic Community (EEC). Created by the Treaties of Rome of March 1957, it owed much to the initiative of Jean Monnet* and began to function in January 1958. Its first members were Belgium, France, the German Federal Republic, Italy, Luxembourg, and The Netherlands. Britain at first refused to join. Its members agreed to develop the already functioning ECSC and Euratom by establishing common policies for agriculture, transport, the movement of capital and labour, and the erection of common external tariffs. Ultimately some form of political integration was also envisaged. In 1961 Macmillan's government in Britain decided to reverse policy and seek membership; but this was delayed until January 1973 following two vetoes by the French. Denmark and Ireland also joined the Community in 1973; Greece in 1981; Spain and Portugal in 1986. Norway planned to be a member but withdrew her application in 1972, as did Greenland in 1985. Austria, Turkey, Malta, and Cyprus have all expressed a wish to join, and the collapse of the East European regimes in 1989 precipitated a debate as to how far the Community could expand. During the 1980s the Common Agricultural Policy (CAP) was absorbing some two-thirds of the Community's budget before subsidy levels were reduced. The European Monetary System (EMS) was established to limit fluctuations in the exchange rates of member states which had joined it; this Britain reluctantly did in October 1990. The Community has an agreement (the Lomé Convention*) with over sixty African, Caribbean, and Pacific countries which eases customs duties and provides aid programmes.

European Free Trade Association (EFTA). It was created on a British initiative by the Stockholm Convention of 1959 as an alternative trade group to the EEC. Membership has consisted of Austria, Britain, Denmark, Norway, Portugal, Sweden, Switzerland, Liechtenstein, Finland, and Iceland. When Britain, Denmark, and Portugal joined the EEC they left EFTA. In 1977 EFTA agreed with the EEC to establish industrial free trade between the two organizations' member countries. This was followed in the 1980s by wider areas of co-operation, and debate as to relationship with the East European democracies.

European Parliament. It meets alternately in Strasbourg and Luxembourg. From 1958 until 1979 it was composed of members drawn from the parliaments/ assemblies of European Community* members. Since then quinquennial elections on a basis of universal suffrage have been held. Treaties signed in 1970, 1975, and 1986 have given the Parliament important powers over budgetary and constitutional matters, assuming, through the Single European Act of 1986, a degree of sovereignty over national parliaments. In the late 1980s strong pressures developed to extend the role of the Parliament *vis-à-vis* the Commission of the Community.

Evatt, Herbert Vere (1894–1965), Australian statesman. Born in New South Wales, he taught at the University of Sydney before developing a law practice which specialized in civil liberties. He was appointed a justice of the Australian High Court in 1931, but in 1940 entered politics, being elected to the federal Parliament for the Labor Party. In October 1941 he became Attorney-General and Minister of External Affairs in the new Labor government of John Curtin*. At the end of the war he played a leading role in the establishment of the United Nations*, eloquently championing the small nations and opposing the power of veto which the five major powers gave themselves on the Security Council. He had remained Minister of External Affairs under Chifley* and was personally responsible for no less than twenty-six amendments to the draft UN Charter. He presided over the UN General Assembly 1948–9. From 1951 to 1960 he was Labor leader in opposition, while at the same time he acted as counsel for some of those indicted in the 'Petrov Affair'*. He was criticized for this at a time when anti-communist hysteria was dividing his party, with the creation of the splinter Democratic Labor Party. He retired from politics in 1960 and served for two years as Chief Justice of New South Wales.

Évian Agreements, a series of agreements negotiated in 1962 at Évian-les-Bains in France. Secret negotiations between the government of President de

Gaulle* and representatives of the provisional government of the Algerian Republic of Ben Bella* began in Switzerland in December 1961 and continued at Évian in March 1962. A cease-fire commission was established and the French government, subject to certain safeguards, agreed to a plebiscite on the future of Algeria. Referenda were held in France (8 April) and Algeria (1 July) and the agreements were ratified by the French National Assembly. They were violently attacked by the extremist Organisation de l'Armée Secrète (OAS)*.

Expressionism, a loose term applied particularly in Germany and Austria-Hungary to the whole Modernist* movement in the arts from *c.*1910 into the 1920s. Munich in particular was a centre of Expressionist art, some of whose giants were, for example: in literature Franz Kafka*; in drama Ernst Toller; in music Arnold Schöenberg; in architecture Erich Mendelsohn; in painting the Blaue Reiter school, including Klee and Kandinsky; in the cinema Robert Wiene (*The Cabinet of Dr Caligari*). Expressionist art seeks to communicate intense emotion through distortion and fragmentation. It was regarded by the Nazis as degenerate, and suppressed.

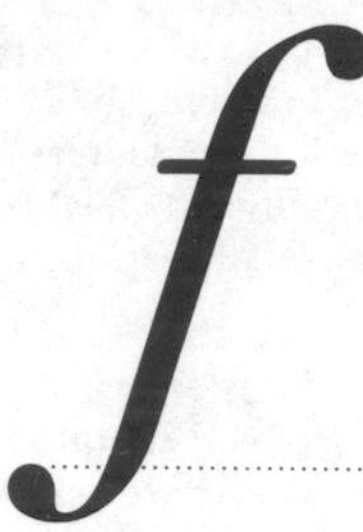

Fabian Society. It was founded in London in 1884 by a group of intellectuals who believed that new political pressures were needed to achieve social reforms. George Bernard Shaw*, Beatrice and Sidney Webb*, Annie Besant*, and Hubert Bland were among its founders. Their early slogan was 'the inevitability of gradualism'. Reforms would be achieved through patient, persistent use of argument and propaganda. They were especially active in London local government. The Society was one of the socialist organizations which helped to form the Labour Representation Committee in 1900. Although it sought inter-party influence, trade union militancy from 1910 until 1926, and the harshness of unemployment in the 1930s, weakened the appeal of Fabian gradualism. However by 1939, with senior members such as Clement Attlee*, himself chairman of the Fabian Research Bureau, the Society's influence revived. It continued after World War II, with regular publication of journals and news-sheets.

Fair Deal, a proposed domestic programme of US President Truman* of January 1949. By it he hoped to advance beyond the New Deal* to introduce measures on civil rights; fair employment practices; education; health, with public health insurance; social security, with extended benefits; support for low-income housing, with price and rent control; and for a farm-subsidy programme. It also proposed guaranteed full employment and an increased minimum wage. A coalition of Republicans and conservative Southern Democrats however blocked most of his measures in Congress. He did manage to secure some advances in housing and social security.

Faisal I (1885–1933) King of Iraq 1921–33. Son of Hussein ibn Ali*, he took part in the Arab Revolt* of 1916, working closely with T. E. Lawrence* and commanding the northern Arab army. In September 1918 he entered Damascus and was proclaimed King of Syria. When however he went to the Paris peace negotiations he found Britain unwilling to support him against the French, who wanted to extend their influence in the Middle East. Back in Damascus he was confirmed as King by a Syrian National Congress, but was forced into exile by the French, the new mandatory power. He was then made King of Iraq by Britain, who held the mandate for that country. Supported by Pan-Arab nationalists throughout the Fertile Crescent, he still managed to maintain British friendship, showing considerable political skill in building up the institutions of a new state. In 1930 he gained full independence for Iraq, followed by membership of the League of Nations.

Faisal II (1935–58), King of Iraq 1939–58. His father King Ghazi was killed when he was 4 and his uncle Abd al-Ilāh* acted as Regent while he grew up. He had been educated in England with his cousin Hussein of Jordan*, and the two planned in 1958 to federate Iraq and Jordan as a counter-balance to the new United Arab

Republic*. This resulted in a pro-Nasser military coup, when he and his uncle were murdered and the Republic of Iraq proclaimed.

Faisal ibn Abd al-Aziz (1905–75), King of Saudi Arabia 1964–75. Son of the first ruler of Saudi Arabia, Abd al-Aziz ibn Saud*, he was appointed Viceroy of the Hejaz in 1926, and in 1934 led a victorious campaign against Yemen. He became effective ruler of Saudi Arabia in 1958, owing to the incompetence and later ill health of his brother King Saud ibn Abd al-Aziz, who was deposed in 1964. As such he had to deal with the main consequences for his country of its immense oil revenues; he encouraged modernization and strong education programmes. Deeply conservative and opposed to all Arab 'revolutionaries', he yet supported Arab claims on Jerusalem and was always opposed to Israel*. After the unsuccessful Arab–Israeli war of 1967 he gave both Jordan and Egypt financial assistance to offset their losses. Worried by Egyptian support for left-wing revolutionaries in neighbouring Yemen, he yet managed to negotiate successfully with Nasser* and became a strong supporter of Sadat*. As the main supplier of US oil, he used his influence to reduce Israel's support from the USA, taking a key role in OPEC* negotiations of December 1973. He was shot by a disgruntled nephew and succeeded by his brother Khalid.

Falange, the, a Spanish political party El Falange Española. Founded in 1933 by José Antonio Primo de Rivera, son of General Primo de Rivera*, its members were equally opposed to the reactionary right and the revolutionary left, their manifesto of 1934 proclaiming opposition to republicanism, party politics, capitalism, Marxism, and the class war. During the Spanish Civil War Franco saw the potential value of the Falange, provided its aims could be made acceptable to monarchists and traditionalists. When Antonio was killed by the Republicans, Franco adopted the movement in April 1937. After the war ended the Grand Council of the Falange acted as a legislative body, but from 1942 its influence declined as Franco adopted more autocratic methods. After World War II the party distanced itself from the Caudillo's policy to restore the monarchy. Its influence has steadily declined.

Falkland Islands (Islas Malvinas), a group of some 100 islands in the south Atlantic, the two main ones having a population of some 2,000 inhabitants, 2,000 British military personnel, and some half-million sheep! Spanish rule ceased in 1806 and in 1820 Argentina claimed to succeed Spain in possession. Britain objected and in 1833 claimed them as a Crown Colony, with a naval squadron occupying them in 1882–3 'for the protection of the seal fisheries'. The Argentinians have disputed British claims since 1833, culminating in the war of 1982. In 1965 the UN invited both sides to negotiate a settlement of their dispute. These discussions were still going on when war broke out. The large British naval and military occupation force was reduced in 1987, following the completion of the Mount Pleasant airport. In 1989 discreet discussions between British and Argentinian diplomats began, in spite of British refusal to concede 'sovereignty'.

Falklands (Malvinas) War, (2 April–14 June 1982). UN attempts to resolve the dispute between Britain and Argentina over the islands having failed, an Argentinian warship was sent by General Galtieri* to land a party of 'scrap-dealers' on South Georgia on 19 March 1982. This was followed by a full-scale invasion of the Falkland Islands on 2 April. Attempts by the UN, Peru, and the USA to secure a peaceful resolution to the conflict failed, and a British task-force of thirty warships, with supporting vessels and 6,000 troops, was assembled and sailed the 13,000 km. to the south Atlantic. South Georgia was easily recaptured on 25 April and raids on East Falkland followed. There was a commando landing on West Falkland on 14 May and an attack on Port Stanley launched on 21 May. The Royal Navy lost two destroyers and two frigates, together with the Cunard container ship *Atlantic Convoy*. On 2 May it sank the Argentinian cruiser *General Belgrano*, although this was away from the area of conflict. The war cost the lives of some 1,000 British and Argentinian servicemen and civilians, ending with the surrender of Port Stanley on 14 June. The British victory resulted in the fall of General Galtieri, but his successor President Alfonsín was unable to persuade Britain to reopen discussions over sovereignty.

FALN (Fuerzas Armadas de Liberación Nacional Puertorriqueña). Its origins can be traced to the violent Nationalist Party in Puerto Rico in the 1930s, which provoked rioting and assassinations against the USA. In 1950 an attempt was made to assassinate President Truman, and in 1954 five US Congressmen were wounded in the Chamber of the US House of Representatives. There were bomb outrages throughout the 1970s, when the title FALN was first used. In 1975 an outrage on Wall Street, New York, killed three people and injured over forty. FALN operated mainly on the US mainland, particularly in Chicago and New York, and does not appear to have gathered much support on Puerto Rico itself. FBI intelligence resulted in mass arrests in 1980 and 1985, after which its activities seemed to decline.

Farouk (1920–65), King of Egypt 1936–52. Educated in England, his early popularity subsided when schemes for land reform collapsed due to endemic corruption. An admirer of Mussolini and hence sympathetic to the Axis powers*, he was obliged by the British in 1942 to accept the Wafd leader Nahhas Pasha* as his Prime Minister. After the war he supported a series of anti-British governments, but his interest in politics declined as he surrounded himself with a corrupt and effete court. Forced into exile in 1952, he spent his last years mostly in Italy.

Fascism, an ideology of the 20th century. The term first appeared in Italy after World War I, where a Fascist movement was led by Mussolini* in 1919. It is derived from the Italian *Fascismo*, itself coming from the Latin *Fasces*, a bundle of rods with a projecting axe, insignia of authority of the consuls of ancient Rome. Arising from the political chaos of post-war Italy, it was less a coherent ideology than a demand for dictatorial government which would personify the authority of the state as supreme source of law, order, and national life. Opposed equally to Marxism and to 19th-century values of 'liberty, equality, and fraternity', it called for the subordination of the individual, whose duty was 'to believe, to obey, to combat', whereby the 'strong will prevail over the weak, the resolute over the irresolute'. In socio-economic terms the Italian Fascist Party was conservative, gaining the support of the military, the landowners, industrialists, and the Catholic Church, and putting forward a doctrine of corporatism*. Its popular appeal lay in promises of extensive public works schemes to reduce unemployment, together with the vision of a rebirth of the ancient Roman Empire. Fascist or neo-Fascist governments emerged in a number of countries between the wars, for example, Romania, Hungary, and Spain, with Nazism in Germany having much in common. As the Cold War developed after 1945, aspects of Fascism began to appeal to an increasing number of right-wing parties in Europe and in Central and South America, where many neo-Fascist regimes prevailed.

Fatah, al- (Arabic; 'Victory'), a militant Palestinian organization. It was founded in Kuwait by Yasir Arafat* in 1958 to fight for the restoration of Palestine to the Arabs. Its first paramilitary operation was in 1964 against an Israeli water-pump. In 1969 it assumed the leadership of the Palestine Liberation Organization*, its guerrilla units being based and organized in Jordan. When they were expelled from there in the fighting 1970–1 an extremist splinter group Black September emerged, who were responsible for the Munich Olympic Games* murders. Al-Fatah now withdrew to Lebanon, from where it continued terrorist activities against Israel until the latter's invasion of 1982, when it was expelled from those areas not controlled by Syria. A year later, in May 1983, there was a split in its ranks in Syrian-held Tripoli, from which it was expelled to Tunis. Since then however it has re-established its presence in Lebanon and, to a much greater extent, in the occupied West Bank and Gaza, organizing a youth movement from which developed the *Intifadah** of December 1987. Arafat remains its supreme commander, aided by Khalil Wazir until his assassination by Israeli commandos in 1988. It has remained the principal group within the PLO, having accepted Arafat's decision to substitute diplomacy for military confrontation against Israel.

Fatima, a small Portuguese town. Here in 1917 three shepherd girls saw a vision of a lady and continued to do so over many months, the lady declaring herself to be 'Our Lady of the Rosary'. She asked for a chapel to be built to which people of all nations could come to do penance for war. Thousands visit the chapel annually, including Pope Paul VI and in 1982 Pope John Paul II, when an assassination attempt was made against him.

Faulkner, Arthur Brian Deane, Baron Faulkner of Downpatrick (1921–77), Prime Minister of Northern Ireland 1971–2. Elder son of a successful Belfast businessman and industrialist, he was educated in Dublin and Belfast and worked in his father's business through World War II. He was a Unionist Member of Parliament at Stormont* (1949–72), Minister for Home Affairs (1959–63 and 1971–2), and Prime Minister 1971–2. His twelve months in office saw sectarian violence reach a new height. He introduced internment in August 1971, which only caused a hardening of Catholic opinion. In March 1972 direct rule was imposed from Westminster. He remained active, seeking to establish a Council of Ireland. However, his negotiations lost him support in his own party and he resigned as leader 1 January 1974, retiring from politics in 1976.

Federal Bureau of Investigation (FBI), the investigative branch of the US Department of Justice. Established by Attorney-General Charles J. Bonaparte in 1908, it was reorganized in 1924, when J. Edgar Hoover* was appointed director, and it was given wider powers to investigate violations of federal laws, with offices established in all major US cities and liaison offices in larger foreign capitals. Hoover successfully led the 1930s drive against criminal syndicates operating across the USA. During World War II the FBI began spying activities against Nazi sympathizers in the USA and South America, co-operating with the CIA* when this was formed after the war. By then it was international communism which was the target. The later excesses of Hoover, in particular his harassment of political dissidents and radicals such as Martin Luther King*, brought its counter-intelligence activities into disrepute. From 1968 the office of director became a presidential appointment on the advice and consent of the Senate. The Bureau was roundly criticized by the Senate in investigations of 1975–6, but continued to employ a large staff of agents throughout the USA and around the world.

Federal Reserve System, a US banking system. From the ending of the Bank of the United States in 1832 the monetary system of the United States was subject to repeated crises until, early in the 20th century, a national Monetary Commission was established. In 1912 this recommended a new single central bank. The Democrats feared this would strengthen the grip of Wall Street on the economy, and when they came to office under President W. Wilson* in 1913, Secretary of State William J. Bryan* devised the Federal Reserve System of twelve federal reserve banks in different parts of the country: Boston, New York, Philadelphia, Richmond, Atlanta, Dallas, Kansas City, St Louis, Chicago, Cleveland, Minneapolis, and San Francisco. They would be 'bankers' banks', owned by private banks, but with power to issue federal reserve notes as currency. The system would be supervised by a Federal Reserve Board consisting of the Secretary of the Treasury, a Controller of the Currency, and five members of a Board of Governors appointed by the President. It began to operate in 1914 and did much to stabilize banking and the economy, although thousands of small banks continued to operate and the Federal Reserve System was unable to prevent the 1929 Stock Market Crash* turning into the Great Depression*. The latter precipitated a reorganization, as part of the New Deal*, with the Banking Act of 1933. This reorganized the central Board of Governors, who would have control over interest rates, reserve requirements, and all activities of the twelve federal reserve banks. The successes of the US post-war economy owe much to the skills of the Federal Reserve System.

Feminism, a movement demanding social and economic, as well as political, equality with men. It has been a powerful force throughout the 20th century. The 19th century had seen women gain some levels of equal opportunity in education and the professions, and in the 1920s writers such as Virginia Woolf of the Bloomsbury Group* began to advocate full social emancipation. Only after World War II, however, did feminism advance from being a fashionable cult to a popular movement, especially in the USA, where Women's Liberation* in the 1960s attacked the all-pervading sexism of Western society with 'its psychological and biological suppression' of the 'second sex'. Socialist feminism criticizes the capitalist system for its demand that women play a specific role, in terms of child-bearing and employment. Radical feminists have argued for a totally separate female society, in which lesbian relationships are the norm and reproduction continued via sperm banks.

Fíanna Fáil (Gaelic; 'soldiers of destiny'), Irish political party. It was founded in 1926 by Eamon de Valera* with the aim of creating a united republican Ireland, politically and economically independent of Britain. It was opposed to the Anglo-Irish Treaty of 1921, which had recognized Northern Ireland. In 1932, under de Valera it won a general election, replacing Fine Gael* in government. It dominated Irish politics for much of the rest of the century. In 1973 it lost to an alliance of the Fine Gael and the Labour Party under Liam Cosgrave*, but was back in power under Charles Haughey* 1979–81, 1982, and again 1987.

Fifth column, a term coined during the Spanish Civil War. Four columns were advancing on Madrid; the fifth was described as consisting of collaborationists within the city. During World War II the term was applied to those in Europe willing and anxious to collaborate with advancing Nazi armies, for example in Norway.

Fifth French Republic. It was created by General de Gaulle* in 1958 and he became its first President. The President is elected for seven years, nominates his Prime Minister, may rule by decree and dissolve the Assemblée, and appeal to the people through referenda. The 1958 Constitution also created the French Community* of autonomous colonies, but this collapsed within two years when independent states were created. The Republic survived the tension of the Algerian War and saw steady growth of French prosperity during the 1970s and 1980s. Its critics have argued against the autocratic powers of the presidency and the reduction of influence of the Assemblée.

Figueiredo, João Baptista de (1918–), President of Brazil 1979–85. After a career in the Brazilian army, in which he rose to the rank of general, he was one of the planners of the 1964 military coup. He was chief of national intelligence until 1979, when the army junta appointed him President. It quickly emerged that he intended to restore democratic government. His regime saw the withdrawal of military political involvement, allowed a restoration of political parties, and redistributed some 47,000 acres of land to landless peasants. But social discontent gathered pace and Tancredo Neves emerged as leader of opposition groups. In 1985 Neves was the first civilian to be elected President of Brazil after twenty-one years of military rule.

Figueres (Ferrer), José (1906–90), President of Costa Rica 1953–8, 1970–4. Educated in Mexico and at the Massachusetts Institute of Technology, he became a coffee plantation owner, but also a mild socialist, being banished in 1942 for criticizing the government. He remained in Mexico until 1944, when he returned to take an active part in Costa Rican politics. He was elected President in 1953, when he instituted many social reforms and nationalized the banks. He was visiting professor at Harvard 1963–4. In his second term of office he faced considerable economic problems, but nevertheless established the Combined Institute of Social Assistance as part of his 'war on poverty'.

Fiji, a group of some 330 islands in the Melanesian archipelago, the largest being Vit Levu. The islands had become a British Crown Colony in 1874, when British settlers immediately began to import Indians to work on sugar plantations on an indenture system. Under British rule some advances were made in education provision, and craft centres and a teacher training college were established. Fiji was an important supply station for US forces in World War II, and this stimulated demands for self-government. These began to be met in 1948, but full independence was not gained until 1970, with a bicameral Parliament, House of Representatives and Senate, together with a powerful advisory body, the Great Council of Chiefs, presided over by the Minister of Foreign Affairs. By the 1950s 50.5 per cent of the population was of Indian descent and they dominated commercial life, while Fijians retained ownership of most of the land and control of the armed forces. The election of a government with an Indian majority in 1987 produced ethnic tensions and a military coup to restore Fijian control. As a result, Fiji withdrew from the Commonwealth of Nations, but civilian rule with guaranteed Melanesian control was restored in January 1990.

Film, the dominant medium of the 20th century for mass communication, entertainment, and artistic expression. In 1914 film meant family photographs and 'the moving pictures', which had been furthest developed in France, its birthplace, but had recently arrived in the Los Angeles suburb of Hollywood. Cinema developed rapidly as an art-form after World War I in Germany (Laszlo Nagy), the Soviet Union (Eisenstein), and France (René Clair). In the USA some classics were filmed during the early 1920s, particularly the comedies of Buster Keaton and Charlie Chaplin, but the attraction of a mass market, with its 21,000 movie houses, meant that the tycoons, with their big studios and complex organizations, soon concentrated on mass escapist entertainment. They were to continue to do so with 'talkies' in the 1930s and even after television had begun to replace the picture house, as it did rapidly after World War II. War stimulates a need for entertainment, but also for propaganda films, which continued to be shown on television networks for the next forty years. War also stimulates the need for training and education, and film steadily developed this function in school, college, and later through Britain's Open University. During the 1950s Hollywood suffered from anti-communist witch-hunts (spearheaded by Ronald

Reagan*) and the challenge of television. Its response to the latter, with colour and the wide screen, only partially held cinema audiences. It was in Italy, Sweden, France, Japan, and to a lesser extent Britain that high-quality cinema was concentrated in the first decades after the war. In India educational film (produced in no less than twelve languages) was balanced against mass entertainment cinema for a population not yet enslaved to the television screen or the video cassette. Satellite television transmission began in the 1970s, which further boosted Hollywood; film material was beamed across the world, with appropriate languages dubbed in, as for example the 'soap-opera' *Dallas*. It is arguable that the communications revolution which television represents precipitated a new political awareness. The mass political protests of 1989 starting in China and spreading to Eastern Europe were undoubtedly inspired by what television film had been able to communicate. Meanwhile the family photograph continued, now often on video, and individual masters of the still studio photograph came and went through the century.

Fine Gael (Gaelic; 'United Ireland'), Irish political party. Founded in 1922 as Cumann na nGaedheal, it changed its name in 1933. It originated among Sinn Fein* supporters of the Anglo-Irish Treaty of December 1921, which had created the Irish Free State. Its leader from then until his retirement in 1944 was William T. Cosgrave*. In 1948 it replaced Fíanna Fáil in power, as the dominant element in a coalition government led by John Costello. It was this government which declared Ireland an independent republic in 1949. Since then Fine Gael has been intermittently in power, but has required coalition support to remain so. It has always advocated the concept of a united Ireland, but to be achieved peacefully.

Finland, a Baltic state. It had been annexed by Russia in 1807 as a grand duchy. Attempts to impose the Russian language and military conscription were unpopular, and the Russian Revolution of 1917 offered the opportunity for national assertion. Independence was achieved in 1919 under Marshal Mannerheim* and a democratic, republican constitution introduced. In 1920 Finland joined the League of Nations, which achieved one of its few successes in resolving the Åland Islands* dispute. After the Nazi–Soviet Pact of 1939 the country was invaded in the Finnish–Russian War* (1939–40). Surrender entailed a considerable loss of territory (Karelia and Petsamo). When Germany invaded the Soviet Union in 1941 the Finns sought to regain these territories by fighting on the Axis side. They again capitulated to the Soviet Union in 1944, burdened with a reparations* bill. In March 1945 Finland declared war against Germany. After the war President Juho Paasikivi (1946–56) established good relations with the Soviet Union, adopting a policy of neutrality in international affairs. Under Social Democrat governments this policy has been maintained, although an economic link with Norway and Sweden was formed in 1956 (the Nordic Council*). In 1961 Finland became an associate member of EFTA*. In 1975 it hosted the Helsinki Conference* under President Kekkonen.

Finnish–Russian War (1939–40) (the Winter War). The Finnish government under President Mannerheim* had rejected Russian demands for bases and frontier revision. In November 1939 Soviet armies attacked on three fronts. At first the Finns' superior skill in manœuvring on skis on frozen lakes, across the Gulf of Finland, and in the forests of their country kept the Soviet forces at bay. After fifteen weeks of fierce fighting, however, Soviet troops breached the 'Mannerheim Line' and on 12 March 1940 Finland was forced to accept peace on Stalin's terms, ceding its eastern territories and the port of Viipuri (Vyborg).

Fisher, Andrew (1862–1928), Prime Minister of Australia 1908–9, 1910–13, 1914–15. Having been a member of the Queensland Legislative Assembly, he was an active proponent of federation and became leader of the Labor group in the House of Representatives in 1907. His governments strengthened social welfare and created the Commonwealth Bank. Prime Minister at the beginning of World War I, he clashed with the bellicose W. M. Hughes* and found the post too stressful. He resigned in 1915 and was Australian High Commissioner in London 1916–21.

Fisher, John Arbuthnot, 1st Baron of Kilverstone (1841–1920), British First Sea Lord 1904–10, 1914–15. Born in Ceylon, he joined the Royal Navy in 1854 and rose steadily through its ranks. He became an early enthusiast for the submarine and by 1899 was Commander-in-Chief in the Mediterranean. As First Sea Lord he was directly responsible for the political decision to strengthen the navy by the implementation of the *Dreadnought* building programme in response to German naval expansion. He supported Winston Churchill's plan for the Gallipoli* expedition, but resented the withdrawal of ships from the North Sea for use in the Dardanelles, since he planned instead a landing in Pomerania. His relations with Churchill became strained, and in May 1915 he resigned. While First Sea Lord he had commissioned the first oil-fired battleships.

Fisher, Sir Norman Fenwick Warren (1879–1948), British civil servant. Born in London and educated at Winchester and Oxford, he entered the British Civil Service in 1903, working with the Inland Revenue. Warren Fisher moved to the Treasury in 1919, as Permanent Secretary and head of the Civil Service, an office he held until his retirement in 1939. As such he was responsible for all recruitment and discipline of civil servants. He insisted that at all times they should show 'integrity, fearlessness and independence of thought and utterance', and be 'utterly impervious to ministerial political pressures'. He was responsible for re-creating a service of the highest calibre, after a period of low morale during World War I.

Fiume (Rijeka), a north Adriatic port. It had been a principal port of the Austro-Hungarian Empire and in 1919 both Italy and the emerging kingdom of Yugoslavia claimed it. A plan to make it a free city was abandoned after D'Annunzio* seized it. In 1924 Mussolini's Italy incorporated most of the city, leaving the small harbour of Susak to Yugoslavia. In 1947 it was ceded to Yugoslavia by the Treaty of Paris*.

FLN *see* FRONT DE LIBÉRATION NATIONALE.

Foch, Ferdinand (1851–1929), marshal of France. Born in the Pyrenees, son of a civil servant, he became a professional soldier, was commissioned, and fought in the Franco-German war 1870–1. He became an artillery expert, and in 1914 he fought in the Battle of the Marne* and was then co-ordinator of Allied forces protecting the Channel ports. He commanded the French troops in the Battle of the Somme (July 1916). He was famous for insisting that constant attack was the sole recipe for victory; by now he was a marshal of France. Mutual dislike and lack of Anglo-French co-operation between the generals had prevented effective concerted action until the German offensive of March 1918. Clemenceau then persuaded the British to accept him as Allied Commander-in-Chief. His July offensive was successful, and in August he began the attack which was to lead to the German surrender at Compiègne on 11 November. He failed to persuade the Paris Peace Conference* to impose tougher terms on Germany and retired.

Food and Agriculture Organization (FAO), a specialized agency of the UN. It was established in October 1945, with a headquarters in Rome. It is one of the largest and has been one of the most successful UN agencies. Through its World Food Programme it has collected and disseminated facts and statistics, given information and technical advice on improvements to food distribution, and provided important technical advice to Third World countries on increasing agricultural production, fisheries, and forestry.

Forbes, George William (1869–1947), Prime Minister of New Zealand 1930–5. A successful farmer, he first entered Parliament in 1908 as a member of the Liberal Party. When this declined after World War I he assisted Joseph Ward* in forming the United Party, which won the election of 1928. He succeeded Ward as leader of the party and Prime Minister in 1930, and won the election of 1931 in coalition with J. G. Coates* and the Reform Party*. As Prime Minister he attempted to meet the Depression by balancing the budget and paying benefits only to men working on labour-intensive public works schemes, for example hydroelectric installations. Although this policy of retrenchment helped him in the 1931 election, he was swept from office in 1935 by the Labour Party on a policy of expansion. Next year he succeeded in amalgamating the United and Reform parties to form the National Party*. He resigned from party leadership in 1938.

Ford, Gerald Rudolph (1913–), President of the USA 1974–7. Born in Nebraska, he graduated at Michigan University and Yale Law School before serving in the US navy. He built a successful law practice in Michigan, which he represented in Congress 1948–73. In 1973 President Nixon chose him to succeed Spiro Agnew as Vice-President, on the latter's disgrace for financial malpractice. On Nixon's own disgrace and resignation he at once became President. He continued Nixon's attempts to control inflation with some success, although unemployment continued to rise. In 1975 North Vietnam successfully annexed South Vietnam, the US involvement ending, with an airlift of some 237,000 troops and refugees. Egypt and Israel were helped to settle a territorial dispute by Secretary of State Henry Kissinger*. Ford won Republican nomination in 1976, but was defeated by the Democrat candidate Jimmy Carter*. After leaving office he returned to work at Michigan University and has held various business appointments.

Ford, Henry (1863–1947), US industrialist. Son of Irish immigrants, he left school in Michigan at 14 and was apprenticed as a machinist in Detroit. He worked for the Edison Company and then formed the Detroit Automobile Company. In 1903 he founded the Ford Motor Company in Detroit, which produced the classic Model T car in 1908. Adapting the mass-production techniques of the conveyor-belt and assembly-line to car production, by 1913 he was able to make great numbers of cars very cheaply. At a time when the average wage in manufacturing was $11 a week, he was paying his employees $5 a day and turning out one 'Tin Lizzie' every three minutes. In World War I he became a leading producer of aeroplanes, tanks, ambulances, and submarine chasers. In the early 1920s one car in two throughout the world was a Ford Model T. World War II saw him once more converting his factories to the production of war material. Among his philanthropic legacies is the Ford Foundation (established in 1936), the largest philanthropic trust in the world.

Foster, William Zebulon (1881–1961), US politician. Born in Massachusetts and apprenticed to a sculptor, he went to sea for three years and then worked in the chemical industry, in mining, lumber farming, and the building industry. He joined the

Industrial Workers of the World* in 1909, as well as the American Federation of Labor*, leading a successful Chicago packing-house wage movement in 1918. He took part in a bloody national steel strike in 1919, when he was denounced in the Red Scare*. In 1921 he joined the newly formed US Communist Party, having founded a Trade Union Educational League. He ran as Communist presidential candidate in 1924, 1928, and 1932. In 1945, after discrediting Earl Browder, the party's wartime leader, he became party chairman. In 1948 he was indicted for subversive activity, but there was no trial. He remained steadfastly loyal to Stalin's leadership and he died in Moscow.

Four Modernizations. The need to modernize Chinese agriculture, industry, national defence, and science and technology had been implied in a speech by Mao Zedong* in 1963, but during the Cultural Revolution* ideology came to be considered more important than economic development. After Mao's death (1976), the defeat of the Gang of Four*, and the emergence of Deng Xiaoping* as the key figure in the Communist Party, these Four Modernizations began to take priority. Training of scientists, engineers, and managers, and the reform of agriculture by the 'responsibility system' (i.e. the transfer of management power from the commune* to the individual), are key examples.

Fourth French Republic (1946–58). In 1944 General de Gaulle established a provisional French government, while protracted debate took place for a new constitution. This was finally agreed in December 1946, de Gaulle having lost his case for a strong presidential republic. A National Assembly and a second chamber, the Council of the Republic, were established, and many Third Republic politicians reappeared, as did many of its structural weaknesses. No less than twenty-three governments were formed within its twelve years. It suffered the humiliation of the Dien Bien Phu* disaster and the outbreak of the Algerian crisis. Its most memorable Prime Minister was perhaps Pierre Mendès-France*, while Robert Schuman's* work as Foreign Minister 1948–53 was largely responsible for the establishment of the ECSC*. In 1958, with civil war threatened by the OAS* over the Algerian crisis, the Fourth Republic collapsed.

Fourteen Points, peace programme by US President Woodrow Wilson*. In his address to the US Congress on 8 January 1918 the President put this programme forward: freedom of the seas, equality of trade conditions, reduction of armaments, adjustment of colonial claims, evacuation by Germany of Russian territory and Belgium, the return to France of Alsace-Lorraine, recognition of nationalist aspirations in Eastern and Central Europe, freedom for subject peoples in the Turkish Empire, independence for Poland, the establishment of a 'general association of nations'. Accepted with some reluctance by the Allies, they became the basis for peace negotiations at the Paris Peace Conference*, although France and Britain demanded a lot more from Germany.

Franchet d'Esperey, Louis Felix (1858–1942), marshal of France. Born in Algeria, he was commissioned into the French army and served in Indo-China and in China. In 1914 he was commanding general of the French 5th Army, and in September halted the German advance in the first Battle of the Marne*. He was less successful in the second battle in March 1918, and was then sent to Salonika, from where he launched a campaign against Bulgaria. He was preparing to advance into Central Europe when the war ended. He served in North Africa after the war, and on retirement entered politics, but he refused to join Marshal Pétain in his establishment of the Vichy* regime.

Francis Joseph (1830–1916), Habsburg Emperor of Austria 1848–1916, King of Hungary 1867–1916. He succeeded to the throne of Austria amid the revolutions of 1848, and this made him fundamentally conservative. In 1867, with the formation of the Dual Monarchy, he became also King of Hungary. He opposed all federalist demands from his Slav peoples, believing in the virtues of an efficient, centralized bureaucracy. He suffered many personal tragedies: his brother Maximilian was shot in Mexico; his son committed suicide; his wife was assassinated in 1898; and his nephew was assassinated at Sarajevo*. This last event only strengthened his determination to resist Slav nationalism in Bosnia, which he had incorporated into the Empire in 1908. He was succeeded by his great nephew Charles*.

Franco (Bahamonde), Francisco (1892–1975), Spanish general and head of state 1939–75. Born in Galicia into a naval family, he graduated at the Toledo Military Academy in 1910. He fought against Abd al-Krim* and in 1927 was principal of the Saragossa Military Academy. A monarchist, he rose rapidly in his profession until Alfonso XIII* abdicated in 1931, when he became temporarily out of favour. But by 1935 he was Chief of the General Staff and in 1936 Governor of the Canary Islands. Elections in February 1936 returned a more left-wing government and the army prepared to revolt. At first he hesitated to join the conspiracy, but in July he led troops from Morocco into Spain to establish a Nationalist government at Burgos. After three years of savage civil war he was victorious and became dictator. In 1937 he had adopted the Falange* and banned all political opposition. During World War II he remained neutral, although sympathizing with the Axis* powers. His government was ostracized by the UN until, with the coming of the Cold War, his hostility towards Communism somewhat restored him to favour. Under pressure from the Catholic hierarchy his domestic policy became slightly more liberal, although the Catalan language remained banned. In 1969 he named Prince Juan Carlos, grandson of Alfonso XIII, not only as his successor, but as heir to the reconstituted Spanish throne. On his death Spain returned to a democratic system of government under a constitutional monarchy.

Franks, Oliver Shewell, Baron Franks of Headington (1905–   ), British diplomat and scholar. Born and educated in Bristol and at the Queen's College, Oxford, he became a Fellow of his college 1927–37 and then Professor of Philosophy at Glasgow University. A temporary civil servant during World War II, he became provost of his college 1946–8 and then British ambassador to the USA 1948–52. During the years 1964–6 he chaired an important Commission on the University of Oxford which made some far-reaching proposals. He has sat on numerous government committees, including one on official secrets and one on the Falkland Islands. Fellow of several Oxford and Cambridge colleges, he was also provost of Worcester College, Oxford, 1962–76.

Fraser, John Malcolm (1930–   ), Prime Minister of Australia 1975–83. Born and educated in Melbourne and at Oxford, he was elected to the federal Parliament in 1955 as a young Liberal. When Robert Menzies retired in 1966, he joined Harold Holt's* government as Minister for the Army, at a time of Australia's controversial Vietnam involvement. He went on to be Minister for Education and Science 1968–9 and Minister of Defence 1969–71 under Holt's successor John Gorton (1968–71). In

March 1975 he became Liberal leader and his Liberal/Country coalition won the election of December. He advocated increased state rights at home, including the emancipation of the Northern Territory*, and greater Australian involvement with its Asiatic neighbours abroad. In 1983 he lost the election to Bob Hawke* and the Labor Party.

Fraser, Peter (1884–1950), Prime Minister of New Zealand 1940–9. He had emigrated from Scotland in 1910 and immediately became involved in New Zealand dock politics, which resulted in a violent 'Red Fed' confrontation in 1912. He was imprisoned in 1916 for 'sedition', that is, opposition to conscription. On release he joined the Labour Party and was elected to Parliament in 1918. Becoming less radical, he rose to be deputy leader of the party in 1933 and joined the government of Michael Savage* in 1935, succeeding him in 1940. Much respected by Churchill, he worked closely with the latter and with President Roosevelt on Pacific strategies, as well as with Evatt*, the Australian Minister of External Affairs. He supported the latter at the San Francisco* UN Conference in his championship of the small nations. After the war he organized a gift of £10m. from the New Zealand people to Britain to assist post-war reconstruction. He held a life-long commitment to the need for state patronage of the arts and education, and the modern New Zealand education system is perhaps his most enduring monument.

Free French, the organization of exiled French men and women in World War II. It was founded in June 1940 and led by General de Gaulle*. Its headquarters were in London, where, apart from organizing forces to take part in military campaigns, and co-operating with the French Resistance*, it also constituted a pressure group that strove to represent French interests. In 1941 its French National Committee was formed and this eventually developed into a provisional government for liberated France. The Free French army in French Equatorial Africa, led by General Leclerc*, linked up with the British forces in Tripoli in 1943 after completing an epic 2,400 km. march across the Sahara from Lake Chad. A provisional Free French government was established in Algiers, moving to Paris in 1944.

Frei (Montalva), Eduardo (1911–82), President of Chile 1964–70. From a wealthy Catholic family, he graduated in law at the Chile Catholic University in 1932, becoming a leader of an anti-Fascist Catholic movement, the Falange Nacional. Professor of Labour Law at his university, he entered politics in 1945. In 1958 he helped to form the Partido Demócratica Cristiano (PDC). As leader of this party he won the 1964 presidential election on the ticket 'revolution in liberty'. The programmes he initiated included extension of education and social welfare, agrarian reform, and the 'Chileanization' of the copper industry, whose controlling interests had until then been held by US companies. He was a severe critic of the US Alliance for Progress*. His failure to check inflation and labour unrest, or to redistribute wealth, turned many of his supporters against him. He retired in 1970, but remained a strong critic of the Pinochet* regime.

FRELIMO (Frente de Libertaçao de Moçambique), African resistance movement founded in the early 1960s. In 1963 it sent recruits for guerrilla training in Algeria and Egypt, one of whom was Samora Machel*. Operations against Portuguese troops began in 1964, led by Eduardo Mondlane. By 1968 Machel and Mondlane could claim to control one-fifth of the country. A Portuguese resettlement programme (*aldeamentos*), together with improved social provision and public-works schemes, failed to satisfy the guerrillas, who were being armed with supplies from China,

Czechoslovakia, and the Soviet Union. Brutal Portuguese counter-terrorism made conciliation ever more difficult, but in 1974 the new Portuguese Prime Minister Mario Soares met Machel in Zambia. After negotiations lasting June–September the Lusaka Accords were signed, agreeing on a cease-fire and independence with effect from July 1975. This precipitated racial violence, with a right-wing White colonial 'Independent Front for the Continuation of Western Civilization' staging a short revolution. Many of the latter's members moved to South Africa, where they used their wealth to finance continued civil war within Mozambique.

French Community (1958–61), a short-lived association of France, its overseas territories and departments, and various emerging African republics formerly part of the French Empire. It was established by de Gaulle, and its members were allowed considerable autonomy, although denied control of higher education, currency, defence, or foreign affairs. The West African territory of Guinea* immediately rejected the scheme and became independent. Hostility elsewhere led to a revision of the Constitution in 1960 and by 1961 it had collapsed, although most members retained close, if informal, links with metropolitan France.

French Empire. The 19th century witnessed a rapid growth of the Empire. The conquest of Algeria began in 1830, while in the Far East Indo-China, New Caledonia, and French Polynesia were added. In the 'Scramble for Africa' Tunisia became a protectorate and by 1914 French Somaliland, Sahara (French West-Africa), Morocco, Dahomey, Senegal, Guinea, Ivory Coast, Congo, and Madagascar had been gained. After World War I Togoland and the Cameroons, former German colonies, became French mandates*, as did Syria and Lebanon from the Ottoman Empire. Defeat in World War II and short-lived post-war governments prevented urgent reforms. In Indo-China the VietMinh* established a Vietnamese republic in 1945, which France refused to recognize, putting forward the alternative of a French Union*. Open warfare 1946–54 ended with French capitulation at Dien Bien Phu* and the independence of Cambodia, Laos, and Vietnam, soon to be racked by further war. In Algeria almost the entire French army failed to quell an Arab rising in 1954. By 1958 de Gaulle had recognized that Algerian independence was inevitable, and it followed in 1962. The abortive concept of a French Union* of 1946 was followed by the idea of a French Community* in 1958. This itself had collapsed by 1961, when the process of decolonization was virtually complete, except in the Pacific.

French Equatorial Africa (1908–58). Formed through the efforts of the Franco-Italian empire-builder Savorgnan de Brazza (1852–1905), it was proclaimed in 1908 and centrally administered from Brazzaville. It included the former French colonies of Chad, Congo, Gabon, and Ubangi Shari (Central African Republic). To these was attached in 1920 the mandated territory of Cameroon*. In 1958 its constituents became autonomous republics within the French Community*. Next year member states formed a loose association called the Union of Central African Republics and in 1960 each became independent, although still loosely linked as part of the Brazzaville Bloc*.

French Foreign Legion, a French volunteer force consisting mostly of foreigners. Formed in Algiers in 1831, it fought in numerous 19th-century wars and in both world wars. In 1962 it was transferred to mainland France. No questions are asked about the origin or the past of its recruits, whose oath binds them absolutely to the regiment. Its motto is 'legio patria nostra' ('the legion is our fatherland').

French Indo-China. In 1893 France had linked Cambodia (colony 1884), Laos (protectorate 1893), Annam (protectorate 1884), Tongking (protectorate 1884), and Cochin-China (colony 1867), forming the Indo-Chinese Union. The royal houses of Laos, Cambodia, and Vietnam (Tongking and Annam) were retained within a federal system, with the Governor-General in Hanoi controlling finance and defence. Cochin-China, with its capital of Saigon, was administered directly by a French prefect. The French educational system was developed and a French university established in Hanoi. In 1905 the Association for the Modernization of Vietnam had been formed by Phan Boi Chau*, whose writings did much to develop nationalist movements such as the Vietnam Revolution Party (1925), the Revolutionary Party of Young Vietnam (1925), a more right-wing Vietnamese Nationalist Party (1927), and the Communist Party of Indo-China (1930), founded by Ho Chi Minh*. In that year the Vietnamese National Party was obliterated, following a small military revolt in Tongking. In the same year on 1 May there were open revolts among the peasantry in Cochin-China and Annam, where a local soviet was established for a short time. French retaliation in 1931 was massive, some 10,000 being killed and 50,000 deported. Although the economy flourished in the 1930s, with rubber plantations and rice production expanding, there was little industrialization. Most of the peasantry remained in poverty and the Communist Party gained in strength. In September 1940 the Japanese obtained military and commercial concessions from the Vichy French administration, with free use of ports and airfields, and the Japanese became a target of nationalist organizations. In 1943 Ho Chi Minh returned, to begin recruiting for the VietMinh*. On 9 March 1945 the Japanese ambassador Matsumoto gave Governor-General Decroux an ultimatum, which he ignored. Next day 750 French officials were imprisoned, of whom 400 died. Cambodia, Laos, and Vietnam proclaimed themselves independent. On 19 August Ho's VietMinh entered Hanoi and the Emperor Bao Dai* abdicated. The Communist Party of Indo-China was dissolved, to be replaced by the Lao Dong (Workers' Party of Vietnam). On 12 September British and French troops landed in Saigon. Violence broke out, there was a general strike, and thousands of Vietnamese were imprisoned. By now the VietMinh were strongly based in Cochin-China and General Leclerc*, who landed in October, initiated a ruthless policy of repression. In Hanoi the VietMinh government was at this stage broadly based, a majority being non-Communist (socialist, Catholic, etc.). On 6 March 1946 Ho signed a treaty whereby Vietnam, as 'a free state', would form part of an Indo-Chinese Federation (Vietnam, Laos, and Cambodia) within the French Union*. Vietnam agreed 'amicably to receive the French army' of not more than 15,000 Frenchmen. While Ho was in Paris for an inconclusive conference on this treaty, the French High Commissioner in Saigon proclaimed Cochin-China a separate republic. Ho returned to preside over an elected National Assembly which rejected this. It confirmed that the Democratic Republic of Vietnam was 'one and indivisible', the north (Tongking), the centre (Annam), and the south (Cochin-China). On 23 November French units arriving in Haiphong claimed they were being attacked. The city was bombarded and perhaps 20,000 died. The French Indo-China War* really dates from this incident. On 20 December French troops occupied Ho's residence in Hanoi and he fled, issuing a national appeal, 'never shall we be enslaved'. General Giap* had perhaps 60,000 men under arms when war broke out. In 1949 ex-Emperor Bao Dai* was invited to return to Saigon as head of state within 'the Union of Indo-China', but war in the north was to last until 1954.

French Indo-China War (1946–54). In November 1946 Haiphong was bombarded by French troops and in December there were popular uprisings in Hanoi,

Hue, and the principal towns of Tongking and north Annam. French troops occupied Hanoi, from which Ho Chi Minh* withdrew into the north Annam mountains—Viet Bac. French political strategy was to unite all anti-Communist nationalists under ex-Emperor Bao Dai*, while militarily preparing for a major assault on the VietMinh* mountain stronghold. In October 1947 they launched 'the greatest military action in French colonial history'—but it achieved nothing. Heavy tanks and artillery destroyed towns and villages, but the VietMinh guerrillas vanished into the jungle. From 1948 French control outside the Red River delta was minimal, while inside the area VietMinh steadily infiltrated the villages. In 1951 General Giap*, now aided by China, launched a frontal assault on the delta, but he was defeated. Next year General Salan's massive counter-attack also failed. By May 1953 his successor General Navarre had a force of half a million French and colonial troops, some of whom were deployed into central Annam and the Mekong delta in Cochin-China, where the VietMinh had infiltrated. In the north he launched a series of paratroop attacks, but they failed. By now the VietMinh was in alliance with the Pathet Lao* in north-east Laos, and his decision to build a massive bridgehead at the strategic point of Dien Bien Phu* proved a disaster. His troops were besieged and capitulated, thus ending the war and enabling Geneva negotiations to take place.

French Polynesia, a collection of some 130 Polynesian islands in five archipelagos: the Society Islands, the largest being Tahiti; Tuamotu; the Îles Gambier (Mangareva); the Marquesas; the Îles Tubuai (Austral Islands). Administered as the French colony of Oceania from 1914 until 1940, they elected to support General de Gaulle and the Free French* during World War II. After the war a strong movement for autonomy was led by Pouvanoa a Oopa, who was arrested in 1960 and exiled until 1971. In 1963 the French began a programme of nuclear testing on the island of Mururoa, but international pressure obliged them to abandon atmospheric in favour of underground tests here and on neighbouring Fangataufa. In 1983 a team of scientists claimed that uninhabited Mururoa was being left as 'a source of radioactive pollution for thousands of years' and that the explosions could damage the porous coral rock, releasing radioactivity into the sea and contaminating the food chain. This claim was refuted. As a French Overseas Territory it acquired a new constitution in 1977, which provided some political autonomy, but French control was retained over finance, foreign affairs, defence, and justice. In 1984 the islands gained further autonomy and their own flag, but pressure for full independence continued to build up, as did opposition to the testing programme on Mururoa, dockers' riots being suppressed by the military in November 1987.

French Union, a reorganization in 1946 of the French Empire* by the government of the Fourth Republic*, in order to implement the Brazzaville Declaration*. It proved unacceptable to both African and Indo-Chinese nationalists and was to be replaced in 1958 by the French Community*.

French West Africa. In the late 19th century France, determined to extend its interests inland from the West African coast, had to use considerable military force to overcome local Islamic African states. In 1895 the formation of Afrique Occidentale Française was proclaimed, although the process of pacification lasted another twenty years. It included what became the republics of Mauritania, Senegal, Mali, Burkina Faso, Guinea, the Côte d'Ivoire, Niger, and Benin. The German colony Togo was captured in World War I and partitioned as a mandate between France and Britain in 1920. French West Africa supported the Vichy* government 1940–2; it then transferred allegiance to the Free French*. In 1958 its members became autonomous republics within the French Community*, except Guinea, which opted, through a referendum, for immediate independence. All members gained full independence in 1960.

Freud, Sigmund (1856–1939), founder of psychoanalysis. After long clinical experience, by 1915 he had reached his theory of repression from the conscious to the unconscious, and of infant sexuality. He went on to develop techniques of psychoanalysis and to identify various neuroses, paving the way for later psychotherapy techniques. He lived and worked in Vienna until 1938, when he took refuge in London. By then he had earned an international reputation. Such concepts as id, ego and superego, regression, identification, and sublimation fundamentally altered the everyday perceptions of 20th-century man, and his work deeply affected literary and artistic culture. At the same time professionally he remains a highly controversial figure in psychiatric medicine.

Freyberg, Bernard Cyril, 1st Baron (1889–1963), New Zealand general. He emigrated to New Zealand with his parents in 1892 and saw distinguished service in World War I, being awarded the VC. A professional soldier in the British army between the wars, he returned to New Zealand in 1939 to command an Expeditionary Force. He was Commander-in-Chief of the unsuccessful Commonwealth expedition to Greece and Crete (1941) and went on to take an active part in the North African and Italian campaigns, in command of the New Zealand Division. From 1945 until 1952 he was Governor-General of New Zealand.

Front de Libération Nationale (FLN), Algerian radical Muslim independence movement and political party. Formed in 1954 by a fusion of three earlier nationalist groups, it was the political expression of the ALN (Armée de Libération Nationale), which fought the Algerian War* 1954–62. In spite of differences of opinion between military, political, and religious leaders, the movement hung together through the war and brought its imprisoned leader Ben Bella* to power as the first President of the Algerian Republic in 1962, following the Évian Agreements*. The principal policies of the party were independence, economic development in a socialist state, non- alignment, and brotherly love with other Arab states. It remained the only party within Algeria until 1989.

Frontier disputes, Latin America. Partly because so much of Central and South America remained unexplored at the beginning of the 20th century, there were a number of fierce frontier disputes through the century. A long-standing dispute between Peru, Brazil, and Bolivia was settled by President Leguía of Peru in 1910. A more intractable quarrel between Peru and Chile, left over from the War of the Pacific (1879–83), involved the provinces of Tacna and Arica with their rich mineral deposits. Leguía again reached a settlement in 1929, Tacna going to Peru and Arica with its nitrate deposits to Chile, Peru receiving an indemnity of $6m. The Leticia dispute was between Peru and Colombia over territory on the Upper Amazon, ceded to Colombia by a treaty of 1924, which Peru then challenged in 1932. This dispute was resolved by the League of Nations, which awarded the area to Colombia. The Chaco War* (1932–5) originated from a frontier dispute between Paraguay and Bolivia dating from the early 19th century. In 1941 Peru defeated Ecuador in another war over the Upper Amazon. At the Chapultepec Conference in Mexico City of 1945, twenty Latin American republics pledged themselves to avoid aggression against each other. This did not, however, prevent numerous further border disputes arising, for example Honduras and El Salvador in 1969 and Colombia and Venezuela in 1988.

Fuad I (1868–1936), King of Egypt 1923–36. Younger son of Khedive Ismail, who had sold the Egyptian Suez Canal shares to Britain in 1875, he became Sultan of Egypt in 1917, having been a keen supporter of University Education for Egypt. He was established as King by the British on 15 March 1923. His entourage was corrupt and his relations with both the British and the nationalists, represented by the Wafd Party*, were uneasy.

Fuchs, Klaus (1911–), physicist and spy. Born in Germany, he joined the German Communist Party while a student in 1932. He did doctoral research in physics at Bristol and Edinburgh, was interned early in World War II, and then released to become a member of the team developing the atomic bomb at Los Alamos, USA. He had been granted British citizenship, and on his return to Britain worked at the Atomic Energy Research Establishment at Harwell. From 1943 he was passing information about the atomic bomb and plans for a hydrogen bomb* to Soviet secret agents. In 1950 he was arrested on evidence obtained from confessed Communist agents in the USA. Fuchs pleaded guilty and was imprisoned. He was released in 1959 and became director of the Institute of Nuclear Physics in the German Democratic Republic, and also worked in the Soviet Union.

Fulani, a Muslim people of West Africa found in Cameroon, Chad, Guinea, Mali, Niger, and Nigeria. They probably migrated from west Sudan from the 13th century, bringing Islam with them. They established many kingdoms, such as Macina, Adamawa, and the Empire of Sokoto, which stretched far south into Yoruba* country and which was broken up by the British in 1900. Although they were originally a nomadic pastoral people speaking Fula, there are today many sedentary rural and urban Fulani, especially in Northern Nigeria, many of whom have adopted the Hausa* language. The social structure of the pastoral Fulani remains egalitarian, their cattle being used for dairy produce rather than for meat. Fula is still spoken through a vast area of West Africa.

Fulbright, James William (1905–) US Senator. Born in Montana, he was educated at Arkansas University and at Oxford, and went on to become a distinguished lawyer specializing in antitrust litigation. Elected to Congress in 1943, he moved to the Senate in 1945, where he remained until 1974. An opponent of isolationism as well as of McCarthyism*, he was chairman of the Senate Foreign Relations Committee from 1959 until 1974. He was an active critic of US policies during the 1960s and 1970s, criticizing the 1965 intervention in the Dominican Republic*, attacking particularly US involvement in the Vietnam War, and urging that Congress should have more control over the President's powers to make war. A Rhodes Scholar himself, in 1946 he sponsored the Fulbright Act, which provided federal funds for the exchange of students and teachers between the USA and other countries.

Fundamentalist churches, Protestant churches stressing particularly the literal truth of the Bible. Fundamentalism developed during the 1920s in opposition to the then 'modern' theological teachings. It has gained many adherents in the USA, especially among Baptist* groups, and is particularly noted for its total opposition to the theory of evolution, as shown in the J. T. Scopes* trial in Tennessee. Since World War II a number of fundamentalist preachers in the USA have achieved fame, and in some cases notoriety, through television, as leaders of the so-called 'Moral Majority' movement. As is the case with Islamic fundamentalism*, membership of fundamentalist churches is steadily increasing.

Futurism, an aesthetic movement which began in Paris in 1909 and particularly flourished in Italy. It aimed to revitalize literature, music, and the visual arts, rejecting the past and advocating emotional involvement in the dynamics of modern life; its painters were fascinated by movement and speed. Umberto Boccioni was a leading painter and sculptor and Antonio Sant'Elia an architect whose idea that buildings should be essentially impermanent greatly influenced 20th-century architecture. Futurists dominated early post-Revolution art in the Soviet Union, for example Vladimir Mayakovsky. Futurism 'blurred the frontiers' between different arts and developed the concept of art as a 'sensational event'.

g

Gabon, a west-coast country of Central Africa. In 1910 it became part of French Equatorial Africa*, the French exploiting its rare woods, gold, diamonds, manganese, and other minerals including oil. The colony became autonomous within the French Community* in 1958 and independent in 1960. With its rich natural resources it has had one of the fastest economic growth rates in Africa. After early political instability there was considerable support for the presidency of Omar Bongo, first elected in 1961 and re-elected four times since. In November 1990 Bongo's Gabonese Democratic Party won a general election, following a decision to restore multi-party politics.

Gaitskell, Hugh Todd Naylor (1906–63), British politician. Born in London and educated at Winchester and New College, Oxford, he taught economics at London University until he entered Parliament in 1945. He held several government posts dealing with economic affairs 1945–51, including the Chancellorship of the Exchequer 1950–1 at a time when the Korean War was necessitating unpopular fiscal measures. He was leader of the Labour Party 1955–63. He represented the moderate right wing of the party as opposed to the 'Bevanites', believing strongly in the welfare legislation of 1945–51, but also in the need for a balance between private and state finance. He was particularly vigorous in his opposition to the government over the Suez Crisis*. A strong supporter of NATO*, he resisted the unilateralists within his own party, whom he defeated in 1961, two years before his sudden death. He was a declared opponent to the proposed British membership of the European Common Market.

Gallipoli campaign, World War I (April 1915–January 1916). Following the failure of the naval Dardanelles* venture to force a passage to the Black Sea, it was decided to land troops, whose task would be to advance all the way to Constantinople. On 25 April a French corps landed on the Asia Minor coast, British and Australian troops under General Ian Hamilton landed at Helles, at the tip of the Gallipoli peninsula, and Australian and New Zealand troops, under General William Birdwood, landed at Ari Burnu (Anzac Cove). Security had been so lax that the Turks under the German General Liman von Sanders were expecting them, and offered strong resistance. British and Anzac* troops desperately tried to advance up the steep hillsides, on the ridge of which Mustafa Kemal* commanded a strong contingent armed with German guns. On 6 August a further landing was made at Suvla Bay, to the west. This was a much more suitable site, but the Allies did not press their advantage. After enjoying two days' swimming they found Mustafa had moved into the hills above them. After intense and bitter fighting a trench-line was established around the peninsula similar to that on the Western Front. During December and January 1916 the Allies withdrew, with minimum casualties. Some 36,000 Allied troops were killed, including

8,587 Australians. A further 82,000 were wounded or made prisoner. Winston Churchill*, who was largely responsible for the idea of the campaign, was blamed for its failure.

Galtieri, Leopold Fortunato (1926–), President of Argentina 1981–2. Born in Buenos Aires, he attended the Argentine Military College and was commissioned in 1945. By 1967 he held the rank of colonel and was subdirector of the Military Engineering Training School. He reached the rank of lieutenant-general in 1979 and was Commander-in-Chief of the army 1979–82 and a member of the junta which appointed him President in 1981. His decision to invade the Falkland Islands* was in part in frustration over the continued failure of UN talks, but more specifically as a gamble to win national support for his regime. When the gamble failed and his forces were defeated, opposition groups successfully re-established democracy in Argentina. He was sentenced to twelve years in prison 'for negligence' and for 'starting and losing' the Falklands War.

Gambia, The. In 1843 the British declared the area around the West African town of Bathurst a British colony and in 1893 extended their control up the River Gambia into the interior, which became a British protectorate. In 1954 the colony and protectorate were united under a new constitution granting a substantial element of home rule. The Gambia became an independent member of the Commonwealth in 1965 and a republic in 1970, with Sir Dawda Kairaba Jawara the country's first President, re-elected in 1982. In 1982 also the Gambia and Senegal formed the Confederation of Senegambia for defence, economic, and foreign policy purposes, but this had lapsed by 1989.

Gamelin, Maurice Gustave (1872–1978), French general. A graduate of the French Military Academy of St-Cyr, he had become Chief of Staff to General Joffre* by 1914. As such he helped to plan the successful first Battle of the Marne*. He served with distinction through the rest of the war, holding various appointments. In 1940 he was appointed Commander-in-Chief of the Allied forces, and he was totally unprepared for the German thrust through the Ardennes which resulted in disaster for his forces. In mid-May he was replaced by General Weygand. He was tried at Riom in 1943 for having caused France's defeat. He was imprisoned, but released by the Allies in 1945.

Gandhi, Indira (1917–1984), Prime Minister of India 1966–77, 1979–84. The daughter of Jawaharlal Nehru*, she was educated in India and in England. In 1939 she joined the Indian National Congress* Party and spent over a year in prison for her wartime activities. Her first years in politics were spent as an aide to her father and in 1942 she married a member of the Congress Party, Feroze Gandhi. She served as president of Congress 1959–60. She became Minister for Broadcasting and Information in 1964 in the cabinet of Lal Bahadur Shastri*, whom she succeeded as Prime Minister in 1966. In her first years in office she was engaged in a protracted struggle with the older leadership of the Congress, but she won the support of the left wing and by 1970 was fully in control. After the successful Indo-Pakistan War* of 1971 her popularity stood high, but it waned during the 1970s. When threatened with a court case for illegal electoral activities, she declared a state of emergency (1975–7), suppressed the Press, and imprisoned her political rivals, governing India dictatorially and assisted by favourites such as her younger son Sanjay. After her defeat in the 1977 elections by Morarji Desai* and the Janata Party, her career seemed finished; but in 1979 her faction of the Congress, the Congress (I) Party, was re-elected to power and she ruled as Prime Minister until her assassination in 1984 by a Sikh extremist. Her son Rajiv Gandhi* succeeded her as Prime Minister.

Gandhi, Mohandas Karamchand (1869–1948), Indian national and spiritual leader. Born into a family of Hindu Bania (merchant) caste in the state of Porbandar in west India, he was educated in India and England, where he was called to the English Bar. He practised law briefly in Bombay before moving to South Africa in 1893, where he became a successful lawyer. During the years 1907–14 he developed his technique of satyagraha* (truth force), leading a campaign of protest against the Transvaal government's policy of discrimination against Indian settlers. He returned to India in 1915, where he joined the Indian National Congress* and organized campaigns for workers' and peasants' rights (1916–18). Subsequently he led nationalists in a series of confrontations with the British Raj*, including the agitation against the Rowlatt Act* of 1919. From 1920 he dominated the Indian National Congress, supporting the Khilafat movement* and initiating the decision to promote a movement of non-co-operation with the British, suffering frequent imprisonment. The Salt March* (1930) was followed by further campaigns, including that of civil disobedience 1931–4, individual satyagraha 1940–1, and that of civil disobedience and the 'Quit India' movement 1942–5. As independence for India drew near, he co-operated with the British, in spite of his opposition to the partition of the subcontinent. In political terms Gandhi's main achievement was to turn the small, upper middle-class Indian National Congress into a mass movement, by adopting a political style calculated to appeal to the ordinary Hindu (using symbols such as the loin-cloth and the spinning wheel), and by creating a network of alliances with political brokers at all levels. Imprisoned 1922, 1930, 1933, and 1942–4, he regularly used the hunger-strike as part of his civil disobedience tactics. In 1931 he took part in the London Round Table Conference*, but only briefly. In social and economic terms, he stressed simplicity and self-reliance, encouraging Indian villages to boycott British goods. At the same time he stressed the need to elevate the status of the Untouchables (harijans) and for communal harmony, always deploring violence. In intellectual terms his emphasis was upon the force of truth and non-violence (ahimsa), in the struggle against evil. His acceptance of partition and concern over the treatment of Muslims in India made him enemies among extremist Hindus. One of these, Nathuram Godse, assassinated him in Delhi. Widely revered before and after his death, he was known as the Mahatma (Great Soul).

Gandhi, Rajiv (1944–91), Prime Minister of India 1984–9. Son of Indira Gandhi*, he was educated in India and England, qualifying as an engineer and gaining a civilian pilot's licence. After the death of his younger brother Sanjay, he was persuaded to enter politics, becoming president of the Congress Party in 1983 and succeeding his mother in October 1984 as Prime Minister. Regarded as a non-abrasive personality, he won early support for his domestic and foreign policies, but his government was accused of corruption and he failed to win the 1989 general election.

Gang of Four. During the Cultural Revolution* four radical Chinese leaders rose to prominence: Jiang Qing* (Mao Zedong's* fourth wife), Wang Hongwen, Yao Wenyuan, and Zhang Chungqiao, with a power-base in Shanghai. After the Tenth Party Congress of 1973 they occupied powerful positions in the Politburo, which they retained until after Mao's death. In October 1976 they were denounced for allegedly plotting to seize power and attacked by Chairman Hua Guofeng* in the Eleventh Party Congress of August 1977. In 1980 they were found guilty of 'plotting against the state', being blamed for the excesses of the Cultural Revolution.

Garvey, Marcus Moziah (1887–1940), Jamaican Black leader. Born and educated in Jamaica, he travelled to London in 1912, and in 1914, back in Jamaica, he founded the Universal Negro Improvement and Conservation Association (UNICA), which aimed to encourage racial pride and Black unity with a slogan 'Africa for the Africans at Home and Abroad'. Failing to attract a following, in 1916 he moved to Harlem, New York, promising repatriation of US Blacks to a new Black-governed republic in Africa (to be created out of former German colonies). By 1919 UNICA claimed a membership of 2 million, with a newspaper *Negro World*, a Negro Factories Corporation, and a shipping line *Black Star Line*. In 1920 Garvey hosted an international convention with a parade of 50,000 through the streets of Harlem. He established four branches of UNICA in South Africa in 1921, which were to encourage growth of a Black movement there in the 1930s. Deeply resented by William Du Bois*, Garvey's followers clashed with more moderate Blacks in the 1920s. He himself, although personally honest and sincere, mismanaged his movement's finances, and he was convicted of attempted fraud in 1923. After serving part of a prison sentence, he was deported to Jamaica, and he died in obscurity in London.

GATT *see* GENERAL AGREEMENT ON TRADE AND TARIFFS.

Gay activism, a movement which began on the west coast of the USA in the mid-1960s. It demanded greater tolerance of homosexual relationships and an end to landlord and employment discrimination against homosexual couples. It made greatest headway in San Francisco and New York, in both of which cities it became politically significant. At the same time attitudes towards homosexuals became more tolerant in other Western countries, as in Britain, for example, where homosexual acts between males ceased to be criminal offences in 1967. In the 1980s, however, with the spread of AIDS*, a wave of antipathy swept both the USA and Britain, where anti-homosexual legislation was enacted.

Gaza Strip, a narrow strip of the Mediterranean coast around the town of Gaza. Between 1920 and 1947 the strip was part of the territory of Palestine mandated to Britain. It was allocated by the UN Commission of 1947 to Arabs, and was occupied by Egyptian troops and administered by Egypt until 1956. In 1955 Israeli troops had raided the strip and left thirty-six Egyptians dead, and in the Suez War (1956) it was quickly occupied by the Israeli army. In 1957 Israel handed it over to a UN task-force, and for the next ten years it was administered by Egypt, receiving a 'constitution' from President Nasser* in 1962. In 1967 the Israeli army quickly reoccupied it (Six-Day War*). Since then it has been held as 'occupied territory'. Some half-million Palestine refugees had crowded into the area in the years of Egyptian control, establishing eight refugee camps. In December 1987, the *Intifadah** movement began, partly as a protest against the appalling conditions within these camps.

Gdansk (Danzig), a great medieval port at the mouth of the River Vistula. From 1793 to 1919 it was incorporated into East Prussia. The Treaty of Versailles made it a 'Free City', giving Poland access to the sea by a 'Polish Corridor'*. It was an attempt by its Nazi gauleiter, Albert Forster, to reunite it with Germany which unleashed World War II. After its capture by the Red Army in 1945 its German population was expelled. It was here in its great shipyards in 1980 that the Polish trade union Solidarity* staged its first strikes, which ultimately led to the collapse of the Communist regime in Poland.

Geddes Axe (1922), British economic proposals. Sir Eric Geddes, director-general of munitions and military railways during World War I, was asked by Lloyd George in 1921 to chair a committee to make recommendations for reducing public expenditure. His report, dubbed 'The Geddes Axe', recommended savings of £75m., with deep cuts in military spending, the axing of a wide range of government posts, and reductions in salary for police, teachers, and others. The National Insurance scheme was to be limited, some government departments to be dissolved, including transport, and the newly begun post-14 continuation education scheme abandoned. The report caused an outcry and not all the recommendations were accepted. In the end the budget reduction in 1922 amounted to £64m. Education was hard hit by the Axe, but the main economies were in the armed services.

Geisel, Ernesto (1907–), President of Brazil 1974–9. Born and educated in Brazil, he had begun a career in the Brazilian army when in 1930 as a young officer he helped Getúlio Vargas* to seize power. He gave enthusiastic support to the latter's *Estado Novo* for the next fifteen years. From 1945 onwards however his career centred on the army, and in 1964 he formed part of the military junta which took over the government of Brazil. In 1974, as a general, he was appointed President. He began a process of liberalization and democratization and sought to develop domestic industries of mining, agriculture, and forestry. Faced by opposition from his military colleagues, however, he abandoned ideas for a restoration of civilian government, but he supported General Figueiredo* as his successor.

General Agreement on Tariffs and Trade (GATT), a specialized agency of the UN. It began as an agreement between twenty-three nations in 1948 to promote international trade by removing trade barriers, laying down maximum tariff rates, and providing a forum for the discussion of trading policies. It was seen at first as a temporary expedient, but by 1960 it had become a permanent specialized agency with a Secretariat and an International Trade Center, which issues full world-trade statistics. Based in Geneva, it had ninety-six nation members by 1990. Its basic aim remains the liberalization of world trade, the ultimate elimination of quotas, and the establishment of uniform customs regulations in order to raise global living standards. In 1965 it was agreed that developing nations would not be expected to offer reciprocity in negotiations with developed countries. In spite of all its efforts, a 'rising tide' of protectionism was detected as a feature of world trade in the late 1980s, with particular tensions developing between the USA, Japan, and the European Community. Regular international conferences are organized, for example the 'Uruguay Round', begun in 1986 at a conference at Punta del Este by the Trade Negotiations Committee. This four-year series of meetings, dominated by agricultural issues, was suspended in deadlock in December 1990.

General Assembly, United Nations. It consists of five delegates from each of the 159 members of the UN. It is responsible for the UN budget and may make recommendations on any UN issue. It holds a regular session each year, but emergency sessions may be convened at any time, if peace is threatened. Two such sessions were held in 1956 over the Soviet intervention in Hungary and the Anglo-French venture at Suez. In the early years the USA and Western powers could always muster a majority, but as more Third World states emerged, the balance shifted to favour the less-developed nations. The Assembly has hence been particularly opposed to any remnants of colonialism.

General Motors, the world's largest industrial corporation. Founded as General Motors Company in 1908 by William C. Durant in Flint, Mich., as an amalgamation of several automobile companies, its early products were the Buick and Cadillac motor-cars, and Reliance and Rapid trucks. The Company was incorporated in

1916 and expanded rapidly during World War I. During the 1920s it pioneered the marketing concept that the automobile should reflect status, power, and sex appeal, as an extension of the owner's personality. It took over subsidiaries overseas—Vauxhall in England, Opel in Germany—and diversified into smaller cars, diesel engines, glass, refrigerators, data-processing, and aerospace, marketing satellites at $5,000m. each. It pioneered high-technology manufacturing processes and consistently invested heavily in research in its laboratories and colleges, including the University of Michigan-Flint. It early accepted the principle of unionization of its work-force, working closely with the United Automobile Workers, founded in 1935.

General Strike, British (4–12 May 1926). After World War I the British coal-mining industry faced severe problems, with overseas competition, declining reserves, and ageing equipment. There was pressure from the National Mineworkers' Union for the industry to be nationalized. This was rejected by a Royal Commission led by Herbert Samuel, in March 1926, which also recommended lower wages, but improved working conditions. The mine-owners demanded longer hours, as well as lower wages, and the Miners' leader A. J. Cook called for the support of other unions—transport, printers, engineers, builders, industrial workers—which the TUC* endorsed. Baldwin's cabinet was split between those who favoured negotiation and those for confrontation. The latter, of whom Winston Churchill* was an enthusiast, prevailed and a General Strike followed. The government used troops and volunteers to maintain basic transport, gas, and electricity, and Churchill edited a special *Gazette* newspaper. After nine days the TUC accepted defeat, but the Miners stayed on strike until November, families suffering considerable privation. In July 1927 Baldwin's government passed a Trade Disputes Act, making general strikes illegal and union affiliation to the Labour Party a matter of individual 'opting-in'. This act was to be repealed in 1946, the year that the coal industry was nationalized.

Genetic engineering, laboratory modification of hereditary features. Following the isolation of DNA by Watson and Crick in 1953, a series of discoveries during the 1960s and 1970s led to a greater understanding of genetic functions and of genetic transcription and translation. The term genetic engineering spans a whole range of bio-medical and microbiological activity for cloning and gene manipulation: the use of sperm banks and artificial insemination; the development of new seed strains with disease resistance; the creation of totally new organisms, the first being a virus patented in the USA in 1986 as a vaccine against rabies. In agriculture patenting became a delicate issue: does the new strain belong to the country of origin or to the research company? Meanwhile the potential moral dilemmas for bio-technologists 'playing God' have been increasingly debated.

Geneva Agreements (20 July 1954), a series of agreements on French Indo-China. They followed a two-month conference convened by the wartime Allies to settle the future of both Indo-China and Korea. This was attended by the foreign ministers of France, Britain, the USA, the Soviet Union, and the People's Republic of China, together with representatives from Laos, Cambodia, North Vietnam (the Democratic Republic of Vietnam), and South Vietnam (the Republic of Vietnam). All troops were to be withdrawn from Laos* and Cambodia*, where elections were to be held. In Vietnam a cease-fire line was to be drawn along the 17th parallel; but this line was in no way to be seen as a political frontier and there was to be a general election before July 1956. This never took place and the two regimes, the Republic of Vietnam under Ngo Dinh Diem* in Saigon and the Democratic Republic under Ho Chi Minh* in

Hanoi, both claimed to represent the whole nation. Only in 1975 were they to be united. The Agreements mark the end of effective French control in Indo-China. On Korea the Geneva Conference made little progress and the Panmunjom talks to end the Korean War* continued.

Geneva Conventions, a series of international agreements on the treatment of victims of war. The first, in 1864, the result of the work of Henri Dunant, founder of the Red Cross, laid down basic rules for the treatment of wounded soldiers and prisoners of war, as well as for the protection of medical personnel. It was amended and extended in the second convention of 1906, which insisted that facilities for treatment of sick and wounded must be available. World War I led to the third convention in 1929, by which the USA and forty-six other nations accepted further rules on the treatment and rights of prisoners of war. Because of the failure of some belligerents to abide by these during World War II, a fourth convention in 1949 extended and codified existing provisions for victims: the sick and wounded; shipwrecked sailors; prisoners of war; and civilians in territory occupied by an enemy. By no means all of these were abided by in later 20th-century wars.

Genro, Japanese elder statesmen. The Meiji Constitution of 1889 was written and introduced by a group of senior counsellors of the Emperor. These and their successors were given the status of *genro*, or elder statesman, and as such attended cabinet meetings and wielded much influence. One of their last survivors was Kimmochi Saionji (1849–1940). The *genro* were an important factor in Japanese development in the early 20th century, but their influence declined in the 1920s and the institution was abolished in 1940.

Gentile, Giovanni (1875–1944), Italian philosopher and politician. Educated at Palermo University, in 1918 he became Professor of History at Rome University. He early on joined the Fascist Party, for whom he tried to produce a philosophical justification. He served in Mussolini's first government as Minister of Education and was later Minister of Science and Arts and also president of the Florence Academy of Sciences and Arts. He was shot by Italian partisans.

George V (1865–1936), King of Great Britain and Ireland (from 1920 Northern Ireland) and Dependencies overseas, Emperor of India, 1910–36. Son of Edward VII, he served in the Royal Navy 1877–92. A strong constitutionalist, in 1911 he insisted that a general election precede any reform of the House of Lords, but his promise to create peers led to the passage of the Parliament Act. In 1914 he brought together party leaders at the Buckingham Palace Conference to discuss Irish Home Rule*, thus possibly preventing civil war. His acceptance of Stanley Baldwin* as Conservative Prime Minister in 1923 contained an element of personal choice. The Silver Jubilee celebrations of 1935 revealed the extent of his popularity.

George VI (1895–1952), King of Great Britain and Northern Ireland and Dependencies overseas 1936–52, Emperor of India 1936–47. He served in the Royal Navy 1909–17 and then in the newly formed Royal Air Force 1918–19. He succeeded his brother Edward VIII after the Abdication Crisis*. His preference for Halifax* as Prime Minister in 1940 rather than Winston Churchill had no effect, but he strongly supported Churchill throughout the war. Likewise he gave his support to Clement Attlee* and his government in the policy of granting Indian independence. He and his wife Elizabeth Bowes-Lyon will be remembered for sustaining public morale during the German bombing offensive 1940–1. As King he had overcome a serious speech impediment and made regular and successful radio broadcasts. He was succeeded by his elder daughter Elizabeth II*.

German Democratic Republic. It emerged in 1949 from the Soviet zone of occupied Germany. Its frontier with Poland on the Oder–Neisse line, agreed at the Potsdam Conference*, was confirmed by the Treaty of Zgorzelec in 1950. In the first five years the Republic had to pay heavy reparations* to the Soviet Union, and in 1953 Soviet troops were used to put down protest riots. In 1954 it was proclaimed a sovereign state and in 1955 became a founder member of the Warsaw Pact*, forming the National People's Army and hosting both nuclear and conventional arms from the Soviet Union. Walter Ulbricht* was general secretary of the Communist Socialist Unity Party 1946–71 and chairman of the Council of State 1960–71. Admission to the UN in 1973, was followed by universal recognition as a sovereign state. Although economic recovery from the war was slower than in the West, East Germany succeeded in establishing a stronger industrial base than most of its fellow members of the Warsaw Pact. But its highly bureaucratic, centralized systems of control steadily atrophied, corruption spread down from the top, and its secret police, the Stasi, became ever more ruthless. Early in 1988 demonstrations began demanding easier travel and civil rights, and by September 1989 a series of demonstrations was taking place, mostly in Leipzig and Berlin, demanding political freedom for a reform group New Forum, which was strongly backed by the Lutheran Church. Mass protests in November led to the opening of the Berlin Wall* 9–10 November and the surrender of the monopoly of power by the Socialist Unity Party. As the level of corruption within that party was revealed, the Politburo was forced to resign. On 4 December the Stasi offices in Leipzig were stormed. New parties, including a Christian Democrat and Christian Social Union, emerged, and moves were made towards reunification with the Federal Republic. The first free elections since 1933 were held in March 1990 and a coalition government under Lothar de Maizière pledged to achieve union by December 1990. In fact this was achieved on 3 October, when the Republic ceased to exist, being absorbed by the Federal Republic*.

German Second Empire (Reich). It was formed by Bismarck in 1871 as a union of twenty-five German states under the Hohenzollern King of Prussia, now Emperor William I. (The First Reich had been the Holy Roman Empire, which ended in 1806.) An alliance was formed with the Austro-Hungarian Empire in 1879 and German economic investment took place in south-east Europe. By 1900 the growth of German industry was making it the greatest industrial power in Europe, and inevitably the search for new markets led to tension with the other colonial powers. The expansion of the German navy under Admiral von Tirpitz* led to rivalry with Britain, while competition with France in Africa led to two crises over Morocco. The assassination at Sarajevo* caught the Empire unawares. After some debate it was decided that the 1879 alliance with Austria-Hungary must be honoured, even if this meant war against Russia and France. During World War I most German colonies were conquered and the Versailles Treaty* stripped Germany of its overseas empire, which became mandated territories administered by the victorious powers on behalf of the League of Nations. On 29 September 1918 Ludendorff* advised the Kaiser that military victory seemed impossible. The liberal Prince Max of Baden was invited to form a government (1 October), which for the first time included socialists. On 29 October the fleet in Kiel mutinied and on 7 November revolution broke out in Munich. On 9 November Prince Max announced that the Emperor had abdicated and that he was handing over the Chancellorship to Friedrich Ebert*. The Empire was at an end.

German Third Empire (Reich). The Weimar Republic* collapsed in 1934, when Hitler announced the Third Reich. This proved to be one of the most radical reversals of democracy. Germany became a national rather than a federal state; non-Aryans and opponents of the regime were removed from the administration and all political parties other than the National Socialists liquidated; the judicial system was made subservient to the Nazi regime, with secret trials meting out summary executions. Anti-Semitism* was formalized by the Nuremberg Laws* and concentration camps* established for dissidents. Both Protestant and Catholic churches were attacked and the Hitler Youth* movement formed to indoctrinate the young. Most industrial workers were won over by the rapid end to unemployment through rearmament and other public spending; the small farmer on the other hand found himself more securely tied to the land. A four-year plan in 1936 aimed at self-sufficiency of the economy in the event of war. In 1935 the Saar Basin had been restored following a plebiscite, and compulsory military service was then introduced. In 1936 the Rhineland was reoccupied in contravention of the Locarno Treaty*. Anschluss* with Austria in 1938 was followed by the annexation of German Sudetenland*. During the spring of 1939 demands were made on the Polish 'Corridor'* to link West and East Prussia, and the Nazi–Soviet Pact* arranged. On 1 September 1939 German forces invaded Poland without a declaration of war, and on 3 September Britain and France declared themselves at war with Germany. During 1940–1 military occupation of most of continental Europe occurred. By December 1941, when the USA declared war, the Empire was at its height: Austria, German Poland, and the protectorates of Bohemia and Moravia, together with Luxembourg, were incorporated into Greater Germany; Norway, Denmark, The Netherlands, Belgium, half of France, the Baltic states, much of Russia, Greece, and Yugoslavia (which was broken up) were occupied; Italy, Hungary, Romania, Bulgaria, Finland, Slovakia, Vichy France, with North Africa, were 'allied'. Total mobilization was introduced in 1942, with forced labour recruited from occupied countries. Armaments production was increased in spite of heavy bombing of industrial and civilian targets. Under Himmler* the SS had immense power. After 1943 the German armies were fighting rearguard actions and by May 1945 the Empire lay in ruins.

Germany, Federal Republic of. Until October 1990 it consisted of the eleven *Länder* (states) of West Germany: Baden-Württemberg, Bavaria (the largest), Bremen, Hamburg, Hesse, Lower Saxony, North Rhine Westphalia (the most wealthy), Rhine Palatinate, Saarland, West Berlin. With unification on 3 October five more *Länder* were added: Brandenburg, Micklenburg, Saxony, Saxony-Anhalt, and Thuringia. Each *Land* has wide powers over domestic policies. The Republic was created in 1949 from the French and US/British zones of occupation and became a sovereign state in 1955, when ambassadors were exchanged with world powers including the Soviet Union. Konrad Adenauer*, its first Chancellor, always refused to recognize the German Democratic Republic*. The challenge of rebuilding shattered cities and of absorbing millions of refugees was successfully met, as was that of re-creating systems of social welfare and health provision. The federal Diet or Bundestag is elected by universal suffrage, using an elaborate system of proportional representation. The Federal Council or Bundesrat consists of members of *Länder* governments. The Federal Republic joined NATO in 1955, when both army and air force were reconstituted. In 1957 it signed the Treaties of Rome*, becoming a founder member of the EEC* in 1958. In that year a crisis developed, when the Soviet Union demanded the withdrawal of Western troops, and again in 1961, when the Kremlin authorized the erection of the Berlin Wall*. In 1971 the policy of *Ostpolitik** resulted in treaties with the Soviet Union and Poland, one of mutual recognition with the Democratic Republic in 1972, a treaty with Czechoslovakia in 1973, and membership of the United Nations. Although the pace of economic growth slackened, the economy remained one of the strongest in the world, under a stable democratic regime. In June 1990 so-called 'two-plus-four' talks began with the GDR, the USA, the USSR, France, and Britain, and agreement was reached on economic and monetary union with the GDR, which took place on 1 July. Later in the month in Stavropol President Gorbachev* conceded that a united Germany could be a member of NATO* and a Treaty of Unification was signed on 31 August. On 12 September the two-plus-four talks ended with a Treaty on the Final Settlement. With reunification Berlin again became the capital and in elections on 14 October the CDU won control of four of the five new *Länder*, only Brandenburg voting for the SDP.

Gestapo, the Nazi secret police or *Geheime Staatspolizei*. In 1933 Hermann Goering reorganized the Prussian plain-clothed political police as the Gestapo. In 1934 control passed to Himmler*, who headed the SS* and who had already restructured the police in other states. The Gestapo became a national force, a subdivision of the SS. In 1939 it was merged with the SD or *Sicherheitsdienst* (Security Service), the intelligence branch of the SS, in a Reich Security Central Office under Reinhard Heydrich*. The powers of the Gestapo were immense: any person suspected of disloyalty could be summarily executed. The Gestapo controlled the concentration camps and extended its activities not only into occupied territories but also into many allied countries.

Ghana, a West African state. Britain had established the colony of the Gold Coast in 1874. Wars against the Asante* followed in 1896 and 1900, after which the Asante region was incorporated into the colony. After 1920 economic growth, based on mining and cocoa-farming, combined with high standards of education, produced a sophisticated people soon demanding home rule. Achimota College, one of the first secondary schools for Africa, was founded in Accra in 1925. Following World War II there were serious riots in Accra (1948) leading to constitutional discussions. In 1957 the Gold Coast and British Togoland were combined to become the independent Republic of Ghana, under the leadership of Kwame Nkrumah*, the first British African colony to be granted independence. Rapid modernization, including the completion of a huge hydroelectric scheme on the River Volta, followed. However, economic problems and resentment over political repression and mismanagement led to the overthrow of Nkrumah by the army in 1966. In 1969 a return to civilian government under Kofi Busia failed and further military rule under General Acheampong followed. After a new coup, a group of junior officers under Flight-Lieutenant Jerry Rawlings* took power in 1979, executing three former heads of government and five other senior officers, and installing civilian rule. When this failed, Rawlings again seized power (December 1982), suspending the Constitution and establishing a Provisional National Defence Council. During the 1980s with IMF* support Ghana regained some economic and political stability.

Gheorghiu-Dej, Gheorghe (1901–65), President of Romania 1961–5. Born in Birlap, he worked as a tram-driver, joining the Romanian Communist Party in 1929. He was imprisoned in Dej for the eleven years 1933–44, during which time he managed to build a secret reputation. He emerged to form a short-lived coalition with King Michael in 1945. When Romania became a Communist republic in 1948 he was its virtual ruler, although not formally President until 1961. He negotiated the withdrawal of all Soviet troops in 1958 and was a close ally of Khrushchev*, advocating 'peaceful coexistence'.

Giap, Nguyen Vo (1912–), Vietnamese general and statesman. Born in central Vietnam and educated at Hue and the French University of Hanoi, he joined the

Revolutionary Party of Young Vietnam in 1926 while still a student. Arrested in 1930, he gained parole and in 1933 joined the Communist Party of Indo-China. He took a doctorate in economics in 1937 and then became a history teacher until arrested in 1939. He escaped to China, but his wife died in prison and his sister was guillotined. Back in Vietnam he began organizing resistance to the Japanese in 1941, linking with Ho Chi Minh in 1943, when they formed the VietMinh*. As commander of the latter 1946–54 he directed the struggle against the French, opposing three distinguished French soldiers, Marshal de Lattre de Tassigny, General Salan, and General Navarre. His victory against the latter at Dien Bien Phu* was decisive. He became Minister of Defence of the Democratic Republic of Vietnam and Commander-in-Chief of its armed forces, which he steadily built up in the years 1955–65. He personally directed the Tet Offensive* of 1968 and the fighting which continued until the capture of Saigon in 1975. He was Minister of Defence and deputy Prime Minister of the Socialist Republic of Vietnam from 1976 until he retired in 1982. As a military commander he had successfully defeated the armies of three great powers, Japan, France, and the USA.

Gibraltar. Having passed to Britain in 1713 it was important as a Royal Navy base in both world wars. It remains a British dependency, with self-government introduced in 1964. In 1967 its inhabitants—Italians, Portuguese, Maltese, and Spanish in origin—voted in a referendum to remain British. In 1969 the border was closed by Spain, which claims possession of Gibraltar. Talks opened in Madrid in 1977 and restrictions were eased in 1980. Following an agreement signed in Brussels in 1984, the border was reopened in 1985.

Gierek, Edward (1913–), Polish statesman, first secretary of the Communist Party 1970–80. Born in Austrian Silesia, where his coal-miner father died in a disaster, in 1923 he emigrated with his mother to France, where he joined the French Communist Party. During World War II he operated in Belgium among groups of Polish underground Resistance fighters. He returned in 1948 to Silesia, now part of Poland, and was appointed to the Polish Communist Party Politburo. In 1970 he succeeded Gomulka* as first secretary, pledged to a 're- evaluation of economic policies'. In 1980, when the Solidarity* protests began, he himself was ousted and replaced by Stanislaw Kania. He was expelled from the party and interned 1981–2.

Gil Robles (y Quinoñes), José María (1898–1980), Spanish politician. Born in Salamanca and a strong Catholic, he trained and practised as a lawyer and in 1931 became leader of the Catholic Acción Party in the new Spanish Republic. In 1932 he broadened this to form the Confederación Española de Derechas Autónomas (CEDA). When this joined the government in 1934 there were left-wing riots, as CEDA was thought to favour restoration of the monarchy. Nevertheless he held office for a short while as Minister of War in 1935. CEDA was increasingly conservative and, following the 1936 election, was the largest single party in the Cortes, in which the left-wing Popular Front had a majority. Many of his supporters now moved to the more right-wing Falange*. He did not actively support the Nationalists in the Civil War, but lived in exile in Portugal 1936–53 and 1962–4. He returned to politics in 1975, helping to form a Spanish Christian Democrat Party.

Giolitti, Giovanni (1842–1928), Prime Minister of Italy 1892–3, 1903–5, 1906–9, 1911–14, 1920–1. From Piedmont and educated in Turin, he dominated Italian politics for twenty years, having first been elected to the Italian Parliament in 1882. A skilled conciliator and liberal, he introduced universal male suffrage for those over 30 in 1912 and did not oppose the growth of trade unionism. His government fought the successful Italo-Turkish war of 1911–12, gaining the Dodecanese and Libya; but he argued against Italy joining the Anglo-French–Russian *entente*, his sympathies lying more with the Central powers. He resigned from office in 1914 over a general strike. From then on his political position moved left and he formulated a programme of state socialism. Faced however in 1920 with massive unemployment and communist agitation on the left and nationalist support for the illegal activities of D'Annunzio* on the right, his government proved impotent and collapsed. Italy was now in a state of virtual civil war for a year, with rioting in most northern cities. In October 1922 the King established Mussolini* as Prime Minister. The ageing Giolitti at first gave his support, but later became critical of Mussolini's dictatorial government.

Giscard d'Estaing, Valéry (1926–   ), President of France 1974–81. Born in Koblenz, where his father was a director of finance, he was educated in Paris and then himself became an inspector of finances before entering politics. He was elected as a Gaullist Deputy in 1956, and in 1962 became Minister of Finance under Prime Minister Pompidou*, whom he succeeded as President in 1974. His coalition of 'Gaullists' and 'Giscardists' worked smoothly for a while, but tension developed as the ambitions of the Gaullist leader Jacques Chirac grew. Giscard's reputation was damaged by allegations of diamond gifts from Jean Bedel Bokassa*, whose regime in the Central African Republic he helped to topple in 1979. He was defeated in the presidential election of 1981 by François Mitterrand*.

Glubb, Sir John Bagot (Pasha) (1897–1986), British soldier. After serving as a young officer through World War I he was appointed as a liaison, political officer in the newly created Iraq. In 1932 he was transferred to Transjordan*, where he took command of the Arab Legion*, building it into a formidable military force. He created a fortification system on the Palestine border with a mobile, highly disciplined 'desert patrol' always available. In World War II his troops took part in operations in Iraq, Syria, and North Africa. He was dismissed in 1956 by King Hussein of Jordan, and returned to England, becoming a lieutenant-general.

Goa, a small coastal state of India. From 1510 it was ruled as a Portuguese colony until 1961 when, together with Daman and the island of Diu, it became a Union Territory, following a brief military action. It became a state of the Union in May 1987, with three elected Members of Parliament.

Goebbels, Joseph Paul (1897–1945). A graduate of Heidelberg University, where he gained a Ph.D. in 1920, he had been rejected by the army because of a club foot, of which he was very self-conscious. He joined the Nazi Party and in 1926 attached himself to Hitler*. He spent the next four years building up the party in Berlin. He founded a newspaper, *Der Angriff* ('The Attack'), and exploited his considerable gifts of oratory and manipulation of the masses to further the Nazi cause. His brilliantly staged parades and mass meetings helped Hitler to power. In 1933 Hitler appointed him Enlightenment and Propaganda Minister, with full control over the Press, radio, and all aspects of culture. After Germany's defeat at Stalingrad he was given special powers as Reich Commissioner for Total Mobilization. Faced with the advancing Soviet armies, he committed suicide in Berlin with Hitler, first killing his wife and six children.

Goering, Hermann Wilhelm (1893–1946), German Nazi politician. Born in Bavaria, he gained the highest award for bravery in the air during World War I. After a brief spell at Munich University he joined the Nazi Party in 1922, commanded their Brownshirt* military organization, and fled the country after being wounded in Hitler's unsuccessful Munich *putsch** in 1923. After living in Sweden he returned to Germany to be elected to the Reichstag in 1928, becoming its president (Speaker) in 1932. On the establishment of the Third Reich he became Prime Minister of Prussia and Reich Air Minister. During the next five years he steadily built up the *Luftwaffe*, while also carrying out diplomatic missions, after being appointed in 1937 Minister for Foreign Affairs. In 1938 he became Hitler's deputy. During the early years of the war he increasingly interfered with military and air policies and this, together with his known and increasing dependency on narcotics, gained enemies. By 1943 he had been deprived of virtually all authority and in 1945 he was dismissed. He surrendered to US forces in Austria in May 1945, after an unauthorized last-minute attempt to make peace. Sentenced to death at the Nuremberg Trials*, he committed suicide in his cell by swallowing poison.

Goldman, Emma (1869–1940), international anarchist. Born in Lithuania, she emigrated to the USA in 1889, landing with $5 in her pocket. As a seamstress she quickly became involved with anarchist socialists in Rochester, NY. In 1893 she was imprisoned in New York City for 'incitement to riot', and then co-operated with her life-long companion Alexander Berkman in anarchist activities, the latter being imprisoned for attempted assassination. In 1919, at the time of the Red Scare*, she was again imprisoned with Berkman for criticizing conscription, and was then deported to Russia. Disillusioned by Soviet Russia, where she was shocked by police methods and by anti-Semitism, she moved to England, where she was given support by such literary intellectuals as H. G. Wells, Rebecca West, and Bertrand Russell*. She travelled and lectured in Canada, France, and Spain, and died in Toronto still a convinced anarchist.

Gold standard, a currency system in which the basic monetary unit of a country was defined in terms of a fixed quantity of gold. Paper money was convertible into gold on demand; gold could be freely imported and exported; and exchange rates between countries were determined by their currency values in gold. In 1821 Britain had been the first country to introduce an official gold standard. It was followed some fifty years later by France, Germany, and the USA. By 1900 all major countries had adopted the gold standard. Its main advantage was thought to be that any country's trade deficit would be automatically corrected. Most countries were unable to maintain the gold standard during World War I. Britain returned to it in 1925, but abandoned it again in 1931 because of the Great Depression*; other countries were soon obliged to follow its example.

Gómez, Juan Vicente (1864–1935), President of Venezuela 1908–35. A full-blooded Indian from the Andes, he joined the private army of Cipriano Castro in 1899, and in 1908 seized power and announced himself President of Venezuela. During his twenty-seven-year rule he established an absolute dictatorship. The foreign investment that he attracted enabled him to build extensive railroads, highways, and other public works. Rich petroleum discoveries in the Lake Maracaibo Basin in 1918 provided a budgetary surplus which not only enabled him to pay off foreign debt, but also assured him a favourable reputation abroad. Because of the brutal nature of his regime, however, this reputation was not shared at home, where he became the richest man in Latin America. When he died in office in 1935 the city of Caracas marched in celebration.

Gomulka, Wladyslaw (1905–82), Polish statesman. Born in what was Austrian Silesia, he became an active trade unionist and joined the Polish Communist Party on its foundation in 1919. He was imprisoned many times in the years 1920–39. A distinguished member of the Polish Resistance forces, he became secretary-general of the underground Polish Workers' Party during the years of German occupation. A member of the post-war government of National Unity under Boleslaw Bierut*, he was made responsible for those territories annexed from Germany. He was dismissed, however, in 1948 for 'nationalist' fervour unacceptable to Moscow. He was imprisoned 1951–5. In August 1956, after the Twentieth Soviet Party Congress*, he was restored to the party. In October, at a time when Polish and Soviet frontier troops had exchanged fire in the wake of riots in Poznań and a workers' trial, he was appointed first secretary of the Polish party, on the intervention of Khrushchev*. He helped to sustain a degree of post-Stalin liberalism, allowing greater personal and religious liberty, and slowing down farm collectivization. He resigned in 1970, following popular disturbances in Gdansk* and elsewhere, being replaced by Edward Gierek*.

Good Neighbour policy, a term used to describe the US Latin America policy of Secretary of State Cordell Hull* during the first term of President F. D. Roosevelt*. It was implemented by the withdrawal of US marines from such countries as Haiti and Nicaragua and the abrogation of the Platt Amendment, which had given the US government a quasi-protectorate over Cuba. A Conference in Montevideo in 1933 declared that 'no state has the right to intervene in the internal or external affairs of another'.

Gorbachev, Mikhail Sergeyevich (1931–), Executive President of the Soviet Union 1990– . Born in Privolnoye in the north Caucasus of peasant family, he was educated at Moscow University in law and then at the Stavropol Agricultural Institute. He began work in 1946 as a machine operator and joined the Communist Party in 1952. He quickly rose through its ranks, becoming organizer of the Stavropol City Party and a member of the Stavropol Territorial Production Board of Collective and State Farms. In 1978 he was appointed Secretary for Agriculture in the Soviet Union and in 1980 was appointed to the Politburo. In 1985 he succeeded Konstantin Chernenko as general secretary of the Communist Party of the Soviet Union. He was a member of the Supreme Soviet of the Russian Federation 1979–90. His efforts to carry out *perestroika* (economic and social restructuring of Soviet society) led to a gradual liberalization, and to the introduction of high technology in a number of fields other than the space industry. However, it met with greater resistance from the bureaucratic hierarchy than he had hoped. Together with his Foreign Minister Eduard Shevardnadze (1928–), he negotiated the 1988 INF Treaty* with the USA, to reduce nuclear forces in Europe. On the domestic front he encouraged a greater degree of *glasnost*, or openness and accountability, so that for the first time the Russian people were told of the enormities of the crimes against humanity perpetrated by the Stalinist regime. He struggled to reduce the excesses of alcohol abuse, the inefficiency, and the corruption which were stifling the Russian economy. During 1989 he faced increasing tensions as the satellite states of Eastern Europe collapsed and the Berlin Wall* was opened. These tensions increased during 1990, as the constituent republics of the Soviet Union one after another sought greater independence, and ethnic violence grew in the south, where Islamic Fundamentalism* appealed to many. The largest of these republics, that of Russia*, elected the radical Boris Yeltsin as its President and was insistent on gaining greater sovereignty. Gorbachev's policies were strongly criticized within the party by conservative opponents, while some senior officers in the army seemed restless. Nevertheless the CPSU re-elected him its general secretary and an extraordinary meeting of the Congress of People's Deputies in March 1990 elected him executive President under a revised constitution. His presidential Council was charged with the task of facilitating rapid reform of the Soviet economy. At the same time he accepted that the political monopoly of the CPSU was ended. In August 1990 he negotiated with Chancellor Kohl that a united Germany could remain in a reformed NATO.

Gottwald, Klement (1896–1953), President of Czechoslovakia 1948–53. Born in Moravia, he fought in the Austrian Imperial Army in World War I and became a founder member of the Czechoslovak Communist Party in 1921 and its general secretary in 1927. Following the Munich Agreement* of 1938, he went to Moscow, where he remained during the war. He returned to Prague in 1944 and was Prime Minister in a coalition government with Beneš 1946–8. After the Communist coup of 1948 he succeeded Beneš* as President. He dominated the country through purges, forced labour camps, and show trials, culminating in the Slansky Trial* and execution of his leading Communist rivals in 1952. He acquiesced in Stalin's plan of reducing Czechoslovakia's industries to satellite-status within the Comecon* economy.

Gowon, Yakubu (1934–) head of state of Nigeria 1966–74. A graduate of Zaria Government College, he joined the Nigerian army, attended Sandhurst and Camberley Staff College, and was a colonel by 1966. At the time of the July 1966 coup he was invited to lead a new government, which proceeded to divide the country into twelve federated states, to replace the four federal regions. The eastern Ibo region rejected the Constitution and declared itself the state of Biafra, under General Ojukwu*. After the Biafran War*, Gowon helped to reconcile the defeated Ibo people. He was largely responsible for the creation of the Economic Community of West African States*. By 1975 he was emerging as an international figure, but he was deposed by the army in July 1975. He studied at the University of Warwick and returned to Nigeria in 1983.

Gramsci, Antonio (1891–1937), Italian political theorist. Born in Sardinia and educated at Cagliari and in Turin, as a student he became increasingly interested in socialist theory. During the Turin general strike of 1920 he argued that the workers, instead of striking, should seize and run the factories, as a starting-point for a new Communist society. In that year he helped to found the Italian Communist Party. He was arrested in 1926 and imprisoned by Mussolini for eleven years, which caused his death. His *Prison Notebooks*, published in 1947, became influential during the 1970s and 1980s, seeking to harmonize Marxist historical materialism with more traditional metaphysical philosophy.

Graziani, Rudolfo (1882–1955), Italian general. A professional soldier, he served in Eritrea before World War I, when he fought on the Macedonian front. In 1935 he became Governor of Italian Somaliland and, after his success in the war against Ethiopia*, he was made Viceroy of Italian East Africa. In 1939 he commanded the Italian forces in Libya which in September 1940 attacked Egypt. He reached Sidi Barrâni, where he was totally routed by the British under Wavell*. He resigned his command in 1941, but joined Mussolini's rump government 1943–5 as Defence Minister. Arrested in 1945 for collaboration with the Germans after the Italian armistice of 1943, he was imprisoned, but released in 1950. He became active in the Italian neo-Fascist Party, but to no effect.

Greater Antilles, a chain of mostly large islands running roughly east–west in the Caribbean. They consist of Cuba*, Jamaica*, Hispaniola (shared by Haiti* and the Dominican Republic*), Puerto Rico*, and a few offshore and small islands, notably the Cayman Islands (a British colony and tax-haven), Curaçao, and Aruba, part of the Netherlands Antilles, which have retained the status of self-governing colonies.

Greater East Asian Co-prosperity Sphere, an attempt by Japan to achieve political and economic supremacy throughout its vast Empire, following the conquests of 1940–2. At first a number of nationalist groups, particularly in Burma and Indonesia, co-operated with the Japanese in the belief that this would enable them to become free from European colonialism. In Indo-China the Vichy French authorities themselves collaborated, while the nationalist movement the VietMinh* turned more towards Communism. In Malaya the Communist Malayan People's Anti-Japanese Army was opposed to occupation from the start, as was the Philippine Hukbalahap movement*. In Burma and Indonesia early co-operation gradually turned to opposition, as the use of forced labour and the requisitioning of supplies produced resentment. By the end of the war the sphere had become an object of hatred and ridicule, often referred to as a Sphere of Co-poverty and Co-suffering.

Great Leap Forward. In 1958 Mao Zedong* threw all his authority behind an attempt to galvanize the Chinese economy. He wanted the development of 'backyard' steel furnaces in the countryside and higher production quotas in the factories, calling upon people's patriotism and enthusiasm for socialist ideals. In agriculture communes* became almost universal. Massive increases in the quantity of production were announced, but quality and distribution posed serious problems. Disastrous harvests and poor products tended to discredit the Leap, and its most important advocate Mao suffered something of a setback until the mid-1960s. The Cultural Revolution* can be seen partly as his attempt to reintroduce radical, socialist policies.

Great Purge *see* YEZHOVSHCHINA.

Greece. In 1862 a Danish prince, William, had been installed as King, taking the title George I of the Hellenes (1863–1913). He was succeeded by his son George II. A military coup established a republic 1924–35. George II was restored by Ioannis Metaxas*, but was to flee into exile in 1941. In 1940 Italian troops invaded from Albania. They were defeated, but in April 1941 German forces attacked from Yugoslavia. A force of some 60,000 Commonwealth troops from North Africa was landed, but it was heavily outnumbered by the advancing Germans, whom it tried to hold at Mount Olympus. Falling back to Crete, 80 per cent of the force was successfully evacuated. With Greece now occupied by the Germans, ELAS* Resistance fighters fought determinedly. After 1944 bitter fighting developed between these and royalists seeking to restore the monarchy. ELAS, under Markos Vafiades, continued a civil war until 1949, when they were defeated by Marshal Papagos* with US aid. The British had restored George II in 1946, and, with the help of the Marshall Aid* programme, recovery and reconstruction began in 1949, with Marshal Papagos becoming civilian Prime Minister in 1952. In 1967 the military coup of the 'Greek Colonels'* took place, with Constantine II* fleeing to Rome, the monarchy being abolished in 1973. A civilian republic was established in 1974. In 1981 Greece joined the European Community and in that year, following a general election, Andreas Papandreou became the first socialist Prime Minister, remaining in office until 1989.

'Greek Colonels', a term used to describe the military regime in Greece 1967–74. Its two strongest men, George Papadopoulos and Stylianos Pattakos, were in fact generals. The former was Prime Minister and Minister of Defence and the latter Minister of the Interior. Their regime was exceptionally oppressive, all political opponents being imprisoned and torture extensively used. Censorship of the arts and the media was tyrannical. In 1973 Papadopoulos proclaimed Greece a republic and himself its President. Student riots in Athens in November 1973 were brutally suppressed.

Papadopoulos resigned in favour of his colleague General Gizikis. The latter promoted a bid in Cyprus to depose Makarios* and proclaim *enosis*. Greece was on the verge of civil war and in July 1974 Gizikis called on Constantine Karamanlis* to return from exile and restore constitutional government. Twenty officers of the regime were placed on trial and Papadopoulos and Pattakos were both imprisoned for life.

Green movement, emergence of political organizations/pressure groups devoted to ecological and environmental conservation. They had been preceded in the 1960s by such groups as the beatniks, the hippies, the yippies, all of whom rejected the affluent consumer society and sought alternative, simpler life-styles. During the 1970s a number of 'green' political parties emerged, for example in France in 1974 (Les Verts) and in Britain in 1975 (Ecology Party). The oil crisis of 1973 helped to boost ecological awareness, which looked for energy conservation, cleaner air and water, greater controls over food production, improved public transport, recycling of waste. Such an awareness was no prerogative of the political left; for example the President of the German Christian Democrats Herbert Gruhl in 1975 published a best-seller *A Planet Plundered.* Other Green parties emerged, in Switzerland 1978, Italy 1980, Ireland 1981, Austria 1982. In West Germany, however, the 1979 NATO 'dual-track' decision to install medium-range nuclear missiles helped to precipitate the Greens to the left. In January 1980, with the SDP in disarray, a new radical federal party Die Grünen was formed, whose policies were to be centred around the proposition 'people are more important than money'. It embraced both the anti-nuclear peace movement and the ecological arguments, together with women's and gay rights and sympathy for the Third World. In 1983 it gained 5.6 per cent of the federal poll and was thus represented in the Bundestag, until support collapsed in 1990. Yet the ecological question remained as urgent and pressing a global challenge as ever.

Greenpeace, an international pressure group whose Council is now based in Amsterdam. It was founded in 1971 in British Columbia to organize protests against US nuclear testing at Amchitka Island, Alaska. It developed rapidly during the next two decades into an international movement, highly skilled in the use of the media and in fund-raising, seeking to prevent environmental abuse and to preserve endangered species. It has had numerous clashes with authority, always adopting direct, non-violent action. It has consistently campaigned against British governments for the dumping of toxic and radioactive waste; in 1985 its ship *Rainbow Warrior** was sunk by French intelligence agents in New Zealand.

Green Revolution, term used since the 1960s to describe the greatly increased agricultural productivity of developing countries, particularly in India, Pakistan, Thailand, and Indonesia. With research and development sponsored by the UN Food and Agriculture Organization*, improved soil fertility, pest control, increased mechanization, and new seed strains (especially wheat and rice) all contributed to a massive increase in food production. Concern began to develop however on pollutant effects of chemicals, together with their ever increasing costs. The UN International Rice Research Institute in the Philippines produced a rice variety IR-36 in 1976, which became one of the world's most widely grown crops in the 1980s.

Grenada, an island state in the Lesser Antilles*. First colonized by the French, it became a British colony in 1763. Universal adult suffrage was granted in 1950, when the United Labour Party, led by the trade union organizer Matthew Gairy, emerged. It became a member of the West Indies Federation* 1958–62. Following the break-up of the latter, the various Windward Islands sought separate independence.

This was gained by Grenada in 1974, when Gairy became Prime Minister. His government became corrupt and he was increasingly autocratic. He was deposed in a relatively bloodless coup in 1979 by Maurice Bishop*, leader of the left-wing New Jewel Movement, who proclaimed the People's Revolutionary Government (PRG). Bishop encouraged a policy of non-alignment, but he was overthrown in September 1983 and killed by army troops commanded by General Austin, who established a Revolutionary Military Council. Military intervention by the USA followed on 25 October, allegedly to pre-empt a Cuban plot, although few Cubans were ever captured. US troops left the island in December 1983, and a general election held in December 1984 was won by the veteran politician Herbert Blaize of the New National Party. Blaize died in office in December 1989, and Nicholas Braithwaite of the National Democratic Conference succeeded him.

Grey, Edward, 1st Viscount Grey of Falloden (1862–1933), British Foreign Secretary 1905–16. From a great Whig family, he was born in London and educated at Winchester and Balliol College, Oxford. He entered Parliament in 1885 and served under Lord Rosebery. During the Boer War he was a strong Liberal Imperialist. As Foreign Secretary he was responsible for the Triple *Entente* in 1907, bringing France, Britain, and Russia together, and for negotiating the end of the Balkan War 1912–13. In 1914 he successfully persuaded a reluctant British cabinet to go to war, because Germany had violated Belgian neutrality. He was British ambassador to the USA after the war and an enthusiast for the League of Nations.

Griffith, Arthur (1872–1922), Irish statesman. Born in Dublin he became a compositor and then a journalist, founding the *United Irishman* in 1899 and a publication *Sinn Fein* in 1906. His initial goal was an independent Irish parliament under a dual monarchy of England and Ireland. In 1902 he had founded an organization Cumann na nGaedheal, which took as its watch word 'Sinn Fein'* ('we ourselves'). This was to become an official political party in 1922 and adopt the name Fine Gael in 1933. Following the emergence of the militant Ulster Volunteer movement* in 1912, Sinn Fein also became a militant political force, and Griffith took part in gun-running for the Irish Volunteers in 1914. He opposed Irish participation in World War I, but did not take part in the Easter Rising* of 1916. Imprisoned by the British 1916–18, he was elected to the Westminster Parliament in 1918, whereupon he was instrumental in forming the unofficial Irish Parliament the Dáil Éireann in 1919, being Vice-President of the rebel Irish Republic and its effective leader in the absence of de Valera*. He led the Irish delegation which signed the Anglo-Irish Treaty in December 1921, establishing the Irish Free State, and was President of the official Dáil Éireann from January 1922 until his sudden death, having parted with de Valera, who opposed the Treaty.

Grimond, Joseph (Jo), Baron of Firth in the County of Orkney (1913–), British politician. Born in St Andrews, Scotland, he was educated at Eton and Oxford and became a successful lawyer, being called to the Bar in 1937. He served in the army in World War II and entered politics as a Liberal in 1950. He became leader of his party in 1956. While opposed to socialism, he was an advocate of radical policies, including expansion of education and 'co-partnership in industry' between management and work-force. He fiercely attacked British policies over the Suez Crisis* in 1956, and his party made some headway under his leadership, winning twelve seats in 1966. He retired a year later, partly for health reasons.

Guam, an island territory of the USA, being the largest of the Mariana Islands in Micronesia*. It was occupied by the US navy in 1898 and ceded to the USA at the end of the Spanish–American War. Captured by the Japanese in December 1941, it was regained after bitter fighting in July 1944, becoming a base for bombing operations against the Japanese mainland. Since 1970 the US Governor of Guam has been elected by universal suffrage, and since 1972 Guam has sent a non-voting delegate to the US House of Representatives. The island is a massive military base and centre for US strategic intelligence.

Guandong (Kwantung) **Army.** Japan had occupied Lushen (Port Arthur) in 1905, following the Russo-Japanese War. It became the administrative centre of the south Manchurian peninsula known as Guandong (Kwantung). Among senior officers of the army based there were many who were determined fully to occupy Manchuria, over which Japan gained further concessions in 1920. In 1928 Guandong Army officers were involved in the assassination of Marshal Chang Tso-lin, warlord of Manchuria, and on 18 September 1931 they staged the Mukden Incident, resulting in occupation and the creation of the militarist state of Manchukuo*. In July 1936, the Army Command in Tokyo having persuaded the government to sign the Anti-Comintern Pact*, General Tojo*, in command of the Guandong Army, launched a campaign to separate all Mongolia from Soviet influence; he created the autonomous republic of Inner Mongolia*, based on Kalgan (Zhangiakou). The Marco Polo Bridge Incident* of 1937 precipitated war against China, as a move to eliminate any threat to Japan's rear in the event of war with the Soviet Union. There were border clashes in 1938 and the Guandong Army was greatly strengthened during 1941, but war with the Soviet Union failed to materialize until August 1945. By then the Army's crack divisions had been moved to other fronts and, when Soviet troops invaded, it quickly disintegrated.

Guatemala, a Central American state. Its population is largely descended from Maya Indians. Having won independence from Spain in 1821 and declared itself an independent republic in 1839, over a century of increasingly despotic rule by dictators followed. In 1944, however, Juan José Arévalo became President and he began a programme of social reform. He was followed by Jacobo Arbenz Guzmán* in 1951. His reforms included expropriation of large hacienda estates for the landless peasantry. The biggest landowner was the US United Fruit Company*, which in 1954, through the CIA, backed a military coup of US-armed Guatemalan exiles. Ten years of disorder followed until the election of Julio César Méndez Montenegro as President in 1966 on a moderate platform. But military intervention recurred, and during the 1970s and early 1980s violent suppression and violations of human rights occurred. In 1982 a guerrilla movement of largely Indian landless peasantry began operations. In 1985 under a new constitution civilian elections were restored, and President Vinicio Cerezo was elected. He ended a dispute over Belize* and restored diplomatic relations with Britain, but he faced major problems. He needed to end guerrilla activity without alienating the military; to restore good relations with the USA; to improve international trade; and to solve Guatemala's economic problems with its vast external debt. Peace talks were held in Oslo in April 1990 between various groups, including guerrilla representatives.

Guernica, the ancient Basque capital, with some 5,000 inhabitants. Under its oak tree in the centre traditional Basque liberties are confirmed. On 27 April 1937 it was heavily bombed by German planes of the Condor Legion*. Picasso's painting was brought back to Barcelona on the restoration of the monarchy in 1975.

Guevara, Ernesto **'Che'** (1928–67), South American revolutionary. An Argentinian by birth, with wealthy parents, he qualified as a doctor and then joined the left-wing regime of Arbenz Guzmán* in Guatemala*. When this was overthrown in

1954 he fled to Mexico. Here he met Fidel Castro* and helped him to prepare the guerrilla force which landed in Cuba in 1956. Shortly after Castro's victory he was given a cabinet post as Minister for Industries. As such he was to play a major role in the transfer of Cuba's traditional ties from the USA to the Communist bloc. A guerrilla warfare strategist rather than an administrator, he moved to Bolivia in 1967 to try to persuade Bolivian peasants and tin-miners to take up arms against the military government. The attempt ended in failure, and Guevara was captured and executed. His refusal to commit himself either to capitalism or to orthodox Communism made him an archetypal hero-figure for radical students of the 1960s and early 1970s.

Guild Socialism, a short-lived British labour movement. Founded in 1906 by Samuel Hobson, it called for revolutionary change in the organization of British industry. Like the syndicalists* in France, it wished workers to be given control of industry, organized through monopolistic guilds authorized by the state. After World War I it established a National Guild League, but this split when many members joined the newly formed Communist Party. Ideological differences followed and the movement had ended by 1923.

Guinea. A West African state conquered by the French in the late 19th century, it was made part of French West Africa*, and remained a French colony until 1958. In that year a referendum rejected membership of the French Community* and Ahmed Sékou Touré* became first President. His presidency was characterized by severe unrest, numerous attempted coups, repression, and almost complete isolation from the outside world. However, before his death in March 1984 a degree of liberalization and stability seemed to be emerging. In April 1984 a military government established the regime of President Lansana Conté, with bauxite mining and aluminium production boosting the economy.

Guinea-Bissau. A West African state, formerly Portuguese Guinea. Its boundaries had been fixed in 1886 by an agreement with France. During the 1960s the struggle against colonial rule intensified, led by Amilcar Cabral*, and in 1974 Portugal formally recognized its independence. In 1977 an unsuccessful attempt was made to unite with Cape Verde*. In 1980 a military coup established a Revolutionary Council which in 1984 became the Council of State, when civilian rule was restored. A new constitution created a Peoples' Assembly elected from the single party (PAIGC), with General João Bernardo Vieira as President.

Gulf Crisis (1990). In July 1990 Iraq accused Kuwait of breaking the OPEC agreement by over-producing oil, and of 'stealing' oil from the Rumalla oilfield worth $2,400m.; it demanded the cancellation of an alleged Kuwaiti debt of $10,000m. An emergency meeting of OPEC on 26 July in Geneva seemed to bring accord. Iraq however now demanded a long-term lease on the islands of Bubiyan and Warba, in order to give access to oil terminals. On 2 August Iraqi troops overran Kuwait, the Emir and his al-Sabah* family escaping. On 28 August Kuwait was proclaimed the nineteenth province of Iraq. The UN Security Council unanimously imposed economic sanctions and a multinational force was sent to the Gulf and Saudi Arabia to protect the latter's oilfields (Operation Desert Shield). Early in November there was a shift in US and British policy. Although there was evidence that sanctions had been effective, it was decided to threaten Saddam Hussein* with a military reoccupation of Kuwait. US forces were greatly increased, to half a million, and Iraq responded by mobilizing reserves on 23 November. On 29 November the UN Security Council agreed to a proposed deadline of 15 January 1991 for Iraqi withdrawal. Intense diplomatic activity followed, by France, Jordan, the Soviet Union, Yasir Arafat*, and others, but to no avail. Hussein was persuaded to release European and US nationals, who had been held virtual hostages, on 6 December, when some hope of conciliation appeared. In spite however of more and more frenzied diplomatic activity the deadline passed, and on 17 January the USA began the Gulf War with a massive air attack.

Guomindang (Kuomintang), the Chinese National People's Party. After the overthrow of the Qing dynasty in 1912, Song Jiaoren and Sun Yixian* (Sun Yat-sen) merged several revolutionary organizations into a republican party to be run along democratic principles. Suppressed in 1913 by Yuan Shikai*, it was reformed in 1920 with Soviet advice. In 1923 it established links with the Comintern* in Moscow and in 1924 adopted the 'Three Principles'*: nationalism, democracy, and social reform. The Communist Party of China (CCP) had been founded in 1921, and many of its members joined the Guomindang. Sun Yixian died in 1925 and Jiang Jiehi* (Chiang Kai-shek) emerged as leader of the Guomindang and Commander-in-Chief of its military arm, the National Revolutionary Army*. In April 1927 he embarked on a process of 'purging' his party of Communists, a ruthless exercise in which thousands died. After the successful conclusion of the Northern Expedition* the Guomindang became internationally recognized, establishing the National government in Nanjing. The NAR organized a sequence of campaigns 1930–4 precipitating the Long March*: but following the Xi'an Incident* the Guomindang was temporarily reconciled, by the United Front*, with the Communists to fight the Japanese. By 1946, when civil war with the Communists recommenced, the party was riddled with corruption and, in spite of massive US aid, its forces were unable to stem the advance of Communist armies. In 1949 the Guomindang withdrew to Taiwan, where it remained the ruling party for the Republic of China.

Gurkhas, the dominant race of Nepal*. As Hindus they had been driven from Rajputana by Muslim invaders, and throughout the 19th and 20th centuries provided regiments of soldiers, first for the British Indian army and after 1947 for the British army. Their military record in both world wars and after was highly distinguished. However, the decline in the number of British overseas possessions resulted in a reduction in the number of Gurkha regiments, with consequent loss of earnings for Nepalese families.

Guyana, a co-operative republic on the northern coast of South America. A British colony from 1831, its sugar plantations were adversely affected by the abolition of slavery, planters importing over 200,000 indentured Indians, as well as Chinese, to work their estates. The 19th century also saw boundary problems with neighbour colonies. Falling sugar prices in 1928 caused Britain to tighten its hold over the Crown Colony, halting moves towards self-government. During World War II the lease of military and naval bases to the USA helped to stimulate the economy, and when the franchise was extended (1943 and 1945), two politicians emerged, Cheddi Jagan, an Indian dentist, and Forbes Burnham*, a Black barrister. At first they co-operated, forming the People's Progressive Party in 1950, dedicated to independence. But their supporters later split, largely along ethnic lines, Indian and Afro-American, Burnham forming the People's National Congress (PNC). Jagan won the first elections following adult suffrage in 1953, but racial violence developed and Britain sent troops in 1962. Under a system of proportional representation Burnham became Premier in 1963, and when independence was granted in 1966 Prime Minister. In 1970 Guyana became a co-operative republic within the Commonwealth. During the 1970s a more radical party emerged, the Working People's Alliance, whose leader was assassinated in 1980. When Burnham died in 1985 he was succeeded as President by Desmond Hoyte, the PNC winning the election of December 1985. Talks with Jagan's PPP collapsed and Hoyte's government embarked on a programme of restoring good relations with the USA by a 'rolling back of co-operative socialism'. His alleged 'sell-out' antagonized the left, but was probably inevitable in the face of economic decline, enabling Hoyte to negotiate an IMF* loan.

Guyane (Guyane Français), a French settlement on the northern coast of South America since the 17th century. It was for long notorious for its convict settlement of Devil's Island, which only closed in 1947. In 1946 it became an Overseas Territory of France, sending Deputies to the Assemblée in Paris. Separatist demands in the 1970s were firmly suppressed, with trade union leaders and separatist agitators being imprisoned. Meanwhile, in the centre of the country a European town was growing around the European Space Agency's Ariane satellite project, the base being protected by units of the Foreign Legion*. Although separatists could only win 3 per cent of the electorate's vote in the 1980s, their agitation was compounded by the problem of refugees from neighbouring Surinam*.

Haakon VII (1872–1957), King of Norway 1905–57. Formerly Prince Charles of Denmark, he was elected by the Norwegian Storting (Parliament) to the throne in 1905, when Norway and Sweden separated. In April 1940 he was driven out by the German invasion of Norway. Refusing the suggestion of the government of Vidkun Quisling* to abdicate, he lived in London as constitutional head of the government-in-exile. He returned to Norway in 1945. He dispensed with much of the regal pomp attached to monarchy, being known as 'the People's King'. He was succeeded by his son Olav VI.

Haganah, a Jewish defence force in Palestine. It was first established before World War I as a secret organization to defend Jewish settlements from Arab attacks. After 1920 it was accepted by the British authorities as a supplementary police force, coming under the control of the Histadrut, the General Federation of Jewish Labour. During the period of Arab–Jewish unrest 1936–9 it acquired a general staff, as a paramilitary arm of the Jewish Agency*, being advised by Orde Wingate*. During the war years it contributed units to the British 8th Army, but was also involved in organizing illegal Jewish immigration from Europe. It condemned the terrorist activities of the Stern Gang* and Irgun*, and in 1947, when the British prepared to leave Palestine, it took on the defence of Jewish Palestine against Arab troops. Some 60,000 strong, it formed the nucleus of the army of the new state of Israel.

Haig, Douglas, 1st Earl Haig of Bemersyde (1861–1928), British Field Marshal. Born in Scotland, he was educated at Clifton College and Oxford and was commissioned into the Royal Hussars in 1895. He served in Egypt, South Africa, and India, where he was Chief of Staff 1909–11. He commanded the 1st Army Corps in Flanders at Ypres* and Loos* in World War I and in December 1915 succeeded Sir John French as Commander-in-Chief. His unimaginative strategy of attrition on the Western Front, while he was prepared to accept huge casualties on the Somme* in 1916 and at Passchendaele* in 1917, was much criticized. His conduct of the final campaign, however, September–November 1918, in collaboration with Marshal Foch*, ended the war more quickly than many had expected. After the war he devoted himself to working for ex-servicemen, founding the British Legion, with its annual 'Poppy Day' appeal.

Haile Selassie (1892–1975), Emperor of Ethiopia 1930–74. A great nephew of Emperor Menelik II, he was baptized Ras Tafari Makonnen as a Coptic Christian and in 1917 named by a Council of Notables Regent to the Empress Waizeru Zauditu, his aunt and heir apparent. The degenerate Lij Yasu had been deposed in 1916 for being pro-German, and Tafari defeated him in a series of campaigns and then exiled him. During his years as Regent he took Ethiopia into the League of Nations, and was crowned King in 1928 and then Emperor in 1930. This represented a triumph for the Coptic Church, as against its Muslim rivals. His new constitution of 1931 proved largely abortive, with Parliament never gaining significant powers. From 1935 his personal rule was interrupted by war with Italy and occupation, which lasted until 1941. During these years he lived in exile in Britain. After his restoration he received massive aid from Britain, the USA, and South Africa, and during the 1960s he succeeded in making Addis Ababa a centre of diplomatic activity and the headquarters of the OAU*. However, he was never in close touch with the social problems of his country. In 1974 he was deposed by a committee of left-wing officers, but allowed to live in his palace until his death.

Haiti, the western third of the Caribbean island of Hispaniola. In 1859 it became a republic independent of the Dominican Republic*, its anarchic history exacerbated by mulatto–Black racial hostility. The USA, which had large investments on the island, feared that Germany might intervene, and landed marines in 1915. These were not withdrawn until 1934, while fiscal control continued until 1947. In 1946 the Black liberal Dusmarsais Estimé had been elected President, the first Black politician to emerge. His policies alienated the landowners and the military, who overthrew him in 1950. One of Estimé's ministers, a Black nationalist intellectual, Dr François Duvalier*, was elected President in 1957. He determined to end mulatto political dominance, to reduce the power of the army, and the influence of the Catholic Church and the USA. He did this by building his private army of National Security Volunteers (Tontons Macoutes), who were empowered to act as ruthlessly as necessary to keep him in power. The country was dominated by him 1957–71 and by his son Jean Claude 1971–86. Following violent anti-government riots in June 1984, there was a general state of revolt through 1985, before the USA flew Jean Claude out to France in February 1986. The Tonton Macoutes were disbanded and freedom of expression allowed by the US-backed government of General Henri Namphy. A new constitution was approved in 1987 and elections took place in 1988, but they were followed by a series of military coups and a regrettable return to violence under General Prosper Avril. Strikes and violence against his regime resulted in his resignation. Judge Ertha Pascal-Trouillot became interim President, until elections in December 1990 brought a dissident Roman Catholic priest, Father Jean-Bertrand Aristide, into office.

Haj Amin al-Husseini (1897–1974), Grand Mufti of Jerusalem. The office of mufti was officially abolished by the Turkish Republic in 1924, but in 1922 the Supreme Muslim Council in Jerusalem, with Britain's approval, had elected Husseini, an ex-Turkish army officer, to be their President and Mufti of Jerusalem. Bitterly resenting the Balfour Declaration*, he did nothing to discourage open rioting in Jerusalem in 1929, and in 1936 headed the revolt of the Arab Higher Committee. When that was suppressed he fled into exile, at first to Iraq and then to Germany (1939–45). After the war he returned to Palestine and installed himself in Gaza, where he formed an 'All Palestinian Government', which recruited guerrillas (fedayeen) for raids into the new state of Israel. From 1956 onwards his influence declined as that of the al-Fatah* developed, and he retired into Egypt.

Haldane, Richard Burdon, 1st Viscount Haldane of Cloan (1856–1928), British statesman. Born in Scotland, he was educated in Germany, where he studied philosophy as well as law. A successful lawyer, he became a QC in 1879 and was a Liberal MP 1885–1911. As Secretary for War 1905–12, he showed great organizational skill in reforming the British army. Recognizing the growing danger of German militarism, he used his knowledge of the German army to develop a British army capable of modern warfare. A small expeditionary force, ready for instant action, was formed, with a Territorial Force as reserve. An Imperial General Staff would be responsible for all military strategic planning. He was sent to Berlin in 1912 to secure a reduction in naval armaments, but failed. He then became Lord Chancellor, but Asquith* dismissed him in 1915 following a vicious, unjustified Press campaign that he was pro-German. After the war he supported the Labour Party and was again Lord Chancellor in 1924. His great public concern at home was higher education, especially in science and technology.

Halder, Franz (1884–1972), German general. A professional soldier, by 1938 he had reached the rank of general and was German Chief of Staff. As such he was responsible for the planning of the blitzkrieg* campaigns of 1939 and 1940. His early dislike of Hitler changed to acquiescence, after the latter's controversial military decisions early in the war seemed to be proved right by events. However, he opposed Hitler's decision to strike against Stalingrad in 1942 and was dismissed. After the July Plot of 1944* he was sent to a concentration camp and was freed in 1945, when he retired.

Halifax, Edward Frederick Lindley Wood, 3rd Viscount and 1st Earl of (1881–1959), British statesman. From a wealthy Yorkshire family, he was educated at Eton and Oxford, being first elected to Parliament in 1910 as a Conservative. He joined Bonar Law's government in 1922 and from 1925 until 1931 he was Governor-General and Viceroy of India (as Lord Irwin). He was closely involved and in many ways sympathetic to that country's struggle for independence. However, he favoured moving towards dominion status rather than the complete independence advocated by Gandhi, whom he imprisoned after the Salt March*. An advocate of appeasement, he became Foreign Secretary in 1938 after Eden resigned. He accepted *de facto* the Anschluss* of Austria and the dismemberment of Czechoslovakia after the Munich Agreement*. In 1939 he refused an invitation to Moscow, thus losing the chance of an agreement with the Soviet Union, and leaving the door open for Hitler and Stalin to draw up the Nazi–Soviet Pact*. He supported Chamberlain's view in September 1939 that there was no alternative to war and in May 1940 was invited by George VI to form a government, but he refused, serving instead under Winston Churchill. He was a successful ambassador to the USA for five years 1941–6. He was always an ardent High Churchman.

Halifax Explosion (Canada). On 6 December 1917 a French munition ship, SS *Mont Blanc*, exploded in Halifax harbour after a collision. Much of the north of the city was destroyed, with 2,000 people killed, 9,000 injured, and over 25,000 left homeless. A relief commission was established which formally completed its work in 1975. It was the biggest man-made disaster up to that date.

Halsey, William Frederick (1882–1959), US admiral. A graduate of the US Naval Academy in 1904, he served in World War I and then rose steadily to the rank of admiral. In 1941, when he was commander of the Pacific fleet aircraft carriers, he and his fleet were fortunately out of harbour when the Japanese attacked Pearl Harbor*. In 1942 he led a spectacular raid against the Marshall and Gilbert Islands, and during the campaign of the Solomon Islands he was given command of the South Pacific area. In October 1944, as commander of the US 3rd Fleet at the battle of Leyte Gulf*, his force sank a number of Japanese aircraft carriers. In 1945 he led the seaborne bombing offensive against Japanese cities. He retired from the navy in 1947.

Hammarskjöld, Dag Hjalmar Agne Carl (1905–61), secretary-general of the UN 1953–61. Born at Jönköping in Sweden, he was educated at Stockholm University, where he later became Professor of Economics 1933–6. He then entered government service and was deputy Foreign Minister 1951–3 and a strong supporter of the UN. He was elected as successor to Trygve Lie in 1953, and re-elected in 1957. Under him the UN established an emergency force to help maintain order in the Middle East, after the Suez Crisis*; and UN observation forces were sent to Laos and the Lebanon. He initiated and directed the UN's involvement in the Congo 1960–1, making controversial use of Article 99 of the UN Charter, which he believed allowed the secretary-general to exercise initiatives independent of the Security Council or the General Assembly. In doing so he incurred the wrath of the Soviet Union, while his policies in the Congo were also contrary to international mining interests. His death in an aeroplane crash over Northern Rhodesia (Zambia) was thought by many not to have been an accident.

Hankey, Maurice Pascal Alars, 1st Baron Hankey (1877–1963), British civil servant. Born in Biarritz, he was educated at Rugby School and was commissioned into the Royal Marines in 1895. He transferred to Naval Intelligence and then became assistant secretary of the Committee of Imperial Defence in 1908 and secretary in 1912. He worked closely with a sequence of prime ministers, all of whom trusted his absolute integrity and relied on his remarkable memory. In 1916 he created the Cabinet Secretariat which serviced Lloyd George's* war cabinet and was linked with the Departments of State by a series of committees, a structure of government which broadly survived into peacetime. He retired in 1938, but during World War II briefly held office under Chamberlain and Churchill until 1942, and was then chairman of the Cabinet Scientific Advisory Committee.

Hara Takashi (1856–1921), Prime Minister of Japan 1918–22. A successful businessman, he became president of the Seiyukai (Friends of Constitutional Government) Party in 1914, a strong advocate of government by political party rather than by interest groups dominated by the *genro**. He was the first commoner to hold the post of Prime Minister. During his period in office the electorate was doubled (1919), but his attempts to build links with the business community brought him under suspicion of corruption. He failed either to prevent post-war industrial unrest or to stop military intervention in the Russian Civil War*. He planned to end the military administrations in Korea and Taiwan and to reduce the influence of the Guandong Army*, but he was assassinated by a right-wing fanatic.

Harlem Renaissance, a US cultural phenomenon. Before World War I New Orleans had been the leading centre of Black culture, especially in music. In 1916 Marcus Garvey* arrived in New York with his determination to stimulate ethnic pride among Blacks. In Harlem, on Manhattan Island, a decade of remarkable cultural achievement followed, the first jazz concert dating from 1917 and a series of literary publications appearing through the 1920s. In 1925 Alain Locke edited *The New Negro*, an influential anthology, and novels and poetry were characterized by a strong interest in African culture, as well as protests against racial bigotry. Langston Hughes (*Weary Blues*), James Weldon Johnson (*God's Trombones*), and Claude McKay (*Home to Harlem*) were leading writers, while painting also flourished and Harlem became a fashionable venue for New York intellectuals. With the Great Depression*, following the 1929 Stock Market Crash, the movement collapsed.

Harriman, William Averell (1891–1986), US diplomat. Son of a wealthy railroad magnate, he himself worked for the Union Pacific Company from 1915. His father was a friend of F. D. Roosevelt* and in 1934 Harriman joined the President's 'National Recovery' programme and became chairman of a Federal Business Advisory Council. In 1940 he was sent to London to expedite the Lend-Lease* arrangements, and was then US ambassador to Moscow (1943–6), attending all wartime conferences. After the war he was special US envoy for European recovery. Never an enthusiast for the Cold War, he held numerous posts for Democratic presidents, including Assistant Secretary of State for J. F. Kennedy. He himself failed to win presidential nomination. He helped to negotiate the 1963 Nuclear Test-Ban Treaty* and led the US delegation to Paris 1968–9, seeking an end to the Vietnam War*. Following his retirement in 1969 his advice continued to be sought after such long experience of foreign affairs.

Hasan, Muhammad Abdille Sayyid (1864–1920), Somali nationalist leader. After a visit to Mecca he joined the Salihiya, a militant, puritanical Islamic fraternity. He believed that Christian colonization was destructive to the Islamic faith and in 1899 proclaimed a jihad (holy war) on all colonial powers in Somalia (French, British, and Italian). Between 1900 and 1904 four expeditions by the colonial powers failed to defeat him. After a truce he resumed war again, allying with Lij Yasu of Ethiopia*. Known by the British as 'the Mad Mullah', he was killed by a combined colonial attack in 1920.

Hasan al-Banna (1906–49). Founder of the Muslim Brotherhood*. Born and educated in Egypt, as a pious Muslim he became a schoolteacher near Suez. Here in March 1928, with six Arab workers from a British camp, he founded a Society of Muslim Brothers, seeking a rejuvenation of Islam. This was to grow into the powerful Muslim Brotherhood, which by 1939 had attracted thousands of Egyptians from all classes. He could not support the fanaticism which developed within the movement, but his piety, frugal life, and sincere, unsophisticated teaching gained him wide sympathy. He was assassinated by government agents in February 1949.

Hashemite dynasty, Arab descendants of the Prophet Muhammad, among whom were the 19th-century emirs of Mecca under the Turks. The last of these was Hussein ibn Ali*, who became King of the Hejaz, but who was ousted by Abd al-Aziz ibn Saud* in 1924. His oldest son Ali briefly replaced him before also being forced into exile by Saud. His other sons Abdullah* and Faisal* became kings of Transjordan and Iraq. The Iraq branch ended in 1958 with the murder of Faisal II*, but the dynasty continues in Jordan.

Hassan II (1929–   ), King of Morocco 1961–   . He succeeded his father Mohammed V (1927–61), becoming the supreme civil and religious authority. Educated in France, as Crown Prince and Commander-in-Chief of the army he personally directed rescue operations at the Agadir earthquake of 1960. There were serious riots in Casablanca in 1965 and he survived coups against his allegedly authoritarian regime 1971, 1972, and 1973, after which he gained in popularity and granted some political liberalization. He has seen himself as belonging to both the Arab world and the African states south of the Sahara, while at the same time keeping close relations with France. In 1972 he was chairman of the OAU* heads of states, while in 1974, 1981, and again since 1986 he has hosted meetings trying to solve the Arab–Israeli impasse. Problems relating to the Western Sahara have however left him increasingly isolated, causing border tension with Algeria and Libya and forcing him to withdraw from the OAU in 1984. In 1975 he had promoted a 'Green March' of some 100,000 unarmed Moroccans into Spanish Sahara, claiming it for Morocco*; but the claim had been consistently challenged by guerrillas of Polisario*, whose government-in-exile was itself elected to the OAU in 1984. In February 1989 he hosted a meeting in Marrakesh when the Maghreb Union* was established.

Haughey, Charles James (1925–), Prime Minister (Taoiseach) of the Republic of Ireland 1979–81, 1982, 1987– . Son of an officer of the IRA who later moved to the Irish Free State Army, he was born and educated in Dublin. Called to the Irish Bar in 1949, and having made a fortune in real estate, he entered politics in 1957. He was elected to the Dáil for the Fíanna Fáil* Party. He held minor offices, but in 1970 was tried and acquitted of trying to smuggle arms for the IRA. In 1979 he succeeded Jack Lynch as party leader, winning the general election. A controversial if charismatic figure, his reputation was marred at that time by financial scandals within his party. He proved a formidable leader of the opposition 1982–7, and in 1990 successfully held office as President of the Council of the European Community.

Hausa, a language group which has given its name to a major group of West African peoples. From the 14th century a series of states developed in Northern Nigeria with fortified towns from which to trade. These achieved high standards in the arts. There were seven main Nigerian Hausa states: Biram, Daura, Gobir, Kano, Katsina, Rano, and Zaria. Islam was introduced from Mali in the 15th century, and in the early 19th century the states were incorporated into a Fulani* Islamic empire, the emirs of each being responsible to Sokoto. These emirates together with Borno were taken over by the British in 1900 to form the protectorate of Northern Nigeria. They retained much of their traditional life-style through the 20th century. The Hausa language is spoken widely in Niger, Nigeria, Chad, and Ghana, and in many towns from Dakar and Agadir to Kinshasa and Khartoum.

Havel, Vaclav (1936–), President of Czechoslovakia 1989– . Born in Prague to a family of 'bourgeois descent', he was excluded from the university. Instead he worked as a laboratory assistant and a taxi-driver, attending evening classes at Prague Technical University. He then obtained entry into a theatrical academy, qualifying in 1967 and becoming a stage-hand and technician. He had already begun serious writing in 1961 with *Autostop*, followed by plays such as *Garden Party* (1963), *The Message* (1965), and *Difficult Understanding* (1968). From then on publication or performance of his work was forbidden as subversive. He managed, however, to continue to write and to publish abroad, including the essay *An Open Letter to Gustav Husák*. He formed a Committee for the Defence of the Unjustly Persecuted (VONS) and signed Charter 77*. He was imprisoned January–May 1977, held on house arrest 1977–9, and imprisoned again 1979–83 for alleged sedition. In 1988 he helped to revive an independent weekly *Lidove Noviny*. Arrested again in January 1989, he was imprisoned for alleged 'incitement and obstruction'. Released in May, he was by now the leading spokesman for political liberalization. In November he helped to form Civic Forum. He was chosen President by the Czech Assembly when Husák resigned in December 1989 and elected by popular vote in 1990.

Hawaii, a chain of Pacific islands in east Polynesia*. They were named the Sandwich Islands by Captain Cook in 1778, and US settlers began to arrive from the 1820s. By 1893 there were demands for US annexation and the settlers were powerful enough to overthrow the Polynesian monarchy under Queen Liliuokalani. Lying between California and Japan and the Philippines, where US commercial activity was developing, they were a valuable coaling station and in 1898 were annexed by the USA, becoming an 'organized territory' in 1900. Thereafter the chief US naval base was built in Pearl Harbor, which after World War II became the largest nuclear arsenal in the world. Hawaii became the fiftieth state of the USA in 1959.

Hawke, Robert James Lee (1929–), Prime Minister of Australia 1983– . Educated at the universities of Western Australia and Oxford, he became a highly successful officer in the Australian Council of Trade Unions 1958–70, having joined the Australian Labor Party*. He was president of the party 1973–8 and was elected to Parliament in 1980. He became leader of the parliamentary party in 1983, and went on to win the election in March of that year; he was described as having 'a charismatic, flamboyant personality'. His government survived, winning the March 1990 election, although losing some support. He faced several government scandals, problems with the USA over its testing programme for the Strategic Defence Initiative*, and severe economic difficulties. He sought to ease the latter by improving trade with Britain, Japan, and New Zealand. He has been strongly critical of French policies in New Caledonia.

Hawley–Smoot Act (USA) (1930), US legislation against imported goods and materials. Drafted before the Stock Market Crash* of October 1929, it was endorsed by President Hoover in the belief that it would help the hard-pressed farmers if increased tariffs were imposed. The Act aroused deep resentment abroad. Within two years twenty-five countries had established retaliatory tariffs, and foreign trade, already declining, slumped even further, thus helping to precipitate the Great Depression*.

Haya de la Torre, Victor Raúl (1895–1979), Peruvian statesman. From a wealthy Peruvian family, he became involved in student political activity in Lima, promoting night schools for the working population. He was deported in 1923 for organizing a demonstration against the Leguía* regime. In exile in Mexico he founded the Alianza Popular Revolucionaria Americana (APRA*), which from 1924 became the spearhead of radical dissent in Peru. After the Leguía regime fell in 1930, he urged his APRA followers to overthrow the army-backed conservative oligarchy. He stood for President in 1931, but ballots were rigged and Colonel Sánchez Cerro was proclaimed victor. Haya was imprisoned 1931–3. After Cerro's assassination he was released, and lived in hiding in Peru for ten years 1935–45, becoming widely known in South America through his writings. In 1945 the Aprista Party took the name Partido del Pueblo (People's Party) and supported José Luis Bustamente as president. During the latter's presidency (1945–8), Haya exerted great influence, but when Bustamente was overthrown by the army in 1948 he took refuge in the Colombian embassy in Lima, and then went into exile in Mexico until 1957. He contested the 1962 election for President, but the army intervened and in 1963 Terry Belaúnde* was declared the winner. In 1979 Haya drafted a new constitution for Peru which restored parliamentary democracy. He died however before his party finally came to power under President García in 1985.

Hayek, Friedrich August von (1899–), Austrian political philosopher. Born in Vienna, he was educated there and in the USA. He taught in Vienna until 1931, when he moved to the London School of Economics, where he taught 1931–50. He then became Professor of Social and Moral Science at Chicago University. In his *The Road to Serfdom* (1945), he denounced totalitarian regimes of both left and right. As a strong advocate of the case for a free economic market as the best safeguard of the individual's liberty, in *The Constitution of Liberty* (1961) he seeks to define a constitution within a capitalist state which will provide maximum liberty without licence or anarchy. This can best be produced by the self-regulating mechanisms of the market.

Heath, Edward Richard George (1916–), British Prime Minister 1970–4. Born in Kent and educated there and at Balliol College, Oxford, he distinguished himself in student politics as a reforming Conservative, hostile to Franco, Mussolini, and Hitler. He served in the Royal Artillery 1940–6 and was first elected to Parliament in 1950, supporting in his maiden speech the idea of European unity. He was made responsible by Macmillan for the first negotiations 1961–3 for Britain's entry into the EEC*, vetoed by President de Gaulle. He became Conservative Party leader in 1965. Elected Prime Minister in 1970, he successfully reopened European negotiations, Britain joining the Community 1 January 1973. Meanwhile the troubles worsened in Northern Ireland, with Brian Faulkner* resigning and direct rule being imposed from London. In domestic affairs his ministry was beset with difficulties. Serious problems of inflation and balance of payments were exacerbated in 1973 by the oil price increases imposed by OPEC*. An unpopular Industrial Relations Act and attempts to restrain wage rises led to a national coal strike in 1972 and strikes in the coal, power, and transport industries during the winter 1973–4. He called an election to try to strengthen his position, but was defeated, becoming a somewhat isolated figure even within his own party. Harold Wilson became Prime Minister. In February 1975 Heath was defeated by Margaret Thatcher as leader of the Conservative Party. During the 1980s he opposed many of her policies, pressing once again for greater participation in Europe. In 1990 he celebrated forty years as a Member of the House of Commons.

Hejaz, the, Red Sea coastal plain of the Arabian peninsula, containing the two holy cities of Medina and Mecca. The 'Pilgrim Railway' from Damascus to Medina had been opened in 1908, and it was this railway that T. E. Lawrence* targeted during the Arab Revolt*, after which the Sherif of Mecca, Hussein ibn Ali*, proclaimed himself King of the Hejaz. He was ousted in 1924 by Abd al-Aziz ibn Saud*, already Emir of Nejd and al-Hasa in central and north-east Arabia, creating the Kingdom of Saudi Arabia. Into the Hejaz come thousands of Muslim pilgrims each year; in July 1987 there were serious riots between Iranian pilgrims and Saudi police, some 400 pilgrims being killed and 650 injured.

Hejaz railway, a railway constructed by German engineers from Damascus to Medina 1900–8 for the purpose of carrying pilgrims to the holy cities. A branch line to Haifa and thence to Jerusalem and Beersheba was built in 1905. During the Arab Rising of 1916 T. E. Lawrence* and Arab guerrillas repeatedly destroyed the line between Ma'an and Medina, thus preventing the movement of Turkish troops. Prolonged negotiations stretching into the 1970s, between the successor states of Syria, Jordan, and Saudi Arabia, failed to reach agreement, and the full length of the line was never reopened, pilgrims moving by air and road rather than rail. In Jordan the line was reopened from Amman to the port of Al Aqabah.

Helsinki Conference (1973–5), a meeting in July 1973 of representatives of thirty-five nations in Helsinki on a Soviet initiative, in order to negotiate a European Agreement on Security and Co-operation, with particular reference to post-1945 frontiers. They met again in Geneva and then over the three days 30 July–1 August 1975 their leaders met in Helsinki to produce the Helsinki Final Act, consisting of three main sections: (1) proposals to prevent accidental confrontations between the East–West power blocs, with a special section on the Mediterranean; (2) proposals for economic and technological co-operation; (3) reaffirmation of human rights, agreeing to respect 'freedom of thought, conscience, religion and belief'. Follow-up meetings were proposed, the first being held in Belgrade in October 1977. Meanwhile in the Soviet

Union a number of dissidents in 1976 formed the Helsinki Human Rights Group, including Yuri Orlov and Andrei Sakharov*, the Group protesting particularly against travel restrictions and prison conditions. Orlov was charged with anti-Soviet agitation and imprisoned along with Alexander Ginsburg and Anatoly Shcharansky. Sakharov was to be placed under house arrest. In 1977 Charter 77* was established in Czechoslovakia, aimed also at the preservation of human rights.

Herrin Massacre (USA) (22 June 1922). It occurred in Herrin, Ill., where employers had attempted to break a strike in a local coal-mine by importing non-union men. Striking miners forced these to stop working, promised them a safe conduct, marched them from the mine, and then opened fire on them, killing twenty-five. A grand jury returned 214 indictments for murder and related offences, but local feeling was such that there were never any convictions.

Hertzog, James Barry Munuik (1866–1942), Prime Minister of the Union of South Africa 1924–9, 1933–9. A successful farmer from the Orange Free State, he was a brilliant guerrilla leader in the Boer War 1899–1902, but in 1910 he joined the first Union of South Africa cabinet. Opposed to the South Africa Party of Louis Botha*, in 1914 he formed the National Party*, aiming at South African independence and opposing support for Britain in World War I. As Prime Minister 1924–9, in a coalition with the South African Labour Party, he made Afrikaans an official language, instituted the first Union flag, and was a protectionist. In 1933 he formed a coalition with J. C. Smuts* and in 1934 they formed the United Party. In racial affairs he was a segregationist, but opposed rigid apartheid*. Although he won the election of 1938, his opposition to joining Britain in World War II split his party and he resigned, retiring from politics in 1940.

Hess, Walter Richard Rudolf (1894–1987), Nazi leader. During World War I he fought at Ypres and then transferred to the German Air Corps. He met Hitler in 1920, when studying at Munich University. He became his political secretary and shared imprisonment with him after the Munich *putsch** of 1923. In 1934 he became Hitler's deputy as party leader and Minister of State. In 1941, secretly and it seems of his own volition, he parachuted into Scotland in order to negotiate peace between Britain and Germany. He was imprisoned by the British for the rest of the war and then for life by the Allies at the Nuremberg Trials*. From 1966 he was the sole inmate of Spandau Prison in Berlin, where he died, allegedly committing suicide.

Heydrich, Reinhard (1904–42), German Nazi police officer. Born in Halle, he served in the German navy until 1931, when he joined the SS* and the Nazi Party. In 1932 he helped Himmler* form the SD and rose rapidly to the rank of *Gruppenführer* (general of police). Within the International Police Commission he developed an elaborate overseas espionage system before 1939. As Reich Protector of Bohemia and Moravia, his bestial regime earned him the names 'the Beast' and 'the Hangman of Europe'. He was assassinated by Czech nationalists in Prague in 1942. The Germans retaliated by a reign of terror, in order to paralyse all opposition, exterminating the village of Lidice*.

Heyerdahl, Thor (1914–), anthropologist and explorer. Born in Larvik, Norway, he was educated at Oslo University and became a research anthropologist, doing field work in Polynesia, Peru, and British Columbia. After serving with Free Norway forces in World War II, he set out to prove his theory that there had been far more cultural contacts across the oceans in prehistory than previously accepted. His voyage on the balsa log-raft *Kon Tiki* in 1947 took him from Callao in Peru to Raroia in Polynesia. In 1953 he clearly established links between Ecuador and the Galapagos Islands. In 1969 he crossed the Atlantic from Morocco to Central America in an Egyptian reed-boat *Ra*, and in 1977–8 his reed-boat *Tigris* proved that ancient Sumeria could have traded by sea with the Indian subcontinent and Africa. Although ethnologists would not accept theories of mass migration, the existence of some cultural links has been increasingly accepted. He has been an active world federalist, and his work greatly stimulated world-wide interest in prehistory and anthropology.

Hieu, Nguyen Van (1918–), Vietnamese politician. Born in the far south of Vietnam, he became a journalist and took part in the VietMinh* risings of 1945. Opposed to the French, he was a victim of Ngo Dinh Diem's* anti-communist purge of 1958, and joined the National Liberation Front* when it was formed in 1960. Throughout the Vietnam War he was an active spokesman for the NLF cause and became well known in the West through his writings and tours abroad, where he tried to put the NLF case. He was appointed Minister of Culture in the Socialist Republic of Vietnam in 1976.

Himmler, Heinrich (1900–45), German Nazi police chief. Born in Munich, son of a schoolmaster, he briefly served in the army 1918–19 before enrolling at the Munich Technical College. Here he became interested in politics and joined the Nazi Party. He worked in a nitrate factory and then in 1928 became a poultry farmer in Bavaria. In 1929 Hitler appointed him leader of his bodyguard police the SS*. In 1932 he founded a security service, the SD, which amassed material on personal lives of all politicians, to be used for blackmail. His forces carried out the murders of the Night of the Long Knives* in 1934 and in 1936 he became chief of all the police services, including the Gestapo*. From 1940 he extended his field of repression into Greater Germany and the occupied countries. From 1943 he was Interior Minister and commander of the reserve army. Although personally nauseated by the sight of blood, he was responsible for putting into practice the programme of genocide of the Jews. He ruthlessly put down the conspiracy against Hitler in the July Plot* of 1944, but soon afterwards was himself secretly negotiating German surrender to the Allies, hoping to save himself thereby. Hitler expelled him from the party and Himmler tried to escape. He was caught by British troops near Bremen and committed suicide by swallowing poison.

Hindenburg, Paul von (1847–1934), German general and President of the Weimar Republic 1925–34. Born in Prussia, he was commissioned into the Prussian Guards and fought at Königgrätz and in the Franco-Prussian War (1870–1). He retired in 1911, but was recalled to active service at the outbreak of war in 1914 and crushed the Russians at Tannenberg* (August) and Masurian Lakes (September). In 1916 he became Chief of the General Staff. With his aide Ludendorff* he established a virtual military dictatorship. They negotiated the Treaty of Brest-Litovsk* with the Bolsheviks and then launched the second Battle of the Marne* in March 1918. After the failure of this offensive he advised the Kaiser to sue for peace. After the war he came to tolerate the Weimar Republic and in 1925 became its second President. Re-elected in 1932, he did not oppose the rise of the National Socialists, and on the advice of von Papen* appointed Hitler Chancellor in January 1933.

Hinduism, the oldest world religion, to which most Indians adhere. It has a rich and tolerant range of gods, but all are manifestations of Brahma, the supreme being of the Hindu pantheon, who is united in the Trinity with Vishnu, the all-preserver, and

Siva, the mystic, miracle-working deity. Into Brahma, the ultimate world-spirit, the reincarnated, purified individual spirit is finally absorbed. The oldest extant scriptures are the 8th-century BC Sanskrit texts the *Upanishads*, from which the systems of Hindu law and social structure are derived, including the caste system. Of later texts the *Mahabharata* remains massively influential, within which is the *Bhagavadgita* (*c*.300 BC), where Krishna, the lord of creation, teaches self-mastery through the Yoga. The greatest modern Hindu reformer was the Mahatma Gandhi*, with his teaching of satyagraha (truth force) and swaraj (self-rule), and his concern for the harijans, the lowest Untouchable caste. Twentieth-century India, a secular state, has had to seek some reconciliation between the age-old caste system and the fashionable doctrines of human rights; inevitable tensions have been the result. Although the religion teaches the sanctity of all life, militant Hinduism has been responsible for numerous acts of violence, particularly against Muslims in both India itself and Bangladesh.

Hindu Mahasabha, a Hindu communal organization. It was first established in the Punjab before 1914 and became active during the 1920s under the leadership of Pandit Mohan Malaviya and Lal Raypat Rai. While stressing non-violence it campaigned for social reform, but also for the reconversion of Hindus from Islam. Its attitude towards Hindu–Muslim relations strained its relations with the Indian National Congress* and in 1937 it broke away and became a separate, right-wing political party under V. D. Sarvarkar. After independence the party declined in influence as the Jana Sangh (see JANATA PARTY) became the leading exponent of Hindu communal ideas.

Hirohito (1901–89), Emperor of Japan 1926–89. Eldest son of Crown Prince Yoshihito, he was appointed Regent in 1921 and, after surviving an assassination attempt, succeeded to the throne in 1926, initiating the Showa* era. Although not approving of military expansion, he failed to exercise his full technical sovereignty, allowing the political triumph of Tojo* and the militarists. He continued to follow advice not to weaken the throne by becoming involved in politics until 1945 when, convinced of the need to end World War II, he intervened to force the armed services to accept defeat. Saved from trial as a war criminal by MacArthur*, he renounced his divinity, but remained as a constitutional monarch under the 1947 Constitution. In private life he was a marine biologist, and from 1947 until his death he adhered strictly to democratic principles. In 1971 he visited Britain, when his Knighthood of the Garter was restored to him.

Hiroshima, a Japanese city in southern Honshu. Undamaged by the US bombing campaign 1944–5, it was chosen as the target for the first atomic bomb attack on 6 August 1945. This resulted in the virtual obliteration of the city centre, some 80,000 immediate deaths, another 60,000 dying within a year, and radiation effects continuing for decades. The attack, together with that on Nagasaki, led to Japan's unconditional surrender and the end of World War II.

Hiss case (1949–50), a legal case in the USA. State Department official Alger Hiss (1904–) was charged with perjury for having denied on oath that he had passed secret documents to Whittaker Chambers, a self-confessed Communist Party courier. Hiss maintained his innocence. In his first trial there was a hung jury, but in the second he was found guilty. At both trials high government officials testified on his behalf. The defence challenged Chambers's sanity and alleged that the FBI had tampered with evidence to obtain a conviction. Hiss was sentenced to five years in prison. The trial symbolizes the fears aroused at that time by the Cold War. Hiss was released in 1954 and returned to private life as a lawyer. Although he was readmitted to the

Massachusetts Bar in 1975, close investigation of the case still leaves the question of his guilt unresolved.

Hitler, Adolf (1889–1945) German dictator. Born in Austria, he lived with his widowed mother in Vienna and Munich, taking casual work as a house-painter. In 1914 he volunteered for the Bavarian army and became a corporal. He twice won the Iron Cross for bravery and was temporarily blinded by a gas attack. After demobilization he became involved in various anti-Bolshevik activities and joined the German Workers' Party as its seventh member. Back in Munich he met Ernst Röhm and in 1921 ousted Anton Dexter as leader of the party, now calling itself the National Socialist Workers' Party. In Vienna before the war he had imbibed the prevailing anti-Semitism* and this, together with tirades against the Treaty of Versailles* and against Marxism, fell on fertile ground in a Germany humiliated by defeat. In 1923 he staged the abortive Munich 'beer-hall' *putsch**. During the months he shared in prison with Rudolf Hess he dictated *Mein Kampf*, a political manifesto in which he spelt out Germany's need to rearm, strive for economic self-sufficiency, suppress trade unionism and Communism, and exterminate the Jewish minority. He differed with Röhm over the function of the *Sturmabteilung* (Brownshirts*), and in 1926 formed his personal force the *Schutzstaffel* (SS*). The Great Depression, which began in 1929, brought him a flood of adherents, so that, aided by violence against political enemies, his National Socialist Party (Nazi) flourished. After the failure of three successive Chancellors, including von Papen*, President Hindenburg* appointed him as Chancellor, with von Papen his deputy, the National Socialists having a majority in the Reichstag. Following the Reichstag fire (see WEIMAR REPUBLIC), he outlawed all other political parties, and in June 1934 eliminated his rivals, including Röhm, in the 'Night of the Long Knives'*. On the death of Hindenburg in August 1934 he proclaimed the Republic at an end and the establishment of the Third Reich*, with himself as President and 'Führer of the People'. He began rearmament in contravention of the Versailles Treaty, reoccupied the Rhineland in 1936, and took the first steps towards expanding the Reich: the Anschluss* with Austria in 1938 and the piecemeal acquisition of Czechoslovakia, beginning with German Sudetenland*. In the summer of 1939 he concluded the Nazi–Soviet Pact of non-aggression in order to invade Poland, but broke this when he attacked the Soviet Union in June 1941. His invasion of Poland had precipitated World War II. Against the advice of his military experts he pursued 'intuitive' tactics, which at first won massive victories in Poland and west Europe. In December 1941 he took personal military control of the armed forces now besieging Leningrad and Moscow. As the tide of war turned against him, he intensified the mass assassination of the Jewish Holocaust*. He escaped the July Plot* to kill him and undertook a vicious purge of all those involved. In 1945, as the Soviet army entered Berlin, he went through a marriage ceremony with his mistress Eva Braun. All evidence suggests that both then committed suicide and had their bodies cremated in his underground bunker.

Hitler Youth. In 1931 Baldur von Schirach* was appointed Youth Leader of the Nazi Party and in 1933 he established the Hitler Youth. In 1936 Hitler outlawed all other youth organizations and announced that all German non-Jewish boys and girls should join the Jungvolk (Young Folk) at the age of 10, when they would be trained in out-of-school activities, including sport and camping, and also receive political education. At 14 the boys would enter the Hitler Youth proper, where they would be subject to semi-military discipline, with tough out-of-doors activities and heavy Nazi propaganda, and girls would join the League of German Maidens, where they would learn motherhood and domestic duties. At 18 they would join the armed forces or the labour service. By 1938 7.7 million had been recruited, but efforts to enrol every boy and girl were failing, so that in March 1939 a conscription order was issued. There was continuous tension between the Hitler Youth and the schools, many of which tried to retain a liberal tradition, and this tension continued through the war.

Hoare, Samuel John Gurney, 2nd Baronet and 1st Viscount Templewood of Chelsea (1880–1959), British statesman. Born in London and educated at Harrow and New College, Oxford, he was first elected to Parliament in 1910 for Chelsea and retained his seat until 1944. He served on military missions to Russia and Italy during World War I, and after the war, as Air Minister, did much to build up the new RAF. Secretary of State for India 1931–5, he devised and defended the Government of India Act* of 1935, which was only partially to be implemented. When appointed Foreign Secretary in 1935 he faced the problem of Italian aggression in Ethiopia. With Pierre Laval*, his French opposite number, the **Hoare–Laval Pact** was devised, which would have partitioned Ethiopia. Public outcry led Prime Minister Baldwin to repudiate the plan and Hoare resigned. The Italians then occupied the whole country. In 1937 he became Home Secretary under Chamberlain (where he proved a notable penal reformer) and was one of the supporters of the Munich Agreement*, for which he was later denounced. He was ambassador to Spain 1940–4.

Ho Chi Minh (1890–1969), President of the Democratic Republic of Vietnam 1945–69. Born Nguyen Tat Thanh of peasant stock in central Annam, he became a schoolteacher in 1907, before travelling to Europe as a ship's steward. In 1918 he was in Paris writing articles for socialist newspapers and helping to found the French Communist Party. From Paris he went to Moscow and then to south China in 1924, where he recruited Vietnamese exiles to Communism. In 1930 he was back in Vietnam, where he formed the Communist Party of Indo-China, which was fiercely repressed after uprisings 1930–1, when he fled to Hong Kong. Here he was arrested by the British on a charge of sedition, but was released and returned in secret to Vietnam. After a failed conspiracy he fled to China where, with fellow Vietnamese exiles, he founded (1941) the VietMinh* movement. He was imprisoned by Jiang Jiehi* (Chiang Kai-shek) 1941–2. In 1943 he again returned in secret to Vietnam, where he began to recruit VietMinh guerrilla fighters and to organize resistance against the Japanese, adopting the name Ho Chi Minh ('He who enlightens'). In August 1945 the Emperor Bao Dai* was persuaded to abdicate and the Democratic Republic of Vietnam was proclaimed on 2 September with great jubilation. This was never recognized by the French, who now sought to re-establish control, General Leclerc moving up from Saigon to Hanoi. The French Indo-China War* (1946–54) was fought mostly in the north and culminated with General Giap's* victory at Dien Bien Phu*, leaving Ho in virtual control of North Vietnam; but by the Geneva Agreements* he had to concede control south of the 17th parallel. In 1960 the regime of Ngo Dinh Diem* in South Vietnam was so unpopular that the NLF* was formed, whose military arm, the Vietcong, began to conduct a guerrilla war, and which appealed to Ho for assistance. By 1963 US advisers in Saigon were giving more and more assistance to the South Vietnamese regime. On his side, by skilful diplomacy Ho had kept the support of both the giant communist powers the Soviet Union and China, and both gave him assistance. In 1965 in its fight with the Vietcong US intervention became open, with heavy bombing of North Vietnam. Ho's five-year plan 1961–5 had greatly increased industrial production and enabled his regime to survive the bombing onslaught. Known as 'Uncle Ho', he kept throughout to a policy of unification, being always prepared to work with non-communists as well as party members. He lived to see the beginning of the long process of negotiation in

Paris which was finally to achieve his aim. In 1975 Saigon was renamed Ho Chi Minh City.

Holland, Sir Sidney George (1893–1961), Prime Minister of New Zealand 1949–57. After service in World War I he became an engineer. In 1935 he followed his father into the House of Representatives and in 1940 became leader of the National Party*. A staunch supporter of New Zealand's part in World War II, he became Prime Minister in 1949. He was closely involved with the decision to send troops to fight in Korea, in the tough handling of a waterfront strike in 1951, and in the decision to abolish the Legislative Council, the Upper House of the New Zealand Parliament. He supported US strategies for the Pacific, New Zealand's membership of the ANZUS* and SEATO* pacts, and Britain's action in the Suez Crisis*; but, in the tradition of pragmatic conservatism, he strengthened much of the social welfare legislation which had been introduced under Labour. He was an advocate of the need to strengthen ties with Japan. He retired in September 1957 due to ill health.

Holocaust, the. Conventionally this ordeal of the Jewish people has been divided into two periods, 1933–41, 1941–5. During the first various anti-Semitic measures were taken in Germany and Austria, and to a lesser extent in Romania and Hungary under right-wing regimes. In Germany, after the 1935 Nuremberg Laws, Jews lost citizenship rights, the right to hold public office, practise professions, intermarry with a German, or use public education. Their property and businesses were registered and sometimes sequestrated. Continual acts of violence were perpetrated against them, and propaganda encouraged 'true' Germans to hate and fear them. As intended, the result was mass emigration, halving the half-million German and Austrian Jewish population. The second phase from 1941 involved forced labour, mass shootings, and concentration camps, many of which in Eastern Europe were equipped with gas chambers for the Nazi 'final solution' of mass extermination. It is estimated that between 4 and 5 million Jews died in these camps. Perhaps a further million died in ghettos by starvation and disease and over a million were shot by mobile killing squads (*Einsatzgruppen*). Out of a population of 3 million Jews in Poland, less than half a million remained in 1945, while Romania, Hungary, and Lithuania also suffered grievously.

Holt, Harold (1908–67), Prime Minister of Australia 1966–7. A successful Melbourne lawyer, he was first elected to the federal Parliament in 1935, supporting the United Australian Party*. In 1940 he joined Robert Menzies's* government and was Minister of Labour and National Service for a brief spell before the government's collapse. After the war he helped Menzies to build his new Liberal Party, which came to power in 1949. He served as Minister of Labour 1949–58 and Minister of Immigration 1949–56, when he relaxed the White Australia policy in favour of Asian students. From 1958 until 1966 he was federal Treasurer, adopting a tight anti-inflation policy. He succeeded Menzies in 1966, supporting US policies in Vietnam and winning a general election in that year. He died tragically when swimming, after less than two years in office, at a time of Australia's controversial involvement in the Vietnam War.

Holyoake, Sir Keith Jacka (1904–83), Prime Minister of New Zealand 1957, 1960–72. A successful farmer from north of Wellington, he entered New Zealand politics in 1932 as a member of the Reform Party* and spokesman for the farmers' interests. In 1936 his party amalgamated with the United Party to form the National Party. He was deputy Prime Minister and Minister of Agriculture under Sidney

Holland* 1949–57. When the latter had to retire in 1957 through ill health, he briefly took over, but lost the election in December. In 1960, however, he was successful, remaining in office until 1972. During those years he became a leading Commonwealth opponent of racism in Rhodesia and South Africa, but also supported US involvement in Vietnam and the participation of New Zealand troops there. As a farmer he was bitterly resentful of British attempts in the 1960s to enter the EEC*, since this threatened to end a century-old market for dairy and meat exports. At the same time he canvassed alternative markets in Australia and Asia. He became Governor-General of New Zealand after his retirement from politics in February 1972, being succeeded as Prime Minister by John R. Marshall. The latter failed to prevent a slide in the standing of the National Party, which was defeated by Labour led by Norman Kirk* in December.

Home Guard, a British volunteer force in World War II. In 1908 the Territorial Force, a home defence organization, had been created, which became the Territorial Army in 1921 and which in 1939 was mobilized. In 1940, after the fall of France, a new force was created, the Local Defence Volunteers, who were to guard especially against parachute landings by German troops. In the absence of arms they equipped themselves with old swords, pikes, and pitchforks. The name was changed to Home Guard in July 1940, and early volunteers were mostly ex-officers and men from World War I, too old to serve, together with boys of 17 and 18. A rifle, bayonet, and tin hat were issued, and by 1941 basic military uniform. The force was accepted as part of the armed forces of the Crown and subject to military discipline, although there was no pay. In 1942 enrolment became compulsory for certain sections of the civilian population. Early in the summer of 1943 its numbers reached 2 million. It was disbanded on 1 November 1944. It was never put to the test, but it did much to keep up British morale in the early years of the war.

Home Rule, Irish. In 1870 Isaac Butt had founded an association aimed at the repeal of the Act of Union of 1801 and the re-establishment of an Irish Parliament. There were to be four Irish Home Rule Bills presented to the House of Commons. The first two failed. The Third Bill was passed by Parliament in 1912 and provoked violent opposition in Ulster among the Protestant majority there. Its operation was postponed when the war broke out in 1914. It left unresolved the question as to how much of Ulster would be excluded from the Act. When the war ended the political situation in Ireland was greatly changed. The Easter Rising* and the sweeping majority for Sinn Fein* in the election of December 1918 were followed by unrest and guerrilla warfare. Lloyd George* was Prime Minister when the Fourth Home Rule Bill was introduced in the Westminster Parliament in 1920. It provided for parliaments in both Dublin and Belfast, linked by a Federal Council of Ireland. It was never fully implemented, since Sinn Fein had already established its own unofficial Parliament, the Dáil Éireann, although the Northern Ireland Parliament at Stormont* was established. In December 1921 the Anglo-Irish Treaty recognized the *de facto* situation, with the Dáil approving its proposals by sixty-four votes to fifty-seven. The Irish Free State was established, with dominion status; in 1949 it would become the Republic of Ireland. The majority group in the Dáil had voted for peace if only partial independence; the minority group, led by de Valera, wanted immediate full independence, and civil war followed.

Honduras, a Central American republic. A Spanish colony until 1821, it became a republic in 1838. An uninterrupted succession of dictator 'caudillos' dominated the remainder of the 19th century. Vast tracts of land had been given to the

United Fruit Company* for banana and other fruit cultivation, and in 1910 US adventurers installed a puppet President Manuella Bonilla. Improvement in the political process came slowly in the 20th century, with military dictators continuing to be more prominent than civilian presidents. The election in 1957 of Ramón Villeda Morales gave hope for the future. This optimism however proved premature, as the Honduran army overthrew him in 1963, before he could implement the reform programme he had pushed through Congress. Military entrenchment was further solidified as Honduras fought a border war with El Salvador in 1969. In 1982 a new US-backed constitution aimed to increase democratic activity, and under it Roberto Suazo was elected President; he came under intense pressure from the army, led by General Gustavo Álvarez. As a condition for US support, Suazo acted as a channel for US aid to 'Contra' rebels in Nicaragua*. In 1985 a new President, José Azcona, was elected. He removed General Álvarez, threatened to stop Contra activity, and supported the peace plan for Central America of Oscar Arias of Costa Rica*. Honduras, however, remained economically dependent upon the USA, and in November 1989 Rafael Callejas was elected President with US support.

Honecker, Erich (1912–   ), head of state of the German Democratic Republic 1976–89. Born in Saarland, he joined the Youth Section of the Communist Party of Germany and was active in resistance to Hitler until 1935, when he was placed in a concentration camp, being released by Soviet troops in 1945. In 1946 he became chairman of the Free German Youth in the Soviet-occupied zone of Germany. He rose in office within the party until elected to the Politburo in 1956 in the German Democratic Republic. In 1974 he became chairman of the Defence Council and in 1976 head of state. In 1989 he lost all his offices and was accused of gross corruption, but spared trial owing to ill health.

Hong Kong, British Crown Colony on the south coast of China, consisting of Hong Kong Island, Kowloon, and the New Territories (comprising some mainland territory and many small islands). Hong Kong was occupied in 1841 and in 1898 its hinterland was extended when the New Territories were leased from China for ninety-nine years. It at once grew rapidly as a trading centre, attracting Europeans and Chinese. After two weeks' fighting it surrendered to the Japanese on 25 December 1941, its remaining European colonists being confined to prison camps. After it was reoccupied by the British in 1945, there was an influx of refugees and capital, especially from Shanghai, following the Communist victory in China. The UN embargo on trade with China during the Korean War stimulated the development of industry and further growth of its financial institutions. During the 1970s and early 1980s it became one of the world's most important economic and business centres, with a population rising to some 5½ million in 1985. With the end of the lease on the New Territories approaching, Britain agreed in 1984 to transfer sovereignty of the entire colony to China in 1997, with China undertaking not to alter the existing economic and social structure for fifty years. Hong Kong's administration is by Governor, Executive Council (by appointment), and Legislative Council (by appointment and election). Although the lack of a strong elective element had been acceptable to the majority of the wealthier business class, a growing demand for democratic government was already arising before the incidents of Tiananmen Square*, after which business confidence lapsed and there was a considerable exodus from the colony. There was pressure for Britain to introduce more democratic government before 1997, and criticism of the immigration legislation of 1981, the British Nationality Act, whereby the 3 million Hong Kong citizens with British passports were to be excluded from entry to Britain.

Hoover, John Edgar (1895–1972), director of the Federal Bureau of Investigation* 1924–72. A graduate of George Washington University in 1916, he at once entered the US Department of Justice and in 1921 the Federal Bureau of Investigation. He was appointed director in 1924, with the task of raising standards after a period of disrepute. He achieved this by vigorous selection and the training of personnel, establishing the FBI National Academy to train agents from all over the USA. He also established a scientific crime-detection laboratory, which pioneered techniques in forensic science. During the 1930s his widely publicized entrapment of certain criminals, while not ending syndicate crime, earned the FBI a reputation for integrity. After World War II his increasingly authoritarian style and persistent interest in the sex lives of various public figures, for example Robert Kennedy, clouded his supposed political impartiality. Although he retained a high international reputation, his antipathy to Black activists and to the civil rights movement earned him widespread criticism. He died in office.

Hopkins, Harry Lloyd (1890–1946), US presidential adviser. Born in Iowa, as a young man he came to New York as a social worker. When Franklin D. Roosevelt* was Governor of New York state 1928–33 he made Hopkins his adviser on social and welfare policies, a position he retained when Roosevelt became President and throughout the years of the New Deal*. His record in public social service was perhaps without equal: head of New York's Emergency Relief Administration (1931); of the Federal Emergency Relief Administration (1933); of the Works Progress Administration (1935); and then Secretary of Commerce (1938–40). He was Roosevelt's manager when he ran for a third term as President in 1940. Before and after the USA entered World War II he served as Roosevelt's untitled second-in-command. He played a pivotal role in the San Francisco Conference* of 1945 which launched the Charter of the UN, and in the last 'Big Three' conference at Potsdam. He was a close friend of Winston Churchill.

Horthy de Nagybanya, Nikolaus Miklós (1868–1957), Regent of Hungary 1920–44. He graduated from the Imperial Naval Academy of Fiume and became an ADC of the Emperor Francis Joseph. By the close of World War I he was Commander-in-Chief of the Imperial Fleet with the rank of admiral, returning to Hungary in 1918. The following year he organized an army to overthrow Béla Kun's* Communist regime, entering Budapest in triumph as Administrator of the Realm. In January 1920 the Hungarian Parliament voted to restore the monarchy, with Horthy as Regent. This post he retained, but thwarted all efforts of the deposed Charles* to regain his Habsburg throne. Charles died in 1922 and his son the Archduke Otto accepted the fact that Horthy would not have him back. Determined to maintain a quasi-feudal social order, Horthy's rule became increasingly dictatorial during the 1930s. In 1940 he won the Vienna Award* which restored much of Transylvania to Hungary from Romania, and he declared war on the Soviet Union as an ally of Germany. But he was on bad terms personally with Hitler and in 1944 tried unsuccessfully to negotiate a separate peace with the Allies. German troops invaded Hungary and he was imprisoned in Germany. He was released by the Allies in 1945 and died in exile in Portugal.

Houphouët-Boigny, Félix (1905–), President of the Côte d'Ivoire* 1960– . Son of a chief in the French colony of Ivory Coast, he graduated in medicine in Paris before returning to the family farm in 1940. In 1944 he was co-founder of the Syndicat Agricole Africain, formed to protect Africans against European agriculturalists. He represented the Ivory Coast in the French Assemblée (1945–59) and in 1958 was appointed as Minister of Health in Paris; but he resigned when his country was offered self-government within the French Community*. In 1946 he had formed the Parti Démocratique de la Côte d'Ivoire. At first allied with the Communist Party, he broke with it in 1950 and co-operated with the French to build up his country's economy. With full independence in 1960 he became President of the new Republic and has been re-elected five times. A great admirer of de Gaulle, he has kept close links with France, and his international policies have been recognizably moderate.

Hoxha, Enver (1908–85), first secretary of the Albanian Communist Party 1945–85. Born near the Greek border of Albania, he was educated and lived for a while in France, where he joined the French Communist Party. He returned to Albania in 1936, and when Italy invaded his country he became leader of the Resistance movement. Operating from the mountains, he steadily built up a National Liberation Army, which by November 1944 had control of most of the country. He formed a provisional government of the Albanian Republic in 1945, and ruled the country with a ruthless hand until his death. He remained a staunch Stalinist, breaking with Khrushchev* in 1956 and building close links with Mao Zedong* in China. By the time of his death Albania was diplomatically isolated from all the rest of Europe and the USA.

Hua Guofeng (Hua Kuo-feng) (1920–), Chinese Premier 1976–80. Born in Shanxi province, he joined the Long March* when he was 14, serving for twelve years under Zhu De* in the 8th Route Army. He became Deputy Governor of Hunan province and survived the Cultural Revolution*, to become a member of the Politburo (1973) and to succeed Zhou Enlai* as Premier. On the death of Mao Zedong* he defeated the Gang of Four* in October 1979, becoming chairman of the Central Committee. His rival for power was Deng Xiaoping*, and he resigned as Premier in 1980 and as chairman in 1981 to be replaced by Deng's supporters.

Huai-Hai campaign, a decisive campaign in the Chinese Civil War, fought in the valley of the River Huang (Huai) in the Shandong and Jiangsu provinces, November 1948–January 1949. In September 1948 Chen Li* had swept south and captured Jinan (Chinan) with its garrison of 80,000 Nationalist troops under General Tu Li-ming, who then withdrew to defend the railway town of Xuzhou (Suchow). A series of engagements followed. 6–22 November Chen Yi engaged Huang Po-tao around the railway town of Nienchuang, suffering 100,000 casualties, but nevertheless defeating Huang, who committed suicide. Xuzhou fell on 1 December and Tu withdrew and regrouped his three army groups, with tanks and artillery, to the south-west around Yungchung. Battle lasted from 6 December to 6 January, when Tu was himself captured, in disguise, and some 327,000 Nationalist prisoners taken. Meanwhile the communist general Liu Bocheng now moved against the beleaguered Nationalist forces, attacking what had been Tu's stronghold of Ch'inglungchi to the north-east. It fell on 22 January, leaving the way open for an advance on Nanjing and Shanghai, which fell in the spring. The Chinese Communists had conquered China. Nationalist losses in this campaign amounted to over half a million.

Huerta, Victoriano (1854–1916), President of Mexico 1913–14. A professional soldier in the Mexican army, in 1910 he joined Francisco Madero in his bid to oust Porfirio Díaz. Appointed commander of the new federal forces in 1912, the next year he led a coup against Madero, who was murdered. As President he instituted a ruthless dictatorship in which torture and assassination became commonplace. Defeated by Carranza*, he fled to Spain and then to the USA. An attempted return to power in 1915 ended unsuccessfully, when he was arrested by US agents while trying to cross the US–Mexican border. He died in a US prison.

Huggins, Godfrey (Martin), Viscount Malvern (1883–1971), Prime Minister of Southern Rhodesia 1933–53 and of the Central African Federation 1953–6. As a doctor he served in World War I, having gone out to Rhodesia in 1911. He entered politics in 1924. As leader, first of the Reform Party and then of the United Party, he was Prime Minister of Southern Rhodesia for twenty years, giving strong support to Britain during World War II, his country providing valuable training facilities for the Royal Air Force. He was convinced of the need for economic integration with neighbouring Northern Rhodesia and Nyasaland, and was an enthusiastic architect of the Central African Federation*, of which he was first Prime Minister, retiring to his farm in 1956. Having at first supported racial segregation, he came to believe in the inevitability of integration, particularly as industrialization proceeded.

Hughes, Charles Evans (1862–1948), US Secretary of State 1921–5; Chief Justice of Supreme Court* 1930–41. Born in New York state, he became a successful member of a New York city law firm and in 1906 was elected state Governor. From 1908 until 1916 he served as a member of the federal Supreme Court, but resigned to fight the presidential election of 1916. Defeated by the Democrat incumbent Woodrow Wilson, he became Secretary of State under Presidents Harding and Coolidge, when he urged US entry into the League of Nations, negotiated a separate peace treaty with Germany, and hosted the Washington Conference* 1921–2, which achieved some naval limitation. His career was crowned by his years as Chief Justice, during which he enhanced the efficiency of the federal court system, and gave firm support to the freedom against state actions guaranteed to the citizen under the First Amendment. He was largely responsible for defeating a plan of President Roosevelt* in 1937 to 'pack' the Court by adding to it extra liberal judges to counter sitting members over 70 years of age who refused to retire. At the same time however his court supported a number of New Deal* proposals, such as the Social Security Act.

Hughes, William Morris (1864–1952), Prime Minister of Australia 1915–23. He had emigrated from Wales in 1884 and was first elected to the federal House of Representatives in 1901 as a Labor member. He sat in Parliament for over fifty years, although representing various parties. A controversial figure, he was strongly pro-British and violently anti-German in World War I, having succeeded Fisher in 1915. In November 1916 the Labor group in Parliament split over the issue of conscription* and Hughes left it, forming a coalition government and a Nationalist Party. This itself split in 1923 when the Bruce*–Page* coalition was formed, since the Country Party* would not serve with him. In 1931, in the international crisis of the Great Depression*, he helped to form the United Australia Party (UAP) and served under J. A. Lyons* in various offices. Strongly racist, he had given much offence to non-European nations at the Paris Peace Conference and remained staunch for White Australia. Briefly leader of the UAP 1941–3, his demagogic approach was uncongenial to Robert Menzies*. When the latter formed his new Liberal Party in 1944 it absorbed the UAP, but when it came to office in 1949 Hughes was left on the back benches.

Hukbalahap movement (People's Anti-Japanese Army), a left-wing nationalist movement in the Philippines. Before World War II tenant organizations were developing, opposed to the great estate owners with their close links with the USA. Their leader on Luzon (the largest Philippine island) was Luis Taruc*, who during the Japanese occupation mobilized these organizations and formed the People's Anti-Japanese Army. In the years 1942–5 they carried out increasingly daring and successful jungle operations, and by the end of the war virtually controlled central

Luzon, the 'breadbasket of the Philippines'. Known now as the Huks, they were reluctant to give up their arms, and when Taruc was refused a seat in the Congress although elected (April 1946), they retreated to the jungle again and began terrorist activities against the great estates as the People's Liberation Army. By 1950 there was more or less open war against the landlord élite. In that year their headquarters in Manila was raided and many leaders arrested. At the same time, with the outbreak of the Korean War and the expansion of US bases in the Philippines, the Huks were seen as a major Communist threat. The USA supported President Roxas* in his determination to destroy the movement. Ruthless action using sophisticated US weaponry had largely achieved this by 1954, when Taruc surrendered, although the Communist New People's Army* was to emerge in the 1970s.

Hull, Cordell (1871–1955), US Secretary of State 1933–44. Born in Tennessee, he became a successful lawyer and entered state politics as a Democrat in 1901. He was then elected to Congress and served in the Senate 1931–3. He resigned to become Secretary of State for F. D. Roosevelt*. He early established the Good Neighbour* policy with Latin America and sought to revive world trade by getting Congress to pass the Reciprocal Trade Agreements Act (March 1934), which was to be a forerunner of GATT*. Because of his efforts to improve relations with Latin America, he created there a climate favourable to the USA when it entered World War II in 1941. He had been a staunch supporter of China in its war against Japan. As soon as war was declared, he began work on creating a post-war peace-keeping body, initiating a Moscow Conference for Foreign Ministers in 1943 from which was to develop the plan for a United Nations. He was never actively involved in the day-to-day planning and decision-making of the war, President Roosevelt allegedly finding him 'too cautious'. He retired in 1944 in ill health.

Humphrey, Hubert H. (1911–78), US Senator 1949–65, 1971–8, Vice-President 1964–9. From a modest family in South Dakota, he trained as a pharmacist and held a number of jobs including that of army air force instructor before entering local politics as a Democrat representing the farming interest. He became Mayor of Minneapolis and was then, in 1948, elected to the Senate, where he developed a reputation for brilliant debating skills as well as for the highest integrity. A presidential candidate in 1960, he became a strong supporter of J. F. Kennedy* and was appointed Vice-President in 1964, after the latter's death. As such he supported US involvement in the Vietnam War. Defeated in the presidential election of 1968, he returned to the Senate, where he served until he died, regarded as one of its 'giants' of the 20th century.

Hundred Flowers campaign (1956–7). Drawing its name from the Chinese Classics, 'let a hundred flowers bloom and a hundred schools of thought contend', the campaign followed the denunciation of Stalin in the Soviet Union. It was launched by Mao Zedong*, who argued that self-criticism would benefit China's development. When criticism of the Communist Party and other institutions began to appear in the Press and on posters, Mao and the party changed course. Critics were attacked as 'rightists' and deprived of citizenship rights.

Hungarian Revolution (23 October–4 November 1956). The de-Stalinization process which was begun in February 1956 at the Twentieth Congress* of the Soviet Communist Party created an atmosphere of hope and patriotism in Hungary. The hated party leader Mátyás Rákosi* was persuaded by Khrushchev to resign in July, being replaced by the almost as repressive Ernó Geró. There were initial demonstrations in Budapest and Soviet tanks appeared in the city. A huge demonstration of students and workers on 23 October was supported by Hungarian soldiers, when the massive statue of Stalin was torn down. Next day on 24 October two former victims of Rákosi, Imre Nagy* and János Kádár*, formed a coalition government, with Nagy Prime Minister. In response to demands for greater democratization he legalized political parties and also released Cardinal Mindszenty*, who had been imprisoned since 1948. The government now announced that it wished to follow Austria and make Hungary a neutral state, withdrawing from the Warsaw Pact. János Kádár at this point withdrew, going to east Hungary, from where he returned with Soviet tanks on 4 November. Budapest was shelled and resistance in the city was soon overcome. Nagy was replaced by Kádár, while some 190,000 Hungarians fled into exile. The Soviet Union reneged on its pledge of safe conduct, handing Nagy and other prominent figures over to the new regime, which executed them in secret.

Hungary. When Austria established the Dual Monarchy of the Austro-Hungarian Empire* in 1867, this was first and foremost an alliance of Hungarian Magyars and Austrian Germans against the Slav nationalities. Defeat in World War I led to the loss of three-fifths of Hungarian territory and two-thirds of its population. The democratic Republic of Károlyi* was followed briefly by the Communist regime of Béla Kun*. Dictatorship under Horthy* was established in 1920 and lasted until 1944. In the later 1930s Horthy allied himself with the Axis powers and in return gained parts of Slovakia and Ruthenia in 1938 by the First Vienna Award*. In March 1939 the remainder of Ruthenia followed. By the Second Vienna Award of 1940 he won three-quarters of Transylvania* from a reluctant Romania. In 1941 he declared war against the Soviet Union, following the German invasion in June. He formally joined the Axis and in December declared war against the USA and Britain. Two-thirds of the Hungarian army died on the Eastern Front. Horthy unsuccessfully sought to make peace early in 1944 and was imprisoned by the Germans. In September Hungary was invaded by Soviet and Romanian troops, and it dutifully then declared war against Germany. The war cost two-thirds of the national wealth, and reparations* of £75m. to the Soviet Union followed, with defeat bringing not only the loss of territorial gains since 1938, but also the creation of the communist 'Republic of Workers and Working Peasants', with Rákosi* first secretary of the party, a rigid Stalinist. By 1953 18 per cent of arable land had been collectivized, but without popular support, Imre Nagy* briefly replacing Rákosi as Premier that year. Following the crisis of the Hungarian Revolution* in 1956, the long regime of János Kádár* introduced an element of liberalization within the Soviet bloc, but with rigid police surveillance continuing. Kádár retired from being party general secretary in May 1988 and in July 1988 a Hungarian Democratic Forum was founded. Other parties followed, including a revived Social Democratic Party and a Smallholders' Party. In June 1989 a rehabilitated Nagy was disinterred and given a massive state funeral in Budapest. The State Office for Church Affairs was closed and diplomatic relations with the Vatican resumed. In October 1989 the Hungarian Socialist Workers' Party (the Communist Party) was restructured by its reformist wing into the Hungarian Socialist Party. At the same time multi-party politics were allowed to operate officially. Elections were held in April 1990, which the Hungarian Democratic Forum, led by Jozsef Antall, won. Arpad Göncz, who had been imprisoned 1956–62, was elected interim President. Tension developed with Romania when 70,000 ethnic Hungarians living in Transylvania demonstrated in Bucharest in March.

Hunger marches, a series of marches to London between the wars. Following World War I, unemployment became an immediate problem, concentrated in 'depressed areas' of dense population and declining industry. The first march was from Glasgow to London in October 1922, supported by the strongly socialist workers of Clydeside. There was another in January 1929. The biggest was that of October 1932 when, with nearly 3 million unemployed, 3,000 men and women marched from South Wales, Glasgow, and north-east England, presenting a petition to Parliament with over 1 million signatures protesting against means-tested relief. The so-called 'Jarrow Crusade' of 200 men from Jarrow in October 1936 was specifically a protest against unemployment in Jarrow, then running at 80 per cent of the work-force. It was headed by the Labour MP for Jarrow, Ellen Wilkinson.

Hussein, Saddam (1937–), President of Iraq 1979– . Born in Tikrit, he was educated at the Universities of Cairo and Baghdad. Known then as Saddam Hussein Takridi, he joined the Ba'ath Socialist Party* in 1957 and escaped to Syria and then Egypt, following a coup against Prime Minister Abdul Karim Kassem*. He returned to Iraq*, but was imprisoned in 1964 for opposing Abdul Salam Aref. He took part in the July 1968 revolution led by General Bakr, as assistant secretary of the Ba'ath Party and a prominent member of its 'civilian' wing. In 1969, following the execution of fifty-one prominent Iraqis on charges of espionage, he became deputy chairman of the Revolution Command Council. In 1976 he was granted the rank of general and in 1979 succeeded Ahmed Bakr as chairman of the RCC and President of Iraq. Soon after launching the Iran–Iraq War* he ruthlessly purged the Ba'ath Party in Iraq, orchestrating an elaborate personality cult in order to win popular Iraqi support. He was provided with lavish technical aid and expertise from the West to fight the war, being then strongly opposed to Islamic fundamentalism* and to Kurdish independence. In August 1990 he ordered the occupation of Kuwait* and in January 1991 was at war with UN forces to defend this. In order to win and maintain popular Arab support through the Middle East and Africa, he moved steadily towards ever greater deference to Islam, in spite of the secularism of his Ba'ath state.

Hussein ibn Ali (1856–1931) (Sherif Hussein), Arab political leader and King of the Hejaz*. A member of the Hashemite* family, he was made Sherif of Mecca under Turkish rule in 1908. In June 1916 he began the Arab Revolt*, calling himself 'King of the Arab countries', a title many would not accept, including Abd al-Aziz ibn Saud*, now Emir of Nejd and al-Hasa. He refused to ratify the Versailles Peace Settlement, arguing that he had been betrayed by France and Britain, who imposed mandate regimes on Syria, Palestine, and Iraq, leaving him only as King of the Hejaz. The caliphate of Islam had ended with the Ottoman Empire and in 1924 he proclaimed himself Caliph. This gave his old rival Ibn Saud an excuse to invade the Hejaz. The cities of Jedda, Medina, and Mecca were lost and he abdicated, being given asylum in Cyprus. One son was installed by the British in Iraq and the other in Transjordan.

Hussein ibn Talal (1935–), King of Jordan 1953– . Grandson of Abdullah ibn Hussein*, he succeeded his father Talal, who was deposed in 1952 as insane. Socially conservative and strongly backed by his army (the Arab Legion*), in the early years of his reign he met continued opposition from pro-Egyptian factions. Although backed and armed by the USA, his regime suffered defeat in the Arab–Israeli war of 1967, when he lost the West Bank* to Israel. PLO* guerrillas now based themselves in Jordan, and during 1971 fighting developed between these and his army. Generously funded by Saudi Arabia, he successfully reasserted his authority, although strongly attacked by many of his Arab neighbours as 'the butcher of Palestinian resistance'. From 1979 onwards his policies became more flexible. During 1985 he was reconciled with Yasir Arafat* and was at the centre of an ongoing peace initiative for the Middle East. In 1988 he renounced all Jordanian claims on the West Bank. During the Gulf Crisis* of 1990 he strenuously sought to prevent the outbreak of war.

Hutu (Bahutu), an agricultural people found in both Rwanda* and Burundi*. They originally displaced earlier Pygmy hunters but were themselves conquered by Nilotic warrior-pastoralists, the Tutsi*, in the 15th century. Under the latter they became a vassal people paying tribute to the Tutsi nobility by labour and produce. Although there was some cultural integration over the centuries, bitter animosity survived, erupting in violence in the mid-20th century. In 1959 in Rwanda the ruling Tutsi were expelled but in Burundi in 1972 a similar revolt against the Tutsi failed, with over 100,000 Hutu killed.

Hu Yaobang (1915–89), Chinese politician. Born in Hunan province, from 1927 he was engaged in youth work in the Jiangxi Soviet*, joining the Communist Party in 1933. He took part in the Long March* and became a close ally of Deng Xiaoping*, with a distinguished record in the war against Japan 1937–45, serving for a while with the Soviet Red Army. Early in the days of the Peoples' Republic he again worked on youth programmes and served on a commission for the elimination of illiteracy. Criticized during the Cultural Revolution*, he was sent to a re-education centre, but was rehabilitated in 1973. Again dismissed by the Gang of Four* in 1976, he became general secretary of the CCP Central Committee in 1979 and replaced Hua Guofeng* as chairman in 1981, still a protégé of Deng Xiaoping. His was a pragmatic approach, eschewing the rigidity of Maoist ideology and strongly encouraging free enterprise and a free market. During the beginnings of student unrest in 1987 he was dismissed for 'mistakes on major issues of political policy', being replaced by Zhao Ziyang. He remained a popular figure, however, and his death in 1989 sparked off the student unrest which was to culminate in the Tiananmen Square Massacre*.

Huyuh Tan Phat (1913–   ), Vietnamese politician. While studying architecture at Hanoi University in the 1930s he became a student activist against the French. He edited an anti-Japanese newspaper during World War II, after which he became secretary-general of the Vietnam Democratic Party. He joined the VietMinh* as a non-Communist and, after the French Indo-China War, opposed Ngo Dinh Diem* in Saigon, joining the National Liberation Front* in 1960 and serving as its general secretary 1964–6. He became President of the Provisional Revolutionary Government in 1969, although still a non-Communist, and since 1975 has held various government offices.

Hyderabad. Formerly a princely state in India. Following an agreement with Lord Curzon in 1902, the nizams of Hyderabad became rulers of the largest princely state in British India. During the 1920s and 1930s they continually resisted demands from the state Congress for more responsible government, and in 1947 the Nizam declared Hyderabad an independent state. The state Congress, supported by the Indian National Congress*, started civil resistance, and a period of political chaos was ended in September 1948 by the Indian army. The Nizam then accepted pressure that Hyderabad join the Union, with himself as Princely Governor. In 1956 it was subdivided between Andhra Pradesh, Mysore, and Bombay, itself ceasing to exist, and with the Nizam becoming a plain citizen of India.

Hydrogen bomb, a device to utilize a thermonuclear reaction in order to produce an explosion. The first such bomb was exploded by the USA on Enewetak Atoll on 1 November 1952. Its force was equivalent to 14 megatons of TNT, that is, more than the total of bombs dropped by both sides in World War II. The radioactive 'fall-out', particularly of radioactive strontium, was greater than expected and far greater than for 'ordinary' atomic bombs. The Russians tested a bomb in August 1953, the British on 15 May 1957, the Chinese in 1967, and the French on 25 August 1968. Later US bombs tested in 1954 on Bikini Atoll in the Marshall Islands made the island uninhabitable for the foreseeable future. The destructive power of the bomb horrified Winston Churchill and prompted a campaign for nuclear disarmament*; but bombs were steadily stockpiled through the Cold War over the next forty years.

i

Ibáñez (del Campo), Carlos (1877–1960), President of Chile 1927–31, 1952–8. Of wealthy parents, he had a successful career in the army and took part in 1924 in a military coup against President Arturo Alessandri*. When the latter was restored, he briefly became his Minister of War before seizing power himself in 1925. Elected President in 1927, he at once imprisoned or exiled his political opponents, but could not prevent Chile's economy from collapsing in the Great Depression*. In 1931 he went into exile, but returned in 1937 to lead a series of unsuccessful neo-Fascist uprisings. In 1952, on a highly reactionary ticket, he won the presidential election. To everyone's surprise his regime was conciliatory, encouraging economic growth and seeking to end corruption. It won the support of the more progressive middle class and of the trade unions, promulgating a Labour Code. Nevertheless it failed to restore the economy and he was succeeded by Jorge Alessandri*.

Ibarruri, Dolores (1895–1989), Spanish politician. Born in poverty near Bilbao, she worked as a seamstress and cook and joined the Socialist Party in 1917. Next year she began to write in a miners' newspaper, assuming the pseudonym 'La Pasionara', by which she was always afterwards known. She had married a miner, Julian Ruiz, in 1915, and he persuaded her to join the Communist Party soon after it was founded in 1920. She began to win a national reputation, being imprisoned on numerous occasions, twice by the Republic after it was founded in 1931. In 1934 she founded an Anti-Fascist Women's Group and in 1936 was elected to the Cortes. During the Civil War she used her rhetorical skills on the radio to exhort the Republican troops, becoming known internationally. She escaped to Moscow in 1939, where she lived as the recognized leader of exiled Spanish Communists. She returned to Spain in 1977, taking an active part again in Bilbao politics.

Ibo (Igbo). A people of the forest belt of Eastern Nigeria in the Niger delta, living in forest clearings and cultivating the yam. Their village structure, with its Council of Heads of Family, was seriously eroded early in the 20th century by British administration, which sought to impose chiefs through whom the policy of 'indirect rule'* could be sustained. The Ibo responded quickly to Christianity and to Western culture and by mid-20th century were a dominant element of society throughout Nigeria. Jealousy of their power and influence was the main factor in the 1966 disturbances, which in turn resulted in a bid for independence in the Biafran War*.

Iceland. It had acquired limited autonomy from Denmark in 1874 and independence in 1918, although it shared its king with Denmark until 1943. It was occupied by the British in 1940 and by the USA in 1941. After a plebiscite it became an independent republic in 1944. It joined the UN and NATO, and has since engaged in disputes with Britain and West Germany over fishing limits. Widespread opposition to the presence of US military bases developed in the 1970s, President Finnbogadóttir being first elected in 1980 on this issue. She has since been twice re-elected. Iceland became a nuclear-free zone (although still a member of NATO) in 1985. Its governments have tended to be coalitions between Social Democrats (SDP), People's Alliance (PA), and the Progressive Party (PP). In 1986 it hosted the USA–USSR summit which began a cautious process towards disarmament.

Ikeda, Hayato (1899–1965), Prime Minister of Japan 1960–4. He entered the government tax service and rose by 1945 to become head of the Japanese National Tax Bureau. Having served as Vice-Minister of Finance in the Yoshida* cabinet of 1947, he was elected to the House of Representatives in 1949 and became successively Minister of Finance and Minister of International Trade. As Prime Minister he devoted himself to sustaining Japanese economic growth through a broadening of international trading connections.

Illich, Ivan (1926–), social and educational writer. Born in Vienna and educated in Rome and Salzburg, he emigrated to the USA in 1951 and in 1961 he helped to found the Center for International Documentation in Mexico, where he worked until 1976. In 1986 he became Professor of Humanities and Sciences at Pennsylvania State University. His critique of society, in such books as *Deschooling Society* (1971), *Energy and Equity* (1973), and *Medical Nemesis* (1975), made him one of the most influential thinkers of the 1970s. The emphasis was on the need for people to be more actively involved in decisions affecting their lives, in such matters as religion, education, health, design. Demystification and deprofessionalization were features of his approach.

IMF *see* INTERNATIONAL MONETARY FUND.

Imjin River, Battle of the (22–5 April 1951), a defensive and inconclusive battle in the Korean War. It was fought between Chinese/North Korean forces and a contingent of British troops serving under UN command, cut off on Hill 235 on the 38th parallel. Public opinion in Britain was aroused, the action being seen at the time as a 'heroic defence of Western values' against Chinese Communist aggression. It was described in Parliament as 'one of the most glorious actions in which the Gloucestershire Regiment has taken part'. British casualties, killed, wounded, and missing, amounted to 1,074.

Immigration Acts (UK). In 1905 an Aliens Act passed Parliament aimed at restricting immigration into Britain of the poorer victims of anti-Jewish pogroms* in Russia, Poland, and south-east Europe. It required all potential immigrants to show that they could support themselves and their dependants; but it did not apply to immigrants from the British Empire. These had free access until after World War II. During the later 1940s and the 1950s there was a large influx of labour from the British West Indies, from India, and from Pakistan; but by the early 1960s racial problems began to develop as the post-war economic boom was ending and unemployment beginning to recur. In response to popular pressure the Conservative government enacted a Commonwealth Immigration Act in 1962, which required all Commonwealth immigrants to have a 'special skill' which would be useful to Britain. It was tightened by the Labour government in 1968 and further extended in 1971, when a single system for aliens and Commonwealth citizens was introduced. This legislation was bitterly resented in Australia and New Zealand, which had always felt close links with 'home', and it did much to loosen the bonds of association within the Commonwealth*. The British

Nationality Act of 1981 further restricted movement, being aimed particularly at the 3 million Hong Kong citizens holding British passports.

Immigration Laws (USA). Although the Statue of Liberty was erected in 1886, welcoming the world's 'homeless' and 'wretched refuse', already in 1882 immigration of Chinese labour had been checked, while in 1887 the anti-immigration American Protective Association was formed. In 1891, by an Undesirable Persons Act, all immigrants had to be screened on Ellis Island*. Japanese immigration was also unpopular, and in 1908 it was much reduced. By then the biggest waves ever of mass immigration from Europe were taking place, some 16 million between 1890 and 1917, mostly from Central and south-east Europe. Three Bills imposing tests on immigrants had failed in 1896, 1913, and 1915, before a Bill passed Congress, over President Wilson's veto, in 1917, imposing a literacy test. In 1920 Congress passed an act to deport aliens and anarchists and in 1921 the first Quota Law, fixing the number of immigrants for each nation. This was stiffened in 1924 and became properly effective. It was not applicable to Canada nor at first to Latin America, while Asian immigrants were virtually excluded. The Great Depression*, however, had a great effect on immigration: in 1932 only 35,570 immigrants arrived. Immediately after World War II special directives empowered admission of displaced persons from Europe and again in 1953 a refugee Relief Act gave visas to some 186,000 escapees from communist regimes above quota. In 1952 the McCarran–Walter Act* updated and codified the 1924 legislation, while in 1965 an Immigration and Nation Act revised McCarran–Walter, introducing greater flexibility into the system, putting all quota countries on an equal footing, with 20,000 maximum from any one country and a total of 170,000 per annum. Since then there has been a steady increase of migrant labour, in a largely uncontrolled fashion, moving into California and the south-west from Mexico.

Immigration Policies (Australia and New Zealand). Immigration patterns and control into Australia and New Zealand have been closely parallel. During the later 19th century immigration was mostly from the British Isles, stimulated by gold (Australia 1850s and New Zealand 1860s) and in the 1880s by refrigeration of shipping, which greatly expanded the meat trade. In 1855 Victoria enacted a Bill against Chinese, and in 1881 one followed in New Zealand. At the turn of the century both countries extended their restrictions to other Asiatic peoples: Thais, Polynesians, as well as people from elsewhere in the British Empire—India and Malaya. Although there was no restriction on European immigrants, neither country proved particularly attractive to German, Italian, or Scandinavian immigrants, unlike the USA. Through the 20th century control of immigration was largely discretionary, in the hands of customs officials. Both countries operated assisted passage immigration from Britain and Europe in the years 1947–65, when a total of 1½ million entered Australia, resulting in a powerful cultural stimulus. During the 1960s tighter entry and landing controls were introduced and Australia introduced a quota system, the criterion being 'skills and availability of jobs'. Britain passed a Commonwealth Immigrants Act in 1972, which applied to all Commonwealth members and which was deeply resented in both countries. Australia reciprocated by ending British citizen rights. At the same time immigrant quotas were reduced.

New Zealand repealed its 1881 anti-Chinese legislation in 1944, but both countries effectively restricted non-White immigration until the mid-1960s, when it began to ease, and a low-key transformation began, Australia allowing a quota of non-White immigrants and ending all racist restrictions in 1973. After this both countries admitted more and more Asiatic students, while there was a marked increase in Asiatics in low-paid service employment, especially in the inner cities. Neither country however has yet moved appreciably to becoming a multi-cultural society.

Imperial conferences, meetings in London of the prime ministers of the United Kingdom and the British dominions. The first was held in 1911. These conferences were successors to earlier periodic colonial conferences begun in 1887 at Queen Victoria's Golden Jubilee. In 1907 it had been agreed to adopt the Canadian term of 'dominion' for all British 'White' colonies. In 1918 an Imperial War Conference discussed both war strategies and the supply of raw materials. Representatives from India attended the 1921 Conference, when it became clear that the dominions were determined to gain total independence over external as well as internal affairs. The 1923 Conference was attended for the first time by the Irish Free State and the issue of imperial preference was discussed at length, as well as disarmament. The Conference of 1926 discussed a formula of Lord Balfour* which was to result in 1931 in the Statute of Westminster*. The 1930 Conference was dominated by the world economic crisis and was followed by the Ottawa Conference in 1932. The final Conference followed the coronation of George VI in 1937, and discussed the worsening international situation. After World War II regular Commonwealth conferences were to be held.

Imperialism, extension of a country's influence over others less powerful. The acquisition of colonies to build an empire, in order to extend the global influence of the state, was widely advocated in Western Europe on the eve of World War I, with Britain, France, Belgium, The Netherlands, Portugal, and Germany the leading imperialist powers. For a short while after the war such imperialist ambitions continued, especially in France and Britain, who sought to absorb much of the Middle East into their empires under the Sykes–Picot Agreement*. Since 1923, however, the story has been one of steady attrition and dismantlement of these empires through decolonization*, so that by 1970 the term imperialism was being used, mostly by Marxists, to describe the activities of international capitalism. This was how Lenin* had already defined it in his book *Imperialism* (1916) and in 1965 Kwame Nkrumah* attacked the multinational corporations* in his book *Neo-colonialism: The Last Stages of Imperialism*. Within the Soviet Union critics of the regime attacked the extent to which Russians had imposed imperialist policies upon the non-Russian republics.

Incident of 26 February 1936, an attempted military coup in Japan. Young extremists of the Imperial Way faction (Kodo-ha) had been active within the Japanese army since the late 1920s, intent on using violent means to overthrow conservative civilian governments and to set Japan on a course of military expansion, particularly in China. Their activities culminated in an attempted coup, in which several prominent politicians were brutally murdered (the Prime Minister Okada Keisuke* only escaping through a case of mistaken identity) and much of central Tokyo seized. The revolt was put down on 29 February and most of its leaders executed, after which leadership of the military expansionist cause passed to the more moderate Control faction (Tosei-ha).

Independent Labour Party (ILP) (UK). It was founded in 1893 at Bradford, Yorks., under the leadership of Keir Hardie, as a socialist organization. Ramsay MacDonald* and Philip Snowden* were both early members. In 1900 it helped to form the Labour Representation Committee, which in 1906 became the Labour Party, with which the ILP remained in uneasy alliance until 1932. In the later 1930s its sympathy for Communism, its pacifism, and its theoretical approach to politics were regarded as electoral liabilities by Labour politicians, who were becoming increasingly pragmatic. Its chairman from 1926 was James Maxton*, MP for Bridgeton, and it became increasingly centred on Glasgow. After his death in 1946 its representation in Parliament declined from three in 1945 to nil in 1950. However, it had been a significant body in Labour's early years, especially in local government and the discussion of foreign affairs. It helped to bring many middle-class idealists into the Labour Party and also developed a newspaper press.

India, British. By 1914 British India consisted of thirteen provinces administered by the Indian Civil Service, whose British senior officers were specially selected and trained, backed by a police force and the Indian army*, both also staffed with British officers. In addition there were some 700 principalities, or princely states, covering two-fifths of the land. These varied in size from great kingdoms such as Hyderabad*, Mysore, and Kashmir* to tiny tribal fiefdoms; but all retained considerable autocratic autonomy, while accepting the Emperor George V and his Viceroy. The Indian judiciary had made considerable progress in Indianization and was already challenging the racist assumptions of the British Raj*.

The development of Western-style higher education, begun with the establishment in 1817 of the Hindu College, Calcutta, had produced a middle-class attracted to the ideologies of nationalism and liberal democracy. Anxious to ensure co-operation of politicized Indians, the British government had provided for limited association of representative Indians within the legislatures by the Councils Act (see INDIA ACTS) of 1909. In 1917 it promised 'progressive realization of responsible government' and in 1919, by the Government of India Act (see INDIA ACTS), transferred some responsibilities to elected ministers in the provinces. Agitation against a Bill for suppression of sedition led to the notorious massacre at Amritsar*. This in turn resulted in greater support for the campaigns for non-co-operation and civil disobedience led by Mohandas Gandhi*. The Simon Commission* 1927–30 recommended increases in responsible government and resulted in the Round Table Conferences* and the Government of India Act (see INDIA ACTS) of 1935. This had not been implemented before the outbreak of World War II, India's participation in which was criticized by Gandhi. His arrest and imprisonment, along with that of many other leaders, led to a critical situation in 1942, when Japanese troops reached the frontier. Even before the end of the war the British opened negotiations which were to lead to independence and partition*.

The Muslim League*, founded in 1905, co-operated with the Indian National Congress* from 1916 and supported the Khilafat* agitation (1919–24). Under the leadership of M. A. Jinnah* it gained the support of the majority of Muslims and from 1937 was emphasizing Muslim separatist aspirations. From 1940 it was demanding a separate Muslim homeland, which was to be achieved with the partition of India and creation in 1947 of the state of Pakistan.

India, Union of (or Bharat), a federal republic consisting of twenty-five states and seven Union Territories as follows: *states*: Andhra Pradesh, Arunachal Pradesh, Assam*, Bihar, Goa, Gujarat, Haryana, Himachal Pradesh, Jammu & Kashmir*, Karnataka, Kerala, Madhya Pradesh, Maharashtra, Manipur, Meghalaya, Mizoram, Nagaland, Orissa, Punjab, Rajasthan, Sikkim*, Tamil Nadu, Tripura, Uttar Pradesh, West Bengal. *Territories*: Andaman and Nicobar Islands, Chandigarh, Dadra and Nagar Haveli, Delhi, Daman and Diu, Lakshadweep, Pondicherry. It adopted its Constitution in 1950, opting to remain a member of the Commonwealth although a republic. The federal Constitution was influenced by both British and US practice and provided for a President, elected by an electoral college, advised by a Council of State which includes a cabinet of ministers from the majority parliamentary party; and a

Parliament or central legislature consisting of an Upper House or Council of States (Rajya Sabha), elected from state legislatures, and a Lower House or House of the People (Lok Sabha), elected through universal adult suffrage. As in the USA each state Constitution mirrors the federal Constitution, although some states opted for bicameral legislatures and some for a single Assembly. Each state was to have a High Court, with the federal Supreme Court acting as the highest court of appeal. The Constitution recognized fifteen regional languages.

During its forty years' history the Union has seen remarkable growth in standards of literacy, in agricultural production resulting from the green revolution*, and in industrial production; but the GNP has only just matched the steady rise in population (15 per cent p.a.). Following the British exodus in 1947 the princely states all acceded to the Union, although pressure was used upon Travancore-Cochin and Hyderabad*, while Jammu and Kashmir only formally joined in 1957. The French voluntarily surrendered their few possessions in India, while the Portuguese territories were integrated through military action. Since independence India has had three wars with Pakistan and one with China, and the relationship with Sri Lanka* remains strained, owing to Indian Tamil support for the Sri Lankan Tamils' movement for autonomy. The Sikh demand for autonomy remains an intractable problem in the Punjab. India's first Prime Minister Jawaharlal Nehru* initiated a policy of planned economic growth and non-alignment. This was substantially followed by his successors Lal Shastri*, Indira Gandhi*, and Rajiv Gandhi*. When the latter lost the elections of 1989 he was succeeded by V. P. Singh of the Janata Party*, who formed an anti-Congress National Front coalition. This split over the issue of Untouchables, and at a time when militant Hindus were fighting Muslims, Chandra Shekhar became Prime Minister of a minority government in November 1990.

India Acts. There were four major British parliamentary Acts leading to Indian independence. (1) **The Indian Councils Act** (1909). This allowed Indians a first share in the work of the legislative councils. (2) **The Government of India Act** (1919). This established a bicameral legislative Parliament for all British India, but without the power to restrain the Viceroy's executive, and in the provinces created elected assemblies which, on the principle of dyarchy*, would concern themselves with 'transferred' matters. (3) **The Government of India Act** (1935). This gave the provincial assemblies full responsible government, being implemented in 1937. It also removed Aden and Burma from British India. Thirdly it proposed that the Indian Empire be transformed into a federal dominion, which would have included the princely states, while retaining the principle of dyarchy for some aspects of government. This section of the Act was never implemented. (4) **The Indian Independence Act** (1947). This was passed through the British Parliament without a division 10–18 July in order to meet Lord Mountbatten's deadline of 15 August for the partition* of the subcontinent. The Act advised princely states to seek integration into one or other of the successor states. A further brief act enabled a republic to remain a member of the Commonwealth of Nations, whose head is the British monarch.

Indian army (British). By 1902, when Kitchener became Commander-in-Chief in India, it was established practice that most regiments of the British army stationed a battalion in India. In 1903 the three armies of Madras, Bombay, and Bengal were united and the Indian army proclaimed by the Viceroy Lord Curzon. Indian troops had served regularly overseas, for example in Egypt and the Sudan. The officers for this army holding the King's Commission were British; they were assisted by Sepoy officers holding the Viceroy's Commission. Kitchener reorganized this army and established a staff college which would recruit and train more Indian officers, and from 1917 Indians became eligible for the full King's Commission. The Indian army served in Iran, Mesopotamia, Palestine, and Egypt as well as on the Western Front during World War I. In 1922 it was again reorganized, with cavalry regiments being reduced and an Auxiliary (Territorial) Force created. During the 1920s it steadily recruited and trained more Indian personnel into specialist units such as engineers and signals. All British officers leaving Sandhurst were still expected to do a stint in India, but the number of British battalions stationed there was steadily reduced. By World War II, under a British commander-in-chief, there were Indian officers of all ranks. The army again served with distinction, in North Africa, Italy, and Burma. Between 1945 and 1947 it loyally endeavoured to maintain peace during the years of civil turmoil. In July 1947 it was reorganized into the Indian and Pakistani armies, British officers being repatriated.

Indian army, (national). Of the three Indian armed services under a civilian Minister of Defence, by far the largest is the army, which is the largest volunteer army in the world. Its strength in peacetime has been of some 1.1 million men, organized into twenty-five divisions under five regional commands, with a full range of training establishments. It is a direct heir of the Indian army (British)*; an important early decision was to abolish the office of commander-in-chief. As with its predecessor, an important role has been that of the maintenance of internal security; at the same time it has been engaged in numerous border disputes in Kashmir with both China and Pakistan.

Indirect rule, the policy adopted by Britain to administer its vast Empire of over 600 million people. It was developed in India during the 19th century, where two-thirds of the country was ruled by native princes. It came to be applied elsewhere, in Malaya and in Africa, as colonies developed. The most influential proponent of the theory was Frederick Lugard*. He expounded it in his book *Dual Mandate of British Tropical Africa.* In monarchical societies, such as Malaya and much of East Africa and Muslim West Africa, this policy of boosting the power and prestige of sultan/emir/*kabaka*/*mwami*/paramount chief, and so on was relatively successful. In other more democratic societies, however, such as that of the Nigerian Ibo*, the creation and imposition of a 'chief' caused bitter resentment.

Indo-Chinese War (20 October–22 November 1962). During the summer of 1962 a series of vituperative notes passed between the governments of China and India, and there was a series of border clashes along the frontier in the Himalaya region. China claimed that the McMahon* decision of 1914 had been unfair. On 20 October two Chinese offensives were launched, one across the McMahon frontier into the North-East Frontier Agency Territory and one into Ladakh in Indian Kashmir. The former advanced rapidly, threatening Assam. The town of Bomdilla fell on 19 November and two days later China announced a cease-fire. The NEFA was evacuated back to the original frontier, but a Ladakh salient was retained. It is probable that this was the main objective and that the NEFA campaign was a deliberate diversion. Some of the border areas remain disputed.

Indonesia. The Republic of Indonesia was formally proclaimed in Jakarta (formerly Batavia) on 17 August 1945 by the Partai Nasionalis Indonesia, whose leader was Sukarno*. Four years of war, the Indonesian Revolution*, were to follow before the government of The Netherlands conceded defeat. Formal transfer of all the Dutch East Indies* (except West New Guinea) took place on 27 December 1949. The early plan for a federation was replaced in 1950 by that of a unitary state over the thousands of islands through nearly 4,000 miles of the Java Sea. Of the major islands of Sumatra, Kalimantan, Sulawesi, and Java, the last was to dominate under President Sukarno, a dominance nearly resulting in civil war with Sumatra in 1956. Disputes with the Dutch over West New Guinea were ended in 1963, when it was reluctantly ceded as Irian Jaya; but the Konfrontasi* with Malaysia to gain control over the whole island of Borneo was a failure. East Timor* was to be formally integrated into Indonesia in 1976, after a brief civil war. Elections to a Constituent Assembly had first been held in 1955, with four main parties: the two Muslim parties of the Masyumi and the Nahdatal Vlama, the PNI, and the PKI*. Sukarno was declared head of state, and, when the Assembly failed to reach agreement, he drew up his own constitution. In 1965 a group of dissident army officers staged an unsuccessful coup, which gave General Suharto* the chance to annihilate the PKI. In 1967 Sukarno was deposed and Suharto became President (re-elected for a fifth term 1988). Although repressive against both left-wing extremists and Islamic fundamentalism*, his regime has seen an improvement in agricultural yield and a commitment by Indonesia both to UN peace-keeping missions and to ASEAN*. By 1990 the pace of deforestation was causing increasing international concern.

Indonesian Revolution (1945–9). Following the Japanese surrender in August 1945 Dutch colonists began to return, seeking to re-establish control over what they still regarded as the Dutch East Indies*. Fighting soon broke out between forces of the PNI (Partai Nasionalis Indonesia), led by Sukarno*, and the colonists, notably in East Java. In November 1945 the British negotiated a truce, and it was agreed to form a United States of Indonesia linked with The Netherlands. But the basis of the link was never clear and Dutch and PNI could not agree. In July 1947 the Dutch launched a 'police action' against the PNI, when control was re-established over most of the Dutch estates. The brutalities of this action, however, resulted in a UN debate, and the USA negotiated another truce, December 1947. In September 1948 there was a revolt at Madiun in East Java by the communist PKI* against the PNI, for allegedly deserting the cause of the nation and making concessions to the Dutch. Sukarno took firm action against the PKI, its leaders being killed or imprisoned; but the Dutch used the incident to justify a new major offensive, bombing Djokjakarta, the PNI capital, which they captured. Sukarno and his cabinet were imprisoned. This second 'police action' was again criticized internationally and the government of The Netherlands agreed to a Conference at The Hague, August 1949, when the independence of Indonesia was accepted.

Indo-Pakistan Wars, conflicts between India and Pakistan: October 1947–January 1949; April–June 1965, September 1965–January 1966; December 1971. The first conflict followed partition* of India and concerned Kashmir*. The main cause of the 1965–6 wars was again an attempt by Pakistan to assist Muslim opponents of Indian rule in Kashmir. Fighting spread to the Punjab, which was the scene of major tank battles. A UN cease-fire was accepted and, by the Tashkent Declaration of 11 January 1966, a troop withdrawal was agreed. There was a brief intervention by Indian troops in East Pakistan in December 1971 as well as renewed frontier fighting in Kashmir, when Bangladesh was seeking independence.

Industrialization, establishing industrial production (primary and secondary) through mechanized, standardized processes. Both Engels and Marx had based much of their condemnation of its evils, with its dehumanizing of the worker, together with its pollution of the environment, on what they saw in England. Yet industrialization steadily accelerated throughout the world during the later 19th century and the 20th century, equally in communist states, for example the Soviet Union, China, North Korea, Vietnam, as in capitalist economies, for example Japan, Brazil, India, South Korea. Its processes, using a wide variety of machines, do not apply only to factory production, but also to agricultural production, whereby ever larger and more efficient machinery replaces human labour on ever larger farms. Twentieth-century industrialization need no longer involve the levels of pollution criticized by Engels, Dickens, and others, but depopulation of the countryside, the erosion of traditional social bonds, the fluctuating demands of production lines, and the adoption of increasingly automated systems, all continue to have a dehumanizing effect.

Industrial Workers of the World (IWW), a radical US labour organization. Popularly known as 'the Wobblies', it was founded in 1905 as an amalgamation of forty-three separate labour groups all opposed to the AFL*, which at that time refused membership to unskilled workers. In 1908 it split, the more radical half under (Big Bill) W. D. Hayward adopting policies of sabotage as well as strikes. Its tactics led to numerous arrests. It opposed US entry into World War I and many members were imprisoned for this opposition. It won a number of victories in the mining and lumber industries of the West, but its membership declined during the 1920s.

Information technology, a term which covers a wide range of skills for the storage, retrieval, manipulation, and dissemination of information. Information science developed as a university discipline from the early 1960s, while courses to master basic IT skills within the school curriculum soon followed. Information can be stored in a variety of ways, book, film, microfilm, videotape, and so on; but its storage on tape and disc in digital form capable of being read by the computer was revolutionary, in that it enabled vast quantities of data to be organized and analysed extremely rapidly, to levels of accuracy beyond human capabilities. Computer technology was then applied to a whole range of activities, from guiding missile systems to operating a superstore check-out.

INF Treaty (Intermediate-range Nuclear Forces) (1988). An agreement was signed in Washington on 8 December 1987 between President Reagan* and General Secretary Mikhail Gorbachev*, which aimed to eliminate all Intermediate Nuclear Missiles such as US Pershing and cruise and Soviet SS-20s: 436 US warheads would be traded for 1,565 Soviet warheads. There was considerable debate during the first half of 1988, centring on whether provision for verification was adequate, but eventually the agreement was accepted by both the US Senate and the Supreme Soviet and the ratified treaty signed in Moscow on 1 June 1988.

Inönü, Ismet (1884–1973), Turkish Prime Minister 1923–37, 1961–5; President of Turkey 1938–50. Born in İzmir, he was commissioned into the Ottoman army in 1903. He served against the Greeks during the Turkish war of independence, being in command at Chanak* at the time of crisis in 1922. He was chosen by Atatürk* as the first Prime Minister of the new Republic and led the Turkish delegation to the Lausanne Conference* in 1923. In 1938 he succeeded as President, being sympathetic to the Allied cause during World War II and liberalizing Turkish society. He remained President until the Democrat victory of 1950. He continued as leader of the Republican People's Party and served again as Prime Minister in the aftermath of the military coup of 1960.

International brigades, international groups of volunteers in the Spanish Civil War. Their members were mostly Communists and they came from Britain, France, the USA, as well as the Soviet Union, to assist the Republic against Franco's

Nationalists. A majority were working people, but there were also writers and intellectuals such as the English poet W. H. Auden and writer George Orwell. At no time were there more than 20,000 in the brigades. They fought mainly in the defence of Madrid and in the Battle of the River Ebro (1938).

International Court of Justice, the judicial court of the UN which replaced the Cour Permanente de Justice in 1945 and meets at The Hague. In 1948 the General Assembly of the UN accepted the Universal Declaration of Human Rights*. By 1966 two international covenants, that on civil and political rights and that on economic, social, and cultural rights, had been drawn up, based on the Declaration, and promulgated. Appeals to the Court in terms of the covenants are enforceable only if the nation concerned has previously agreed to be bound by its decision.

International Development Association, a specialized agency of the UN. It was founded in 1960 to augment the work of the World Bank*. Its aim has been to assist with loans the Least Developed Countries (LDCs) in Africa and Asia, with an emphasis on intermediate technology in the rural situation.

International Labour Organization, specialized agency of the UN. It was established by the Treaty of Versailles in 1919 as an agency of the League of Nations* to improve labour and living conditions throughout the world, by working towards an international code of labour law and practice. Its aim was to provide an international forum for labour demands and to recommend to governments constitutional means to answer these, through law and conciliation. It has been affiliated to the UN since 1946, and there are 135 member nations. It has become increasingly concerned with human rights and the provision of technical assistance to developing nations. It has also continually sought to reduce child labour in the less-developed countries, whether in factory, field, or mines.

International Monetary Fund (IMF), specialized agency of the UN. It was proposed at the Bretton Woods Conference* and constituted in 1946, designed to stabilize international trade. Member countries subscribe funds in accordance with their wealth; these provide a reserve on which they may draw on certain conditions, to meet obligations during periods of economic difficulty.

International Socialism. Nineteenth-century Socialism* believed that in the final class struggle the proletariat, which 'had nothing to lose but its chains', could only overcome bourgeois capitalism if united internationally. In 1864 the London Working Men's Association had founded the **First International,** which had collapsed in 1876. The **Second International** had been founded in 1889, with headquarters in Brussels and, by 1914, with members from all European countries, from Japan, Canada, and the USA. It did not survive the outbreak of World War I, its plan for an international general strike being swamped by a wave of nationalism in all countries. The **Third International,** usually known as the Communist International or Comintern* (1919), was founded by Lenin* to promote world revolution and a world Communist state. It drew up Twenty-One Points of doctrine to be accepted by all seeking membership. This resulted in splits between Communist parties, which accepted them, and socialist parties, which did not. The Comintern, increasingly an instrument of Stalinist foreign policy, was dissolved in 1943. The **Fourth International** was founded by Trotsky* and his followers in 1938 in opposition to Stalin. After Trotsky's murder in 1940 it was controlled by two Belgian Communists, but broke up in disagreement in 1953. In 1951 a Socialist International was founded, with headquarters in London with a membership of over sixty democratic socialist parties. It has strongly supported the UN, adopting a 'World Plan for Mutual Aid'. Its influence, however, has not been great.

Intifadah (protest uprising). Sparked off by an incident on 8 December 1987 and anonymously led, there was a continued series of incidents in some fifty towns of the West Bank* and Gaza Strip* through early 1988—stone-throwing, tyre-burning, and window-smashing. Israeli troops retaliated with considerable violence, some 300 Palestinians being killed during 1988. Although it was supported by the official PLO led by Yasir Arafat*, a more militant Islamic Resistance Movement or Hamas, hostile to the PLO, emerged claiming to be organizer. The heavy Israeli response to the *Intifadah* provoked international criticism and, before the Gulf Crisis*, had helped to convince the USA of the need for an international conference on the Middle East.

Invergordon Mutiny (1931). In August 1931 the National government* proposed severe if uneven pay cuts for the Royal Navy: 7 per cent for admirals, 3.7 per cent for lieutenant- commanders, 13.6 per cent for unmarried able-seamen. Ratings of the British Atlantic Fleet on Cromarty Firth, Scotland, refused to go on duty. The cuts were slightly revised and the mutiny ended; the ringleaders were discharged. Nevertheless, at the thought of the Royal Navy in mutiny foreign holders of sterling took alarm. On 14 September an Act suspending the gold standard* was rushed through Parliament, but the value of the Pound fell by more than a quarter.

Iqbal, Muhammad (1876–1938), Indian political philosopher. Educated at Cambridge, he returned to India to practise law and was appointed a Professor of Philosophy. In a set of poems published in 1924 he stressed the need for a renewal of Islam, and his lectures *Reconstruction of Religious Thought in Islam* (1928) were highly influential. He took an active part in politics in the Punjab and was president of the Muslim League* in 1930, when he advanced the idea of a separate Muslim homeland in north-west India, the beginning of the concept of Pakistan.

IRA *see* IRISH REPUBLICAN ARMY.

Iran. This ancient country, now the Islamic Republic of Iran and formerly Persia, was ruled by a series of dynasties over 3,000 years. It adopted Islam, but was never part of the Ottoman Empire. In 1914 it was ruled by the Qajar* dynasty. Oil concessions had been granted to Britain in 1901 and in 1909 the Anglo-Persian Oil Company (later BP) was founded. The Anglo-Russian *Entente* of 1907 had allocated north Iran as a Russian 'sphere of influence' and the south to Britain. In 1918, after the Russian Revolution*, British troops invaded Russia from Iran. At the end of this 'War of Intervention' an Iranian officer, Reza Khan, seized power (1921). In 1925, backed by the British, he deposed the last Qajar Shah and proclaimed himself as Reza Shah Pahlavi*. Iranian nationalism increasingly resented the British presence and in World War II the Shah favoured the Axis powers of Germany and Italy. In August 1941 Iran was occupied by British and Soviet troops, it being a major route for sending supplies to the Soviet Union. Reza Shah was obliged to abdicate in September in favour of his son Muhammad Reza Shah Pahlavi*. British and Soviet troops withdrew in 1946, but strong anti-British feelings survived. The new Shah's position was precarious. He was shot and wounded in 1951, after which he was obliged to accept Mohammed Musaddiq* as his Prime Minister, when the Anglo-Iranian Oil Company was nationalized. In 1953 the Shah dismissed Musaddiq, concentrating more power into his own hands. Between 1960 and 1963 some moves were made towards liberalizing politics, together with attempts at land reform and more industrialization. Urbanization was rapid and in 1963 there was widespread urban unrest. This was suppressed by the army with the help of SAVAC, the intelligence service. After 1963 no political opposition was allowed, political prisoners being tried by military tribunals. In 1975 the Shah formed the National Resurrection Party and Iran became officially a one-party, secular state, women having been emancipated in 1963. Its governing élite had become increasingly wealthy from oil revenue. Student demonstrations took place in 1977, spreading to riots in Qum and Tehran, which resulted in martial law. Ayatollah Khomeini* co-ordinated rebellion from exile in France. In January 1979 the Shah fled and Khomeini returned, proclaiming Iran an Islamic republic to be administered by a Revolutionary Committee of Shia clergy. In November 1979 the US embassy was besieged, resulting in the Iran Hostage Crisis*. Extremist groups fought pitched battles in Tehran and the elected President Abol Hassan Bani-Sadr was denounced by Khomeini. He escaped, but over 300 'counter-revoluionaries' were executed and the new President assassinated (August 1981). By now the Iran–Iraq War* had begun; it was to claim millions of young Iranians. In 1989 a vitriolic campaign against the novelist Salman Rushdie brought the aged Khomeini back into the centre of the Islamic world. Following the death of Khomeini in 1989 and a confused power-struggle, Hasheni Rafsanjani* was elected President.

Iran Hostage Crisis (4 November 1979–20 January 1980). Followers of the Ayatollah Khomeini* were alleging that the USA was involved in military plots to restore the Shah Muhammad Reza Pahlavi*. In November 1979 they seized the US embassy in Tehran, taking sixty-six US citizens hostage. All efforts of President Carter* to free the hostages failed, including economic measures and an abortive helicopter rescue bid. The crisis dragged on through the American presidential election until 20 January 1981, when Algeria successfully mediated and the hostages were freed. It seriously weakened Carter's bid for re-election.

Iran–Iraq War (1980–8). When Saddam Hussein* came to power he acted vigorously against Shia Muslims, whom he regarded as a threat to the Ba'ath* secular state of Iraq. In April 1980 an ayatollah was executed, an act which Ayatollah Khomeini* described as 'an affront to Islam'. Saddam went on to abrogate the 1975 Algiers Agreement, which had granted some 518 sq. km. of border to the north of the Shatt al-Arab waterway to Iran in return for an end to Iranian military assistance to Kurdish insurgents in Iraq. What began as a border dispute quickly developed into full-scale war. Iraq demanded not only a revision of the Iraq–Iran border, but also the return of three islands in the Straits of Hormuz seized by Iran in 1971, and the granting of autonomy to minorities in Iran. Iraqi armour gained some initial successes in September 1980 around Abadan, but Iran resisted strongly and in May 1982 counter-attacked, recovering the port of Khorramshahr. Proclaimed by Khomeini as a holy war or jihad, it was fought by the regular armies and by hundreds of thousands of conscripts ruthlessly used as 'cannon-fodder'. In August 1982 there were unsuccessful attacks on both Baghdad and Basra. By 1985 Iraq, with help from the USSR, France, and the USA, had gained air superiority and Iraqi planes bombed a partially constructed nuclear power-station in Bushehr, as well as civilian targets. In return Iran fired shells and rockets into Basra and Baghdad. In 1987 the war entered a new phase, with Iran increasing hostilities against commercial shipping in the Persian Gulf. The USA and other nations sent in naval escorts and a number of incidents followed. The UN Security Council unanimously agreed to call for a cease-fire, to which Iraq agreed but not Iran. During 1988, however, skilful negotiation by the secretary-general Pérez de Cuéllar* resulted in Iran's agreement to an armistice in July. A peace settlement was finally agreed in August 1990. The war is estimated to have cost at least 1½ million lives.

Iraq, part of the Ottoman Empire from 1638 until 1918, the three ancient Mesopotamian provinces of Mosul, Baghdad, and Basra were occupied by Britain in 1918. France had wanted control of Mosul, which was also claimed by Turkey, but all three were made a British mandate* by the League of Nations in 1920. When Faisal* ibn Hussein was rejected by the French as King of Syria, he was installed by the British here instead, in August 1921, and proclaimed King of Iraq. With generous help and the political skill of Nuri al-Said*, he succeeded in welding the area together, in spite of Kurdish revolts in the north. Oil was discovered in 1927 and a pipe-line constructed to carry it to the Mediterranean. The reign of Faisal I's son Ghazi (1933–9) saw the growth of increasing anti-British nationalism. When Ghazi was killed in an accident his young son Faisal II* succeeded. In 1941 a pro-German nationalist, Rashid Ali*, was Prime Minister, and British troops occupied the country, remaining until 1947. The Regent Abd al-Ilāh* and pro-British Prime Minister Nuri al-Said now succeeded in keeping radical extremists at bay until July 1958, when a revolution, led by Brigadier Kassem*, created a left-wing republic. Kassem was replaced by the Ba'ath* socialist President Aref (February 1963), who in 1964 quarrelled with his party and established himself as virtual dictator. He improved relations with Egypt, nationalized the oil industry, and moved further to secularize the state. Aref died suddenly in a helicopter crash in 1966; it took two years of internal struggle before his successor, the leader of the Socialist Ba'ath Renaissance Party, Hasan al-Bakr (1968–79), was established as President. He fought against the Kurds in the north, but also achieved rapid economic modernization of both agriculture and industry in two five-year plans 1971–5, 1976–80. There were massive irrigation projects financed by oil revenues. He signed a treaty of alliance with the Soviet Union (1972), supported the aims of the PLO*, and joined Syria and Egypt in the unsuccessful Yom Kippur War* (1973). Vice-President Saddam Hussein* succeeded Bakr in July 1979. Iraq feared the spread of the Islamic revolution of Iran, with whom Hussein was quickly involved in a long and costly war. Damage to oil installations led to a loss of revenue and to austerity measures, while renewed Kurdish pressure for autonomy developed in the north. The Iraqi National Assembly allegedly granted some autonomy in May 1986, but government forces in the mountains were in fact fighting Kurdish rebels with chemical weapons. Following the ending of the Iran–Iraq War a dispute with Kuwait resulted in the Gulf Crisis* of August 1990, which deteriorated into war in January 1991.

Ireland, Republic of. After years of intermittent fighting the Anglo-Irish Treaty of December 1921, concluded by Lloyd George* with the Sinn Fein* leaders, gave separate dominion status to a new Irish Free State. There would be a Governor-General and an Executive Council with a President from within the elected Dáil Éireann. Irish republicans led by de Valera* rejected the agreement, and fought a civil war against the Irish Free State forces until defeated in 1923. In 1926 de Valera founded his Fíanna Fáil* Party and in 1932 became President of the Council. In 1937 he introduced a new constitution which abolished the office of Governor-General, thus ending the powers of the British Crown. The new sovereign state of Ireland would have an elected President and be known as Eire. In 1938 an agreement ended British occupation of Irish naval bases and the country remained neutral in World War II. Ireland left the Commonwealth and was recognized as an independent republic in 1949, when its present Constitution was established, with an elected President as head of state, but a Prime Minister (Taoiseach) as leader of the majority within the Dáil. De Valera held that office 1951–4, 1957–9; then he became President. He was succeeded as Taoiseach by Sean Lemass (1959–66) and Jack Lynch (1966–73). In 1973 Ireland joined the European Community and a Fine Gael–Labour coalition led by Liam Cosgrave* came to power. Subsequent governments have been controlled alternately by the Fíanna Fáil and the Fine Gael–Labour coalition. In November 1985 Ireland signed the Anglo-Irish Accord (the Hillsborough Agreement), giving the Republic a consultative role in Northern Ireland. Regular meetings followed, but tension with Britain over its role in Ulster remained, especially over the issue of extradition of terrorist suspects. During the 1980s Ireland's economy sharply declined, resulting in unemployment and renewed emigration.

Irgun (Hebrew: 'Irgun Zvai Leumi'), an underground Zionist* organization founded in 1931. Aimed at first at Palestinian Arabs, from 1939 onwards it targeted British mandate forces as well, claiming over 200 acts of terrorism, and organizing illegal immigration. Under its leader Menachem Begin* it blew up the King David Hotel on 22 July 1946, killing ninety-one people, and in the Arab–Israeli fighting 1947–8 carried out raids against Arabs, most notably at Dir Yassin on 9 April 1948, when all 254 inhabitants were killed. It was disbanded in September 1948.

Irigoyen, Hipólito (1850–1933), President of Argentina 1916–22, 1928–30. After a career as a rancher, lawyer, and teacher, he entered politics in 1896 as leader of the Radical Party. Having won his fight for a secret ballot, he was elected President in 1916. He kept Argentina out of World War I and actively supported organized labour until a series of strikes in 1918 and 1919 threatened economic paralysis. He then turned against the trade union movement with the same enthusiasm that he had shown previously in supporting it. Having survived his first term, however, he was re-elected in 1928 at a difficult time in Argentine history. Corruption, continued labour unrest, and large budget deficits were all exacerbated by the Great Depression*. The Argentine military stepped in to overthrow Irigoyen in 1930.

Irish Republican Army (IRA). Its origins go back to the Fenian Brotherhood created in the USA in 1858; it was revived by Sinn Fein* in 1919 as a nationalist armed force recruited from Irish Volunteers. Its first commander was Michael Collins*, with Sean McBride as Chief of Staff. Collins signed the Anglo-Irish Treaty of December 1921, for which he was regarded as a traitor and assassinated. In June 1922 the IRA shot the Chief of the Imperial Staff Sir Henry Wilson outside his home in London. It waged war in Ireland until 1924, when de Valera took an oath of allegiance, after which it became a terrorist organization, outlawed by the Irish government. Since its establishment the IRA has been able to rely on support from sympathizers in the Irish-American community in the USA. There were bomb explosions in England during 1939, one in August in Coventry killing five people. During World War II hundreds of members were interned without trial in Ireland and in England. In 1956 violence erupted in Northern Ireland, with the IRA performing a number of border raids. Following violence against civil rights demonstrators from 1968 by both IRA and Ulster Unionists, the IRA split into the Official and the Provisional wings in 1971. The Provisional IRA (PIRA) and the Irish National Liberation Army staged demonstrations, hunger-strikes, assassinations, and bombings in both Northern Ireland and Britain; these included the Birmingham bombing of 1974 and the attack on the British cabinet at Brighton in 1984. In September 1981 nine IRA prisoners died in a hunger-strike, demanding status as political prisoners and regarding themselves as prisoners of war.

Iron Guard (Garda de Fier), Romanian Fascist party. It was founded in 1927 by Cornelieu Zelca Codreanu as a strongly nationalistic, anti-Semitic organization.

During the 1930s it was responsible for a number of terrorist atrocities. By 1937 it could command 17 per cent of the electorate. King Carol* attempted to suppress it in 1938, but it reappeared after his abdication. It never gained the support it expected from the German Nazi Party.

Irredentism (derived from the Italian *Italia irredenta*, 'unredeemed Italy'). This Italian patriotic movement, formed after 1866, aimed at the incorporation of all lands where Italian was the dominant language, into the newly united Italy. These were mainly in the south Tyrol and the northern Adriatic coast, which had remained in the Austro-Hungarian Empire. Irredentism was the main motive for Italy's entry into World War I in 1915. The Paris Peace Conference satisfied most irredentist claims, one exception being that for Fiume (Rijeka), which D'Annunzio* then seized in 1919.

Islam (Arabic; 'surrender to the will of Allah'), a major world religion, preached by the prophet Muhammad. Its tenets are to be found in the Qur'ān (believed by Muslims to be the final revelation which came to the prophet from God) and the *Sunna* (Muhammad's words and actions recorded by his companions, but not written by him), while the political and social framework is provided by the shariah (the legal code of Islam). With some 800 million believers, the appeal of Islam lies in its racial tolerance, the simplicity of its worship, and the strength of its moral code. With the collapse of the Ottoman Empire*, the caliphate (spiritual leadership of the Muslim community) in Constantinople was discontinued in 1924. In the Republic of Turkey*, the successor state, a process of secularization* and modernization began, those regulations of the Qur'ān regarded as outmoded, for example polygamy and the subjection of women, being abandoned. Elsewhere, however, opposed to such Westernization, there were other movements loosely referred to as Islamic fundamentalism*, which steadily gathered support throughout the Muslim world. The two major groupings of Islam are **Sunni,** comprising the main community in most Muslim countries, and the **Shiites,** centred in Iran and southern Iraq. Both adhere to the same body of tenets, but differ in community organization, in theology, and in legal practice. Central to these differences is the Shiite belief that only the direct descendants of the prophet Muhammad may adopt the title and role of imam (religious leader), while the Sunni choose the latter by consensus. Tensions between Sunni and Shiite have been a major cause of social and political unrest in Middle Eastern countries during the 20th century.

Islamic fundamentalism, a loose term to describe the various manifestations of a revival of Islam during the 20th century. Much of the Islamic world had been colonized by European powers during the 19th century: Indonesia, Malaysia, India, Iran, Iraq, Syria, Egypt, and Africa north of the Sahara. Decolonization* gave a powerful stimulus to Islamic revival, with Europeanization increasingly rejected. The death of Gordon at Khartoum in 1885 perhaps marks a turning-point, in that Mahdism* became one of its early manifestations, together with the Salafiyya movement of Muhammad Abduh (1849–1905), whose leader in East Africa was Muhammad Abdille Sayyid Hasan*. In British India Maulana al-Mawdudi (1903–1980) founded Jama' at-i-Islami (Islamic Society) and became highly influential in Pakistan, while in Egypt Hasan al-Banna* founded the Muslim Brotherhood, one of whose members, Sayyid Qutb, provided a fundamental theological (Sunni) and ideological basis for the new radicalism in *Islam and the Problems of Civilization* (1960). The Wahabi* movement in Arabia resulted in the creation of Saudi Arabia after World War I, firmly in the hands of fundamentalists, certainly up till the time of the Gulf Crisis* in 1990. In Iran the deposition of the Shah by Ayatollah Khomeini* in 1979 resulted in a restoration of the power of the (Shiite) imams and a reversal of the trend towards Europeanization and of women's liberation*. In the Maghreb an Islamic Salvation Front emerged in the 1980s in Algeria, legalized in 1989 and claiming 3 million members. Central to all revivalism has been the demand for Islamic shariah law, with traditional Islamic penalties to be enforced for such acts as theft, adultery, fornication, and consumption of alcohol. Politically the Islamic revival has been exploited by such leaders as Qadhafi* in Libya, General Zia* in Pakistan, and by the opposition Mujaheddin* in Afghanistan.

Isolationism, a persistent factor in US politics, whose origins go back to Presidents Washington and Monroe, advocating the avoidance of all alliances or participation in world affairs outside the USA. It foiled Woodrow Wilson* in his attempt to take the USA into the League of Nations*, and during the 1930s was responsible for the Neutrality Acts*. Since US entry into World War II it has been much less vociferous, but by no means an irrelevance. It has advocated political and military withdrawal from overseas bases and the establishment of a 'fortress America', protected by military systems such as the Strategic Defense Initiative*.

Israel. The UN Palestine Commission of 1947 recommended the creation of two successor states, Jewish and Arab, the latter to consist of the West Bank* territory linked to an area north of Sinai including Gaza on the coast. In the early months of 1948 Arab–Jewish fighting took place, Irgun* massacres driving out Arab refugees. On 14 May 1948 the state of Israel was proclaimed, whereupon Britain declared her mandate* at an end. Troops from Egypt, Transjordan, and Syria immediately entered Palestine, and Transjordan occupied the Arab West Bank. The new state of Israel promptly enlarged its territory by occupying the remaining lands allocated to the Arabs, apart from the Gaza Strip*. A UN mediator, Count Bernadotte*, was assassinated, but his successor, Ralph Bunche*, had achieved an uneasy armistice by July 1949, whereupon Israel established a series of defended settlements along the new frontiers. In 1955 Israeli troops raided the Egyptian Gaza Strip, and President Nasser blockaded Israel's new port of Elat at the head of the Gulf of Aqabah. This contributed to the Israeli decision to launch the Sinai campaign*, which precipitated the Suez Crisis*. After this UN forces were established in Sinai. In May 1967 Israel launched a 'preventive war' against alleged Jordanian and Egyptian military preparations. Within six days Israel had occupied the West Bank, Old Jerusalem*, the Golan Heights, the Gaza Strip*, and Sinai back to the Suez Canal, which was closed. Acts of terrorism and counter-terrorism continued until 1973, when Egyptian and Syrian armies unsuccessfully launched a new attack, the Yom Kippur* or October War. Pressure from OPEC* countries, especially upon the USA, led to a new UN peace, followed by President Sadat's 'Initiative', which resulted in the Camp David Accord* of 1978 and the end of the state of war between Egypt and Israel. The latter agreed to evacuate Sinai. Tension continued, however, along the borders of Jordan, Lebanon, and Syria, culminating in full-scale occupation of south Lebanon June 1982–June 1985. In December 1987 the Arab *Intifadah** erupted in Gaza and the occupied West Bank. Israel's fierce reaction, in which hundreds of young Arabs were killed, shocked international opinion, especially US Jewry.

The pioneer settlers from before 1914 had begun to establish kibbutzim, whose idealistic farming methods proved highly successful. Jewish immigration, mostly from Europe, raised the population from 700,000 in 1948 to 4.2 million in 1985; many of these settled in the annexed territory of the West Bank. Politically the socialist traditions of early Zionism survived under the first prime ministers, but since 1974 the Labour Party* has been more and more strongly challenged by the right-wing Likud*, which is supported by a growing number of fiercely orthodox Jews. Economically Israel has successfully survived the Arab Boycott* and high inflation, partly because of continued and extensive financial aid from the USA.

Italian campaign, World War II (1943–5). Following the North African campaigns Montgomery* and Patton* prepared to invade Italy. The landing on Sicily was launched from Malta, in July 1943, and by the end of the month both the island's principal cities, Palermo and Catania, were captured. On the mainland German forces moved down into north Italy and took over Rome's airfields. Kesselring* withdrew his troops from Sicily and British and US forces crossed to Reggio on the mainland. Mussolini had been obliged by the King to resign and he was imprisoned. On 8 September an armistice was agreed, ending hostilities between the Anglo-American forces and the new government of Badoglio*. On 9 September the US 5th Army was landed at Salerno, where desperate fighting to hold the bridgeheads took place, the struggle developing into a great battle for Naples. On 12 September Mussolini was abducted by German paratroops. Meanwhile a third surprise Allied landing on the 'heel' of Italy captured the ports of Taranto and Brindisi. Montgomery and these forces advanced through southern Italy, captured Foggia on 28 September, and established a line north of Naples for the winter. On 11 October Italy declared war on Germany, and from now on a large and well-organized partisan force harassed Kesselring's forces, which were receiving reinforcements from the north. Kesselring took a stand at Monte Cassino, site of the ancient monastery of St Benedict. The Allies decided to bypass this, landing 50,000 men at Anzio, south of Rome, on 22 January 1944. Fierce fighting again took place around the bridgehead, and lasted until May. Cassino was heavily bombed and was finally captured by Polish troops on 18 May. The main advance was now resumed. Rome fell on 4 June, followed by Florence, after bitter fighting in August. The Germans consolidated in the River Po valley, along the so-called Gothic line, and fought a hard battle through the autumn and winter months. In April 1945 the Allies launched their final attacks, Milan falling on 29 April, and on 2 May General Alexander accepted the surrender of the whole German army group serving in northern Italy and southern Austria.

Italo-Ethiopian War (1935–6). In 1896 Italian forces had been defeated by the Ethiopian Emperor Menelek in the Battle of Adowa. Mussolini was determined to avenge this and to incorporate Ethiopia* into Eritrea and Italian Somaliland as Italian East Africa. In December 1934 there was a clash between Italians and Ethiopians at Walwal. During 1935 further border incidents occurred, over which the League of Nations sought to mediate. On 2 October the Italian army invaded from north and east. The Ethiopians mustered 40,000 men equipped with spears and ancient rifles. They were helpless against Italian aircraft, tanks, and poison gas. The League imposed limited sanctions and France and Britain tried unsuccessfully to mediate with the Hoare*–Laval Plan. In April 1936 a pitched battle at Marchew was followed by the capture of Addis Ababa by Marshal Badoglio* on 5 May, who became Governor of Abyssinia. Guerrilla fighting lasted throughout the next five years. In 1940 the Italians occupied British Somaliland, but in 1941 Commonwealth troops, aided by Ethiopian nationalists, evicted the Italians from Eritrea, Somaliland, and Ethiopia in a four-month campaign. Emperor Haile Selassie* was reinstalled in Addis Ababa.

Italy. During the Turko-Italian war of 1911–12 Italy occupied the Dodecanese* and conquered Tripolitania in Libya*. By 1914 it had occupied most of

Libya, which in 1939 it was to declare an integral part of the country. In 1915 Italy entered World War I in support of the Allies, hoping to gain those areas of the Austro-Hungarian Empire which were predominantly Italian-speaking. It suffered the military disaster of Caporetto* in 1917, but at the Paris Peace Conference* gained Trieste and south Tyrol*. The Fascist dictator Mussolini* determined to avenge the defeat of Adowa of 1896 and to combine Ethiopia* with Somaliland and Eritrea. By 1936 Italian East Africa had been established. In World War II Mussolini allied himself with Hitler, occupying Albania (1939) and invading Greece in 1940. In 1943, after the Allies had conquered Italy's North African and East African Empire and had launched an invasion of the mainland, Mussolini was deposed and Italy declared war on Germany. In 1946 King Victor Emanuel III* abdicated in favour of his son Umberto, who ruled for three weeks before a referendum voted for a republic. Following the Paris Peace Treaty* of 1947, Italy saw sustained economic growth, but also political instability, characterized by frequent changes of government. The Italian Communist Party under Togliatti* successfully adjusted to democracy, but there were terrorist kidnappings and outrages during the 1970s, by the 'Red Brigade', notably the murder of former Prime Minister Aldo Moro*. Post-war governments have mostly been formed from elaborate coalitions, but the Christian Democrats have tended to command the largest share. Their leader Giulio Andreotti has been Prime Minister 1972–3, 1976–9, 1989– , and also Foreign Minister 1983–9. During 1987 a determined effort was made to eliminate organized crime, with a massive trial of the Mafia* in Palermo. Italy has been an enthusiastic member of the European Community.

Iwo Jima, Battle of, World War II. The tiny island of Iwo Jima lies some 1,200 km. south of Tokyo in the Bonin archipelago. It had a major air base for the Japanese in World War II. Early in February 1945 US marines were landed to capture the airfield. Three weeks of intense fighting resulted in over 20,000 marine casualties, including some 5,000 killed. The island was then used to mount the bombing offensive against Japanese cities.

Jackson, Jesse (1941–), US civil rights leader. Born in South Carolina of a poor family, he won scholarships to study sociology and theology, and did postgraduate studies at Chicago before being ordained a Baptist minister. While still a student he became an assistant of Martin Luther King*, taking part in the march on Selma, Ala., in 1965. In 1967 King appointed him director of Operation Breadbasket, the economic arm of his Southern Christian Leadership Conference concerned with poverty in the South. Later he was founder of Operation PUSH (People United to Serve Humanity), before seeking Democrat nomination for President in 1984 and again in 1988. Described as 'the most charismatic Black leader since King', he gained considerable support from both White and Black voters in the 1988 primary elections, promoting the idea of a 'rainbow coalition'.

Jackson, Robert Houghwout (1892–1954), Justice of US Supreme Court. Born in Pennsylvania, he was early articled to a law firm, pleading his first case when still a minor and being admitted to the Pennsylvania Bar when he was 21. He worked in the New York Justice Department and was a supporter of Franklin D. Roosevelt* when he was both Governor of New York and President. He was appointed Solicitor-General and then Attorney-General, when he acted for the President in New Deal* legislation. In 1941 he was appointed to the Supreme Court, where he showed himself a keen supporter of judicial independence, antitrust legislation, and public order. He took leave from the Court to act as chief US prosecutor for the Nuremberg War Criminal Trials* 1945–6.

Jajce Congress. In November 1943 Tito's Communist partisans* captured the Bosnian town of Jajce from the Germans. By now the partisans controlled most of the mountainous regions of Yugoslavia and they convened a Congress of the Anti-Fascist National Liberation Committee. On 29 November the Congress resolved to create a Federal Republic of Yugoslavia, with Tito Marshal of Yugoslavia. By November 1945 the rival chetnik* forces of Mihailovich* were in disarray and the Republic was officially proclaimed in Belgrade, 29 November becoming a national holiday.

Jamaica. A British colony from 1655, its sugar industry declined after the abolition of slavery and it suffered a major 'Negro' revolt in 1865. It was a Crown Colony from 1866, and representative government was gradually introduced from 1884. There was widespread rioting during the 1930s, caused by racial tension and economic depression. In 1938 two rival political parties were formed, the Jamaica Labour Party (JLP) of Alexander Bustamente*, which was strongly anti-Communist, and the People's National Party (PNP), a mildly socialist party led by Norman Manley*. In 1944 self-government, based on universal suffrage, was granted, and economic recovery followed World War II. In 1958 Jamaica became a founding member of the Federation of the West Indies*. When this collapsed in 1962 the Jamaican Labour Party negotiated independence as a dominion within the Commonwealth. Administrations have alternated between the JLP and PNP, whose leader Michael Manley*, son of Norman, introduced many social reforms in the 1970s. In 1980 the JLP returned to power under Edward Seaga*. His conservative government supported US policies in the Caribbean, but it failed to reverse economic decline, exacerbated by the decision of two US bauxite mining companies to move their operations to Australia. In 1989 the PNP returned to power under Michael Manley again as Prime Minister, an enthusiastic supporter of CARICOM*.

Janata Party, an Indian political party. It originally consisted of a grouping of mostly right-wing Hindu political parties, based upon the Jana Sangh (People's Party), founded 1951. The so-called Janata Front was a broad coalition formed to oppose Indira Gandhi* in 1975. The Janata Party proper was formed in 1977 under Morarji Desai*, with Chandra Shekhar its president. After Desai's government collapsed in 1979, a splinter right-wing party, Bharatiya Janata, was formed, but in 1988 V. P. Singh formed Janata Dal as an anti-Congress National Front. This won many states in elections early in 1990, but split over the issue of Untouchables, Singh being replaced by Chandra Shekhar as Prime Minister.

Japan, a constitutional monarchy under the House of Yamato, whose emperors renounced all political power under the Constitution of 1947. This established a new Diet* of a House of Representatives and a House of Councillors, candidates for the latter coming from the forty-seven prefectures. Japan's newly industrialized economy had brought victory in the Sino-Japanese War (1894–5) and the Russo-Japanese War (1904–5), establishing the country as the dominant power in north-east Asia. Early political leaders had all been men of wealth, usually with military connections. The first professional politician to gain power was Hara Takashi*, to be followed by the moderate Kato Takaaki*. The Great Depression* had very important results for Japan, in that it gave discontented military leaders an opportunity to criticize what they considered to be corrupt government dominated by the *zaibatsu**, whose markets had now collapsed. Already in 1928 officers of the Guandong Army* had precipitated events in Manchuria which led to its occupation in 1933, a triumph for the expansionist-militarist factions. Crazed extremists from the latter staged a military revolt on 26 February 1936 which was suppressed, but a year later the Guandong Army successfully precipitated war in the rest of China. In 1940 Japan occupied parts of northern Indo-China and negotiated with Vichy France a joint protectorate of the colony. The USA now sought to support China by imposing an oil embargo. This had crippling effects. Japan's eye turned south to the oil-rich Dutch East Indies, and the argument was put forward that through a Greater East Asian Co-prosperity Sphere* Japan could obtain the raw materials needed, especially oil and rubber, in return for liberation from European colonial masters. In October 1941 the aristocratic Konoe Fumimaro* was toppled by his War Minister Tojo Hideki*, and war began in December. From December 1941 until June 1942 Japan was triumphant. Following the Battle of Midway Island*, however, a slow but steady reversal took place. By July 1944 Japan was clearly losing the war and Koiso Kuniaki replaced Tojo, only himself to be replaced in 1945 by Admiral Suzuki Kantaro. By now the cities were being systematically destroyed by US fire-bombs and the Emperor sought means to end the war, asking the Soviet Union to act as mediator. The USA and Britain however ignored this information, and pressed ahead with plans to drop atomic bombs, having concluded that the cost of lives of an amphibious landing on the main islands would have been over 1 million men, so fiercely were the Japanese defending themselves. Following the formal surrender on 2 September, General MacArthur* established SCAP* and began the process of demilitarization. The first intent was to break up the *zaibatsu* as 'warmongers', but in the event only twenty-eight of the 1,200 concerned were dissolved, the remainder providing the means for economic recovery. The Japanese Constitution* of 1947 established democratic political activity, and the close links with the USA 1951–60 helped to bring prosperity. Although anti-American feeling developed in 1960 and student riots and bomb outrages by the Japanese 'Red Army' followed (1968–9), the growth of the economy steadily continued. The Liberal Democratic Party* survived financial scandals under Noboru Takeshita and Sosuke Uno to win the election of February 1990 under Prime Minister Toshiki Kaifu. By the 1980s Japanese investors had controlling shares in many multinational corporations*, as well as heavy investment in US government stock. Meanwhile Japanese manufacturing companies, particularly automobiles and electronics, had come to dominate world markets.

Japan, occupation of (1945–52). After Japan's unconditional surrender on 2 September 1945, it came under the control of the Allied forces of occupation led by General Douglas MacArthur* in his capacity as Supreme Commander of the Allied powers, who established the organization of SCAP*. Although technically backed by an eleven-nation Far Eastern Commission and a Four-Power Council (Britain, China, USA, and the Soviet Union), the occupation was entirely dominated by the USA, with policy in the hands of MacArthur and, after his dismissal in April 1951, of his successor General Matthew Ridgway*. US occupation policy had two main components, the demilitarization of the country and the establishment of democratic institutions and ideals. The first objective was achieved through the complete demobilization of the armed forces and the destruction of their installations, backed up by the peace clause of the new Constitution*. The second was much more difficult, and although the Constitution was in operation before the occupation was formally ended in 1952, the impact of US-inspired reforms on Japanese socio-political institutions is still debated.

Japanese Constitution (1947). Drafted under US influence, emphasis in the new Constitution of Japan was placed on ending militarism and extending individual liberties. It was finally adopted on 3 May 1947, leaving the Emperor as head of state, but stripping him of governing power and vesting all legislative authority in a bicameral Diet*. Executive power is vested in the cabinet, which is headed by the Prime Minister and is responsible to the Diet. The Constitution specifically renounces war, but it has been interpreted as allowing self-defence, which has led to the creation of the Self-Defence Forces. However, defence expenditure remains low.

Japan–United States Security Treaty (1951). This was negotiated as part of the package of arrangements attending the formal return of Japanese independence following World War II. It established the USA as the effective arbiter of Japan's defence interests, granting it a large military presence on Japanese territory. Renewal of the treaty in 1960, as a Mutual Security Treaty with revised terms, produced a major political crisis in Japan. When renewal next came due in 1970, both countries agreed to a process of automatic extension, on the condition that revocation could be achieved on one year's notice by either.

Jazz, a musical term which began to be used in the Southern states of the USA in the years 1913–15. It was preceded by 'ragtime' *c.*1890–1917 and 'the Blues', whose craze began *c.*1900. 'Dixieland' jazz flourished in New Orleans and moved to Chicago c.1917, that city becoming the centre of jazz for the next decade. Early musicians were Jelly Roll Morton, Scott Joplin, and Louis Armstrong. Apart from 'swing' there are three characteristics of jazz: repetition, syncopation, and improvisation. In the 1920s jazz moved to New York, Paris, and London, and the decade has been described as 'the Jazz Age'. The 1930s was the decade of big bands, often referred to as the 'swing era'. Since the 1960s, with such fashions as 'cool jazz' and 'free jazz', it has become increasingly intellectual. Although 'pop groups' became dominant, nevertheless jazz continued to flourish with such groups as the Modern Jazz Quartet.

Jehovah's Witnesses, a fundamentalist Christian sect. In 1872 Charles Taze Russell founded the International Bible Students' Association in Pittsburg, USA. The Association flourished, stressing the fundamental powers of Yahweh or Jehovah, whose agent was Jesus Christ. In 1913 the term Jehovah's Witnesses was coined by Russell's successor Judge Joseph Rutherford (1869–1942), who built up a national organization based in New York. He equipped his members with portable phonographs which played his sermons. Witnesses were based on a Kingdom Hall and each was assigned a territory to cover, operating part-time and developing the 'front-porch assault'. The movement spread to Europe after World War I, Witnesses predicting the establishment of God's Kingdom following Armageddon. Several dates have been identified for the latter, including 1874, 1914, and 1990. By 1990, 10 million copies of the Witness magazine *Watchtower* were being printed in eighty languages.

Jellicoe, John Rushworth, 1st Earl (1859–1935), British admiral. Born in Southampton, he joined the Royal Navy when he was 13 in 1872 and saw service all over the world, rising to the rank of admiral by 1914, when he was appointed Commander-in-Chief of the Grand Fleet. In June 1916 he fought the Battle of Jutland*, in which he suffered heavier casualties than the Germans, after which he was criticized for excessive caution. He became First Sea Lord, being succeeded by Beatty* as Commander-in-Chief. He had some influence on strategic planning and implemented the convoy system* introduced by Lloyd George*, but was dismissed from office by the latter in December 1917. After the war he served as Governor-General of New Zealand 1920–4.

Jenkins, Roy Harris, Baron Jenkins of Hillhead (1920–), British statesman. Born in south Wales, son of a miner who had entered Parliament and been parliamentary private secretary to Clement Attlee, he won a scholarship to Balliol College, Oxford, and then served in the Royal Artillery during World War II. He was first elected to Parliament in 1948 as a Labour MP, and was author of a number of distinguished biographies and historical works during the 1950s and 1960s. In 1964 he accepted office as Air Minister (1964–5), and later became Home Secretary (1965–7) and Chancellor of the Exchequer (1967–70). In both the latter posts he is considered to have been highly successful, overseeing a number of social reforms including legalization of abortion and homosexuality, and presenting a series of skilful budgets which eliminated a large balance-of-payments deficit. Briefly Home Secretary again in 1974, he resigned to become president of the European Commission 1976–81. On the right wing of the Labour Party, he resigned in 1981 to form the short-lived Social Democratic Party, for whom he sat in Parliament until retiring to the House of Lords in 1988. He became Chancellor of Oxford University in 1987.

Jerusalem. A holy city for each of the three monotheistic religions, Judaism, Christianity, and Islam, containing the Muslim Dome of the Rock, the Jewish Wailing Wall and ancient synagogues, and for Christians the ruins of the Temple in which Jesus worshipped, together with the Church of the Holy Sepulchre. It was liberated from the Turks in December 1917 by General Allenby*, who promised it to the Arabs. It became the capital of the British mandate government and the scene of Arab–Jewish fighting 1929 and 1936–9, and of Zionist* terrorist attacks 1940–7. In 1947 the UN recommended that it become an international city, but the 1949 armistice gave west Jerusalem to Israel and the Old City and east Jerusalem to Transjordan, who had occupied the adjacent West Bank*. In 1967 Israel annexed the whole city and in 1980 declared it the capital of Israel, although in practice Tel Aviv remained the administrative centre. The Israeli claim is deeply resented by the Muslim world.

Jewish Agency, an organization established by Britain, with the support of the League of Nations, to represent the Jewish community in Palestine. It was formally constituted in 1929, although it had in fact been operating since 1920. Half its members came from outside Palestine, nominated by the World Zionist Organization. It became responsible not only for establishing kibbutzim settlements, but also for Jewish immigration and investment. Its defence force Haganah* took on an increasingly military role against Arabs. The latter strongly resented Britain's refusal to establish a similar Arab Agency. In 1948 it was members of the Agency who formed the embryonic government of the new state of Israel.

Jewish Brigade, (World War II). Formed in 1944 after long British objections, it numbered some 5,000 men and saw service in Egypt, north Italy, and north-west Europe. Its members helped to smuggle out many survivors of the Holocaust*. It was officially disbanded in 1946, and many of its operations were taken over by the Haganah*.

Jewish Legion, a military organization in World War I. It was formed in 1917 as a result of Zionist pressure to accelerate a Jewish national home in Palestine. One battalion was recruited in England, another in the USA, and others in Egypt and Palestine, joining Allenby as he advanced into Ottoman Turkey. It was disbanded in 1921, but many members helped later to form the Haganah*.

Jiang Jiehi (Chiang Kai-shek) (1887–1975) President of the Republic of China 1928–45; President of the Republic of China (Taiwan) 1950–75. Son of a village merchant, he attended the Chinese Imperial Military Academy and later lived in Japan, where he received military training. He returned to China in 1911 as a supporter of Sun Yixian*. In 1923 he established a military academy with Soviet help in Guangzhou, from which he was to recruit his favoured officers. In 1926 he became Commander-in-Chief of the National Revolutionary Army*, leading the successful Northern Expedition* 1926–8. In 1927 he ruthlessly suppressed trade union and Communist organizations, purging the Guomindang of Communists with mass arrests and executions. His Nationalist* government, established in Nanjing in 1928, lasted until 1937 and succeeded in unifying much of China, while losing Manchuria to the Japanese. Financial reforms were carried out, and communications and education improved. A New Life Movement tried to reassert traditional Confucian values to combat communist ideas. His government was constantly at war—with provincial warlords*, with the Communists in their rural bases, and with the Japanese. In 1936 he was kidnapped at Xi'an* and agreed to co-operate with the Communists against the Japanese. Nanjing was lost in 1937 and he lost control of the coastal regions and most of the major cities early in the conflict, retiring from Nanjing to Chongqing (Chungking). Encouraged by his wife Soong Mei-ling, his nationalist forces received massive aid from the USA by airlift and via the Burma Road, but corruption and inefficiency prevailed. Under US patronage he attended the Cairo Conference* with Churchill and Roosevelt in 1943 and won a place as a permanent member of the Security Council of the United Nations in 1945. In that year talks with Mao Zedong* failed to provide a basis for agreement and in the ensuing Civil War* his forces were gradually worn down until he was forced to resign as President and evacuate his remaining forces to Taiwan. Here he continued as President of the Republic of China until his death. Throughout these years he hoped in vain to lead a campaign to 'roll back communism' on the mainland and was deeply resentful of President Nixon's* decision to recognize the People's Republic of China in 1972.

Jiang Qing (1914–), wife of Mao Zedong* and member of Gang of Four*. A minor film actress in Shanghai, in 1937 she fled from the Japanese to Chongqing, became a Communist, and moved to Yan'an, where she became a drama teacher. Here she met Mao Zedong and married him, agreeing however to stay out of politics. From 1963 onwards she was increasingly critical of the Beijing Opera, seeking a return to popular art forms in opera and ballet. This criticism gradually escalated as the full-scale Cultural Revolution* developed, when she gave her full support to the Red Guards. Her influence declined as the Revolution waned, but she re-emerged in 1974 as the aged Mao's spokesperson. On his death she and the other members of the Gang of Four were criticized and arrested. She was expelled from the Chinese Communist Party (1977) and later sentenced to death, this being commuted to life imprisonment.

Jiangxi Soviet. From 1927 onwards it was the determination of the Guomindang* to exterminate Communism in China. Under continuous harassment many Communists moved into the countryside, maintaining their strength through guerrilla warfare in remote mountain regions. A group led by Mao Zedong* established itself in 1931 on the Hunan–Jiangxi border; it merged with another group led by Zhu De*, and the First National Congress of the Chinese Soviet Republic of Jiangxi was held in November 1931. Four 'encirclement' campaigns by the National Revolutionary Army* of the Guomindang (at that time advised by German officers) were thwarted by guerrilla tactics between December 1931 and early 1933; but a fifth, beginning in October 1933, forced the evacuation of the Soviet, and the commencement of the 'Long March'* a year later. Many Communist policies, including land reform, were first tried out in the Jiangxi Soviet, which, at its height, had a population of some 9 million.

Jibouti, a small East African republic with a population of some half-million. During the 1880s France occupied the enclave around the port of Jibouti, which was rapidly developing. In 1892 this was named the colony of French Somaliland, its importance arising from its strategic position on the Gulf of Aden. In 1958 it was declared to be the Territory of the Afars and the Issas within the French Community*; in 1977 it was granted total independence as the Republic of Jibouti under President Hassan Gouled Aptidon, re-elected in 1981 and 1987. Famine and wars inland have produced many economic problems, refugees arriving in large numbers. Under a mutual defence agreement France has both aircraft and troops stationed in the country.

Jinnah, Muhammad Ali (1876–1948), first Governor-General of Pakistan 1947–8. From a prosperous merchant family, he was educated in Bombay and England, where he was called to the Bar. In 1906 he entered politics as a strong supporter of the moderates in the Indian National Congress* and proponent of Hindu–Muslim unity. In 1916 he was one of the principal architects of the Congress–League Lucknow Pact, in which Congress conceded that Muslims should always have adequate representation. He did not support the Simon Commission* in 1928, but attended the Round Table

Conferences* in London 1931–3. By now doubts had arisen over the possible future for Muslims in an independent India. Returning to India in 1934, he was elected president of the Muslim League*, which he led in the 1937 elections. Although personally believing in Hindu–Muslim accord, he became increasingly convinced of the need for separate homelands, devoting his energies to extending the hold of the Muslim League over all Muslims in British India. In 1946–7 his determined rejection of attempts to find a compromise on the status of the proposed Muslim homeland of Pakistan led to the partition of India and the creation of an independent state. He became its first Governor-General and President of its Constituent Assembly until his death.

Joffre, Joseph Jacques Césaire (1852–1931), marshal of France. Born in the Pyrenees, son of a cooper, he joined the army in 1870, was commissioned, and served in the French Empire with distinction. By 1911 he was Chief of the General Staff and devised Plan xvii to meet a German invasion. Its effectiveness was proved in the first Battle of the Marne* in September 1914, which frustrated German hopes of a quick victory. As Commander-in-Chief of all French forces (1915), he took responsibility for French unpreparedness at Verdun* and the failure of the Battle of the Somme*. He became president of the Allied War Council and then served in the French War Ministry.

John XXIII (1881–1963), Pope 1958–63. Son of a peasant, Angelo Giuseppe Roncalli, he was ordained in 1903 and served as a war chaplain in World War I. He then held various posts in the Vatican until being elected Pope. During his pontificate he made energetic efforts to liberalize Roman Catholic policy, especially on social questions. Particularly notable were his encyclicals *Mater et magistro*, on the need to help the poor, and *Pacem in terris*, on the need for international peace. In 1962 he summoned the Second Ecumenical Vatican Council* to revitalize the Church. Unity of all Christians was his ultimate goal. This highly popular figure was succeeded by Paul VI*.

John Paul II (1920–   ), Pope 1978–   . Born Karol Wojtyla and educated in Cracow in Poland, he was obliged to work in a stone quarry during the German occupation. Ordained priest in 1946, he became a parish priest, with a keen interest in sport and youth activities. By 1967 he was a cardinal. When Pope John Paul I died after only thirty-three days in office, he was elected as the first non-Italian pope since 1522. No pope has travelled as extensively as he, visiting the USA, numerous Latin American countries, Africa, Japan, Indonesia, Pakistan, the Philippines, as well as England, Ireland, Spain, Portugal, France, Germany, and his native Poland. Here his visits did much to strengthen the resolve of a deeply Catholic nation to overthrow the Communist regime in 1989. He was shot in St Peter's Square in May 1981, possibly as part of an international conspiracy. On his visit to England in 1982 he took part in a service in Canterbury Cathedral, the first pope to enter the building since the Reformation. He deplored the Falklands War*, which followed soon after. While he has supported certain aspects of 'Liberation Theology'*, his moral stance has remained sternly conservative, in that he opposes abortion, birth control, extra-marital sex, and homosexual practices.

Johnson, Lyndon B. (1908–73), President of the USA 1963–9. Born in Texas, he had a variety of jobs as a young man before going to Washington in 1931 as an administrative assistant. In 1935 he returned to Texas as director of the National Youth Administration in Texas, part of the New Deal* programme. In 1937 he was elected to Congress for the Democrats. Here he was a strong supporter of President Roosevelt's policies. After naval war service in 1948 he moved to the Senate. In 1951 he became the Democrat 'whip', displaying remarkable talents for achieving consensus by astute 'wheeling and dealing'. In 1960 he was elected Vice-President to J. F. Kennedy, and when the latter was assassinated he was immediately sworn in as President. He acted decisively to restore confidence, and pressed Congress to pass the former President's legislation, especially the civil rights proposals. He won a sweeping victory in 1964 with the largest presidential majority ever, with Hubert Humphrey* as Vice-President. The administration introduced an ambitious programme of social and economic reform. It took his considerable negotiating skills to persuade Congress to support his measures, which included medical aid for the aged (medicare*) through a health insurance scheme, housing and urban development, increased spending on education, and federal projects for conservation. In spite of these achievements, urban tension increased. There were serious race riots in many cities and Martin Luther King* and Malcolm X* were assassinated. The USA's increasing involvement in the Vietnam War* overshadowed all domestic reform and led him on an increasingly unpopular course involving conscription and high casualties. By 1968 this had forced Johnson to announce that he would not seek re-election.

Jordan, or the Hashemite Kingdom of Jordan. In 1920, following the defeat of the Ottoman Turks, the League of Nations allocated the mandate* of Palestine to Britain. In July of that year Abdullah ibn Hussein* appeared and tried to raise an army against the French in Syria who had expelled his brother Faisal. Britain dissuaded him, inviting him instead to be Emir of the area east of the Jordan, to be known as Transjordan. He would be heavily subsidized and provided with British advisers. The British mandate expired in 1946 and Transjordan became a kingdom. In 1949 it changed its name to the Hashemite Kingdom of Jordan, following the conquest of the West Bank*. King Hussein ibn Talal*, who succeeded in 1952, met strong anti-British feeling and was obliged to dismiss John Glubb*, his military adviser, in 1956. In 1958 he briefly formed a federal union with his cousin Faisal II* of Iraq. Jordan offered full citizenship to all West Bank Palestinians, but there was tension, and the offer was rejected by the Palestine Liberation Organization*, formed in 1964 and dedicated to achieving an independent Arab state. Its members now organized al-Fatah* guerrilla raiders (fedayeen) against Israel. Jordan lost the West Bank in the Six-Day War of 1967, whereupon Palestinian fedayeen established themselves on the east bank of the Jordan. Hostility between these and the Jordanian government built up into civil war September 1970–July 1971, after which the PLO was ejected. During the Yom Kippur War* (1973) Jordan sent tanks to aid Syria on the Golan Heights, but there was no fighting along the Jordan. In 1974 Hussein agreed at an Arab summit in Morocco to a resolution that the PLO was the 'sole legitimate representative of the Palestinian people', after which some improvement took place in his relations with his Arab neighbours. Jordan supported Iraq in the Iran–Iraq War* and has itself been heavily financed by Saudi Arabia. In 1987 it hosted the Arab League* summit in Amman which readmitted Egypt and which pressed for an international peace conference on the Middle East. During the Gulf Crisis* of 1990 its economy suffered severely as a result of sanctions against Iraq.

Judaism, one of the world's major religions. With an estimated world population of Jews of 13 million, it is centred in the USA (5.7 million), in Israel (3.5 million), and in the Soviet Union (1.5 million). Its beliefs are enshrined in the Hebrew Scriptures (the Torah) and in its oral traditions (the Mishnah and the Talmud). Over the centuries Jews had spread around the Mediterranean (the Sephardi), and into Holland, Germany, Poland, the Baltic states, and Eastern Europe, whence millions fled to the

USA during the 19th century. Legal limitations to Jews were abolished in most European countries during that century, when three strands of Judaism emerged: Reform, Orthodox, and Conservative. The Reform movement sought to reconcile Judaism with contemporary Europe, among other things allowing the language of the country to be used in the synagogue. Orthodox Judaism was a reaction to the liberal Reformists, seeking to reject secular culture and to preserve ancient practices; it included supporters of Hasidism, an austere, mystical group who stressed the development of a personal spiritual life and attacked every manifestation of modernity. Between these emerged Conservative Judaism, whose Positive Historicism aimed to harmonize Jewish tradition with modern knowledge. When Zionism* emerged in the late 19th century, both Reform and Conservative Judaism supported it, while a majority of Orthodox for long were suspicious of the concept of a Jewish state, which was nevertheless created in 1948. While preserving some traditions of the Jewish community, most Jews outside Israel have continued their assimilation into the population of their country. But in Israel itself fundamentalist Orthodox Judaism has steadily gained adherents and influence, holding the political balance in the Knesset.

July Plot (20 July 1944), a plot to assassinate Adolf Hitler. Disenchanted by the Nazi regime, an increasing number of senior army officers believed that Hitler had to be assassinated and an alternative government formed to negotiate peace. Plans were made in 1943 and a number of unsuccessful attempts made before that of July 1944. The plot was carried out by Count Berthold von Stauffenberg, a colonel on Hitler's staff. He left a bomb at Hitler's headquarters at Rastenburg in East Prussia. The bomb exploded, killing four people, but not Hitler. Stauffenberg, believing he had succeeded, flew to Berlin, where the plotters, led by Field Marshal von Witzleben and General von Beck, established a government under the Resistance leader Carl Goerdeler, Mayor of Leipzig, and planned to seize Supreme Command Headquarters. Before this, however, news came that Hitler had survived. A counter-move resulted in the arrest of some 200 plotters, including Stauffenberg, who was strangled. Some 150 alleged conspirators, including Beck, Witzleben, and Goerdeler, were shot, hanged, or strangled, including thirteen other generals and two ambassadors. Fifteen other leading figures, including Field Marshal Rommel*, were obliged to commit suicide. The regime used the occasion to execute many prominent protesters such as Dietrich Bonhoeffer*, as well as many younger liberals who were working for the German Resistance.

Jumayyil, Amin (1942–), President of Lebanon 1982–8. Younger son of Pierre Jumayyil*, he became President on the assassination of his brother. Leader of the Maronite* Christians, his support came from them and from Sunni Muslims. He remained suspicious of the Druzes* and hostile to all Shia factions, as well as to extremists within the Maronite Phalangists. After the failure of two reconciliation conferences in Switzerland 1983–4, he succeeded in forming a coalition government with both Sunni and Druzes. In 1987 he reluctantly accepted Syrian intervention in Lebanon and the need to end the tradition of allocating political office according to religion. He managed to survive his full period of office until September 1988, even though these years saw some of the bitterest civil unrest in Lebanon, especially Beirut.

Jumayyil, Pierre (1905–84), a leading Maronite* Christian in Lebanon. In 1936 he founded the Phalange Lebanese Party. This developed a paramilitary wing aimed at protecting Christian Lebanon against Pan-Arabism. In 1958 he led resistance to pro-Nasser rebels and then held office in many governments in the 1960s in alliance with Camille Chamoun*. He spoke out against the policy of support for PLO* guerrillas and led the Phalange militia during the civil war 1975–6. He was a member of Parliament 1960–84.

Jumblat, Kamal (1917–77), Lebanese politician. As the wealthy leader of the Druze* community, he was active in Lebanese politics from 1947. In 1958 he sided with the Muslims against the Maronites*, and as Minister of the Interior he supported the PLO*, encouraging them to operate from Lebanon. During the civil war 1975–6 he aimed for a Muslim coalition, and was assassinated in 1977.

Jung, Carl Gustav (1885–1961), Swiss psychiatrist. After working in an asylum for the insane, he developed a theory of the unconscious, and then worked closely with Freud 1906–13. His clinical experience led him to define neurosis as 'the suffering of the soul which has not discovered its meaning'. After working with Pueblo Indians in America he developed the theory of the collective unconscious, in which are lodged fundamental archetypes. A deeply religious man, his analysis of the human psyche, his theory of the libido, his stress on myth and the interdependence of Logos and Eros, (the cognitive and the affective), all profoundly influenced 20th- century culture—religion, education, anthropology, and the arts.

Justo, Agustín Pedro (1876–1943), President of Argentina 1933–8. A professional soldier, he had reached the rank of general by 1930, when he participated in the conservative military coup which overthrew President Hipólito Irigoyen*. As Commander-in-Chief of the army, he was elected President in November 1931. Faced by the need for political and economic reconstruction, with high unemployment and economic decline caused by the Great Depression*, his regime was highly autocratic. He outlawed the Communist Party in 1936 and supported closer links with Britain and Pan-American co-operation. He lost the 1937 election to Roberto Ortiz, having been instrumental as an arbitrator in ending the Chaco War*. A supporter of the Allies in World War II, he enlisted in the Brazilian army in 1942 and was killed.

Jutland, Battle of (31 May 1916), the only major sea battle in World War I. It began between two forces of battle-cruisers, the British under Beatty* and the German under von Hipper. Suffering heavy losses, Beatty sailed to join the main British North Sea fleet of some 150 vessels under Jellicoe*. At 6 p.m. this engaged the German High Seas fleet of ninety-nine vessels under Scheer. Firing was at long range, approximately 14 km. As the Germans headed for home in the night, they collided with the British fleet, several ships sinking in the ensuing chaos. Both sides claimed victory. The British lost fourteen ships, including three battle-cruisers, and with 6,100 casualties; the Germans lost eleven ships, including one battleship and one battle-cruiser, with 2,550 casualties. The British retained control of the North Sea, the German fleet remaining inside the Baltic for the rest of the war.

JVP (Janatha Vimukthi Peramuna), a banned political party in Sri Lanka. Support for this People's Liberation Front, with its Marxist ideology, has come from extremist anti-Tamil Sinhalese. By mid-1988 it had been responsible for over 500 assassinations, including that of the chairman of the United National Party. Its terrorist activities were concentrated in the south of the island, and a counter-offensive by security forces in 1989 led to the death or capture of its leadership. Meanwhile attempts were made to bring the party into the democratic political process.

Kádár, János (1912–89), Prime Minister of Hungary 1956–8, 1961–5; President 1988–9. Born at Kapoly, he trained as an instrument-maker and in 1932 joined the illegal Communist Party (the Hungarian Workers' Party) in Budapest, being often imprisoned. During World War II he helped to organize Resistance groups and in 1945 was appointed Deputy Chief of Police and then Minister of the Interior in 1949. Imprisoned during the Rákosi years 1951–4, he was released shortly before Rákosi retired and in October 1956 was appointed to succeed him as general secretary of the party. He supported the short-lived government of Imre Nagy during the Hungarian Revolution* of that month, proclaiming that 'the Communist Party had degenerated into perpetuating despotism and national slavery'. Fearing that Nagy was going too far, however, he left Budapest and on 4 November returned with Soviet tanks to set up a reign of terror in Hungary. He twice served as premier, decentralizing the economy and introducing some liberalization. In 1988 he gave up the post of general secretary, becoming President until May 1989. He died in July 1989 shortly before the regime collapsed.

Kafka, Franz (1883–1924), Czech writer. Born in Prague into a middle-class Jewish merchant family, after university he continued to live in the family home, where his love/hate relationship with a dominating father powerfully influenced his writing. He was a profoundly unhappy person, and a sense of desolation and despair pervades his writing, where powerless characters, searching for identity, feel alienated and unable to find security and a sense of purpose. By day Kafka worked as an insurance agent, while writing through the night and dying of tuberculosis. Most of his work was published posthumously and he was really only discovered after 1945. Both *The Trial* and *The Castle* anticipate the totalitarian, faceless bureaucracies which were to become an increasing feature of 20th-century life.

Kalinin, Mikhail Ivanovich (1875–1946), head of state of the Soviet Union 1919–46. Born at Tver (now Kalinin) on the Upper Volga, of a peasant family, he moved to St Petersburg to work in a metal factory and joined the Russian Social Democratic (Workers') Party on its formation in 1898. He was arrested numerous times, joined the Bolshevik* faction in 1903, took part in the 1905 Revolution, and helped to found the party newspaper *Pravda*. In the 'October' Revolution (1917) he was appointed Mayor of Petrograd. In March 1919 he became head of state in Soviet Russia, a role he continued to play until a few months before he died. The Soviet Union was created in 1922, when four republics agreed on the new Constitution, and his title changed several times, becoming that of Chairman of the Praesidium of the Supreme Soviet in 1938. A member of the Politburo, he always supported Stalin and managed to survive the period of the Great Purge*, remaining a solid figurehead for nearly thirty years.

Kamenev, Lev Borisovich (1883–1936), Soviet politician. He joined the Russian Social Democratic (Workers') Party when a young man in 1901, and sided with the Bolshevik* faction when the party split in 1903. He became a member of the Duma, but was exiled to Siberia for opposing Russian entry into World War I. He returned in 1917 and presided over the second All-Russian Congress of Soviets. After the 'October' Revolution he became chairman of the Central Executive Committee of the Soviets and also chairman of the Moscow Soviet. He became a member of the Politburo and a deputy chairman of the Council of People's Commissars. On Lenin's death Kamenev, Stalin*, and Zinoviev* formed a triumvirate to exclude Kamenev's brother-in-law Trotsky* from power. Later, however, Kamenev began to oppose Stalin, and in 1935 he was sentenced to five years' imprisonment. Next year he was retried in the first of Stalin's show trials, accused of complicity in the murder of Sergey Kirov, who had been assassinated in December 1934. Kamenev was shot; later evidence revealed that Kirov had been killed on Stalin's orders. His trial marked the beginning of Stalin's Great Purge*.

Kamikaze (Japanese; 'divine wind'), the term used to describe Japanese aircraft loaded with explosives and suicidally crashed on a target by the pilot. At first volunteers were used, but the practice later became compulsory. It was first used against US warships in November 1944 in the Leyte Gulf and again in January 1945 against the landings on Luzon in the Philippines. In May 1945 over 300 kamikaze pilots died in one action off Okinawa*, when these tactics came closest to repelling an Allied invasion.

Kapp putsch (March 1920). On 13 March 1920 Wolfgang Kapp, a right-wing Prussian landowner, seized Berlin, aiming to overthrow the Weimar Republic* and restore the monarchy. He was supported by Ludendorff* and backed by General von Luttwitz and his 'Freikorps', which the government had been trying to disband. The German officer corps ignored him, and the *putsch* failed when Berlin workers went on strike and civil servants refused to obey him. Kapp and Luttwitz fled on 17 March. Their symbol had been the swastika.

Karamanlis, Constantine (1907–), Prime Minister of Greece 1955–63, 1974–80; President 1980–5. Born in Macedonia, he qualified and practised as a lawyer until entering Parliament in 1935 for the Populist Party. He became Minister of Public Works under Metaxas*. After the war he formed the right-wing National Radical Union Party and worked closely with American aid officials. On the death of General Papagos he became Prime Minister. During his first ten years in office he sought unsuccessfully to bring Greece into the EEC and was challenged by the *enosis* movement in Cyprus. He lived in Paris during the regime of the 'Greek Colonels'*. When he returned he was re-elected Prime Minister and then was President of the Republic.

Károlyi, Mihály, Count (1875–1955), Hungarian statesman. Born in Budapest of an aristocratic Hungarian family, he entered Parliament in 1905. Of liberal views, he favoured a less pro-German policy for the Austro-Hungarian Empire* and equal rights for all nations within it. There was no hope of achieving this until the Empire collapsed in November 1918, when Hungary proclaimed itself a republic with Károlyi as its President. When, in March 1920, he learned that Hungary must cede territory to Romania, Czechoslovakia, and Yugoslavia, he resigned. The Communist regime of Béla Kun* followed, and he went into exile. He returned to Hungary in 1946, and served for three years as a diplomat, but he again went into exile in 1949 with the establishment of a totalitarian Communist regime.

Kasavubu, Joseph (1910–69), President of the Republic of Congo 1960–5. Educated in Roman Catholic mission schools, he became a teacher and then a civil servant, and in the early 1950s a member of undercover nationalist associations to free the Congo of Belgians. In 1955 he became president of Abako (Alliance des Bakongo), a cultural association of his Bakongo people, and turned it into a powerful political organization. On independence in 1960 he became head of state. His Abako Party formed a coalition with the Mouvement National Congolais of Patrice Lumumba*, whom he ousted early in 1961. In 1965 he himself was deposed by General Mobutu* in a bloodless military coup. He retired to live on a farm.

Kashmir. In 1914 the princely state of Jammu and Kashmir, with its pleasant hill climate, was a highly popular resort for British administrators. In 1947 the Hindu Maharaja Sir Hari Singh would have preferred it to become an independent state, but, faced with insurgent Muslim border tribesmen, he declared his accession to India and called for military aid from the Indian army. The majority of the population being Muslim, Pakistan challenged this decision of the Maharaja and sporadic fighting took place between Indian and Pakistani troops from October 1947 until January 1949, when a UN Commission established a demarcation line allocating Azad Kashmir to Pakistan and the remainder of the state to India. The Maharaja abdicated in 1951 in favour of his son, but the state Assembly declared the state a republic, and in January 1957 it was formally integrated within the Union of India as the state of Jammu and Kashmir. Border disputes over the McMahon Line* began with China in the 1950s culminating in the Indo-Chinese War* of October 1962. There were further border conflicts with Pakistan in the Indo-Pakistan Wars* of 1965 and 1971, since when there have been periodic demands for a UN-supervised plebiscite, together with outbreaks of violence.

Kassem, Abdul Karim (1914–63), Iraqi politician. Son of a carpenter, he joined the Iraqi army and rose to the rank of brigadier. Strongly nationalist and anti-British, he was a leader in the 1958 Iraqi revolution, becoming Prime Minister of the new Republic of Iraq. He survived an assassination attempt, but his rule became increasingly repressive. In 1961 a Kurdish revolt tied down much of his loyal military support and in February 1963 a dissident army group, supporting more extreme Ba'ath* rivals, seized power; Kassem committed suicide.

Kato Takaaki (also called Komei) (1860–1926), Prime Minister of Japan 1924–5. His early career was with the Mitsubishi *Zaibatsu**, which always gave him its full support. In 1887 he entered politics, being appointed ambassador to London in 1908. He became Foreign Minister 1913–15, but resigned following the presentation to China of the Twenty-One Demands*. He became chairman of the Kenseikai (Constitutionalist) Party and was leader of the opposition until 1924, when his party came to power. During his two years in office he introduced universal male suffrage, reduced the size and influence of the army, and lessened the power of the House of Peers. At the same time, however, he sponsored an anti-subversive Peace Preservation Law to balance the possibly destabilizing effects of manhood suffrage, and introduced military training into high school. His government was described as 'the most democratic' in pre-war Japan.

Katyn Massacre. In April 1943 the German army claimed to have discovered a mass grave of some 4,500 Polish officers in Katyn Forest near Smolensk. The International Red Cross was asked to investigate and it was assumed by the Polish government of General Sikorski* in London that they were part of a group of some 15,000 Polish prisoners who had disappeared in 1940, and whose fate remained unknown. Stalin denied any Soviet involvement and used the incident to break off diplomatic relations with the Sikorski government. The Soviet Union continued to deny any involvement in the massacre until April 1990, when it was confirmed that the assumption of Western historians that the officers had been killed in the early days of close Nazi–Soviet collaboration was correct.

Kaunda, Kenneth (1924–), President of Zambia 1964– . A trained teacher, he was a headmaster and then a welfare officer in the Rhodesian copper mines. In 1949 he joined the African National Congress (ANC)* and led opposition to the Central African Federation*, instituting a campaign of 'positive non-violent action'. For this he was imprisoned by the British. Released in 1960, he was elected president of the newly formed United National Independence Party (UNIP), which by 1964 had become the leading party. He was appointed Prime Minister of Northern Rhodesia in January 1964, and when independence was granted he was elected first President of Zambia. During his presidency education was expanded, and the government made efforts to diversify the economy, to release its dependence on copper. Ethnic differences, the Rhodesian and Angolan conflicts, and the collapse of copper prices engendered unrest and political violence. This led him to assume autocratic powers in 1972 and to grant a new constitution in 1973 which gave his party UNIP a privileged position. In spite of these difficulties he was re-elected President in 1978, 1983, and for the sixth time in 1988. A staunch supporter of the Commonwealth, he has also taken a strong line in demanding international sanctions against South Africa.

Kautsky, Karl (1854–1938), political theorist. Born in Prague, he joined the Austrian Social Democrats when a student in Vienna. As a young man he met both Engels and Marx, whose thinking greatly influenced him, as did that of Darwin. He was a dominant figure of the Second International (see INTERNATIONAL SOCIALISM) and a founding member of the German Social Democrat Party in 1900. Although he defended Marxist orthodoxy against the 'revisionism' of Bernstein*, he was a bitter opponent of Bolshevik extremism, which he regarded as Lenin's imperialism. After World War I he edited and published official documents for the Weimar Republic and from 1924 lived and worked in Vienna, considered the leading Social Democrat theorist in Europe. At the time of the Austrian Anschluss* he fled to Holland, where he died.

Kefauver, Carey Estes (1903–63), US Senator 1949–63. Born in Tennessee, he became a successful lawyer and State Commissioner of Finance and Taxation, before being elected as a Democrat to Congress in 1939 and to the Senate in 1949. He came to national prominence in 1951, when chairman of a Senate committee investigating organized crime, whose hearings were the first ever to be televised. The committee exposed nation-wide gambling and crime syndicates which had infiltrated legitimate business and many areas of local politics. Evidence of corruption among federal tax officials led to several dismissals and to the resignation of the Commissioner of Internal Revenue. Kefauver won the Democratic Party's nomination for Vice-President in 1956, when Eisenhower was re-elected.

Keller, Helen (1880–1968), educator of the deaf and dumb. At the age of 19 months she lost her sight and hearing and became mute. Her parents obtained the services of Anne Sullivan, herself partially sighted, to tutor her. Helen learned to speak by pressing her fingers on Anne's larynx; she learned the alphabet by the feel of letters in the palm of her hand. She attended the Horace Mann School for the Deaf in Boston and learned braille. In 1904 she graduated at Radcliffe College, Cambridge, Mass. She devoted the rest of her life to the promotion of education for the blind, deaf, and dumb. Her work liberated thousands, enabling them to have full and worthwhile lives.

Kellogg–Briand Pact, or Pact of Paris (August 1928), an international peace pact. This grew out of a proposal in April 1927 by the French statesman Aristide Briand* to the USA that the two countries negotiate a treaty outlawing war between themselves. The US Secretary of State Frank B. Kellogg countered with a suggestion of a multinational treaty of the same character. A nine-nation conference, which included Germany and the USSR, met in Paris and in August 1928 issued a convention renouncing war. Sixty-two nations eventually signed and ratified the Pact, but its failure to provide measures of enforcement nullified its contribution to international order.

Kemal, Mustafa *see* ATATÜRK.

Kennedy, John Fitzgerald (1917–63), President of the USA 1961–3. The second of nine children in the powerful and wealthy Kennedy family of Boston, he graduated from Harvard in 1940 and then joined the US navy, being severely wounded in the Solomon Islands campaign. Kennedy never lost an election, entering Congress as a Democrat in 1946 and in 1952 running for the Senate, with his brother Robert* organizing his campaign. In January 1960 he began his campaign to win the presidency, taking his defeated rival Lyndon Johnson* as his Vice-President and defeating Richard Nixon* in a close election. From the time of his inaugural address ('ask not what your country can do for you—ask what you can do for your country'), he brought a new spirit of hope and enthusiasm to the office. Although Congress gave support to his foreign aid proposals and space programme, however, it was reluctant to accept his domestic programme, known as the 'New Frontier' proposals for civil rights and social reform. In foreign affairs he recovered from the disaster of the 'Bay of Pigs'* to resist Khrushchev* over Berlin in August 1961 and again over the Cuban Missile Crisis* in 1962. He helped to secure a Nuclear Test-Ban Treaty* in 1963, having established his Alliance for Progress* programme for Latin America in 1961. He became increasingly involved in Vietnam, dispatching more and more 'military advisers' and then US troops into combat readiness there. In November 1963 he was shot while visiting Dallas, Tex. The Warren* Commission, appointed by his successor Lyndon B. Johnson, concluded that he had been killed by Lee Harvey Oswald. John Kennedy had married the wealthy Jacqueline Lee Bouvier, and at their 'court' in the White House they entertained artists, musicians, and intellectuals, who themselves were responsible for the 'myths' which surrounded his presidency. In terms of achievements there is little to record, Kennedy perhaps having too easily been persuaded of the Communist global threat, to which he responded with the Vietnam intervention.

Kennedy, Robert (1925–68), US politician. From the wealthy Massachusetts Kennedy family, he was a successful lawyer and campaign manager for his brother John F. Kennedy* during the 1950s and for the presidential election of 1960. Appointed Attorney-General in 1961, he acted vigorously against racial discrimination and organized crime. Following his brother's assassination he was elected to the US Senate (November 1964), where he strongly criticized the war policy in Vietnam. In 1968 he campaigned for the presidency, and had won a series of primary elections as candidate for the Democrats, when he was assassinated in California by a Palestinian immigrant, Sirhan Bishare Sirhan, who claimed that Kennedy was an active supporter of Zionism*.

Kent State University Shootings. On 4 May 1970 some 500 students at Kent State University, Ohio, were involved in a demonstration against US involvement in

Vietnam and Cambodia. The university summoned the State National Guard, who opened fire on the demonstrators, killing four and wounding many more. The incident produced a shock-wave of sympathy throughout the USA.

Kenya. In 1888 the British East African Company secured lease of a coastal strip from the Sultan of Zanzibar* and in 1896 the British East Africa protectorate was formed, when thousands of Indians were brought in to build railways. The British Crown Colony of Kenya was created in 1920. By then a great area of the 'White Highlands' had been reserved for White settlement, while 'Native Reserves' were established to separate the two communities. During the 1920s there was both considerable immigration of farmers from Britain and the development of African political movements demanding a greater share in the government of the country. Kikuyu* nationalism developed steadily, with Jomo Kenyatta* one of its early leaders. From the resultant tension grew the Kenya Africa Union and the militant Mau-Mau* movement of 1952–7. An election of 1961 led to the two African political parties, the Kenya African Union (KANU) and the Kenya African Democratic Union (KADU), joining the government. Independence was achieved in 1963, and in the following year Kenya became a republic, with Kenyatta as President. Under him the country remained generally stable and prosperous, and after his death in 1978 his successor Daniel Moi* mounted a successful four-year development plan. Falling coffee and tea prices in 1982 led to unrest, culminating in a bloody attempted coup. Elections in 1983 saw the return of comparative stability, with Moi still President, but of an increasingly corrupt and autocratic regime.

Kenyatta, Jomo (1893–1978), Prime Minister of Kenya 1962–4, President 1964–78. Son of a Kikuyu* farmer, he early became an orphan. From his uncle and grandfather he learned the customs of the Kikuyu, to which he always remained faithful, even though baptized a Christian at a Church of Scotland mission school. In 1916 he became involved in political protest in Nairobi against loss of land to British ex-servicemen, and the latter's use of forced labour. In 1929 and 1931 he went to London, unsuccessfully to press the case for restoration of Kikuyu land. He stayed in Europe for fifteen years, travelling, studying anthropology at the London School of Economics, writing, and teaching. In 1945 he attended the Pan-African* Conference in Manchester before returning to Kenya, where he became a leader of the Kenya African Union. Arrested in 1952, he was convicted of having 'managed' the Mau-Mau* movement. Seven years' 'hard labour' were followed by detention, where he became a symbol of Kenyan nationalism. Released in 1961, against the advice of the Governor, he was elected to the new Legislative Council, and became Prime Minister in 1962 and President of the independent Republic of Kenya in 1964. His conciliatory policies, encouraging White Kenyans to remain ('we are going to forget the past and look to the future'), resulted in considerable economic and social stability. He retained immense popularity in Kenya, and was held in high regard as a statesman throughout the world.

Kerensky, Alexander Feodorovich (1881–1970), Russian Prime Minister July–October 1917. Born in Simbirsk, son of a schoolmaster, he studied law at St Petersburg, where he became interested in politics. He was elected to the Fourth Duma (1912–17) as a moderate Social Democrat. In 1917, after the Tsar's abdication in March (February, old style), he was made Minister of War in the provisional government* of Prince Lvov, succeeding him as Premier four months later. Determined to continue the war against Germany, he failed to implement promised agrarian reforms, while the Russian economy continued to collapse. His government was overthrown by the

Bolsheviks* in the 'October' Revolution. He escaped to Paris, where he continued active propaganda against the Soviet regime until 1940. He then went to the USA, where he taught.

Kerr, Sir John Robert (1914–), Governor-General of Australia 1974–7. Born in Sydney, where he was educated, he was called to the Australian Bar before serving in World War II, at the end of which he was appointed organizing secretary of the South Pacific Commission* 1946–7. He then resumed his legal career, rising to become Chief Justice of the New South Wales Supreme Court. As Governor-General he was called upon to handle the crisis in the administration of Gough Whitlam* in 1975.

Kesselring, Albrecht (1885–1960), German field marshal. Born in Bavaria, he joined the German Imperial Army in 1904. After two years with the artillery on the Western Front 1914–16, he was transferred to the General Staff, on which he served until 1926, when he was appointed to the German High Command, specializing in artillery and the development of the dive-bomber. Having learned to fly, he became Chief of Air Staff, and commanded the bombing offensives over Poland, The Netherlands, and France in 1939–40. In fighting the Battle of Britain he was allegedly hampered by the interference of Hitler and Goering. In April 1942 he was made Commander-in-Chief of the South. His bombing offensive against Malta* failed and he likewise failed to prevent the Allies from landing at and winning the Battle of Salerno in September 1943. He remained in command in Italy, however, until being transferred to the Western Front early in 1945. He was condemned to death by a British military tribunal in Venice in 1947 for atrocities against Italian partisans, but the sentence was commuted to life imprisonment. He was released in 1952.

Keynes, John Maynard, 1st Baron Keynes (1883–1946), British economist. Born and educated in Cambridge, he became a life Fellow of King's College there in 1912, after a short spell in the India Office of the Civil Service. During World War I he worked at the Treasury, whom he represented at the Paris Peace Conference*. He achieved prominence when, in *The Economic Consequences of the Peace* (1919), he criticized the vindictive policy of German reparations imposed by the Conference, which, he argued, would do severe damage to the international economy. He remained a strong critic of the economic policies being followed by Britain and the USA during the 1920s, including the return to the gold standard* in 1925 at pre-war parity. He regarded free trade, still popular among some Liberals, as an unwanted 'Victorian relic', as, equally, was an economic policy which left all to market forces. He did not support state socialism, but did argue that governments had greater financial responsibilities than to balance budgets. He collaborated with Lloyd George in 1929 to argue that in a depression they should increase rather than decrease expenditure, in order to 'prime the pump' of economic activity. His thinking influenced President Roosevelt's New Deal*, and his advocacy of full employment influenced the Beveridge Report* and governments in the immediate post-war years. Keynes believed that when necessary governments should consider deficit financing, relying on international arrangements to right the balance 'in the long run'. This doctrine was adopted by the Bretton Woods Conference* of 1944, which was to result in the establishment of the International Monetary Fund* and the World Bank*. He helped to negotiate a loan for Britain from the United States in 1945. 'Keynesian Economics' underpinned the establishment of the British Welfare State*, but their validity came to be questioned and in part abandoned in the 1970s and 1980s. Married to a Russian ballerina, he was a great patron of the arts and the first chairman of the British Arts Council in 1945.

KGB. Formed in the Soviet Union in 1953 as a Committee of State Security in succession to the NKVD*, it was to be responsible for external espionage, internal counter-intelligence, and the prevention/detection of internal 'crimes against the state'. The most famous chairman was Yuri Andropov (1967–82), who went on to become Soviet leader 1982–4. He made KGB operations more sophisticated, especially against internal dissidents. With the onset of *glasnost* some operations of the KGB have become more public.

Khalid ibn Abd al-Aziz (1913–82), King of Saudi Arabia 1975–82. He succeeded his brother Faisal when the latter was assassinated in 1975. Before becoming King he concerned himself particularly with the Bedouin and with schemes for desert reclamation. A moderate, he supported the Camp David Accord* and was a much respected conciliator. He was succeeded by his younger brother Fahd ibn Abd al-Aziz.

Khama, Sir Seretse (1921–80), President of Botswana 1966–80. He was educated in South Africa and Britain, where his marriage to an Englishwoman for a while prevented him from returning to his native Bechuanaland and succeeding to his chiefdom. In 1956, however, he returned and formed the Bechuanaland Democratic Party, which won the elections of 1965. In that year his country gained independence as Botswana, and he became its first President. A strong believer in multi-racial democracy, he strengthened the economy of Botswana and achieved universal free education.

Khilafat movement, an Indian Muslim movement. It aimed to arouse world opinion against the harsh treatment of the Ottoman Empire after World War I and specifically against the treatment of the Ottoman Sultan and Caliph (khalifa) of Islam. It began in 1919 under the leadership of two brothers, Muhammad Ali (1878–1931) and Shaukat Ali (1873–1938), assuming a political character in alliance with the Indian National Congress*. In May 1920 it adopted a non-co-operation programme. The movement gained considerable support from Muslim Indians, but subsided after the abolition of the caliphate by Atatürk* in 1924.

Khmer Rouge (Red Cambodians). The Communist Party of Indo-China had been founded in 1930 by Ho Chi Minh and it appealed to many French-educated young Marxist Cambodians. During World War II the 'Khmer Minh' were in close touch with Ho Chi Minh in opposing Japanese occupation. In 1960 the Communist Party of Kampuchea was founded, Pol Pot* becoming secretary in 1963. It became more and more involved in organizing peasant guerrilla insurgents, which were given the name Khmer Rouge. In 1967 there was something of a peasant rebellion near Samlaut, ruthlessly suppressed by Sihanouk's* Prime Minister Lon Nol*, with perhaps 10,000 killed. Undaunted, however, the young Revolutionary Army of Kampuchea began what was to be a virtual civil war 1968–75. They allied themselves with both the North Vietnamese and the NLF of South Vietnam, helping to facilitate the movement of supplies along the Ho Chi Minh route. In March 1969 the US government first sanctioned mass bombing against the Khmer Rouge (including napalm) and then a US/Saigon invasion. Surviving all this, they increasingly dominated the Cambodian countryside and in April 1975 captured Phnom Penh. The Khmer Rouge regime under Pol Pot which followed (1975–9) was prepared to use the most brutal methods in its efforts to achieve a self-sufficient socialist economy, many thousands of town-dwellers being forcibly relocated on work-sites in the countryside. Relations with its former ally, but traditional enemy, Vietnam (now the Socialist Republic of Vietnam), deteriorated, and the Khmer Rouge regime fell in January 1979. In exile it retreated into the Thai mountains, from which it organized border incursions. In 1982 it joined the Coalition Government of Democratic Kampuchea (CGDK) of Prince Sihanouk, while retaining strong support in much of the Cambodian countryside, where there was increasing infiltration. During 1990 the USA withdrew support for the CGDK, while China ended its military support. Increasingly isolated, in December representatives were taking part in the Paris International Conference on Cambodia*, Vietnam having withdrawn all troops.

Khomeini, Ruholla (1900–89), Iranian ayatollah. Son and grandson of Shia leaders, he was acclaimed an ayatollah ('token of God') in 1950. In 1963 he protested against the Shah's Europeanization policies, especially the emancipation of women. He was arrested in Qum and exiled. He joined the Shia community in the Iraqi pilgrimage city of Najaf, but was expelled by Saddam Hussein* in 1978, settling in Paris. During that year there were riots throughout Iran and Khomeini called on the Iranian army to rebel. The Shah fled in January 1979 and in February 3 million people welcomed Khomeini's return to Tehran. For twelve months he ruled Iran, imposing strict Islamic puritanism and fiercely attacking the USA, whose embassy was besieged. In February 1980 the first President of the Islamic Republic Bani-Sadr was elected, but when he offended Khomeini he was expelled. Towards the close of the Iran–Iraq War*, moderate pragmatists, led by Hojatolislam Rafsanjani*, re-emerged, anxious to achieve peace and restore relations with the USA and Europe. They succeeded, in that in July 1988 Khomeini accepted a cease-fire in the Iran–Iraq War ('I have drunk poison'); but his Guardian Council sitting in Qum steadily rejected any proposals of the Islamic Assembly which it regarded as in breach of Islamic law (Shariah). In 1989 Khomeini launched a violent attack against the British author Salman Rushdie for his novel *Satanic Verses*.

Khrushchev, Nikita Sergeyevich (1894–1971), Soviet statesman. Born in the Ukraine, son of a miner, he himself became a miner. He fought in the Civil War with the Red Army 1919–21, after which he attended technical school and, as an active member of the Communist Party, rose steadily in its ranks, first in Kiev and then in Moscow. Here he became involved in the construction of the Moscow Metro. He was Prime Minister of the Ukraine 1943–7, appointed by Stalin to reorganize Soviet agriculture, enlarging state farms, which replaced many collectives devastated by the war. On the death of Stalin he became first secretary of the Communist Party (1953–64) and Chairman of the Council of Ministers (1958–64). In a historic speech at the Twentieth Party Congress* in February 1956 he denounced Stalin for his many crimes against the party and for his cult of personality. At home he attempted to tackle the problem of food supply by arranging for the cultivation of the virgin lands of Kazakhstan. He continued the programme of partial decentralization, and introduced widespread changes in regional economic administration. He restored some legality to police procedure and closed many prison camps. In foreign affairs he subdued the Hungarian Revolution*, but restored Gomulka* to power in Poland. He met Kennedy in a summit conference* in Vienna in 1961 and in 1962 he came close to global war with the USA in the Cuban Missile Crisis*, but agreed to the withdrawal of Soviet missiles. His ideological feud with Mao Zedong* threatened Sino-Soviet war, all Soviet technicians being withdrawn from China. His policy of peaceful coexistence did noticeably ease international tensions. He was dismissed from office in 1964, his colleagues fearful of war with China, but also critical of his failure to raise agricultural production as much as hoped. He lived in retirement outside Moscow.

Kibbutz, a co-operative venture in Israel, usually agricultural, but sometimes industrial. It is owned collectively by its members, rather than by constituent families, whose children are educated collectively. Kibbutzim were first introduced in 1909 as socialist, egalitarian Jewish settlements within Turkish Palestine. As tension developed with Arabs, they took on responsibility for their own defence. Since the establishment of Israel kibbutzim have attracted many student volunteers from Europe and the USA. Their number has grown to over 300.

Kiesinger, Kurt Georg (1904–88), Chancellor of the German Federal Republic 1966–9. Born in Tübingen, he was educated there and in Berlin, where he practised law. He joined the Nazi Party, but refused membership of the Guild of National Socialist lawyers in 1938. During World War II he worked in radio propaganda. In 1945 he was interned by US troops, but cleared by a de-Nazification court. He entered politics as a founding member of the Christian Democrat Party and served in the Bundestag 1949–58. He was then Minister- President (Prime Minister) of Baden-Württemberg 1958–63. In 1966 he replaced Erhard* as Chancellor, being strongly pro-Western in sympathy. The German economy recovered in the later 1960s and he made tentative moves towards an *Ostpolitik**. But he lost the support of the SPD in the Bundestag and was succeeded by their leader Willy Brandt*.

Kikuyu (Gikuyu). A people of the highlands of Kenya whose economy was for long based on millet cultivation and on cattle. Over the centuries they developed strong militarist traditions, cattle raids being a part of the life-style. Their traditional grazing lands attracted European farmers in the early 20th century and the Kikuyu were foremost in the Kenya protest movement which developed in the late 1920s. Their leader Jomo Kenyatta* described their egalitarian system of government (age groups under the guidance of a Council of Elders) in his book *Facing Mount Kenya.*

Kim il-Sung (1912–), Premier of the Democratic People's Republic of North Korea 1948–72; President 1972– . Of peasant stock, his father was an active Resistance fighter against the Japanese, and while still at school Kim was often engaged on liaison missions with guerrilla units. He formed a Marxist Young Communists' League in 1927 and was imprisoned. Released in 1930, he joined the Korean Communist Party, his ideology being deeply nationalist and concerned as a first priority with the needs of the peasantry. He spent the next fifteen years as a guerrilla fighter, mostly in Japanese Manchuria. His Korean People's Revolutionary Army adopted increasingly bold tactics and in 1936 he formed an Association for the Restoration of the Fatherland, with a ten-point programme including land reform and abolition of sex discrimination. After 1941 he was in close touch with the Soviet army, in which he held the rank of major, and in 1945 his partisan troops joined with Soviet units in the liberation of northern Korea. As did most returning Korean exiles, he strongly opposed the idea of UN trusteeship, pressing for a united Korean government. The conflict which developed along the 38th parallel made this increasingly unattainable, and in February 1946 he formed the North Korean Interim People's Committee, which set about radical land reform, breaking the power of the landed class and nationalizing industry. At the same time the North Korean Workers' Party was formed, with Kim as chairman. In 1948 Soviet troops withdrew and the Democratic People's Republic of North Korea was formed, with Kim as Premier as well as Commander-in-Chief of the Korean People's Army, now reinforced by Korean units returning from the Chinese Civil War. His decision to launch this army against South Korea in June 1950 led to the Korean War*. During the 1950s he organized agricultural co-operatives and sought to diversify industry and trade within the Communist bloc. He successfully survived an attempted army coup in 1958 and managed to maintain independence from both the Soviet and Chinese regimes, in accordance with his ideology of *juche* or autonomy. In 1972 he was proclaimed President for life, and in 1984 his son was declared his heir.

King, Martin Luther, Jr. (1929–68), Baptist minister and US civil rights leader. Born in Atlanta, son of a Baptist minister, he graduated at Morehouse College, Atlanta, and went on to take a theology Ph.D. at Boston University. In his studies he had been much influenced by the thinking of Mohandas Gandhi*. In 1954 he was appointed pastor of the Dexter Avenue Baptist Church in Montgomery and at once became involved in local civil rights agitation. He became leader of the bus protest which began in Montgomery on 1 December 1955, and which lasted over a year, during which time his home was bombed. King now founded the Southern Christian Leadership Conference and, operating from Atlanta, he became a national leader for civil rights. Imprisoned in Birmingham along with hundreds of others for taking part in a march, he was released in time for the March on Washington of 28 August 1963, when he spoke from the Lincoln Memorial ('I have a dream') to a crowd of over 200,000. President Kennedy was committed to civil rights legislation, but blocked by Congress. His assassination, and a rising tide of national sympathy for King's cause, enabled President Johnson* to achieve the Civil Rights Act of 1964. King was awarded a Nobel* Peace Prize, and continued to campaign in the South, leading a march on Selma, Ala., in 1965. Here, blocked by state troopers, he knelt in prayer and withdrew, to the chagrin of his more militant 'Black power' supporters. In 1966 he moved north to tackle racism in Chicago, where he met the opposition of the powerful Mayor Daley. By 1967 he was campaigning against the Vietnam War and planning a Poor People's March on Washington. Early in 1968 he visited Memphis to support a sanitation workers' strike and was shot by a White assassin on 4 April, dying in the belief that one day 'racial and economic justice', would prevail in the USA. In 1986 the US Congress voted the third Monday of January as Martin Luther King Day.

King, William Lyon Mackenzie (1874–1950), Prime Minister of Canada 1921–6, 1926–30, 1935–48. Born in Ontario, he studied law at Toronto before becoming a civil servant. He first became an MP in 1908, serving briefly as a Minister of Labour. In 1911 he lost his seat and for ten years was a labour consultant for the Rockefeller* Foundation. Chosen as leader of the Liberal Party in 1919, he re-entered Parliament as Prime Minister in 1921. In 1926, following a constitutional crisis with Governor-General Lord Byng, he established that the latter's office be confined to state ceremony. A man with great political skills, he drew support from both French-Canadian and Western Progressives, as well as from urban English Canada, always asserting Canada's role as an independent nation for whom 'Parliament will decide'. An admirer of F. D. Roosevelt, he made trade agreements with the USA, while his National Housing Act of 1938 and other social legislation owed a debt to New Deal* thinking. He had the ability to choose strong ministers and senior civil servants and then allow them to run their departments or regions without interference. He personally signed the Order-in-Council bringing Canada into World War II, and his wartime Prices and Trade Board was a highly efficient instrument in running the economy. Conscription being politically unpopular, it was only introduced for overseas service in 1944, when King's position was unassailable. A close friend of Winston Churchill, he hosted the conferences in Quebec in 1943 and 1944, and in his later years was a powerful supporter of the UN.

King–Crane Commission. In March 1919 the Paris Peace Conference* sent a Commission of Inquiry to the Middle East. Britain, France, and Italy refused to join, so that it became a US commission under Henry C. King and Charles R. Crane. Their report, which foresaw many of the problems which subsequently arose, criticized British and French policies as imperialist. It was also anti-Zionist*, although in later years the USA was strongly to support the Zionist cause.

Kiribati, a Pacific islands state in east Micronesia*, consisting of some thirty-three islands, including Ocean Island (Banaba), with its now exhausted phosphate deposits, and Kiritimati (Christmas Island, scene of the British hydrogen bomb test of 1957). In the 19th century many of these islands, together with those of Tuvalu*, were named the Gilbert Islands, becoming a British protectorate in 1892. In 1916 they were renamed the Crown Colony of the Gilbert and Ellice Islands. They were occupied by Japan in 1942, when the whole population of Ocean Island was deported. They were liberated by US marines in 1943 after fierce fighting. In 1974 the Polynesian Ellice Islanders voted to secede, becoming the independent state of Tuvalu in 1978. In 1979 the remaining islands of the colony gained independence as Kiribati with a unicameral Assembly, which elects the President. It is a member of the Commonwealth of Nations and of the South Pacific Forum*. Ieremia Tabai, President since 1979, was re-elected in 1987.

Kirk, Norman (1923–74), Prime Minister of New Zealand 1972–4. Son of a cabinet-maker on the South Island, he worked on the New Zealand railways, joined the Labour Party, and took part in local government, before entering Parliament in 1957. In 1965 he successfully challenged A. H. Nordmeyer as party leader, and after two defeats led Labour to a landslide victory in December 1972. As Prime Minister he strongly condemned French nuclear tests in the Pacific, and embarked on a programme of social reform which was cut short by his sudden death, not before he had honoured his election promise of withdrawing from the Vietnam War and from SEATO*. He was succeeded briefly by Wallace Rowling, but his government lost popularity and was defeated in the 1975 election.

Kirov, Sergey Mironovich (1886–1934), Russian Communist revolutionary. Born in Urzhum in the Vyatka province, he moved to St Petersburg, where he became involved as a young man in political activity, being arrested several times and at 19 joining the Bolsheviks* in 1905. He took part in the Revolution that year and after the 'October' Revolution of 1917 was sent to the Caucasus, where he organized victory in the Civil War and became head of the Communist Party in Azerbaijan. He helped to create the Transcaucasian Soviet Federated Socialist Republic in 1922, one of the first four republics to form the USSR. In 1926 he was moved to Leningrad and became a member of the Politburo. Kirov at this stage seems to have been a close ally of Stalin, and remained so until the Seventeenth Party Congress in 1934, when he criticized his increasingly arbitrary personal rule. He was shot in December at the party headquarters by a young man, Leonid Nikolayev. His death resulted in the execution of hundreds of Leningrad citizens and the deportation of thousands more. In addition Stalin used it as an excuse to eliminate his supposed 'rivals' in the Great Purge* which followed. It has since become almost certain that Stalin himself ordered his assassination.

Kishi Nobusuke (1896–1987), Prime Minister of Japan 1957–60. A graduate of Tokyo University, he entered the Civil Service, reaching high positions and serving in Manchukuo*. In 1941 he joined the cabinet of Tojo Hideki*, but was increasingly opposed to his policies as the war continued. Imprisoned in 1945, he was released without trial. Elected to the new House of Representatives (1953), he was one of the architects of the new Liberal Democratic Party*, of which he became leader (1955), becoming Prime Minister in 1957. His policies aimed to ease tension with neighbouring Asian countries, while encouraging the US–Japanese link. In domestic matters he was conservative, especially in education and over law and order. He resigned following a riot in the Diet building, allegedly over his revised Japan–USA Security Treaty*.

Kissinger, Henry Alfred (1923–), US Secretary of State 1973–7. Born in Germany of Jewish parents, he came to the USA in 1938 and studied at City College, New York. During World War II he served in the US army in Europe and later with the US military government in Germany, before returning to complete his studies at Harvard. Here in 1962 he became Professor of Government. As such, he became a government consultant, and in 1968 a campaign adviser for Richard Nixon*, who then appointed him Special Adviser on National Security and then head of the National Security Council 1969–75. He was Secretary of State for Nixon's last eleven months in office and throughout President Ford's* term. Kissinger repudiated the Dulles* strategy of 'massive retaliation', believing in the need for pragmatic and flexible responses. Although he feared a 'missile gap' and pressed for massive spending on nuclear weapons, he was at the same time largely responsible for the Strategic Arms Limitation Treaty* (SALT) of 1969. In addition, he helped to achieve a resolution of the Indo-Pakistan War (1971), *rapprochement* with Communist China (1972), which the USA now recognized for the first time, and above all resolution of the Vietnam War. He had at first accepted the case for intervention, supporting the bombing offensive against Cambodia*, but he changed his views, and, after prolonged negotiations, he reached agreement (see PARIS PEACE ACCORDS, January 1973) for the withdrawal of US troops. Later in that year he helped to resolve the Arab–Israeli Yom Kippur War* and restored US diplomatic relations with Egypt. He managed to remain reasonably untainted by the Watergate scandal*, continuing in office under the new President. As head of an international consultancy agency, he exerted considerable influence on US affairs during the 1980s.

Kita Ikki (d. 1937), Japanese revolutionary. A former socialist and member of the nationalist Kokuryukai (Black Dragon Society), Kita played a key role in the upsurge of violent right-wing militarism in Japan in the 1930s, inspiring young dissidents with his call for a revolutionary regime, headed by the military, which would nationalize wealth, sweep away existing political structures, and prepare Japan for leadership over all Asia. He was executed in 1937 for alleged involvement in the Incident of 26 February 1936*.

Kitchener, Horatio Herbert, 1st Earl of Khartoum and Broome (1850–1916), British general and Secretary for War 1914–16. Born in Ireland, he was commissioned into the Royal Engineers in 1871 and saw wide service before commanding the Anglo-Egyptian army which conquered the Sudan in 1898. In the Boer War (1899–1902) his policy of destroying Boer farmhouses and placing non-combatants in concentration camps earned him criticism, but the peace terms at Vereeniging owed much to him. Commander-in-Chief in India 1902–9 and in Egypt 1911–14, he was appointed Secretary of State for War on 5 August 1914. Unlike many of his cabinet colleagues, he foresaw a long struggle and campaigned urgently for volunteers. Set-backs on the Western Front, blunders over the supply of artillery shells, and his advice to abandon the Gallipoli campaign, which was a disaster, all damaged his reputation. He was drowned in 1916 when on his way to Russia.

Kohl, Helmut (1930–), Chancellor of the Federal Republic of Germany 1983– . Born in Ludwigshafen, he studied at Frankfurt and Heidelberg, where he was involved in student politics for the CDU (Christian Democratic* Union Party). After working in industrial relations he entered politics full-time and was elected to the state Parliament of Rhineland Palatinate in 1959 and was chairman of the state's CDU 1966–76. In 1969 he was elected state Minister-President (Prime Minister), and served 1969–76. He was elected to the Bundestag in 1976, when he challenged Helmut Schmidt* for the chancellorship. In 1982 he succeeded, on a parliamentary vote, which was confirmed in the election of 1983. Attacked from the right by Franz Josef Strauss*, his early years as Chancellor were much criticized for alleged failure to take positive action. Even in December 1989 this seemingly rather remote figure was far from popular on the streets of Berlin as the Wall fell. However, having established excellent working relationships with President Bush, he adopted a highly positive and forceful stance towards the ailing German Democratic Republic*. He took part personally in elections and offered a currency unification for July 1990 on a 1 : 1 basis. In the same month he successfully negotiated with President Gorbachev for a united Germany to remain within NATO*. In September he agreed that the cost of Soviet troops in the East would be met by the Federal Republic ('the price of German unity'), and in December 1990 was re-elected Chancellor.

Kolchak, Alexander Vasileyvich (1874–1920), Russian admiral and counter-revolutionary. Born in the Crimea, of Tartar descent, he joined the Russian Imperial Navy and rose steadily in rank. In 1905 he commanded Port Arthur in the unsuccessful Russo-Japanese War (1904–5), after which he helped to reform the Russian navy. Later he explored a possible route between European Russia and the Far East, and during World War I he commanded the Black Sea fleet. After the 'October' Revolution of 1917 he became War Minister in a government set up in Omsk in western Siberia by anti-Bolshevik forces. In October 1918 he proclaimed himself supreme ruler of Russia. With Denikin* he led the 'White' Russian* forces, clearing the Bolsheviks from Siberia. There were, however, continued defections from among his supporters and ultimately he was betrayed, captured, and shot by the Red Army at Irkutsk in February 1920.

Konfrontasi (1963–6), diplomatic and military confrontation between Indonesia and Malaysia. It was precipitated by the formation of Malaysia* in 1963, which President Sukarno* opposed, hoping to rally nationalist and anti-colonist forces. Asserting that the Federation of Malaysia was part of a British plot against Indonesia, he launched a guerrilla war in the Borneo territories of Sarawak* and Sabah* in April 1963, relying on support from local Chinese Communists. His 'confrontation' policy, however, only served to increase support for the new federal arrangements within the Malaysian states. Malaysian troops were assisted by British, Australian, and New Zealand contingents. It also led to increased disaffection in the Indonesian army, which ultimately contributed to Sukarno's downfall. His successor General Suharto* ended the Konfrontasi in 1966.

Koniev, Ivan Stepanovich (1897–1973), marshal of the Soviet Union. He served in the army during World War I and in 1917 joined the Russian Bolshevik* Party. He then joined the Red Army, fighting in the Civil War, and made a successful career as a professional soldier. Having escaped Stalin's purge of the Red Army, he commanded several army groups in World War II. In 1945 his 1st Ukrainian Army Group advanced through Poland and Silesia and took a major part in the capture of

Berlin. After the war he became Commander-in-Chief of Soviet land forces 1946–55 and then Commander-in-Chief of Warsaw Pact forces until his retirement.

Konoe Fumimaro (1891–1945), Prime Minister of Japan 1937–9, 1940–1. He entered politics after World War I and, as a member of the Upper House of the Diet, emerged as a leading advocate of popularly based parliamentary democracy and opponent of the domination of government by the military. As Prime Minister he strove unsuccessfully to control the political situation and to prevent war with the USA; but in October 1941 he was forced out of office by the militarists, led by his War Minister Tojo Hideki*. He committed suicide in December 1945, when summoned to answer charges of war crimes.

Korea, a peninsula 600 miles long in north-east Asia. After centuries of seclusion it had been opened to Western trade in 1876 and, with its rich mineral resources, annexed by Japan in 1910. Japanese colonial policy was to maintain traditional landholdings (75 per cent of the population were peasant tenant farmers), while developing mineral exploitation and heavy industry. A railway system second only to Japan's own was built, while levels of literacy grew steadily. A Central Advisory Council of aristocrats and businessmen advised the Governor-General, but no Korean held high administrative office. The last armed revolt against the Japanese was suppressed in 1919, after which an exiled Korean provisional government was formed in China. In 1924 a Korean Communist Party began underground guerrilla activity. From 1937 onward the Korean peasantry was increasingly mobilized to work in mines and industry, the country being held in the tight grip of a ruthless Police Bureau. In 1943 Roosevelt, Churchill, and Jiang Jiehi* (Chiang Kai-shek) pledged that 'in due course Korea shall become free and independent'. On 10 August 1945 Soviet troops entered Korea with units of the Korean People's Revolutionary Army from Manchuria under Kim il-Sung*. On 8 September units of the US army landed in the south, and for a few brief weeks the whole country experienced the euphoria of liberation. Japanese forces surrendered in two separate ceremonies and an arbitrary line, the 38th parallel, marked where the two liberating armies met. This ran through provinces and even cities, and was at first seen as a temporary expedient. In fact, after the failure of a Joint Commission in May 1946, it was to become an armed frontier along which open fighting quickly developed, claiming some 100,000 lives even before the outbreak of war in 1950. In 1948 the two separate regimes of North and South Korea were proclaimed.

Korea, North (officially the People's Democratic Republic of Korea). After Soviet troops entered Korea in August 1945 a left-wing People's Republic of Korea was proclaimed among much rejoicing. But with the arrival of US troops this was replaced in the south by a military administration. In December 1945 a Four-Power Trusteeship Commission was established in Moscow. This was unpopular in both north and south, and in February 1946 a Provisional People's Committee for North Korea was formed under Kim il-Sung*. In February 1947 the latter announced that elections would be organized throughout Korea for a Supreme People's Assembly. These were duly held in the north during August 1948, UN-sponsored elections having taken place in May in the South. The People's Democratic Republic was proclaimed on 9 September, with Kim il-Sung as Premier. While claiming to be the lawful government of all Korea, it in fact only controlled north of the 38th parallel. During the Korean War* most of its cities were destroyed by bombing. An intense programme of rebuilding was followed by a series of economic plans, three, five, and seven years, implemented since 1954. Secondary and university education expanded, industrial production (iron and steel, shipbuilding, mining, electrical, and textiles) greatly increased, and agriculture was modernized, with some 3,000 co-operative units formed. Rapid urbanization resulted in a shortage of farm workers. A heavy emphasis on military expenditure was justified by the strongly nationalist ideology of *juche* or autonomy (political, economic, and military) within the Communist bloc. In 1985 a series of North–South economic talks began. Proposals for 'normalization of relations' with South Korea were supported by Kim il-Sung and in September 1990 there were meetings between prime ministers, and also proposals for restoring diplomatic relations with Japan.

Korea, South (officially the Republic of Korea). On 6 September 1945 Korean nationalists proclaimed a People's Republic of Korea, but when US troops landed they refused to recognize this, Lieutenant-General John Hodge arguing that military government in the two zones, US and Soviet, was 'the only government'. In February 1946, however, he agreed to a Representative Democratic Council being formed in the south to advise him. Its chairman was Syngman Rhee*. In September 1946 the UN General Assembly accepted a US proposal for elections to a Korean National Assembly. These were in fact only held in the South (May 1948) and the Republic was inaugurated on 15 August with Rhee as President. In December (1948) the United Nations declared the Republic the only lawful government in Korea, the USSR having declared in October the North's People's Democratic Republic as the only lawful government. Following the Korean War* South Korea faced the problems of assimilating millions of refugees and rebuilding the economy. With generous aid from the USA and the UN, power facilities were restored, many new manufacturing industries established, and some land reforms initiated. By 1960 however the economy was stagnating, unemployment and inflation damaging the reputation of the government. Its increasing brutality and corruption led to its overthrow. After the failure of a second civilian government, the army, led by General Chung Hee Park*, seized power in 1961. Park proclaimed himself President in 1963 and organized a successful economic programme, from which South Korea emerged as a strong industrial power. His repressive policies, however, led to his downfall in 1979. His successor General Doo Hwan Chun continued his policies until forced partially to liberalize them after widespread student unrest in 1987. A referendum was held and a new constitution proclaimed. Under this President Roh Tae Woo was elected, and his party, the Democratic Justice Party, in 1990 amalgamated with the Reunification Democratic Party and New Democratic Republican Party to form a new Democratic Liberal Party. This supported a policy seeking reunion with the North and established diplomatic relations with the Soviet Union.

Korean War (1950–3), a war fought along the Korean peninsula. Its origins go back to the liberation of Korea after World War II and the emergence of North* and South Korea*. In the north Kim il-Sung* restructured society, breaking the power of the landed class, and nationalizing industry. The Soviet army was content to remain in the shadows. In the south US forces faced opposing factions: right-wing nationalists led by Syngman Rhee* and left-wing groups looking to the guerrilla hero of the north Kim il-Sung. In October there were serious peasant uprisings in the south, mainly against the police, which the US military had inherited from the Japanese. The policy of the US commander Lieutenant-General John Hodge was to make the south 'a bulwark against northern communism and southern rebellion', while confirming the social and economic positions of the landed and business classes. By 1948 the two respective leaders, Syngman Rhee and Kim il-Sung, were implacably opposed. The latter saw himself as the potential leader of a reunited Korea, the former as the leader of a coalition with

Jiang Jiehi* (Chiang Kai-shek) which would 'roll back' communism in Asia. In December 1948 Soviet troops withdrew from North Korea, leaving behind the Korean People's Army, reinforced by Korean units returning from the Chinese Civil War. In June 1949 US troops were withdrawn from the south. They were replaced by the Korean Constabulary, recruited from the Police Bureau and armed by the USA. In January 1950 the American Secretary of State Dean Acheson* refused to be committed to the defence of South Korea, whereupon Rhee threatened unilaterally to invade North Korea. The latter argued that its invasion of the south on 25 June was a pre-emptive strike against Rhee. It was seen by General MacArthur in Japan as Soviet-inspired aggression and as such had to be resisted. North Korean troops were welcomed by the peasantry and within weeks had reached Mokpo in the far south. In July MacArthur sent US aid, including tanks, which were landed at Pusan, where a bridgehead was established. At the United Nations the Security Council supported the US action and, with the Soviet representative absent, asked member nations to support South Korea. On 15 September MacArthur landed forces at Inchon, while also counter- attacking from Pusan. Heavy bombing, using napalm for the first time, obliterated villages thought to be sympathetic to the north, and a rapid advance recaptured the capital Seoul. UN troops then crossed the 38th parallel and rapidly drove through North Korea all the way to the Yalu River, near the Chinese frontier. By October China was threatening to intervene unless they withdrew, and on 27 November 1950 Chinese troops first engaged UN forces, who were forced to retreat by superior numbers. In January 1951 Seoul was again lost, after which the Chinese appear largely to have withdrawn. Having regained a position slightly north of the 38th parallel, MacArthur urged a full offensive, including the bombing and invasion of Manchuria, but in April he was dismissed by President Truman*. In July peace talks were begun in the village of Panmunjom. It was here that 159 plenary sessions were held before a final armistice was signed on 27 July 1953, with the battle line almost along the 38th parallel accepted as boundary between the two Koreas. In spite of numerous attempts, including over 400 meetings of the UN Korean Military Armistice Commission, set up in 1953, this armistice has never been followed by a peace treaty. Casualties in the war were high, with some 140,000 US dead and perhaps a total of 4 million lives lost.

Košice Incident (26 June 1941). Four days after the German attack on the Soviet Union, the town of Košice in eastern Slovakia was bombed by three planes. The town had been ceded to Hungary by the Munich Agreement*. The planes were in fact German, but it was asserted that they were Russian and that it 'was an act of unprovoked aggression by the USSR'. The incident was stage-managed in order to secure Hungary's collaboration in the Russian campaign. Hungary duly declared war.

Kosygin, Alexei Nikolayevich (1904–81), Soviet Prime Minister 1964–80. Born and educated in St Petersburg, he worked as a textile operative in Leningrad (as it became) and joined the Communist Party in 1927. He became an expert in industrial economics and was Mayor of Leningrad 1938–9. In 1939 he was appointed Commissar for the Textile Industry, which he remained through the war. He held a number of posts in the years 1948–60 and in 1960 he became chairman of the State Economic Planning Commission. In October 1964 he succeeded Khrushchev as Chairman of the Council of Ministers (i.e. Prime Minister). During his period of office he worked closely with both Brezhnev* and Gromyko and achieved a notable diplomatic success in bringing the Indo-Pakistan War to an end in 1966. His domestic aim was to decentralize control of industry and to increase production of consumer goods. He was largely unsuccessful. Ill health obliged him to retire in 1980.

Kotahitanga movement, a movement to create a separate Maori* Parliament in New Zealand and ultimately Maori home rule. It had been founded at Waitangi in 1892 and there were meetings each year until 1902. At the meeting of 1900 the young Apirana Ngata* urged that the New Zealand Parliament was the sole source of legal redress for Maori, although delegates could legitimately scrutinize legislation to see if it 'was fair to both sides, pakeha and Maori'. The Kotahitanga Parliaments were succeeded by the Kingitanga Parliaments, which met regularly until the 1920s, an extremist element led by Rua Kenana being suppressed by militant police action in 1916. The Parliaments discussed land grievances and the identity of the Maori people, issues taken up by such Young Maoris as Ngata and Peter Buck*.

Kreisky, Bruno (1911–90), Chancellor of Austria 1970–83. Born in Vienna into a wealthy Jewish family, he became a Doctor of Law there and also a socialist. He was briefly imprisoned in 1936. Persecuted by the Gestapo for his Jewish origins, he escaped to Sweden, where he lived and worked 1938–50, joining the Austrian legation there in 1946. After the war, as a strong Social Democrat, he served in Parliament and was Foreign Minister 1959–66, becoming Chancellor in 1970. His policy of 'active neutrality' eased relations with his neighbours Yugoslavia and Czechoslovakia. At the same time his domestic aim was to pursue a life-long commitment to a Welfare State* which would provide full employment, wage and price restraint, minimum inflation, and technological innovation. As a distinguished Jew he supported the Arab cause in Palestine, as well as the cause of all less-developed nations in the Third World.

Krupps of Essen. The Krupp family first began arms production at Essen in the 17th century. In 1810 a major factory was built and the family began to diversify into mining, shipbuilding, and other industries. During the 1890s Krupps pioneered steel production for the German navy and manufactured massive breech-loading artillery guns, and it had a virtual monopoly of arms production for World War I, employing 80,000 men. From 1920 it further diversified into manufacturing and into agricultural machinery. Gustav Krupp (1869–1950) was an ardent supporter of Hitler, and undertook the rearmament programme, which began in 1934, with a work-force of 120,000, and yet further diversification into hotels, banking, and so on. His son Alfried Krupp served four years' imprisonment as a war criminal, having employed concentration camp forced labour and having been responsible for the murder in a camp of Robert Rothschild, who had refused to comply with his demands. Under an Allied decree of 1953 the company was ordered to dispose of 75 per cent of its value. As, however, there were no buyers, Alfried was able to restore his and the family's position, now working within the ECSC* and gaining international contracts, including within the Soviet Union. Alfried's son Arndt did not wish to inherit and on his father's death in 1967 the Krupp firm became a public corporation, Alfried Krupp von Bohlen. It remained one of the largest enterprises within the European Community.

Kubitschek (de Oliveira), Juscelino (1902–76), President of Brazil 1956–61. After graduating in medicine in 1927 he worked in Paris, London, and Vienna, before joining the Medical Corps of his home state of Minas Gerais. In 1934 he entered politics, serving in the federal Chamber of Deputies. He became Governor of Minas Gerais in 1951 and was then elected President in 1955 on a programme of 'power, transportation, and food'. Determined to diversify the economy and reduce unemployment, he embarked on a massive programme of public works, including the creation of the new capital city of Brasilia. Economic prosperity followed, but at the cost of high inflation. Brazil's national debt rose to $4 billion, while its population soared to over 60 million. Renominated for President in 1964, he was forced into exile by a military junta. He returned to Brazil in 1967 to become a banker.

Ku Klux Klan, a secret society in the USA. Originally founded in 1866 to maintain White supremacy in the Southern states, it became famous for its white robes and hoods, spreading fear among Blacks to prevent them voting. The Klan reappeared in Georgia in 1915 and during the 1920s spread north and into the mid-West and Oregon, being responsible for bombings, whippings, and lynchings, and winning additional support from anti-Catholic, anti-Semitic, and anti-Union groups. At its height in the mid-1920s it boasted 4 million members, including high federal and state officials, but it also aroused intense opposition. A series of scandals and internecine rivalries sent it into rapid decline, but there was a new resurgence in the 1960s as a protest against civil rights legislation. It survived into the 1970s at the local level in the Southern states, often allied with neo-Nazi groups of the far right.

Kulak (Russian; 'fist'). The term originally applied to money-lenders, merchants, and anyone who was acquisitive. It came specifically to apply to wealthy peasants who, as a result of agrarian reforms by Stolypin in 1906, acquired relatively large farms and were financially able to employ labour. As a new element in rural Russia they were intended to create a stable middle class and a conservative political force. Following the period of Lenin's New Economic Policy*, they increasingly seemed a threat to a Communist state, and Stalin's* collectivization policy from 1928 inevitably aroused their opposition. By 1934 the great majority of their farms had been collectivized and the kulaks annihilated.

Kun, Béla (1886–1938), Hungarian Communist leader. Son of a Jewish lawyer, he himself graduated as a lawyer at Kolozsvar before World War I. During this he was captured on the Russian front and joined the Bolsheviks. He returned to Hungary and in March 1919 he persuaded the Hungarian Communists and social democrats to form a coalition government. This proceeded to inaugurate wholesale communization of property, which was bitterly resented in the rural areas. His army overran Slovakia, but promised Soviet help was not forthcoming and in August 1919 a Romanian army, which supported Admiral Horthy*, who was himself organizing a counter-revolutionary force, entered Budapest, and Kun fled to Vienna. He then went to Moscow, but reappeared in Vienna again in 1928 and was briefly imprisoned. At that time he was president of the Comintern* in Moscow, and he remained a member of the international executive until his disappearance in 1937 in Stalin's Great Purge*. He was rehabilitated in 1958. In 1990 it was revealed in Moscow that he had been shot in August 1938 as a 'Trotskyist'.

Kurdistan, a mountainous area divided between Turkey, Iran, and Iraq, with small areas in Syria and the Soviet Union. It is inhabited by islamic Kurds. Kurdish nationalism had developed in the Ottoman Empire in the late 19th century. After World War I the Treaty of Sèvres* promised an independent Kurdistan; but this never materialized, largely because Britain wanted the oilfields of Mosul under its mandate* for Iraq. Since then there have been intermittent Kurdish revolts: in Iraq 1922–32, 1958–74, 1975, 1984–8 and 1990– ; in Turkey 1944–5, 1978–9, 1984; in Iran, where in the years 1944–5 a short-lived Kurdish Republic of Mahabad was formed with Soviet help. The prolonged, armed struggle in Iraq 1958–74 resulted in a plan for limited autonomy, but fighting was resumed in 1975 when the Kurdish leader Mustapha Bazarni was forced into exile. During the Iran–Iraq War* Kurdish guerrillas in Iraq, aided by Iran, made considerable military progress. Their two parties the Patriotic

Union of Kurdistan and the Democratic Party of Kurdistan fused, but the forces of Saddam Hussein* retaliated in 1988 by resorting to chemical warfare, even while Baghdad was claiming that autonomy was being conceded. The Gulf Crisis* of 1990 was to precipitate a new Kurdish revolt.

Kursk, Battle of, World War II (5–15 July 1943). Kursk had been liberated by the Red Army on 8 February 1943, following victory at Stalingrad. In June Hitler ordered the elimination of this Soviet salient, with its important rail junction, in 'Operation Citadel', which was to be a pincer movement from north and south. Under Field Marshal Walter Model, he concentrated 2,700 tanks, together with assault guns, supported by 1,000 aircraft. They were confronted by Marshal Zhukov's* Tank Army, backed by five infantry armies. Many of the larger German tanks became stuck in the mud and hundreds were immobilized by mines. The Russians had more guns, tanks, and aircraft, and when they counter-attacked, the Germans were forced to retreat, losing some 70,000 men, 1,500 tanks, and most of their aircraft. The Soviet victory, the largest tank battle in history, ensured that the German army would never regain the initiative on the Eastern Front.

Kut, Siege of, World War I (December 1915–April 1916). Kut al-Amara is on the River Tigris. It was captured by British and Indian troops advancing from Basra in September 1915 and garrisoned by this imperial force. General Townshend advanced towards Baghdad, but was defeated at Ctesiphon and fell back to Kut, which was then besieged by Turkish forces on 7 December. Badly organized relief forces failed to break through and the garrison capitulated on 29 April 1916, after a four-month siege. Ten thousand prisoners were marched across the desert, two-thirds dying on the way, while some 23,000 troops of the relieving forces were also lost. The defeat severely weakened Britain's prestige as an imperial power, although Kut al-Amara was recaptured in February 1917.

Kuwait, a small independent state, with a busy port, at the head of the Persian Gulf, ruled by a hereditary emir of the al-Sabah* family. Although nominally under Ottoman suzerainty until 1918, it had become *de facto* a British protectorate in 1899. Oil was discovered in 1936 and after World War II Kuwait became one of the world's largest oil producers, with banking also a key activity. In 1961 an Iraqi claim was warded off with British military assistance and settled in 1963 by Arab mediation. Kuwait's defensive pact with Britain lapsed in 1971, after which it tried to pursue a policy of neutralism. Its massive wealth was channelled into modernization programmes and the development of education. Several of its tankers were victims of the Iran–Iraq War*, after which Iraq accused it of 'stealing' its oil and of ignoring OPEC agreements. It was invaded on 2 August 1990. Thousands of foreign workers from Egypt, Bangladesh, and elsewhere escaped and the country was stripped of its wealth, before the UN campaign for its liberation began in January 1991.

Ky, Nguyen Cao (1930–), South Vietnamese politician. Educated at the French Colonial Academy, he trained as an air-force pilot under US guidance, becoming Commander of the South Vietnamese air force in 1963. He took part in the overthrow of President Ngo Dinh Diem* in that year and was Premier of the military government set up, and then Vice-President under Van Thieu* (1967–71) during the Vietnam War. He was criticized for incompetence and corruption and escaped to the USA in 1975.

l

Labor Party (Australia). In the 1880s and 1890s various state Labor groups emerged, though none was ever strongly influenced by Marxist doctrines. In 1901 Australia was described as the country of 'socialisme sans doctrine', although protection of industry and extension of social welfare were both early policies within the Labor group in the new federal Parliament. This formed governments in 1904, 1908–9, 1910–13, 1914–16. When the controversial W. M. Hughes* split the group in November 1916, a national Australian Labor Party was formed, amalgamating the various Labor groups and gaining a constitution. As with the British Labour Party, Labor split in 1931 over policies for the Great Depression*, but the party returned to power in 1941, governing Australia through the dark days of war. A new split took place after the war, with a virulent anti-communist Catholic wing forming the Democratic Labor Party. Labor returned to office in 1972 under Gough Whitlam* and again in 1983 under Bob Hawke*. Pragmatic and with little ideological baggage, the party has tended to favour egalitarian policies, to foster trade unions, and to support the underdog. Until the 1950s split it had traditionally received the Catholic vote, with its Irish origins.

Labour Party (Israel). A Jewish Labour Party, Mapai, was founded in Palestine among Jewish immigrants in the 1920s, its early members being Ben-Zvi*, Levi Eshkol*, and Ben-Gurion*. In 1968 Mapai combined with two other social democrat groups formally to constitute the Israel Labour Party, under Levi Eshkol. He was succeeded as Prime Minister by Golda Meir*, but since 1974 the party has lost its monopoly of Israeli support. In 1977 it was in opposition and in 1984 under Shimon Peres formed a coalition with Likud*.

Labour Party (New Zealand). Individual Labour politicians stood for Parliament in New Zealand from the 1890s and gave their support to a number of key items of social legislation of the Liberal government of Richard Seddon, for example women's suffrage and industrial conciliation and arbitration. It was not however until 1916 that a national Labour Party was organized, gaining its supporters from both farm workers and the growing urban proletariat. The first Labour government under Michael Savage* and Peter Fraser* (1935–49) stimulated economic recovery by a public works programme, state subsidies for marketing primary produce and for railways, and nationalization of the Reserve Bank. At the same time it introduced an extensive social-security system, a state health system, and a state mortgage system. Some of this legislation was during World War II, for which conscription was successfully introduced and an Expeditionary Force sent to the Mediterranean. The party supported the US strategy for the South Pacific, and after the war a policy of collective security. In 1957 it regained power under Walter Nash*, but its majority was slender and New Zealand was facing its worst economic recession since 1931; hence Nash's government 1957–60 was a cautious one. The party was back in power 1972–5, when it withdrew New Zealand from SEATO*, and again in 1984. By then it had reversed its earlier support for a nuclear strategy; against international criticism, its policy of rejection of all nuclear power, whether as energy or in warheads, was successfully implemented. The party has always had a pragmatic approach, with little Marxist ideology to hinder it. It has gained from the increasing urbanization of New Zealand since World War I and has achieved one of the highest levels of health, education, and social welfare in any Western democracy.

Labour Party (UK). In 1900 a Labour Representation Committee was formed, which in 1906 succeeded in winning twenty-nine seats in the House of Commons and changed its name to the Labour Party, though still a loose federation of trade unions and socialist societies. In 1918 it adopted a constitution drawn up by Sidney Webb*. Its main aims were a national minimum wage, democratic control of industry, reform of national finance, and 'surplus wealth for the common good'. Although it was to be in power several times in the 20th century these aims were to be only partially realized. By 1920 membership was over 4 million and it became a major force in British municipal politics as well as gaining office with the Liberals in 1924 and 1929–31. The party strongly supported war in 1939, and through its leaders Attlee*, Bevin*, and Morrison* played a major role in Winston Churchill's government 1940–5. In 1945 it gained an overall majority in Parliament and extended the welfare legislation begun during the war. It also pursued policies of nationalization, full employment, and colonial freedom. It was in power under Harold Wilson* 1964–70, when much social legislation was enacted, and 1974–9, when it faced grave financial and economic problems. During the 1970s and early 1980s left-wing activists pressed for a number of procedural changes, for example in the election of leader and of candidates for Parliament. From the right wing a group of senior members split to form the Social Democratic Party* in 1981. Labour has always favoured military disarmament and in 1986 adopted a policy of unilateral nuclear disarmament, which it later abandoned. After its defeat in 1987 it embarked on a major policy review, which recommended a policy of pragmatism and the further abandonment of its 1918 Constitution. On balance, revisionism and pragmatism have been more dominant than ideological extremism, and it has been resolutely non-Marxist.

La Follette, Robert Marion (1855–1925), US Senator. Born and educated in Wisconsin, he became a successful lawyer and was elected to the House of Representatives in 1880. He became increasingly critical of corruption within the Republican Party, emerging as leader of a progressive group demanding reforms. He served as a successful, reforming Governor of Wisconsin from 1900 to 1906, when he was elected to the US Senate. Here he carried on his campaigns for more open and less corrupt government and for greater control over the huge multinational corporations* which were emerging. He opposed US entry into World War I. In 1924 he ran as a Progressive Party candidate for President, supported by much of organized labour, farm groups, socialists, and dissident Republicans. He polled 5 million votes, but only carried his home state.

Landfall. A New Zealand literary magazine edited by the poet Charles Brasch, son of German Jewish refugees. It was first published in 1947 and for the next twenty years it moulded literary and artistic opinion in New Zealand, gaining a state subsidy from the New Zealand Literary Fund.

Lange, David Russell (1942–   ), Prime Minister of New Zealand 1984–9. Son of a doctor and educated in Auckland, he became a successful lawyer before entering politics in support of the Labour Party. Elected to Parliament in 1977, he became deputy leader of the party in 1979 and then its leader. He had developed a particular concern for Pacific Island affairs, and after becoming Prime Minister implemented with firm determination Labour's non-nuclear policy, banning nuclear-powered and nuclear-armed ships and expressing outrage over the *Rainbow Warrior** affair. His government reduced income tax, implemented a goods and service tax, created a Ministry for Women, and introduced a Bill of Rights. He resigned in August 1989, being succeeded by Geoffrey Palmer.

Lansbury, George (1859–1940), British statesman. Born in Suffolk of a poor family, he grew up and worked in London's East End before emigrating to Australia. He returned in 1885 and joined the Liberal Party. He entered local politics, joined the Labour Party, and was elected MP in 1910. Always an idealist, he supported the suffragette movement, opposed entry into World War I, and, as Mayor of Poplar 1919–20, was imprisoned rather than reduce unemployment relief benefits. In 1919 he founded and edited the Labour newspaper the *Daily Herald.* Refusing to join the National government* in 1931, he became leader of the rump of the Labour Party (1931–5). His rejection of sanctions against Italy, following that country's invasion of Ethiopia in 1935, alienated his colleagues, and he resigned, to be succeeded by Clement Attlee. He remained a convinced pacifist, but his integrity and idealism won him wide respect both in Parliament and in the country.

Lansdowne, Henry Charles Keith Petty-Fitzmaurice, 5th Marquess of (1845–1927), British statesman. From a wealthy aristocratic family with lands in Ireland, he was educated at Eton and Oxford and held various political offices under Gladstone. An opponent of Irish Home Rule, he became a Conservative and served as Foreign Secretary 1900–5, negotiating the *Entente Cordiale* with France. From 1906 to 1915 he led the Conservative opposition in the House of Lords. He was the author of a famous letter to the *Daily Telegraph* in November 1917 advocating a negotiated peace with Germany.

Laos. The ancient kingdom of Lanxang had gradually broken up and come under Thai rule before becoming part of French Indo-China*. In 1914 it was ruled by French officials in the capital Vientiane and from Hanoi, but with the monarchy in the person of King Sisavang Vong (1904–59) remaining in the holy city of Luang Prabang. The French had established their educational system and by 1939 a generation of highly educated Laotians, many of whom had studied in Paris, was beginning to demand independence. After 1940 Vichy French officials continued to rule, subject to oversight by the Japanese, who allowed Thailand to retake an area in the west. Resistance to Japan developed, and in March 1945 Japan declared French colonial rule ended. In a confused situation there were both anti-Japanese and anti-French movements, until French troops returned in 1946 and pacified the country, confirming Sisavang Vong as King, and in 1949 granting self-rule within the French Union*. The radical nationalist group the Pathet Lao* under Prince Souphanouvong*, which emerged in 1950, demanded immediate independence. They allied themselves with the VietMinh* and, although the French granted full independence in 1953, had established virtual control in the north-east provinces by 1954 and the Geneva Agreements*. The next twenty-one years were a time of struggle between the Pathet Lao, which came increasingly to be dominated by the Communist Party (the People's Revolutionary Party of Laos), founded in 1955, the neutralists led by Souvanna Phouma*, and right-wing militarists led for a while by Phoui Sananikone. Elections were held in 1955 and in March 1956

Souvanna Phouma formed a government in alliance with the Pathet Lao. This collapsed in 1958 and Sananikone denounced the Geneva Agreements, but was superseded by a US-backed military government led by General Phoumi Nosavan. He was deposed by a group of young officers and Souvanna was restored, only to be again deposed. In 1962, after long talks, a truce to the civil confusion was established and Souvanna formed another coalition government with his half-brother Souphanouvong of the Pathet Lao. A declaration of neutrality was accepted by all the great powers. Unfortunately this accord soon collapsed, and as the Vietnam War escalated Laos became more and more involved. The Pathet Lao facilitated VietMinh supplies using the 'Ho Chi Minh Trail', whereupon intensive US bombing followed. In 1971 US and South Vietnamese troops (estimated at 22,000) invaded and campaigned. Souvanna remained nominal Prime Minister, while a succession of US-backed officers tried to prevent the steady growth of Pathet Lao support. In 1973 a cease-fire ended US bombing and a new agreement between the two half-brothers followed. In April 1975 Pathet Lao troops were welcomed through town after town, entering Vientiane in August. They proclaimed a 'People's Revolutionary Administration'; in December the monarchy was abolished and the People's Democratic Republic formed under Kaysone Phomvihane*, who remained in power for the next fifteen years. In 1989 there began some relaxation of his regime, and moves to restructure the economy. A rebel United Front for the National Liberation of the Lao People claimed Thai support and during 1990 appeared to be gaining peasant support.

Largo Caballero, Francisco (1869–1946), Prime Minister of the Spanish Republic 1936–7. Born in Madrid, he became a plasterer and a trade union official. As a socialist he was imprisoned for life in 1917 for organizing a general strike, but released on his election to the Cortes in 1918. After the abdication of Alfonso XIII* he joined the government of the Republic as Minister of Labour. When this collapsed he was imprisoned again for supporting an abortive Catalan uprising, but acquitted and released in 1935. He was leader of the Popular Front*, which won the elections of February 1936, but did not become Prime Minister until September 1936, two months after the outbreak of Civil War, when he headed a coalition of Socialists, Republicans, and Communists. He resigned in May 1937 following a Communist take-over in Barcelona and went into exile in France. He was arrested by the Germans and spent the war years in Dachau concentration camp, which he survived, dying in exile.

Lateran Treaties (11 February 1929), agreements between Mussolini's government and Pope Pius XI. They aimed to regularize the relations between the Vatican and the Italian government, which had been strained since 1870, when the Papal States had been incorporated into a united Italy. A treaty, or Concordat, recognized Roman Catholicism as the sole religion of the state, and the Vatican City as a fully independent state under Papal sovereignty. A further treaty, or convention, recognized that the loss of the Papal States had involved great loss of papal income, and the Vatican received in cash and securities a large sum in settlement of its claims.

Latin American Integration Association (LAIA). This was formed by the Treaty of Montevideo in 1980. It replaced an earlier, more ambitious organization, the **Latin American Free Trade Association** (LAFTA), which had been formed in 1960 by an earlier Treaty of Montevideo and which had envisaged the eventual establishment of a Latin American Common Market. Little progress however had been made by 1980, and this at the expense of the poorer members. All the republics of South America joined the new Association, with observers from Central America and the Caribbean.

With a secretariat in Montevideo, it made some progress during the 1980s at tariff-cutting.

Latvia, a Baltic republic. Annexed by Russia in the 18th century, it was a land of vast estates owned by German families, the so-called Baltic Barons, within the tsarist empire. In April 1918 Latvian nationalists proclaimed independence. After the Russian Civil War, fought between Latvians, Germans, and Bolsheviks, international recognition was gained in 1921 and the Constitution of the Republic agreed in 1922. During the period 1922–40 seaports and industry declined with the loss of the Russian market, but agriculture flourished. In 1934, with the Great Depression* at its worst, a neo-Fascist regime was formed by Karlis Ulmanis. He vainly tried to win Hitler's support, but was sacrificed in the Nazi–Soviet Pact of 1939. The Red Army occupied it in June 1940, but the German army took Riga on 1 July 1941 and was welcomed. Reoccupied by the Red Army in October 1944, it became a constituent republic of the Soviet Union. Latvian nationalism never died, and in May 1990 a newly elected Supreme Soviet passed a resolution demanding independence from the Soviet Union, based on the Constitution of 1922. Negotiations for a Baltic confederation of republics were begun in Moscow.

Laurier, Sir Wilfred (1841–1919), Prime Minister of Canada 1896–1911. He was the first French-Canadian Prime Minister, but, as leader of the Liberal Party, was a strong opponent of separatism along religious/linguistic lines. His government was popular, with a strong policy of immigration, but collapsed in 1911 on the issue of trade with the USA. In 1914, as leader of the opposition, he supported Canada's part in World War I, but opposed the introduction of conscription in 1917.

Lausanne, Treaty of (24 July 1923). This replaced the earlier Treaty of Sèvres*, following the successes of Mustafa Kemal*. Smyrna (Izmir) was recognized as Turkish, and Greece surrendered eastern Thrace, including Adrianople (Edirne). Italy was confirmed in the Dodecanese and Britain in Cyprus. All the Aegean islands except Imbros and Tenedos would go to Greece. Turkey accepted that Palestine and Syria were to be mandated to Britain and France. The Dardanelles were to be demilitarized and open to shipping, as supervised by a League of Nations Commission. As a result over 1 million Greeks were to leave Turkey and some 350,000 Turks to leave Greece.

Laval, Pierre (1883–1945), French Prime Minister 1931–2, 1935–6, and of the Vichy regime 1942–4. Born in the Auvergne, he qualified and practised as a lawyer before entering politics as a young socialist Deputy in 1903. He strongly criticized France's entry into World War I, narrowly escaping imprisonment. Gradually moving to the right, as Prime Minister 1934–6 he was also Foreign Minister; as such he was co-author with Samuel Hoare* of the Plan to partition Ethiopia between Italy and Ethiopia. The plan was disowned by Britain's Prime Minister Baldwin and shocked many as a cynical appeasement of Mussolini, and he resigned. After France's defeat in 1940 he joined Pétain's regime as deputy Prime Minister. At this stage he advocated a policy of neutralism, refusing military co-operation. Pétain basically disliked him and he was dismissed. In 1942 he was obliged however to reinstate him as Prime Minister when Germany occupied all France, making Laval virtual dictator. He now advocated support for Hitler, drafting labour for Germany, authorizing a French Fascist militia, and instituting a rule of terror against Resistance fighters and Jews. He was distrusted by the Germans, who arrested him in 1944; but he escaped to Spain. He returned to France in 1945, and was arrested, tried, and shot at Fresnes.

Law, Andrew Bonar (1858–1923), British Prime Minister 1922–3. Born in Canada, he was educated in Scotland, and was a successful iron merchant before entering politics. Elected to Parliament as a Conservative in 1900, he was described, perhaps surprisingly, as 'an outstanding speaker'. He became leader of the Conservatives in 1911 and supported Ulster's resistance to Home Rule*. A tariff reformer, he joined Asquith's coalition as Colonial Secretary and continued under Lloyd George* as Chancellor of the Exchequer and member of the war cabinet. He was Lord Privy Seal 1919–21. In 1922 the Conservatives rejected Lloyd George as leader of the coalition government, and he was appointed Prime Minister. His health had already been failing and he resigned in May 1923, dying of cancer shortly afterwards.

Lawrence, Thomas Edward (1888–1935), British soldier and author. Born in Cornwall, he grew up and was educated in Oxford, from where he worked as an archaeologist in the Near East before World War I. In 1915 he was invited to join the Arab Bureau in Cairo and he played a part in the negotiations with Sherif Hussein ibn Ali*, and his son Faisal*, to enter the war against the Ottomans in return for Allied recognition of Hussein as King of the Hejaz. He took an active part in the Arab Revolt* which followed, working with guerrilla forces sabotaging the Hejaz railway*, and taking Aqabah in July 1917. In November 1917 he was captured and severely tortured before escaping and taking part in the capture of Damascus in October 1918. After the war he argued for Britain to honour its promises, urging support for Arab claims in Syria against the French. In 1921 he joined Churchill's new Middle Eastern Department as adviser, and helped to plan the Middle East settlement of that year. He then withdrew from public life, first enlisting in the ranks of the RAF under the name of John Hume Ross. In 1923 he joined the Tank Corps as T. E. Shaw, which name he retained. He returned to the RAF, in which he served 1925–35, engaged in research on fast-rescue motor boats. He was killed in a motor cycle accident. His account of the Arab Revolt, *The Seven Pillars of Wisdom* (1926), became a classic.

League of Nations, organization for international co-operation. It was established at the Paris Peace Conference*, which adopted its Constitution or Covenant. This embodied the principles of collective security, arbitration of international disputes, reduction of armaments, and open diplomacy. It also established a number of specialist agencies. The US Senate, which had refused to accept the Treaty of Versailles, dissociated itself from the Covenant. Germany was admitted to the League in 1926 and the Soviet Union in 1934. With its headquarters in Geneva, it accomplished much of value in post-war economic reconstruction, negotiating international loans for a number of countries. Its International Labour Organization* did much to improve working conditions, especially in colonial societies. Yet the League failed in its prime purpose, through the refusal of member nations to put international interests before national ones. It was powerless in the face of Japanese, Italian, or German expansionism. A number of nations left the League, including Brazil in 1926, Japan and Germany in 1933, and Italy in 1937. In 1945 it was replaced by the United Nations.

Lebanon, an east Mediterranean state. In 1914 Beirut and Mount Lebanon were nominally under Turkish rule, although since 1860 they had been *de facto* under French protection. In 1915 the area was placed under Turkish military rule. In 1918 the French High Commission in Beirut declared the coastal strip from Tyre in the south to Tripoli in the north, together with the Lebanon mountains, to be the new state of Greater Lebanon. In 1920 the League of Nations gave France the mandate for this and for Syria. In 1926 it was named the Lebanese Republic and granted considerable autonomy. During the 1930s a tradition emerged for the President to be a Maronite*, the Prime Minister a Sunni Muslim, and the Speaker of the Chamber of Deputies a Shia Muslim, the Chamber to have Christians and Muslims in the ratio 6 : 5. In 1941 Lebanon was occupied by Free French forces who announced independence. However, when a nationalist anti-French President was elected in 1943 he was arrested and only released after pressure from Britain and the USA. In 1946 British and French troops withdrew, and from then until the 1970s Lebanon maintained an uneasy stability, Beirut* becoming the financial centre for all the Middle East. It managed to stay neutral in the Arab–Israeli crisis of 1948, but a split appeared over the Suez Crisis*, when Muslims supported Egypt and the Maronite Christians France and Britain. President Camille Chamoun* only averted civil war (1958) by calling upon US assistance. Although Lebanon again avoided active involvement in the wars of 1967 and 1973, increasing numbers of Palestinian refugees entered the country, creating camps around Beirut and other cities. PLO guerrillas began to operate from these, especially after they were ejected from Jordan (1970–1). This caused intermittent retaliation from Israel, but was also the main cause of civil strife: Maronite Phalangists wanted the PLO eradicated, but some Muslims wanted to give them protection. When a bus-load of Palestinian refugees was massacred in April 1975, civil war erupted, with the Druzes* siding with Muslims and the PLO, as traditional enemies of the Maronite Phalangists. In June 1976 Syria invaded and fierce fighting followed until October, when much of Beirut was destroyed. Lebanon was now effectively divided: Syria held most of the east of the country; Muslims held both Tripoli in the north and much of southern Lebanon; Beirut was divided between Christian East Beirut and Muslim West Beirut, the latter split between Sunni and the increasingly extremist Shia. Israel bombed PLO centres in 1977 and in March 1978 occupied south Lebanon, being replaced in June by a UN peace-keeping force. Israel resumed bombing in 1980 and in 1982 again invaded, severely damaging Tyre and Sidon. Palestinian camps around Beirut were besieged and the city was bombed and shelled. In August 1982 the PLO officially withdrew from Beirut, but a Phalangist massacre of men, women, and children in two camps Chabra and Chatila (September 1982) brought a renewed UN peace effort. A multi-nation force (US, British, French, and Italian) failed, and there were inconclusive conferences in Switzerland 1983–4. Fierce fighting in Beirut through 1984 resulted in a 500-yd. wide 'Green Line' of devastation between the Christian and Muslim halves of the city. In the same year President Jumayyil* succeeded in forming a coalition government under Rashid Karami (assassinated 1987). Israeli occupation of south Lebanon ended in June 1985, when Israel installed a Maronite South Lebanon Army (SLA). There were now numerous militia groups on the streets of the cities: Maronites, split between hard-line Phalangists and moderate Chamounists (National Liberation Party); Druzes (Progressive Socialist Party); Shia Amal Muslims and two more extreme Shia groups, the Iranian-backed Hezbollah and the al-Jihad al-Islami. In February 1987 Syrian troops moved into West Beirut to try to obtain release of some twenty-six hostages held by militia groups, and in July a pro-Syrian Unification and Liberation Front emerged. Israel still backed the SLA and there were over twenty Israeli air raids during 1988. In March 1989 the Maronite Christian General Aoun launched an all-out war against occupying Syrian troops and fighting lasted all summer. In October Arab League arbitration resulted in the Taif Accord*, but very soon afterwards the new President Mouawad was assassinated. He was succeeded by President Elias Hrawi. He seemed at first to be impotent, as General Aoun refused to recognize him. However, fighting soon broke out between the latter's supporters and the Phalangist Lebanese Front led by Samir Geaga. On 3 April 1990 Geaga accepted the Taif Accord and recognized Elias

Hrawi. Syria, having supported US policies in the Gulf Crisis*, felt secure enough to impose itself, and on 11 October Syrian forces surrounded Aoun's palace. Aoun gained asylum in the French embassy and fierce fighting resulted in some 800 casualties. The Green Line was dismantled and a frail peace established.

Lebensraum ('Living-room'), a term coined by the German political scientist Karl Haushofer in the 1870s. He argued the inevitability of territorial expansion for Germany in order to maintain a balance between urban and rural living, on which its moral health depended. It became a central tenet of Nazi propaganda during the 1920s and 1930s and was used to justify the absorption of Bohemia, Moravia, and western Poland in 1939.

Leclerc, de Hauteclocque, Philippe (1902–47), French general. From a French aristocratic family in Picardy, he graduated at the French Military Academy of St-Cyr in 1924. By 1939 he held the rank of captain. He joined the Free French and was sent by de Gaulle to the Cameroons. In 1943 he brought a motorized column up the Congo and across the Sahara to join the advancing 8th Army at Tripoli. He was next given command of the 2nd Armoured Division for the Normandy landings and in August 1944 dashed for Paris, which he liberated. In the autumn of 1945 he was sent to Indo-China, landing at Saigon, where he took highly repressive measures against the insurgents. He was killed in an aircraft crash

Le Corbusier, Charles-Édouard Jeanneret (1887–1966), Swiss architect. He was an early exponent of ferro-concrete, and in his *Towards a New Architecture* (1927) he put forward the idea of mass-produced housing, based on a modular skeleton, together with the concept of pedestrian segregation. Although influenced by Cubism*, his buildings adopted an increasingly massive and 'brutal' style, which he realized in the *Unité d'Habitation* of Marseilles and in Chandigarh, the capital city of the Punjab in India. Immensely influential, in lesser hands his concepts resulted in some town-planning disasters, as in Sheffield and other British cities, many of which were being demolished by the end of the century.

Le Duc, Tho (1911–90), Vietnamese politician. One of the founders of the Communist Party of Indo-China in 1930, he was imprisoned by the French but escaped to China. On his return to Indo-China in 1941 he helped to found the VietMinh*, an anti-Japanese movement, which after the war led resistance against the French. He was elected a VietMinh delegate for South Vietnam in 1949, at the same time becoming leader of the Vietnam Workers' Party (Communist). From 1955 he served on the Politburo of the latter and in 1966 led resistance to US intervention in Vietnam. In 1968 he unsuccessfully negotiated with William Harrison in Paris for a cessation of US bomber offensives, being spokesman for the North Vietnamese delegation to the Paris Peace Conference. In 1973 the conference achieved a cease-fire, for which, with Henry Kissinger* the American negotiator, he was offered the Nobel* Peace Prize, which he refused. He remained a member of the Politburo of the Republic of Vietnam until his resignation in 1987.

Lee, John Alexander (1891–1982), novelist and dissident member of the New Zealand Labour Party. Born in Dunedin, after primary school he worked as a labourer and served time in Borstal. Both severely wounded and decorated in World War I, he joined the newly formed Labour Party, describing himself as 'having come out of the gutter'. He was a Member of Parliament 1922–8 and 1931–43. As a novelist he identified with those whom he regarded as the victims of the social and political system. When

Labour won the election of 1935 he failed to gain a cabinet post, becoming instead Under-Secretary to the Minister of Finance. He put forward a radical low-cost housing plan, but after the 1938 election became a fierce critic of the government for its alleged 'loss of nerve'. He was expelled from the Labour Party in March 1940, whereupon he set up his own Democratic Labour Party. This failed to take off and he was not re-elected in 1943, after which he became a book publisher.

Lee Kuan Yew (1923–), Prime Minister of Singapore 1959–90. Born into a wealthy Singapore Chinese family, he was educated at Cambridge and called to the English Bar. On returning to Singapore, he was a trade union advocate before entering politics. In 1955 he formed the People's Action Party, a democratic, socialist organization, which under his leadership has dominated Singapore politics ever since. He took part in negotiations in London for self-government at the time of the Malayan Emergency* and in 1959 formed his first government, with a policy of greater industrialization. He led Singapore as a component state of the newly formed Federation of Malaysia* in 1963, and since 1965 as a fully independent republic. His policies developed along increasingly autocratic lines, centred on the basis of one-party rule. Tight government planning, and a hard-working population supported by an extensive social welfare system, within a free market economy, resulted in one of the world's most successful economies. In November 1990 he announced his retirement.

Leguía y Salcedo, Augusto Bernardino (1863–1932), Prime Minister of Peru 1903–8; President 1908–12, 1919–30. A member of a wealthy Spanish colonial family in Peru, he had a successful career in business before entering politics. He was first elected Prime Minister in 1903, and during his first term in office he settled a long-standing frontier dispute with Bolivia and Brazil, introduced administrative reforms, and improved the public health system. He was reinstated as President by the army in 1919, introducing a new constitution in 1920. He chose however largely to ignore this, governing by increasingly dictatorial methods and alienating liberal public opinion. Although this second term saw rapid industrialization, it was adversely affected by the Great Depression* and fall in commodity prices. Criticized for the settlement of 1929, which ended the long Tacna–Arica frontier dispute with Chile, he steadily lost popularity and in 1930 was forced from office by the army.

Lend-Lease Act, US legislation. In March 1941 the US Congress passed an act allowing President Roosevelt* to lend or lease equipment and supplies to any state whose defence was considered vital to the security of the USA. Some 60 per cent of the shipments went to Britain and Commonwealth countries as a loan, in return for British-owned military and naval bases leased to the USA and for the accommodation of US troops, especially after the US entry into war in December 1941. About 20 per cent went to the Soviet Union. Altogether thirty-eight nations received Lend-Lease aid, including China. President Truman* ended Lend-Lease immediately the war ended in September 1945, to the dismay of many recipients, especially the Soviet Union.

Lenin, Vladimir Ilich (1870–1924), Russian revolutionary leader and founder of the Soviet Union. Born in Simbirsk, son of a senior civil servant (his real name was Vladimir Ilich Ulyanov), he was expelled from Kazan University for participation in student riots. His older brother was executed in 1887 for an attempt on the Tsar's life and this had a profound effect on him. He moved to St Petersburg, where he graduated with a gold medal in law as an external student. In 1895 he formed the Union for the Struggle for the Liberation of the Working Class and was exiled to Siberia. Released in 1900, he lived in Geneva, Paris, and London, where he set up the Iskra Group 'for the preservation of orthodox Marxism'. Here he also took part in the crucial debate in 1903 of the Second Party Congress of the Social Democrats, when Bolshevik* and Menshevik* factions split. His *What is to be Done?* (1902) became a basic text for left-wing activists. As leader of the Bolshevik faction he took a prominent part in socialist propaganda and organization in the years before World War I. He returned to Russia in 1905 and helped to organize the St Petersburg Soviet. He had to escape and was then in Prague and Cracow, forming in 1912 the 'Central Committee of the Social Democratic Party' of a dozen hand-picked Bolsheviks, and founding the newspaper *Pravda*. He was in Switzerland when war broke out, publishing *Imperialism*, a Marxist explanation for the war, and *State and Revolution*, in which he foresaw a proletarian revolution in Russia. In March 1917 he returned there via Germany in the famous sealed train and was in Petrograd in April. In July, however, he was expelled by the provisional government*, and went to Finland. He returned again in October, and led the rising of the Petrograd Soviet in the 'October' Russian Revolution*. As chairman of the Council of People's Commissars he became virtual dictator of the new state. Once in power he emphasized more and more the need for a strong Central Committee of the party, arguing the case for 'democratic centralism', from which was to develop the huge bureaucratic monolith of the Soviet Union. He took Russia out of the war with Germany at Brest-Litovsk* and successfully resisted the counter-revolutionary forces in the Civil War* 1918–21. His initial economic policy of war communism, which included nationalization of major industries and banks, and control of agriculture, was an emergency policy demanded by civil war. It was followed by his New Economic Policy* (NEP), permitting private production and trading in agriculture. It came too late to avert terrible famine 1921–3. He did not live to see the marked recovery as agricultural and industrial production increased, having suffered a stroke in 1922 from which he died. Lenin's own outlook and character deeply affected the form the revolution took. He set an example of personal austerity and of political impersonality and ruthless determination. Later Communists throughout the world continued to look to his writings (forty-five volumes!) for inspiration.

Leningrad, Siege of, World War II (1941–4). The German army had hoped to capture Leningrad during the 1941 campaign against the Soviet Union, but as a result of slow progress in the Baltic area it was September before the German/Finnish forces surrounded the city. Resistance was so fierce that the German High Command abandoned efforts to storm the garrison, relying instead on a blockade. This was to last nearly 900 days. As few preparations had been made, and as evacuation of the population had been ruled out by the Soviet government, there may have been over 1 million civilian deaths in the siege, caused by starvation, cold, and disease, as well as by enemy action. Over 100,000 bombs were dropped on the city, and between 150,000 and 200,000 shells fired at it. Soviet counter-attacks began early in 1943, but it was nearly a year later (27 January 1944) before the siege was completely lifted.

Leopold III (1901–83), King of the Belgians 1934–51. He was educated at Eton and at Ghent University and succeeded his father in 1934. When the Germans invaded in May 1940 he took command of the Belgian army, but was hopelessly outnumbered and on 28 May ordered a cease-fire. This order came to be the centre of a long controversy, the Belgian Prime Minister at once calling on Belgium to continue to resist. Leopold was imprisoned by the Germans and liberated in May 1945, going to Switzerland and declaring that he had never been a collaborator. He refused to abdicate and a commission investigated his actions, which it justified. In 1950 a referendum was held and a small majority voted for his return; but there were serious riots and Belgium was near civil war. He was finally persuaded to abdicate in favour of his son Baudouin in July 1951.

Lesotho, a southern African state surrounded by the Republic of South Africa. Founded by Moshoeshoe I in 1832, it was placed under direct control of the British government in 1884, with its paramount chief titular head of what the British named Basutoland. When the Union of South Africa was formed in 1910 Basutoland came under the jurisdiction of the British High Commissioner in South Africa, much against the will of many South Africans. In 1966 it became independent and was renamed Lesotho, as a constitutional monarchy, Constantine Mottohei Moshoeshoe II, paramount chief of the Sotho people since 1940, becoming King. In 1986 a military coup disbanded the National Assembly, the King to rule through a Council of Ministers and a College of Chiefs. Moshoeshoe was deposed in November 1990 and replaced by his son Letsie II.

Lesser Antilles, numerous small islands in the Caribbean grouped into the Leeward and Windward Islands, which were all colonized during the 17th century by Spanish, British, French, Dutch, and Danish settlers. They include: the US Virgin Islands*, about fifty small islands bought by the USA in 1917 from Denmark for strategic reasons and since developed for tourism; the British Virgin Islands*, a smaller group which retain colonial status; Montserrat, Martinique, and Guadeloupe, overseas dependencies of France; and a string of now independent Commonwealth states which were colonized by the British and which became independent after unsuccessful attempts at federation into the Federation of the West Indies* (1958–62).

Lévesque, René (1922–87), Premier of Quebec 1976–85. Having served as a war correspondent with the US army in World War II, he became a well-known radio and television journalist in the years 1946–60. He then entered the Quebec legislature as a Liberal, holding a number of posts, including Minister of Natural Resources, when he nationalized Hydro-Québec. Frustrated with federal government policies, he left the Liberal Party and founded Parti Québécois* in 1968, becoming Premier of Quebec in 1976. His government was less radical than some had feared, although strongly supporting French language rights. It refused to accede to Pierre Trudeau's* new Constitution of 1982, which defined the powers of provincial governments, but lost popularity when public-sector salaries were reduced. He retired in 1985 and was succeeded as Premier by his old Liberal rival Robert Bourassa, his aim of independence having been thwarted. In 1987 Bourassa indicated willingness to accept the Constitution at the Meech Lake conference.

Lewis, John Llewellyn (1880–1969), US labour leader and president of the United Mine Workers of America 1920–60. Son of Welsh miner immigrants in Iowa, he began work as a miner aged 15. In 1905 he became legal representative of the UMWA and in 1920 its president. A leading member of the American Federation of Labor* (AFL), he successfully organized mass-production workers into unions. This resulted in a clash with AFL policy in 1935, and in 1936 all such unions, including his miners, were expelled from the AFL; they formed the Congress of Industrial Organizations, with Lewis as president. During the next four years he led a number of militant and bitter CIO strikes in such industries as steel, automobiles, tyres, and electrical products. In 1940, in protest against Roosevelt's third-term nomination, he resigned from the CIO (which was to be reconciled with the AFL in 1955) and in 1942 withdrew the miners, whose president he remained. A strong personality, he would defy any authority in the interests of his members. No communist, he defied the Taft–Hartley Act* (1947) by refusing to declare on oath that he was not.

Leyte Gulf, Battle of, World War II (October 1944). In the campaign to recover the Philippines, US forces under General MacArthur* landed on the island of Leyte on 20 October 1944. On 23 October four Japanese naval forces were converging, to attack US transports. In a series of scattered engagements forty Japanese ships were sunk, forty-six were damaged, and 405 planes destroyed. US losses were one light carrier, two escort carriers, and three destroyers. Following this major US victory, the Japanese fleet withdrew from Philippine waters.

Liang Qichao (Liang Chi-chao) (1873–1929), Chinese political theorist. As a leading Chinese scholar he had endeavoured in the 1890s to create a modern Chinese school system and to reform the 2,000-year-old Civil Service exam. When the Empress Dowager halted all reform he went abroad, but his weekly articles still appeared in the Chinese Press. He returned to China in 1912 as a liberal republican opposed both to the restoration of a monarchy and to the radical policies of Sun Yixian*. After World War I he travelled extensively in Europe and India, returning to China to spend his later years as a university academic, seeking means to achieve an intellectual and cultural regeneration in China, while eschewing the increasingly influential doctrines of Marxism.

Liaquat Ali Khan, Nawabzada (1895–1951), first Prime Minister of Pakistan 1947–51. Son of a wealthy landowner, he was educated at Oxford, and practised as a barrister. He entered politics in 1923, becoming a Muslim leader in the Legislative Council of the United Provinces. From 1933 he was right-hand man to Jinnah* in the Muslim League*, and in 1940, as deputy leader of the League, was elected to India's Central Legislative Assembly. In 1946 he was appointed a member of the Viceroy's Executive Council and Finance Minister of the interim government. When Pakistan was created he became its Prime Minister and Minister of Defence. As such he accepted the UN ruling on Kashmir*. On the death of Jinnah in 1948 he became the most powerful politician in Pakistan. Deeply committed to membership of the Commonwealth of Nations and to improved relations with India, he concluded the Delhi Pact of 1950; he was murdered by extremists wanting war with India.

Liberal Democratic Party (Japan). Political alignments were slow to coalesce in post-war Japan, but in 1955 the Progressive Party and the old Liberal Party combined to form the Liberal Democratic Party, which has succeeded in holding power ever since. Early leaders were Kishi Nobusuke* and his brother Sato Eisaku*. Tanaka Kakuei* was forced to resign in 1974 over a bribery scandal; four leaders followed within eight years as the party's fortunes waned, before recovery under the forceful leadership of Nakasone Yasuhiro*, president of the party and Prime Minister 1982–7. The party developed close links with business and interest groups such as fisheries and agriculture. A key feature has been its structure of internal factions, less concerned with policy than with patronage, electoral funding, and competition for leadership. The party has appealed to a wide range of the electorate, although a Japanese Socialist Party has gradually gained more support.

Liberal Party (Australia). There were numerous state Liberal parties before 1901 and many early federal politicians such as Alfred Deakin called themselves Liberals, often allying with Labor politicians to form what was termed a Fusion Party within the federal Parliament. When W. M. Hughes* split the Australian Labor ranks over conscription, many Liberals joined him to form a Nationalist Party. This itself split in 1923, but the Bruce*–Page* government of 1923–9 contained many Liberals, as did the United Australia Party founded by J. A. Lyons* in 1931. It was not however until 1943, when the UAP had only twelve members in Parliament, that Robert Menzies* decided to build a national Liberal Party. In the immediate post-war years, when Communism seemed a major threat throughout South-East Asia, he was able to construct a strong right-wing party. He allied himself with the Country Party* and this Liberal/Country coalition won fifty-five seats in the election of 1949. Since then it has alternated in office with Labor. Strongly committed to free enterprise, it has tempered its earlier fiercely anti-Communist stance and promoted closer links with its Asiatic neighbours.

Liberal Party (Canada). The party traces its origins back to the early 19th century and first formed a federal government in 1873. Under Sir Wilfred Laurier it had established its main principles of encouraging immigration, opening the West, introducing social legislation, and encouraging English/French unity and greater autonomy for Canada within the British Empire. With a strong base in Quebec, it held office through most of the 20th century until the Quebec issue divided it in the 1980s. Its leaders Mackenzie King*, Louis St Laurent*, Lester Pearson*, and Pierre Trudeau* have all shown considerable political skills. When in power the party has steadily built up a federal welfare system by introducing pensions, family allowances, unemployment benefits, and a medicare* system which provides comprehensive health care. Such a system has inevitably strengthened the role of the federal government as against those of the provinces, and one aim of the Liberal government's 1982 *Charter of Rights and Freedoms* was to protect provincial rights against federal encroachment. In recent years the party has become increasingly wary of the influence of US capital on the Canadian economy.

Liberal Party (New Zealand). As with the British Liberal Party, the greatest days of that of New Zealand had been in the years before World War I, especially under Richard Seddon 1893–1906, when factory acts, trade union legislation, women's suffrage, and old-age pensions had all been achieved. Although it formed a wartime coalition (1915–19) with W. F. Massey*, the party lost support after the war either to Labour or to Massey's right-wing Reform Party* of North Island farmers. The party leader Joseph Ward* tried to attract the latter, and in 1928 restyled his party as the United Party, winning the election of that year. During the Great Depression* of the early 1930s the two parties drew closer and amalgamated in 1936 as the National Party* of New Zealand, which was to alternate in power with Labour after World War II.

Liberal Party (UK). It had emerged in the mid-19th century as successor to the Whig Party and was the major alternative to the Conservatives until 1918. Its government 1905–15 implemented a large programme of social legislation on which later Welfare State legislation was to be partly based. After the outbreak of war in 1914 it formed coalitions with the Conservatives, first under Asquith and then under Lloyd George, but was badly split at the time of the Coupon Election* in December 1918. It was in coalition with Labour in 1924 and 1929–31, but split again in 1932 over the Ottawa Agreements*, John Simon* leading the National Liberals, who stayed in the government, and Herbert Samuel the Liberal Nationals, who resigned. The former remained in government until 1945. After World War II it was an opposition party of declining fortune, forming a Lib–Lab pact with Labour 1977–8 and then an Alliance (1983–7) with the Social Democratic Party*. In 1988 it merged with that party to form the Social and Liberal Democrats. It had been especially strong in Scotland, Wales, and south-western England, and its continued decline has been partly explained by a lessening of church/chapel issues in society.

Liberation Theology, a movement within the Roman Catholic Church. It began to emerge in Latin America following a Bishops' Conference in Colombia in 1968. It seeks to apply religious faith by active political/social activity on behalf of the poor and oppressed. Liberation theologians stress that the Bible can only be fully understood within the context of poverty and that, especially in Latin America, the Church is a church for and of the poor. To build this Church local *comunidades de base*, with thirty or so members, would meet both to study the Scriptures and to try to meet immediate needs of food, water, and sewage. Many such groups sprang up during the 1970s and a leading supporter was Archbishop Oscar Romero, murdered in El Salvador* in 1980. Because of their overwhelming concern for the poor as against wealthy élites, they have been accused of Marxism and of fostering social revolt. Official support from the Vatican has been no more than luke-warm.

Liberia, the oldest independent republic in Africa (1847). It owes its origin to the American Colonization Society, which resettled African slaves from the USA. The beginning of prosperity was in the 1920s, when the Firestone Rubber Company planted huge areas and provided a permanent and stable market for rubber. W. V. S. Tubman* was President from 1944 until his death in 1971. With a decline in rubber prices as a result of plastic alternatives, the economy suffered during the 1970s. In a bloody revolution in 1980 President Tolbert was assassinated, political parties were banned, and a People's Redemption Council, a military government, under Master-Sergeant Samuel Doe was installed. The latter was then named President and Commander-in-Chief. In 1990 he was overthrown in a bloody civil war by his former ally Charles Taylor and the National Patriotic Front.

Libya. With a predominantly Arab population, Libya (the provinces of Tripolitania and Cyrenaica) had been under Ottoman rule from 1551 until 1911, when it was conquered by Italy. Effective Italian occupation was largely confined to the coastal regions and Tripolitania and Cyrenaica were treated as separate colonies until 1939, when they were linked. The scene of fierce fighting in World War II 1940–3, the country was placed under British military rule in January 1943. Italy renounced all claims in 1947 and the country was declared independent by a UN resolution in 1951—a federal state of three provinces, Tripolitania, Cyrenaica, and the Fezzan, an inland desert region to which France had until then laid claim. The Emir of Cyrenaica, Muhammad Idris al-Sanussi, was proclaimed King Idris I. In return for considerable economic and technical aid both Britain and the USA were granted concessions for military bases. The discovery of oil in 1959 and its exploitation from 1961 greatly reduced Libyan dependence on foreign aid. In 1963 the federal system was abolished and women enfranchised, moves which were by no means universally popular. In September 1969 Idris was deposed by a group of young officers, who created a radical socialist state to be ruled by a Revolutionary Command Council under the charismatic leadership of Muammar al-Qadhafi*. The Anglo-Libyan treaty of friendship of 1953 was terminated and British and US bases closed. Libya joined Egypt and Syria in adopting a strong anti-Israeli policy. In 1973 Qadhafi nationalized the oil industry, using oil wealth to build up military capability, Libyans being encouraged to replace foreigners throughout the economy. Libya broke with Egypt over the Camp David Accord* (1978) and began to interfere in the affairs of Tunisia, Chad, and Morocco. It provided a base for a number of extremist terrorist groups, and relations with Europe and more especially the

USA steadily deteriorated, with military incidents 1981, 1986, 1988. In 1989 it joined the Maghreb Union* and President Qadhafi condemned the Iraqi occupation of Kuwait in August 1990, adopting a neutral stance.

Li Dazhao (Li Ta-chao), 1888–1927, co-founder of the Chinese Communist Party. Born of a peasant family, he gained entrance to Tientsin University and later studied in Japan. He returned to China in 1916 and became Professor of History and Librarian of Beijing University, where he employed young Mao Zedong* as assistant. From 1917 he was lecturing on Marxism, and the May Fourth Movement* stimulated him into merging his study groups into the infant Chinese Communist Party (1921). He visited the Soviet Union and by 1926 he was in virtual control of the Guomindang in Beijing and recognized as the leading Chinese Communist intellectual. In April 1927 Beijing was occupied by the warlord Chang Tso-lin. Li took refuge in the Soviet embassy, but he was seized and hanged. As a student of politics his main interest was in the relationship between the concept of democracy and socialist theory.

Lidice, a mining village in Bohemia. Five days after Reinhard Heydrich* was assassinated in Prague it was destroyed by German security forces in retaliation. All its men were killed and women and children sent to concentration camps. It remains a symbol of Resistance heroism in Czechoslovakia.

Lie, Trygve Halvdan (1896–1968), first secretary-general of the UN 1946–53. Born and educated in Oslo, after university he quickly became a successful barrister and in 1919 joined the Norwegian Labour Party (later the Social Democrats). He became its legal adviser and was elected to the Storting (Parliament). From 1935 he held a series of ministerial posts, and became Minister of Foreign Affairs in the exiled Norwegian government in London in 1941. He was elected to his UN post as a compromise candidate. When forces of the Republic of North Korea crossed the border into South Korea in 1950 he took the initiative by sending UN forces to restore peace. In doing so he lost the support of the Soviet bloc within the UN, and by 1952 his position had become very difficult. He had been re-elected to a further three years of office in 1951, but he decided to retire in 1953, returning to Norwegian politics.

Liebknecht, Karl (1871–1919), German politician. Son of Wilhelm Liebknecht (1826–1900), who had been an associate of Karl Marx, he graduated from Berlin University and became a brilliant lawyer. Elected to the Imperial Reichstag as a socialist, he was expelled in 1916 and sentenced to four years' penal servitude for 'attempted high treason'. He was released in October 1918 and then helped to form the German Communist Party in December. He led the Spartakist* rising with Rosa Luxemburg, and was arrested and shot by his guard in January 1919.

Likud, an Israeli political coalition of right-wing parties. Its roots go back to Russian Zionism* of the 1920s, and two of its leaders have been Menachem Begin* and Yitzhak Shamir*. Essentially a coalition within the Knesset (Parliament), it formed a government in 1984 with the Israeli Labour Party under Shimon Peres. This collapsed in March 1990, when the Likud remained in power, now supported by extreme Orthodox groups.

Lin Baio (Lin Piao) (1907–71), Chinese military leader. Of a modest family background, from central China, he joined the Socialist Youth League when at school and in 1925 enrolled at the Whampoa Military Academy, where he graduated with high honours before joining Jiang Jiehi's* (Chiang Kai-shek) Northern Expedition* in July 1926. On the split between the Guomindang* and the Communists in 1927, he joined the Chinese Communist Party and became one of the military commanders of the Jiangxi Soviet* against the Nationalists. He commanded the 1st Army Corps in the Long March* to Yan'an, where at 28 he became head of the Red Army Academy. He was severely wounded in the Sino-Japanese War, going to Moscow for medical treatment and returning to China in 1942. During the Chinese Civil War his capture of Manchuria from the Nationalists was achieved through patient wooing of the peasantry, building up an army of a million men, capturing Mukden (now Shenyang) in October 1948, and then moving down to take Beijing in January 1949, and to participate in the conquest of south China. He was created a marshal (1955) and Defence Minister (1959). He supported Mao in the Cultural Revolution*, becoming recognized as his successor in 1969. His policy of politicizing the People's Liberation Army* was seen however as an increasing threat to the CCP. Mao seems to have sensed danger, and Lin was alleged to have organized a plot among senior military officers, in a desperate move to avoid being purged. He was killed in an air crash in September 1971 while fleeing to the Soviet Union.

Lindbergh, Charles Augustus (1902–74), US aviator. Son of a US Congressman from Minnesota, after university he entered the US Army Air Corps Reserve in 1925. In 1927 he won $25,000 for being the first person to fly the Atlantic solo, making him a folk-hero in the USA. Because of his world-wide fame the kidnapping and murder of his 2-year-old son in 1932 resulted in massive media coverage, from which he fled to Europe. Here he was impressed by what he saw of Nazi Germany, while warning of its growing air power. He criticized Franklin D. Roosevelt's* open support for the Allies 1939–41, but once war was declared he took an active part in the Pacific. He later became a technical consultant for various airlines.

Lithuania, a Baltic republic. It had been a vast state during the Middle Ages, at one time stretching from the Baltic to the Black Sea. By 1917 however it had all been absorbed by tsarist Russia, although a nationalist movement had emerged during the 1880s. In March 1918 Kaiser William II recognized Lithuanian independence and a German king was elected by the Lithuanian Assembly. In November 1918 he was deposed and a republic proclaimed. Bolshevik troops from Russia now invaded and a short Russo-Lithuanian war was ended in March 1920 with the Treaty of Moscow, which recognized the independence of the Republic, which incorporated the German-speaking port of Memel*. It failed in its bid to incorporate Vilna, which went to Poland. At first a democratic republic, with a strong Social Democratic Party, its politics began to polarize, and in 1926 its President Antanas Smetona established a neo-Fascist dictatorship. In October 1939 a Soviet–Lithuanian Mutual Assistance Pact permitted Lithuania to take Vilna, on condition that it was garrisoned by Soviet troops, Memel having been lost to the Germans in March. In July 1940 the Lithuanian Assembly voted for incorporation into the Soviet Union as a constituent republic. It was occupied by German troops 1941–4, when its large Jewish population was almost wiped out. Reoccupied by the Red Army in 1944, the Republic has been the scene of considerable nationalist agitation since. In 1956 there were serious riots, which were suppressed. In March 1990 a unilateral declaration of independence was made. The Soviet Union responded by an economic blockade, cutting oil and gas supplies. This, together with mediation by France and West Germany, led in May to an agreement to enter negotiations with President Gorbachev* on a possible confederation of Baltic states.

Li Tsung-jen (1890–?), Chinese soldier and Nationalist politician. Of peasant stock, he became a young graduate of the Kwanzi Military Academy in 1913. He rose to become a warlord*, who by 1924 controlled all the southern province of Guangxi

(Kwanzi). In 1926 he joined forces with Jiang Jiehi* (Chiang Kai-shek), his troops forming the 7th Army of the National Revolutionary Army* and taking part in the Northern Expedition*. After 1937 he commanded the 5th War Area against the Japanese. In 1948 he became Vice-President of the National Government*, fighting unsuccessful campaigns against the People's Liberation Army*, withdrawing from Nanjing and Hangzhou to the Nationalist capital Chongqing (Chungking). He was briefly President of the National government, January 1949–March 1950, conducting unsuccessful negotiations with the Communist leadership. In 1950 ill health took him to the USA and Jiang Jiehi resumed the presidency, now in Taiwan. In 1965 he went to live in the People's Republic.

Little Entente (1920–38). In August 1920 the Czech Foreign Minister Edvard Beneš* concluded treaties with both Romania and the Kingdom of Serbs, Croats, and Slovenes (later termed Yugoslavia), establishing an *Entente*. Its aim was to achieve a common foreign policy which would prevent both the extension of German influence and the restoration of the Habsburgs to the throne of Hungary. France supported the *Entente*, concluding treaties with each member. In 1929 it pledged itself against both Bolshevik and Magyar (i.e. Hungarian) aggression in the Danube Basin, while also seeking the promotion of Danube trade. During the 1930s its members gradually grew apart. Romania under Carol II* leaned towards Hitler's Third Reich, Czechoslovakia signed a non-aggression pact with the Soviet Union (1935), while in February 1934 Romania and Yugoslavia joined Greece and Turkey in the Balkan Pact*. In 1937 Yugoslavia and Romania were unwilling to give Czechoslovakia a pledge of military assistance against possible German aggression, and when the Sudetenland* was annexed in September 1938 the *Entente* collapsed.

Little Rock, the capital city of the US state of Arkansas. It achieved notoriety when in 1957 the Governor Orval Faubus called out the National Guard to prevent Black children from entering segregated White schools. A federal court injunction required the guards to be removed, and President Eisenhower sent federal troops to secure the authority of the federal government, which incidentally led to educational desegregation. After this incident segregation in US schooling rapidly declined.

Litvinov, Maxim Maximovich (1876–1951), Soviet Foreign Commissar 1930–9. Born of a Jewish family in Russian Poland, he joined the Russian Social Democratic (Workers') Party on its foundation in 1898. Always a supporter of Lenin, he was soon imprisoned, but escaped in 1902 and then acted abroad as an arms agent for the 1905 Revolution. He returned to Russia but was deported and went to France and then London, where he married. After the 'October' Revolution in 1917 he was appointed the first Bolshevik representative in London, but was deported in 1918. He then worked in the Soviet Foreign Office and from 1926 was virtually in control of Soviet foreign policy, although not appointed Foreign Commissar until 1930. He headed the delegation to the disarmament conference of the League of Nations 1927–9, negotiated diplomatic relations with the USA, and signed the Kellogg–Briand Pact* of 1928. He was a strong advocate of collective security against the Axis powers and was dismissed by Stalin before he signed the Nazi–Soviet Pact in 1939. He was Soviet ambassador to the USA 1941–3.

Liu Shaoqi (Lui Shao-chi) (1898–1974?), Chinese statesman. Born of a peasant family in Hunan province, he attended the same school as Mao Zedong. In 1920 he went to Shanghai and joined the Socialist Youth League, and then went to Moscow, returning in 1922 to join the newly formed Chinese Communist Party. He served as a

Communist trade union organizer in Guangzhou (Canton) and Shanghai before becoming a member of the Central Committee of the CCP in 1927 and one of its chief theoreticians. After the Long March* he was appointed a Political Commissar and during the Civil War he organized guerrilla forces in north China. On the establishment of the People's Republic in 1949 he was appointed chief vice-chairman of the party. In 1959 he succeeded Mao as Chairman of the People's Republic but remained second to Mao within the party. During the Cultural Revolution* he was fiercely criticized by Red Guards as a 'renegade, traitor, and scab', having disagreed with Mao on a number of issues. In 1968 he was stripped of office and disappeared the next year. His death was announced in 1974 and he was posthumously rehabilitated in 1980.

Liverpool and London Riots (1981). During the early 1980s street rioting returned to Britain, reminiscent of the 18th and early 19th centuries, and the first time since the anti-Fascist riots of the 1930s. Most riots were sparked by race hatred, fanned by rising unemployment. There was street violence in St Pauls, Bristol, in April 1980, but in 1981 severe rioting occurred in Brixton, London (April), and Toxteth, Liverpool (July), with other outbreaks at Moss Side, Manchester, and Southall, London. The Toxteth riots were described by the Home Secretary as of 'extraordinary ferocity' against the police, who for the first time used CS gas and who were henceforth trained for riot procedures and equipped with riot gear. Lord Scarman, a retired judge, was asked to report on the Brixton riot. His recommendations on community policing did something to improve race relations, but social deprivation, especially among Black youth, remained a potential source of violence, which became an increasing feature of Britain in the 1980s. In March 1990 there was severe anti-Poll Tax rioting in Trafalgar Square, London.

Lleras Camargo, Alberto (1906–), President of Colombia 1945–6, 1958–62. Born and educated in Bogotá, he worked as a journalist there 1925–30 before entering politics, joining the Liberal Party. He became chairman of the Colombia House of Representatives in 1932 and held various government posts in the 1930s. He became ambassador to the USA and represented Colombia at the San Francisco Conference* which founded the UN in 1945. He held office as President of Colombia for a year 1945–6 and was then secretary-general of the OAS* 1947–54. In 1957 he helped to organize a National Front of Liberal and Conservative parties which succeeded in removing the military junta led by General Rojas. As President he sought to re-establish democracy in Colombia and to improve relations with the USA, being an enthusiastic supporter of the Alliance for Progress*.

Lloyd George, David, 1st Earl Lloyd-George of Dwyfor (1863–1945) British Prime Minister 1916–22. He was born in Manchester of Welsh parents; his father died when he was 1 and he was brought up by a shoemaker uncle in North Wales. He became a successful lawyer and in 1890 was first elected as a Liberal MP for Caernarvon Boroughs, which he represented until 1945. He strongly denounced the Boer War, for which he was vilified, but in 1905 entered Campbell-Bannerman's* government as President of the Board of Trade. In 1908 he succeeded Asquith as Chancellor of the Exchequer. His budget of 1909, challenged by the House of Lords, led to the Parliament Act of 1911 and provided an opportunity for his unique oratorical skills. He was responsible for the National Insurance Act of 1911, protecting some of the poorest in the community from the hazards of ill health. Created Minister of Munitions in 1915, by his administrative drive he ended the shell shortage on the Western Front. In December 1916 he became Prime Minister, replacing Asquith and forming a powerful coalition government. His war cabinet was unique and highly efficient, and his powers of leadership were responsible for Britain's steady recovery from near defeat. He galvanized the Admiralty into accepting the convoy system* against U-boat attacks. His tactics at the Coupon Election*, however, permanently weakened the Liberal Party. At the Paris Peace Conference*, in spite of tough talk in the election, he strove for moderation, fearing the consequences of French vindictiveness against Germany. In 1921 he reached agreement with Sinn Fein* over Irish self-government. At home, the Conservatives, disliking his individualistic style of government and his contempt for the 'honours' system, left the coalition and forced him to resign as Prime Minister in October 1922. The Liberal Party, already split between the followers of Asquith and those who remained loyal to Lloyd George, was overtaken in the 1920s by the Labour Party. In 1929 he collaborated with Keynes* in a programme to combat unemployment. In the 1930s he opposed the National government over the Ottawa Agreements*, leaving the Liberal Party to form his own Independent Liberal Party. He had some sympathy with German grievances, but opposed the Munich Agreement* and supported Britain's entry into the war in 1939.

Locarno, Treaties of (1925), a series of international agreements. Signed on 1 December, they sought to guarantee the boundaries of Germany, France, and Belgium as specified by the Treaty of Versailles. The main architects of the conference were Austen Chamberlain*, Aristide Briand*, and Gustav Stresemann* for Germany. The latter refused to accept Germany's eastern frontiers with Poland and Czechoslovakia as unalterable, but agreed that alterations must come peacefully, following negotiation. He was therefore prepared to sign arbitration conventions with both Poland and Czechoslovakia as well as with France and Belgium. France signed Treaties of Mutual Guarantee with Poland and Czechoslovakia. The main Locarno Treaty, however, was that confirming the inviolability of the German frontiers with France and Belgium and the demilitarized zone of the Rhineland. In the 'spirit of Locarno', Germany was invited to join the League of Nations*. In 1936 Hitler was to denounce the main treaty when he occupied the demilitarized Rhineland.

Lockheed scandal. By 1971 the US Lockheed Aircraft Corporation was in severe financial trouble and it embarked on a policy of allocating $24m. as bribe money to win foreign orders. In February 1976 revelations to a US Senate Subcommittee led to official action in Japan, The Netherlands, and Italy. In Japan payments had been made to ensure that All Nippon Airways bought Lockheed aircraft and ex-Prime Minister Tanaka Kakuei* was arrested (later bailed) for receiving 500m. yen in violation of currency regulations. In The Netherlands the husband of Queen Juliana, Prince Bernhard, Inspector-General of the Armed Forces, was discredited for receiving $1.1m.; while in Italy two former defence ministers were named as having purchased Lockheed aircraft for the Italian air force and eight people were arrested.

Lodge, Henry Cabot (1850–1924), US Senator 1893–1924. From a long-established and respected Massachusetts family, he studied and taught at Harvard until entering state politics. In 1893 he was elected to the Senate for Massachusetts, becoming a dominant figure in senatorial politics. A close friend of Theodore Roosevelt, he supported the gold standard* and high protective tariffs. A conservative Republican, he came to specialize in foreign affairs, as chairman of the Foreign Relations Committee. He led the fight against US membership of the League of Nations, being a bitter foe of Woodrow Wilson and of the Versailles Peace Treaty*. He was a US representative at the Washington Conference* 1921–2 on arms limitation.

Logical positivism, a system of philosophy. It emerged in Vienna during the 1920s and early 1930s, under such thinkers as Rudolf Carnap. It continued in the USA when many leading German and Austrian philosophers emigrated there after Hitler came to power. The system asserts the meaninglessness of metaphysics, conceiving philosophy as consisting purely of analysis conducted by logic. Although as a school of philosophy it lost support, at a popular level it reflected the secular beliefs of the Western world in the middle years of the 20th century.

Lomé Conventions, trade agreements replacing the earlier Yaoundé Convention between eighteen African states and the European Economic Community of 1969. Lomé I (1975) was between forty-six African, Caribbean, and Pacific states and the EEC for technical co-operation and development aid. The developing countries received free access for their products into the markets of the EEC, plus aid and investment. Lomé II was signed in 1979 and provided over £3,000m. of aid in the years 1980–5. It was renewed in 1984 (Lomé III) between sixty states and the EEC. Sixty-eight countries signed Lomé IV in December 1989.

London Treaties (1915). There were two somewhat disreputable secret agreements made by Britain in London at the beginning of World War I. The Constantinople Agreement* with Russia (April 1915) would have given Constantinople to Russia but for the Revolution. The London treaty of 26 April between Britain, France, and Italy promised confirmation of Libya and the Dodecanese, seized by Italy 1911–12, together with extensive territory in the Tyrol and along the Dalmatian coast, if Italy would enter the war against Austria-Hungary. Italy duly declared war on 24 May, but, although some of the promised gains were granted at the Paris Peace Conference*, by no means all were. This was to prompt D'Annunzio's* Fiume venture and to give Mussolini valid excuse for resentment.

Long, Huey Pierce (1893–1935), US Senator 1931–5. Born in Louisiana of humble background, he early established himself as a successful lawyer. He then entered state politics, building himself a strong if ruthless political machine, and rising through various political offices in Louisiana to become Governor 1928–31 and then Senator. As both Governor and Senator he used dictatorial methods to modernize his state. Known as 'Kingfisher', he persuaded the state legislature to pass a series of welfare measures, especially for education, while he had roads, bridges, and public buildings constructed by private contractors, fighting the vested interests of the public utility companies as well as the oil companies. Despite initial support, Long quickly turned against Franklin D. Roosevelt, seeing the New Deal* as a rival programme to his own 'Share our Wealth' programme for the redistribution of income. His charismatic, if neo-Fascist, appeal came close to dividing the Democratic Party, but he was assassinated before he could run for the presidency.

Long March (1934–5). By 1934 the Jiangxi Soviet* was close to collapse after repeated attacks by the Guomindang* army. In October a force of some 100,000 evacuated the area, Mao Zedong* taking over the leadership in January 1935, backed by Zhu De*. For nine months they struggled through mountainous terrain cut by several major rivers, averaging 50 miles a day. On 20 October 1935 Mao and some 6,000 survivors reached Yan'an* in the province of Shanxi, having marched some 9,600 km. (6,000 miles). Other groups arrived later, in all about 30,000 surviving the journey. During the next twelve months they were joined by other 'Route Armies'.

Lon Nol (1913–85), President of the Khmer Republic (Cambodia*) 1972–5. Educated in France, he rose to become a provincial governor under the French administration of Cambodia and, after World War II, a general in the Cambodian army. He led campaigns against Communist guerrillas 1946–54 and early in the 1960s against the Khmer Rouge*, who were gaining control of northern Cambodia. He was Prime Minister under Prince Sihanouk* 1966–7, when he ruthlessly suppressed a peasant revolt in Samlaut. He was opposed to Sihanouk's policy of neutrality and in 1970 led the coup which deposed him. He established close ties with US and South Vietnamese forces in Saigon, allowing them to operate in Cambodia. On 10 March 1972 he took full power as President of the Khmer Republic, but was flown out by US helicopters in April 1975 and lived in California.

Loos, Battle of, World War I (September–October 1915), a British offensive on the Western Front*. The attack across a flat rain-sodden plain was ordered by the Commander-in-Chief Sir John French, against the advice of commanders in the field. It was to be in conjunction with an attack by the French 3rd Army to the east. It began on 21 September, when chlorine gas was used, which blew back into the British trenches. It lasted until 13 October and achieved nothing except 'useless slaughter of infantry', some 60,392 British being killed for 20,000 Germans.

López (Michelson), Alfonso (1913–), President of Colombia 1974–8. Son of a former President of Colombia, he spent many years in graduate and postgraduate law studies in the USA and Europe before returning to Colombia in 1958 to practise law and to enter politics. He criticized the arrangement of 1957 between the Liberal and Conservative parties (the so-called National Front) whereby they would alternate in power, but in 1968 he joined the Liberal Party. He was Minister for Foreign Affairs 1968–70 and was elected President in 1974 with a landslide victory. While he sought improved relations with the Soviet Union, his economic policies were those of an advocate of the free market. Government subsidies were eliminated and unemployment allowed to rise. This precipitated peasant unrest and guerrilla activity. He was accused of corrupt practices with drug-dealers, but was elected leader of the Liberal Party in 1982. In that year he failed in a bid to be re-elected President.

López Mateos Adolfo (1910–69), President of Mexico 1958–64. Born in the year of the Mexican Revolution, of a mestizo family, his father was a dentist. After attending university and graduating in law, he became a librarian and university teacher of Spanish American literature, before entering politics in 1946. He was elected to the Mexican Senate and took part in a diplomatic mission to the United Nations. He became secretary-general of Mexico's only significant party, the Partido Revolucionario Institucional, and was a successful Secretary of Labour under President Cortines (1952–8). When President himself he took a firm line with railway strikers and Communist agitators. Nevertheless, still relatively youthful, he retained immense popularity, resuming the earlier policies of the Revolution, of nationalization (public utilities) and of creating co-operative peasant-run *ejidos*, distributing some 45 million acres of land. The Mexican population explosion had almost nullified the earlier drive to eliminate illiteracy, and his government's largest single budget item was on education expansion (particularly adult) and the building of rural schools. In foreign affairs he established close relations with the USA, being on excellent terms with President Kennedy.

López Portillo (y Pacheco), José (1920–), President of Mexico 1976–82. Born in Mexico City, he graduated in law and became a Law Professor at the National University of Mexico and adviser to President Echeverría*. On his election as President he placed more emphasis on expanding industry and oil production and less on land redistribution, but he gave strong support to population control. By increasing the size of the Mexican Chamber of Deputies he aimed to encourage opposition parties to participate in the political process, which had been dominated by the Partido Revolucionario Institucional (PRI) since 1929. By the end of his period in office there was widespread evidence of corruption and financial maladministration. He was accused of defrauding the state-owned oil agency PEMEX.

Ludendorff, Erich (1865–1937), German general. Born in German Poland, he was commissioned into the Imperial Army in 1882 and in 1906 joined the General Staff, where he worked on the Schlieffen Plan* for possible war in the west. After the capture of Liège in 1914 he joined Hindenburg's* staff on the Eastern Front, largely planning the victory of Tannenberg*. As Quartermaster-General 1916–18 he exercised enormous power in Germany, directing the war effort. He and Hindenburg forced the resignation of Chancellor Bethmann Hollweg* and together established a virtual military dictatorship. He planned the offensive of March 1918, but when the Allied counter-offensive developed in August, he urged the Emperor to make peace. In October 1918 he fled to Sweden. He returned to try to overthrow the Weimar Republic in the Kapp *putsch**. He joined Hitler in the abortive Munich *putsch** of 1923 and sat in the Reichstag as a National Socialist 1924–8, but only won 1 per cent of the votes in the presidential elections of 1925. In 1929 he broke with Hitler and became a querulous propagandist for total war and of Aryan racist dogma, writing pamphlets accusing 'supranational powers'—Roman Catholics, Jews, Freemasons—of a common plot against Germany.

Lugard, Frederick Dealtry, 1st Baron (1858–1945), British colonial administrator. Born in Madras, India, he was educated in England and commissioned into the army in 1878. After serving in Afghanistan, the Sudan, and Burma, he joined the British East Africa Company, and in 1890 was posted to Uganda, persuading the British government in 1894 to assume a protectorate over the country. He then moved to Nigeria to work for the Royal Niger Company, and forestalled French and German efforts to occupy the north. By 1903 he had defeated the Fulani* and established the Northern Nigerian protectorate. He was Governor of Hong Kong 1907–12, but then returned to Nigeria, where he served 1912–19, amalgamating the northern and southern protectorates into a single colony. It was then that he developed the doctrine of indirect rule* which was to be widely applied through the British Empire, believing that the colonial administration should exercise its control through traditional native chiefdoms and institutions. He worked for the League of Nations 1922–36, being particularly involved in the administration of mandates*.

Lumumba, Patrice Emergy (1925–61), Prime Minister of the Congo Republic 1960. Born in central Zaïre, he was a postal clerk until 1957, when he founded the influential MNC (Mouvement National Congolais), to bring together radical nationalists. In 1958 he attended a Pan-African Conference in Accra, where he met Nkrumah* and Nasser*. He was accused of instigating public violence and imprisoned by the Belgians, but released to take part in the Brussels Conference (January 1960) on the Congo. He became Prime Minister and Minister of Defence when the Congo became independent in June 1960. Sections of the army mutinied, Belgian troops returned, and Katanga (Shaba) province declared its independence (Congo Crisis*). Lumumba appealed to the UN, which sent a peace-keeping force. President Kasavubu*, his rival in power, dismissed him and shortly afterwards he was put under arrest by Colonel Mobutu*. He escaped but was recaptured and handed over to Katangese troops of Tshombe*, who murdered him in January 1961. He is regarded as a national hero and martyr by many in Zaïre, and a university in Moscow was named after him.

Lusaka Manifesto (14–16 April 1969). Following a two-day meeting in Lusaka, Zambia, of representatives from East and Central African states, a Manifesto was issued on behalf of the peoples of southern Africa. It called for an end to apartheid* in South Africa, a solution to the Rhodesia crisis (see ZIMBABWE), and independence for Angola, Mozambique, and Namibia: 'we do not accept that any one group within a society has the right to rule any society without the continuing consent of all the citizens.' It criticized the UN for failing to implement its agreed policies.

Lusitania, a British transatlantic liner. It was torpedoed on 7 May 1915 off the Irish coast without warning by a German submarine, with the loss of 1,198 lives. The sinking, which took 128 US lives, created intense indignation throughout the USA, which until then had accepted Woodrow Wilson's* policy of neutrality. Wilson sent a strong protest, but Germany refused to accept responsibility, claiming that the ship had been carrying war materials, and no reparations settlement was reached. Two years later, following Germany's resumption of unrestricted submarine warfare in February 1917, the USA severed diplomatic relations and entered the war on the side of the Allies.

Luthuli, Albert John (1898–1967), South African political leader. A Zulu by birth, he trained and taught as a teacher before becoming chief of the Abasemakholweni people in 1936 and member of the Native Representative Council until its dissolution in 1946. In 1956 he was elected president of the African National Congress* (ANC). He became universally known as a leader of non-violent opposition to apartheid*. In 1956 he was charged with high treason. The charges were dropped, but he suffered frequent arrests and harassment by the South African government, until in 1959 it banished him to his village. In 1960 the government banned the ANC. In 1961 Luthuli was awarded the Nobel* Peace Prize and next year published his book *Let My People Go*, which rejects militancy. He died in 1967, 'falling in front of a train'.

Luxembourg, Grand Duchy of. Established as an independent state in 1815, it was invaded by German troops in 1914 and 1940 and was incorporated into Germany 1942–5. With close economic links with Belgium, it established a customs union with The Netherlands and Belgium in 1948 which developed into the Benelux Economic Union of 1960. Under the Grand Duke there is an elected Chamber of Deputies. The Duchy is the centre of massive iron-ore mining and steel production. It is the home of the European Court and the Secretariat of the European Parliament*.

Luxemburg, Rosa (1871–1919), founder of the German Communist Party. Born in Russian Poland, she was educated in Warsaw and Zurich. An internationalist socialist, she took part in the anti-Tsar riots in Poland in 1905. With Karl Liebknecht* she argued the cause of 'revolution through the mass-strike', against the moderates in the German Social Democrat Party. From 1910 she organized a series of Sunday demonstrations in Berlin. In 1915 she formed the Spartakist* Group, dedicated to international peace. Although imprisoned from then until 1918, she smuggled out letters criticizing Lenin for imposing a dictatorship in Russia. Released from prison in November 1918, she at once helped to found the German Communist Party in December, but was captured by the right-wing Freikorps in Berlin and shot during the Spartakist revolt in January 1919.

Lyautey, Louis Hubert (1854–1934), marshal of France. Born at Nancy, he graduated from the French Military Academy of St-Cyr in 1873 and spent most of his professional life in the French colonial service. After spells in Algeria, Madagascar, and Indo-China, he went to Morocco in 1912 as Resident-General. His first action was to replace the Sultan with his 'more compliant' brother Moulay Yusuf. He was recalled to Paris as Minister of War in 1916, but returned to Morocco in 1917 and served there until retirement in 1925. While respecting Islamic culture and the role of the Sultan, his rule was still a paternalistic one, seeking to train administrators who would provide fair and just government. As such it was resented by younger nationalists. During his years in Morocco major engineering and irrigation projects were completed.

Lyons, Joseph Aloysius (1879–1939), Prime Minister of Australia 1932–9. He was a teacher in Tasmania until elected to the Tasmanian Assembly in 1909 as a Labor politician. He was Tasmanian Premier 1923–9 and was then elected to the federal Parliament as a Labor member. His party split in 1931 during the Depression crisis. He joined a group of Nationalists led by W. M. Hughes to form a United Australia Party, which won the election in December. Lyons was Prime Minister until his sudden death in April 1939. Coming to office at a time of deep depression, he saw the revival of the Australian economy as gold and wool prices recovered and industry expanded. This latter was stimulated by a three-year defence programme which he launched in 1934.

m

Macao (Macau). A territory in south China first seized by Portugal in 1511 and consisting of some 16 sq. km. with an estimated population of half a million. In 1887 it was recognized by China as a Portuguese dependency and it became an Overseas Province of Portugal in 1961. In 1974 it became an autonomous Chinese territory under Portuguese Constitutional Laws, with an elected Legislative Assembly. It is due to be handed totally to the Chinese People's Republic in 1999. Tourism and the textile industry are the two major sources of income.

MacArthur, Douglas (1880–1964), US general. Born in Little Rock, Ark., he was commissioned into the US army at West Point in 1903. He served as an aide in Tokyo and in Mexico before World War I, when he distinguished himself as a divisional commander in France. By 1930 he had reached the rank of general and became Chief of Staff of the US army until 1935. He was then seconded to the Philippines to build up a Filipino defence force, retiring from the army in 1937 but staying on in the Philippines until 1941. He was then recalled to the army by President Roosevelt* and led the defence against the Japanese invaders until March 1942, when he was ordered to Australia as Commander of Allied Forces in the South-West Pacific. He commanded the Allied counter-attack against Japan (July 1942–January 1943) in the Papuan campaign in New Guinea. From here (1943–4) his troops advanced towards the Philippines, which were recaptured in the spring of 1945. By now he was commander of all US army forces in the Pacific and received the Japanese surrender on board the USS *Missouri* on 2 September 1945. As commander of the Allied forces of occupation in Japan he took an active role in many reforms, including the drafting of the new Japanese Constitution*. Appointed commander of the UN troops in the Korean War*, he led them into North Korea in October 1950, but was forced to retreat when the Chinese entered the war. In 1951 he resumed the offensive, but tension arose with President Truman*, who believed that MacArthur was prepared to risk a full-scale atomic war. He was dismissed in April 1951. He failed to obtain nomination for the presidential election of 1952, which was to be won by Eisenhower.

McCarran Act (1951), US internal security legislation. Passed over President Truman's veto, it required the registration of all Communist organizations and individuals, prohibited the employment of Communists in defence work, and denied entry to the USA to anyone who had belonged to a Communist or Fascist organization. It arose out of the fear, whipped on by McCarthyism*, of a Communist conspiracy, or Red Menace, against the USA. In 1965 the Supreme Court ruled that an individual could refuse to admit being a Communist by claiming the constitutional privilege enshrined in the Fifth Amendment against self-incrimination.

McCarran–Walter Act (1952), a codification of US immigration law*. Passed over President Truman's veto, it maintained the basic 1924 quota system, with an annual quota total of 154,658 persons, but with some significant changes. At a time when McCarthyism* gripped the country, it gave the Attorney-General powers to refuse admission to any 'subversive' and to deport any member of a 'Communist or Communist-front' organization. Immigration quotas for eastern and south-eastern Europe were reduced. On the more positive side, it provided for selective immigration on the basis of skills, and it granted an annual quota total of 2,000 persons per annum from eight Pacific nations previously banned, with Japan being allocated 185.

McCarthy, Joseph Raymond (1908–57), US Senator 1947–57. Born in rural Winsconsin, he was a farm worker and store-keeper until articled as a lawyer in 1929. He built a successful law practice, entered local politics, and was then elected to the US Senate in 1946, after claiming a distinguished record of war service. He quickly launched a campaign alleging that there was a large-scale Communist plot to infiltrate the government at the highest level. In February 1950 he announced that he had evidence of fifty-seven 'card-carrying communists' in the State Department and some 205 'sympathizers'. Despite the conclusions of a Senate investigating committee under Millard Tydings that such charges were fraudulent, McCarthy continued to make repeated attacks on the administration, the military, and public figures, his demagogic methods of anti-intellectualism and social envy having an early popular appeal. As Communism triumphed in China, the 'Red Menace' was seen as a potent threat and the term 'McCarthyism' became synonymous with the 'witch-hunt' that gripped the USA 1950–4. In 1953, as chairman of the Senate Permanent Subcommittee on Investigations, McCarthy conducted a series of televised hearings, where his vicious questioning and unsubstantiated accusations destroyed the reputations of many of his victims. At length he went too far, attacking the army, and in December 1954 the Senate censured him for his conduct. After the 1954 election, with the Democrats again in control of Congress, his influence rapidly declined.

MacDonald, James Ramsay (1866–1937), British Prime Minister 1924, 1929–35. Born in poverty in Lossiemouth, Scotland, he worked in London from 1886, doing clerical jobs and attending night-school. He became private secretary to a Liberal MP and in 1894 joined the Independent Labour Party*, working now as a journalist. Largely responsible for the formation of the Labour Representation Committee in 1900, he was its secretary until 1912, being himself elected to Parliament in 1906. In 1911 he became leader of the Parliamentary Labour Party as a notable theorist of socialism. At the outbreak of war in 1914 his belief in negotiation, not war, with Germany made him unpopular and he resigned. He was re-elected to Parliament in 1922 and became leader of the opposition, and then Prime Minister January–November 1924. This ministry, the first time Labour had formed a government, relied on Liberal support, which was withdrawn when he failed to prosecute a communist, J. R. Campbell, for alleged incitement to mutiny. His second government (1929–31) broke down in August 1931 through cabinet divisions over proposals to reduce unemployment benefits. He continued as Prime Minister of a National government* until 1935, when he resigned through failing health. He continued however to serve in Baldwin's* cabinet. Throughout his years in office MacDonald was closely involved in international affairs and disarmament negotiations. Like many others he did not discern the menace of Nazism, hoping always for collective security through the League of Nations*, of which he was a passionate supporter. Criticized in his day for having insufficient regard for socialism and for being 'seduced by social success', in retrospect he is often considered a politician of skill and vision.

Macedonia. This ancient land had been part of the Ottoman Empire since the 15th century, but a nationalist movement IMRO (Internal Macedonian Revolutionary Organization) was founded in 1893, aiming at independence. After a confused period of fighting, the Treaty of Bucharest (August 1913) partitioned it between Serbia and Greece. It was the scene of fierce fighting through World War I. An Allied force landed at Salonika in October 1915 and, in September 1918 under General Franchet d'Esperey*, advanced against Bulgaria. After the war the Treaty of Neuilly* confirmed that northern Macedonia would become southern Serbia, while southern Macedonia and Salonika remained Greek. IMRO terrorists between the wars were supported by Bulgaria, from which they operated against both Yugoslavia and Greece. During World War II Bulgaria occupied Macedonia, but the Treaty of Paris of 1947 confirmed that of Neuilly, 'southern Serbia' becoming the Republic of Macedonia within the Federal Republic of Yugoslavia. Some half-million Macedonians continued to live in Bulgaria. The IMRO aim of 'Macedonia for the Macedonians' had become a forlorn dream, but the ancient Macedonian language and a cultural revival have been encouraged in the universities of both Skopje and Salonika. In elections in December 1990 the anti-Communist Democratic Party for Macedonian Unity was the largest in a hung Parliament, while racial tension against Albanians increased.

Machel, Samora Moises (1930–86), President of the Republic of Mozambique 1975–86. From a poor family, he became a hospital nurse and then in 1963 went to Tanzania, where he joined Eduardo Mondlane, a fellow nationalist. He trained as a guerrilla in Algeria and in 1964 returned to Mozambique as a member of FRELIMO* (Frente de Libertaçao de Moçambique). In 1966 he became its Commander-in-Chief and in 1970 succeeded Mondlane as its President when the latter was assassinated. He led the FRELIMO War against the Portuguese (1964–74) and became the first president of the People's Republic of Mozambique. A Marxist, he nationalized multinational companies and allowed his country to be used as a base for nationalist guerrillas from Rhodesia and South Africa. Nevertheless, his politics became increasingly pragmatic, accepting Portuguese aid and contact with South Africa, notably with the Nkomati Accord* (1984). He died in an aircraft crash.

McLuhan, Marshall (1911–80), communications theorist. His studies at the University of Toronto of mass electronic communications and their impact on the human psyche made him one of the most influential thinkers of the 1950s and 1960s. Popularizing the concept of the 'global village', his books include *Gutenberg Galaxy* (1962) and *Understanding Media: The Extensions of Man* (1964), with its view that 'the Medium is the Message, because it is the medium that shapes and controls the scale and form of human association and action'.

McMahon Line, a boundary line dividing Tibet and India. It was marked out by the British representatives led by Sir Henry McMahon at the Simla Conference of 1914 between Britain, China, and Tibet. The Chinese refused to ratify the agreement, and after the reassertion of control by China over Tibet in 1950, boundary disputes arose between India and China culminating in the Indo-Chinese War* of 1962.

Macmillan, Maurice Harold, 1st Earl of Stockton (1894–1987), British Prime Minister 1957–63. Born in London, a member of the Macmillan publishing family, he was educated at Eton and Oxford, serving throughout World War I. He was three times wounded and was appalled at the wastage of life in that war and the failures of post-war governments to succeed in honouring the promise to 'make Britain a land fit for heroes to live in'. He was first elected to Parliament in 1924 for Stockton-on-Tees, and regarded himself as being on the radical wing of the Conservative Party. During the 1930s he was critical of the policy of appeasement and of economic policy, being influenced by the ideas of J. M. Keynes*. In 1940 he joined the government of Winston Churchill* and in 1942 became Minister for North Africa with cabinet rank; he remained responsible for British policy in the Mediterranean until the end of the war. In 1951 as Housing Minister he was responsible for the largest local authority building programme yet seen in Britain. He became Minister of Defence in 1954 under Churchill and then Foreign Minister and Chancellor of the Exchequer under Eden*, whom he succeeded after the Suez Crisis*, comfortably winning the 1959 election. During his years in office people in Britain began to taste the fruits of affluence, most accepting his election slogan 'Our people have never had it so good.' In 1958 life peerages were introduced. However, the new affluence also brought balance-of-payments difficulties, and unemployment began to reappear. This fanned race tensions with post-war immigrants, and in 1962 the Commonwealth Immigration Act was passed. Macmillan enjoyed good relations with President Kennedy, supporting him over the Cuba Missile Crisis* and reaching the Nassau Agreement* for British submarines to be armed with US missiles. The government's bid to enter the European Economic Community was vetoed in 1963 by President de Gaulle*. Macmillan accepted the need for decolonization in both the West Indies and southern Africa, where he made his 'Wind of Change'* speech in February 1960. He was closely involved in the negotiations which led to the Nuclear Test-Ban Treaty* between the USA, the USSR, and Britain in July 1963. His health failed him at the end of that year and he resigned. In 1984 he entered the House of Lords, where he strongly criticized the economic and social policies of Mrs Thatcher.

McNamara, Robert (1916–), US Secretary of Defense 1961–8, President of World Bank 1968–81. Born in San Francisco, he graduated at the University of California in 1937 and during World War II taught at the Harvard Business School. Joining the Ford Motor Company, he rose rapidly to become its president, then immediately resigned to join the administration of John F. Kennedy*. Repudiating the Dulles* belief in 'massive retaliation', he advocated 'counter-insurgency' and 'flexible responses' as a means to combat world communism. He successfully gained control over Pentagon spending, axing obsolete weapons and applying strict cost-accounting methods. He visited Vietnam 1962, 1964, and 1966, when he sought to boost the US-backed Saigon government against the National Liberation Front*. While supporting the early bombing offensive against North Vietnam in 1965, he came to repudiate full-scale military involvement, resigning from the Johnson administration in February 1968. He was appointed president of the World Bank*, where his main concern was to ease the burden of debt repayment incurred by developing countries, many of whose economies were devastated by high oil prices in the 1970s.

Madagascar, a large island state in the southern Indian Ocean. France established it as a protectorate in 1890, not without meeting considerable resistance. After 1945 it became an Overseas Territory of the French republic, sending Deputies to Paris. In 1958 it became an autonomous republic of the French Community* and gained full independence in 1960 as the Malagasy Republic. In 1975 it changed its name back to Madagascar. Severe economic difficulties caused recurrent political problems through the 1970s and 1980s, including social unrest and frequent changes of government. President Admiral Didier Ratsiraka was re-elected in 1982 and again in March 1989, being 'guided' by a Supreme Revolutionary Council.

Mafia, the, an international secret society. Its origins can be said to date from the period 1806–15 in Sicily, when under British pressure attempts were made to break up the huge estates of the Sicilian feudal aristocracy. Their disbanded private armies often became brigands and proceeded to offer an alternative system of control and justice beyond the purview of the official authorities. In the late 19th century many Sicilians emigrated to the USA and the Mafia became established in New York and Chicago. In the 1920s the Italian Fascist government brought many Mafia leaders to trial, but some escaped to the USA, where they were active during the Prohibition* era. Since World War II Mafia activities have become world-wide, increasingly centred on the drug trade. In Italy periodic much publicized trials have failed to eradicate the problem of Mafia control, particularly in the south and in Sicily.

Maghreb Union, a trading union among the states of the area of north-west Africa known as the Maghreb. In February 1989 representatives of Algeria, Tunisia, Morocco, Libya, and Mauritania signed a trade agreement in Marrakesh. This fulfilled a dream of many early Arab nationalists and was achieved in spite of many unresolved problems between the participating states. At its second meeting, in Algiers in July 1990, it supported the *Intifadah* in Israel.

Maginot Line, a series of defensive fortifications in France. Begun in 1929, it stretched along the eastern frontier from Switzerland and then west to Luxembourg. Named after the Minister of War André Maginot, it was built because French military theorists believed that defence would predominate in the next war, and because it reduced the need for soldiers. Partly because of objections from the Belgians, who were afraid they would be left in an exposed position, the line was never extended along the Franco-Belgian frontier to the coast. Consequently it could be outflanked, as indeed happened in the spring of 1940.

Mahdism. Muhammad Ahmad al-Mahdi, or' divinely guided one', had died at Omdurman in 1885 soon after his conquest of Khartoum and the death of General Gordon. In a matter of only four years (1881–5) he had created a religious empire throughout the Upper Nile which his successor Khalifar Abd Allah and son Abd ar-Rahman (d. 1959) consolidated. During the 20th century Mahdism has fired Islamic nationalism throughout the Sudan and much of Africa, with its emphasis on an austere adoption of the basic teachings of the Qur'ān and strong opposition towards Christianity and Europeanization.

Makarios III (1913–77), Archbishop of Cyprus and President of Cyprus 1960–77. Born Mihail Christodoulou Mouskos in Paphos on Cyprus, son of a peasant, he became a monastic novice. After studying in Athens and in the USA, however, he decided to become a priest and was ordained in 1946. He became a bishop and then in 1950 was elected Archbishop of the Orthodox Church of Cyprus. A sincere believer in *enosis* (union with Greece), he was thought by the British to be involved with the terrorist organization EOKA* and he was deported to the Seychelles in 1956. He now renounced *enosis*, accepting the British offer of Cypriot independence within the Commonwealth. He returned to become President in 1960. In July 1974 a group of Greek officers of the Cypriot National Guard attacked his palace, forcing him into exile and proclaiming *enosis*. The coup failed. Makarios returned from London and was reinstated, but he could not reassert control over what had become Turkish Cyprus. He remained in office until his death.

Malan, Daniel François (1874–1959), Prime Minister of the Union of South Africa 1948–54. Born in Cape Colony he became a minister of the Dutch Reformed Church from 1905 until 1915, when he began to become involved in politics and founded the newspaper *Die Burger*. An ardent Afrikaner nationalist, his political thinking was dominated by desire for secession from Britain, republicanism, and racial segregation. He served under Hertzog* 1924–33, but when the latter formed a coalition with Smuts* in 1934 he resigned and formed the Purified National Party. On Hertzog's resignation in 1940 he became leader of the opposition, and the reunited National Party, of which he was now leader, swept into power in 1948 on the promise that apartheid* legislation would be introduced. When in 1952 this was declared unconstitutional by the Supreme Court, he introduced legislation making Parliament the ultimate High Court in South Africa. He retired in 1954 to be succeeded by Johannes Strijdom*.

Malatesta, Enrico (1853–1932), Italian anarchist. Heir to his ancient family's wealth, he studied at Naples, but was expelled for anarchism and gave away his fortune. A friend of the Russian anarchist Bakunin, before World War I he was constantly in prison and escaping. He founded the London Anarchist Club and numerous reviews in various countries. From 1919 he lived in Italy, unmolested by Mussolini, who had been attracted to his revolutionary idealism.

Malawi, a land-locked country in Central Africa formally known as Nyasaland. British colonial administration was instituted when Sir H. H. Johnston proclaimed the Shire Highlands a British protectorate in 1889. This became British Central Africa in 1891 and then Nyasaland 1907–64. Unwillingly a member of the Central African Federation* 1953–63, it gained its independence in 1964 as Malawi, with Dr Hastings Banda* as first Prime Minister. He became President in 1966, when the country became a republic, and President for life in 1971. It is a one-party state governed by the Malawi Congress Party; an opposition group, the Malawi Freedom Movement, has been consistently repressed. Malawi is a member of the SADCC*.

Malaya, a peninsula in South-East Asia, since 1963 part of Malaysia. By 1914 Malaya was under British control, with heavy investment in expanding rubber plantations and tin mines. It consisted of (1) the *Straits Settlements* of Malacca, Penang, Labuan, and Singapore, with a Governor in Singapore; (2) four *federated Malay states*, each with its own ruler, Perak, Selangor, Negri Sembilan, and Pahang, with a British Resident-General in Kuala Lumpur; (3) three *unfederated states* of Johore, Kelantan, and Trengganu, again with their own rulers, but loosely under British protection. Kedah and Perlis were to join this group in 1923 and 1930.

In 1932 a reorganization established that the Governor in Singapore be concurrently High Commissioner for all nine Malay states, in addition to the Straits Settlements. Chinese and Tamil Indians had increasingly been introduced to work the mines and plantations. By 1939 the Chinese equalled the Malay population, with Indians 10 per cent of the total. The ideas of Communism had already influenced many Chinese, serious strikes occurring 1936 and 1937. After June 1941 the Malayan Communist Party became more co-operative, so that rubber and tin could assist the Soviet Union's war production. Following the entry of Japan into World War II in December 1941 some 200 Communists were given a crash course in guerrilla fighting at a special training school in Singapore. These were to form a cadre for resisting the Japanese after they had occupied the country in February 1942. This cadre steadily built up the Malayan

People's Anti-Japanese Union (later Army). After the war, however, this force, the MPAJA, went into hiding as British control was reimposed. In 1946 a new Governor, Sir Edward Gent, was sent to establish a Malayan Union (1946–8). In 1948 this was succeeded by a Federation of Malaya, excluding Singapore. These moves were seen as revived imperialism and resulted in the emergence of a United Malays National Organization*, demanding self-government of Malaya by Malays. The largely Chinese Malayan Communist Party now took to the jungle in opposition to both the British and the UMNO. Malaya was in a state of emergency for the next ten years.

Early British attempts to cope with the Malayan Emergency* were insensitive and incompetent, ex-Palestine police officers being brought east with fairly disastrous results. In 1952 Churchill appointed General Templer* as High Commissioner, and in three years the latter succeeded in winning 'the hearts and minds' of both Chinese and Malay, so that, when independence was gained in 1957, it was as an integrated multi-racial society. The threat of Communism had been successfully overcome by skilful political as well as military action.

Malayan campaign, World War II (December 1941–February 1942). Through an agreement with the Vichy French administration the Japanese had established military bases in Vietnam as early as July 1941. At the same time they made an agreement with Field Marshal Pibul Songgram* of Thailand for free passage of troops through Thailand, which had been allied with Japan since 1939. Thus on 8 December 1941 Japanese forces crossed into northern Malaya without hindrance, while their aircraft bombed Singapore. Kuala Lumpur fell 11 January and British, Indian, and Australian troops fell back to Singapore, where they were mostly to be captured in February after the fall of Singapore*. During the retreat a guerrilla resistance force was organized to conduct sabotage, operating behind enemy lines. This was the Malayan People's Anti-Japanese Army (MPAJA), rapidly recruited and trained for jungle warfare by the British and consisting largely of Chinese, most of whom were Communists. In May 1944 Allied troops began the gradual reconquest of Burma and were preparing to invade Malaya when Japan surrendered. The British regained Singapore on 3 September 1945.

Malayan Emergency, a crisis in Malaya* 1948–60. Resentment by the minority Chinese of Malay political dominance of the new Federation of Malaya (1948) was exploited by the (mainly Chinese) Communist guerrillas who had fought against the Japanese. They initiated a series of attacks on planters and other estate owners, which flared up into a full-scale guerrilla jungle war. Led by Chin Peng and supported by its own supply network (the Min Yuen), members of the Malayan Races Liberation Army caused severe disruption in the early years of the campaign. During General Templer's* period as High Commissioner (1952–4) the insurgents were gradually defeated by troops from Britain and the Commonwealth (Australia, New Zealand, East Africa, and Fiji), through the use of new jungle tactics and of helicopters, and through the disruption of their supply network. Templer created a series of 'new villages' which could be well protected from attack; this helped to maintain the loyalty of the Malay and Indian population. Skilful use by the British of local leaders in government committees facilitated the peaceful transition to independence in 1957. By then the insurrection had been all but beaten, although the emergency was not officially ended until 1960.

Malaysia. Created by Britain in September 1963 as one of the richest countries of the region, it is divided into West Malaysia, along the Malay Peninsula, and

East Malaysia (Sarawak* and Sabah*). It is a federal constitutional monarchy, whose head of state (Yang di-Pertuan Agong) is elected every five years from among the nine hereditary state rulers. The federal Parliament consists of a House of Representatives (Dewan Rakyat), elected on universal suffrage, and a Senate (Dewan Negara), whose members are partly appointees of the Prime Minister and partly representatives of the state legislatures. Each state has its own constitution and legislative assembly. Within two years of its establishment Malaysia faced two crises—pressure for secession by Singapore*, and the Konfrontasi* of Indonesia over Sarawak and Sabah. In 1969 inequalities between the politically dominant Malays and economically dominant Chinese resulted in riots in Kuala Lumpur, and parliamentary government was suspended until 1971. As a result there was a major restructuring of political and social institutions designed to ensure Malay predominance, the New Economic Policy being launched to increase the Malay stake (*bumiputra*) in the economy. In more recent years racial tensions have arisen, with Islamic fundamentalists* desecrating Hindu temples, and Chinese boat people* demanding asylum. The largest single political party has remained the United Malays National Organization* created by Tunku Rahman*. Since 1955 this has ruled in an uneasy alliance with the Malaysian Chinese Association (MCA) as a coalition, the National Front, since 1981 under Prime Minister Dr Mahathir bin Mohamed.

Malcolm X (1925–1965), US civil rights leader. Born Malcolm Little in Omaha, he later took the name Al Hajj Malik Shabazz, but was always known as Malcolm X. When he was a boy, his home was burned down by the Ku Klux Klan*. Later he moved to Boston, where he worked as a waiter and spent seven years in prison for burglary. Here he was converted to Islam. He became a follower in 1952 of the ascetic Elijah Muhammad, founder of the Black Muslims. He was sent to New York, where he worked in Harlem and as a brilliant speaker became nationally known for his part in the civil rights movement*. At first a strong advocate of Black separatism, he was prepared to condone violence as a means of self-defence. In 1964 he quarrelled with Muhammad and left the Black Muslims. He made a pilgrimage to Mecca, as a result of which he modified his views on separatism, becoming an advocate of world brotherhood and forming the Organization of Afro-American Unity. Hostility developed between his followers and the Black Muslims, who shot him at a rally of his followers. His *Autobiography* (1965) remained a best-seller, its ideology appealing especially to Black young people.

Maldives, Republic of, a coral islands state in the Indian Ocean. From 1887 until 1952 the islands were a sultanate under British protection. Maldivian demands for self-government and constitutional reform had begun during the 1930s and internal self-government was granted in 1948. Full independence was achieved in 1965 and in 1968 the sultanate was abolished and a republic declared. There are no political parties and President Maumoun Abdul Gayoon was first elected in 1978. In 1988 with Indian help he suppressed an attempted coup. The Republic became a full member of the Commonwealth in 1985.

Malenkov, Georgy Maksimilianovich (1902–88), Soviet Prime Minister 1953–5. Born in Orenburg, when he was 17 he joined Trotsky's Red Army and fought in the Civil War in Turkistan. He joined the Communist Party in 1920 and was always a close associate of Stalin. He rose through the party ranks and was deeply involved in the Terror of the Great Purge* 1934–8. During World War II he served on the Defence Council and became a member of the Politburo in 1946 and a deputy Prime Minister.

On Stalin's death in 1953 he became Chairman of the Council of Ministers (Prime Minister), working at first amicably with Khrushchev*, the party secretary. His policies of reducing arms production and improving standards of living were opposed by many die-hards, and during an internal party struggle in 1955 he was deposed and in 1957 expelled from the party. He was however allowed to live quietly in retirement.

Mali a vast land-locked West African state. Freed from Moroccan rule at the end of the 18th century, it was colonized by France in the 19th century as part of French West Africa*. In 1946 it became an Overseas Territory of France. It was proclaimed the Soudan Republic in 1958, an autonomous state within the French Community*. In 1959 it united with Senegal as the Federation of Mali, but this did not work, and Senegal withdrew in 1960, when Soudan became the independent Republic of Mali. A military government took over in 1969 under Lieutenant Moussa Traoré. Civilian government was restored in 1974 and General Traoré elected President. He was re-elected in 1985. A brief border war broke out with Burkina Faso* in 1985. Prolonged drought, locusts, and torrential rain (August 1988) have all caused problems.

Malinowski, Bronislaw Kaspar (1884–1942), Polish anthropologist. Born in Cracow, he studied folk-psychology at Leipzig and then taught at the London School of Economics from 1922. His methodology for a series of studies of the culture of the Trobiand Islands off New Guinea established ethnographic and anthropological studies on an entirely new level, even though his utilitarian theory of culture, that it 'is no more than the means to satisfy basic human needs', never won total support.

Malta, an island state of the central Mediterranean. It was annexed to the British Crown in 1814. A Legislative Assembly was established in 1921 but there were spasmodic riots against colonialism throughout the 1920s; in 1933 the Constitution was suspended. The island was severely bombed during World War II, especially 1941–2, 14,000 tons of bombs destroying some 35,000 buildings. George VI awarded the island the George Cross in 1942. After the war Britain did not at first envisage decolonization, only a measure of autonomy, naval considerations being still considered paramount. There was rioting in 1956, with integration within the United Kingdom being discussed. Eventually in 1964 independence was granted, the first government being led by Dr Giorgio Olivier. Dominic Mintoff (b. 1916), leader of the Labour Party, dominated Maltese politics from 1971, when he became Prime Minister, until his retirement in 1984. He negotiated an agreement with NATO*, which could continue to use naval bases, but all British land forces were to be withdrawn in 1979. In 1989 a 'neutrality and non-alignment' amendment to the Constitution forbade any foreign military base or nuclear-carrying ship. In 1990 Malta was seeking membership of the European Community.

Manchukuo, Japanese-occupied Manchuria. On September 1931 a detachment of the Japanese Guandong Army*, stationed in Manchuria under treaty rights, used an allegedly Chinese-inspired railway explosion to occupy the city of **Mukden** (now Shenyang). Within five months Japanese forces had extended their power over all Manchuria. The last Emperor of China, Pu Yi*, was established as a puppet ruler, but effective power lay with the Guandong Army and Japanese officials. Japan extended the railroad system, linking it with that of Korea, and developed mineral extraction; but continuous guerrilla warfare by Chinese resistance forces never ceased. In 1945 Soviet forces invaded and the Japanese were expelled, the Soviet Union removing large quantities of industrial equipment. Chinese Communist forces also moved in, linking up with guerrillas, and Manchuria became a Communist stronghold.

Mandate, a form of international trusteeship, devised in 1919 by the League of Nations. It was a device for the administration of those colonial territories of Germany, and provinces of the Ottoman Empire, which were to be assigned to the victorious powers following World War I. Marking an innovation in international law, the mandated territories were to be supervised by the League's Permanent Mandates Commission. The latter, however, had no means of enforcing its will on the mandatory power, which was responsible for the administration, welfare, and development of the 'native population' until the latter was considered ready for self-government. Britain and France received the largest share of mandates, which in Africa at least were to be treated as if they were colonies. In 1946 the Mandates Commission was replaced at the UN by the International Trusteeship Council.

Mandela, Nelson Rolihlahla (1918–), South African nationalist leader. Son of a chief, he studied and practised as a lawyer in Johannesburg until 1958, when he became a member of the Executive Committee of the African National Congress* (ANC) and a member of its militant subsidiary, the Spear of the Nation. He was banned from South Africa 1953–5, but in 1956 was among those charged in a mass treason trial which lasted until 1961. All were acquitted. He continued his opposition to apartheid* by campaigning for a free, multiracial, and democratic society. He was arrested in 1962 and imprisoned for five years. Before this sentence expired, he was charged under the Suppression of Communism Act; after a memorable trial (October 1963–June 1964), in which he conducted his own defence, he was sentenced to life imprisonment. His authority as a moderate leader of Black South Africans did not diminish, though his absence from the political scene enabled a more militant generation of leaders to emerge. Offered a conditional release by the South African government, Mandela refused to compromise over the issue of apartheid. His wife Winnie Mandela continued to be politically active. His long imprisonment became a matter of international concern. In 1989 he was moved to a clinic and had meetings with government officials including the new President de Klerk*; and in 1990 he was liberated. In May 1990 he took part in preliminary discussions on the possible dismantlement of apartheid.

Manhattan Project, code name for the development of the atomic bomb during World War II. Both German and Allied scientists were working towards an atomic bomb, to the urgent concern of Albert Einstein*. A large research establishment was created at Oak Ridge, Tenn., in August 1942, where scientists from Canada, the USA, and Britain, together with refugee scientists from Italy, Germany, and Austria, worked. These included the Italian Enrico Fermi, who had discovered the atomic chain reaction in 1934. A test site was selected in New Mexico, where a test bomb was exploded on 17 July 1945. Precautions against radiation were minimal, resulting in cancer development in many of the site workers.

Maniu, Iuliu (1873–1953), Prime Minister of Romania 1928–33. Born in Transylvania, then part of the Austro-Hungarian Empire, he was elected to the Hungarian Parliament in 1906, where he urged the minority rights of Romanians living in his home area. He served in the Imperial Army, but in May 1918 he helped to organize a successful mutiny of Romanian troops. In December an *ad hoc* Romanian Council of Transylvania voted for union with Romania. He formed a National Peasant Party and was intermittently Prime Minister during the years of the Great Depression*. He did not welcome Carol II's* return to Romania in 1930, although he was briefly his Prime Minister 1932–3. In 1937 he formed an unsuccessful alliance with the Fascist Iron Guard*, to try to gain political control from the King. After the latter resigned in 1940 he at first supported Antonescu*, but later became leader of a Resistance movement, which in August 1944 organized a coup which toppled Antonescu and brought Romania into the war against Germany. Although he had been allied with the Soviet Union, he was a strong anti-Communist and in 1947 was imprisoned by the new regime, accused of espionage. He died in prison.

Manley, Michael Norman (1923–), Prime Minister of Jamaica 1972–80, 1988– . Son of Norman Manley*, he was educated in Jamaica and at the London School of Economics. He worked in London with the BBC 1950–1 and retained a close interest in journalism. He returned to Jamaica to become a trade union organizer, holding various posts in the sugar-workers' union, becoming its president in 1984. In 1962, on Jamaica gaining independence, he became a Senator until 1967, when he was elected to the House of Representatives. He succeeded his father as leader of the People's National Party and leader of the opposition in February 1969. He won the 1972 election and became Prime Minister and Minister of External Affairs, supporting policies of non-alignment and making strong efforts to rally support against South African apartheid*. His mildly socialist policies alienated the USA and there were CIA* attempts to 'destabilize' his regime. He won re-election, however, in 1976, and when economic problems developed resisted demands from the IMF* for austerity; he claimed he would seek 'an alternative economic strategy'. Street violence and splits within his party followed and he lost heavily in the election of October 1980. While again leader of the opposition he wrote a history of cricket and rebuilt the PNP to win the October 1988 election.

Manley, Norman Washington (1893–1969), Premier of Jamaica 1959–62. Son of a planter, he was educated at Jamaica College and Jesus College, Oxford. He served in World War I and was called to the Bar in 1922. He became a KC in 1932, when he was described as 'the foremost lawyer of the day'. In 1938 he entered Jamaican politics, founding the People's National Party (PNP). In the same year he successfully defended his cousin and political rival Alexander Bustamente* on charges of sedition. He was a member of the Jamaican House of Representatives from 1949, Chief Minister of Jamaica 1955–9, and Premier 1959–62. He was an enthusiastic supporter of the West Indies Federation*. In 1962 his party lost at the polls to Bustamente and he became leader of the opposition. In February 1969 he retired in favour of his son Michael, dying in the same year.

Mannerheim, Carl Gustav Emil, Baron von (1867–1951), President of Finland 1944–6. Commissioned into the tsarist Imperial Army in 1889, he served with distinction in the Russo-Japanese War 1904–5. During World War I he rose to the rank of general, fighting German troops on the Eastern Front. At the end of 1918 Finnish Bolsheviks occupied Helsinki. Mannerheim drove them out with a force of Finnish nationalists, going on to attack all Russian soldiers in Finland. Fighting against Bolsheviks continued until 1921, when the new Soviet Union recognized Finland's independence. He retired from active service, but in 1930 was appointed chief of the Finnish Defence Council. He planned a line of defensive fortifications, the Mannerheim Line, across the Karelian isthmus, within 30 km. of Leningrad. When World War II began the Soviet Union demanded access to certain Finnish ports. This was refused and the Finnish–Russian War* followed. When Germany attacked the Soviet Union in 1941 Mannerheim negotiated an alliance and German troops used Finland as a basis for the attack on Leningrad*. By 1944, however, Soviet troops were advancing. He was appointed President and in March, faced with defeat, he signed an armistice with the

Soviet Union, which demanded reparations*. In March 1945 he brought Finland into the war against Germany. Failing health led him to retire in March 1946.

Maoism, a political theory developed from the writings of Mao Zedong*. Mao's puritanical idealism saw 'moral incentive' as replacing the profit motive. He placed great emphasis on peasant communities, believing it possible for a people to transform itself in a revolutionary direction without adopting a programme of industrialization. Maoism has inspired many Third World Communist parties, as it does the Sendero Luminoso* in Peru. In China it was overturned after the defeat of the Gang of Four*. Its most notorious distortion was in Cambodia*, where the entire bourgeoisie was eliminated by Khmer Rouge fanatics.

Maori. The Maori peoples of New Zealand (Aotearoa), who live mostly on the North Island, had steadily lost their land and been deprived of their heritage through the 19th century. Their population had fallen from an estimated 100,000 in 1769, when Captain Cook first visited the island, to about 40,000 in 1914, at which time the Young Maori Party* was developing. Since then there has been a steady recovery and renewal. Much of this in the early days was the achievement of Sir Maui Pomare, the first Maori Health Officer, appointed in 1900, and of Sir Peter Buck*. Sir Apirana Ngata*, of the Young Maori Party, saw through Parliament in 1929 a Maori Land Settlement Scheme, and the Ratana/Labour Alliance of 1935 achieved many reforms for Maori, including the secret ballot in 1937. By the 1970s the Maori population had risen to 300,000. Although many have been urbanized, there has been a growing awareness of the cultural heritage and a determination to survive alongside, rather than be assimilated by, the predominant European (pakeha) life-style. One of the most important figures in this recovery of Maori self-esteem was Te Puea* (1883–1952), with her determination 'to make the Maori a people once again'. Maori units fought in both world wars. After the second the important Social and Economic Advancement Act was passed by Parliament, at which time (1948) the term 'Native' was officially replaced by that of 'Maori'. In 1962 a Maori Welfare Act passed Parliament from which resulted the Maori Council*. In 1867 four Maori parliamentary electorates had been created for male Maori—Western, Northern, Eastern, and Southern and in 1976 Maori electors became eligible to vote in either a Maori or a general electorate. In 1979 the Waitangi Tribunal* was established to protect Maori rights.

Maori Council (New Zealand). It was established under the Maori Welfare Act of 1962 and at the national level consists of three delegates from each of the seven Maori land districts. At regional or district level a series of Maori executive committees was established, and below these local Maori community committees. Charged with a concern for the social, economic, spiritual, and cultural well-being of the Maori, the Council makes submissions to parliamentary select committees. It has affected legislation on a wide area of Maori affairs, as well as on land and fisheries. Its grass-roots connections have helped to stimulate Pan-Maori consciousness.

Maori Land Settlement Scheme. A Native Land Act of 1862 had legalized private land transactions between a Maori and a European (pakeha) immigrant. In 1894 a Native Land Settlement Act had tried to impose tighter control over crown lands, but by means of bribery and intimidation the Maori had lost most of their best land by 1914. After World War I the Young Maori Party* pressed for reform and in 1929 the Minister of Native Affairs, Apirana Ngata*, drew up a Land Development Scheme, whereby government subsidy would enable an increased area of Maori land to be brought under cultivation. This was steadily implemented from 1931 onwards.

Mao Zedong (1893–1976), founder of the People's Republic of China. Son of a Hunan peasant farmer, his schooling was interrupted in 1911 when he volunteered to take part in the Revolution*. In 1918 he went to Beijing, where he obtained work in the University Library under Li Dazhao*, whose ideas greatly influenced him. He supported the May Fourth movement*, but, disillusioned with Western liberalism, he went to Shanghai in June 1921 as a founding member of the Chinese Communist Party*. He returned to Hunan to build a party branch, already committed to fundamental egalitarianism. In 1927 he fled to the mountains and with Zhou Enlai* and Zhu De* he established a base on the border of Hunan, which became the Jiangxi Soviet*. Beginning in 1930, the army of the National government in Nanjing embarked on five successive campaigns to crush the Communists, and against this onslaught Mao and Zhu De developed their successful guerrilla tactics. Nevertheless by 1934 the Communists felt the need to retreat, some 100,000 embarking on the Long March* and perhaps a total of 30,000 reaching their new base of Yan'an. In 1936 Zhou Enlai successfully negotiated the United Front* at Xi'an and the long struggle against the Japanese followed, during most of which time Mao remained in Yan'an developing his ideology and building up the party, while his guerrilla forces under Zhu De harassed the enemy. In 1945 US envoys tried to persuade Mao and Jiang Jiehi* (Chiang Kai-shek) to form a coalition government. The talks failed and civil war developed. On 1 October 1949 Mao took office as Chairman of the People's Republic of China. He at once concentrated on land reform. Holdings of landlords were confiscated under an Agrarian Reform Law and schools established or reorganized. Fully committed to the Korean War*, which he saw as a clear struggle between East and West, he used it internally to strengthen political control and to denounce 'American imperialism'. After the war he sought to woo the intellectuals with his Hundred Flowers campaign* and pioneered efforts to galvanize the economy with the Great Leap Forward*. He travelled to Moscow in 1957, where he made an agreement for China to receive nuclear technological data. Two years later, however, the Soviet Union scrapped that agreement as relations between the two powers deteriorated. In April of that year (1959) Mao retired from the post of Chairman of the Republic. He re-emerged in 1966 to launch the Cultural Revolution*, when mass rallies of Red Guard young people shouted frenzied adoration of 'Chairman Mao'. It is claimed that 350 million copies of his red book *Quotations from Chairman Mao Zedong* were published in 1967–8. In the struggle for succession he eliminated his once trusted disciple Lin Baio*, and he gave his support to his wife and the Gang of Four* in the last months of his life. He accepted the 'tactical necessity' of reconciliation with 'American imperialism' as brought about in 1972, but to the last claimed that 'strategically' he 'despised' it.

Maquis (from Corsican Italian word *macchia*, 'undergrowth'), French Resistance movement*. After the fall of France in June 1940, it carried on the fight against Nazi occupation. Supported by the French Communist Party, but not centrally controlled, its membership rose 1943–4 and constituted a considerable hindrance to the German rear after the Allies landed in 1944. By then its various groups were being co-ordinated into the Forces Françaises de l'Intérieur. Considerable tension arose, however, between its Communist members and the liberating Free French forces backed by the SOE (see RESISTANCE MOVEMENTS).

March on Rome. In October 1922 Italy seemed threatened by civil war, with strikes and riots in all the major northern cities. King Victor Emanuel III invited Mussolini, leader of the Fascists, to be Prime Minister and form a government. He travelled to Rome from Milan by train and on 30 October announced a government.

Next day 25,000 Blackshirts* arrived by train to take part in a ceremonial parade. There was no 'March on Rome' as such.

Marco Polo Bridge Incident. During the night of 7 July 1937 a small Japanese marauding military force near the Marco Polo Bridge outside Beijing demanded entry into the walled town of Wan-ping to search for a missing soldier. The Chinese Nationalist garrison refused entry and a shot was fired. Fighting continued through the night. Negotiations followed, but both sides refused to make concessions. The incident marks the beginning of the Sino-Japanese War*.

Marcos, Ferdinand Edralin (1917–89), President of the Philippines 1965–86. Of a wealthy Philippine family, he was a successful lawyer before World War II. He then joined the Philippine army, and was captured by the Japanese but escaped. He acted as a US intelligence officer, later claiming that he was an active anti-Japanese partisan. He was a special assistant to President Roxas* 1946–7 and was elected to the Philippine Congress in 1949, becoming leader of the Senate in 1963. Two years later he was elected President. He initially achieved some success as a reformer and identified closely with the USA; but after his election to a second term he became increasingly involved in campaigns against nationalist guerrilla forces and the Communist New People's Army*. In 1972 he declared martial law and then assumed dictatorial powers. Although martial law was lifted in 1981 and some moves were made towards restoration of democracy, hostility towards him intensified after the murder of the opposition leader Benigno Aquino. US support for his increasingly corrupt and brutal regime began to wane as he failed to achieve any kind of consensus. In February 1986 he was forced to leave the country, following his attempts to retain power after an election disputed by Corazón Aquino*. He fled to the USA, allegedly with many millions of dollars made from his years in office.

Market Socialism, a term first coined during the 1930s, seeking to reconcile social ownership of means of production, through workers' co-operatives, with a free market mechanism for distribution. It was adopted in Yugoslavia under Tito* and was attractive to many in the Soviet Union and in Eastern Europe in the early 1990s as an alternative to a totally free market economy.

Marne, Battles of the, World War I (1914, 1918). The first battle marked the climax of the German plan to destroy French forces before Russian mobilization was complete. German troops were already within 25 km. of Paris (the French government had retreated to Bordeaux) when on 4 September Joffre* launched a counter-attack. On 6 September the Germans crossed the Marne in order to outflank the French, but by 12 September they were in retreat. They moved north to the River Aisne where they dug in, setting the pattern for trench warfare for the next four years. Joffre's success represents one of the decisive battles in history. On 21 March 1918 the Germans launched a strong offensive north of the Somme. This first offensive failed to break the Allied line and did not reach its objective of Amiens. On 15 July, however, Ludendorff* renewed the attack further east. His troops crossed the Marne east of Château-Thierry and again Paris was threatened; but on 17 July Marshal Foch* ordered his counter- attack, successfully using tanks. This began the Allied advance of British, French, and US forces which was to culminate in the armistice of 11 November.

Maronites, an Eastern-rite Christian community. Centred on Antioch and dating back to the 4th century, in the 12th century they gave their allegiance to the Latin (Western) Church. Some 1.5 million adherents live in Lebanon, where under the French mandate 1920–46 they gained a privileged position. After Lebanon's independence they formed the dominant religious/political group, a convention being devised that the President would always be a Maronite, the Prime Minister a Sunni Muslim, and the Speaker of Parliament a Shia Muslim. Traditional rivals of the Maronites are the Druzes*. In 1936 Pierre Jumayyil*, a wealthy Maronite, formed a right-wing Phalange Party aimed at protecting Maronite Lebanon from the rising tide of Pan-Arabism, as well as from the Druzes. By the Taif Accord* of October 1989 they lost their built-in political majority over the Muslim population of Lebanon.

Marshall, George Catlett (1880–1959), Chief of Staff of US army 1939–45; US Secretary of State 1947–9; US Secretary of Defense 1950–1. Born of long-established Virginian families, he graduated at the Virginia Military Institute in 1901 and had a distinguished record in World War I with the US 1st Army on the Western Front. Rising rapidly in rank between the wars, he was appointed Chief of Staff on 1 September 1939 to an army of some 200,000. When he left office in 1945 this had risen to 8½ million. He attended all the major wartime conferences and was a strong advocate of an Allied landing on the Normandy coast. In November 1945 he was asked by President Truman to attempt to mediate in the renewed Chinese Civil War*, but had to report that US aid to the Guomindang was being dissipated by wholesale corruption and that he had failed to achieve a pacification. As Secretary of State he initiated massive aid to Greece and Turkey and was the author of the European Recovery or Marshall Plan*. Following the collapse of his plan to aid Eastern Europe, because of Soviet hostility to it, he helped to create NATO* and supported the firm line taken by the Western powers over the Soviet blockade of West Berlin. Ill health obliged him to resign in 1949, but he returned to office in 1950 when war in Korea broke out. He remained a distinguished government consultant throughout the 1950s. He is the only general ever to have received a Nobel* Peace Prize.

Marshall Islands, Republic of the, a group of some thirty-three small islands in Micronesia which were German colonies until 1920. Captured by Japan in 1914, they were then mandated to Japan until captured by US forces in 1944. In 1946 they became part of the US Trust Territory of the Pacific Islands*, being granted a series of island councils to run their home affairs. On 21 October 1986 the Republic of the Marshall Islands came into being, with the USA continuing to provide defence and paying compensation for its extensive military presence.

Marshall Plan (European Recovery Program). Passed by Congress in 1948 as the Foreign Assistance Act to aid European Recovery after World War II, it was the achievement of Secretary of State George Marshall*. The Plan invited the European nations to outline their requirements for economic recovery in order that material and financial aid could be used most effectively. The Soviet Union refused to participate and put pressure on its East European allies to do likewise. Aid was to be provided by direct grants and by loans. To help administer the plan, the Organization for European Economic Co-operation was set up, eventually consisting of eighteen European nations including West Germany, together with the USA and Canada. Between 1948 and 1951 some $13.5 billion was distributed. The Marshall Plan greatly contributed to the economic recovery of Europe, restoring industrial and agricultural production, and bolstering international trade. In 1951 the Plan's implementation was transferred to the Mutual Security Program, which continued to operate until 1956. In 1961 the OEEC was superseded by the Organization for Economic Co-operation and Development*.

Martov, Julius (1873–1923), Russian revolutionary and leader of the Menshevik Party. Born Yuliy Osipovich Tsederbaum, as a young man he joined the Russian Social Democrats and worked as a journalist. He supported Lenin* in 1900, when he formed his Iskra group, publishing a newspaper *Iskra*, which campaigned for 'orthodox Marxism' and the 'Liberation of Labour'. But in 1903 he split with Lenin, refusing to join his Bolsheviks. This was at a meeting of the party in London, when he disagreed over the extent to which the party should be controlled by a small élite cadre. He and his Menshevik supporters favoured a mass labour party. He did not however support the Menshevik provisional government* of February 1917, and after the 'October' Revolution supported Lenin against the 'White' armies. Nevertheless he became the official leader of the Menshevik Party and in 1920 left Russia as head of a Menshevik delegation in Berlin. He remained there, continuing his work as a journalist.

Marxism, the body of thought derived from the writings of Karl Marx. As a young man he developed Hegelian thought, seeing human nature itself as not immutable but subject to change within historical time, as a result of social and economic conditions. In his main work *Das Kapital* (1867), he developed his theory of 'the historical essence of man'. History itself is propelled by material forces, but because the existing system of capitalism 'of necessity' is a system of exploitation of human nature, eventually and inevitably that system will collapse, when a classless proletariat will inherit the earth and man will be free and restored to his dignity. By mid-20th century, Marxist thought had developed into a vast body of theory, much of it not mutually compatible. In the Soviet Union Marxism-Leninism developed, which claimed to combine Marx's analysis of capitalism with Lenin's* theory of democratic centralism. This has itself been subject to revisionism, as Stalinism has been replaced by a more liberal polycentric ideology. Not only has Marxism provided socialist and communist politics with an ideology, it has also profoundly influenced all 20th-century historiography and economic theory.

Masaryk, Jan (1886–1948), Czechoslovak statesman. Born in Prague, son of Tómaš Masaryk*, he was educated in Prague and Vienna and served for a while in the Austro-Hungarian army. In 1907 he emigrated to the USA, but returned to Prague in 1918 and became a personal assistant to Edward Beneš*, attending the Paris Peace Conference*, where the new state was recognized. As ambassador to Britain 1925–38, he resigned in protest at his country's betrayal at Munich. On the liberation of Czechoslovakia by the Allies in 1945 he became Foreign Minister, and was dismayed at the Soviet veto of Czechoslovak acceptance of Marshall Aid*. At the request of President Beneš he remained in his post after the Communist coup of February 1948, but he either committed suicide or was murdered three weeks later.

Masaryk, Tómaš Garrigue (1850–1937), President of Czechoslovakia 1918–35. Son of a coachman on a Habsburg estate, his ability was early recognized. After attending the universities of Leipzig and Vienna he became Professor of Philosophy at Prague University 1882–1914, where he defended Slav and Semitic minorities. He revealed that the government was forging documents to discredit nationalist leaders in the Habsburg Empire and he escaped to London. Here he taught at King's College and campaigned with Beneš* for an independent Czechoslovakia. In 1917 he went to Russia to recruit a Czech Legion from prisoners of war and then he went to the USA to gain support. Czech independence was proclaimed in Prague in November 1918, when he was elected President. He returned in triumph in December, having won the full backing of President Wilson. He favoured friendly relations with

Germany and Austria, and was a strong supporter of the League of Nations. He was twice re-elected, but in 1935 felt that the rising Nazi menace needed a younger president, and resigned.

Massey, Vincent (1887–1967), first Canadian-born Governor-General of Canada 1952–9. After teaching at Toronto University he served in World War I, before becoming president of the family firm of Massey-Harris. In 1926 he was appointed first Canadian Minister to the USA, and was Canadian High Commissioner in London 1935–46. After the war he was chairman of an influential Royal Commission on the Arts and Sciences in Canada, which resulted in considerable extension of federal support. He served as Governor-General with some distinction.

Massey, William Ferguson (1856–1925), Prime Minister of New Zealand 1912–25. He had emigrated from Ireland in 1870 and become a successful farmer near Auckland, acquiring huge areas of Maori land. He first entered Parliament in 1894 as a Conservative and in 1905 founded the Reform Party* of New Zealand farmers. Campaigning for freehold tenure and free enterprise, he won the election of 1912. His government forcefully suppressed a docks strike led by militant 'Red Feds', using farmer constables, 'Massey's Cossacks'. When Britain declared war in August 1914 he gave New Zealand's fullest support, being a strong imperialist. Some 80,000 volunteers sailed to the Middle East as part of the Anzac* army, and in 1916 he introduced conscription, having formed a coalition with the Liberals under Joseph Ward*. He represented New Zealand at the Paris Peace Conference*, after which he split with Ward on policies of resettlement. His Reform Party won the election of 1919, but his post-war government faced economic problems, with a fall in meat and wool prices. He died in office 10 May 1925.

Matteotti, Giacomo (1885–1924), Italian politician. From a wealthy family in northern Italy, he became a socialist, and was elected as such to the Italian Chamber of Deputies. In 1921 he began to organize a United Socialist Party. He openly accused Mussolini and his Blackshirts* of winning the 1924 parliamentary elections by force, giving examples of attacks on individuals and the smashing of opposition newspaper offices. Within a week he was found murdered, his death provoking widespread rioting. How far Mussolini was personally responsible is not known, but in January 1925 he accepted responsibility for the crime. There were arrests, the perpetrators receiving token sentences. Mussolini used the Matteotti Affair to impose Press censorship and to ban all socialists. Matteotti Brigades were formed in 1943 by the Italian anti-Fascist partisans.

Mau Mau (1952–9), a term used to describe the nationalist rebellion in Kenya. The Kikuyu Central Association (KCA) had been formed in 1924, demanding restoration of or compensation for expropriated land, expansion of education, and the removal of commercial restrictions on Africans in Kenya. In 1940 it was declared illegal. In 1944 the more broadly based Kenya African Union (KAU) was formed. Frustrated at the lack of progress, an underground movement began to develop from 1950, imposing strong oaths of loyalty on its members. Acts of violence against Europeans—persons and property—followed. This movement was dubbed 'Mau Mau'. In 1952 a state of emergency was declared and some 200 prominent KAU members detained, including teachers, clergy, and journalists. Lacking leadership, hundreds of less prominent nationalists disappeared into the forest as 'freedom fighters'. Special detention centres were established, and suspects arrested and subjected to a 'rehabilitation process'. This allowed the use of 'compelling' force to obtain a confession, but not 'overwhelming' force. In 1958 at Hola Camp, from a group of eighty-eight hard-core detainees, eleven died and many were seriously injured during interrogation. Political protest in Britain against this 'Hola incident' resulted in the decision (1959) to call a Constitutional Conference resulting in Kenyan independence. Mau Mau has been termed a 'peasant revolution'. Official casualties up to 1957 were ninety-five Europeans and 1,920 Africans killed by Mau Mau and 11,503 Kikuyu killed by defence forces, including a 'loyalist' home guard. The origins of the term are not known; it may be related to the Kenya rift-valley Mau Escarpment.

Mauritania, a West African republic. French penetration of the West African interior which forms part of the Sahara desert resulted in it becoming a French protectorate in 1903. In 1920 it was made a territory of French West Africa*. It became an autonomous republic in 1958 within the French Community* and fully independent in 1960. Following the Spanish withdrawal from the Western Sahara in 1976, Mauritania and Morocco divided between them the southern part of this territory, known as Tiris el-Gherbia. Bitter war ensued, and in 1979 Mauritania relinquished all claims, although the Polisario* struggle continued. The country's first President, Moktar Ould Daddah, was replaced in 1978 by a military government, the Military Committee for National Salvation (CMSN), which elects a President and appoints a Council of Ministers. In 1989 some 40,000 Black Mauritanians were expelled to Senegal* following ethnic violence in both countries which deteriorated into a border war through 1990.

Mauritius, an island state in the Indian Ocean. The British occupied it in 1810 and under their rule massive Indian immigration took place, for an economy based upon sugar-cane production. In 1967 it became self-governing and in 1968 a fully independent state within the Commonwealth. Politically it has maintained stability as a multi-cultural state. Sugar remained the principal crop and a drop in world prices in the early 1980s resulted in the fall of Sir Seewoosagur Ramgoolam's Mauritian Labour Party and its replacement by the Mouvement Militant Mauriçien, with Aneerood Jugnauth as Prime Minister. In 1983 the Labour Party formed a coalition with Jugnauth's new party the Mauritian Socialist Party. This alliance remained in power through the 1980s with Jugnauth Prime Minister. In 1965 the atoll Diego García was detached and its population forcibly removed. It was then leased to the USA to provide a military base in the Indian Ocean. This action resulted in deep resentment among Mauritians.

Maxton, James (1885–1946), British politician. Born and educated in Glasgow, he joined the Conservative Club at the university. However, in 1904 a speech at the Club by Philip Snowden* inspired him to join the Independent Labour Party*. He became a teacher, but in his spare time began to make a name for himself as an orator and a rebel. In 1916, during World War I, he urged Glasgow workers to strike, and he was imprisoned for a year for sedition. He represented Clydeside in Parliament from 1922 until he died, becoming chairman of the ILP in 1926. In 1932 he broke with the Labour Party, considering it to have betrayed socialism. An advocate of left-wing 'direct-action' socialism and always a pacifist, he remained one of the most respected and popular members of the House of Commons.

May Fourth Movement (1919), an ideological debate in China. On 4 May 1919 there were student protests in Beijing against the Treaty of Versailles* confirmation of Japan taking over German concessions in the Shandong peninsula. This stimulated 'a New Culture Movement', in which Chinese intellectuals grappled with

Marxism* and liberalism in their search for reforms. Socialist ideas became popular and the movement played a major role in the revival of the Guomindang* and the creation of the Chinese Communist Party*.

Mboya, Tom (1926–69), Kenyan politician. Born on a sisal plantation, from the Luo people, he won a scholarship to Ruskin College, Oxford, and then served as treasurer of the Kenya Africa Union. He was the chief trade union organizer in Kenya in the late 1950s. In 1960, as one of the leaders of the Kenya Independence Movement, he attended a conference in London on the future of Kenya, where he won a high reputation for his moderation, helping to obtain a draft constitution which would give Africans political supremacy. He became secretary-general of the newly formed Kenya African National Union (KANU), and after independence served in various government posts including Minister of Labour and Minister of Justice. Widely regarded as heir to Kenyatta*, he was assassinated in Nairobi in July 1969.

Meanjin, an Australian literary quarterly begun in 1940 and for some twenty years the focus for work of younger poets and short-story writers, as well as a forum for critical assessment of Australian art and literature in the post-war period. It featured much of the earlier work of Judith Wright and A. D. Hope, and helped to precipitate a new Australian cultural identity.

Medical and surgical advances. The 19th century had made some fundamental advances in understanding and beginning to control a number of world diseases, for example rabies, bubonic plague, typhoid, diphtheria, and malaria. X-rays had been discovered and the make-up of the blood understood. The early 20th century saw the conquest of syphilis by salvarsan, the development of blood transfusion, the discovery of insulin, and the development of the electrocardiogram (1903) and electroencephalogram (1929). Sulphonamide drugs (1935) proved effective against a number of infections; penicillin was mass-produced from 1941; and after World War II the pharmaceutical industry* developed a whole range of new drugs. Plastic surgery developed, stimulated by the challenge of war casualties, and anaesthetics made startling advances. After World War II surgery began to make very rapid advances, particularly in the fields of eye and heart surgery, organ transplant, and bone replacement. Laser surgery enabled incredibly delicate operations to be performed, for example on the eye. Pharmaceutical research resulted in successful treatment of a number of diseases: leprosy, pneumonia, tuberculosis, yellow fever, meningitis. Life expectancy advanced throughout the world, helped by increasingly sophisticated electronic scanning machines. But only the most developed nations could afford the high costs of sophisticated surgical technology. Preventive medicine through improved diet, better hygiene, and a purer environment were everywhere accepted as the main means by which to improve a nation's health. Nevertheless the century witnessed the death through starvation of many millions, especially in eastern Africa; influenza remained incurable (an epidemic killed 15 million in 1918–19); cancers remained stubbornly resistant to treatment; and HIV (Human Immunodeficiency Virus) threatened many millions with AIDS*.

Medicare. Both the USA and Canada have introduced medicare schemes, providing medical care and assistance. In the USA a scheme was introduced to Congress by President Johnson* in July 1965 as one of a series of reforms for his Great Society. Funded by compulsory health insurance, it meets a percentage of the cost of medical treatment and hospital care for persons aged 65 or over. The scheme also includes Medicaid for sick people with absolutely no personal resources. The American Medical

Association had staunchly opposed any more general provision as 'socialist medicine', having successfully resisted *any* medicare scheme from its first proposal by President Truman in 1947. In Canada a national hospital insurance scheme was introduced in 1961 and a full medical insurance scheme in Saskatchewan in 1962. Following the Federal Medical Care Act of 1968 Saskatchewan and British Columbia both became eligible for federal reimbursement of 50 per cent of the cost of a medicare plan. Other provinces quickly followed, providing Canada with a comprehensive, universal health service akin to the British National Health Service.

Meir, Golda (1898–1978), Prime Minister of Israel 1969–74. Born Golda Mabovitch in Kiev, Russia, she emigrated with her parents to the USA in 1907. She married Morris Mayerson, and they changed their name to Meir when they settled in Palestine in 1921 to live and work on a kibbutz. A strong socialist and member of the Jewish Labour Party (Mapai), she worked for several organizations before becoming (1946–8) head of the Political Department of the Jewish Agency*. As such she was involved in secret negotiations leading to the foundation of Israel, when she was appointed the first ambassador to the Soviet Union. From 1949 to 1956 she was Minister of Labour and 1956–66 Foreign Minister. In 1966 she left the government to become secretary-general of the Mapai Party and in 1967 helped to merge it with two left-wing parties to form the Israel Labour Party*. She was Prime Minister of a coalition government 1969–74, but she distrusted the forceful policies of General Dayan*, who left her government after the relative failure of the Yom Kippur War*. She was heavily criticized for her policies at this time and in March 1974 unexpectedly resigned, being succeeded by General Rabin.

Melanesia, a loose term for an island area of the south-west Pacific in a 3,500-mile arc north-east of Australia. It includes Irian Jaya and Papua New Guinea, the Solomon Islands, Vanuatu, New Caledonia, and Fiji. The name, 'Black Islands', refers to the dark skins of the islanders, some of whom during the 19th century suffered grievously from European exploitation. The upheavals of World War II strongly affected most of the islands, particularly New Guinea, and the pace of Europeanization, with demands for self-government, was greatly accelerated after the war. Almost all the islands, with the exception of New Caledonia*, are now self-governing, and there has been some resurgence of Melanesian culture, with its strong kinship ties.

Mellon, Andrew William (1855–1937), US Secretary of the Treasury 1921–32. Son of a wealthy Pittsburgh banker, he joined his father's bank and by 1920 was one of the richest men in the USA. Appointed by President Harding in 1921, he served under his two Republican successors, Presidents Coolidge and Hoover. He opposed high expenditure by the federal government, working to reduce taxes. He presided over the boom of the 1920s, convinced that government was an extension of big business, to be run on business lines. His tax-cuts purposely helped the rich to aid their investments which, he felt, would in time bring employment and other benefits to the less well-off. In the early 1930s, in face of accelerating depression, he had no policy other than of increasing retrenchment. He donated his considerable art collection, together with funds, to establish the US National Gallery of Art (1937), in Washington.

Memel (Klaipeda). This Baltic port and hinterland had been part of East Prussia, but was assigned to the newly independent Lithuania* in 1919 by the Treaty of Versailles. By 1935 the National Socialist Party had a majority in the Memel Diet and in 1938 they won 90 per cent of the seats. In March 1939 it was incorporated into East Prussia by Hitler. It was conquered by Red Army troops during World War II and reassigned to the Socialist Republic of Lithuania, its German population being expelled and its name changed to Klaipeda.

Mencken, Henry Louis (1880–1956), US journalist and writer. Born and educated in Baltimore, he became a police reporter and then joined the staff of the *Baltimore Sun* (1906), on which he worked for most of the rest of his life. After World War I he founded and edited the *American Mercury*, which had a powerful influence on public opinion. During the boom years of the 1920s he satirized the 'new rich' and fashionable society. Cynical of democracy and of racial equality, he was nevertheless a perceptive literary critic, patronizing D. H. Lawrence, Ford Madox Ford, and Sinclair Lewis, among others. His popularity began to decline during the 1930s and 1940s, when in particular his admiration for German 'high culture' was no longer fashionable.

Menderes, Adnan (1899–1961), Prime Minister of Turkey 1950–60. Born in İzmir, he graduated and practised as a lawyer there and entered politics in 1932, a cautious critic of President Atatürk*. In 1945 he founded the Democratic Party, winning the election of 1950. As Prime Minister, he brought Turkey into NATO and accepted the London decision of 1959 that Cyprus be independent. By then, however, the Turkish economy was in difficulties and there were riots. Menderes took dictatorial powers, but was deposed by the army, charged with breaking the Constitution. He was hanged in September 1961 'for treason'.

Mendes, Chico (1944–88), Brazilian trade union leader. Born in Xapuri in the Brazilian state of Acre, he became president of the Brazilian Rural Workers' Union, founder of the Union of Forest People, and a leader of the National Rubber Tappers' Association. As such he led the Brazilian rubber tappers in erecting blockades (*empates*) against machines sent in to clear the Brazilian forest for cattle ranches. He fought for the preservation of the forests of Brazil against those wanting to develop the mahogany trade and successfully blocked the building of a highway through the forest in his home state. He was awarded the Global 500 Prize by the UN Environmental Programme in 1987. There were many attempts on his life before his murder in December 1988, probably on the instigation of the right-wing Brazilian Landowners' Association, the UDR.

Mendès-France, Pierre (1907–82), Prime Minister of France 1954–5. Born and educated in Paris, where he was a brilliant law student, he practised law and entered politics as a Radical Socialist, being elected Deputy in 1932. He was an Economics Minister in the government of Léon Blum* in 1936. He was imprisoned by the Vichy government, but escaped to London to join the exiled Free French* government of de Gaulle. After the war he was critical of France's colonial policy in Indo-China and became Prime Minister after the disaster there of Dien Bien Phu*, promising that France would pull out. He honoured this pledge at the Geneva Conference, rejected the plan for a European Defence Community*, and prepared Tunisia* for independence. While supporting claims for ultimate independence in Algeria too, he nevertheless sent troop reinforcements there under the uncompromising Governor Jacques Soustelle, and this offended his left-wing supporters. This and an austere economic policy led to his downfall in February 1955. He served in the government of Guy Mollett* (1956), but was unhappy with the presidential constitution of the Fifth Republic* created by de Gaulle in 1958. He resigned from the Radical Party in 1959, after which he never had an effective power-base. He became increasingly opposed to de Gaulle's autocratic use of his powers and supported the bid by François Mitterrand* to oppose him in 1965. He retired from political life in 1973.

Menon, Vengalil Krishnan Krishna (1897–1974), Indian politician. Born in Calicut, he was educated in Madras and at the London School of Economics, becoming a history teacher in England before being called to the English Bar. In 1929 he became secretary of the India League in London and an active lobbyist for independence. A close friend of Jawaharlal Nehru*, he was High Commissioner in London 1947–52 and then India's representative at the UN (1953–62), where he helped to negotiate for peace in both Korea and Indo-China. A strong neutralist, he criticized Britain's action over the Suez Crisis*. As Minister of Defence 1957–62 he helped to develop India's arms industry, but was forced to resign over his country's defeat in the Indo-Chinese War*. He left the Congress Party but continued to sit in Parliament until his death.

Menzies, Sir Robert (1894–1978), Prime Minister of Australia 1939–41, 1949–66. Son of a store-keeper in Victoria, he became a successful barrister before entering the Victoria Parliament 1928–34. He was elected to the federal Parliament in 1934 as a member of the United Australia Party, and served as Attorney-General under J. A. Lyons, on whose sudden death in April 1939 he became Prime Minister. He became unpopular with his colleagues, however, particularly over the issue of conscription, and he was deposed in August 1941, enabling Labor to come to power and remain there throughout the war. Menzies was fiercely anti-Communist at this time and built his new Liberal Party* on what he perceived as a Communist threat to Australia. As Prime Minister he strongly supported US strategies in the Pacific, while also initiating the Colombo Plan* to preserve poorer Commonwealth nations from communism. His government saw massive industrial expansion in the 1950s, helped by a programme of assisted immigration which had been initiated by J. J. Curtin*. Menzies supported British action over Suez in 1956 and US policies in Vietnam. In 1965 he succeeded Churchill in the archaic office of Warden of the Cinque Ports in southern England.

Meredith Incident, an episode in the US civil rights struggle. In September 1962 a federal court order required the University of Mississippi to allow a Black man, James Meredith, to enrol as a student. The Governor, Ross Barnett, made defiant statements, thus encouraging thousands of White segregationists to attack the federal marshals assigned to protect Meredith as he entered the university. President Kennedy* sent in National Guardsmen and US army troops to restore peace and enforce the court's orders, two people dying in the riots. Meredith received his degree in August 1963, as the university's first Black student. In 1966 he led a civil rights march on Jackson, Miss.

Mers el-Kebir, a former French naval base in the Gulf of Oran, Algeria, finally evacuated in 1968. Shortly after the French armistice of 1940 a British fleet arrived with orders to prevent the French fleet falling into German or Italian hands. When the French refused to scuttle or be escorted to the West Indies, the British bombarded, damaging three capital ships and killing over 1,000 French seamen. In retaliation the French bombed Gibraltar. The event caused deep resentment in France and lost General de Gaulle's Free French* much support.

Mesopotamia campaign, World War I (1914–18). In 1913 Britain had acquired the Abadan oilfield of Persia (Iran), and when war broke out in 1914 it was concerned to protect both the oilfield and the route to India. When the Ottoman Empire joined the war in November 1914, British and Indian troops occupied Basra in Mesopotamia (Iraq). They began to advance towards Baghdad, but were halted and suffered the disaster of Kut al-Amara*. General Sir Frederick Maude recaptured Kut in

February 1917, entering Baghdad on 11 March. One contingent of British troops reached the oilfields at Baku in May 1918, and occupied it until September, when the Turks reoccupied it. A second contingent moved up the River Euphrates to capture Ramadi in September 1917, while a third moved up the River Tigris as far as Tikrit (July 1918) before advancing on Mosul. Meanwhile, from Egypt General Sir Edmund Allenby was driving north into Palestine, aided by Arab partisans. In December 1917 Jerusalem was occupied, from where Allenby moved north towards Damascus which, following the Battle of Megiddo, was occupied in October. Following the armistice of Mudros* on 30 October British troops briefly reoccupied Baku, to use it as a base in the Russian Civil War* and to deprive the Bolsheviks of its oil. It was to be evacuated in August 1919. Having occupied all Mesopotamia, the British for a while contemplated creating a single dominion which would have consisted of Palestine, Transjordan, Iraq, and Iran, thus linking Egypt with India, and providing a bulwark against Bolshevism.

Metaxas, Ioannis (1871–1941), Greek general and statesman. Born on Ithaka, he was commissioned into the Greek army in 1870. He fought the Turks in the Balkan War 1912–13 and became Chief of Staff of the Greek army in 1913. In 1915 he was exiled, disapproving of Greece's entry into World War I. He returned in 1920 and, after leading an unsuccessful coup, was exiled again 1923–4. A strong monarchist, he held several ministerial posts from 1928 to 1936. In that year, and with royal approval, Parliament was dissolved and he ruled as dictator until 1941. Under him corruption was reduced and social services developed. In 1940 he led a united country against the Italian invaders. He died in January 1941 and in April the Germans occupied his country.

Methodist Church, a Protestant Church originating in the late 18th century. After the death of John Wesley in 1791 it rapidly became a distinct church separate from the established Church of England and gaining particular support in Wales. It suffered many secessions during the 19th century, including the Methodist New Connection, the Primitive Methodist Church, and the Bible Christians in Britain, while in the USA it divided into numerous groups over the issue of slavery. In the 20th century various Methodist bodies came together again, notably by the merging of the Wesleyan Methodists, the Primitive Methodists, and the United Methodists into the Methodist Church in Britain in 1932. In 1939 three Methodist churches were united as the Methodist Church of the USA. The World Methodist Council, founded in 1881, provides a link between the 40 million Methodists in the world, many of whom are in Africa, as a result of missionary activities. Attempts to reunite the Methodist Church in Britain with Anglicanism have so far all failed, in spite of the ecumenical movement*.

Mexican Revolution (1910–40). The roots of this prolonged revolution lie in the conflicts and tensions generated by demographic, economic, and social changes during the long presidency of Porfirio Díaz. His regime (1876–1911) became increasingly centralized and dictatorial, favouring a privileged minority and failing to incorporate either the growing urban middle-class or Labour groups into national politics. In November 1910 Francisco Madero, leader of the anti-re-electionist movement, received an enthusiastic response to his call to arms to overthrow the dictator. Díaz resigned in May 1911, and Madero was elected President; but he failed to satisfy either his radical friends or his Porfirian enemies. He was assassinated in a counter-revolutionary coup led by General Victoriano Huerta* in 1913. Huerta was defeated by an arms embargo and diplomatic hostility from the USA, and by a coalition of revolutionary factions led by Emiliano Zapata*, Pancho Villa*, Venustiano Carranza*, and Alvero Obregón*. The victorious revolutionaries then split: the

Constitutionalists, who pressed for a democratic constitution, led by Carranza and Obregón; and the Conventionists, who wished to implement the radical proposals of a Convention at Aguascalientes in 1914, led by Zapata and Villa. The civil war which followed was protracted and bitter. Villa and Zapata were defeated and in February 1917 the US-inspired Constitution of Mexico (still valid) was promulgated. Carranza became President; however, he failed to implement much of the Constitution and in 1920 was assassinated. Mexico's new leaders Obregón and Calles faced the difficult task during the 1920s of economic regeneration and the re-establishment of central authority, which had almost broken down. They were hampered by strong opposition from the Catholic Church. Tension culminated in the so-called War of the Cristeros (1928–30), when tens of thousands of Christian peasants arose in protest against the new 'godless' state, and were finally defeated at the Battle of Reforma. It was not until President Cárdenas* came to office in 1934 that sufficient stability had been achieved for many of the original ideals of the Revolution to be implemented, including nationalization of railways and oil, and land redistribution of the hacienda estates. When Ávila Camacho* was elected President in 1940 a period of consolidation and reconciliation marked the end of the Revolution and the beginning of a period of industrial development.

Mexico, a federal republic whose present constitution dates from 1917. This vast country of some 2 million sq. km. consists of thirty-one states: Aguascalientes, Baja California Norte, Baja California Sur, Campeche, Chiapas, Chihuahua, Coahuila, Colima, Durango, Guanajuato, Guerrero, Hidalgo, Jalisco, México, Michoacán, Morelos, Nayarit, Nuevo León, Oaxaca, Puebla, Querétaro, Quintana Roo, San Luis Potosí, Sinaloa, Sonora, Tabasco, Tamaulipas, Tlaxcala, Veracruz, Yucatán, Zacatecas. The long dictatorship of Porfirio Díaz ended in 1911 with the Mexican Revolution*, which was to last until 1940. The 1917 Constitution envisaged a strongly interventionist federal government which would seek to promote social, economic, and cultural well-being. All land and minerals were regarded as a national asset, although private property and contracts were protected; organized labour would have a right to strike; there would be a comprehensive system of social security and public health provision; education would be free, compulsory, and secular, the Roman Catholic Church losing its control. In 1929 the Partido Revolucionario Nacional was formed, which has held power throughout the 20th century, changing its name in 1946 to Partido Revolucionario Institucional (PRI). Under President Miguel Alemán* (1946–52) the process of reconciliation begun by his predecessor Ávila Camacho* was continued. Since then democratic governments have continued to follow moderate policies, while seeking further to modernize the economy, boosted by oil revenues. The presidency of Miguel de la Madrid Hurtado (1982–88) was faced by a fall in oil prices, a massive national debt, one of the fastest growing birth rates, and the largest gap between rich and poor in the world. In addition, the earthquake of 1985 in Mexico City killed 20,000 people and did some $3,700m. of damage. The 1988 elections were again won by the PRI, with Carlos Salinas de Gortari taking office as President. An austerity programme initiated by his predecessor was continued, as was Mexico's involvement in seeking a pacification for Central America. Yet, in spite of massive foreign debt, Mexico continued to enjoy the advantages of a strong manufacturing base, self-sufficiency in oil and natural gas, and large capital investment in a modernized agricultural system. In May 1990 President Salinas privatized state-run banks.

MI5 and MI6 (Military Intelligence). The Security Service MI5 was founded in 1909 and covers internal security and counter-intelligence on British territory. MI6, the Secret Intelligence Service, was formed in 1912. It covers all areas outside the United

Kingdom and overseas dependencies. Operations of MI6 during World War II, in co-operation with Resistance movements, mostly Communist-inspired, contributed considerably to the outcome of the war. Since then, however, it has suffered through disclosures that some of its employees, notably intellectuals recruited in Cambridge in the 1930s, such as Philby, Burgess, Maclean, and Blunt, were in fact double agents acting for the Soviet Union. There have been consistent, but largely unsuccessful, efforts to make the two services publicly accountable. Strong evidence emerged in the 1980s of MI5's extra-constitutional role during the years of the Labour governments 1974–9, which it was seeking to destabilize; but no disciplinary action appears to have followed.

Micronesia, the regional name for the widely spread tiny islands of the Pacific north of Melanesia* and west of Polynesia*. Many of them were under Japanese mandate 1920–44, having been German colonies. The largest island is Guam*, a major US military base in the Marianas. The rest of the Marianas, the Marshall Islands, and the Carolines were all placed within the US Trust Territory of the Pacific* in 1946. Micronesia was the scene of extended nuclear testing in the years 1946–63. The USA exploded sixty-six atmospheric test weapons from Bikini and Enewetak. The test of 1 March 1954 not only made Bikini permanently uninhabitable but has resulted in extensive cancer tumours and deformities in the neighboring island of Rongelap. In 1986 the Marianas north of Guam became the self-governing Commonwealth of the Northern Marianas, in political union with the USA; at the same time the Republic of the Marshall Islands* was formed. The Federated States of Micronesia, consisting of Yap, Truk, Pohnpei (formerly Ponape), and Kosrae (Caroline Islands), were granted a constitution by the USA in 1978, and in 1986 they gained full self-government, with their own citizenship, but with the USA retaining links in 'free association' and providing defence.

Midway Island, Battle of, World War II (4 June 1942). The USA had a naval/air base on this tiny island 1,500 miles west of Hawaii, which the Japanese planned to seize by an amphibious operation. US intelligence learned of the plan and a US naval force attacked. The battle was fought by carriers some 250 miles apart and by US land planes from the base. Four Japanese carriers, each with a complement of planes and some 1,500 men on board, were sunk, together with three transports carrying 6,000 men. One US carrier was lost. The battle was decisive in that it turned the naval balance in the Pacific towards the USA.

Mihailovich, Draza (1893–1946), Yugoslav chetnik* leader. Born in south Serbia, he became a professional soldier, fighting in World War I. During the 1930s he made a special study of guerrilla warfare, and by 1941 he held the rank of colonel. After the invasion of Yugoslavia by German troops in 1941, he organized royalist Resistance forces into bands of chetniks*, whose relations with the Communist partisans of Tito were uneasy. The exiled Peter II was obliged by the British to withdraw his support, when chetnik forces were alleged to be collaborating with the Germans against Tito in 1944. But Mihailovich continued to fight on the borders of Bosnia and Serbia. In March 1946, after Tito had gained power, he was tried and shot for 'collaboration and war crimes'.

Miki Takeo (1907–   ), Prime Minister of Japan 1974–6. Having studied in California he entered politics in 1937. He publicly opposed Japan's participation in World War II, after which, as a member of the Liberal Democratic Party*, he held ten cabinet posts. In 1974 he became critical of the financial methods of the party and of Tanaka Kakuei*, whom he replaced as Prime Minister. He pushed for an inquiry into the Lockheed scandal*, which electorally damaged his party in the election of 1976, when he resigned.

Mikoyan, Anastas Ivanovich (1895–1978), President of the Praesidium of the Supreme Soviet 1964–5. Born in Armenia, he was educated at a theological seminary for the priesthood, but when he was 20 he became interested in politics and joined the Bolsheviks*. He took part in the revolutions of 1917 and in the Russian Civil War in the Caucasus. He then worked in Tbilisi and Baku, and in 1921 became head of the party organization in Nijni Novgorod and then north Caucasus 1922–6. From 1926 onwards he held various posts in the field of trade, being a supporter of Stalin in the power-struggle 1924–7. He became a member of the Politburo in 1935 and a deputy Prime Minister in 1937. After Khrushchev's denunciation of Stalin in 1956, with which he associated himself, he became his negotiator in Soviet relations with discontented East European states. An advocate of 'collective leadership', he held the office of President of the Praesidium for the year 1964–5, before his retirement.

Military mechanization, the development of tank warfare. In 1914 Kaiser William II rode into war with his cavalry accompanied by his generals, and horse cavalry took part throughout World War I. Cavalry officers always retained status over their infantry colleagues. Early British tanks on the Somme in 1916 proved inadequately armoured and too cumbersome. By 1918 a new lighter 'whippet' tank had been developed, which helped finally to break the stalemate on the Western Front*. It was to be another decade in Britain before horse cavalry slowly began to be mechanized, the first regiments being the 11th Hussars and 12th Lancers in 1928. The process was far from complete, however, by 1939, when the Royal Armoured Corps was formed. In Britain it was General J. F. C. Fuller who successfully argued for the tank, and in France General Charles de Gaulle*. With rearmament, the German army developed a theory of armoured warfare where the role of infantry was to support the tank, a theory successfully put into practice with the early blitzkriegs* of 1939 and 1940. Britain in 1939 held the opposite view, that the rôle of tanks was to support the infantry. It was not until 1942 that British tank strategy was reversed.

Milner, Alfred, 1st Viscount (1854–1925), British statesman. Born in Germany and educated there and at Oxford, he became a journalist. Later, as private secretary to G. J. Goschen, Chancellor of the Exchequer, he developed a capacity for finance which he applied in Egypt (1889–92) and as chairman of the Board of Inland Revenue. As High Commissioner for South Africa (1897–1907), his policies precipitated the Boer War, but he made preparations for peace by schemes of reorganization and reform which helped to give stability to the new South Africa. His 'kindergarten' of young imperialists were a notable reforming group. He joined Lloyd George's* war cabinet in 1916 and was Secretary of State for the Colonies 1919–21. He supported the War of Intervention in Russia and had a strong belief in Britain's imperial mission to the 'coloured peoples of the world'. At the same time he was an advocate for the education and preparation of these people for eventual self-government. Together with Churchill he supported Curzon's* idea of creating a Middle East dominion, while urging that all the dominions should unite to form a federal Commonwealth. He quarrelled with Lloyd George over policies in Egypt and resigned in 1921, in many ways, as one of the last imperialists, a disillusioned man.

Mindszenty, József (1892–1975), Primate of Hungary 1945–8. Born József Pehm, he was ordained priest in 1915, becoming Bishop of Veszprém in 1944. A staunch anti-Nazi, he was arrested by the Germans in 1944, when German troops invaded

Hungary. When their puppet government collapsed in January 1945 he was released and made Archbishop of Hungary. After visiting the USA, however, he was arrested by the new Communist government and charged with treason and currency offences. At his trial in 1948 he expressed his hostility to Communism and his support for the Habsburg claims to the throne of Hungary. He was sentenced to penal servitude for life, but this was later commuted to house detention. At the time of the Hungarian Revolution* of 1956 he was released by the Nagy government. When Soviet forces entered Budapest, however, he took refuge in the US embassy. He stayed there until 1971, by which time the Hungarian government and the Vatican had become reconciled. He was persuaded to go to Vienna and then to Rome.

Minorities treaties. Because of the racial and ethnic tangles in Eastern Europe and the Balkans, the League of Nations in 1920 obliged new or re-established states to sign agreements to give minorities equal civil and political privileges, including the use of language and religious freedom. The Baltic republics also voluntarily accepted the same obligation. Breaches of the treaties were referred to the International Court of Justice. Unfortunately there was no effective means of enforcing on minorities their obligations to majorities. During the 1930s, with the growth of nationalism, but also of anti-Semitism, the treaties were increasingly ignored. After World War II the UN Commission on Human Rights again tried to grapple with this problem on a global front, not however with much greater success.

Miranda v. Arizona, a US Supreme Court* ruling. In 1966, by a five to four majority under Chief Justice Warren*, the US Supreme Court overturned an Arizona court's decision regarding the conviction of Ernesto Miranda. The Court ruled that statements made in custody are inadmissible unless the suspect has been clearly informed: (1) of the right to remain silent; (2) that anything said could be used as evidence against him/her; (3) of his/her right to consult an attorney; (4) of his/her right to have an attorney present during interrogation. The US law enforcement lobby campaigned for a reversal of the judgement under Chief Justice Burger*, but his Court, while limiting the scope of safeguards, stopped short of a reversal.

MIRV, a ballistic missile that carries several nuclear warheads, which can be released at different speeds on different trajectories (Multiple Independently-targeted Re-entry Vehicle). They were developed in the USA during the 1970s for land intercontinental and submarine launching. Soviet MIRV technology developed during the 1980s, allegedly overtaking its US rival.

Mitterrand, François (1916–   ), President of France 1981–   . Born near Bordeaux, son of a stationmaster, he was brought up a Catholic. He studied law in Paris at the time of Blum's Popular Front government, with which he had much sympathy, becoming a socialist. Captured by German troops in 1940, he escaped and became a prominent member of the Resistance, going to London in 1943 on a secret mission. In 1946 he was elected a Deputy in the Constituent Assembly and served in all the governments of the Fourth Republic. Seeking to build a coalition between all French parties of the left—Radical, Socialist, and Communist—he founded the Federation of Democratic and Socialist Left in 1965. Standing against President de Gaulle, he won 7 million votes in the election. He stood unsuccessfully again against Giscard d'Estaing* in 1974. In 1981 he defeated Giscard, who was seeking re-election. Early measures to decentralize the government, raise basic wages, increase social benefits, and nationalize key industries were followed by an economic crisis and a reversal of some policies. He was re-elected in 1988. A committed supporter of both nuclear power and a nuclear bomb for France, he has advocated a strong foreign policy, but one that should take account of the massive changes in Eastern Europe in 1989. In the European Community he was often opposed by Mrs Thatcher for wanting to move more quickly towards political integration than did the British Prime Minister. At the same time he disagreed with the latter on the significance of the French Revolution, which he commemorated at enormous expense in 1989.

Mobutu, Sese Seko (1930–), President of Zaïre 1965– . He joined the Force Publique (Belgian Congo army) in 1949, rising to the rank of colonel. In 1960 he attended the Brussels Conference on the Congo. On independence he was appointed Chief of Staff of the new Congolese army and then Commander-in-Chief. With the support of Belgium and the USA he ousted Lumumba* in January 1961. Following years of chaos and civil war, in November 1965 he deposed President Kasavubu* and announced himself President. While maintaining order by harsh military discipline, he appointed civilians as ministers and struggled to give stability to the economy. As part of the programme of 'authenticity' he changed the name of the country from Congo to Zaïre in 1971. While Mobutu remained in power for over two decades and achieved some degree of political reform, his regime was persistently shaken by violent opposition, including rebel invasions from Angola in 1977 and 1978. He encouraged mineral exploitation and hydroelectric schemes by foreign companies. In April 1990 he announced that the single-party system would end, after 'a transitional period'.

Modernism, a term loosely and broadly used by later generations to describe the remarkable artistic upheaval of the early 20th century. Activated by a need to break away from received conventions of the later 19th century, in painting this meant going beyond Impressionism to Post-Impressionism*, Cubism*, Surrealism*, and Dadaism; in literature it meant creating new verse and prose forms, accepting the insights of the psychologists Freud* and Jung*; in sculpture it meant discarding the long Greco-Roman tradition of sculpting the human form, by returning to basic, natural forms; in music it meant searching for new modes, including a twelve-note scale; in architecture it meant replacing the pompous and eclectic with clean line and utilitarian purpose, as promulgated at the Bauhaus*. Many of the achievements of Modernism were for long rejected by middle-class taste, for whom 'modern art' remained a pejorative term. In the years following World War II the term 'Post-Modern' became current, but no coherent 'Post-Modern' aesthetic ever emerged.

Moi, Daniel arap (1924–), President of Kenya 1978– . Educated in mission schools, he was a teacher in Kenya 1945–57. He then entered politics and was appointed to the Kenya Legislative Council 1957–63. After independence he held various government offices and was Vice-President 1967–78, when he succeeded Jomo Kenyatta*. He announced a purge of the allegedly corrupt Kenyan bureaucracy and imposed a stern regime, allowing little criticism. The so-called Mwakenya Conspiracy of 1986–7 reflected growing resentment at his authoritarian methods, which were increasingly criticized internationally. He was returned unopposed as President in 1988, and played an active part in seeking an end to civil war in Mozambique. In 1990 the murder of his Foreign Minister Rodert Ouko further weakened his regime.

Moldavia, a Soviet socialist republic. Bessarabia was an area lying between the rivers Pruth and Dniester, with a majority of its population speaking Romanian. It was assigned to Russia at the Congress of Vienna in 1815, but in February 1918 a special council proclaimed union with Romania. This was confirmed at the Paris Peace Conference*. In June 1940 the Soviet Union demanded its return and King Carol II* reluctantly agreed. In June 1941 Romanian forces recaptured it, but the Red Army returned in March 1944 and the Moldavian Socialist Republic was created and incorporated into the Soviet Union, some parts of Bessarabia being assigned to the Republic of the Ukraine. Pro-Romanian sentiment survived, however, and in 1990, with the Soviet Union racked by resurgent nationalisms, there were strong pressures for independence.

Moley, Raymond Charles (1886–1975), US journalist, academic, and Assistant Secretary of State 1933. Born in Ohio, he studied and taught at Columbia University, being Professor of Public Law 1928–54. In 1928 he joined Franklin D. Roosevelt's team working for Roosevelt's election as Governor of New York, and in 1932 he helped Rexford G. Tugwell and Adolph A. Berle to form the 'Brain Trust' which advised Roosevelt for his presidential campaign and early months in office. He wrote many of Roosevelt's speeches and coined the phrase 'New Deal'*, advocating close co-operation between government and business to revive the US economy in the interests of 'the forgotten man at the bottom of the economic pyramid'. He soon came to regard Roosevelt as too radical, and abandoned active politics for the academic life and journalism, endorsing Republicans for President, including Richard Nixon.

Mollet, Guy (1905–75), Prime Minister of France 1956–7. Son of a textile worker, he trained as a teacher and taught in Arras, becoming a Resistance leader in World War II. Three times arrested and tortured by the Gestapo, he managed to avoid imprisonment and in 1945 entered politics, being elected to the Constituent Assembly. He became secretary-general of the Socialist Party (1946–69) and served in several governments of the Fourth Republic*. As Prime Minister he was persuaded to enter secret negotiations with Israel and Britain at the time of the Suez Crisis*, and also to send reinforcements to Algeria, which worsened the crisis there. Both these actions aroused strong criticism. He led the opposition to the Gaullists in the Fifth Republic, retiring from politics in 1971.

Molotov, Vyacheslav Mikhailovich (1890–1986), Soviet leader. Born near Kazan on the River Volga into a wealthy family, he joined the Social Democrat Party when he was 16 and still a student. He was at once exiled by the tsarist regime, but returned to Russia in 1912 and played a prominent part in the establishment of the Communist newspaper *Pravda*, becoming its editor. He supported Lenin* in his opposition to the provisional government* of February–October 1917, being a member of the Petrograd Soviet. After the 'October' Russian Revolution* he rose rapidly in the party. Working closely with Lenin and later with Stalin, he was responsible for the nationalization of Russian factories and workshops. For the next forty years he remained at the heart of the Soviet political élite. He took a leading part in the liquidation of the Mensheviks (see BOLSHEVIK), becoming a member of the Politburo, and in 1926 helped defeat the Zinoviev* opposition. He was chairman of the Council of People's Commissars 1930–40, and Commissar for Foreign Affairs 1939–49 and 1953–6. In 1939 he was the signatory for the Soviet Union of the Nazi–Soviet Pact and, after Hitler's invasion, he signed the Anglo-Soviet Treaty of 1942. At Yalta* and Potsdam* in 1945 he was Stalin's closest adviser. He took a truculent part in periodic meetings of a Council of Ministers, the foreign ministers of the four occupying powers of Germany, 1945–7. This functioned until 1948, when plans developed for a federal constitution for the three Western zones, when he supported the idea of a blockade of Berlin and the rejection of Marshall Aid*. Surviving Stalin, he was expelled from all his posts by Khrushchev*, who appointed him ambassador to the Mongolian People's Republic. In

1960 he became the Soviet delegate to the International Atomic Energy Agency in Vienna, but soon afterwards was expelled from the party. He retired, but in 1984 was rehabilitated within the party, two years before his death.

Monash, Sir John (1865–1931), Australian general. From a distinguished Jewish family in Melbourne, where he was educated, he practised as an engineer before World War I. He served with distinction at Gallipoli*, in command of a brigade, and then went on to command an Australian division on the Western Front 1916–18. In August 1918 he took command of the Australian Corps and helped to launch the final offensive against the Germans. Described as 'the finest Jewish soldier since Masséna', he reached a higher rank as an officer in the British army than any previous Jew. After the war he was responsible for organizing the repatriation and demobilization of the Australian armed forces. He became Vice-Chancellor of Melbourne University and Victoria's second university is named after him.

Monetarism, an economic theory which regards control of the money supply as crucial to control of inflation. It rejects Keynesian* policies of government attempts to reflate an economy through public spending, since this normally increases the money supply. The Chicago economist Milton Friedman became highly influential during the 1970s and monetarism had become a fashionable theory by the end of the decade. The first Thatcher* administration imposed monetarist policies, but with unemployment rising to over 3 million these were largely abandoned, even though inflation in Britain rose again to over 10 per cent by 1990.

Mongolia. Outer Mongolia was a Chinese province until 1911. After a confused period, Chinese warlords were driven out with Russian help and it emerged in 1924 as the **Mongolian People's Republic** in close alliance with the Soviet Union, and with the Communist Mongolian People's Revolutionary Party as the sole political party. Its independence was confirmed in 1945 by a plebiscite. In 1950 a Sino-Soviet treaty guaranteed its independence, and a new boundary agreement with China was agreed in 1987. Its vast area of 1,567,000 sq. km. has a rising population of under 2 million, the official language being Mongolian. There has been a series of five-year plans since 1950. The MPRP won 85 per cent of the votes in the first multi-party elections held in July 1990. The autonomous state of **Inner Mongolia** in China was established under Japanese direction 1937–8. It was liberated by Soviet forces in August 1945, but has remained within China, Chinese outnumbering Mongolians by 10 : 1.

Monnet, Jean (1888–1979), French economist and administrator. Born in Cognac, he became a civil servant 1915–19, when he developed a high reputation for finance. He was an adviser for the League of Nations* and an advocate of economic co-operation between the USA and Europe. In December 1945 de Gaulle* appointed him to devise a recovery plan for France. In 1947 he became Commissioner-General of this Monnet Plan, which in five years enabled France to outstrip its pre-war production. He was an early advocate of European Union, and it was his drive and enthusiasm which helped the Schuman* Plan to succeed; he became the first president of the ECSC* 1952–5. His ultimate aim was to see a United States of Europe, and he became a critic of de Gaulle's policies towards the European Community, welcoming Britain's entry in 1973.

Monroe Doctrine, a US foreign policy declaration, warning European powers against intervention in the American hemisphere and disclaiming any intention of the USA to take part in the political affairs of Europe. It was put forward by President Monroe in 1823, but was infrequently invoked during the 19th century. During the early

20th century, however, it developed into a policy whereby the USA regarded itself as the policeman of both North and South America, and this consistently complicated relations with Latin American countries.

Montagu–Chelmsford Proposals (1918), constitutional proposals for British India. They were made in the Report on Indian Constitutional Reforms by Edwin Montagu (1879–1924), Secretary of State for India, and Lord Chelmsford (1868–1933), Governor-General of India 1916–21. They followed the promise of responsible government made in 1917, and envisaged a gradual progress towards devolution of power and Indian self-government. The Government of India Act* (1919) which followed went some way towards their implementation; but it was regarded by Indian nationalists as inadequate and as a measure to shore up imperial authority by containing opposition.

Montenegro, one of six constituent republics of Yugoslavia, the only southern Slav nation to remain outside the Ottoman Empire. Under Nicholas Petrovic Njegos, who assumed the title of king in 1910, it engaged in the Balkan Wars of 1912 and 1913 and greatly extended its territory. It was occupied in 1916 by Austrian troops, when Nicholas fled to France. In 1919 a Grand Association declared the monarchy ended and that it would accept the Corfu Pact* and become part of the Kingdom of Serbs, Croats, and Slovenes, being absorbed into Serbia. It was invaded by German and Italian troops in 1941 and a bitter Resistance movement developed, spearheaded by Communists. In 1946 it accepted the federal Constitution of Marshal Tito*, but a strong spirit of independence survived, being voiced by Milovan Djilas* in the 1950s and reasserting itself in 1989. Nevertheless in elections held in December 1990 the Communists (LCY) held power.

Montessori, Maria (1870–1952), Italian educationalist, the first woman doctor to graduate in Italy (1896). Her work as a paediatrician with mentally handicapped children led her to formulate lasting theories on nursery and infant education. Children learn in a 'prepared environment', needing to be provided with a wide range of graded materials with which to develop skills through self-correction and self-education. Together with that of Piaget*, her influence on educational thought and practice went well beyond the Montessori schools established through Europe and the USA. In England during the 1950s and 1960s, through the encouragement of educational administrators such as Alec Clegg, a complete revolution took place in the classrooms of nursery and infant schools.

Montgomery, Bernard Law, 1st Viscount Montgomery of Alamein (1887–1976), British Field Marshal. Educated at St Paul's School, London, and Sandhurst, he was commissioned into the army in 1908. He served with distinction on the Western Front in World War I and held various commands between the wars. In World War II he commanded the 8th Army 1942–4 in the North African* and Italian campaigns*. He led his troops at El Alamein*, one of the decisive battles of the war, which enabled the Allies to begin the advance that removed the Germans from North Africa. In 1944 he commanded the British Commonwealth armies in Normandy with considerable success. His idea of an attack on Arnhem failed, but he played a major role in defeating the German counter-offensive in the Ardennes* and he formally received the German surrender on Lüneberg Heath 4 May 1945. He held various senior military posts after the war including Chief of the Imperial General Staff. He had the ability to arouse intense personal loyalty from the troops under his command, but was also undiplomatic and sometimes a political embarrassment.

Montreux Convention (July 1936). In 1936 Turkey claimed that Mussolini was seeking to use Italy's bases in the Dodecanese* to launch imperialist attacks on Asia Minor. An international conference was held in Montreux, where Turkey obtained revision of the Lausanne Treaty*, permitting fortification along the Dardanelles and Bosporus. A new Convention was agreed upon for tonnage through the Straits, criticized at times by the Soviet Union but still valid.

Moore, Henry (1898–1986), British sculptor. He was born and educated in Castleford, Yorks., and was trained in Leeds and at the Royal College of Art. As a young man he was greatly influenced both by ancient Egyptian art and by Mexican and African sculpture, with their strong sense of mass and form, and close relationship with materials. Deeply humanist, he wished increasingly to relate man to his environment, and found increasing significance in natural forms, whether stone, bone, or wood, while always 'seeking nature's principles of form and rhythm'. Having been commissioned for works throughout the world, most of his later work was in bronze. He was described shortly before his death as 'the greatest living Englishman'.

Morgenthau Plan (September 1944), a plan for post-war Germany. Henry Morgenthau was US Secretary to the Treasury, and at the Quebec Conference between President Roosevelt and Winston Churchill he presented a plan to deindustrialize Germany after its defeat. This 'pastoralization' plan was quickly rejected as impracticable, but it was leaked and may have helped to stiffen German resistance in the later months of World War II.

Morínigo, Higinio (1897–1985), President of Paraguay 1940–8. A professional soldier, he was a senior officer during the Chaco War* and subsequently Minister of War. When President José Félix Estigarribia was killed in an air crash, Morínigo became President. He suspended the constitution and established a harsh military dictatorship, banning all political activity as well as trade unions. Housing and public health were however improved during his time in office. Morínigo had sympathized with the Axis powers during World War II, and after the war pressure was exerted on him to liberalize his regime. In 1946 political opponents were allowed to return and to enter a more democratic cabinet. When these were ousted on charges of treason (1947), civil war broke out and a broadly based opposition, led by the liberal Colorado Party, forced Morínigo's retirement and exile to Argentina in 1948. He returned briefly to Paraguay in 1956 but then lived in Argentina.

Mormon Church, or Church of Jesus Christ of Latter-day Saints. It was founded in 1830 in New York state by Joseph Smith, and was based on the revelation to him of the Book of Mormon, a mixture of religious concepts and mythical history, including the belief that American Indians were the lost tribe of Israel. After many hardships, the Mormons created a self-contained economy centred around Salt Lake City, although continued difficulties over the issue of polygamy prevented Utah becoming a state of the USA until 1896. Mormons dominate the administration of the state and have also brought their faith to Britain and to Europe during the 20th century.

Moro, Aldo (1916–78), Italian Prime Minister 1963–8, 1974–6. Born in south Italy, he studied law at Bari University, where he became a prominent anti-Fascist as a student. He practised as a lawyer, and in 1946 was a member of the Constituent Assembly devising a new Italian constitution. He sat in Parliament as a Christian Democrat and, as Minister of Justice 1955–7, considerably reformed the Italian prison system. He was Foreign Minister in several governments between 1965 and 1974 as well as being Prime Minister. He was kidnapped in March 1978 by the terrorist group the Red Brigade, who demanded the release of imprisoned terrorists for his return. The government's refusal to accede led to his murder in May 1978.

Morocco, an independent sultanate since the Middle Ages. Early in the 20th century there were tribal disturbances against the Sultan, while both France and Spain were seeking to establish protectorates. This they did in 1912, with the port of Tangier* to be an 'international zone'. There was violent fighting in the early 1920s by rebels from the Riff mountains led by Abd al-Krim*, first against the Spanish and then against the French. The latter's Resident-General Marshal Lyautey*, however, showed considerable administrative skill and managed to maintain the loyalty of the majority of urban Moroccans. Demands for some autonomy developed during the 1930s. When these were rejected by France in 1934 a nationalist movement, the Istiqlas Party, emerged, demanding full independence. Spanish Morocco was the base from which General Franco launched the Spanish Civil War and Hitler was keen to use it during World War II. Franco rejected his request and in November 1942 Morocco was occupied by an Anglo-American force which then advanced into North Africa. After World War II some moves were made towards self-government in French Morocco, but there were serious anti-French riots 1953–5. At the same time there were clashes between royalists, who wanted to vest power in the Sultan, republicans, and Berber tribesmen seeking independence. In 1956 France agreed to an interim period leading to the establishment of Sultan Mohammed Yusuf as a constitutional monarch, Mohammed V. At the same time Tangier and Spanish Morocco were to be absorbed. King Hassan II* has survived both republican opposition and abortive military coups. Since 1975, however, his claims over the mineral-rich Western Sahara to the south have led to increasing isolation. A convention was signed in 1976, dividing the area between Morocco and Mauritania*, but the latter renounced its claims in 1979; major battles were fought with guerrillas of the Polisario* through 1979 and March 1980. Moroccan troops have made some headway, building a series of desert walls, but international support for the Polisario has increased, the Saharan Arab Democratic Republic having been recognized by most African countries and many other members of the UN.

Morrison, Herbert Stanley, 1st Baron Morrison of Lambeth (1888–1965), British statesman. Born in Lambeth, London, he left school at 13 and became a shop assistant and then a telephone operator. He joined the London Labour Party and became its secretary in 1915, being a strong pacifist during World War I. He became Mayor of Hackney in 1920 and was elected to the London County Council in 1922 and as MP for Hackney in 1923. Morrison was Leader of the LCC 1934–40, when he unified the transport system, and as Minister for Transport 1929–31 he created the London Passenger Transport Board. During these years he was also influential in the creation of a 'Green Belt' around the metropolis. He was briefly a member of MacDonald's* government in 1931 and was Minister of Supply and then Home Secretary in Churchill's coalition government 1940–5, creating the National Fire Service and sitting in the war cabinet. He drafted the programmes for nationalization and the social services for the 1945 election. As deputy Prime Minister to Attlee* 1945–51 he implemented Labour's nationalization policy, whereby 'public ownership' was achieved by the creation of autonomous central boards. He was briefly Foreign Secretary in 1951. He was defeated in 1955 for leadership of the Labour Party by Hugh Gaitskell*, a much younger man.

Mosley, Sir Oswald Ernald, Bt. (1896–1980), British politician. His father was a baronet from a wealthy family, and he was educated at Winchester and

Sandhurst. He served in World War I in the cavalry and then in the Royal Flying Corps. He was a Member of Parliament successively as Conservative (1918–22), Independent (1922–24), and Labour (1924–31). He was briefly a member of MacDonald's* government in 1929, presenting proposals to solve the mounting economic depression. When these were rejected he resigned and formed his own 'progressive socialist' movement, the New Party (1931), which advocated greater state intervention. He now became increasingly racialist and, after a visit to Mussolini, he was calling for a dictatorial system of government, forming the British Union of Fascists in 1932. Its blackshirted followers staged violent anti-Semitic marches and rallies in the East End of London. Mosley was interned during World War II 1940–3. In 1948 he founded the 'Union Movement', whose theme was to be that of European unity.

Mountbatten, Louis Francis Albert Victor Nicholas, 1st Earl Mountbatten of Burma (1900–79), British admiral and administrator. He was a son of Prince Louis of Battenburg*, a grandson of Queen Victoria, who adopted the name Mountbatten in 1917. He joined the Royal Navy and served in World War I as a midshipman in Beatty's flagships. After the war he accompanied the Prince of Wales on two Empire tours. In 1940–1 he commanded a destroyer flotilla which was badly bombed off Norway and later in the battle for Crete. He became Chief of Combined Operations in 1942, always having the confidence of Winston Churchill. He did much for the planning of landings in North Africa, Italy, and Normandy. In October 1943 he was appointed Supreme Allied Commander South-East Asia, where he restored the morale of British and Commonwealth forces. His SEAC* successfully recaptured Burma and made plans for combined operations to take Singapore and Malaya, which were largely invalidated by the fall of Japan in September 1945. In 1947 he became the last Viceroy of India, charged with the transfer of sovereignty from the British Crown. This was promptly effected in August 1947, although marred by inter-communal massacres. At the invitation of the new government of India, he stayed on until 1948 as the first Governor-General. Resuming his naval career, he rose to be Chief of the Defence Staff (1959–65), in which capacity he supervised the merging of the service ministries into a unified Ministry of Defence. Active in retirement, he criticized British defence policies for becoming too reliant on nuclear, as against 'conventional', weapons. He was assassinated in 1979 by the IRA* while on a holiday in Ireland.

Mozambique, a southern African state. As a Portuguese colony its African resistance movements were suppressed throughout the 19th and early 20th centuries. In 1964 the Frente de Libertaçao de Moçambique (FRELIMO) was formed by Eduardo Mondlane and Samora Machel*, and its guerrilla groups steadily gained control. By the mid-1970s Portuguese authority had reached the point of collapse, and in 1975 an independent People's Republic was established under Machel. Support for guerrilla campaigns in Rhodesia and by ANC supporters in South Africa led to repeated military incursions by troops of those countries. The establishment of a stable government was further hindered by the weak state of the economy, based as it was on village democracy within a one-party Marxist state. In 1984 Mozambique and South Africa signed a non-aggression pact, the Nkomati Accord*, but unofficial South African support, funded by Portuguese ex-colonists, continued, with some 10,000 well-armed troops operating within the country. In July 1989 at a Party Congress there was some relaxation of the previous Marxist-Leninist line. Presidents Moi of Kenya and Mugabe sought means of ending the fourteen-year war and in December 1989 President Joaquim Chissano met the rebel leader of the Mozambique National Resistance (MNR), Alfonso

Dhlakama, for talks. Formal peace negotiations in Rome reached a third round in November 1990, when a multi-party constitution came into effect.

Mubarak, Mohamed Hosni (1928–), Egyptian statesman and President of Egypt 1981– . Trained as an air pilot, he became air force Chief of Staff 1969–72 and Chief Commander 1972–5, when his forces took part in the Arab–Israeli war of 1973. As Vice-President to Sadat* (1975), he helped in the mediation of a dispute over the future of Spanish Sahara. Although bitterly opposed to Israeli action in Lebanon in 1982, he remained loyal to the Camp David Accord*. Essentially a pragmatist, his regime has been threatened by both Islamic fundamentalism* and economic stagnation. He has sought to improve relations with other Arab states, while also pressing for an international solution to the Palestine problem. Against much criticism he deployed Egyptian troops against Iraq in the Gulf Crisis* of 1990.

Mudros, armistice of. It was signed on 30 October 1918 on the island of Lemnos, ending the Ottoman Empire's part in World War I. The Dardanelles and Bosporus were to be opened to all and the Turkish army demobilized. The Allies used it as a pretext to occupy Istanbul and parts of Anatolia, thus precipitating the revolt of Mustafa Kemal*.

Mugabe, Robert Gabriel (1924–), Prime Minister of Zimbabwe 1980–7 and President 1987– . Born in Mashonaland and educated at mission school, he went on to university in South Africa and London and then worked as a teacher 1952–60, being in Ghana from 1956. In 1961 he helped to form the Zimbabwe African People's Union (ZAPU) with Joshua Nkomo*. Imprisoned in 1962, he escaped to Tanganyika and in 1963 helped to form ZANU (Zimbabwe African National Union), a more militant party. He was again imprisoned in 1964 for 'subversive speech', during which time he was elected leader of ZANU, much of whose support came from the Shona* people. He was freed in 1975 and, with Nkomo, led the guerrillas against Ian Smith's* regime, the two parties ZAPU and ZANU having combined to form the Patriotic Front*. When the war ended he won a landslide victory in elections held under British supervision in 1980, and became Prime Minister. Tension now developed with Nkomo and his ZAPU supporters, whose power-base was in Matabeleland. Nkomo left the government in 1982 and did not rejoin until 1987, during which years there was severe civil tension within the country. In 1988 Mugabe signed an 'Accord' of friendship with Nkomo. A strong supporter of non-alignment and also of the Commonwealth, Mugabe took an active part in pressing for international sanctions against the regime in South Africa, being openly opposed to British government policy.

Mujaheddin ('fighter in a holy war'). Various Islamic 'dedicated fighter' groups have emerged during the later 20th century, some inspired by Sunni fundamentalists of Wahabi* Saudi Arabia, others dedicated to preserving Shiite Islam, as in Iran. Afghan Islam had traditionally been moderate, drawing on that of India and later Pakistan; but more fundamentalist Mujaheddin sects were emerging in the 1970s, a Sunni Mujaheddin based in Peshawar and backed by Saudi Arabia, and an Iranian-inspired Shia group led by Mohammed Karim Khalili. During the decade of Soviet intervention in Afghanistan seven Mujaheddin groups formed a loose alliance, which received generous supplies of US arms via Pakistan. In January 1989 both Sunni and Shia groups held discussions with the Soviet Union, which withdrew from Afghanistan in February, whereupon some 450 Mujaheddin delegates in Pakistan elected Imam Seghbatullah Mujja as President of a government-in-exile; he was opposed by the extreme Sunni fundamentalist group Hizb-i-Islami, led by Gulbuddin

Hekmatyar. As they had been agreed about little else except the need to eject Soviet troops, it was not surprising that the various Afghan Mujaheddin sects soon began to disagree and to lose momentum in their civil war.

Muldoon, Sir Robert (1921–), Prime Minister of New Zealand 1975–84. Born in Auckland, he made his early career in accountancy, becoming president of the Institute of Cost Accountants in 1956. In 1960 he entered politics as a member of the National Party* and was elected to the House of Representatives. He became leader of the party in 1974 and won the general election of 1975, becoming both Prime Minister and Minister of Finance. His years in office were a difficult time for the New Zealand economy. Oil prices had risen steeply in 1973 and the traditional market in Britain for New Zealand farm and dairy produce was reduced by the Britain's entry into the EEC. He wished to provide economic incentives for industrial diversification, but he also imposed freezes on prices and wages, which led to conflict with the trade unions. He served on the Board of Governors of the World Bank 1979–80. In 1984 the National Party lost power to the Labour Party.

Mulroney, Martin Brian (1939–), Prime Minister of Canada 1984– . Son of Irish immigrants, he grew up and was educated in Quebec, where he became involved in student politics. Having built a successful law practice, he became president of the Iron Ore Company of Canada and then turned to national politics, challenging Joseph Clark* for leadership of the Progressive Conservative Party* in 1983. Elected for the first time to the House of Commons, he led the opposition to Pierre Trudeau*. In 1984 he successfully defeated the latter's successor John Turner in the general election, when he showed great skill in debate on television. His government suffered a series of patronage scandals and failed in a bid to reduce old-age pensions, but its significant achievement was the Meech Lake Accord (1987), whereby Quebec agreed to accept the 1982 Constitution if given 'distinct society' status. He was re-elected in 1988 on the issue of closer trade and industrial links with the USA.

Multinational corporations, corporations with at least 25 per cent of their output and activity outside their country of origin. Many have extremely diverse interests in commodity production, manufacture, and distribution, as for example the US corporation ITT. In the USA multi-state corporations and trusts had developed late in the 19th century and one of the earliest multinationals was the United Fruit Company*; but the British, French, Dutch, Germans, and Japanese all developed such corporations, particularly after World War II. Many consider their international influence, for example that of the oil companies, to be beyond the control of particular governments, and that it is an insidious form of imperialism*. In an attempt to regulate their activities the OECD* has endeavoured to draw up a code of conduct.

Munich Agreement (1938), international agreement on the future of Czechoslovakia. Hitler had long demanded 'protection' for the German-speaking Sudetenland* and shown readiness to risk war to attain his end. To avert conflict the British Prime Minister N. Chamberlain* met Hitler at Berchtesgaden on 15 September and again at Bad Godesberg on 23 September, by which time Hitler had extended his demands. He now wanted not only the immediate annexation by Germany of Sudetenland but also that Germans elsewhere in Czechoslovakia should be given the right to join the Third Reich. In a final effort Chamberlain appealed to Mussolini, who organized a conference at Munich where he, Chamberlain, and Hitler were joined by Daladier*, the French Premier. No Czech or Soviet representative was invited. On 29 September an Agreement was signed, Hitler gaining most of what he wanted. On 1

October German troops entered Sudetenland and President Beneš* resigned. Poland was to be allowed to occupy all of the disputed area of Teschen* in Moravia, while the First Vienna Award in November granted the Felvidék region of Ruthenia and Slovakia to Hungary. In March 1939 the rest of Slovakia became an 'independent' client state, while Moravia and Bohemia were occupied by Nazi troops. Germany emerged as the strongest power in Europe and Czechoslovakia had disappeared.

Munich 'beer-hall' putsch. On 8 November 1923 in a beer-hall in Munich a meeting of right-wing politicians, denouncing the Weimar Republic* and calling for the restoration of the Bavarian monarchy, was interrupted by a group of Nazi Party members led by Adolf Hitler*. In a fierce speech Hitler won support for a plan to 'march on Berlin' and there install the right-wing military leader General Ludendorff* as dictator. With a unit of Brownshirts* (SA) he kidnapped the leader of the Bavarian government and declared a revolution. Next day a march on the centre of Munich by some 3,000 Nazis was met by police gunfire, sixteen demonstrators and three policemen being killed and many wounded, including Herman Goering*, in the riot that followed. Many were arrested. Ludendorff was released. Hitler was sentenced to five years in prison, serving only nine months, during which he dictated the first volume of his autobiography and manifesto *Mein Kampf* to his fellow prisoner Rudolf Hess*.

Munich Olympic killings. During the 20th Olympiad at Munich on 5 September 1972, Palestinian terrorists, from the 'Black September' gang, attacked the Israeli team's quarters in the Games village. Two Israelis were killed and nine taken hostage. In efforts to rescue them they were all killed, together with five terrorists. The remaining terrorists were flown to Libya and their release later negotiated. The incident resulted in greater caution by authorities when involved in kidnapping negotiations.

Muñoz Marín, Luis (1898–1980), Governor of Puerto Rico 1948–64. His father was a publisher, and he was educated in and worked in the USA before returning to Puerto Rico in 1926 to edit a newspaper, *La Democracía*, owned by his father. Elected to the Puerto Rican Senate in 1932, he was at that time a strong advocate of independence, and remained so throughout the 1930s. In 1941, however, when Governor Rexford Tugwell was appointed, he rather changed his position, working closely with the Governor on a New Deal* programme for housing and economic development. When Puerto Rico gained the right to elect its own Governor he won an overwhelming victory. As Governor he sought to continue policies initiated by Tugwell, but he retired in 1964. He is remembered as 'the father of modern Puerto Rico'.

Musaddiq, Mohammed (1880–1967), Prime Minister of Iran 1951–3. A wealthy landowner, he joined the first government of Reza Shah*, but then withdrew from politics until 1942. Strongly anti-British, he genuinely believed that if the Anglo-Iranian Oil Company were nationalized all Iranian poverty could be eradicated. Increasingly violent speeches gave him a strong following. Early in 1951 the Iranian Parliament (the Majlis) nationalized the oil industry and, after rioting in Abadan, forced the Shah to appoint him Prime Minister. Granted dictatorial powers by Parliament, he failed to gain the expected oil revenues, foreign technicians having mostly withdrawn. A socialist, he vainly strove for land reforms; but being emotionally unstable, he lost the support of colleagues and was dismissed in 1953. Civil unrest followed, whereupon he was imprisoned and sentenced to solitary confinement.

Muslim Brotherhood. Created by Hasan al-Banna* as a movement to rejuvenate Islam, it sought to impose Islamic law (shariah) upon all social and political activity. Although it expanded rapidly throughout the Arab world, its political influence was largely confined to Egypt. During the 1940s its membership there grew to some one million, including a fanatical youth section. It was banned in 1948, but survived as an underground terrorist organization, being strongly anti-Western, but equally rejecting involvement with the Soviet Union. The Brotherhood was involved in recruiting Arab irregulars to fight in Palestine in 1948, and had some influence in Syria, where, under the name Islamic Liberation Party, it was suppressed by President Asad's* regime. In Egypt it allegedly attempted to assassinate Nasser, and six members were executed for treason. It remains an underground element in Egyptian politics. In the Sudan the Brotherhood, known as the National Islamic Front, has had steadily increasing influence since 1958.

Muslim League, an Indian political party. It was founded in 1905 to represent the separate interests of Indian Muslims, who felt threatened by the prospects of a Hindu majority in any future democratic system. The radical nationalist elements in the League forged a pact with the Indian National Congress* in 1916, on the basis of separate electorates and of reserved seats in provinces with Muslim minorities. A section of the League supported Congress in the various non-cooperation movements of the 1920s and 1930s. In the provincial elections of 1937 the League captured very few Muslim seats, but it succeeded in convincing the Muslim masses that elected Congress ministers were oppressing Muslims. In 1940 at Lahore it put forward the demand for autonomous Muslim homelands. After the war, during the transfer-of-power negotiations, its leader M. A. Jinnah* interpreted this as a demand for an independent Muslim state, Pakistan. He called for a Day of Direct Action in August 1946. Mass rioting followed, whereupon the British and the Congress agreed to partition*. A triumphant League dominated Pakistan politics in the early years of independence, but it gradually disintegrated. In the 1970 elections the Awami League* won in East Pakistan and the Pakistan People's Party in West Pakistan.

Mussolini, Benito (1883–1945), Prime Minister of Italy and Dictator 1922–43. Born in the Romagna, he trained as a teacher and taught for a year, and then in 1902 went to Switzerland, where he worked as a mason and became a socialist. He was expelled for revolutionary activities and returned to Italy. After military service he taught again, and in 1910 founded a weekly socialist newspaper, being imprisoned for extreme anti-clerical views. In 1911 he went to Milan as editor of the socialist paper *Avanti.* Here he was expelled from the Socialist Congress of Milan for supporting war against Austria. By now an influential editor, he founded a newspaper *Il popolo d'Italia*, and when Italy declared war in 1915 he enlisted and was wounded, being demobilized in 1917. He returned to journalism, now attacking both socialism and pacifism. He joined with various groups to form the Fasci d'Italiani di Combattimento in 1919. He linked up with the Blackshirts of D'Annunzio*, whom he much admired. His party began to attract industrialists and won thirty seats in the 1921 elections. During 1921–2 it organized bands of anti-Communists throughout urban Italy, strikes prevailing in most large cities. In October 1922 King Victor Emanuel III* invited him to form a government. Violence and intimidation helped him to win the election of 1924, but he took the opportunity of the Matteotti* Affair to silence all opposition. Italy became a single constituency in which 400 candidates nominated by the Fascist Grand Council were voted for. Trade unions were suppressed, industrial matters being settled by a National Council of Corporations. An extensive public-works programme was initiated, and Church/state relations improved with the Lateran Treaty* in 1929. His quest for a new Italian empire led to the annexation of Ethiopia* (1936) and Albania* (1939). He was to be less successful in trying to conquer Greece. Hitler, one of Mussolini's early admirers and imitators, became his ally and then a resented senior partner in the Axis*. Having entered World War II at the most favourable moment (1940), Mussolini was nevertheless unable to avoid a series of military defeats, in East and North Africa. He was deposed by hitherto acquiescent colleagues in July 1943, but was rescued by German paratroopers and established a puppet government in the small north Italian town of Salo. In 1945 he was captured by Italian partisans and shot. His body was brought to Milan and displayed amongst rejoicing.

Mutesa II, Sir Edward Frederick (1924–69), last *Kabaka* of Buganda and President of Uganda* 1963–6. He assumed the office of *Kabaka* in 1939 after a regency, aged 18. Progressive in spirit, he nevertheless backed the British government in suppressing Buganda nationalist risings in 1945 and 1949. In 1953, however, fearing the loss of Buganda's independence, he claimed the right of his kingdom to secede from the Ugandan protectorate. This was denied him by the Uganda High Court, and he was deported to Britain. In 1955 he returned as a constitutional monarch and in 1963 was elected first President of independent Uganda. He disagreed with the left-wing policies of Prime Minister Milton Obote*, who deposed him in 1966. He returned to Britain to live in exile until his death.

Muzorewa, Abel Tendekayi (1925–), bishop of the United Methodist Church and Prime Minister of Rhodesia 1979. Educated in Methodist schools in Southern Rhodesia, he went on to study in the USA, where he also taught for some years. He returned to Africa in 1963 as Director of Youth Work and in 1968 became a bishop in the United Methodist Church. He became involved in the Rhodesian political struggle, forming the moderate United Africa National Council and attending the Geneva Conference of 1976. In 1978 he joined the Transitional Executive Council to prepare for Rhodesian majority rule, and when his party won the election of 1979 he became Prime Minister. In the following elections of 1980, when White influence had largely disappeared, his party won only three seats. He was imprisoned 1983–4, accused of 'subversion', after which he travelled in the USA, returning to Zimbabwe in November 1986, having withdrawn from politics.

Myanmar *see* BURMA.

My Lai Massacre (16 March 1968), an incident in the Vietnam War* when a US unit commanded by Lieutenant William Calley killed 109 unarmed inhabitants of the village of My Lai who had allegedly been harbouring Vietcong guerrillas. After a four-month court martial Calley was sentenced to life imprisonment on 29 March 1971, but his superior officer Captain E. Medina was acquitted. Calley served a short portion of his sentence, but the affair was used by opponents of the Vietnam War in the USA to argue that the war was morally debasing as well as militarily disastrous.

n

NAACP *see* NATIONAL ASSOCIATION FOR THE ADVANCEMENT OF COLORED PEOPLE.

Nagy, Imre (1896–1958), Hungarian Prime Minister 1953–5, 1956. Born in south Hungary, he served in the Imperial Army during World War I and was captured on the Russian front. He became a Communist and took part there in the Revolution and Civil War. In 1921 he returned to Hungary, but had to escape in 1928, going first to Vienna and then to Moscow, where he studied agriculture. He returned to Hungary again in 1944, and was Minister for Agriculture 1945–6, responsible for major land reform. He was briefly Minister of the Interior, but Prime Minister Rákosi* removed him as ideologically unsound. In 1953 he himself replaced Rákosi as Prime Minister. His liberal regime relaxed Press censorship, reduced the rate of farm collectivization, and encouraged manufacturing industry. In February 1955, however, he was forced from office by his rival Rákosi, accused of 'Titoism'. He was expelled from the party, but retained his popularity. In October 1956 Rákosi retired to Moscow. Nagy was restored to favour and on 24 October became Prime Minister again with János Kádár* the new first secretary of the party. Their coalition government only accelerated the pace of the Hungarian Revolution*. Kádár left Budapest, to return with Soviet tanks. Nagy took refuge in the Yugoslav embassy but he was seized, imprisoned, and shot on 18 June 1958. He was rehabilitated in 1989.

Nahhas Pasha, Mustafa al- (1879–1965), Egyptian politician. One of the founders of the Wafd* Party, he was Prime Minister in 1928 and 1930 and again in 1936, when he negotiated the Anglo-Egyptian Treaty. Dismissed by Farouk* in 1937, he was reimposed by the British in 1942. He helped to form the Arab League* in 1944, when he was again dismissed by Farouk. In 1950 he formed a new Wafd government and proclaimed the Sudan to be part of Egypt. He initiated a strong anti-British campaign, but fell from office in 1952, accused of corruption. He was never tried.

Nakasone Yasuhiro (1918–), Prime Minister of Japan 1982–7. He was first elected to the Japanese House of Representatives in 1947, and as a leading member of the Liberal Democratic Party* he held intermittent cabinet posts from the 1960s. He succeeded Suzuki Zenko* as Prime Minister and President of the LDP in October 1982, and held power until his resignation in 1987, despite opposition to his strong style of leadership and the damage done to his party by the revelations of the Lockheed scandal*. His domestic policies were based on a package of administrative, fiscal, and educational reforms, while internationally he was committed to close ties with the USA and to greater involvement of Japan in world affairs.

Namibia. In 1878 Britain annexed Walvis Bay in South West Africa and incorporated it into Cape Colony; in 1884 the remainder of South West Africa became a

German protectorate. In 1910 Walvis Bay was placed under Cape Province, where it remains. In 1915 South West Africa was conquered by South African troops, and in 1920 South Africa was given the mandate* by the League of Nations. After World War II South Africa refused to place it under UN trusteeship, but in 1967 the UN established a Council for South West Africa and changed the name of the Territory to Namibia. In 1971 the International Court of Justice ruled that South Africa's presence was illegal. After negotiations, South Africa agreed to a multiracial Advisory Council, and complex negotiations continued throughout the next two decades, while SWAPO*, formed in 1959, conducted guerrilla campaigns. SWAPO operated from Angola, which South Africa invaded. A Geneva Protocol was finally signed on 5 August 1988. Elections were held in November 1989 and the country became formally independent in March 1990.

Nanjing Massacre, a mass atrocity in China. On 13 December 1937 Japanese troops entered Nanjing (Nanking). The massacre of an estimated 200,000 followed—men, women, and children being raped, tortured, and killed over the next two months, including some 7,000 women refugees. Over one-third of Nanjing's houses were burned and protests by refugee organizations failed to prevent the massacre.

Nansen, Fridjof (1861–1930), Norwegian explorer and politician. Born near Oslo, he led expeditions to cross Greenland (1888) and to reach 86° 13′ N. in 1896, becoming a national hero. In 1905 he entered politics to work for Norway's independence from Sweden, and he became an ardent supporter of the concept of a League of Nations*. He worked with the latter when it was established in 1920, being particularly concerned with relief work for refugees from Russia and Greece. It is estimated that he helped to feed and rehabilitate some 10 million refugees. In 1926 he became Lord Rector of St Andrews University, but he refused to stand for Norway's premiership.

Naoroji, Dadabhai (1825–1917), Indian politician. Born in Bombay, he had been a Professor of Mathematics before going to Britain, where he was the first Indian to be elected to the British House of Commons, representing Finsbury 1892–5. He had helped to found the Indian National Congress*, being president on three ocassions, and his book *Poverty and Un-British Rule in India* (1901) stimulated Indian nationalism. He remained an influential critic of British imperialism until his death.

NASA (National Aeronautics and Space Administration). Established in 1958 in the wake of the Soviet 'Sputnik' flight, it was a development of the National Advisory Committee for Aeronautics, created in 1915. NASA was generously funded throughout the 1960s in order to realize President Kennedy's* aim to put a man on the moon by 1969. In July of that year it achieved this through the Apollo Program, going on to a number of unmanned space flights and to placing 'spy' satellites, together with weather and communications satellites, into the earth's orbit. Its programme to develop a reusable space shuttle suffered a severe setback in January 1986 with the disaster of a *Challenger* shuttle explosion, investigations into which revealed both malpractices and inefficiencies, with slack quality control. The resumed shuttle programme authorized by President Reagan was to be devoted to scientific research, with private corporations developing commercial space-launchers. The US Department of Defense remained responsible for military research and development in outer space. A permanent space laboratory, already achieved by the Soviet Union, remained a NASA ambition for the 1990s.

Nash, Sir Walter (1882–1968), Prime Minister of New Zealand 1957–60. Born in Worcestershire, he worked as a clerk in Birmingham until emigrating to New Zealand in 1909, where he became a commercial traveller. A life-long Christian Socialist, he joined the Labour Party in 1916 and was a member of its executive 1919–60. He was first elected to Parliament in 1929 and became the leading spokesman for the moderate wing of the party. With Peter Fraser* he organized the party on a national basis and formulated policies which won the election of 1935, when he became Finance Minister. He played a major role in piloting through Parliament in 1938 a programme of child allowances and national health, establishing the most extensive system of social security in the world at that time. He became leader of the opposition in 1950 and led Labour to a narrow victory in 1957. His period in office was one of financial stringency, although he did introduce further social reforms. He was defeated in the election of 1960.

Nassau Agreement. In December 1962 Prime Minister Macmillan* and President Kennedy* met at Nassau in the Bahamas for defence discussions. Kennedy agreed to arm British Polaris submarines, operating under NATO command, with nuclear missiles. The agreement was resented by President de Gaulle*, who argued that Britain was not genuinely interested in European integration. Four weeks later he vetoed the first British application to join the EEC.

Nasser, Gamal Abd al- (1918–70), President of Egypt 1954–70. Educated at Cairo Military Academy, where he became an instructor, from 1942 onwards he was secretly encouraging mildly socialist and republican ideas among cadets and junior officers. In 1948 he formed the secret Free Officers' Movement, which gained momentum as a result of Egypt's defeat in Israel that year. In 1952, with eighty-nine Free Officers, including General Neguib*, he achieved an almost bloodless coup, forcing the abdication of King Farouk*. Nasser became Minister of the Interior in the new Republic, succeeding Neguib on 17 November 1954. He proclaimed Egypt a one-party state, suppressing Islamic extremists such as the Muslim Brotherhood* and securing some land reform. A treaty for the withdrawal of British troops from the Suez Canal area had already been concluded (October 1954), and this was to be completed in June 1956. His greatest project was the High Dam at Aswan*, and when he failed to get Anglo-American financial backing, he turned instead to the Soviet Union, with whom he had already made arms deals; at the same time he nationalized the Suez Canal. This precipitated the Suez Crisis*, out of which he emerged as the leading statesman of the Arab world. He negotiated the short- lived UAR* with Syria and established links with Yemen and Iraq. For a while he was regarded as the patron and protector of socialist independence movements throughout Africa, but the popularity of 'Nasserism' began to fade. At home more extreme left-wing groups criticized him and forced him to take an increasingly hostile attitude to Israel. When the latter retaliated by launching the Six-Day War* his air force was destroyed and his army left in disarray. On 9 June 1967 he resigned, but reconsidered the decision next day. For the next three years he was increasingly criticized by the PLO*, while seeking to re-establish links with the USA. In 1970 he tried to mediate in the civil war in Jordan, but died suddenly of a heart attack, one year before his greatest achievement, the Aswan Dam, was opened.

National Association for the Advancement of Colored People (NAACP). Created in 1900 by a group of young Blacks led by W. E. B. Du Bois and concerned Whites, its main activities before World War II were in fund-raising to finance civil rights litigation and to improve public awareness. After the war its activity increased, and its victory in *Brown* v. *Board of Education of Topeka** was crucial in the ensuing civil rights movement. From its headquarters in Baltimore it has continued to press for affirmative action* in favour of 'colored people'.

National government of China (1928–49). Following the split between the Guomindang and the Communists and the purge of the latter in 1927, the Northern Expedition* was resumed and the major warlords defeated. On 28 October 1928 a National government of the Republic of China was proclaimed in Nanjing, with Jiang Jiehi (Chiang Kai-shek) as President. A series of campaigns was at once launched against the Communists, who by 1934 were so squeezed as to have to embark on the Long March*. Meanwhile the Nanjing government did much to modernize the country, expanding public education, improving road and rail transport, and encouraging industry. These policies benefited the cities, but the mass of the peasantry remained illiterate, a prey to disease, and increasingly subject to conscription. Manchuria was one of the richest provinces, where Japan already had extensive concessions. The decision of Japan to take it over as Manchukuo* dealt a major blow to the National government, which in December 1936 also suffered the humiliation of the Xi'an Incident*. During the early months of the Sino-Japanese War, which developed in 1937, the government lost the cream of its army and air force and retreated to Chongqing (Chungking). Although the war dragged on there was a stalemate in the years 1939–43, while guerrilla units effectively harassed the Japanese. After the fall of Burma, bitter conflicts arose between Jiang and the US government, whose General Stilwell* demanded that the Chinese army be reformed and corruption eradicated. By 1944 civil war between the Nationalists and the Communists already threatened, although it was delayed until the final Japanese defeat when the National government and the CCP embarked on policies of confrontation. This led to the government's defeat and its retreat to Taiwan.

National governments, a term used to describe coalition government in Britain 1931–40. In August 1931 a financial crisis led to a split within the Labour government, nine ministers resigning rather than accept cuts in unemployment benefits. The Liberal leader Herbert Samuel suggested that the Prime Minister, MacDonald, create 'a government of national salvation', by inviting Conservatives and Liberals to replace them, and the first National government was formed on 24 August. An emergency budget was introduced, which increased taxes and proposed to reduce both benefits and public-sector salaries. When naval ratings staged the 'Invergordon Mutiny'* there was further financial panic, and sterling fell by 25 per cent. Britain abandoned the gold standard* and proposed to adopt a policy of protection. Samuel himself was to withdraw over this abolition of free trade in 1932, while the Labour Party split, regarding those who supported MacDonald as traitors to socialism. In October MacDonald won a general election and formed a second National government. He went on to achieve the Ottawa Agreement* over Empire trade, but resigned as Prime Minister in 1935, to be succeeded by Baldwin*, who won a further general election (the last until 1945). The latter formed a National government (1935–7), but it was effectively a Conservative government, even though MacDonald himself stayed a member. That of Neville Chamberlain (1937–40) was even more so.

National guards, a term used in many American countries to describe a reserve militia system for national defence. The US National Guard goes back to colonial days, when all able-bodied men were expected to take part in anti-Indian defence. In 1792 a Militia Act empowered each state to enrol such numbers of free White able-bodied citizens between the ages of 18 and 45 as to maintain law and order. After the Civil War, state-controlled militia or National Guards were regularly used to break strikes, and in 1914 a National Guard Act declared that they would form part of the US army in time of war or national emergency. The 1933 National Guard Status Act established the National Guard as a Federal Reserve Force. Formations were required to conduct drill and training classes at least one night a week, for which they would receive one day's regular army pay. In addition they had to attend two weeks' summer camp with pay. The Guard played a crucial role in the early days of World War II and the Korean War. In 1965 Robert McNamara* achieved some reorganization of the National Guard, since when it has been called upon to maintain public order during race riots and protests, and also at times of such emergencies as floods, earthquakes, and so on.

National Health Service (UK). The National Health Service (NHS) Act of 1946 established a comprehensive health service in England and Wales with effect from July 1948. Separate acts were passed for Scotland and Northern Ireland. It was to be funded partly from National Insurance but mainly from taxation. The Bill had been fiercely opposed by many family doctors and hospital consultants, but the skill of Aneurin Bevan*, Minister of Health, had enabled it to become law, the great majority of hospitals of that time, as well as general practitioners, coming within the scheme, to be administered by regional health authorities. During the 1960s a massive hospital building scheme was instituted, but rising costs steadily beset the service, partly a result of its own success in helping to prolong expectation of life. Bevan's aim was for the Service to be totally free, and he resigned in April 1951 when Hugh Gaitskell* instituted prescription charges of 1 shilling (5p, to rise to £2.80 in 1990) and charges on dentures and spectacles. In 1989–90 a massive, highly unpopular reorganization was imposed by Mrs Thatcher's government in an effort to grapple with ever-rising costs. New Zealand, Australia, Canada, as well as most European countries have all developed a health service, some more comprehensive than others.

National Insurance (or social insurance), state insurance financed by compulsory contributions from employee and employer. Pioneered in Germany by Bismarck, schemes were introduced in other European countries including Britain, and also in New Zealand, before World War I, for state assistance in sickness, old age, accident, and unemployment (for certain categories of employee). In Britain, as a result of the Beveridge Report* (1942), National Insurance was extended to all adults in employment after World War II. In the USA, as part of the New Deal*, a federal insurance scheme was introduced by the Social Security Act of 1935, but this did not include medical care. The majority of US citizens, however, rely more on private insurance schemes for health, accident, pension, and other provision. The Universal Declaration of Human Rights* lists access to a social security scheme as a basic human right, but many less-developed nations are unable to afford this.

Nationalism, the sentiment of attachment to a nation in terms of territory, language, customs, and culture. The Napoleonic attempt early in the 19th century to impose a common European state collapsed, and European nationalism was an increasing factor before 1914, some regarding it as the main cause of World War I. It continued to play a part in European politics between the wars, although German and Soviet imperialist ambitions were more significant. After World War II the movement towards a united Europe somewhat weakened nationalist ardour in Western Europe, although it remained a strong latent factor, surfacing in such events as the Falklands War*. Before, during, and after World War II it was the emergence of strong nationalist feelings in Asia which powerfully contributed to the process of decolonization*, resulting in such states as Pakistan and India, as well as in the successful establishment of Communist regimes in countries such as China and Vietnam. Since 1985 nationalism has once again become an increasingly potent force in Central and Eastern Europe, as well as in the republics of the Soviet Union.

National Liberation Front (NLF), a Vietnamese resistance movement. Formed 'somewhere in the south' in the Mekong delta area of South Vietnam, its first aim was the overthrow of Ngo Dinh Diem*, and beyond this the unification of all Vietnam. It quickly absorbed members of the VietMinh* working in the south and developed a complex power structure linked with its allies in Hanoi. In 1962 a first Congress was held and a People's Revolutionary Party, offspring of the Vietnam Communist Party, Lao Dong (Workers' Party), was founded. Its members rapidly infiltrated through all society; villages, towns, and into Saigon itself. A Foreign Relations Committee established links with all Communist countries and several neutral ones, sending representatives to the UN and in 1968 to the Paris peace talks. Increasingly sophisticated propaganda machinery—film, radio, pamphlet, poster—was created. Meanwhile the Liberation Army, or Vietcong, steadily recruited volunteers. In the early days its main weapons were those of the terrorist—ambush and assassination. By 1963, when Diem was murdered, the NLF, from its secret headquarters somewhere in Pathet Lao* country in Laos, had gained control over the majority of the 2,500 villages of the south. Increasingly the Vietcong became engaged in more formal confrontations with the South Vietnamese army; with the Americanization of the war from 1964, it became militarily more closely allied with General Giap's* forces in the north, from whom came its supplies along the Ho Chi Minh route.

National Party (New Zealand). In 1936 the United Party (restyled Liberal Party*) and the Reform Party* of New Zealand amalgamated in opposition to Labour and came to power after World War II as the National Party. It has formed governments 1949–57, 1960–72, 1975–84, 1990– . Ideologically it has been right of centre, being particularly sensitive to the rural vote, but basically pragmatic in its social and economic policies.

National Party (South Africa). It was originally founded by General J. B. M. Hertzog* in 1914, after his secession from Louis Botha's* South Africa Party. In 1924 it became the Nationalist–Labour Alliance under Hertzog. In 1934 the latter joined Smuts* in a coalition, the United Party. In opposition to this D. F. Malan* founded his Purified National Party, which in 1939 was reunited with Hertzog, emerging as the Afrikaner-dominated National Party, which swept to power in 1948, pledged to introduce apartheid*. It has held uninterrupted power since then, although attempts by Presidents P. K. Botha* and de Klerk* to meet the twin threats, of domestic unrest and international criticism, by the easing of apartheid led to defections to extreme right-wing groups.

National People's Congress (NCP), constitutionally the highest state organ of power in China. The first Congress was elected in 1954 in accordance with the Constitution of that year. The Congress takes its procedures from the French Assemblée Nationale. The seventh Congress was elected in 1988.

National Revolutionary Army (NRA), a Chinese army. With civil war raging in northern China between warlords*, Sun Yixian* organized an army to pacify the region. He died in Beijing in March 1925, and in August Jiang Jiehi* (Chiang Kai-shek) was appointed Commander-in-Chief of this National Revolutionary Army of some 85,000 troops. Following a split between the Guomindang and the Communist Party he used it first to attempt to suppress the latter before going on successfully to complete the Northern Expedition* in 1928, when the Nationalist government* of the Republic of China was proclaimed in Nanjing. Its army retained the title National Revolutionary Army, with Jiang Jiehi its generalissimo.

National Trust. The National Trust for England and Wales was founded in 1895 by Octavia Hill, Sir Robert Hunter, and Canon Rawnsley to preserve for the people countryside and properties of outstanding beauty or historical interest. It was given statutory approval by the National Trust Act of 1907. Its role was strengthened by the later Acts of 1919, 1937, and 1939. Under the 'Country House Scheme' owners were enabled to continue to live in houses given to the Trust, but these can only be accepted if they are endowed sufficiently to make them self-supporting. The Trust has acquired many miles of coastline in Dyfed and Cornwall. It is administered by a central council and over 100 local committees. There is a separate Trust for Scotland. During the 20th century many other countries established similar bodies to protect their heritage, for example the Republic of Ireland, Australia, and the USA.

Native American Church (peyotism), a religious movement among North American Indians in both Canada and the USA. The peyote or peyotl is part of a cactus which produces hallucinogenic effects and was used in Aztec days in Mexico. Having spread steadily north from Mexico, peyote rites vary, but they include all-Saturday-night ceremonies of prayer, contemplation, and sacramental eating of peyote, followed by a Sunday-morning communion breakfast. Jesus is regarded as an intercessor with God, the 'Great Spirit', and Bible readings form part of the ceremonies. The religion enjoins family care, brotherly love, and avoidance of alcohol. Persecuted through the 19th century, the Church was given official acceptance in Oklahoma in 1918 and by 1960 in eleven other US states.

NATO *see* NORTH ATLANTIC TREATY ORGANIZATION.

Nauru, a Pacific island state lying west of Kiribati, most of whose inhabitants are Polynesians. In 1914 it was part of the German Marshall Islands protectorate. A British mining company had begun to extract one of the world's richest deposits of phosphate of lime in 1906, and in 1920 the island was mandated by the League of Nations to Britain, Australia, and New Zealand. It was occupied by Japan 1942–5, after which it became a UN Trust Territory, again jointly under Britain, New Zealand, and Australia, the latter providing the administration. In 1968 it became an independent state and member of the Commonwealth.

Naxalite movement, part of the activities of the Communist parties* of India. It was named after the village of Naxalabi in the Himalayan foothills of West Bengal, where it first began in 1967. The theoretician and founder of the movement, Charu Majumdar, a veteran Communist, broke away from the Communist Party of India (Marxist) and established the Communist Party of India (Marxist-Leninist). This CPI (M-L) first organized armed risings of landless agricultural labourers, especially in eastern India. Later it developed (1968) into an urban guerrilla movement, especially in Calcutta. Its programme of strikes, riots, and murders reached a peak in 1971, being suppressed by 1977 with considerable violence. The CPI (M-L) split into several factions, some of which adopted a policy of participating in constitutional politics. In October 1990 forty-seven passengers were killed in a train ambushed by a Naxalite group in Andhra Pradesh.

Nazi Party (Nationalsozialistische Deutsche Arbeiterpartei or National Socialist German Workers' Party). It was founded in 1919 as the German Workers' Party by a Munich locksmith, Anton Drexler. It adopted its new name in 1920 and was taken over by Hitler* in 1921. In so far as it had a coherent programme this consisted essentially of opposition to democracy in general and to the Weimar Republic* in particular. It promulgated so-called theories of 'the purity of the Aryan race' and hence of anti-Semitism, allied to the Prussian military tradition and to an extreme sense of nationalism. It was able to fan the humiliation felt by many Germans over the Treaty of Versailles*. It drew its ideology from the racist theories of the 19th-century French writer the Comte de Gobineau (1816–62), from the national fervour of Heinrich von Treitschke (1834–96), from the superman theories of Friedrich Nietzsche (1844–1900), as well as from the 19th-century romanticization of the German *Volk*. It shared many of the features of Fascism*, and was given dogmatic expression in Hitler's *Mein Kampf* (1925). The success of the party in the early 1930s is explained by the widespread desperation of Germans over the failure of the Weimar Republic governments to solve economic problems during the Great Depression*, and by a growing fear of Bolshevik power and influence. Through Hitler's oratory the party offered Germany new hope. In the period leading up to war aspects of the ideology found adherents in many countries, while Nazi systems were imposed on occupied Europe 1938–45. The Nazi Party was disbanded in 1945 and its revival officially forbidden by the Constitution of the German Federal Republic.

Nazi–Soviet Pact (23 August 1939), a military agreement negotiated by Molotov* and von Ribbentrop* and signed in Moscow. Germany and the Soviet Union renounced war between their two countries and pledged neutrality by either party if the other were attacked by a third party. Each promised not to join any grouping of powers which was 'directly or indirectly aimed at the other party'. The Pact contained secret protocols whereby Poland was to be divided between the two signatories, and the Soviet Union was to have a free hand to deal with Finland, the Baltic states, and Bessarabia (see MOLDAVIA).

Ndebele, an ethnic group of the Ngoni* people. Following Zulu* persecution in the early 19th century, they had trekked into south-western Zimbabwe under their leader Mzilikazi during the *Mfecane* (Time of Troubles); here they forcefully displaced Shona* inhabitants to create Matabeleland. In 1893 their King Lobengula was defeated by the British and they reluctantly accepted British control. They form about 13 per cent of the Zimbabwean population and Joshua Nkomo* has been their advocate through much of the 20th century.

Neguib, Mohammed (1901–84), President of Egypt 1952–4. Born in Khartoum, he joined the Egyptian army, rising to the rank of general and distinguishing himself in the fighting against Israel in 1948. In July 1952 he overthrew King Farouk with the help of Colonel Nasser's* dissident Free Officers' Movement. Egypt was proclaimed a republic in June 1953, with Neguib its first President. He was largely responsible for the Anglo-Egyptian agreement on self-determination in the Sudan and was anxious to allow a multi-party democracy to develop in Egypt. Younger officers of the army, however, led by Nasser, felt him too conservative. In November 1954 he was obliged to resign, being placed under house arrest until 1960, after which he retired into private life.

Nehru, Jawaharlal (named 'Pandit', 'teacher') (1889–1964), first Prime Minister of the Union of India 1947–64. Son of Motilal Nehru*, he was educated in

England at Harrow School and Cambridge. He returned to India in 1912 to practise law. He became actively involved in politics in 1919 following the Amritsar Massacre*. Attaching himself to Mohandas Gandhi*, he conducted numerous campaigns of civil disobedience, being first imprisoned in 1921 and altogether spending nine years in British prisons. A member of the Indian National Congress*, whose president he was in 1929, his conviction that the future of India lay in its becoming an industrialized society brought him into some conflict with Gandhi's ideal of a society centred on self-sufficient villages. On his release from prison in 1945 he took an active part in the negotiations leading to partition*, becoming Prime Minister and Minister of Foreign Affairs in 1947. He was immediately faced with the first Indo-Pakistan War* over Kashmir, which resulted in a massive influx of Hindu refugees from Pakistan. His government also faced the challenges of the integration of the princely states (sometimes by coercion) and of the first Communist government in Kerala (1956–9). It was also committed to a series of five-year economic plans to underpin the new state. In 1961 India annexed the Portuguese colony of Goa*. In foreign affairs he adopted a policy of non-alignment, being a strong supporter of the UN and involved in mediation over Korea (1951) and Indo-China (1954); but in 1962 he sought Western aid with the outbreak of the Indo-Chinese War*.

Nehru, Motilal (1861–1931), Indian political leader. Together with C. R. Das (1870–1925), he organized the Swaraj* Party in 1922. This aimed to participate in Indian Legislative Councils, but to oppose the British by obstructive tactics from within, as an alternative to Gandhi's* policies of non-co-operation. In 1928 he chaired an All Parties' Committee, which produced the Nehru Report. This set out a proposed Indian constitution, with dominion status for India.

Nemieri, Jaafar al- (1930–), President of Sudan 1971–85. He graduated at Khartoum Military College in 1952, being a keen Muslim. He was always an admirer of Gamal Nasser* in Egypt. A spokesman for younger officers, he led the military coup which overthrew the government of General Abboud* in 1969. He became head of a Revolutionary Council and in 1971 was elected President by a plebiscite, whereupon he proclaimed the Sudan Socialist Union a one-party state. In 1972 he granted limited autonomy to the south, accepting the right of schools in the southern regions to teach Christianity, and in English not Arabic. He supported President Sadat's peace policies and as president of the OAU* in 1978 sought to keep Africa non-aligned to either superpower. His early policies were making Sudan a major food producer, but his imposition of Islamic law on the south in 1983 precipitated a renewal of civil war and mass famine. By April 1985 a deteriorating economy, and an increasingly corrupt and authoritarian regime, led to his overthrow, while he was abroad in Egypt. He continued to live there in exile.

Nepal, a Himalayan kingdom. It had been conquered by the Gurkhas in the 18th century, and in the 19th century effective rule passed from the royal family to a family of hereditary prime ministers, the Ranas, who co-operated closely with the British, Gurkhas being recruited into both the British and Indian armies. Growing internal dissatisfaction led in 1950 to a coup, when the Rana family was ousted and royal powers reaffirmed under the King, Tribhuvan (1951–5). His successor King Mahendra (1955–72) experimented with a more democratic form of government, but tension arose in 1960, Crown v. cabinet, and in 1962 a new constitution abolished political parties and restored monarchic rule, together with a system of panchayat—elected village and district councils. This continued under his son King

Birendra Bir Bikram (1972–), but, following pro-democracy demonstrations and mass arrests from 1989 onwards, the King granted a new constitution in November 1990, establishing a bicameral Parliament.

Netherlands, The. This constitutional monarchy remained staunchly neutral during World War I, giving refuge to Kaiser William II* in 1918. In spite of industrial and agricultural expansion, it suffered economic difficulties during the Great Depression*, but had introduced universal suffrage and much social legislation by 1939. In May 1940 it was occupied by German troops, welcomed by a large fifth column* of National Socialists. Queen Wilhelmina* and her government moved to London. Many Jews were deported to concentration camps, but German attempts at Nazification failed, largely as a result of the determined opposition of the churches. A courageous Resistance movement developed and in September 1944 Allied forces began to arrive. Until World War II The Netherlands was the third largest colonial power, controlling the Dutch East Indies, various West Indian islands, and Guiana in South America. The Japanese invaded the East Indian islands in 1942 and installed Sukarno* in a puppet government for all Indonesia. In 1945 he declared independence, and four bitter years of war followed. Dutch Guiana received self-government as Surinam* in 1954 and independence in 1975, but Curaçao and other Antilles islands remained linked to The Netherlands. The last major possession, West Irian, became Indonesian in 1963. The Netherlands joined the Benelux* union in 1947, and then NATO, the ECSC, and the European Community. Following the long reign of Wilhelmina (1890–1948), her daughter Juliana became Queen. She retired in 1980 and her daughter succeeded as Queen Beatrix.

Neuilly, Treaty of (27 November 1919), the peace treaty for Bulgaria, negotiated at the Paris Peace Conference* and signed at the Château of Neuilly in Paris. Bulgaria had to cede the rich wheat-growing area of southern Dobrudja to Romania and western Thrace to Greece, thus losing direct access to the Mediterranean. Small areas were also to go to Yugoslavia, and reparations were to be negotiated. Its army was to be limited to 20,000 men.

Neutrality Acts, US legislation. During 1934–6 a US Senate Committee to investigate the munitions industry, chaired by Gerald P. Nye, revealed high profits, but no evidence that President Wilson had had a financial stake in urging war with Germany in 1917, as was being alleged. The Committee's findings helped to stimulate isolationist feelings, which wanted the USA to remain aloof from global commitments. A series of Acts was passed through Congress 1935–9 which prohibited loans or credits to belligerents and placed embargoes on direct or indirect shipments of arms or munitions. Congress declared it would take no stand on issues of international morality by distinguishing between aggressor and victim nations. The Acts of 1935 and 1936 both affected US policy on the Italian war in Ethiopia. The two Acts in 1937 limited the US response to wars in Spain and China. The Act of 1939 repealed arms embargoes and authorized 'cash and carry' exports to any belligerent power, but it still forbade US ships to carry 'belligerent cargo'. During 1940 President Roosevelt fought for repeal of the Acts on the ground that they encouraged Axis* aggression and endangered US security. They were gradually relaxed before Pearl Harbor* made them irrelevant.

New Caledonia (Nouvelle Calédonie), a group of Melanesian islands, the largest of which is Noumea, lying some 750 miles east of Queensland, Australia. They had been annexed by France in 1853, and were a penal colony 1883–98. The Kanak inhabitants from here and from Vanuatu were imported as virtual slave labour into

Australia in the 19th century, but forcibly repatriated from 1906 as part of the White Australia policy. By 1914 mining had become a major activity, especially of nickel and chrome. Noumea was occupied by the USA 1942–5, becoming the headquarters of the US Armed Forces in the South Pacific under Admiral Halsey*. French administration was resumed in 1946, New Caledonia becoming a French Overseas Territory. In 1949 the South Pacific Commission* was established on Noumea, using the US headquarters buildings. There have been increasing demands for independence, with the formation by the Kanaks of the Kanak Socialist National Liberation Front (FLNKS), opposed by French settlers and 8,000 French troops.

New Deal, the term used to describe President F. D. Roosevelt's* social and economic reforms 1933–9. There are at least two claimants for coining the term, Raymond Moley* and Samuel Rosenman, an FDR adviser 1929–32. It was first used in the 1932 election. The First New Deal Program (1933–5) aimed to restore public confidence and to relieve the plight of some 14 million unemployed. Immediate measures of the 'First Hundred Days' included an Emergency Banking Act (March 1933), an Economy Act (March 1933), and the establishment of a Federal Emergency Relief Administration (March 1933), to be followed in June by the creation of a National Recovery Administration (NRA), which was to concern itself with such issues as child labour, working hours and practices, and collective bargaining. In May 1933 an Agricultural Adjustment Administration (AAA) was established, aiming to limit production of staple crops and to stabilize prices by a policy of federal subsidies. The legislation for this was invalidated by the Supreme Court*, and a new Act was passed in 1938. These measures together ended the immediate crisis. At the same time there was a programme of public-works legislation: (1) The Tennessee Valley Authority was created in May 1933. An independent corporation backed by federal funds, it built dams and hydroelectric installations in seven states. It also took over a project begun in 1916, the Muscle Shoals project in Alabama extracting nitrate, and, in addition to providing cheap electricity, engaged in reforestation to check soil erosion throughout the Tennessee River Basin. (2) A Civilian Conservation Corps (CCC) was established in November 1933. Between 1933 and 1941 it found work for some 2 million in reforestation and other projects. (3) A Civil Works Administration, established in February 1934, employed some 2½ million on a variety of public-works projects.

To try to prevent a repetition of the 1929 financial collapse, a Federal Deposit Insurance Corporation was created to strengthen the banking system, a Securities and Exchange Commission was formed (June 1934) to protect investors against fraud, while a Banking Act of June 1933 had reorganized the Federal Reserve System*. At the same time, to protect 'the forgotten homeowner', legislation had been passed in June 1933 to refinance mortgages which had suffered from the financial crisis. Having begun the process of recovery, Roosevelt's administration moved in the Second New Deal Program (1935–8) to consolidate and to provide greater social protection. (1) In order to improve the efficiency of emergency agencies already created, a co-ordination body, the Works Progress Administration (WPA), was established in 1935 by the Emergency Relief Appropriation Act, whose initiators were Harold Lekes and Harry Hopkins*. Not only did the WPA engage in such public works as the building of roads, bridges, airfields, schools, hospitals, playing fields, and so on, it also ran programmes on adult literacy, the provision of community theatres, and the creation of works of art on public buildings, and it sponsored a National Youth Administration. Some 8½ million people passed through WPA schemes at a cost of some $11,000m. It changed its name to Works Projects Administration in 1939 and was wound up in 1943. (2) The National

Labor Relations Act (Wagner–Connery Act) of 1935, introduced by Robert Wagner, established a National Labor Relations Board with powers to supervise and conduct elections of union representatives, thus hoping to eliminate employer domination within labour unions. (3) A massive Social Security Act, partly responding to Francis Townsend's* pressure group, in August 1935 established old-age and widows' pensions, unemployment benefits, and disability insurance. (4) A Fair Labor Standards Act, introduced by Frances Perkins*, established a minimum wage of 40 cents and limited hours to forty per week. Although Congress had given initial support, by 1937 critics were saying that the New Deal was establishing 'creeping socialism'; while in the Supreme Court it was argued that the Constitution gave the federal government no powers to regulate industry or to legislate on social reform, these being individual states' responsibilities. Roosevelt's plans to 'pack' the Court were defeated, and the pace of legislation had much reduced by the time the President sought a third term in 1940.

New Delhi, capital of the Republic of India. In 1911 Britain decided to move the capital of British India from Calcutta to Delhi, building a new administrative city alongside ancient Delhi. The architects Edwin Lutyens and Herbert Baker designed and built what are now Parliament House and the President's palace, other public buildings, office blocks, and a residential area for Europeans, in a spacious planned environment, blending in with such monuments as the Tomb of Emperor Humayun. The city is in stark contrast to the massed dwellings and buildings of Old Delhi.

New Democratic Party of Canada, formed in 1961 as a democratic socialist party, when the Co-operative Commonwealth Federation* and the Canadian Labour Congress merged. It is committed to universal medicare*, public ownership of essential industries, adequate old-age pensions, and workers' compensation. It has formed administrations in Saskatchewan (1961–4, 1971–82); Manitoba (1969–77, 1981–8); British Columbia (1972–5). In 1987 it led federal opinion polls, but failed to win a majority in the 1988 election. It advocates a nuclear-free Canada and is opposed to NATO rearmament.

New Economic Policy (NEP). By 1921 famine faced Russia, weakened by years of Revolution and Civil War. Lenin's* new policy represented a shift from his earlier 'war communism'. This had been adopted during the Civil War to supply the Red Army and the cities, but had alienated the kulak* peasants. The NEP permitted private enterprise in agriculture, trade, and industry; encouraged foreign investment; and virtually recognized the previously abolished rights of private property. It met with some success, which Lenin did not however live to see, but in 1928 was ended by Stalin's five-year plans and agricultural collectivization*.

Newfoundland, an island province of Canada. In 1914, having rejected proposals to join the Confederation of Canada, Newfoundland was still a British colony, although its Constitution provided internal self-government. In 1932, at the time of the Great Depression*, its economy collapsed; the Constitution was suspended and the island ruled directly by the British Colonial Office. After World War II a campaign developed, led by J. R. Smallwood, to join the Confederation, which it did on 31 March 1949, with Smallwood Liberal Premier 1951–71. The economy has revived since World War II, with the development of mining, forestry, fishing, and hydroelectric installations.

New Guinea, a large island separated from Australia by the Torres Straits. In 1914 the western half formed part of the Dutch East Indies*, while the eastern half was divided between Britain and Germany. After 1920 both these areas were mandated to

Australia by the League of Nations. The latter introduced an element of self-government, but only for European and Australian officials. Heavy fighting took place on the island from January 1942, when the Japanese landed, until September 1945, when their last jungle units surrendered. Since then Papua New Guinea* has emerged, while the western area joined Indonesia in 1963 as West Irian, now Irian Jaya.

Ne Win (1911–), President of Burma 1962–81. A member of the nationalist movement Dobama Asi-ayone from 1936, he served as Chief of Staff in Aung San's* Burma National Army from 1943, defecting with it to the Allied side in 1945. As Commander-in-Chief of the Burmese army after independence in 1947, he led the campaign against insurgent hill tribes and the Karen guerrillas of the Irrawaddy delta. He served briefly as Prime Minister in a caretaker government 1958–60. In 1962 he led a coup against the administration of U Nu*, abolished the parliamentary system, and proclaimed the Socialist Republic of the Union of Burma. He went on to expel 300,000 foreigners, in an attempt to gain control over the Burmese economy, a move which was to have disastrous effects. Constantly harassed by Karen guerrilla disturbances, he succeeded in maintaining his power as a military dictator, establishing working relations with both the USA and China. Although he stepped down as President in 1981, his influence remained considerable.

New People's Army, a Filipino Communist resistance movement. The violent suppression of the Hukbalahap* movement in the 1950s did not end Communist agitation in the Philippines. President Marcos* used its resurgence to justify martial law in 1972. By December 1985 he was claiming that '10,000 innocent civilians' had been killed by Communists, now reorganized as the New People's Army. When Corazón Aquino came to power, she negotiated a sixty-day truce in November 1986. Unfortunately, however, negotiations broke down and guerrilla activity resumed in 1987, especially in Mindanao in the south, where their strength was estimated at 25,000. Security forces against the outlawed organization were active through 1988, in April capturing the Commander-in-Chief Romulo Kintanar, but losing him again in an escape in November. In September 1990 a new cease-fire broke down, and in November one of the Army's intelligence officers was captured.

New Zealand. Self-government of the British colony of New Zealand had been granted in 1856, with the Crown retaining responsibility for defence and foreign affairs through a Governor. Following a series of Anglo-Maori wars, British farming communities, mostly in the North Island, had profited from the invention (1869) of refrigeration of meat. Immigration into the South Island had been stimulated by a gold rush in the 1860s, but Asian immigrants had been excluded since 1881. During the years 1891–1911 Liberal administrations had introduced extensive social legislation, including the enfranchisement of women, 1893. In 1907 the title of dominion was adopted, with a bicameral Parliament: a House of Representatives and a Legislative Council, the latter being abolished in 1950. The Statute of Westminster* was not to be fully adopted until 1947, and in 1949 the British Parliament renounced its control over the New Zealand Constitution.

In 1914 the conservative government of William Massey* was in office. Massey was a strong imperialist and some 10,000 New Zealand volunteers served in the Anzac* Middle East army. In 1916 conscription was adopted. Following World War I New Zealand was given mandatory powers over German West Samoa*. The conservative Reform Party* remained in office through the 1920s under Massey and Coates*, but the farming industry was not as prosperous as before the war. The world's Great

Depression* severely affected the economy, although the Ottawa Conference* (1932) decision to adopt imperial preference helped to restore it. The Labour Party*, organized by Nash* and Fraser*, held power during the years 1935–49, instituting a programme of public works after the war. It has since alternated in office with the National Party*. New Zealand sent troops to the Korean and Vietnam Wars. In 1962 the government of Keith Holyoake* created a new office, that of Parliamentary Commissioner or Ombudsman. That government supported the USA in Vietnam, but in 1972 the Labour government withdrew from SEATO* and brought its troops home. Concern over nuclear pollution has been a growing issue in New Zealand. Prime Minister Holland protested to Britain that he had not been warned of the first hydrogen bomb tested from Christmas Island (Kiritimati) in 1957. This issue culminated in the decision of David Lange's government to adopt a totally non-nuclear policy, excluding nuclear-powered shipping from its ports and withdrawing from ANZUS*. Concern over continued French testing of nuclear weapons was the cause of the *Rainbow Warrior** incident. Since the British accession to the EEC, New Zealand has strengthened its trade links with Australia and its Asian neighbours, after attempts to renegotiate protocol 18 of Britain's Treaty of Accession on dairy and sheep-meat products had largely failed. When the National Party formed a government in 1990 it confirmed the non-nuclear stance of its Labour predecessor. New Zealand retains as overseas territories the Kermadec Islands, Tokelau, and the Ross Dependency. As a multiracial society, New Zealand has, during the later 20th century, achieved increasingly successful integration between the two races, Maori* and European (pakeha).

Ngata, Sir Apirana Turupa (1874–1950), Maori politician and cultural leader. After leaving Te Aute College he became the first Maori to obtain a university degree, qualifying as a lawyer in 1897. He had a great influence on the Kotahitanga movement* and was one of the founders of the Young Maori Party*, entering Parliament in 1905 for the Eastern Maori electorate. He was Minister for Native Affairs 1928–34 and a close ally of Te Puea*. In 1931 he began to inaugurate his Maori Land Development Scheme*, which greatly expanded Maori land under cultivation. He used a Maori Purposes Scheme to finance school construction and was chairman 1928–34 of the Maori Board of Ethnological Research. In 1934 he was forced to resign, being criticized for unconventional methods by the rising Ratana movement*. He remained a close friend of Peter Fraser*, especially during the years of World War II.

Ngoni. An African people consisting of various ethnic groups who for long inhabited southern Africa, from the Great Fish River north into Mozambique. There were two major catastrophes in the 19th century. In the north the establishment of a Zulu* empire in the 1820s resulted in the so-called *Mfecane* (Time of Troubles) and exodus of Ngoni people into south-west Zimbabwe (see NDEBELE), Swaziland, Tanzania, Zambia, and Malawi. In the south the Xhosa or 'Kaffir' wars by British colonists established European farms throughout Cape Colony, depriving the Xhosa* Ngoni of their lands, a process of deracination accelerated after 1886 with the discovery of gold in the Witwatersrand. In the later 20th century the peoples of the so-called Bantustan homelands* are mostly Ngoni, and efforts have been made to keep alive cultural traditions. Ngoni people share similar Bantu languages, for example Xhosa, Zulu, Swazi.

Ngugi, James Thiong'o (1938–   ), Kenyan playwright and novelist. Born in Kenya, he attended a Kikuyu school and Makerere and Leeds universities. He taught in various East African schools and began to publish novels and plays, writing with nostalgia of the pre-European Kenya Highlands, but also becoming a trenchant critic of corruption in Kenyan politics. He taught at the Northwestern University, USA, from 1970, but in 1977, back in Kenya, he was arrested and held in detention without trial for over a year following the performance of his play *Ngaahika ndeenda* ('I will marry when I want'). He lived in exile from Kenya after 1982, publishing novels such as *Weep not, Child* and *Petals of Blood*. In 1989 he wrote *Matigari*, a novel in his native language of Gikuyu. In this he has gone beyond the call for freedom to the struggles of African peoples to achieve social change and social justice for 'the dispossessed majority'.

Nicaragua. The largest Central American country, it gained independence from Spain in 1821. The 20th century opened with the country under the vigorous control of the dictator José Santos Zelaya, who extended Nicaraguan authority over the coastal Mosquito kingdom. The USA, which was apprehensive of his financial dealings with Britain, supported the revolution which overthrew him in 1907. US presence, including two occupations by marines, dominated the country until 1933. In that year it fell under the control of Anastasio Somoza*, who ruled until his assassination in 1956. He was succeeded by his son Luis (1957–63) and then by Luis's brother General Anastasio Somoza Debayle (1967–72, 1974–9). When not officially presidents, the Somoza family installed puppets. In 1961 the three principal left-wing opposition groups formed the Frente Sandinista de Liberacion Nacional (FSLN) (see SANDINO), committed to armed insurgency. It gained increasing support from the landless peasantry and engaged in numerous guerrilla clashes with the National Guard, ending in virtual civil war 1976–9. By then the excesses of the Somoza regime had led proscribed political parties such as the Christian Democrats to form a partnership with the FSLN. It is estimated that some 50,000 people had died in civil strife by the end of the war in June 1979, when the FSLN formed a Provisional Government of Reconstruction with its civilian allies. Daniel Ortega* was appointed co-ordinator of a three-man junta and the Sandinista-led government won widespread early support. The Somoza estates (10 per cent of all the land) were allocated to peasants and co-operative farms encouraged. There was a drive to improve literacy, health, and housing. Gradually, however, the government became more radical. The USA discontinued all aid and it relied increasingly on advice and aid from Eastern-bloc countries. More haciendas were expropriated and mines and forests nationalized. Dispossessed and exiled owners, together with ex-National Guard refugees, formed the Fuerzas Democráticas Nicaragüenses (FDN), which, funded by the CIA*, proceeded to recruit an army of some 10,000 'Contras'. As Contra activities from Honduras and Costa Rica escalated, a state of emergency was proclaimed in March 1982. The Reagan* administration in the USA gave increasing aid to the Contras ('Freedom Fighters'), whose headquarters were based in Miami. It refused to accept that the presidential election of Ortega in 1984 was democratic. In so far as Ortega imprisoned political opponents, closed opposition newspapers and radio stations, and expelled unsympathetic priests, it was right. Apart from funding the Contras the CIA was empowered to engage in numerous attempts at 'destabilizing' the Sandinista regime, such as mining Nicaraguan ports in 1984, an act condemned by the International Court of Justice. In 1986–7 the US administration was seriously embarrassed by the exposure of illegal diversion of money to the Contras from US sale of arms to Iran (the so-called Irangate scandal). Nicaragua refused to accept the Contadora peace plan of June 1986, but agreed to the plan of President Oscar Arias of Costa Rica* on 7 August 1987, which led to the release of 985 political prisoners and direct negotiations between Contra leaders and President Ortega. These broke down, but when President Bush took office in 1989, direct US military funding to the Contras ended. Elections were called in Nicaragua for 1990, with opposition groups heavily financed by the USA. The country's economic problems remained acute as long as 50 per cent of its budget continued to be devoted to military expenditure. In the election the Sandinistas lost to a coalition group, the National Opposition Union, Violeta Chamorro becoming President. Contrary to the fears of many, a peaceful handover was achieved, Contra rebels were gradually demobilized, and the US Congress voted a $300m. loan.

Nicholas II (1868–1918), last Emperor (Tsar) of Russia 1894–1918. The son of Alexander III, on coming to the throne he at once formalized the alliance with France. This replaced Russia's earlier alliance with Germany and Austria-Hungary. In 1904–5 his Far Eastern ambitions ended in disaster when the Japanese sank the Russian fleet sent to fight the Russo-Japanese War. This helped to precipitate the first Russian Revolution of 1905, after which he agreed to an elected Duma and Upper Chamber. His entry into war against Germany and the Habsburgs was popular, but he unwisely took personal command of the armies, leaving government to the Tsarina and the sinister Rasputin*. Government chaos and mismanagement led him to agree to abdicate in March 1917, while at his headquarters. He was later imprisoned and in 1918 moved by the Bolsheviks to Siberia. The fear of counter-revolutionary action led to the murder of Nicholas, the Tsarina Alexandra, and their family at Ekaterinburg (now Sverdlovsk) 16 July 1918.

Niemöller, Martin (1892–1984), German Protestant theologian. He joined the navy and became a U-boat commander during World War I. In 1924 he was ordained priest, working in Berlin. A strong opponent of Communism, at first he welcomed the National Socialists and joined the party. But in 1933 he founded the Pastors' Emergency League to try to help rising discrimination against Christians with a Jewish background. His opposition to attempts to Nazify the Church led to his arrest in 1937. He was released but rearrested by the Gestapo in 1938 and confined in concentration camps until 1945. By now he was a pacifist, and he became president of the World Council of Churches 1961–8.

Niger, a land-locked republic in West Africa. The French first arrived in 1891, but the country was not fully pacified until 1914, being incorporated into French West Africa in 1922. It became an autonomous republic within the French Community* in 1958 and fully independent in 1960, with special arrangements with France covering finance, defence, technical development, and cultural affairs. From 1974 it was governed by a Supreme Military Council, with all political associations banned. In September 1989, under President Ali Saibou, a new constitution was approved by referendum, allowing a single civilian political party (the Mouvement National de la Société de Développement) to organize elections.

Nigeria, a federal republic with the highest population (105 million) of any African country. Abuja is the federal capital and there are twenty-one states: Anambra, Bauchi, Bendel, Benue, Borno, Cross River, Gongola, Imo, Kaduna, Kano, Kwara, Lagos, Niger, Ogun, Ondo, Oyo, Plateau, Rivers, Sokoto, and, since 1989, Akwa Ibom, and Katsina. The island of Lagos was an important centre for the 18th-century slave trade. Explorers worked their way inland, and in 1893 the Royal Niger Company was taken over by the British Colonial Office, which created the Niger Coast protectorate. Following the conquest of the ancient Kingdom of Benin, this became the Southern Nigeria protectorate in 1900, when the protectorate of Northern Nigeria was also proclaimed. In 1906 the colony of Lagos was absorbed into the southern protectorate, and in 1914 the two protectorates were amalgamated to form Nigeria, the largest British colony in Africa. Under its first Governor Frederick Lugard* it was to be administered by a policy of indirect rule*; where emirs or chiefs did not exist they were to be created. In Northern Nigeria Muslim chiefs from the Fulani* maintained a conservative rule over the majority of the country's Hausa* population. In the west the Yoruba* dominated; the Ibo* people were centred in the east.

By 1939 a wealthy and sophisticated Nigerian middle class, based mostly in Lagos, was demanding some form of home rule. During the war many thousands of Nigerians served in the British army in Africa and Burma, and by its end the first Nigerians were being commissioned. After the war anti-colonial pressure grew intense, and in 1954 a constitution created the Federation of Nigeria, to consist of three regions: Northern, Eastern, and Western, together with the UN Trust Territory of the Cameroons, and the federal territory of Lagos. In 1960 this Federation became an independent nation within the Commonwealth, with Nnamdi Azikiwe* first Governor-General and later President, and in 1963 a republic. Following the installation of military rule in 1966, the regions were replaced by twelve states in a new constitution of 1967. The result was the Biafran War*, which had been partly stimulated by the discovery of oil off Port Harcourt. After successfully suppressing the Biafran secession General Gowon* was deposed in 1975. In 1979 the military government organized multi- party elections, but corruption and unrest precipitated further military take-overs in 1983 and 1985, when General Ibrahim Babangida became head of state. In spite of wealth from its oil revenues Nigeria has continued to face deep social and economic problems. Civilian rule was scheduled to be restored in 1992.

'Night of the Long Knives', the name coined by Hitler himself for the weekend of murders in Germany 29–30 June 1934. It followed a secret deal between himself and the SS* units. Precise details remain unknown, but the army is believed to have promised to support Hitler as head of state after Hindenburg's* death in return for the elimination of the more radical Nazi private army the SA (*Sturmabteilung*), or Brownshirts*, led by Ernst Röhm. Hitler announced that seventy-seven people had been summarily executed for alleged conspiracy; but there were arrests by the SS throughout Germany, usually followed by murder. These numbered many hundreds, including some non-party figures. Among those killed was former Chancellor Schleicher*.

Nimitz, Chester William (1885–1966), US admiral. Born in Texas, he graduated at the US naval academy in 1905 and served in submarines during World War I, becoming a strong advocate for submarine warfare. After various surface ship commands and shore appointments, he took over command of the Pacific fleet in 1941, after Pearl Harbor*. From his headquarters there, he deployed his forces to win the Battles of Midway* and Coral Sea*, and subsequently supervised the moves in the Pacific campaign* which led to successful actions off Guadalcanal, the Gilbert Islands, the Marshalls, the Marianas, and in the Leyte Gulf*. To a large extent he was responsible for making the Pacific fleet, weakened by Pearl Harbor, the instrument of Japan's defeat, concentrating on submarine and aircraft attack. After the war he was briefly chief of US naval operations 1945–7.

Niue, a Polynesian island in the Pacific. It was annexed by New Zealand and administered as part of the Cook Islands in 1901. In 1904, however, it had been granted a separate New Zealand resident Commissioner. This arrangement lasted until 1974, when a referendum voted for self-government in association with New Zealand. A twenty-member Legislative Assembly was established, which elects the Prime Minister;

at the same time Niueans would retain their New Zealand citizenship. Niue joined the South Pacific Forum* and adopted a non-nuclear policy in 1978.

Nixon, Richard Milhous (1913–), President of the USA 1969–74. Born in California and educated at Duke University, NC, he was a rising lawyer in 1942, when he joined the US navy. Elected to Congress in 1946, he at once became involved in seeking out 'Un-American Activities'. In 1950 he moved to the Senate and in 1953 became Vice-President to President Eisenhower. He lost narrowly to John F. Kennedy in 1960, when he returned to his law practice, but re-emerged to defeat Hubert Humphrey* as President in 1968. His administration initiated a New Economic Policy (1971) to counteract inflation, which included an unprecedented attempt to control prices and wages in peacetime, as well as a reversal of many of the social policies of President Johnson*. In an attempt to achieve a balance of trade the dollar was twice devalued, in 1971 and 1973. His presidency is best remembered for its achievements in foreign affairs, for which his Secretary of State Henry Kissinger* was at least partly responsible. Having inherited the Vietnam War*, Nixon began by extending it, by invading Cambodia (1970) and Laos (1971), and by saturation bombing. From 1971, however, a policy of gradual withdrawal of US forces began, while negotiations were taking place, ending with the cease-fire accord of Paris in 1973. At the same time support was being given to the West German policy of *Ostpolitik**, with a presidential visit to the Soviet Union achieving agreements on trade, joint scientific and space programmes, and nuclear arms limitation. Recognition was given to the Communist regime of the People's Republic of China as the official government of China, and in February 1972 he paid a visit to China. Although re-elected in 1972, his second term was scarred by the Watergate* scandal. He announced his resignation on 9 August 1974 and was succeeded by his Vice-President Gerald Ford*, who proceeded to grant him a 'pardon', even though he had not been found guilty of any 'crime or misdemeanour'. He was debarred from legal practice, but made a fortune from his memoirs. In 1981 he re-entered Republican politics as an increasingly respected 'elder statesman' and consultant.

Nkomati Accord (1984), a non-aggression pact signed on 16 March between President P. W. Botha* of South Africa and Samora Machel* for Mozambique, on the border between the two countries. It takes its name from the small River Nkomati. The pact stated that neither country would allow 'organizations or individuals which plan or prepare to commit acts of violence, terrorism or aggression'. It effectively curbed ANC* activity for a short while, but MVRM (Mozambique National Resistance Movement) continued to operate. Machel was killed in 1986 and the Accord was increasingly ignored.

Nkomo, Joshua Mqabuko Nyongolo (1917–), Zimbabwean statesman. Educated in South Africa, he was for many years a trade union organizer on the Rhodesia railways, becoming secretary-general of the Rhodesian Railways African Employees' Association. He was president of the Rhodesian branch of the African National Congress* from 1957 until 1959, when it was banned. In 1960 he founded the National Democratic Party. When this was banned he created the Zimbabwe African People's Union (ZAPU). He was twice detained in the years 1962–4, and then imprisoned for ten years 1964–74. On release he travelled widely to promote the nationalist cause, and was in Geneva in October 1976 to meet Ian Smith*. His ZAPU, mainly supported by the Ndebele* in south-western Zimbabwe, allied uneasily with Mugabe's ZANU as the Patriotic Front*, guerrilla operations being mounted from neighbouring Zambia. He lost the 1980 election to Mugabe, who appointed him Minister of Home Affairs. In 1982 he left the government, and Zimbabwe was near to civil war until his reconciliation in 1987, when he rejoined the government.

Nkrumah, Kwame Francis Nwia Kofi (1909–72), Prime Minister and President of Ghana 1957–66. Educated at Achimota College, in the USA, and at the London School of Economics, he lived and worked in London before and during World War II. He was a key organizer of the Manchester Pan-African Congress of 1945. He returned to the Gold Coast in 1947 as general secretary of the United Gold Coast Convention, an African nationalist party founded by J. B. Danquah*. In 1949 he founded the Convention People's Party (CPP), a more radical organization, and led a series of strikes and boycotts for self-government. After a short imprisonment by the British for sedition, he was appointed Prime Minister and led his country to independence as Ghana (1957), the first British African colony to achieve this. His style of government was autocratic, but in his first years he was immensely popular with his policy of Africanization. In 1964 he was declared President for life, but economic pressures were resulting in political unrest. In 1966, while he was on a visit to Hanoi, a military coup deposed him. He took refuge in Guinea, where Sékou Touré* made him 'Co-president'. An outstanding African nationalist and a firm believer in Pan-Africanism*, he died in exile in Romania.

NKVD (Soviet People's Commissariat for Internal Affairs), a secret police agency. In 1934 OGPU (see CHEKA) was reorganized and this agency established. It was the main instrument for Stalin's purges in the years 1934–8. In 1946 it was itself reorganized by Beria* and merged with the MVD and MGB. The MVD (Ministry of Internal Affairs) had controlled all police forces in the Soviet republics and administered the prison camps, while the MGB (Ministry of State Security) had overseen all police forces in satellite countries, eliminating anti-Communist opposition. After the fall of Beria the NKVD was placed under the KGB*.

NLF *see* NATIONAL LIBERATION FRONT.

Nobel Prizes, annual prizes awarded since 1901 and funded from the fortune of the Swedish scientist Alfred Nobel (1833–96). They are in Physics, Chemistry, Medicine and Physiology, Literature, and Peace.

Non-alignment, a policy which arose during the 1950s and 1960s among states anxious not to align themselves with either the USA or the USSR. The Bandung Conference* of 1955 aimed to develop non-alignment in South-East Asia, and at a later Belgrade Conference in 1961 thirty-one states, including many Middle East and newly liberated African states, took part. The non-aligned states had some influence on world politics through the UN during the 1970s, but their coherence declined as the tensions of the Cold War also declined during the later 1980s.

Non-proliferation Treaty (1968), agreement on nuclear weapons. On 1 July 1968 Britain, the USA, and the Soviet Union signed an agreement for the non-proliferation of nuclear weapons, endorsed by fifty-nine other states. It was agreed that with effect from 1970 the three signatories would give no assistance to other states in obtaining or producing nuclear explosives. Since then Israel, South Africa, India, Pakistan, and possibly other states have developed nuclear capacity.

NORAD (North American Air Defence Command) was established in September 1957 'to integrate air reconnaissance and defence systems' of Canada and the

USA, with a combat centre in the Cheyenne Mountains and some 16,000 Canadian personnel in a command of 120,000. A line of fifty tracking, warning, and control stations, the DEW line, had already been established across the northern Arctic from Greenland to Alaska, for which Canadian personnel became responsible in 1959, as they were also for a Mid-Canada Line which existed 1954–65. In 1963 surface-to-air missiles were first introduced with nuclear warheads, much against the wish of many Canadians and with fierce opposition in Parliament. In 1968 NORAD was renewed, with none of the bitter opposition of five years before, and in 1981 it changed its name to North American Aerospace Defence Command (NAADC).

Nordic Council. It was founded in 1952 as an assembly of sixteen members from the Parliaments of each of the five Scandinavian countries, Denmark, Finland, Iceland, Norway, and Sweden. It meets annually and makes recommendations to the respective governments on matters of common interest. It was developed in 1972 by the formation of the Nordic Council of Ministers under a Helsinski Convention of 1962. This consists of ministers from the five countries, meeting regularly for 'consultation and co-operation on matters of common interest' in such fields as scientific research and development, education, social welfare, and the arts. The Council funds a number of projects and its formal declarations are binding on member governments. Meanwhile the older Nordic Council continues to meet in an advisory role.

Normandy campaign, World War II (June–September 1944). Five beaches had been designated for the Allied invasion of Normandy, code-name 'Operation Overland', for which General Eisenhower* was Supreme Commander. All the beaches, code-names Utah, Omaha, Gold, Juno, and Sword, had been reconnoitred by commandos. On 6 June, D-Day, five separate groups landed. British and Canadian troops fought across the eastern beaches, the Americans the western. Four beaches were taken easily, but there was fierce German resistance at Omaha. A bombing offensive had destroyed bridges across the Seine and the Loire, preventing German reinforcements from reaching forward units. At the height of the fighting Rommel*, who commanded the western defences, was wounded and was recalled. Meanwhile old ships were towed across the Channel and sunk to provide more sheltered anchorages. On D-Day+14 two vast steel-and-concrete artificial harbours, code-name Mulberry, were towed across the Channel, one being sunk in a freak storm. The other was established at Arromanches on beach Gold, to provide the harbour for the campaign; at the same time a series of pipe-lines, code-name Pluto, began to be laid across the Channel, to supply the thousands of vehicles now being landed. Having taken Omaha, US forces under General Bradley* immediately cut off the Cotentin Peninsula (18 June) and accepted the surrender of Cherbourg. The British attacked towards Caen, securing it after heavy fighting (9 July) and then advancing towards Falaise. US troops broke through the German defences to capture the vital communications centre of Saint-Lô. The Germans launched a counter-attack, but were caught between the US and British armies in the 'Falaise Gap' and lost 60,000 men in fierce fighting. Field Marshal Model, transferred from the Eastern Front, was unable to stem Patton's advance, which now swept across France, while Montgomery moved up the English Channel. Paris was liberated by General Leclerc* on 26 August and Brussels on 3 September. By 5 September more than 2 million men, 4 million tonnes of supplies, and some 450,000 vehicles had been landed, at the cost of some 224,000 Allied casualties.

North African campaigns, World War II (June 1940–May 1943). When Italy declared war in June 1940 General Wavell* had some 36,000 Commonwealth troops in

Cairo at his disposal. In September Italian troops under General Graziani* crossed the Egyptian frontier and advanced to Sidi Barrani. In December Wavell counter-attacked with a large force of tanks, which surrounded Sidi Barrani, capturing some 120,000 prisoners. Wavell went on to take Tobruk*, Benghazi, and all of Cyrenaica. In July 1940 the Italians had occupied parts of the Sudan and British Somaliland, but in January 1941 the British counter-attacked, and on 6 April Ethiopia and all of Italian East Africa surrendered, thus opening the way for Allied supplies and reinforcements to reach the Army of the Nile. In March 1941 General Rommel* arrived, and in April Wavell dispatched Commonwealth forces to help Greece. The British were obliged to withdraw, leaving Tobruk besieged. Under a new British commander, General Auchinleck*, an offensive, 'Operation Crusader', was planned. At first successful, the campaign swung back and forth across the desert, both German and British tank casualties being high. Tobruk fell in June 1942 and the British took up a defensive position at El Alamein* in July. From there in October the reinforced 8th Army of 230,000 men and 1,230 tanks, now under General Montgomery*, launched their attack. Rommel's supply position was weak and he steadily fell back towards Tunisia. Meanwhile 'Operation Torch' was launched; an amphibious landing of US and British troops on 8 November under General Eisenhower* near Casablanca on the Atlantic and Oran and Algiers in the Mediterranean. Vichy* French troops of Admiral Darlan* at first resisted, but after three days acquiesced. From Fort-Lamy in Chad, General Leclerc* brought his contingent of Free French to link up with the 8th Army at Tripoli. From November 1942 until May 1943 German armies, although reinforced, were being squeezed between the 8th Army advancing from the east and the Allied forces advancing from the west. On 7 May Tunis surrendered. Some 250,000 prisoners were taken, although Rommel skilfully succeeded in withdrawing the best troops of the Afrika Korps to Sicily.

North Atlantic Treaty Organization (NATO). Founded in 1949, it was established to counter the perceived military threat from Soviet power in Eastern Europe. Its original members were: Belgium, Canada, Denmark, France, Great Britain, Iceland, Italy, Luxembourg, The Netherlands, Norway, Portugal, and the USA. Greece and Turkey joined in 1952, West Germany in 1955, and Spain in 1982. In 1966 France withdrew its forces from the NATO Military Committee, though remaining a nominal member of the NATO Council. The organization was inevitably dominated by the USA, although during 1989 tensions developed over the issue of short-range nuclear missiles based in Germany. At its meeting in 1990 it faced the problem of the disappearance of a Soviet military threat, while at the same time it was agreed that a unified Germany should remain a member. A debate began as to its possible future role, the 'hand of friendship' being offered to East European countries 'which were our adversaries in the Cold War'.

Northern Expedition (1926–8), a military campaign in China. Before he died in March 1925 Sun Yixian* had hoped to organize a campaign to pacify northern China. In August 1925 Jiang Jiehi* (Chiang Kai-shek) was appointed Commander-in-Chief of the National Revolutionary Army*, and from September 1926 to March 1927 he successfully advanced north from Guangzhou (Canton), Nanjing falling in March 1927. At this point Jiang, as leader of the Guomindang, broke with the Chinese Communist Party, and the massacre of Shanghai* followed in April. In 1928 Jiang resumed his expedition against the northern warlords*. Beijing fell on 8 June 1928, its warlord Chang Hsueh-liang* pledging support for the National government, which was set up in Nanjing.

Northern Ireland, the six north-eastern counties of Ulster. It was established as a self-governing province of the United Kingdom by the Government of Ireland Act (1920), as a result of pressure from the Ulster Volunteers* and from its predominantly Protestant population. A Parliament with a built-in Protestant majority was to be established in Stormont Castle*. Discrimination against the largely working-class Catholics (about one-third of the population), particularly in jobs and housing, resulted in an eruption of violence in 1968, with demands for civil rights. Paramilitary groupings, such as the Provisional Irish Republican Army* (IRA), the militant, 'loyalist' Ulster Defence Association (UDA), and the Ulster Defence Force (UDF), clashed more and more violently. In 1969 extra British troops were sent to the province, at first welcomed by beleaguered Catholic families, and have remained ever since, with fatalities reaching some 3,000 by 1990. Following 'Bloody Sunday' (30 January 1972), when troops fired on a Catholic demonstration in Londonderry, killing several civilians, the British government suspended the Northern Irish Constitution, and dissolved the Stormont government, imposing direct rule from London. A more representative Northern Ireland Assembly was elected in 1973, but it soon collapsed. There was to be a non-sectarian Northern Ireland Executive, but leaders such as the Revd Ian Paisley*, together with the Ulster Workers' Council, forced its dissolution and foiled attempts for power-sharing between both sides. However, since 1979 closer co-operation between the Republic of Ireland and Britain has developed, with the Anglo-Irish Accord (the Hillsborough Agreement), signed in 1985, giving the Republic a consultative role in Northern Ireland. Attempts to organize an agreed permanent system of government have so far failed. At the same time terrorist outrages have continued to be committed by extremists of both sides.

Northern Territory (Australia). The Territory (519,800 sq. miles) was transferred to the Commonwealth from South Australia in 1911, being administered from Darwin as the responsibility of the Ministry for External Affairs. Darwin was severely damaged in early 1942 by Japanese air raids, after which the area was placed under military control, while women and children were evacuated. In 1947 an Administration Act created a Legislative Council, which in 1974 became an elected body of nineteen members. After World War II extensive mining developed, copper, gold, uranium, bauxite, and manganese. In 1978 the Northern Territory Self-Governing Act established the Territory as virtually on a par with the six states. It would have an 'administrator' instead of a 'governor', and a 'chief minister' instead of a 'premier'. As Australian citizens (emancipated since 1948) Aborigines* have full equality, and the Aboriginal Land Rights Act (1978) gave freehold rights to former reserve land. The Territory elects one member to the Federal House of Representatives and has two Senators.

North-West Europe campaign, World War II (September 1944–May 1945). Following the Normandy campaign*, Montgomery's* forces captured Antwerp on 4 September and crossed the Albert canal. The US 1st Army captured Namur and Aachen, while the US 3rd Army moved east and reached the Moselle. Montgomery's attempt to seize the lower Rhine by dropping the 1st Airborne division at Arnhem* ended in disastrous failure. In November the Germans consolidated and launched a counter-attack in the Ardennes*. In January Montgomery's forces pushed forward to the Rhine. In March a massive bombardment of Wesel preceded a successful crossing of the lower Rhine, while French forces moved up the upper Rhine towards Lake Constance. The US 7th Army pushed east towards Munich and the 3rd Army crossed the Rhine at Frankfurt, and swept through central Germany into Bohemia. On 11 April

Montgomery reached the River Elbe. Following the capture of Berlin by the Red Army and the suicide of Hitler, Montgomery received the surrender of the German forces in north-west Europe on Lüneburg Heath on 4 May. Four days later, VE-Day, the war in Europe was declared at an end.

North West Territories (Canada). Created by statute in 1875, the Territories were reduced by the creation of the provinces of Alberta and Saskatchewan in 1905. With a population of 51,384 (some 50 per cent Indian and Innuit) over 3,379,700 sq. km. (1,304,903 sq. miles), they are administered by a Commissioner, appointed since 1966 by the Federal Department of Indian Affairs and Northern Development, an elected Legislative Assembly, and a seven-member cabinet with a government leader. The Territories have one Member of Parliament, an administrative capital at Yellowknife, and five administrative regions. They were divided in 1920 into three districts, Franklin, Keewatin, and Mackenzie, where in 1964 significant oil finds were made. The exploitation of these and the opening of lead, zinc, and other mines are severely threatening the ecological balance of this northern environment.

Norway. Union with Sweden was unilaterally declared dissolved in June 1905, when Haakon VII* was elected King. A Liberal Party government introduced women's suffrage and social reform and maintained neutrality during World War I, when the basic timber industry suffered grievously. A brief Labour government in 1928 was followed by high unemployment, but Labour returned in 1935 and stayed in power until 1940. In World War II the Germans invaded, defeating Norwegian and Anglo-French forces at Narvik in 1940 and imposing the puppet government of Vidkun Quisling*. The latter failed in its efforts to Nazify the schools, but in 1943 introduced concentration camps. In 1945 the monarchy and a new Labour government returned, the economy being greatly helped by Marshall Aid*. Norway joined EFTA* in 1960 and signed up to join the European Community in 1972, but withdrew after a national referendum. In 1973 it signed a free trade agreement with the EEC and since then 64 per cent of its export revenue has come from EEC countries and only 15 per cent from EFTA. Norway was expected to reapply for EC membership before 1992. In 1986 it became clear that acid rain—mostly originating in British power stations, but also from Soviet sources—had done immense damage to Norway's forests and lakes.

Novotný, Anton (1904–75), President of Czechoslovakia 1957–68. Son of a bricklayer, he was trained as a locksmith and worked in an arms factory near Prague. In 1921 he joined the Communist Party and worked for it throughout the inter-war years. He was placed in a concentration camp in 1941, and on liberation in 1945 quickly rose to become first secretary of the party, which he remained until he retired in 1968. On the death of Antonin Zapotocky he succeeded as President. He was a hard-line Stalinist, out of sympathy with Khrushchev's* attempts to liberalize the Soviet bloc. His adherence to Comecon* policies of concentrating on heavy industry led to an economic recession 1961–3 and to student unrest, his reactions to which being too little and too late. In February 1968 he called on the army to occupy Prague, and it refused. He retired, to be replaced by General Svoboda* as President and Alexander Dubček* as first secretary for the brief 'Prague Spring'*. In 1971 he was restored to favour within the party but did not regain office.

NRA (National Rifle Associations), a US pressure-group. In Britain a Rifle Association was formed in 1860 to organize the sport of rifle and pistol shooting. Although it sponsors competitions, its activities are modest compared with those of the US Association. Founded in 1871, it has a membership of some 2 million, and has proved itself a highly effective political lobby against restrictions and control over the possession of firearms, long regarded as a basic freedom in the USA.

NSC (National Steel Corporation), the largest steel corporation in the USA. Founded in 1929 through an amalgamation of a number of steel companies, by Ernest T. Weir, it successfully survived the Great Depression*, for long resisting the unionization of the work-force. During the 1970s it diversified into aluminium and in the 1980s into computers, becoming a holding company known as National Intergroup Inc., half of which is owned by Japanese investors.

Nu, U (1907–   ), Prime Minister of Burma 1948–56, 1956–8, 1960–2. A graduate of Rangoon University, he became a teacher and journalist and helped to lead a student strike in 1936. In 1937 he joined the nationalist movement Dobama Asi-ayone, and was imprisoned by the British in 1940 for alleged sedition. He served in the puppet government of Ba Maw during the Japanese occupation, but broke with it towards the end of the war, helping both to form the Anti-Fascist People's Freedom League and to negotiate independence. He became Prime Minister after Aung San's* murder. During his years in office he sought to build a neutralist state, based on strong Buddhist foundations and providing maximum social welfare. Economic problems, and Communist and ethnic opposition, led to his downfall in 1962. He was imprisoned by Ne Win* 1962–6 and then exiled. He made several attempts to regain office from his exile in Thailand between 1969 and 1974. He returned to Burma in 1980.

Nuclear energy. Although the energy potential of the atom had been grasped early in the 20th century, the first nuclear reactor using uranium 235 was built by Enrico Fermi at the University of Chicago in December 1942. Plans were immediately made 'for the harnessing of the atom for peaceful purposes', while at the same time the Manhattan Project* developed the first atomic bomb. In 1945 a small reactor was operating in Ontario and in 1956 Calder Hall in Cumberland (later renamed Sellafield) became the first commercial nuclear power-station. The USA quickly followed, with a research programme into five possible designs: 'pressurized water', 'boiling water', 'sodium-graphite', 'homogeneous', and 'fast-breeder', and mining of uranium supplies in Zaïre, Canada, and elsewhere was developed. In Europe the Atomic Energy Commission was established in 1957 and in 1962 forecasts were being made of the need for a substantial provision of nuclear energy to supplement fossil fuels. Extensive and rapid development followed in the USA, in Japan, France, Britain, Germany, the Soviet Union, and elsewhere. At the same time, nuclear-powered submarines and surface ships developed, although the latter proved commercially uneconomic. During the 1970s a change of public attitudes towards nuclear energy began, prompted partly by the realization that no satisfactory mechanism for disposal of radioactive waste could be devised. In France and Germany there were public protests against further developments, while the Three Mile Island* accident in 1979 seemed to confirm fears of radioactive hazard. The Chernobyl* disaster of 1986, together with the realization that costs for nuclear power generation were still appreciably higher than for fossil fuels, further polarized opinion. At the same time it became clear that if nuclear power poses a radioactive hazard, the burning of fossil fuels produces both acid rain and an excess of carbon dioxide, precipitating a 'greenhouse effect'.

Nuclear Test-Ban Treaty (1963), agreement not to test nuclear weapons in the atmosphere, in outer space, or under water. It was signed on 5 August 1963 by the USA, the Soviet Union, and Britain. The issue of disarmament had been raised at the Geneva Conference of 1955, and discussions on banning nuclear testing had begun in Geneva in 1958; by then the appalling pollution effects of atmospheric testing had been scientifically established. Between 1963 and 1965 ninety other countries signed the treaty, but not France or China. In spite of the treaty, however, the potential spread of nuclear weapons became a major preoccupation of the superpowers and in 1968 a Non-proliferation* Treaty was negotiated.

Nuffield, William Richard Morris, 1st Viscount (1877–1963), British industrialist and philanthropist. Born in Worcester and educated in a village school in Cowley near Oxford, he set up his own bicycle business in a shed in Cowley when he was 16. In 1902 he designed a motor cycle and by 1912 he was repairing motor cars. He bought a small factory, where the Morris Cowley, the first British car planned for middle-class family use, began production in 1913. After World War I mass-production kept this and later models cheap, and Cowley developed into an industrial suburb of Oxford. Business acumen earned him a vast fortune, much of which he gave away, including £3m. to Oxford University to fund four chairs in medicine. He later established Nuffield College and the Nuffield Foundation, a trust to encourage educational and medical development.

Nujoma, Sam Daniel (1929–), leader of Namibian SWAPO Party and President of Namibia 1990– . Educated at the Finnish Protestant School in South West Africa, he worked on the state railways until 1957. After various clerical jobs, in 1959, with Andimbe Toivo ja Toivo, he founded SWAPO*, was arrested, escaped, and went into exile. He appeared before a UN Committee on South West Africa in 1960. Having established a headquarters for SWAPO in Dar es Salaam he returned home, to be rearrested and forbidden the country. In 1966, his claim against South Africa having been rejected by the International Court of Justice, he now resorted to an armed struggle operating from Angola and Zambia. Toivo was captured and imprisoned in 1968. In 1971 Nujoma again appeared before the UN, which now gave its support to SWAPO. In 1984, after an eighteen-year guerrilla war, he was reunited with Toivo, who was released from prison and became general secretary of the party. With US support, the two men succeeded in reaching an accord with South Africa in 1988 and in organizing an election victory in November 1989. In March 1990 he became first President of independent Namibia.

Nuremberg Laws, Nazi legislation. Each September from 1933 to 1938 Goebbels* organized a giant Party Rally at Nuremberg, with torchlight processions culminating in a speech by Hitler. At the 1935 rally he announced decrees which debased all Jews to second-class citizens, closed the professions to them, and forbade mixed marriages or sexual relations between Jew and non-Jew, 'for the protection of German Blood and German Honour'. The decrees were unanimously adopted by the Reichstag.

Nuremberg Trials, a series of thirteen trials over two years before US, British, French, and Soviet judges: 177 Germans and Austrians were indicted; twenty-five were sentenced to death; twenty to life imprisonment; ninety-seven to shorter terms; and thirty-five acquitted. The charges were conspiracy against peace, crimes against peace, violation of the laws and customs of war, and crimes against humanity. Judgements against the twenty-four Nazi leaders included death sentences, imprisonment, and 'not guilty'; ten were hanged, while Goering and Ley committed suicide. Hess was sentenced to life imprisonment. As a result of the trials several Nazi organizations such as the SS* and the Gestapo* were declared to be criminal bodies.

Nuri al-Said (1888–1958), Prime Minister of Iraq 1930–2, 1938–40, 1941–7, 1949–52, 1953–7, 1958. Born near Baghdad, he joined the Turkish army fighting in the Balkan War of 1912. In 1914 he defected and fled to Egypt. When the Arab Revolt* broke out he became Chief of Staff to Emir Faisal. When the latter became King Faisal I* of Iraq he was his chief adviser and virtual founder of the Iraqi army. He was to form no less than sixteen governments as Prime Minister during the rest of his life. His aim was for a confederation of Iraq, Syria, and Transjordan, linked with Christian Lebanon and a Jewish enclave in Palestine; but the rise of Ba'ath* socialism and mutual antipathies always prevented its achievement. When Faisal II* became King in 1939, he was a close associate of the Regent Abd al-Ilāh* and strongly supported British intervention in Iraq in 1941. He helped to create the Baghdad Pact* (1955), but regarded Nasser's ambitions and Soviet Middle East involvement with apprehension. Being strongly pro-British, he supported the idea of a federation with Jordan after the Suez Crisis*, but his opponents were too strong. After the murder of King Faisal he tried to escape but was caught and killed.

Nyerere, Julius Kambarage (1922–), Prime Minister of Tanganyika 1961, President of Tanganyika and later Tanzania 1962–85. Son of a chief, he was educated at Makerere College and Edinburgh University. He was a teacher until 1954, when he became politically active. He organized the Tanganyika African National Union (TANU), and in 1956 the British administration nominated him as the TANU representative on the Legislative Council. In 1957 he resigned, complaining of too slow progress towards self-government. On Tanganyika's independence in 1961 he was appointed Prime Minister, but again resigned after a month. He spent the next year building up his party, and in 1962 was elected President of the Tanganyika Republic. In 1964, following a revolution in Zanzibar*, he effected union between it and Tanganyika as the Republic of Tanzania. In 1967 he helped create the short-lived East African Community, a common market with Kenya and Uganda. In the Arusha Declaration* (1967) he outlined the socialist policies that were to be adopted in Tanzania. He was a major force in the Organization of African Unity* (OAU) and over a broad range of African politics, especially in relation to Uganda, Zimbabwe, and South Africa. He resigned the presidency in 1985, having been an eloquent supporter of the Commonwealth, but also a proponent of international sanctions against South Africa.

OAS *see* ORGANIZATION OF AMERICAN STATES *or* ORGANISATION DE L'ARMÉE SECRÈTE.

OAU *see* ORGANIZATION FOR AFRICAN UNITY.

Obote, Milton (1924–   ), President of Uganda 1966–71, 1980–5. After attending mission school he held a variety of jobs in Uganda and Kenya as labourer, clerk, and salesman, becoming a founder member of the Kenya Africa Union. He became increasingly politically active, but was regarded by the British as fundamentally moderate. He was a member of the Uganda National Congress 1952–60 and of the Uganda Legislative Council 1957–61. He was appointed first Prime Minister of independent Uganda in 1962, but in 1966 he overthrew Mutesa II* *Kabaka* of Buganda and first President of Uganda, assuming the presidency himself. He was overthrown by Idi Amin* in 1971, but returned from exile in Tanzania in 1980 to resume the presidency. He failed, however, to restore the economy or to stop corruption and tribal violence, and was once again overthrown in 1985, by Yoweri Muscveni, seeking refuge in Zambia.

Obregón, Alvaro (1880–1928), President of Mexico 1920–4. Son of a farm labourer, in 1912 he led a group of armed peasants to support President Francisco Madero in his struggle against General Huerta*. His military skills resulted in quick promotion in the Revolutionary Army. He was responsible for the defeat of Pancho Villa* and dominated the Convention in 1917 which produced the Mexican Constitution. A strong anti-clerical, he was dismayed by what he regarded as Carranza's* reactionary policies, and in April 1920 overthrew his old comrade. He was elected President in December and succeeded in achieving some measure of agrarian, educational, and labour reform, although peonage remained strong. His implementation of the Constitution brought him into bitter conflict with the Catholic Church. After turning the presidency over to his successor Plutarcho Eliás Calles* in 1924, he was elected to a second term in 1928 but, prior to taking office, he was assassinated for his alleged persecution of the Church.

Oceania, a loose term to describe the island states and territories of the Pacific Ocean, from Hawaii in the east to New Zealand and New Guinea in the west. After 1945 the USA was more and more concerned over the defence of this vast area, while at the same time seeing it as a springboard for attacking the USSR if necessary. Hence Fifteen Trident submarines were permanently stationed here, each with twenty-four nuclear warheads. B52 bombers operated from the Marianas, from Guam, and from other islands of Micronesia, while to the east Hawaii, headquarters of the US Pacific

Command, had 110 military installations and one of the largest arsenals of nuclear weapons anywhere. Testing of both nuclear weapons and missiles is conducted by France in French Polynesia* and by the USA in Micronesia. Between 1885 and 1940 the islands of French Polynesia were administered as the colony of Oceania.

Oder–Neisse Line. This frontier, formed by the two rivers, had been that of medieval Poland. As a result of the Potsdam Agreement* in 1945 it was restored to post-war Poland as its frontier with the then Russian-occupied zone of Germany. The German inhabitants of the area now assigned to Poland were expelled and Polish settlers moved in. The frontier was agreed to by the Federal Republic of Germany in 1972 as part of the policy of *Ostpolitik** and confirmed in 1990 when negotiations for the unification of Germany were taking place.

OECD *see* ORGANIZATION FOR ECONOMIC CO- OPERATION AND DEVELOPMENT.

Ohira, Masayoshi (1910–80), Prime Minister of Japan 1978–80. Unlike many Japanese politicians he came from a poor farmer's family. He worked his way through university and was a civil servant from 1936 until 1952, when he entered politics. As a member of the Liberal Democratic Party* he held various cabinet posts, showing great negotiating skills. As Foreign Minister he was responsible for the establishment of diplomatic relations with South Korea and the People's Republic of China. In 1978 he became president of his party and Prime Minister, but died suddenly in the election campaign of 1980. The China–Japan Peace and Friendship Treaty* was his achievement.

Ojukwu, Chukwuemeka Odumegwu (1933–), head of state of Biafra 1967–70. Son of a wealthy Ibo businessman, he was educated at Oxford and then joined the Nigerian army. He had reached the rank of lieutenant-colonel by 1966. Following the January 1966 revolt, he became military Governor of his home Eastern Region. After the September massacre of Ibos in the north, he put forward a scheme for constitutional reform. When this was rejected he succumbed to mounting Ibo pressure for the secession of the Eastern Region, declaring independence in May 1967. He personally led Biafran troops in the Biafran War*, at the end of which he fled to the Côte d'Ivoire. He returned to Nigeria in 1982 and attempted to re-enter politics, his activities leading to imprisonment during 1984. Since then he has lived in retirement.

Okada Keisuke (1868–1952), Japanese admiral and Prime Minister 1934–6. After graduating at naval college, he rose rapidly to become a full admiral (1924) and Commander-in- Chief of the Combined Fleet. In 1927 and again in 1932 he served as Minister for the Navy, before becoming Prime Minister. He narrowly escaped assassination in the incident of 26 February* 1936, after which he resigned, feeling a responsibility for its atrocities. During World War II he was one of the senior officers pressing the Emperor to dismiss the Tojo* regime and negotiate for peace.

Okinawa, one of the Ryuku Islands between Taiwan and Japan. It was captured from the Japanese in World War II by a US assault that lasted April–June 1945. With its bases commanding the approaches to mainland Japan, it was a key objective and was defended by some 120,000 Japanese, almost to the last man, with kamikaze* air attacks inflicting substantial damage on US ships. After the war it was retained under US administration until 1972. It was then returned to Japan, following a vociferous campaign, but with a substantial US military presence remaining.

Olympic Games. These were revived in 1896 in Athens by the enthusiasm of Baron Pierre de Coubertin, and are administered by the International Olympic

Committee in Lausanne. They have been held every four years since, each time in a different city, except during the years of the world wars. Early competitors were all genuinely amateur and the events were mainly track and field athletics. In 1924 a cycle of winter games began at Chamonix, while the range of sports involved has steadily widened, including a full list of team events, swimming, yachting, and archery. The 1936 Games in Berlin were the first to be flavoured by politics, and the 1972 Munich Games* were marred by the tragedy of terrorist murders. The 1980 Moscow Games were boycotted by the USA and many Western nations as a protest against the Soviet invasion of Afghanistan. While standards have steadily risen, national rivalries have resulted in some loss of the early spontaneity among competitors.

Oman, an independent sultanate in south-eastern Arabia. The most powerful state in Arabia during the 19th century, by 1914 the sultans had concluded a treaty with Britain, having lost control of much of their lands (the rich island of Zanzibar had been lost in 1861). There had been traditional tension between the tribes of the hinterland and those of the coast, including the inhabitants of the capital Muscat, many of whom are Indians. In 1939 a new treaty of commerce and navigation was made with Britain, and after World War II oil was discovered, being first exported in 1967. In 1970 the Sultan Said bin Taimur was deposed in a palace coup, being succeeded by his son Qaboos bin Said (1940–). Oman then joined the Arab League* and UN. The policy of Sultan Qaboos, the 14th sultan of the al-Busaid dynasty, has been to encourage Westernization and economic links with Britain and the USA. Internal opposition from the left-wing Popular Front for the Liberation of Oman has been suppressed and Oman successfully survived the tensions of the Iran–Iraq War.

OPEC *see* ORGANIZATION OF PETROLEUM EXPORTING COUNTRIES.

Organisation de l'Armée Secrète (OAS), a secret terrorist organization in Algeria and France. Soon after he became French President (1959), General de Gaulle* began to move to the view that the Algerian War* could not be won. Unhappy with his conduct of the war, a group of senior army officers, led by General Salan*, and of die-hard settlers in Algeria, formed the organization, pledged to keep Algeria French by any means. In April 1961 Salan staged an unsuccessful revolt in Algiers but escaped. When the OAS learned of the secret negotiations leading to the Évian Agreements* there were bomb outrages in Paris and assassination attempts against de Gaulle, whose car was machine-gunned at Petit Chamart in March 1962. Salan was captured in Algiers in April 1962 and the OAS gradually disintegrated, many members being arrested and imprisoned.

Organization for African Unity (OAU). Developed from the Brazzaville Bloc*, it was founded in 1963 for mutual co-operation and elimination of colonialism. The leaders of thirty-two African countries signed its Charter at a Conference in Addis Ababa. All African states, except South Africa, have at one time belonged. There is an annual assembly of heads of state and government, a council of ministers meeting every six months, a general secretariat, and commissions for health, education, social affairs, and for mediation, conciliation, and arbitration. The latter has had only limited success in resolving problems on the continent.

Organization for Economic Co-operation and Development (OECD). This was established in 1961 as a successor to the OEEC, which had administered the Marshall Plan*. Eighteen European countries, together with the USA and Canada, signed the original convention, since when Australia, New Zealand, Iceland, and

Yugoslavia have joined. It is based in Paris; the heads of delegations meet each week, and there is an annual ministerial conference. Its main targets have been liberalization of international trade and finance, together with the co-ordination of economic aid to developing countries. Its annual output of economic analysis and prediction covers a wide area, from capital markets and tax structures, to lumber and air pollution. It also issues an annual evaluation of the economies of member countries.

Organization of American States (OAS), founded at the 9th Pan-American* Conference at Bogotá in 1948, with a Charter designed to achieve 'peace and justice and to promote American solidarity'. By 1990 there were thirty-two member states on a basis of absolute equality, with a General Assembly meeting annually, an elected secretary-general, and three councils: the Permanent Council, the Economic and Social Council, and the Council for Education, Science, and Culture. Perhaps inevitably the OAS was dominated by the USA in its earlier years, but this dominance has lessened. At the same time OAS effectiveness was much reduced by the Central American crisis during the 1980s. Canada became a member in January 1990.

Organization of Petroleum Exporting Countries (OPEC). The first moves to establish closer links between oil-producing countries were made by Venezuela, Iran, Iraq, Kuwait, and Saudi Arabia in 1949. In 1960, following a reduction in oil price by the international oil companies, a conference was held in Baghdad of representatives from these countries, when it was decided to establish a permanent organization. This was formed in Caracas, Venezuela, next year. Other countries later joined: Qatar (1961), Indonesia (1962), Libya (1962), United Arab Emirates (1967), Algeria (1969), Nigeria (1971), Ecuador (1973), and Gabon (1975). OPEC's activities extend through all aspects of oil negotiations, including basic oil price, royalty rates, production quotas, and government profits. In 1973, following a crisis with the oil companies, the price of crude oil was raised by some 200 per cent. This steep increase adversely affected the economies of the less-developed countries, while making some Arab producers extremely rich. The appearance of new, non-OPEC oil producers such as Britain, Norway, and Alaska has somewhat reduced the influence of the organization, but it continued to play a major part in influencing world prices through control of production. The 1990 meeting of OPEC preceded the Iraqi invasion of Kuwait, which deeply divided the organization.

Orlando, Vittorio Emanuele (1860–1952), Prime Minister of Italy 1917–19. Born in Palermo, he was educated there and became Professor of Law at Palermo University. He entered politics just before World War I and supported Italy's entry into the war. He became Minister of Justice in 1916 and, after the Caporetto* disaster, Prime Minister. At the Paris peace talks he clashed with President Wilson over what were considered excessive claims by Italy to former Austrian territory, and he resigned. As social unrest developed in Italy he supported Mussolini's appointment, but after the Matteotti* murder, he resigned from Parliament in protest and left the country (1925). After the fall of Mussolini in 1943 he returned to become leader of the Conservative Democratic Union.

Ortega Saavedra, Daniel (1945–   ), President of Nicaragua 1985–90. Son of a peasant guerrilla fighter, he briefly attended Managua University before becoming an underground member of the Sandinista National Liberation Front (FSLN), quickly becoming in charge of urban units opposed to the Somoza regime. From 1959 onwards he was imprisoned and tortured on a number of occasions before his arrest in 1967. He then spent seven years in prison, but was released in 1974 on a deal between the Somoza government and the FSLN. After a spell in Cuba he returned to Nicaragua to co-ordinate the Sandinista victory of 1979. He was appointed 'co-ordinator' of the Sandinista junta and in 1984 elected President, after which he made strenuous efforts to gain international support for his regime, which continued to struggle against the US-backed Contras.

Orwell, George (1903–50), English novelist. Born in India Eric Arthur Blair, he won a scholarship to Eton and then returned east as a colonial officer in Burma. Disturbed by what he saw of British imperialism, he resigned to work as a journalist under his *nom de plume*, living for a while as a tramp and producing *Down and Out in Paris and London* (1933). Although he fought for the Republicans in Spain, he was unsympathetic to democratic centralism, advocating instead a decentralist, libertarian socialism. His later books *Animal Farm* (1944) and *Nineteen Eighty-Four* (1949) brilliantly warned of the dangers of totalitarianism as practised in Nazi Germany and Stalin's Russia. They had a profound influence on the political consciousness of his contemporaries.

Ostpolitik. This 'Eastern policy' was a reversal of West Germany's refusal to recognize the German Democratic Republic, as propounded by the Hallstein Doctrine (1957). This had asserted that West Germany would sever diplomatic relations with any country (except the Soviet Union) which recognized East German independence. The policy was pursued with particular vigour by Willy Brandt*, both as Foreign Minister and as Chancellor of the Federal Republic. A General Relations Treaty of 1972 normalized relations between the two republics, while treaties between West Germany and Poland and the Soviet Union gave formal recognition to the Oder–Neisse* frontier.

Ottawa Agreements (July–August 1932), a series of agreements on tariffs and trade between Britain and its dominions. They were concluded at an Imperial Economic Conference convened by the Canadian Prime Minister R. B. Bennett* at Ottawa. There would be a system of imperial preference which would provide for quotas of meat, wheat, dairy products, and fruit from the dominions to enter Britain free of duty. In return, tariff benefits would be granted by the dominions to imported British manufactured goods. A year later quotas were applied to Crown Colonies. The economic gains of imperial preference were helpful but not massive. After World War II the system was maintained through GATT*, but its benefits steadily eroded for Britain. With the prospect of British entry into the European Community, the agreements were seen as increasingly dispensable. Although a serious factor in the 1961–3 negotiations, they played little part in the 1971–2 terms of entry, apart from New Zealand dairy products, which were to continue to have some protection.

Ottoman Empire. By the Treaties of London and Bucharest in 1913 the Empire had lost most of its remaining European lands to Greece, Serbia, Bulgaria, Romania, and newly independent Albania, preserving only Constantinople and eastern Thrace. In the later 19th century a movement for more liberal government had produced the Young Turk* revolution of 1908 and the deposition of Sultan Abdul Hamid II. During World War I Britain and France occupied much of what remained of the Middle East section of the Empire, encouraging Arab nationalism and creating, after the war, such successor client states as Transjordan, Syria, Lebanon, and Iraq, as well as promising a Jewish national home in Palestine (1917). The abortive Treaty of Sèvres* attempted to reduce the Empire to only part of Anatolia, together, reluctantly, with Constantinople. Following the Turkish war with Greece the Treaty of Lausanne* reversed much of the earlier treaty, giving Turkey its present frontiers. By then the last

Sultan, Mehmed VI, had been overthrown, the caliphate was abolished, and the new Republic proclaimed under Mustafa Kemal Atatürk*.

Outback (Australia). Originally a colloquial term to describe the inland grazing lands beyond the Great Dividing Range of New South Wales, it was here that vast sheep stations of many thousands of acres had developed in the 19th century. Today the term is used to describe much of inland Australia. Life on sheep stations and cattle ranges has been considerably affected by radio communications and the helicopter. The former enables children to study at home via radio links with teachers hundreds of miles away, while the latter has ended the isolation which had earlier characterized the life-style of the outback.

Pacific Campaigns, World War II (December 1941–September 1945). On 7 December 1941 Japanese aircraft attacked the US naval base of Pearl Harbor* in Hawaii. Japan already had air bases in Vichy-controlled Indo-China and was allied to Thailand. Their land forces quickly occupied Hong Kong, Malaya, Singapore, and Burma. Other forces captured islands across the Pacific, while, following their naval victory in the Battle of the Java Sea (27 February–1 March), they occupied Borneo and the Dutch East Indies*. By April the Philippines were occupied, followed by northern New Guinea. General MacArthur withdrew to Australia, where he organized a counter-attack. The Battle of the Coral Sea* (5–8 May) prevented Japanese landings on southern New Guinea and ended the threat to Australia. It was followed by the decisive Battle of Midway Island*, which shifted the balance of naval power to the USA. In August 1942 US marines landed on Guadalcanal and Tulagi in the Solomon Islands, where fighting raged until February 1943. The remaining Solomon Islands were retaken during the year, Bougainville falling in November, followed by New Britain early in 1944. In November 1943 US marines captured Tarawa (Kiribati), although with heavy casualties. In June 1943 MacArthur had launched his campaign to reoccupy New Guinea, and through 1944 US forces gradually moved back towards the Philippines. In June the Japanese lost some 470 planes in the Battle of the Philippines Sea* and in July the Mariana Islands were taken, from which bombing raids on Tokyo began to be organized. In October the Battle of Leyte Gulf* marked the effective end of Japanese naval power, while on the mainland the Burma campaign* had reopened land communication with China. Manila fell in March 1945 and from April to June US forces gradually captured Okinawa*, at the cost of very high casualties on both sides (5,900 Japanese planes, as well as 1,900 Kamikaze attacks; 763 US planes and 12,000 dead). Plans to invade Japan ended with the decision to drop atomic bombs, which resulted in Japan's surrender and the end of the short-lived Japanese Empire.

Pacific Islands, Trust Territory of the, a group of some 2,000 Pacific islands in Micronesia* administered on behalf of the UN by the USA 1946–86. The islands had mostly been mandated in 1920 to Japan, having been previously part of the German Pacific Empire. Japan wished to exploit and develop them as part of the Japanese Empire. Following the recapture of the US island of Guam* in 1944 the islands were gradually all liberated and placed under US administration. In 1952 the US Congress devised a Code of the Trust Territory, but this came under criticism within the UN Trusteeship Council, and in 1965 a Congress of Micronesia was established, from which developed the Federated States of Micronesia, the Republic of the Marshall Islands*, and the Commonwealth of the Northern Marianas. Full independence for the islands of

Belau (formerly Palau) has been delayed owing to the reluctance of its people to accept a US military base equipped with nuclear weapons.

Pacifism, the belief that all war and the employment of force is unjustifiable. Such a belief has a long history within the monastic traditions of Buddhism and early Christianity, although both the Orthodox and Catholic churches evolved the concept of 'a just war' in their struggle against Islam. In recent centuries it has been held by the Society of Friends and the Plymouth Brethren. The word itself was coined early in the 20th century by those who were arguing that all international disputes should be settled by arbitration. It was then applied to those in World War I who refused conscription. They were usually imprisoned. Since the invention of the hydrogen bomb* an increasing number of people have adopted the pacifist ideal, on the grounds that nuclear war would be too horrendous in its consequences to contemplate.

Paderewski, Ignacy Jan (1860–1941), Polish statesman and concert pianist. Born in Podolia in the Ukraine, son of a landowner, he studied music at Warsaw and Berlin and later at Vienna, becoming well regarded as a concert pianist and composer. A strong Polish patriot, during World War I he worked for the Polish National Council in Paris, giving many fund-raising concerts, and he travelled to the USA to try to win support for an independent Poland. Soon after the end of the war he went to Poznań in German Poland, where he was wildly received, thus ensuring that it would be part of the new Republic of Poland until it was seized by Hitler in 1939. He was briefly Prime Minister under Pilsudski* and attended the Paris Peace Conference* to argue the case for Poland's frontiers. He then retired to California in 1920, where he had acquired an estate. He remained a popular pianist, performing all over the world.

Page, Sir Earle Christmas Grafton (1880–1961), Australian statesman. After serving in World War I he was elected to the federal Parliament as a member of the Country Party*, of which he was leader 1920–39. He formed a coalition with Stanley Bruce in 1923, being deputy Prime Minister until 1929. In 1934 his party joined the government of J. A. Lyons*, under whom he again held office, being leader of an Australian trade delegation to Britain 1936–8. On the sudden death of Lyons in April 1939 he became Prime Minister for a few weeks before being succeeded by Robert Menzies*. During the war years 1941–2 he was invited to join the British war cabinet and after the war joined Robert Menzies's Liberal/Country coalition, being Minister of Health 1949–58, when he introduced a national health plan for Australia.

Pahlavi, Muhammad Reza Shah (1918–80), Shah of Iran 1941–79. He succeeded his father Reza Shah Pahlavi* when he was deposed. After the fall of Musaddiq* in 1953 he used the new oil revenues to finance social and economic development, while sustaining an increasingly repressive regime. He steadily alienated all sections of Iranian society: Islamic zealots rejected his social changes, especially the emancipation of women; he was unsympathetic to the Arab cause; his entourage at court was corrupt and displayed excessive wealth; his secret police SAVAC acted ruthlessly against all left-wing opponents. After severe rioting in 1978 he left Iran in January 1979 'to go on holiday'. He never returned. A sick man, he sought refuge in the USA, Mexico, and elsewhere before being granted asylum in Egypt, where he died.

Pahlavi, Reza Shah (1878–1944), Shah of Iran 1925–41. An officer of the Persian Cossack Brigade, in 1921 he led an army coup which established a military dictatorship under Ahmad Shah. One of his first actions was to conclude a treaty with the new Soviet Union, being successively Minister of War and Prime Minister, before deposing Ahmad and himself becoming Shah. He followed a policy of rapid modernization (army, civil service, legal and educational systems) and economic development, notably through the Trans- Iranian Railway (1927–38). He crushed tribal and other opposition to his policies. In World War II his refusal to expel German nationals led to occupation by Soviet and British forces. He was obliged to abdicate and died in exile in South Africa.

Paisley, the Revd Ian Richard Kyle (1926–), Irish politician. Son of a Baptist minister, after attending the Reformed Presbyterian Theological College in Belfast, he was ordained in 1946. He established his own Church in Ballymena, County Antrim, in 1951, the Free Presbyterian Church of Ulster. This quickly expanded to over thirty churches in Northern Ireland. From the mid-1960s he became leader of extreme Protestant opinion, founding the newspaper the *Protestant Telegraph* in 1966. He was first imprisoned that year for 'unlawful assembly', accusing Britain of supporting unduly the Catholic hierarchy. In 1969 he led a breakaway group from the Ulster Unionist Party*, the Protestant Unionist Party, for which he was elected to the Westminster Parliament in 1970. In 1972 he founded the Democratic Unionist Party, for which he sat in the short-lived Northern Ireland* Assembly 1973–5. In 1979 he was elected to the European Parliament and in 1985 resigned from the House of Commons in protest against the Anglo-Irish Agreement, but he returned in 1986. He has led numerous marches and demonstrations, but in the early 1990s there were occasional signs that, in spite of the rhetoric, he might be willing to seek some form of compromise with the Catholic minority of Northern Ireland.

Pakistan, a federal Islamic republic, consisting of four provinces: Baluchistan, North-West Frontier province, Punjab, and Sind. In addition the state of Azad ('free') Kashmir is economically and administratively linked to Pakistan. At the time of the partition* of the Indian subcontinent the victorious Muslim League* won control of both north-west and north-east India, creating West and East Pakistan (the latter seceding in 1971 as Bangladesh*). At independence a Constituent Assembly was elected, dominated by the Muslim League. Its task was to devise a constitution, and for the interim the federal structure envisaged by the Government of India Act* (1935) was adopted. This established both central and provincial cabinet government responsible to elected assemblies, with wide powers for the Governor-General, the first being Ali Jinnah*. When his Prime Minister Liaqat Ali Khan* was assassinated by an Afghan fanatic in 1951 civil disorder developed, culminating in a state of emergency in 1954. The Assembly finally produced a constitution in 1956, but it was abandoned in 1958, when the country was placed under martial law by Ayub Khan*. Two further constitutions, 1962 (establishing presidential government) and 1973, have been introduced, with military rule consistently alternating with attempts to operate a democratic system. One explanation for the dominance of the military élite may lie in the fact that a high proportion of senior officers in the British Indian army* had come from the north-west of India and were to form a cadre for the new Pakistani army. The general election of 1970 brought Zulfikar Bhutto* to power. His government introduced constitutional, social, and economic reforms, the latter highly necessary after the loss of Bangladesh. It withdrew from the Commonwealth in 1972, following Britain's recognition of the latter. After Bhutto was deposed, military rule resumed under General Zia* until 1988, when the Pakistan People's Party won the election in December and Bhutto's daughter Benazir* became Prime Minister.

Bitterness over Kashmir* has resulted in three Indo-Pakistan wars* and continued tension between the two states, although normalization of relations with India advanced during 1989. With the invasion of Afghanistan by Soviet troops in 1979 President Zia embarked on a massive military build-up, financed by the USA. At the same time Zia faced the need to provide for millions of refugees. To facilitate this and to maintain order he enforced martial law 1981–6. At the time of his death Pakistan was moving towards a return to democracy. However, with ethnic violence following the dismissal of Benazir Bhutto, there were signs of increasing military influence. In November 1990 Mian Mohammad Nawaz Sharif became Prime Minister.

Palestine. Originally the strip of Mediterranean coast between Jaffa and Gaza settled by Philistines, the area known through history as Palestine was fought over by Jews, Syrians, Egyptians, Persians, Greeks, Romans, Muslims, Christian Crusaders, Mongols, and Ottoman Turks, being part of the Ottoman province of Syria 1517–1918. Early in the 20th century the Zionist* movement was arguing that Palestine should become an independent state and a national home for Jews from all over the world. By 1914 some 60,000 Jews were living there as against some half-million Arabs. From 1917 to 1920 the area was under British military rule. The Paris Peace Conference* (1920) agreed to the establishment of a Palestine state, with Britain as a mandatory power. Thus Britain became responsible for a huge area 'Palestine', linking its protectorate of Egypt with the other British mandate of Iraq. However, the largely desert lands east of the Jordan were quickly offered (1921) as an 'emirate' to Abdullah ibn Hussein*, becoming the Kingdom of Transjordan in 1946. British administrators west of the Jordan, based in Jerusalem, had now to try to balance the conflicting promises of the Balfour Declaration* with those made to the Arabs during the Arab Revolt*. There were riots in Jaffa (1920) and Jerusalem (1921), and again in Jerusalem in 1929 at the Wailing Wall, when there were over 200 dead. Arab nationalism was being fanned by the Mufti of Jerusalem, Haj Amin al-Husseini*, resenting increased Jewish immigration and the establishment of the Jewish Agency*. Anti-Semitism in Germany and Poland in the early 1930s increased pressure for Jewish immigration, some 61,000 entering in 1935. In 1936 there was serious rioting. All Arab work and trade stopped and an 'Arab Higher Committee' was formed to co-ordinate protest. Roads were blocked and trains sabotaged. Martial law was imposed, 20,000 British troops brought in, the Supreme Muslim Council in Jerusalem broken up and its members exiled. A Palestinian Conference in London in 1939 failed, but when World War II broke out both sides at first gave support to Britain, although Jewish extremists Irgun* and Stern Gang* soon began activities. In 1945, with thousands of Jews being rescued from concentration camps, a British White Paper rejected any change in immigration policy. This was criticized by the USA, which pressed for an immediate 100,000. In July 1946 British government offices in the King David Hotel were destroyed; Jewish leaders were arrested and deported; a British alternative to partition, the creation of provinces or cantons, was rejected; and in February 1947 Britain agreed to a UN Commission. This recommended partition into a Jewish and an Arab state and the ending of the mandate on 15 May 1948. On 14 May a Jewish state, Israel*, was proclaimed in Tel Aviv. The Arab world refused to recognize this, while Transjordan absorbed the West Bank as far as Jerusalem. Arab–Jewish fighting lasted for a year. The state of Palestine had now ceased to exist, but Palestinians would spend the next forty years seeking a solution to the Arab–Israeli problem.

Palestine Liberation Organization (PLO). It was founded in 1964, largely on Egyptian initiative, as a movement to unite various Palestine Arab groups opposed to Israeli presence in the former territory of Palestine*. It is an 'umbrella' organization for some nine political factions, the largest being al-Fatah* (the Palestine National

Liberation Movement), the Marxist Popular Front for the Liberation of Palestine, and the Democratic Front for the Liberation of Palestine. Its Charter called for a 'democratic and secular Palestine state' and for the 'elimination of Israel'. In 1967 Israeli forces drove Jordan from the West Bank and the PLO became dominated by al-Fatah, which proceeded to organize guerrilla raids by commandos (fedayeen) from Jordan against Israel. In 1969 Yasir Arafat* became its chairman. Increasing tension now developed in Jordan between the PLO, wanting an aggressive anti-Israeli policy, and the King, backed by his army. In September 1970 this erupted into virtual civil war, after which the PLO was forced to leave Jordan. It now became based in Lebanon and Syria. During 1974 a meeting of Arab states in Morocco announced that the PLO was the sole representative of all Palestinians, and its representatives addressed a UN debate. There have been numerous splits and feuds within the movement, with extremist groups such as 'Black September' carrying out assassinations, hijackings, and bomb outrages. The growing strength of the PLO prompted Israel to take action against Palestinian refugee camps in 1981 and to invade Lebanon in 1982, when the PLO officially evacuated Beirut, moving to Tunisia and elsewhere. It remained active in the camps, however, and in other cities. In May 1983 the al-Fatah leadership split and some 5,000 PLO supporters of Arafat were obliged to leave Tripoli in north Lebanon. Arafat rebuilt his power-base and during 1988 persuaded the movement to renounce violence. In November an extraordinary session of the Palestine National Council, its governing body meeting in Algiers, declared the establishment of an independent Palestinian state, which recognized the existence of the state of Israel. In December Arafat again addressed the UN, and the USA began to hold official talks with his representatives. The PLO has been accepted by an increasing number of states as being a government-in-exile.

Palme, Olof Joachim (1927–86), Prime Minister of Sweden 1969–76, 1982–6. Born in Stockholm into a wealthy family, he studied in the USA and then read law at Stockholm University. In 1953 he became personal secretary to Tage Erlander* and was elected to the Swedish Parliament in 1957. As a Social Democrat he held various government offices 1963–9. He succeeded Erlander in 1969 as leader of the Social Democrats and as Prime Minister. In addition to his domestic role he became an active international statesman. In 1977 he led a fact-finding mission to South Africa. He was chairman of an Independent Commission on Disarmament and Security issues, which pressed for a nuclear-free zone in Europe. He was a member of the Brandt Commission*, whose report was published in 1982, and also a member of the UN team which tried to find a solution to the Iran–Iraq War (1980–8). Re-elected in 1985, he was assassinated in a Stockholm street 28 February 1986. His killer was never satisfactorily identified.

Pan-Africanism, a movement for political unity in Africa. This became a positive force with the first Pan-African Conference in London in 1900. An international convention in the USA in 1920 was largely inspired by the Jamaican Marcus Garvey*. A series of further conferences followed. The invasion of Ethiopia by Italy in 1935 produced a strong reaction within Africa, stimulating anti-colonial nationalism, particularly in British West Africa, as well as keeping the ideals of Pan-Africanism alive. The Pan-African Conference in Manchester in 1945 was dominated by Jomo Kenyatta*, Kwame Nkrumah*, and by the 'father of Pan-Africanism', the American W. E. B. Du Bois*. In 1958 a conference of African states was held in Accra, followed by two further conferences in Monrovia in 1959 and 1961. In 1963 in Addis Ababa the OAU* was founded, by which time Pan-Africanism had moved from being an ideal into practical politics.

Pan-Africanist Congress (PAC), a South African movement. A militant offshoot of the African National Congress* (ANC), it was formed in 1959 by Robert Sobukwe. He advocated forceful methods of political pressure such as strikes and boycotts, and sponsored demonstrations such as Sharpeville*, in which sixty-seven Black Africans were killed. The South African government outlawed the PAC as well as the ANC and imprisoned Sobukwe and other leaders. Some PAC members escaped into exile, continuing their campaign under the secretary of the party, Potlako Leballo. By 1990 the PAC chairman Clarence Makwetu was moving away from commitment to 'the armed struggle' and beginning to work with the ANC.

Panama. Despite many nationalist insurrections against Colombia during the 19th century, it only became independent in 1903 as the Republic of Panama. The USA had given help in the so-called 'War of a Thousand Days' in return for a concession to complete the half-built canal across the isthmus, with the lease of a zone around it. The volatile, élite-dominated politics which characterized Panama during the 20th century led to its occupation by US forces in 1908, 1912, 1918, 1941, and 1989. From 1968 until 1981 General Omar Torrijos controlled Panama, working to diversify the economy and to reduce US sovereignty over the Canal Zone, long resented by nationalists. Under his regime the economy flourished, especially in the fields of banking and shipping. Torrijos signed the Panama Canal Treaties (1977) with President Carter and in 1978 announced plans to restore democracy while remaining Commander of the National Guard. In October 1979 US control over the Canal Zone officially ended, but US troops remained and Torrijos was mysteriously killed in an air crash in 1981. In 1983 Panama under President de la Espriella hosted the Contadora Group*, while at the same time accusing the USA of seeking ways to prevent the transfer of the administration of the Canal. Meanwhile in August 1983 Manuel Noriega succeeded Torrijos as Commander of the Panamanian Defence Forces (FDP), as the National Guard was now termed. In spite of accusations of the sale of US national security secrets and of drug-trafficking, he continued to build himself a power-base. In June 1987 the US Senate called on Noriega to step down, but in 1988 he deposed President Del Valle and in May 1989 there were clearly fraudulent and violent elections. Taking a strong anti-USA stance, Noriega had considerable support from the poorer classes, but the business community felt threatened and welcomed the US invasion of 1989–90, when Noriega was removed for trial in Florida and the US-backed Guillermo Endara installed as President. In March 1990 he won a $500m. reconstruction grant from the USA.

Panama Canal. This canal, about 81 km. (51 miles) long, was begun by a French company under Ferdinand de Lesseps in 1882, but not completed until 1914 by the USA at a cost of nearly $400m. In 1903 a treaty had gained for the USA a concession from Panama* for a 16 km.-wide Canal Zone under perpetual control of the US government. After World War II Panamanians became increasingly hostile to this US presence, and following years of negotiation President Carter in 1977 succeeded in obtaining congressional approval of the Panama Canal Treaties, which provide for Panama taking over Canal administration and the USA relinquishing all control of the Canal Zone by 1 January 2000, while assuring the Canal's perpetual neutrality. The regime of General Noriega led to a strengthening of US forces in the Canal Zone and a clear reluctance in the USA to honour the treaty.

Pan-Americanism. The first Pan-American Conference was held in Washington, DC, in 1889, to encourage trade between the twenty-one republics of South, Central, and North America, and to preserve peaceful relations between them.

The seventh Conference at Montevideo in 1933 was important because the USA, in harmony with the Franklin D. Roosevelt 'Good Neighbour' policy, adopted the principle of non-intervention (not always adhered to afterwards!). The Conference at Buenos Aires in 1936 adopted a treaty for the peaceful resolution of conflicts between American states. The Conference at Chapultepec in 1945 agreed on a united defence policy for the signatory nations. At the Conference in Bogotá in 1948 the Organization of American States* (OAS) was established, transforming the *ad hoc* Pan-American system into a formal organization within the framework of the UN.

Panay Incident. On 12 December 1937 Japanese aircraft bombed, machine-gunned, and sank the US gunboat *Panay* as she was sailing down the Chang Jiang (Yangtze) river with refugees from the Nanjing Massacre*. There was an official apology for this 'terrible blunder', but almost certainly it was a deliberately provocative act by the Japanese militarists to dissuade the USA from involvement in China.

Pankhurst, Dame Christabel Harriette (1880–1958), women's suffragette leader. Born in Manchester, she grew up in London and in 1903 persuaded her mother Emmeline* to form the Women's Social and Political Union (WSPU). She worked tirelessly in Lancashire with Annie Kenney when a student, being expelled from a Manchester public meeting for displaying a banner 'Votes for Women'. She graduated from Manchester University in 1906 and in 1907 became secretary of the WSPU. In June 1908 she organized a rally of over half a million supporters in Hyde Park, London, showing great organizational skills. Her campaign became increasingly militant, and she was imprisoned with her mother. She then went to Paris, from where she continued to direct the Union's activities. In 1914 she returned to England and devoted herself to recruiting both men and women for war service. After the war she rejected the idea of maintaining the Union as a force for social change, once political emancipation had been granted. She lived mostly in Canada and the USA, being appointed a DBE in 1936. Her younger sisters, Sylvia and Adela, were both more extreme, and became active socialists.

Pankhurst, Emmeline (1858–1928), women's suffragette leader. Born in Manchester, she married a London barrister and in 1889 founded the Women's Franchise League. An early member of the Independent Labour Party*, but frustrated by lack of progress towards winning the vote, she and her daughter Christabel* founded the Women's Social and Political Union (WSPU) in 1903. After 1906 they adopted increasingly militant tactics. Frequently imprisoned, she used hunger-strike methods to gain temporary release, being sentenced to three years in 1913 for arson. Released after a year, in August 1914 she abandoned her campaign and encouraged women to join the police, the services, or go into industry. She visited the USA to win support for the Allies. She lived in Canada after the war, where she became involved in child welfare. She returned to England in 1926 and joined the Conservative Party shortly before her death.

Papagos, Alexander (1883–1955), Prime Minister of Greece 1952–5. Born in Athens, he was commissioned into the Greek army and took part in the unsuccessful war against Turkey 1919–22. During the early 1930s, by then a general, he became involved in politics, being briefly Minister of War before being appointed by Metaxas* to be Chief of the General Staff with the task of modernizing the Greek army. When this was defeated by the Germans in 1941 he became a prisoner in Germany until 1945. In 1949 he was recalled from retirement to become Commander-in-Chief of the royalist forces in the Greek civil war. He enjoyed close relations with US advisers and his right-wing party the 'Greek Rally' won elections in 1951, when he became Prime Minister. By the time of his death he had become involved in the problems of Cyprus*, which strained relations with Britain and Turkey.

Papandreou, George (1888–1968), Prime Minister of Greece 1944, 1963, 1964–5. Born in Salonika, he trained and practised as a lawyer before entering Greek politics after World War I. As a moderate socialist he served in several governments 1923–35, opposing the regime of Metaxas* and resisting German occupation. He escaped in 1942 and returned to Greece in 1944 to head a brief government. He was always, however, unacceptable to the army, and although his Centre Union Party had a wide following, he never held office for any length of time. He clashed with Constantine II* and resigned in 1965. It was to prevent a new electoral victory of his party in 1967 that the 'Greek Colonels'* seized power. His son Andreas was Prime Minister of Greece 1981–9.

Papen, Franz von (1879–1969), Chancellor of the Weimar Republic 1932. From an aristocratic family, he entered the German Imperial Diplomatic Service, but during World War I was Chief of the Turkish General Staff in Palestine. During the 1920s he joined the Catholic Centre Party* and organized an aristocratic group in Berlin, the Herrenklub, which persuaded Hindenburg* to appoint him Chancellor in 1932. Although at that time opposed to Hitler he lifted the ban on Brownshirts*, but resigned as support for the National Socialists grew. He changed policy, persuading banking and industrial interests to support a National Socialist government, believing that, as his Vice-Chancellor, he would be able to restrain Hitler. In this he proved mistaken. Many of his friends died in the Night of the Long Knives*, but he survived, being ambassador to Austria (1934). He worked for Anschluss* in 1938 and was then ambassador to Turkey 1939–44. He was tried as a war criminal but acquitted. Rearrested under de-Nazification laws, he served three years in a labour camp before release in 1952.

Papua New Guinea, an island republic in Melanesia*. The island of New Guinea* had been the scene of some of the heaviest jungle warfare in World War II. Over 1 million US and Australian troops fought across it, their presence greatly accelerating a process of Europeanization. In 1949 the eastern half of the island, together with those islands on the Bismarck and Solomon archipelagos for which Australia had received the League of Nations mandate* in 1920, were united administratively to form Papua New Guinea, as a UN Trust Territory for which Australia was again responsible. Australia had established legislative councils before World War II, but only Europeans or Australians had sat on these; the first Papua New Guineans were elected to a House of Assembly in 1964. In 1965 a university was founded at Port Moresby, whose first graduates quickly formed a nationalist movement. From 1967 the first political party Pangu Pati was formed, demanding independence. After an election in 1972 self-government was granted (December 1973) and independence was gained in September 1975. Strong regional loyalties have survived and some of the islands have called for secession. There is a defence treaty with Australia, whereby the latter would assist Papua New Guinea in repelling any invasion, for example across its long frontier with Indonesian Irian Jaya. Extensive deforestation has caused international concern, one of its effects being detribalization. Papua New Guinea is a member of the Commonwealth of Nations. Its Prime Minister in 1987, Paias Wingti, leader of the People's Democratic Movement, was a strong critic of French policies in the Pacific. In May 1990 the Pangu Pati government of Rabbie Namaliu faced a declaration of independence from the island of Bougainville.

Paraguay, a south American republic. It is estimated that it had lost 70 per cent of its population in a disastrous war with its neighbours 1864–70. Political turmoil followed into the 20th century. In 1912 Edward Schaerer of the Liberal Party became President (1912–17). He achieved some economic improvement financed by foreign investment, together with social improvement for the peasantry, some of the poorest on the continent. During the 1920s armed conflict developed along the Bolivian border, leading to the Chaco War* 1932–5. Although Paraguay won the war it was a costly victory, with the Chaco failing to produce the expected oil. A dictatorship under Higinio Morínigo* was followed by civil war between the Liberal and Colorado parties. The latter won, but its leader Frederico Chávez was deposed by General Alfredo Stroessner* in May 1954. Stroessner had some support from the USA and gained IMF* assistance in curbing inflation. He purged the Colorado Party to make it an instrument for his personal, autocratic rule. With a fast-growing population there was an urgent need for land reform, and there was some redistribution to landless peasants. A massive hydroelectric scheme was begun in 1973 to stimulate economic growth, but cattle exports to Europe declined with the development of the European Community. From 1985 there was a fall in living standards, while Stroessner's repressive regime lost the support of the USA. By 1987 Storessner could not prevent the re-emergence of political protest and he was deposed in February 1989. Elections brought General Andrés Rodríguez to office as President, himself a wealthy landowner. Eighty per cent of Paraguay's land remained in the hands of 1 per cent of landowners, many of these being foreign corporations.

Paris Peace Accords (January 1973). A series of agreements reached in Paris and signed by France, the USA, the Soviet Union, China, Britain, and representatives of the National Liberation Front* and of North Vietnam. They ended five years of negotiation. In 1968, following the Tet Offensive*, President Johnson of the USA had agreed to end the bombing of North Vietnam in return for an attempt at negotiation. This began in 1969, with representatives from the National Liberation Front and Tho Le Duc* from the VietMinh. After many false starts, including periodic resumptions of air warfare and the bombing of Cambodia and Laos, Henry Kissinger* and Le Duc finally reached an agreement on 27 January 1973. It was repudiated by President Thieu* of South Vietnam, who determined to continue the struggle alone.

Paris Peace Conference (1919–20), conference of delegates of the belligerent powers in World War I, to draw up a series of peace treaties. It first met on 18 January in the Palace of Versailles, and the early sessions were dominated by 'the Big Four'—Clemenceau*, Woodrow Wilson*, Lloyd George*, and Vittorio Orlando*. Bitter disagreement as to the treatment of Germany developed, with the French delegation demanding the harshest possible terms. The main Treaty of Versailles* with Germany was agreed upon in June 1919, although it left details on reparations* to be finalized. The Treaties of St-Germain* and Trianon* involved endless debate upon frontiers, as the new state of Czechoslovakia was established and the Kingdom of Serbs, Croats, and Slovenes was recognized. The Treaty of Sèvres* was to be abrogated and renegotiated at Lausanne*. The Treaty of Neuilly* was agreed by Bulgaria in November 1919. The other achievement of the Conference was the establishment of the League of Nations* and the adoption of its Covenant. Its lawyers set about devising the concept of mandate*. After July 1919, foreign ministers, sitting in a Council of Heads of Delegations, negotiated the details of the treaties, which failed to eliminate a number of potential flash-points, for example Teschen* and Fiume*. Following the Russo-Polish War* Poland's frontiers were to be settled by the separate Treaty of Riga in 1921. Great bitterness was engendered in that the defeated states were denied the chance to negotiate, having to accept peace terms dictated by the Conference.

Paris Peace Treaties (1947). From July 1946 until February 1947, delegates from twenty-one nations met in Paris to decide on peace terms for Germany's five allies in World War II: Bulgaria, Hungary, Finland, Italy, Romania. Italy was obliged to cede most of the the Istrian peninsula, including Fiume (Rajeka), and some Adriatic islands to Yugoslavia, and the Dodecanese to Greece; it had also to accept some minor frontier adjustments and renounce all claims in Africa. Romania regained Transylvania*, but ceded Bessarabia (see MOLDAVIA) and the northern Bukovina* to the Soviet Union; south Dobrudja*, regained from Romania by Bulgaria in 1940, would remain Bulgarian. Trieste was to be a free city, a status it retained until 1954; Hungary would be limited to its frontiers of the Treaty of Trianon*; Finland had to cede Petsamo to the Soviet Union. This settlement proved far more lasting than had that of 1919–20.

Park, Chung Hee (1917–79), Korean statesman and President of the Republic of Korea 1963–79. After having served in the Japanese army 1940–5, he made his career in the army of South Korea. A member of the military junta which seized power in 1961, he became President of the Republic in 1963. In 1971 he banned all anti-government activity and many intellectuals were arrested and killed. He proclaimed martial law in 1972 and was then re-elected twice in 1972 and 1978. His government became increasingly corrupt and authoritarian, and he was assassinated by the head of the South Korean Central Intelligence Agency on 26 October 1979.

Parliament, British. It is the supreme legislature in the United Kingdom*, comprising the sovereign, as head of state, and two Chambers which sit in the Palace of Westminster—the House of Lords and the House of Commons. The Prime Minister, from the majority party in the latter, appoints the cabinet of ministers from either House, which is responsible for formulating policy. Acts of Parliament in draft, known as Bills, each of which has to be 'read' (debated) three times in each House, are referred in the Commons (and occasionally in the Lords) to parliamentary standing or select committees for detailed consideration. The House of Lords is the ultimate British Court of Appeal. The sovereign's powers of government are entirely dependent on the advice of ministers, who in turn are responsible to Parliament. Reform of the House of Lords has been an issue on the political agenda throughout the 20th century. The Parliament Act of 1911 imposed general elections once in five years instead of seven (except in times of emergency) and reduced the power of the Lords to a suspensive veto of two years. An Act in 1949 reduced this to one year, while in 1958 women were admitted to the Lords and the rank of life-peer was created. There were several commitments for more radical reform of the Lords, but none was realized; for example governments in 1922, 1927, 1933, 1945, and 1964 all pledged change, as did the Labour Party's policy review of 1989. Reforms for the election of Members of the House of Commons have continued more successfully. In 1918 the Representation of the People Act added some 8 million to the electorate, including 6 million women over 30. That of 1928 reduced the voting age of women to 21. That of 1948 abolished 'plural voting' (university graduates and some businessmen had had a second vote), while that of 1969 lowered the voting age to 18. Parliamentary constituencies have been steadily adjusted to reflect population shifts through the century, the aim being constituencies of between 40,000 and 80,000. The Single European Act (1986) laid down that decisions of the European Community's institutions should, if necessary, take precedence over those of Parliament. The British parliamentary system was adopted by many European countries and by most countries of the the Commonwealth on gaining independence, but without a hereditary second chamber.

Parsees, followers in India of the 9th-century BC Iranian prophet Zoroaster. They appear to have fled to India to escape Muslim dominance during the 8th–10th centuries AD, settling in the area of Bombay. They later traded successfully with the Portuguese and British, and during the 19th and 20th centuries their wealth helped to industrialize India. Parsee religious practice involves a great many ancient rituals and the recitation of fragments of sacred texts, in worship of the one God, Ormuzd. Such worship enables the soul (*urvan*) to overcome the influence of evil, the work of Ahriman, in the great cosmic battle of good versus evil.

Partai Kommunis Indonesia (PKI). It was formed in 1920 and was early active in trade union activities in the Dutch East Indies*. Severe Dutch repression, especially after abortive uprisings in Banten and west Sumatra 1926–7, led to its virtual eclipse, although it was involved in guerrilla activity against the Japanese during the occupation. In 1948 there was a confrontation with the nationalist PNI party in east Java during the Indonesia Revolution*. This rivalry between the two parties survived into the 1950s, when the PKI became one of the largest Communist parties outside China and the Soviet Union, winning 20 per cent of the vote in the 1955 election. The military was determined to break the party, and an attempted coup in 1965 gave them their opportunity. Estimates of the number of victims vary between 300,000 and a million, but General Suharto's* purge in the years 1966–8 was one of the most ruthless of 20th-century massacres, virtually wiping out the party.

Parti Québécois. Founded in October 1968 from the Mouvement Souveraineté Association and the Ralliement National, by René Lévesque*, its aim was unilaterally to declare Quebec a separate state and French its official language. The separatist movement had received wide publicity in July 1967, when President de Gaulle had exclaimed 'Vive Québec libre' in Montreal, and it gathered pace during the 1970s, weakening the Canadian Liberal Party, which had traditionally had the support of French Canadians. When however it won a majority in the Quebec National Assembly in November 1976, the efficient administration of Lévesque was far from radical. In 1980 a referendum on secession gave a 60 per cent 'non' vote, after which the party began to disintegrate, but it started to revive again after the collapse of Brian Mulroney's* Meech Lake Accord in 1990.

Partisans, the term used by Communist-dominated guerrilla resistance groups in World War II. It had originated in Russia when Napoleon invaded, and Russian partisans inflicted immense damage on German supply lines during the campaigns of 1941–3. There were active partisan groups in Slovakia, Moravia, Albania, Greece, as well as Italy, and in Yugoslavia. Here Tito's National Army of Liberation and Partisans clashed with Mihailovich's chetniks*; operating from the mountains of Montenegro and Bosnia-Hercegovina they linked up with the Red Army as it advanced during 1944.

Partition of India. In March 1942 Stafford Cripps* led an unsuccessful mission to India. In May, with Japanese forces at the frontier, Mohandas Gandhi* called upon the British to 'Quit India' and leave Indians to resolve the problem with Japan. He and Congress leaders were imprisoned and a brief period of guerrilla violence followed, which was quickly suppressed. In April 1944 C. R. Rajagopalachariar* produced a plan to resolve the Hindu–Muslim problem; put simply it was that a referendum be held among Muslim areas as to whether a majority wanted secession into a new homeland. In Britain a new Labour government came to power in 1945, firmly committed to independence. In the 1946 Indian elections the Muslim League* won massive support, and the struggle in India now became one between the League, led by Jinnah*, and the Congress, led by Nehru*, Gandhi, and V. Patel*, who worked tirelessly to persuade Indian National Congress* members to accept the principle of partition. Frustrated by delays, the Muslim League called for 'Direct Action', and announced that 6 August 1946 would be 'Direct Action Day'. This produced massive strikes, but also horrendous atrocities, 4,000 people dying in Calcutta. In London Attlee* announced that independence would be no later than June 1948 and appointed as Viceroy Louis Mountbatten* to replace Wavell*. Mountbatten now proposed that the elected Legislative Assemblies of the thirteen provinces choose whether or not they wanted partition. For Sind, Baluchistan, and the North-West Frontier it was a clear vote for a new, as yet unnamed, state. Both Bengal* and the Punjab* voted to divide themselves; thus Muslim East Bengal would become part of the new Muslim state, separated by thousands of miles from the rest, and so would West Punjab. The Sikh* community would be divided. Mountbatten now announced 15 August 1947 as the day of transfer. There were ten weeks to find a name for the new state—Pakistan; to agree on frontiers, only done with great bitterness; to divide Indian finances between the two new states; to dismantle the Indian army* and create two new armies; and to persuade the princely states to participate in the whole operation. For millions of peasants in the areas of partition there was suffering and loss. It is estimated that some 8 million refugees moved each way into both India and Pakistan, with perhaps 200,000 dying in the process. The two new states were almost immediately at war over Kashmir*. Yet probably only the decision to divide the subcontinent prevented far worse bloodshed in massive civil war.

Pašić, Nikola (1845–1926), Prime Minister of Serbia five times and of the Kingdom of Serbs, Croats, and Slovenes 1921–4 and 1924–6. Born in Zajecar in Serbia, he studied engineering in Belgrade and Zurich, where he became interested in politics. First elected to the Serbian Parliament in 1878, he founded the Radical Party and was exiled 1883–9. He dominated Serbian politics from 1891, when he first became Prime Minister, until his death. Although suspicious of Croats on both political and religious grounds, his ideal was of a 'Greater Serbia' which would include much of Croatia and Dalmatia. In 1917 he skilfully negotiated the Corfu Pact*, which resulted in a union of Serbs, Croats, and Slovenes into the Kingdom which he represented at the Paris Peace Conference* in 1919. He was twice Premier of the new Kingdom, which in 1929 was to adopt the name Yugoslavia, always seeking conciliation within the precariously balanced ethnic groups.

Passchendaele, Battle of, World War I (31 July–10 November 1917), the third Battle of Ypres*. The name of this Belgian village became notorious for the worst horrors of trench warfare during World War I. Haig*, the British Commander-in-Chief, remained convinced, despite the Somme*, that frontal assaults in superior numbers must succeed, his plan being to seize the Passchendaele ridge and then sweep into Belgium. A preliminary bombardment and torrential rain, which lasted all through August, reduced Flanders to a sea of mud, making advance impossible. Only on the final day, 9 November, did the Canadians reach the ruined village of Passchendaele, by when the Allies had suffered some 300,000 casualties. Even this nominal gain was surrendered in the retreat before Ludendorff's* final offensive in April 1918. The Germans first used mustard gas during this three-month-long battle.

Passive resistance. This usually involves a refusal to co-operate with government authorities and/or a defiant, non-violent breach of laws and regulations. During the 20th century it has been a major weapon of many nationalists, among wartime resistance groups, and in social movements. One of the most successful campaigns was that waged by Gandhi* against British rule in India, when widespread and large-scale civil disturbances and protests persuaded the British to make major concessions. Gandhi's example was an inspiration for the civil rights movement* in the USA from the 1950s.

Patagonia. In the later 19th century Chilean and Argentinian rangers began settling in this territory south of the 44th parallel, previously occupied by Indians. They fell into bitter boundary disputes, which were resolved in 1901 when Chile ceded the area to Argentina. Europeans, notably Basques, Welsh, and Scottish, immigrated in the early 20th century. The pastoral economy of the region was challenged after World War II by the discovery of iron ore, petroleum, uranium, and natural gas deposits, all of which began to be exploited. By 1989 the population of its four provinces Chubut, Rio Negro, Santa Cruz, and Tierra del Fuego totalled approximately 1 million.

Patel, Sardar Vallabhbhai (1875–1950), Indian statesman. A successful criminal lawyer, in 1917 he met and was deeply influenced by Mohandas Gandhi*. He became the principal organizer of many civil disobedience campaigns in the 1920s, suffering frequent imprisonment by the British. He held many posts in the Indian National Congress*, including that of president in 1931. He played an important role in the negotiations for the partition of India*. As Indian deputy Prime Minister (1947–50) and with the assistance of V. P. Menon (1894–1966) he was responsible for the integration of 562 of the princely states into the Indian Union. As Minister for Home Affairs, he firmly opposed extremist action by Communists, but was a skilful negotiator in the interests of harmony between Hindus and Muslims in India.

Pathans (or Pashtuns), a Pashto-speaking Muslim people of Pakistan and Afghanistan, more especially the tribesmen of the mountainous region along the north-west frontier of Pakistan. In 1893 an international frontier, the Durand Line, had been established, with about two-thirds of Pathans living in British India and one-third in Afghanistan. In spite of protest by the Red Shirt movement* this line was inherited by Pakistan in 1947, when the British North-West Frontier Province voted to join the new state. Red Shirt supporters of Ghaffar Khan (see RED SHIRTS) continued to press for a Pathan homeland, Pathanistan or Pakhtunistan, and this was a factor in the civil war in Afghanistan 1979–89.

Pathet Lao, a nationalist group in Laos founded in 1950. It organized resistance to French attempts to reimpose control during the French Indo-China War*, using guerrilla methods. By 1953, when France recognized the Kingdom of Laos as independent, it had established itself in the north-east, where it claimed to be the legal government for all Laos. After the Geneva Agreements*, its political wing, the Lao Patriotic Front, took part in various Laotian governments. In 1965 its military wing renamed itself the Lao People's Liberation Army and assisted North Vietnam to use the so-called Ho Chi Minh Trail through its territory. A state of virtual civil war lasted in Laos until 1972, when negotiations began, which resulted in a cease-fire in February 1973, following the Paris Peace Accords*. In 1974, after further negotiations, it joined a Provisional Government of National Unity, with equal numbers from Pathet Lao and from supporters of Phouma Souvanna* in Vientiane. When the right-wing governments in Saigon and Phnom Penh collapsed in April 1975 and US advisers departed, Pathet

Lao troops swept through Laos into Vientiane. In December the Lao People's Democratic Republic was proclaimed, with the Communist People's Party of Laos in control.

Patiño, Simón Iture (1860–1947), Bolivian capitalist. Born in the Cochabamba Valley of Bolivia of mestizo artisan background, he was apprenticed in a silver mine and then in 1894 bought his first share in a tin mine in Oruro. By 1897 he had a controlling interest and in 1900 struck the richest vein of tin ever. His enterprises expanded rapidly and he became a powerful influence on the Bolivian government before World War I. In 1916 he took control of the world's largest tin smelter in Liverpool. When President Bautista Saavedra restructured the tax on tin in 1923 he moved Patiño Mines and Enterprises out of Bolivia to Delaware, USA. By 1924 Patiño Enterprises dominated Bolivian mining, with 50 per cent of production and over 10,000 workers. From now on he lived permanently abroad in the USA and Europe, dying one of the world's richest men. He is still criticized in Bolivia for not investing more in the development of his own country.

Patriotic Front, a Zimbabwean Political Party. From 1974 until 1979 the two rival parties for independence in Zimbabwe, ZANU (Mashona-based) and ZAPU (Matabele-based), worked in uneasy alliance as the Patriotic Front. This split in 1982 when Joshua Nkomo* left the government of Robert Mugabe*. In July 1987 the Constitution of Zimbabwe was altered to end the reservation of parliamentary seats for Whites; prolonged discussions followed to end the period of hostility between the two parties. Finally in December 1987 the two merged under the name ZANU (Patriotic Front).

Patton, George Smith (1885–1945), US general. Born in California, he graduated from the US Military Academy in 1909 and joined the staff of General John Pershing*, with whom he went into Mexico in 1916 and to France in 1917. Between the wars he became an expert on tank warfare, and in World War II commanded a corps in North Africa and then the 7th Army in Sicily. He temporarily lost his command in 1943, after a publicized incident in which he hit a soldier suffering from battle fatigue, but in 1944 he led the US 3rd Army in the Normandy campaign*. His tendency to make rapid military advances, skilfully deploying his tank units, but at times with no regard for support units or allies, became evident in his spectacular sweep through France, across the Rhine, and into Czechoslovakia. As military governor of Bavaria he was criticized and relieved of his command in October 1945 for his leniency toward Nazis. He was killed in a motor accident in December 1945 while commanding the US 15th Army.

Paul VI (1897–1978), Pope 1963–78. Born Giovanni Battista Montini, near Brescia, he was ordained priest in 1920. He spent much of the next thirty years working in the Vatican, until appointed Archbishop of Milan in 1954. Elected Pope to succeed John XXIII* he continued many of the latter's policies, reconvening the Second Vatican Council* 1963–5. Important post-conciliar commissions were set up on liturgical reform, Christian unity, lay participation, and reform of the *Curia.* He was the first pope to travel out of Italy, going to India, Central and South America, and the USA, where he offered to mediate in the Vietnam War. His support of the use of the vernacular in the Mass offended many, but in matters of personal morality he was a traditionalist and suspicious of any innovation which might undermine the authority of the Church.

Paulus, Friedrich von (1890–1957), German field marshal. A professional soldier who served in World War I, he had risen to the rank of field marshal by 1939. As deputy Chief of Staff he planned the German invasion of the Soviet Union, 'Operation

Barbarossa'*. In 1942 he failed to capture Stalingrad, was surrounded, and surrendered in February 1943. In captivity he joined a Soviet-sponsored German organization and publicly advocated the overthrow of the Nazi dictatorship. He lived in East Germany until his death.

Pavelić, Ante (1889–1959), Croatian Fascist leader. Born in Bradina in Bosnia, he trained and practised as a lawyer in Zagreb, entering local politics in 1920. Elected to Parliament in 1927, he criticized the increasing centralization of government and its dominance by Serbs under Alexander I. He fled to Italy in 1929 when the latter assumed dictatorial powers. Here he became leader of the Uštaše* terrorist movement and organized Alexander's murder in 1934. As head of the Independent State of Croatia 1941–5 he was responsible for brutal atrocities against both Jews and Orthodox Serbs. He went into hiding in 1945 and then escaped to Argentina, dying in Madrid.

Paz Estenssoro, Victor (1907–), President of Bolivia 1952–6, 1960–4, 1964, 1985–9. Born in Tarija and educated at the University of Mayor de San Andrés, he worked in finance and banking in the 1930s before becoming a university Professor of Economic History and government economic adviser. In 1941 he entered politics and helped to form the left-wing party Movimiento Nacionalista Revolucionario (MNR), becoming Minister of Finance 1941–4. In 1952 he was elected President. The suffrage was granted to Indians, who also gained grants of arable land on the central plateau, and the tin mines were nationalized. He was re-elected President in 1960, when he reached an understanding with international financiers for the reorganization of the tin industry. Elected a third time in 1964, he was overthrown by the army and went into exile, becoming a university teacher in Peru until 1971, when he returned to Bolivia. Military government ended in Bolivia in 1982, and in 1985 he was re-elected President in succession to Dr Siles Zuazo.

Peace Corps, a US government agency. It was created by President Kennedy* in 1961 with the aim of sending volunteers to developing countries to work in such fields as education, agriculture, health, technology, and so on. It grew from some 900 volunteers working in sixteen countries in 1961 to 10,000 working in fifty-two countries by 1966. The project is parallel to the British Voluntary Service Overseas scheme established in 1958. Volunteers are expected to work for two years, to speak the native language, and to adopt the life-style of their host country. During the 1980s a greater emphasis was placed on higher skills. This resulted in a drop in the number of volunteers, but an increase of receiving countries to sixty-three.

Pearl Harbor, a harbour on the island of Oahu in Hawaii. It is the site of a major US naval base, where a surprise attack by Japanese carrier-borne aircraft on 7 December 1941, delivered without a prior declaration of war, brought the USA into World War II. A total of 188 US aircraft were destroyed, and eight battleships sunk or damaged. The attack was a strategic failure, however, because the crucial element of the US Pacific fleet, its aircraft carriers, was out of the harbour on that day.

Pearson, Lester Bowles (1897–1972), Prime Minister of Canada 1963–8. His studies at Toronto University were interrupted by overseas service in World War I, after which he taught at Toronto until entering the diplomatic service, going to London in 1935 as a first secretary and to the USA in 1945 as Canada's first ambassador. He was a delegate to the founding conference of the UN, and as Canadian Minister for External Affairs (1948–57) helped to establish NATO. He was active in the UN over Palestine, Korea, and Suez in 1956, when he helped to create the UN Emergency Force. In

January 1958 he became leader of the Canadian Liberal Party, then in opposition, and Prime Minister in 1963. Although rocked by scandals, his government greatly strengthened social welfare provision by a 'Canada Assistance Plan', a 'Canada Pension Plan', and 'Medicare'*. It united the armed forces and agreed on a new national flag. During his years in office the Quebec separation movement developed, the Parti Québécois* being founded in 1968, the year he handed the Liberal Party over to Pierre Trudeau and retired.

Peel Commission. Early in 1936 there was serious rioting in Palestine, especially Jerusalem, with Arab attacks against Jewish settlements. The British government set up a Royal Commission under Lord Peel, a former Indian Secretary. The Commission reported in July 1937, recommending partition of Palestine into a Jewish and an Arab state, with Britain retaining direct control over the holy places of Bethlehem, Nazareth, and Jerusalem. The recommendations were rejected by both Arab and Jewish communities, and in the House of Lords. It was the first public proposal of partition as a solution to the Palestine problem.

Peng Dehuai (P'eng Te-huai) (1898–1974), Chinese military leader. As a poor orphan from Hunan province he worked on the land and in a coal-mine before joining the Hunan provincial army in 1918. By 1927 he was a major, but as a socialist he was 'purged' when the Guomindang* and the Communists split. He joined the Chinese Communist Party and became leader of a Chinese Workers' and Peasants' Army of guerrillas against the Nationalists. He joined Mao Zedong* on the Long March and served with distinction under Zhu De* in the war against Japan and later in the Civil War, recapturing Yan'an in 1948. As Commander of the Chinese forces in the Korean War* he signed the Panmunjom armistice in 1953 and next year became Minister of Defence and a member of the Politburo of the CCP. In 1959, however, he was removed from office for criticizing Mao Zedong's great Leap Forward* policy. In 1978 he was rehabilitated.

Peng Pai (P'eng P'ai) (1896–1929), Chinese Communist leader. Born into a wealthy family in the Guangdong province, he enrolled in Tokyo University in 1918. Here he became a socialist, returning to China in 1921 to join the newly formed Chinese Communist Party. He was appointed superintendent of education in Haifeng hsien, and formed a Haifeng Peasants' Association, which embarked on a policy of land expropriation, rent reduction, and the elimination of non-co-operative landlords. The Association was enlarged into a Peasants' Bureau, in which Mao Zedong* was later involved and which successfully organized thousands of peasants in the Guangdong province. When the Communists and the Guomindang* split in 1927 Peng raised a Peasants' Revolutionary Army, but he was captured and executed by the Nationalists.

Pentagon Papers, a 1967 leaked official study of US policy in South-East Asia. Extracts were first sent to the *New York Times* in June 1971 by a former government employee, Daniel Ellsberg, revealing miscalculations, deceptions, unauthorized military offensives, and policy clashes between Robert McNamara* and President Johnson*. Their full publication in forty-seven volumes provoked demands for more open government, and stimulated opposition to the Vietnam War*.

People's Liberation Army (PLA). In August 1945 the armies of the Chinese Communist Party under Zhu De* and Lin Baio* received the surrender of numerous Japanese units in north China. On 1 May 1946 they adopted the title the People's Liberation Army and engaged in three years of civil war against Jiang Jiehi's (Chiang

Kai-shek) National Revolutionary Army*, which now had Japanese co-operation and vast US financial and military support. By the end of 1949 what remained of the Nationalist army had retreated to Taiwan*. During the 1950s the PLA played little part in internal Chinese affairs, while becoming involved in the war in Korea. In 1959 Lin Baio became Defence Minister and by 1966 the army had nuclear capability. As a result of the Red Guards* movement during the Cultural Revolution*, the army was employed to restore order. Lin Baio appears to have been planning a coup by the PLA when he was killed in 1971. During the next decade the PLA, under ageing senior officers, became an increasingly conservative force, and on 4 June 1989 units brought into Beijing committed the massacre of Tiananmen Square*.

Pérez de Cuéllar, Javier (1920–   ), secretary-general of the UN 1982–   . Born in Lima of aristocratic birth, he entered the Peruvian diplomatic service after reading law at the university. He served in France, the UK, Brazil, and the Soviet Union, and in 1971 became Peru's ambassador to the UN. In 1979 he was appointed UN under-secretary for political affairs and in 1981 secretary-general to succeed Kurt Waldheim*. He failed to resolve the Falklands dispute between Britain and Argentina, but had greater success with his efforts to bring peace to Afghanistan and Namibia and to end the Iran–Iraq War. He was unable to resolve the Gulf Crisis* of 1990.

Perkins, Frances (1882–1965), US Secretary of Labor 1933–45. Born and educated in Massachusetts, she took up church social work, working in New York City from 1910 until 1929, when she was appointed State Industrial Commissioner (1929–32) by Governor Franklin D. Roosevelt. In the New Deal*, against bitter business and political opposition, she became Roosevelt's Secretary of Labor, the first woman cabinet member in the USA. She played an active role in minimum wage and maximum hours legislation, and helped to draft the Fair Standards Act (1938). Her influence was particularly evident in the Second New Deal after 1935, when she gained increased responsibilities and much respect for her skilful administration.

Perón (Duarte de), Eva (1919–52), second wife of Juan Perón*. A minor actress, Evita, as she was called, became the mistress of Juan Perón when he was bidding for power in 1945, basing his support on the *descamisados* ('shirtless ones'). In August Perón was imprisoned as a result of a political coup, but Evita rallied his supporters and he was released. On 17 October he and Evita addressed a crowd of 300,000 from the presidential palace. A few days later she married him, and on his election as President she became in effect Minister of Health and Labour. She founded the Eva Perón Foundation, funded by national lottery, which established hospitals, schools, and old people's homes. By 1947 she owned or controlled almost every radio station in Argentina and had closed or banned over 100 newspapers and magazines. A gifted orator, she was a militant champion of women's rights, winning them the vote. Although she knew she was dying of cancer she bid for the vice-presidency in 1951, only to be blocked by the army. Her death at the age of 33 shocked her Perónist followers, who tried unsuccessfully to have her canonized. For twenty years there were disputes over her body, which in 1955 was stolen and kept in Italy for sixteen years, before finally coming to rest in her family tomb.

Perón (Domingo), Juan (1895–1974), President of Argentina 1946–55, 1973–4. Born in a small town on the Argentine Pampas, at 16 he entered military school. He was commissioned and during the 1930s was military attaché in Italy, where he became an admirer of Benito Mussolini. In 1943, with the rank of colonel, he joined a military junta which seized power, and became Minister of War. A counter-coup in 1945 placed him briefly in prison, but he was released with the help of his future wife (*see* PERÓN, EVA) and went on to be elected President in February 1946. Preaching the virtues of 'the Third Position', between Communism and capitalism, he adopted a strongly anti-British and anti-USA stance, seeking to establish Argentina as the leading South American power. He nationalized railroads and utilities, sponsored a wide programme of public works, and established a social-security system. He was fortunate in that Argentina had amassed a large foreign exchange surplus during World War II. His policies favoured the industrial proletariat at the expense of the rural worker. He was re-elected in 1951, but he became increasingly dictatorial, having alienated many of the wealthier sections of Argentine society. In September 1955 he was deposed by the army and went into exile in Spain, where he married his third wife. In Argentina he had left a strong Perónista Party, and when civilian rule was restored in 1972 Perón returned. He was elected President in March 1973 among scenes of wild Jubilation, with his wife Isabel as Vice-President. He died a year later. Isabel vainly tried to hold the office of President, but was deposed by the army in March 1976.

Perónism, a vague ideology based on the ideas and practices of President Juan Perón*, who himself was influenced by Italian Fascism of the 1930s. Social welfare by the establishment of a social-security system, guaranteed minimum wage, protection of the rights of trade unions, concern for the health and safety of industrial workers, public ownership of basic utilities and industries, all form part of its thinking, together with a nationalist fervour and a reluctance to allow political debate. After Perón was exiled to Spain in 1955 the Perónista Party retained the support of millions, who in 1972 brought their hero back into office. By then, however, there were splits appearing left and right, and violence between the wings began to develop. By 1989 the Perónista Party had changed its name to the Judicialist Party. In May their candidate Carlos Menem was elected President and the party won a majority in the Chamber of Deputies.

Pershing, John Joseph (1860–1948), Chief of Staff of the US army 1921–4. Born in Montana, he graduated at West Point in 1886 and served in the later Indian wars, gaining the nickname 'Black Jack'. He served in the Spanish-American War (1898) and in the Philippines. In 1913 he became commander of the 8th Cavalry Brigade, and led a US expedition into Mexico in 1916. In May 1917 he was appointed commander of the American expeditionary forces in France. His talent for organization was largely responsible for the moulding of hastily trained US troops into an army of well-integrated combat troops. He strongly resisted pressure from the British and French to integrate the AEF into their armies. In 1919 he became general of the armies of the USA and in 1921 army Chief of Staff. His assault in September 1918 of the Saint-Mihiel salient marked the beginning of the Allied advance to victory in World War I.

Peru. In 1821 José de San Martín proclaimed it an independent republic. By the end of the 19th century its prosperity was reduced by the loss of its saltpetre mines to Chile in 1884 and the decline of its guano deposits. In 1889 national bankruptcy was proclaimed. By then civilian politics had emerged, with two parties, the Democrats and the Civilians, alternating in office. The latter, led by Augusto Leguía*, held power 1908–30, introducing much progressive legislation and settling the long Tacna–Arica dispute. President Manuel Prado, elected in 1939, aligned Peru with US policies during World War II. After the war the APRA party of Haya de la Torre*, which sought greater participation in politics by Indians, gained increasing influence, although the army was in power from 1948 until Terry Belaúnde* gained office in 1963. In 1968 a left-wing army junta returned, seeking to nationalize US-controlled industries. A more moderate junta succeeded in 1975 and elections were again held in 1979, when Belaúnde was re-elected under a new constitution, largely the work of Haya de la Torre. In the face of severe economic problems Belaúnde succeeded in redemocratizing the country, and in 1985 President Alan García was elected. Confronted by large rescheduling requirements for Peru's foreign debts, his regime imposed an austerity programme and engaged in a campaign against the Sendero Luminoso*. In 1986 270 of the latter's supporters were killed in prison riots and there were student protests against the alleged brutality of García's government. At the same time, a programme of bank nationalization alienated Peru's international creditors. García's APRA Party did badly in the 1990 elections, when the son of Japanese immigrants, Alberto Keinya Fujimori, of the Cambio 90 Party, was elected President. His austerity programme quickly resulted in protests.

Pétain, Henri-Philippe (1856–1951), marshal of France and head of state 1940–4. Commissioned at the French Military Academy of St-Cyr in 1878, he had held numerous military posts before 1914, by which time he had reached the rank of general. He was acclaimed a military hero for halting the German advance at Verdun (1916), and replaced Nivelle as French Commander-in-Chief in 1917, under Generalissimo Foch*. After the war he commanded the French army against Abd el-Krim* and in 1929 became Inspector-General of the Army. He became increasingly a symbol of the conservative elements in French society. He was appointed Minister of War in 1934 and France's first ambassador to Franco in 1939. In 1940 he was recalled to succeed Reynaud* as Prime Minister, and concluded an armistice with Germany. This provided that French armed forces be disbanded and that three-fifths of France be surrendered to German control. The French National Assembly established its seat at Vichy and conferred on him the rank of 'Head of the French State', with power to establish an authoritarian regime which would purge France of 'moral decadence'. He designated Laval* as Vice-Premier and Foreign Minister, but later dismissed him for Darlan*. German forces occupied Vichy France in 1942 and Pétain was forced to reinstate Laval. Now a puppet of Nazi Germany, he was taken there in 1944 and arrested in 1945 by the Allies. He was brought to Paris and tried for treason. The death sentence was commuted by General de Gaulle to life imprisonment on the Île d'Yeu.

Peter II (1923–), King of Yugoslavia 1934–45. He inherited the throne on the assassination of his father Alexander in 1934. His uncle Prince Paul, acting as Regent, was keen to ally with the Axis powers. When the latter invaded in 1941, however, Peter led an unsuccessful revolt and then fled to England. From there he gave his support to the chetnik* resistance fighters under Mihailovich* and was opposed to Tito*. In 1944 he quarrelled with the British government over this issue and went to live in Egypt. The Yugoslav monarchy was abolished in 1946, his estates were confiscated, and he lost his nationality.

Petrov Affair. In April 1954 a Soviet diplomat in Australia, Vladimir Petrov, and his wife, asked for political asylum, which was granted. The Petrovs then proceeded to make allegations of a widespread Communist spy ring in Australia and a Royal Commission was established to investigate. Herbert Evatt* acted as counsel for some of the accused until debarred by the Commission, as it allegedly conflicted with his duties as leader of the opposition in Parliament. The affair weakened the Labor Party, and helped to produce a splinter group, the Democratic Labor Party. Little of substance was in fact revealed by the Petrovs.

Phan Boi Chau (1867–1940), Vietnamese author and resistance leader. Educated in the mandarin tradition, by 1900 he had become a staunch Vietnamese nationalist in French Indo-China, operating for a while in Japan and then in China. In 1914 he was imprisoned for an unsuccessful attempt to assassinate the French Governor-General of Indo-China. While in prison he adopted Marxist ideology and when released in 1917 he resumed protests and resistance against the French. In 1925 he was arrested in Shanghai, brought to Hanoi, and condemned to death; but following riots in the city, he was allowed to retire under surveillance to Hue. His writings were to inspire a whole generation of revolutionaries. In the later years of his life he took a less antagonistic attitude towards the French.

Pharmaceutical industry. The first treatise on pharmacology dates from AD 60, drawing on a long tradition back into prehistory. The vast majority of early drugs were organic compounds extracted from plants and herbs, and occasionally from animals. From 1500 onwards new drugs arrived from America, Asia, and Africa. Such groups as the Pharmaceutical Society of Great Britain (1841) traditionally supervised the dispensation of drugs by individual apothecaries. It was in the late 19th century that the synthesizing and manufacture of organic compounds began, the first being salicin, marketed in 1888 as aspirin, and barbitral, in 1903 marketed as sedative barbiturates. One of the first inorganic compounds to be successfully used as a drug was salvarsan, discovered by Paul Ehrlich as a cure for syphilis (1909). In 1921 insulin was discovered by Sir Frederick Banting and its extraction from the pancreas of pigs, sheep, and oxen began. The real growth of the drug industry, however, begins in 1935, when the antimicrobial qualities of a red dye, prontosil, were noticed, from which sulphanilamide was isolated and some 6,000 derivatives developed. M and B 693, penicillin (and a whole range of semi-synthetic penicillins), streptomycin, and tetracycline followed. The synthesized hormone progesterone led to contraceptive pills, while librium and valium revolutionized mental health treatment. By 1980 drug sales in the USA were some $12,000m. per annum and the world market was growing at 14 per cent per annum. The great majority of new drugs have been discovered through the R. & D. of the great multinational drug companies, Roche in Switzerland; Hoechst in Germany; Glaxo and Distillers in Britain; Dupont, Upjohn, Smith Kline & French in the USA; for example it has been claimed that of the sixty-six most important drugs discovered since 1888, fifty-seven were discovered within the industry, with more and more organic substances being synthesized in the laboratory. At the same time many drug companies have regularly reported annual profits of 50 per cent. Controlled clinical trials have not always proved adequate, especially since side-effects can differ markedly from patient to patient, while addiction can prove another hazard. The thalidomide tragedy 1958–62 proved too that licensing a new drug will not always prove infallible. Promotion of new drugs poses considerable ethical problems; in the USA over 170 million pieces of promotion literature are mailed each year.

Philippines. There are some 7,000 islands in the Philippines group, the largest being Luzon. Nationalists had resisted US occupation under Emilio Aguinaldo*, but in 1902 the country accepted US occupation on the promise that this was to last only until it had been 'trained for self-government'. English was introduced into a US-style school system and in 1907 an elected Assembly was introduced. In 1916, by the US Jones Act, this was expanded into a bicameral Congress (Senate and House of Representatives), but with finance, foreign policy, and defence all reserved to the USA. By now various political parties had emerged, all calling for full independence. This came nearer in 1934 when the US Tydings–McDuffie Act gave full autonomy, subject to US veto, and a promise of total independence in 1946. Manuel Quezón* was elected President. US policy had been to discourage industrialization but to concentrate on the development of large haciendas to export tobacco and other agricultural products. The Great Depression* drove many smaller cultivators into bankruptcy and in 1935 there was a major revolt of landless peasants in central Luzon in which Luis Taruc* was involved. Manila fell to the Japanese in January 1942 and all the islands had capitulated by May; but the Anti-Japanese People's Army (Hukbalahaps*) was highly successful in gaining control in the rural areas. Manila was heavily bombed before being recaptured by MacArthur's forces in February 1945; the liberated country was split between ex-collaborators and ex-guerrillas, many of whom were Communist. In spite of these problems the Republic was inaugurated under its first President Manuel Roxas* on 4 July 1946, subject to the continued presence of US military bases. Having defeated the Huk rebellion, Ramón Magsaysay became President in 1953 but was killed in an air crash in 1957. His short presidency had raised expectations and enthusiasm. It was followed, however, by a series of administrations which proved unable to deal with severe economic problems and regional unrest. By 1972 there was Communist guerrilla insurgency conducted by the New People's Army* in Luzon, and violent campaigns by Muslim separatists, the Moro National Liberation Front, in the southern islands. President Marcos* declared martial law, assuming dictatorial powers. While his regime achieved limited success in dealing with both economic problems and guerrilla activities, this was at great cost in terms of human freedom. After the murder of Benigno Aquino in 1983, resistance to the Marcos regime coalesced behind Aquino's widow Corazón Aquino* and the United Nationalist Democratic Organization. Backed by the Church and with US support Corazón replaced Marcos restoring a fragile democracy.

Philippines campaign, World War II (June 1944–July 1945). In the Battle of the Philippines Sea*, fought in June 1944 while US forces were securing air bases in the Marianas, the Japanese suffered crippling losses. It was followed by the decisive Battle of Leyte Gulf* on 25 October. Five days previously US forces had landed on the island of Leyte (20 October) and the Japanese were driven out by 23 December. General MacArthur* next moved to Luzon, landing north of Manila on 9 January 1945. Three weeks later other US forces landed in the Batangas province and advanced north. The two armies surrounded Manila, which was entered on 4 February, MacArthur having been in touch with Resistance forces within the city. It quickly capitulated, and by July MacArthur could announce that all the Philippine islands were liberated, although detached groups of Japanese, in accordance with their instructions to fight to the last man, were still at large when the war ended.

Philippines Sea, Battle of, World War II (19–21 June 1944), a naval battle fought off Saipan (Belau). As US marines landed (15 June) and fought their way across the island, a naval battle between US and Japanese fleets under Admirals Raymond Spruance and Jisaburo Ozawa developed. On 19 June 243 Japanese planes, including some kamikaze*, and two carriers were lost, with another carrier and 233 planes in the next two days. US losses were 130 aircraft. The battle ensured the capture of Saipan and prepared the way for the reoccupation of the Philippines.

Phomvihane, Kaysone (1920–), Chairman of the Council of Ministers of Laos 1975– . Son of a civil servant, he attended Hanoi University, and in 1944–5 fought against the Japanese as a guerrilla with the VietMinh*. In 1945 he was in Bangkok with a nationalist group, the Free Laos Front. In 1949 he joined the Lao Dong (Vietnam) Communist Party and in 1950 helped to form the Pathet Lao*, becoming

Commander-in-Chief of their forces from 1954. In 1955 he formed the People's Revolutionary Party of Laos (Communist), and was its general secretary. In 1975 he became Chairman of the Council of Ministers (Prime Minister) of the new Republic of Laos.

Piaget, Jean (1896–1980), Swiss psychologist. After study in Zurich and Paris he became director of the Jean-Jacques Rousseau Institute in Geneva. His work on child development revolutionized educational thought and practice. He devised a model to trace development from birth to adulthood through a series of four stages, until both space and time can be comprehended, abstract ideas manipulated, and logical processes followed to conclusions. The so-called child-centred educational methodology developed as a result of the new insights provided by Piaget. The teacher ceased to be 'the transmitter of knowledge' and became instead the 'guide to the child's own discovery of the world'.

Pibul Songgram (1897–1964), field marshal and Prime Minister of Thailand 1938–44, 1948–57. From a farming family, he had a military education in Thailand 1909–14 and later in France 1924–7. In 1932 he took part in the bloodless revolution which ended the absolutism of the monarchy in Thailand, becoming Minister of Defence and deputy Commander-in-Chief of the army. As such he strongly influenced the moderate Premier Phya Bahol, whom he ousted in 1938, ruling in an autocratic, neo-Fascist style until 1944. In 1940 he signed a pact of friendship with Japan, allowing that country military access, and in 1942 declared war against the USA and Britain. His government used ruthless means to suppress the anti-Fascist Free Thai Movement, but it collapsed in July 1944 when he retired. In 1948, however, he returned to office on a wave of anti-Communist feeling and became a strong supporter of the USA, joining the SEATO* pact, which made its headquarters in Bangkok. He was ousted in 1957, his government being attacked as corrupt and inefficient. He went to Japan, where in 1960 he became a Buddhist monk.

Picasso, Pablo Ruiz y (1881–1973), Spanish artist. Son of an art teacher, he showed remarkable talents as a boy and moved from Barcelona to Paris in 1904. Influenced by both African sculpture and child art, his work moved through a whole series of innovative changes, Post-Impressionist, Cubist, realist, neoclassicist, Expressionist, Surrealist; but always centred on the human form. In 1937 he painted *Guernica**, and vowed he would never return to Spain as long as Franco ruled. He lived in Paris throughout World War II. Picasso was not only a painter but a sculptor, a ceramicist, a lithographer, and a printmaker. In his later years he became increasingly interested in the powers of magic. Incredibly prolific, he was the most inventive and influential of all 20th-century artists.

Pilsudski, Joseph Klemens (1867–1935), Polish statesman. Born in Vilna in Russian Poland, he studied at the University of Kharkov, where he became involved in student politics. In 1887 he was exiled to Siberia, where he remained until 1892. He returned to Poland, where he founded an underground newspaper which agitated for Polish independence. When war broke out in 1914 he raised a force of Polish volunteers to join the Austro-Hungarians on the Russian front. German refusal to guarantee the ultimate independence of Poland after the Russian Revolution in February 1917 led him to withdraw his support for Germany, and he was then interned 1917–18. When the new Republic of Poland was proclaimed, he became head of state (1918–22) and Chief of the Army Staff (1918–27). He successfully commanded the Polish forces against the Bolsheviks in the war 1919–20 which established the eastern frontier. In 1926, after a miltary revolt, he assumed the office of Minister of Defence, establishing a virtual dictatorship until his death. In 1934 he tried to guarantee Poland's independence by signing non-aggression pacts with both the Soviet Union and Germany.

'Ping-pong diplomacy' (China). During 1970 it was noted that President Nixon* had ceased to use the term Red China to describe the People's Republic. Early in 1971 a group of US table-tennis players was given a warm reception by Zhou Enlai*, who claimed their visit 'opened a new page in the relations between the Chinese and American peoples'. This was a signal that the US government had been waiting for, and in July Henry Kissinger* visited China, when a visit of the US President was agreed upon. The term has been used to describe the more pragmatic foreign policy adopted by China in the 1970s.

Pinochet, Augusto (1915–), President of Chile 1973–90. A professional soldier who had graduated from the Santiago Military Academy in 1936, he rose to become Commander-in-Chief of Chile's armed forces in September 1973. Eighteen days after his appointment he master-minded a military coup which deposed and killed President Allende*. Pinochet became President of the Council of Chile (a junta of military officers) and imposed harsh military rule, during which some 130,000 people were arrested, many tortured, and thousands never seen again. He was proclaimed President of Chile in 1974, and his free market policies at first reduced inflation and resulted in a short-lived economic boom. In 1978 he announced that a plebiscite had given 75 per cent endorsement of his policies. A new constitution in 1980, again endorsed by a plebiscite, gave him authority to be sworn in for another eight years as President. In 1982, however, the economy began to collapse and his popularity to wane. In 1990 he resigned the presidency but remained Commander-in-Chief of the army.

Pius XII (1876–1958), Pope 1939–58. Born Eugenio Maria Giuseppe Giovanni Pacelli in Rome, he was educated there and ordained priest in 1899. He spent almost all his career in the papal diplomatic service and became Secretary of State for Pius XI. As such he negotiated a Concordat in 1933 with the Nazi government. Its repeated violations by Hitler led to the encyclical *Mit brennender Sorge*, which in 1937 branded Nazism as fundamentally anti-Christian. Nevertheless as Pope he remained politically impartial between the Allied and Axis powers during World War II. At the same time he supervised a programme for the relief of war victims through a Pontifical Aid Commission and made the Vatican City an asylum for many refugees. While unwilling to speak out against the Nazi Holocaust*, he did give assistance and asylum to a number of individual Jews. After the war he inveighed against Communism, threatening its supporters with excommunication, and he concluded accords with both Portugal (1950) and Spain (1955).

PKI *see* PARTAI KOMMUNIS INDONESIA.

Plekhanov, Georgi Valentinovich (1857–1918), Russian revolutionary. As a young man he became interested in Marxist ideas and in 1877 became leader of a populist Russian organization, Land and Liberty. When this turned increasingly to terrorist methods, he formed a splinter group to continue peaceful mass agitation. This developed into the League for the Liberation of Labour, the first Marxist revolutionary organization in Russia, which in 1898 merged with the Social Democratic Workers' Party. In the 1903 split with Lenin* he supported the Mensheviks, but always tried to reunite the party. Having lived in Paris and London, he returned to Russia in 1917, but failed to prevent the Bolsheviks* from seizing power in October.

PLO *see* PALESTINE LIBERATION ORGANIZATION.

Podgorny, Nikolay Viktorovich (1903–83), Chairman of the Supreme Soviet of the USSR 1965–77. Born in Karlovka in the Ukraine, he was educated at the Kiev Institute for Food Technology. He had become interested in politics when at school, and in 1939 became a commissar for food-processing in the Ukraine, a post he retained throughout the difficult war years 1941–5. In 1946 he was appointed to the Ukrainian Council of Ministers and then to the Central Committee of the Communist Party of the Soviet Union 1963–5. He then became Chairman of the Praesidium, ceremonial head of state. He was ousted from this post in 1977 when Brezhnev* assumed the title. He lived in retirement in Moscow.

Pogrom (Russian; 'riot'), a mob attack approved or condoned by authority, frequently against Jews. The first occurred in the Ukraine, following the assassination of Alexander II in 1881. After that there were many pogroms throughout Russia, and Russian Jews began to emigrate to the USA and Western Europe. Anti-Semitic persecution increased in Russia after the 1905 Revolution. Pogroms conducted on a large scale in Germany and Eastern Europe after Hitler came to power ultimately led to the Holocaust*.

Poincaré, Raymond (1860–1934), President of France 1913–20; Prime Minister 1922–4, 1926–9. Born in Bar-le-Duc in Lorraine, he became a successful Paris lawyer before entering politics, being elected Deputy in 1887. In February 1913 he was elected President, believing that his role should be more than that of figurehead. This led him to clash with Prime Minister Clemenceau*, although he supported the latter's demand for tough measures against Germany, demanding stringent reparations. He was the first ex-President to become Prime Minister, his policies being nationalist and conservative. When Germany defaulted on reparations he ordered French troops to occupy the Ruhr*. He could not however sustain this policy and resigned in 1924. Prime Minister again in 1926 of a government of 'National Union', he eased an acute economic crisis by a deflationary policy, balancing the budget, and securing the Franc (1928) at one-fifth of its former value.

Point Four Program, a US aid programme 1950–3. It was so called because it developed from the fourth point in the inaugural speech of President Truman in 1949, in which he undertook to make 'the benefits of America's scientific and industrial progress available for the improvement and growth of under-developed areas'. Each year Congress provided technical assistance for the long-term development of industries, agriculture, health, and education in those developing countries which came under the scheme. The project also encouraged the flow of private investment and in 1953 was merged with other foreign aid programmes.

Poland. A great medieval country, it had been obliterated during the 18th century, being absorbed by Russia, Austria, and Prussia. There were revolutions in 1830, 1848, and 1863, and just before 1914 a resurgence of Polish nationalism. When war broke out many Poles enlisted with the Austro-Hungarian forces against Russia. On 5 November 1916 a Kingdom of Poland was proclaimed by the German and Austrian emperors. Following the 'February' Revolution of 1917 in Russia, three Polish army corps from the Russian army formed the basis of a new Polish army. Germany now seemed the bigger threat, and in July 1917 Pilsudski* and other leading Poles were interned by the Germans. Meanwhile a Polish National Committee was active in Paris and Paderewski* travelled to the USA to win support. In January 1918 President

Wilson gave US backing for the formation of a Polish republic. With the collapse of Austria in October 1918, this was proclaimed in Lublin, and when Pilsudski returned to Warsaw from internment this Republic was confirmed on 11 November, with Pilsudski head of state and Paderewski Prime Minister. Frontiers were still a matter of argument, Germany wishing to retain Poznań and Russia looking to retain the Ukraine. The Treaties of Versailles* (1919) and of Riga (1921), and the League of Nations arbitration over Teschen*, settled these, with Poland having access to the Baltic via a 'Polish Corridor'* separating East and West Prussia. By 1926 democratic politics were polarizing, with a strong Communist Party and extreme right-wing anti-Semitic parties. In 1926 Pilsudski took powers as virtual dictator, being succeeded by Smigly-Ridz*. The existence of the Corridor provided the excuse for the Nazi invasion of September 1939, precipitating World War II and the division of the country between Germany and the Soviet Union. Following the Warsaw Rising* a Polish provisional government was established under Bierut*, with Red Army protection. Two million Germans from Poland and East Prussia (now in Poland) fled to the Western-occupied zones of Germany, while Poles from the east were settled in their place. The frontier with the Soviet Union was to be the Curzon Line* and that in the west the rivers Oder and Neisse (Oder-Neisse Line*). The new government co-operated with Stalin to bring Poland within the Soviet bloc. For a while political opposition was neutralized and in 1952 a Soviet-style constitution was adopted. In 1956, following the de-Stalinization speech at the Twentieth Congress*, Polish workers went on strike to protest against food shortages and other restrictions. Although they were suppressed by military force, Wladyslaw Gomulka* was restored to power and some liberalization followed. Further strikes and demonstrations in 1970 were again forcefully suppressed. The election of John Paul II* in 1976 strengthened the influence of the Catholic Church in the country. In 1980 more strikes, now organized by the free trade union Solidarity*, erupted at the Gdansk Shipyard. Martial law was imposed by General Wojciech Jaruzelski 1981–2, military tribunals continuing to operate even after it had officially ended. In 1987 the government, beset by severe international debt problems, put forward plans for limited decentralization. In January 1989 the banned Solidarity Union was legalized, and in February round-table discussions were held between the Polish United Workers' Party (Communist Party), Solidarity, the Catholic Church, and other groups. Multi-party elections were held June–July, when Jaruzelski was elected President. He then invited Tadeusz Mazowiecki, a moderate supporter of Solidarity, to form a coalition government. In February 1990 the IMF* granted extensive credit, but at the cost of draconian economic policy. Increasing criticism then developed, with Lech Walesa* aiming at the presidency, which he won in November 1990.

Polisario (Popular Front for the Liberation of Saguira Hamra and Rio de Oro), a movement of Saharawi tribesmen in Western Sahara unwilling to be absorbed by Morocco and seeking independence. Fighting took place on an increasing scale after 1975, when Hassan II* sent some 100,000 Moroccans into the area. In 1976 Polisario announced the formation of a government-in-exile, the Sahara Arab Democratic Republic (SADR). In 1979 Mauritania renounced that part of Spanish Sahara it had shared with Morocco in 1976. Morocco then unilaterally tried to annex it. Some 20,000 Polisario guerrillas, operating from and armed by Algeria, fought back, whereupon Moroccan troops responded by building six stretches of 'wall' right across Western Sahara to the Atlantic; a mined sand barrier equipped with US-supplied electronic devices. By 1987 sixty-nine countries had recognized the SADR as a legal government. When it took its seat at the OAU* summit in 1989 Morocco resigned. The UN General

Assembly resolved on Western Sahara's right of self-determination and in 1988 a UN delegation achieved a short truce between SADR and King Hassan. While Saharawi refugees crowded into camps in Algeria, a new Polisario offensive was launched in September 1989. At the same time the UN and OAU made plans for a referendum in Western Sahara.

Polish Corridor. A strip of territory separating East Prussia from the rest of Germany, it was granted to Poland by the Treaty of Versailles* to provide access to the sea. It contained the lower course of the River Vistula except for Danzig (Gdansk*), which was to be a free city. Historically it belonged to Polish Pomerania in the 18th century, but it had been colonized by German landowners. When Hitler's forces annexed the Corridor, Danzig, Poznań, and districts along the Silesian frontier, World War II began.

Politburo, the highest policy-making committee of Communist regimes. The Soviet Politburo was founded, together with the Ogburo (Organizational Bureau), in 1917 by the leading Bolsheviks in order to provide continuous leadership during the Russian Revolution. Both bureaux were then re-formed to control all aspects of Soviet life. Towards the end of Stalin's life the Soviet Politburo's influence lessened as he became increasingly dictatorial, and in 1952 it was abolished. In its place was to be a Praesidium of the Central Committee. In 1966 this body's name reverted to Politburo, as part of the policy of collective leadership, and to be distinguished from the Praesidium of the Supreme Soviet, which is elected by universal suffrage. Other regimes, notably East Germany, China, and the Balkan states, developed similar institutions within their party structure. As regimes such as the Soviet Union entered multi-party politics, so the powers of the Politburo would correspondingly lessen.

Pol Pot (1928–   ), Prime Minister of Cambodia (Kampuchea) 1975–9. Born Saloth Sar, of peasant family, he trained as a Buddhist monk and then worked on a rubber plantation. During World War II he was an active member of the Communist Party of Indo-China, working against the French. In 1949 he was sent to Paris on an engineering scholarship and returned to Phnom Penh in 1953, when he became a teacher. He fled north in 1963 and became secretary of the Communist Party of Cambodia. As leader of the Khmer Rouge* he became Prime Minister of the Democratic Republic of Kampuchea in 1975. Throughout his years in office the CPK was continually being purged of dissidents, as he sought to transform Kampuchea into a vast, self-sufficient, militarist work-site, at the cost of immense suffering, especially for the urban population. His regime collapsed in January 1979, and he withdrew to the Thailand border as Commander-in-Chief of the Khmer Rouge. He was sentenced to death *in absentia* for genocide, by the new People's Republic of Kampuchea.

Polynesia, a loose term to describe the hundreds of islands in an immense area of the Pacific inhabited by Polynesian peoples. It includes Hawaii* and French Polynesia*, but also Samoa (West Samoa* was the first Polynesian state to gain political independence), Tonga*, the Cooks, and Tuvalu*. The people of Fiji* have a Polynesian culture but are racially Melanesian. Easter Island, annexed by Chile 1881, marks the limit far to the south-east, while the Maori* people of New Zealand are of Polynesian descent.

Pompidou, Georges Jean Raymond (1911–74), President of France 1969–74. Born in the Auvergne, he became a schoolmaster before World War II, in which he fought with the Resistance. He joined de Gaulle* in 1944 and became one of his administrative assistants. As such he was chief contact figure with the Algerian rebels, negotiating the Évian Agreements*. Elected a Deputy for the Fifth Republic* in 1958, he was appointed Prime Minister in 1962, when he showed considerable political skill: but de Gaulle dismissed him following the strikes and street violence of May 1968. He announced that he would run as the next President, and when de Gaulle suddenly resigned in 1969 he was elected. In swift and decisive policy changes, he devalued the franc, introduced a price freeze, and lifted France's veto on British membership of the European Community. Suddenly his health gave way with cancer, and he died in April 1974. His name is perpetuated in the Pompidou Centre in Paris.

Pop culture, a term coined in the 1950s to describe mass entertainment in film, art, music, literature, and clothing. At that time it was particularly targeted at the affluent teenage consumer. It opposed both the so-called working-class culture of pre-war Britain and élitist 'high culture'. In the 1960s Pop Art broke through to the latter and powerfully influenced the world of advertising. Equally, the best of pop music, for example The Beatles*, came to be recognized as of intrinsic aesthetic value. Pop fashions in clothing have dominated fashion houses throughout the decades.

Popular Front, a political coalition of left-wing parties, including Communists. Such coalitions became possible as a result of the Comintern* policy decision of 1934, to defend democratic forms of government believed threatened by right-wing Fascism*. In France such an alliance gained power after elections in 1936, under the leadership of Léon Blum*, who implemented a programme of radical reform. In Spain Popular Front governments of Azaña*, Caballero*, and Negrin were in office 1936–9 and fought the Spanish Civil War against Franco and the Nationalists. A Popular Front government was in power in Chile 1938–47.

Portal, Charles Frederick Algernon, 1st Viscount Portal of Hungerford (1893–1971), Marshal of the Royal Air Force. Educated at Winchester and Oxford, he served in World War I, joining the newly formed Royal Flying Corps in July 1915. By 1937 he was an air marshal and Director of Organization at the Air Ministry. In April 1940 he was placed in charge of Bomber Command. The aircraft available had many technical deficiencies, especially in navigation, but by carrying the bombing offensive into Germany they raised British morale, while to some extent disrupting industry, rail junctions, and power plants. Gradually the aeroplanes available to him improved and he pressed for a policy of 'area bombing', to replace that of specific targets, even though he accepted that this would result in high civilian casualties. After the war he became Controller of Atomic Energy in Britain (1945–51).

Portugal. Through most of the 19th century there was political instability, until 1910, when a republic was established. In 1926 there was a military coup led by General Gomes da Costa. He was deposed by General Carmona*, who appointed Dr Salazar* his Finance Minister. In 1932 the latter became Prime Minister and virtual dictator until 1968, strongly supported by the Catholic Church. Portugal supported the Allies in World War I and in World War II was theoretically neutral, while allowing the Allies naval and air bases. After the war Goa*, Diu, and Daman were lost to India, but Macao* in south China was retained. Salazar's policies were largely continued by Marcello Caetano*, until a military coup in 1974. Guinea Bissau* gained independence in 1974 and Angola* and Mozambique* the next year, after bitter guerrilla fighting. After two years of political instability at home, a more stable democracy began to emerge, following the election of Antonio Eames as President in 1976. Moderate coalition governments both of left and right of centre have alternated, all struggling with severe economic problems. The Social Democrat Mario Soares was elected President in 1986, having been Prime Minister since 1983. Portugal joined NATO* in 1949, EFTA* in 1960, and the European Community* in 1986. In May 1987 the Social Democrats retained power with Aníbal Silva as Prime Minister.

Post-Impressionism, a specifically English term to describe the work of that body of painters, mainly French, who immediately succeeded the French Impressionists. It was coined for the 1910 and 1912 exhibitions at the Grafton Gallery in London. It included the work of Gauguin, Cézanne, early Picasso, and the Fauves group of painters such as Matisse and Vlaminck.

Post-industrial society, a loose term, fashionable in the 1980s, to describe those Western societies, particularly Britain and the USA, which had abandoned their lead in heavy industries and manufacturing in favour of computer-based service industries.

Potsdam Conference (17 July–2 August 1945), held in the former Hohenzollern palace at Potsdam outside Berlin. This last World War II Conference was attended by Churchill* (replaced by Attlee* during its course), Stalin*, and Truman*. It implicitly acknowledged Soviet predominance in Eastern Europe by, among other things, accepting Polish and Soviet administration of certain German territories, and agreeing to the transfer of the German population of these territories and other parts of Eastern Europe (over 10 million people) to Germany. It established a Council of Ministers to handle peace treaties, made plans to introduce representative and elective principles of government in Germany, discussed reparations, and decided to outlaw the Nazi Party, to de-monopolize much of German industry, and to decentralize its economy. The final agreement, vaguely worded and tentative, was consistently breached, as Communist and capitalist countries polarized into respective blocs. The Potsdam Declaration of 26 July 1945 demanded from Japan the choice between unconditional surrender or total destruction.

Powell, John Enoch (1912–), British politician. Born in Birmingham, he was educated there and at Trinity College, Cambridge, where he taught classics 1934–8. He became Professor of Greek at Sydney University, and then served in the army in India throughout World War II. In 1950 he entered politics, being elected to the House of Commons as a Conservative. He was Minister of Health 1960–3, during which time he did much to expand the National Health Service. From then onwards he became something of a 'lone wolf' in British politics. He deplored the trend in Britain towards a multi-cultural society, strongly and consistently attacked British membership of the European Community, and, after 1974, when he was elected as an Ulster Unionist MP, for South Down in Northern Ireland, acted as a spokesman for the Protestant majority and opponent of the Anglo-Irish Agreement of 1985. In 1990 he appeared to reconcile himself to Mrs Thatcher's leadership of the Conservative Party.

Prague Spring, 1967–9, a brief period of liberalization in Czechoslovakia. Following an economic recession 1962–3 and student demonstrations, President Novotný* made some concessions to the Slovaks, who were demanding greater autonomy, and dismissed some of his 'hard-line' colleagues. But discontent against his policies of centralization and what was alleged as economic exploitation by other Comecon* countries continued, as did Slovakian demands for more independence. In June 1967 Novotný was openly criticized at a Writers' Congress in Prague and there were widespread student demonstrations in October. In February 1968 Novotný ordered the Czech army to move into Prague and its officers refused, whereupon he resigned. The liberal Slovak Alexander Dubček* replaced him as first secretary. On 5 April a programme of liberalization and reform was introduced, which would have restored multi-party politics and personal freedom. Dubček assured Brezhnev in Moscow that Czechoslovakia would remain within the Warsaw Pact, but in August Soviet troops, together with contingents from Poland, East Germany, Hungary, and Bulgaria, occupied Prague. Dubček abandoned most of his programme and was ousted from office in April 1969, although it was agreed that Slovakia would become an 'autonomous republic'.

Prasad, Rajendra (1884–1963), first President of India 1950–62. From a wealthy Bihar family, he was educated in Calcutta and became a successful lawyer. In 1917 he began working with Mohandas Gandhi*, being imprisoned by the British on many occasions for civil disobedience and 1942–5 for opposition to the British war effort. He supported the conservative wing of the Indian National Congress*, of which he was president 1934, 1939, and 1947. In 1946 he was appointed Minister of Food and Agriculture in the interim government and was president of the Constituent Assembly (1946–9) which devised the Indian Constitution. When the Republic of India was proclaimed, he became President.

Primary elections, preliminary elections in the USA to select candidates. National party conventions began in 1832, nominating a party's candidate for President. At first delegates were chosen by a closed 'caucus' system, that is, senior party members in the state chose the delegates 'behind closed doors' and instructed them how to vote. Beginning with Wisconsin in 1903, however, democratic primary elections steadily replaced 'caucus primaries'. They are held by the state, and the results are legally binding on the delegates. There are both 'open' and 'closed' presidential primaries. In the former any adult voter in a state may take part, regardless of his or her own party preference. In the latter, only those who are registered members of the party may vote. Primary elections are also used for a wide range of local and state elections, but nominating conventions are only held for the offices of President and Vice-President. Since the 1930s unofficial pre-primary endorsing conventions have been held by both Republican and Democrat parties.

Primo de Rivera, Miguel (1870–1930), Spanish dictator 1923–30. His father was Governor-General of the Philippines, which he unsuccessfully defended in 1898. As a professional soldier, Miguel had reached the rank of general in the Spanish army by 1923, when Alfonso XIII* encouraged him to take power as a dictator, in the belief that current disorder in Spain was the product of corruption and inefficiency. His authoritarian regime attempted to unite the nation around the motto 'Country, Religion, Monarchy', which hardly appealed to Communists or socialists, many of whom were imprisoned. In 1927 he successfully ended war in Morocco, but was less successful in suppressing agitation in Catalonia for autonomy. His reliance on the landlord class prevented urgently needed agricultural reform. Although he survived three attempts to remove him, increasingly poor health and a loss of army support led him to resign in 1930, the year of his death. His son Antonio founded the Falange* Party.

Prison camps (Soviet). The tradition of exiling political protesters to Siberia was well established in 19th-century Russia. By a decree of 1919 Lenin maintained it, operating through his police agency Cheka. During Stalin's rule millions were arrested, including peasants who resisted collectivization, Christians, especially Baptists, Jews, intellectual dissidents, and political protesters. Prisoners were passed to GULAG (Main Administration of Corrective Labour Camps), established in 1930 and responsible for administering the forced labour system. The camps were mostly in eastern Siberia and were referred to as the Gulag Archipelago. Estimates of numbers confined by Stalin vary, ranging from 6 million to 15 million. Among the authors whose writings testify to their experiences are Evgenija Ginzburg, *Journey into the Whirlwind* (1975), and Alexander Solzhenitsyn, *One Day in the Life of Ivan Denisovich* (1962) and *The Gulag Archipelago* (1971). After the arrest and execution of Beria* and the de-Stalinization policy of Khrushchev* from 1956, there was a steady decline in the worst excesses of the camps, which were formally replaced by Corrective Labour Colonies. Many distinguished Soviet citizens were still however either exiled to Siberia or detained in 'psychiatric hospitals'. Since 1987, in the wake of the policy of *glasnost*, the release of the majority of such detainees has been ordered.

Privatization, a term coined after World War II to describe a process of denationalization by the sale of a state enterprise through the issue of shares on the stock market. It was first used in 1959 to describe policies in post-war West Germany. In 1977 the British government, under a Labour Prime Minister, James Callaghan*, first looked at 'the possibility of the private sector providing goods and services that are now provided through government enterprise and programs'. In 1979 the Tory Party adopted the policy for its election manifesto and the Thatcher* administrations (1979–90) embarked on an extensive series of privatization exercises including British Telecom, British Airways, Water, and Electricity. In Eastern Europe and the Soviet Union privatization schemes were developed with the collapse of the Communist-dominated economies.

Profumo Affair, a British political scandal. In March 1963 John Profumo, Minister of War, resigned from Parliament after lying to the House of Commons about a relationship with a call-girl, Christine Keeler, who also had a Soviet naval attaché as a lover. A public outcry resulted in an inquiry which concluded that national security had not been at risk. However, the affair weakened the popularity of the Macmillan government and the Prime Minister resigned later in the year.

Progressive Party (Canada). Founded in 1920 to obtain low tarriffs for western farmers, it enjoyed early popularity in the western provinces. In 1921, with sixty-four MPs, it held the balance of power in the federal Parliament, but internal divisions soon weakened the party. Only a small 'ginger' group survived the 1925 election, who then went on to help form the Co-operative Commonwealth Federation* in 1932. Many Progressives, such as Robert Forke, joined forces with Mackenzie King's* Liberals, calling themselves Liberal-Progressives. Others, such as John Bracken, joined the Conservatives, who in 1942 adopted the title Progressive Conservative Party*.

Prohibition era (USA) (1920–33). This was a culmination of the US Temperance Movement, whose Anti-Saloon League had been founded in 1893. It began when the Eighteenth Amendment to the US Constitution was ratified in January 1919. The Volstead Act, passed by Congress in 1919 over President Wilson's veto, enacted that federal law enforcement was to prevent the manufacture, distribution, or sale (but not consumption) of liquor, wines, and beer. Moderates had only sought the banning of liquor sale and had expected each state to act unilaterally. The Amendment came into force in January 1920 and, despite the securing of some 300,000 court convictions, drinking continued. Speakeasies (illegal bars) and bootlegging (illegal distilling of alcohol) flourished. The success of gangsters like Al Capone*, who controlled the supply of alcohol, led to corruption of police and government. After the Wickersham Commission reported in 1931 that the prohibition laws were unenforceable and encouraged public disrespect for the law, the Eighteenth Amendment was repealed by the Twenty-First Amendment. A number of states and counties retained partial prohibition, but by 1966 no state-wide prohibition laws existed.

Propaganda. The *Propaganda Fidei* was established in Rome in 1622 'for the propagation of the Faith'. During the 20th century the term 'propaganda machine' was particularly applied to the attempts in Nazi Germany and Stalin's Soviet Union to win public support, regardless of truth or reason, that of Goebbels* being particularly notorious. Britain in both world wars used increasingly sophisticated methods of 'psychological warfare', while the USA developed 'the Voice of America', with which to fight the Cold War. The concept of 'the enemy' is an important ingredient in all propaganda, for example 'Bolshevik', 'counter-revolutionary', 'Jew', 'yellow devil'. As one enemy becomes an ally so a new one tends to be identified, whether internally by a party machine or externally in international affairs. As state manipulation or control of the media has developed, so has government propaganda become an increasing norm throughout the world in the late 20th century.

Protectionism, the theory that a nation's economic prosperity needs to be protected from foreign competition by the imposition of tariffs. Britain began the 20th century deeply divided on this issue, but with its US and European competitors imposing tariffs the Gladstonian theory of free trade was steadily abandoned after World War I. By 1939 protectionism, together with imperial preference (see OTTAWA AGREEMENTS), dominated world markets and, in retrospect, was considered as much cause as effect of the Great Depression*. After World War II the United Nations endeavoured to establish a policy of tariff reduction with the establishment of GATT*. Following the creation of the European Community with its heavily subsidized agriculture, the issue of protectionism, especially between the USA and the Community, became a crisis issue.

Proust, Marcel (1871–1922), French novelist. Born into a wealthy Parisian family, but with recurrent ill health, he spent most of his life in the capital, where he had access to the changing social scene of the *fin de siècle*. He devoted most of the last twenty years of his life to writing his twelve-volume novel *A la recherche du temps perdu*, in which he chronicles the decadence of the old French aristocracy in favour of a vulgar *nouveau riche* society. After one of the most profound analyses of the human condition beset by jealousy for the beloved, the narrator in the novel at last finds peace of mind. It is the taste again of *la petite madeleine* which confirms for him the existence of time, embedded in the unconscious, to which his involuntary memory carries him back in almost unspeakable bliss. The stylistic brilliance and the psychological insights of the book deeply influenced later 20th-century literature.

Provisional government (Russia). When the imperial troops commanded by Nicholas II refused to accept him as their Tsar, he and his family abdicated on 15 March. This was the so-called 'February' Revolution. The Duma elected a government under Prince Lvov, who had been a prominent agitator for constitutional reform, formed of liberals and Menshevik socialists. It regarded its main tasks as being the continuation of the war and the organization of a Constituent Assembly to decide on a new constitution. In many cities all over Russia soviets of workers' and soldiers' deputies had appeared, that in St Petersburg, now Petrograd, being led from May onwards by Trotsky*. These soviets* were reluctant to support a war effort. Lenin* had returned to Petrograd in April, where he organized his Bolshevik followers. In July Alexander Kerensky* took over as Prime Minister, having earlier been Minister of Justice and Minister of War. He re-exiled Lenin to Finland and imprisoned the royal family. In August General Kornilov, Commander-in-Chief of the army, tried to seize Petrograd and to reinstate the Tsar. He was arrested and imprisoned (he was to escape in December and start to organize a counter-revolution). Kornilov's failed coup strengthened support for the Bolsheviks, who on 6 November seized control of Petrograd, and the government collapsed in the so-called 'October' Revolution. (The Russian calendar at that time was out of step with the Western, hence 'February' and 'October' Revolutions.)

Prussia. By 1871 the Kingdom of Prussia consisted of most of north Germany from the Rhine and the North Sea in the west to the Niemen and the east Baltic. In 1918 it was reduced to becoming one state within the Weimar Republic, East and West Prussia, divided by the Polish Corridor*. It disappeared in 1947, being divided between Poland, the Soviet Union, and the German Democratic Republic. Many thousands of its inhabitants became refugees in the West.

Pueblo Incident. In January 1968 the USS *Pueblo*, an intelligence ship loaded with electronic spy equipment, was captured by North Korean patrol boats. Public opinion in the USA put pressure on President Johnson* to take military action, and there was an escalation of the US military presence in South Korea. In the end, however, Johnson took no action, but instead negotiated the release of the crew who, while in captivity, signed a 'confession' acknowledging the ship's intrusion into North Korean territorial waters. This confession was later challenged, but no action followed. The incident helped to convince President Johnson, already deeply committed in Vietnam, of the need for the USA to wind down its South-East Asian commitment.

Puerto Rico, the Commonwealth of, an island in the Caribbean. In 1898, during the Spanish-American War, it came under US military rule, and was ceded to the USA at the end of the war. In 1917 the US Congress passed the Jones Act, declaring Puerto Ricans to be US citizens, but with an appointed Governor and without the right to elect to the US Congress or Presidency. Since the 1940s, with a decline in the sugar industry, there have been successful efforts to diversify the economy. In 1948 Muñoz Marín* was the first elected Governor, being re-elected three times. In 1952 the USA granted a new constitution, ratified by plebiscite, whereby the island became a self-governing 'Commonwealth'. The dominant political party since then, the Partido Popular Democrático (PPD), has supported the status quo, while urging greater autonomy. Its rival the Partido Nuevo Progresista (PNP) would like the island to become the fifty-first state of the Union. Although a PNP candidate, Romero Barceló, was elected Governor in 1980, the PPD retained control of the Senate and Lower House, and the status quo remained. A small Puerto Rican Independence Party exists (PIP), which wants an independent republic. The violent separatist organization the FALN* has had little support on the island. Each year from 1978 the UN has passed a resolution urging the USA to withdraw from the island pending a plebiscite on its future. With a large tourist industry, the island's economy clearly gains from its links with the USA.

Pugwash, an ongoing international conference on science and world affairs. It was first convened in July 1957 by a wealthy philanthropist Cyrus Eaton, in the village of Pugwash, Nova Scotia, on the initiative of Bertrand Russell* and Albert Einstein*, both of whom were horrified by the developing nuclear arms race. Distinguished scientists from the Soviet Union and Eastern Europe joined others from the USA, Canada, Britain, France, India, and elsewhere to discuss the social responsibilities of science. A continuing Committee was established, to be based in London, and subsequent conferences have been held in many countries, including the Soviet Union, India, Romania, Sweden, and the USA. It is as yet unclear how far Pugwash conferences helped to reduce tensions in the Cold War* and the precise influence which Pugwash members had upon disarmament negotiations.

Punjab, formerly a north-western province of British India. In 1799 the Sikhs* had established a kingdom there but, after their defeat in the Sikh Wars 1845–7, the province remained under British rule until 1947. Muslims constituted more than half the province's population and the Muslim League's capture of Punjabi votes in the 1945–6 election strengthened the demand for a separate Muslim homeland. Growing communal violence pushed Britain and the Congress Party into a reluctant acceptance of partition*. Violence, during which hundreds of thousands died, increased as Hindus and Sikhs crossed to India, and Muslims to Pakistan. The province suffered most in the process of partition, as east Punjab became the state of Punjab (India), and West Punjab a province of Pakistan. In 1960 the Indus Waters Treaty was agreed between India and Pakistan, after a decade of dispute over irrigation canals in West Punjab. Both Indo-Pakistan Wars* 1965 and 1971 involved fierce fighting on the Punjabi frontier. A new Pakistan national capital was built at Islamabad, and Punjabis have been increasingly dominant in Pakistan's national life. In 1966 also the Indian state of Punjab (India) was divided on a religious/linguistic basis to provide a predominantly Punjabi-speaking state of Punjab (with a Sikh majority), and a Hindi-speaking state of Haryana, with Chandigarh as a joint capital. Militant Sikhs, however, continue to campaign for an independent Sikh state.

Pu Yi (P'u-i) (1906–67), last Qing Emperor of China 1908–12. Proclaimed Emperor Hsuan T'ung at the age of 2 by his great aunt the Dowager Empress Cixi, he reigned until the Chinese Revolution*, when he was forced to abdicate. He continued to live in the imperial palace with extensive privileges, serving as a focus for monarchist movements and even experiencing a twelve-day restoration in 1917 by a warlord*. When the Guomindang* occupied Beijing in 1924 they abolished his title, and he sought protection in the Japanese concession of Tianjin. Here he cut off his queue and called himself Henry Pu-Yi. After the Japanese seizure of Manchuria, he was placed at the head of the puppet state of Manchukuo* as the Emperor K'ang Te. Deposed and captured by Soviet troops in 1945, he was exiled to Siberia until 1950, when he was handed over to the People's Republic. After a period of 'political re-education' he was allowed to live out his life as a private citizen.

Qadhafi, Muammar al- (1942–   ), President of Libya 1977–   . Son of a Bedouin farmer, he was educated at the University of Libya and the Military Academy, from which he graduated in 1965. In 1969 he led a military coup which deposed King Idris and made him Commander-in-Chief of the Libyan armed forces, chairman of the Revolutionary Command Council, Prime Minister, and Minister of Defence. He removed British and US military bases, expelled many Jews and Italians, and became president of the Federation of Arab Republics. In 1973 he nationalized outstanding petroleum assets and outlawed alcohol and gambling, being both a strict and devout Muslim and a populist socialist. In 1977 he became President of Libya with the rank of major-general, while still retaining the title colonel and adhering to a strictly modest life-style. His government became involved in a number of incidents with neighbouring states—Egypt, Sudan, Tunisia, Morocco, and most persistently Chad*. Firmly opposed to any negotiation with Israel, he gave asylum to a number of extreme terrorist groups, not only Arab but also such groups as the IRA*. A series of Libyan assassination and terrorist incidents in Europe and the USA led to increasing isolation and tension, including three military forays by US forces in 1981, 1986, and 1988. His dream of a single North African Arab Federation from Egypt to Morocco remained unrealized, although he joined the Maghreb Union* in 1989. His policies appeared to be more conciliatory during 1990.

Qajar dynasty (1796–1925), Iranian ruling family. In 1914 Iran had recently experienced a constitutional revolution establishing a Parliament (the Majlis). Shah Muhammad Ali had been deposed and his son Ahmad had just come of age as the last Qajar ruler. He was deposed by Reza Khan Pahlavi* in 1925, who pronounced himself Shah.

Qatar, an independent Arab sheikhdom in the Persian Gulf, ruled by the al-Thani family. It was under Ottoman suzerainty until 1913 and in 1916 made a treaty with Britain. Sheikh Abdullah ibn Jassim al-Thani agreed that foreign affairs would be conducted by British agents, while Britain would provide naval protection. Oil was discovered in 1939 and exploited from 1949, bringing great wealth. In 1968 Qatar joined a Federation of Arab Emirates, but in 1971 it declared full independence under a constitution by which the Sheikh governs as Prime Minister.

Quebec Liberation Front (Front Libèration du Québec). An extremist terrorist organization for Quebec separatism which operated in Quebec and Montreal from 1963 until its successful suppression in 1970. It began by bomb attacks in mail-boxes in the wealthy English-speaking suburbs of Montreal, and later used robbery, kidnapping, and murder. In October 1970 it kidnapped the British Trade

Commissioner James Cross and a Quebec politician Pierre Laporte. When Prime Minister Trudeau* and the Quebec Premier Bourassa refused negotiation Laporte was murdered. A massive military and police hunt resulted in 400 arrests. On 6 December the whereabouts of Cross was revealed and negotiations allowed the remaining FLQ kidnappers safe passage to Cuba. Laporte's murderers were eventually prosecuted and the movement, having overreached itself, disappeared.

Quemoy (Chin-mentao), an island within the territory of Taiwan but lying only 6 miles from Xiamen (Amoy) on the Chinese mainland. It was regularly used during the 1950s by Nationalist troops as a base for guerrilla raids against the Chinese People's Republic. In 1955 and again in 1958 there were threats of invasion by the latter, but these were countered by a strong US naval presence in support of Taiwan.

Quezón, Manuel Luis (1874–1944), President of the Philippines 1935–42. He was a supporter of Aguinaldo* in the wars against Spain and the USA (1896–1902), and afterwards was elected to the Philippines Assembly. From 1909–16 he served as Resident Commissioner for the Philippines in Washington, his efforts helping to produce the Jones Act of 1916, which extended Philippines autonomy. He sat in the Philippines Senate until 1935, when he was elected the first President of 'The Commonwealth of the Philippine Islands'; he invited General MacArthur* to be his military adviser. In 1942 he left the islands, on Japanese occupation, to go into exile in Australia. He died of tuberculosis in New York.

Quiet Revolution (Quebec), a term given to the changes in Quebec brought about by the Liberal government 1960–6 led by Premier Jean Lesage. It involved major social and educational reforms in favour of the urban population, together with increased government participation in the province's economy. Improved social insurance, a programme of school and hospital building, and nationalization of electric power into Hydro-Québec, were all features of these years. The 'revolution' helped to provide French Canada with a more confident perception of itself and fuelled the demand for separatism. Lesage wanted 'special status' within the Confederation, but René Lévesque*, who formed the Parti Québécois*, demanded independence.

Quisling, Vidkun Abraham Lauritz Jonsson (1887–1945), Prime Minister of Norway 1942–5. Son of a Lutheran pastor, he entered the Norwegian army and had a brilliant early career, being a general staff officer by the age of 24. He entered politics in 1931, and in 1933 founded the Fascist Nasjonal Samling Party. In December 1939 he visited Hitler, whom he much admired, and helped him to prepare the conquest of Norway, which followed in April 1940. He became Prime Minister of the puppet pro-German government and remained in power until 1945, when he was arrested, charged with high treason, and executed. By this time 'Quisling' had become a term for any politician who supported a country's invaders.

r

R 101 disaster. Airships had been steadily built since before World War I, filled with highly inflammable hydrogen. In October 1930 the largest to date was the British R 101, which crashed and burnt out in France on a maiden flight to India, killing all forty-eight passengers. Britain then abandoned airships for flying-boats, 'to link the Empire'. Germany's airship programme ended in 1937 with the crash in the USA of the *Hindenburg*. Airships filled with helium began to reappear after World War II, being thought by many to have a considerable future in a world with declining oil supplies.

Race Relations Act (UK) (1976). This Act, passed by Parliament under a Labour administration, repealed those of 1965 and 1968, and strengthened the law on racial discrimination by extending the 1968 ban to housing, employment, insurance, and credit facilities. It also established a permanent Race Relations Commission to promote equality of opportunity and good relations between different racial groups within Britain. Large numbers of West Indians had emigrated to Britain in the 1950s to meet labour shortages, and in the early 1960s immigrants from India and Pakistan increased dramatically, rising from an average of 7,000 per annum to 50,000 per annum before the first Immigration Act was passed in 1962. General economic problems, particularly increased unemployment, tended to highlight prejudice. In spite of continuing efforts by the Commission, racial tensions continually flared up in a number of inner-city areas, for example Bristol, Toxteth (Liverpool), Brixton (London), and Handsworth (Birmingham).

Racism, a pseudo-scientific ideology of the late 19th century. Its theoretician had been the Comte de Gobineau, whose ideas (*Essay on the Inequality of the Human Races*, of 1853) were given pseudo-biological support from the Social Darwinists*. Its practical manifestations were in European imperialism*, which operated on the agreed assumption that Western culture was racially superior to all others; in segregation legislation in the Southern US States and in South Africa; and in anti-Semitism*. Although the latter had roots much further back in Christian society, it undoubtedly reached a culmination under Nazi rule in Europe. By the mid-20th century racism may have been biologically totally discredited, but it remained embedded in much popular thinking in both Europe and North America.

Radek, Karl (1885–1941), international Communist leader. Born in Austria-Hungary, he joined the German Social Democratic Party and then the Polish Party. In 1905 he took part in anti-Russian riots in Warsaw. He crossed Germany with Lenin after the outbreak of the Russian Revolution and was a close associate of his, taking part in the Brest-Litovsk* peace negotiations. In 1919 he was imprisoned in Germany after taking part in the Spartakist rising*. He returned to Russia in 1922 and became a leading member of the Comintern*. Expelled from the party in 1928 for alleged support for Trotsky*, he was exiled to Siberia, but was later readmitted after recanting. He became a leading figure in Soviet foreign affairs until 1937, when he was accused of treason in the second show trial of the Great Purge* and sentenced to prison. He was released in 1941, but then disappeared.

Radhakrishnan, Sir Sarvepalli (1888–1975), President of India 1962–7. Of a poor Brahmin family, he was educated in Madras and taught philosophy there, at Mysore and Calcutta, before becoming Professor of Eastern Religion and Ethics at Oxford 1936–52. He wrote extensively on Hindu religious and philosophical thought, seeking to reinterpret it for modern times. He also served as India's first ambassador to the Soviet Union 1949–52 and then as Vice- President of India, before succeeding President Rajendra Prasad* in 1962. As a scholar without political affiliations he occupied a rare and detached position in Indian life, stressing the need for India to establish a classless and casteless society.

Rafsanjani, Hojatolislam Hashemi Ali Akbar (1934–), President of Iran 1989– . As a wealthy young politician, he survived the years under the last Shah in Iran and the decade of the Ayatollah Khomeini*. He became Speaker of the Iran Islamic Consultative Parliament (the Majlis) and in the last months of the Iran–Iraq War Commander-in-Chief of the armed forces. As a member of the Islamic Republican Party, he won election in July 1989 as President, following the death of Khomeini. His reputation is that of a skilful negotiator and moderate pragmatist.

Rahman, Abdul Putra, Tunku (Prince) (1903–90), Prime Minister of Malaya 1957–63; Prime Minister of Malaysia 1963–70. Born at Kuala Kedah, one of the sons of the Sultan, he was educated at Cambridge and called to the English Bar. In 1931 he entered the Kedah state Civil Service, where he continued to work through the Japanese occupation. In 1952 he succeeded Dato Onn bin Jafaar as leader of the United Malays National Organization* (UMNO) and played a central role in organizing an alliance with the moderate Malayan Chinese Association (founded in 1949 by Tan Cheng Lock). At the end of the Malayan Emergency* this provided the political base for the achievement of independence. In 1955 he became leader of the Federal Legislative Council and Malaya's first Prime Minister in 1957. In 1963 he successfully presided over the formation of the Federation of Malaysia*, which he led as Prime Minister, successfully securing the support of both the Chinese and Indian peoples by skilful use of committees. The Tunku survived the Indonesian Konfrontasi* and remained in office as leader of the Alliance Party, but in May 1969 he faced ethnic riots in Kuala Lumpur between Chinese and Malays. In this emergency he struggled to restore good community relations but in January 1970 he stood down. Through active political journalism he remained an influential figure in Malaysia during the years of his retirement.

Rahman, Sheikh Mujibur (1920–75), Prime Minister of Bangladesh 1972–5; President 1975. Born in East Bengal, son of a landowner, he was educated at the universities of Calcutta and Dacca, and became a successful lawyer. Co-founder of the Awami League* in 1952, he was arrested on many occasions on the orders of President Ayub Khan*. When the Awami League won the election of 1970 Mujibur demanded independence for East Pakistan. There were riots, savagely suppressed by the Pakistani army, and he was imprisoned for alleged treason. In December 1971 he was released after Indian intervention and became Prime Minister of the new state. A convinced socialist, he found it impossible to operate a democratic system, and assumed dictatorial powers as President in 1975. He was murdered, along with his family, by a group of right-wing senior army officers.

Rainbow Warrior Affair. On 10 July the 'flagship' of Greenpeace* *The Rainbow Warrior* was blown up in Auckland Harbour, New Zealand, killing a photographer. Two French secret service agents were tracked down and arrested. They pleaded guilty to manslaughter and sabotage and were imprisoned for a short while in French Polynesia before returning to France, allegedly for 'medical treatment'. The Prime Minister of New Zealand David Lange* described the affair as 'sordid, state-backed international terrorism'. The affair affected French domestic politics but also increased international concern over policies for nuclear testing in French Polynesia*.

Rajagopalachariar, Chakravarti (1878–1972), Indian politician. Known as Rajaji, he was a successful brahmin lawyer in Madras before entering politics. After World War I he joined the Indian National Congress* and from 1919 was a close associate of the Mahatma Gandhi*, editing his newspaper *Young India*. Imprisoned on several occasions for non-co-operation with the British, he worked in the Congress 1922–42 and also served as Premier of Madras 1937–9. Himself a Hindu, he accepted the need for minority safeguards for Muslims and the creation of the separate state of Pakistan. He was the only Governor-General of India 1948–50, before the latter became a republic, and again served as Chief Minister for Madras 1952–4. In 1959 he left the Congress Party and was one of the founders of the Swatantra Party, having become a critic of the policies of Jawaharlal Nehru*.

Rajk, Laszlo (1909–49), Hungarian politician. Born of a Jewish family and educated in Budapest, he became a Communist while a student at the university and went to Spain to join an international brigade. Here he was wounded. He returned to Budapest and spent the war years working in the Communist Party underground. He emerged in 1944, and in 1946 became Minister of the Interior in Hungary and then Foreign Minister. Suddenly in May 1949 he was arrested and tried, accused of 'treasonable links with Tito'. He was hanged, the evidence having clearly been fabricated, it is presumed on the orders of Stalin; too liberal in his views, he was also a Jew. In 1956 he was rehabilitated by Rákosi*.

Rákosi, Mátyás (1892–1971), Prime Minister of Hungary 1952–3. Born and educated in Budapest, where he became a Communist, he held office in the Béla Kun* government, escaping to Vienna and thence to Moscow in 1920. He returned to Hungary in 1924, but was arrested and tried in 1927 and sentenced to prison. He was released in 1940, as part of an exchange deal, and allowed to return to Moscow. Here he became first secretary of the Hungarian Communist Party, and returned to Hungary in 1944, when a rigidly Stalinist regime was established. He was Prime Minister 1952–3, and continued the real ruler of the country as party 'boss' until opposition to his policies persuaded Moscow that he should resign, to be succeeded by the equally unpopular Ernö Geró. In an attempt to appease his critics he had rehabilitated Laszlo Rajk*, but it was the brutality of his secret police which precipitated the Hungarian Revolution* three months later.

Randolph, Asa Philip (1889–1979), Black US trade unionist and civil rights leader. Son of a Methodist minister, he grew up and was educated in Harlem, New York, attending night schools. In 1912 he founded an employment agency for Black workers, and in 1925 was founding president of the Brotherhood of Sleeping Car

Porters, winning a major contract with the Pullman Company in 1937. He took his union out of the AFL* in 1938 for its failure to fight racial discrimination, joining the CIO. As US involvement in World War II deepened his threat to lead a march on Washington contributed to the end of race restrictions on employment in the defence industries (June 1941). As president of the Non-Violent Civil Disobedience League against Military Segregation his activities helped to persuade President Truman* to end segregation in the armed forces in 1948. In 1955 he helped to reunite the AFL with the CIO, becoming a vice-president of the combined AFL-CIO. In 1963, with Martin Luther King*, he helped to organize the March on Washington for Jobs and Freedom, one of the largest civil rights demonstrations ever held. Soon afterwards ill health forced him to retire.

Rapacki Plan (1957). This was presented to the UN General Assembly by the Polish Foreign Minister Adam Rapacki in October 1957. It proposed a zone in Central Europe consisting of Poland, Czechoslovakia, and the German Democratic Republic, which would be free of both stockpiled nuclear weapons and their means of production, as would be the Federal Republic of Germany. Negotiations would commence for the reunification of Germany. The plan was accepted by Khrushchev with enthusiasm, but rejected by the USA, Britain, and West Germany. They argued that it would leave the Warsaw Pact* with a strategic advantage of superior numbers of men and conventional weapons.

Rapallo, Treaties of. There were two such treaties. The first, in November 1920, tried to settle differences between Italy and the Kingdom of Serbs, Croats, and Slovenes. Italy obtained the Istrian peninsula, while Dalmatia went to the Kingdom (later to be called Yugoslavia). Fiume (Rijeka) was to become a free city, after its illegal occupation by D'Annunzio* in 1919. It was to be annexed by Mussolini in 1924. The second treaty was between Germany and the Soviet Union in April 1922. This was just prior to Lloyd George's* Genoa Conference in May, when twenty-nine European nations met, and which ended in stalemate. By the Rapallo Treaty Germany and the Soviet Union agreed to abandon any financial claims which each might bring against the other following World War I. Secretly, in defiance of the Treaty of Versailles, German soldiers were to be allowed to train in the Soviet Union. Five years of accord between the two countries followed.

Rashid Ali al-Gailani (1892–1965), Prime Minister of Iraq 1933, 1936–8, 1940–1. From a distinguished Iraqi family, he graduated as a lawyer and served as Minister of Justice and of the Interior. A strong nationalist, during the 1930s he was a consistent opponent of General Nuri Said*. After the fall of France in 1940, Pan-Arab opinion in Iraq strongly resented a continued British presence, and argued for an alliance with Germany. Rashid Ali at first resisted this pressure and allowed Britain to reinforce. He was briefly ousted, but reinstated by the Iraqi army, which now openly resisted British forces, April–May 1941. The regent Abd al-Ilah* and Nuri Said fled, and fighting lasted thirty days, after which the Iraqi army capitulated to the British, who expelled Rashid and restored the Regent. Rashid returned in 1958, but was accused of plotting against Kassem* and died in exile.

Rasputin, Grigori Yefimovich (1871–1916), Russian religious fanatic. Born in Tobolsk in Siberia, son of a peasant, he developed mystical healing and hypnotic powers. In 1903 he came to St Petersburg, where his beneficial treatment of the haemophiliac Crown Prince won him a disastrous hold over Tsarina Alexandra, an influence which increased when Nicholas II left the court to command the army in 1915.

He and the Tsarina were thought to be lovers, and together they virtually ruled Russia, dismissing all more liberal ministers. The resulting government inefficiency was responsible in large measure for the failure to get adequate supplies to the Tsar and the army. Rasputin's drunken habits and sexual excesses discredited the court and in December 1916 he was murdered by a group of nobles led by Prince Yusupov.

Rastafarianism, a cult which spread from West Indian society to both the USA and the UK among young Blacks during the 1980s. It sees the (deposed) Emperor of Ethiopia (see HAILE SELASSIE, born Ras Tafari Makonnen) as symbolically the rightful protector of African civilization and the representative of the true message of the Gospels. It helped to give Black minority groups a degree of self-respect in an often hostile and essentially White culture.

Ratana, Tahupotiki Wiremu (1873–1939), a Maori Methodist farmer who in November 1918 had something of a spiritual vision, claiming that he had heard God say that He had 'come back to Aotearoa'* to choose 'the Maori people as his own'. He began to preach and then discovered that he had the gift of healing. From a humble background, he appealed especially to detribalized Maori, often urbanized and unemployed. A Ratana political movement began in 1922, which gradually gathered momentum, displacing the Young Maori Party* in its appeal. At the same time a Ratana Church emerged, with its own hymns and prayers, glorifying Ratana as God's mouthpiece. It was disowned by other churches until a reconciliation in the 1960s. By 1988 it had 36,000 adherents, with the Ratana 'Pa' (home settlement) its spiritual headquarters. Politically the movement was significant from 1935 onwards, when Ratana candidates were elected in alliance with the Labour Party, gaining all four Maori electorates in 1943. The Labour/Ratana alliance from 1935 enabled many Maori grievances to be righted, granting, for example, secret electoral ballot, equal unemployment benefits, housing opportunities, and improved health and educational provision. The Ratana movement remained a significant political factor until into the 1970s. In 1979 a group of younger Maori politicians, impatient with the Labour/Ratana alliance, founded a new party, Mana Motuhake.

Rathenau, Walther (1867–1922), German industrialist and statesman. Born in Berlin of a Jewish family, he inherited and became director of a large electrical company, the AEG, which his father had founded. In 1916 he was invited to take responsibility for Germany's war economy and afterwards became Minister of Reconstruction (1921) and Foreign Minister (1922) in the Weimar Republic. He accepted that Germany must fulfil its obligations under the Treaty of Versailles*, including payment of reparations. In April 1922 he negotiated the Treaty of Rapallo* with the Soviet Union, establishing trade and military links. He was assassinated later in the year by anti-Semitic nationalists.

Rawlings, Jerry John (1947–), Ghanaian statesman. He attended Achimota College and Teshie Military College (1968). He then trained as a jet pilot, but, shocked by corruption among senior military officers, he plotted with thirteen other young officers, and in May 1979 openly challenged them. Arrested and court martialled, he escaped from prison and organized a successful coup, as head of the Armed Forces Revolutionary Council. The Council's courts tried and executed eight top military officers for corruption, including three previous heads of state. After two months Rawlings handed over power to a civilian President, Hilla Limann. However, in December 1981 he re-established himself as Chairman of the Provisional National

Defence Council of Ghana. Since 1985 his government, with the assistance of the IMF*, has restored economic stability in Ghana.

Recruit Cosmos scandal. The Recruit Cosmos Company was a Japanese real estate company. In November 1988 some sixteen Japanese politicians were named as being involved in 'insider trading' in the company's shares, Takumi Veda of the Japanese Socialist Party being the first to resign from the Diet, on 4 November. On 9 November an official of the company was arrested for trying to bribe investigators. In December the Finance Minister Kiichi Miyazawa resigned as being involved, as did the Justice Minister. The scandal eventually brought down the government of Noboru Takeshita during 1989.

Red Guards, militant young supporters of Mao Zedong* during the Chinese Cultural Revolution*. Taking their name from army units organized by Mao in 1927, they provided the popular, paramilitary vanguard of the Cultural Revolution*. Numbering several million, they attacked supposed reactionaries, including the Communist Party establishment, China's cultural heritage, as well as all aspects of Western intellectual and artistic influence. They maintained momentum through mass demonstrations, a constant poster war, harassment of individuals, and attacks on property. There was argument between different groups which itself resulted in violence. As the Cultural Revolution subsided many Red Guards were sent into the countryside for forced 're-education'.

Red Scares (USA), alleged Communist subversion. There were two Red Scares in the USA. In February 1919 workers in Seattle staged a general strike, which the city mayor declared was part of a Bolshevik plot. There were a number of bomb outrages across the country—probably the work of anarchists—and, with rapid inflation, further strikes and May Day demonstrations. The formation of the American Communist and Labour Party in August 1919 confirmed for many Americans that the country was in crisis. Two hundred and fifty Russian immigrants were shipped back to Russia and J. Edgar Hoover*, a young member of the staff of Attorney-General Mitchell Palmer, reported that 'revolution was imminent'. Palmer authorized widespread raids, over 6,000 persons were arrested, and 556 of them deported. Public hysteria built up through early 1920, but when 1 May passed off peacefully, it subsided. Black veterans returning from the war were regarded as suspect, and many of them were victims of Ku Klux Klan* violence. One result of this first scare was the 1924 immigration legislation.

The second Red Scare developed in 1946, when a world-wide Soviet spy-ring was thought to be in place. President Truman* ordered that all federal employees be given 'loyalty' tests and 2,000 resigned. The Alger Hiss* and Rosenberg* trials seemed to confirm the worst fears of the scaremongers, whose leader Joseph McCarthy* dominated the headlines for three years 1950–3. In retrospect it is difficult to know as yet how much truth lay behind this wave of anti-Communist hysteria and how far it was the result of ruthless manipulation.

Red Shirts, a Pathan* movement in the North-West Frontier province of British India. Correctly titled Khudai Khidmatgar (Servants of God), it was formed in 1929 by Abdul Ghaffar Khan (1890–1988), a Muslim follower of Mohandas Gandhi*. It provided the main support for Ghaffar Khan's influence within the province until 1946, during which time it supported the Congress Party rather than the Muslim League. In the debate over the partition* of India, the Red Shirts campaigned for a separate state of Pakhtunistan, rather than absorption into Pakistan, but they were unsuccessful. After 1947 the new government of Pakistan banned the movement and imprisoned Ghaffar Khan for thirty years.

Referendum, a political device whereby an issue is put to an entire electorate, which is asked to answer a specific question. Such a device has been used in Europe on such issues as the fate of Schleswig-Holstein*, and has sometimes been given the alternative name of **plebiscite** (a practice of the ancient Roman Republic). While the referendum is a constitutional requirement in certain countries, for example Switzerland and Australia, it is unknown in the USA and, although used in Britain on such issues as Scottish and Welsh devolution (1979), is considered by lawyers to have no constitutional authority within the UK.

Reform Party (New Zealand), a conservative political grouping formed from various local organizations from 1905 onwards, under the leadership of William Massey*. It won the election of 1911 on a programme demanding reform of land tenure and critical of the trade unions, whom it strongly attacked on coming into office. It lost support in the early 1920s, joined a coalition government 1931–5, and then amalgamated with the United Party* in 1936 to form the National Party*.

Refugees. The 20th century has witnessed the plight of many millions of refugees, fleeing from political, ideological, or religious oppression, for example: Russian Revolution refugees; Armenian and Greek refugees from Turkey; Jewish refugees from Germany and Austria; Spanish refugees from Franco; Hindus and Muslims, following partition of India in 1947 (*c.*18 million with perhaps another 12 million when Bangladesh was created in 1971); East German refugees following World War II (*c.*12 million); Palestinian refugees from Israel in 1948; perhaps 2 million refugees from the Communist revolution in China in 1949; many thousands of refugees following various aspects of decolonization*; the Vietnamese boat people* of the 1970s. In 1921 the League of Nations appointed a High Commissioner for Refugees, the first being the explorer Fridtjof Nansen*, and UN commissioners have grappled with the challenge over the post-war decades. Numerous voluntary agencies have helped to resettle refugees, especially where they have been able to reach an already existing ethnic minority, for example Jews in New York.

Reith, John Charles Walsham, 1st Baron Reith of Stonehaven (1889–1971), director-general of the BBC 1927–38. He was born and educated in Scotland, where he attended the Royal Technical College of Glasgow. During World War I he was engaged in weapons procurement in the USA. In 1922 he became general manager of the BBC, and in 1927 its first director-general. His strongly Calvinistic temperament moulded the early years of broadcasting, with an emphasis on programmes which were educational in the widest sense: classical music, book reviews, news, and drama. His aim was that the BBC should earn respect for its impartiality and sense of responsibility, the more so because of its monopoly of radio broadcasting at that time in Britain. In 1936 he inaugurated British television. In 1940 he was elected MP for Southampton, and was appointed Minister of Works by Winston Churchill. During 1943–4 he was at the Admiralty planning the movement of supplies, war materials, and transport for the invasion of Europe. After the war he was chairman of the National Film Finance Corporation 1948–51 and of the Colonial Development Corporation 1950–9. But in his later years he felt unappreciated and frustrated. He was Lord High Commissioner of the Church of Scotland 1967–8.

Relief agencies. Relief charities for children, the aged, the poor, and the destitute have existed since the Middle Ages, and in the later 19th century there were many such founded, as for example the Salvation Army* and Dr Barnardo's. World War II produced a number of international agencies, one of the first being Oxfam (the

Oxford Committee for Famine Relief), founded in 1942 to raise funds to feed starving children in occupied Greece. It rapidly expanded in order to help provide aid for war refugees. Since 1960 it has developed a world-wide network aimed at improving food production, with associate organizations in the USA, Canada, Australia, and elsewhere, and providing emergency aid for areas struck by natural disaster. Its chain of charity shops manned by volunteers raises millions. A UN agency **Unicef** was established after the war, specifically to work among refugee children, but it has continued to operate over a wide range of children's needs. In Britain the **Save the Children Fund** was also established after World War II to work among refugee children. It has grown into one of the largest international children's charities, operating in fifty countries, with a philosophy of 'working with the poorest'.

Renner, Karl (1870–1950), Chancellor of the first Austrian Republic 1919–20; President of the second Austrian Republic 1945–50. Born in Moravia, he was educated at the University of Vienna and qualified and practised as a lawyer there. In 1907 he was elected to the Austro-Hungarian Reichstag as a Social Democrat, soon becoming party leader. In 1919 he led the delegation to Paris which settled the frontiers of the new Republic, becoming briefly its first Chancellor. After a short spell in prison in 1934, he lived in retirement during the years 1934–45, being then invited to head a provisional government. In December 1945 a second Austrian Republic was proclaimed, with Renner its President. He was acceptable to the Soviet Union, although totally opposed to Communism. His presidency saw a healing between Catholic and Socialist groups in Austria and established firm foundations for the new Republic.

Reparations, compensation for damage done in war by a defeated enemy. They were a condition of the armistice for World War I and part of the Versailles Treaty*. France, which had paid reparations to Germany in 1871, hoped by their means to bankrupt Germany. British civilians had sustained little damage, and so Lloyd George claimed only the cost of war pensions. The US Senate did not ratify the Versailles Treaty, and waived all claims on reparations. A Reparations Commission in 1921 agreed on the sum of 132,000m. marks (£6,500m.) from Germany, a figure which the British economist J. M. Keynes* argued would ruin the interdependent world economies, but one which was less than half that demanded by the French. Hungary, Bulgaria, and Austria were also to pay huge sums, while Turkey, being more or less bankrupt, agreed to an Allied Finance Commission. To enforce its claims France occupied the Ruhr* (1923), precipitating an inflationary crisis in Germany. With Britain and the USA unhappy about reparations, various plans were devised to ease the situation, that is, the Dawes Plan* of 1924 and the Young Plan* of 1929. The Lausanne Pact (1932) substituted a bond issue for the reparation debt, but German repayments were never resumed. After World War II reparations took the form of Allied occupation of Germany and Japan. A European Reparations Commission was again established, but this time it was decided that payment would be in kind, by the transfer of technical productive capital equipment, France and the Soviet Union obtaining the greatest shares, some 15 per cent of German capital equipment going to the Soviet Union. Britain, France, and the USA ended reparation collections in 1952, but Stalin continued to plunder the East German zone. In Japan the USA administered the removal of capital goods, and the Soviet Union seized Japanese assets in Manchuria. Following the San Francisco Conference of 1951, Japan worked out reparations agreements with the various states which had been victims of its aggression. Since 1953 West Germany has paid $37 billion as reparations to Israel for damages suffered by Jews under the Nazi regime.

Republican Party (USA). The present Republican Party was formed in 1854. It won its first presidential election in 1860, with Abraham Lincoln. From then until 1932 it lost only four such contests, two each to Grover Cleveland and Woodrow Wilson*. The opulence of the 'gilded age' at the end of the 19th century contrasted sharply with increasing poverty among immigrants and the urban proletariat. This split the party when Theodore Roosevelt formed his Progressive Party. After World War I the party's isolationist policy brought it back to power, and there were three successive Republican presidencies*. From 1932, however, there were five successive presidential defeats, the party only regaining the White House through the massive popularity of General Eisenhower* in 1952. Under Republican Presidents Nixon*, Ford*, and Reagan* it became strongly associated with military spending and a forceful assertion of US presence world-wide, especially in Central America. Strongly backed by corporate business, it has nevertheless failed to maintain a grip on Congress, which has tended to have a Democrat majority, even when the President has been Republican. Under President Bush* the party had to adjust to a new post-Cold War situation.

Republican presidencies (USA) (1920s). In 1920 the Republican candidate for President was Warren Gamaliel Harding (1865–1923), a Senator from Ohio, whose state political machine was notoriously corrupt. He campaigned on the ticket 'back to normalcy', a popular sentiment after two years of war and many post-war problems. His presidency is remembered for the Teapot Dome scandal* and for the wave of isolationism, as well as a Red Scare* which swept the USA in these years. Although far from competent, his own integrity was never in question. On his sudden death (August 1923) he was succeeded by his Vice-President John Calvin Coolidge (1872–1933). In 1924 he was elected in his own right, serving one term. At a time of reckless spending and considerable corruption, his personal characteristics were of thrift, caution, and honesty. Certain sections of the economy experienced difficulties, especially mining and farming, but he showed little awareness of this or compassion for the victims. He retired into private life, to be succeeded by Herbert Clark Hoover (1874–1964), a successful mining engineer. He had led a relief commission to Europe after World War I and had served as Secretary of Commerce 1921–8, being personally convinced of the need for planned economic policies. When, however, the US economy itself collapsed, following the Stock Market Crash* of October 1929, his administration had no effective policies to stave off depression. With banks closed and factories bankrupt, he was defeated by the Democrat Franklin D. Roosevelt in 1932. After World War II he was again involved in relief work as co- ordinator of the European Food Program.

Resistance movements, World War II, extensive underground movements against both Nazi Germany and Japan. Apart from committing acts of sabotage, their activities involved conveying intelligence by secret radio, publishing underground newspapers, and helping prisoners of war, and in Europe Jews, to escape. Resistance to the Nazi regime was active within Germany from 1934 onwards, at first expressed by both Protestant and Catholic churches, but also from 1939 by groups such as the Roman Catholic student group Weisse Rose, and the Communist Rote Kapelle, which carried out sabotage and espionage for the Soviet Union until betrayed in 1942. Admiral Wilhelm Canaris*, head of German Counter-Intelligence (*Abwehr*), was a key figure until betrayed and hanged after the July Plot* of 1944. In occupied Europe there were often deep divisions between the Communist and non-Communist organizations, notably in France, where the Maquis* was active, as well as in Belgium, Greece, and Yugoslavia. European Communist parties had at first remained passive, but after the Germans invaded the Soviet Union in June 1941, they formed or joined underground groups. Dutch, Danish, and Norwegian Resistance remained unified and worked closely with London, where in 1940 the British Special Operations Executive (SOE) was set up to co-ordinate all subversive activity, both in Europe and the Far East, supplying arms and equipment by secret air drops. In Eastern Europe the long German lines of communication were continually harassed by partisans*, and Polish Resistance was almost certainly the largest and most elaborate in Europe. Some Eastern European groups later turned against the Red Army as it advanced west (1944–5), the Polish Warsaw Rising* being a tragic example of the tensions between Communist and non-Communist forces. In the Far East numerous clandestine operations were carried out, supported by British and American intelligence organizations. The latter were involved in psychological warfare and prisoner of war recovery, as well as intelligence gathering, while the sabotaging of selected installations and communication lines was conducted by native-born, nationalist, and often Communist-inspired guerrillas. In Vietnam and Indonesia the Resistance movements went on to fight decolonization wars against the French and the Dutch. In Malaya* the MPAJA became the core of the Communist resistance during the Emergency of the 1950s, while in the Philippines the Anti-Japanese People's Army formed the post-war Hukbalahap* movement.

Reuther, Walter Philip (1907–70), US trade union leader. Born in West Virginia, he left school at 16 to become an apprentice, but was dismissed for protesting against Sunday work. In 1926 he went to Detroit and became a foreman at the Ford Motor Company. Dismissed in 1932 because of his union activity, he bicycled across Europe and worked in a Ford plant in the Soviet Union. In 1936, back in the USA, he worked for General Motors* and helped to found the United Automobile Workers union, serving as its president 1946–70. He pioneered negotiations for guaranteed employment, wage increases tied to productivity, and welfare provisions for his members. A strong anti-Communist, he was president of the Congress of Industrial Organizations 1952–5 and fought strenuously to rid unions of corruption and racketeers. In 1955 he led the reunion between the CIO and AFL*. Disagreement, however, with George Meany, president of the AFL-CIO, led him to take the UAW out of the organization in 1969. Shortly before his death he formed a short-lived Alliance for Labor Action with the Teamsters*. A strong supporter of civil rights and co-operative movements, he fought strenuously for the right of organized labour to participate in industrial planning. He died in an aeroplane crash.

Revisionism, a term used loosely through the 20th century to describe any attempt to revise accepted ideological or historiographical beliefs. It was used in both the Soviet Union and China to denounce the beliefs and practices of opponents. Between the wars it was applied to those who were arguing the case for German, Hungarian, and Bulgarian participation in World War I. At a later date the term was adopted by those historians critical of the orthodox US version of the origins of the Cold War.

Reynaud, Paul (1878–1966), Prime Minister of France 1940. From southern France, he became a successful Paris lawyer before World War I, in which he was awarded the Croix de Guerre. He entered politics in 1919, a member of the right-wing Democratic Alliance, and held various offices during the 1930s. He was Prime Minister of France for twelve weeks at the time of the crisis in 1940, with Pétain* his deputy. The latter negotiated an armistice with the Germans and Reynaud resigned. He was interned by the Vichy government and imprisoned by the Germans in Austria 1943–5. He held the offices of Finance Minister (1948) and Vice-Premier (1955) in the unstable Fourth

Republic*. Early on he supported de Gaulle in the Fifth Republic, but he quarrelled with the President and retired from politics.

Rhee, Syngman (1871–1965), Korean statesman and President of the Republic of Korea 1948–60. He was an early supporter of Korean independence, and after a spell in prison for alleged revolutionary activity (1897–1904) he went to the USA. In 1919 he formed a Korean government-in-exile with a small group of supporters, and during World War II advised the US government in Washington. He returned to Korea in 1945, and during the US occupation was the leading anti-Communist politician, advocating the reunification of Korea as a first step towards 'rolling back the tide of Communism in Asia'. Elected President of the Republic of Korea (South Korea*) in 1948, he welcomed the subsequent ending of the US occupation, but considered the levels of US military support to his regime to be inadequate. He only reluctantly supported the long armistice negotiations to end the Korean War*, consistently opposing attempts forcefully to repatriate prisoners of war. Having gained a US treaty of friendship, he was re-elected in 1952 and 1956. His government had inherited a ruthless police bureau, and it became more and more corrupt and autocratic. As economic conditions deteriorated, increasing unrest developed. Accusations of electoral 'rigging' and serious riots in 1960 forced him to return to the USA in exile.

Rhineland. The success of the French revolutionary armies in 1794 had brought the left bank of the Rhine to France, but this was ceded to Prussia in 1815, as a bulwark against renewed French expansion. In 1918 the Rhineland was demilitarized under League of Nations supervision, but allowed to remain in the Weimar Republic*. In 1936 Hitler's troops 'remilitarized' the area, meeting with no effective opposition from France or the League. The scene of heavy fighting in 1944, it was recaptured by US troops early in 1945. It now forms part of the Federal Republic as the *Land* (state) of Rheinland-Pfalz (Rhineland Palatinate).

Ribbentrop, Joachim von (1893–1945), German Nazi statesman. From an aristocratic family, he served in the cavalry as an officer during World War I, after which he married the daughter of a wealthy wine merchant. He joined the National Socialist Party in 1932, and became a close associate of Hitler*. He was ambassador in London 1936–8 and then Foreign Minister 1938–45. He personally conducted the negotiations with many states destined to become Hitler's victims. He regarded the Nazi–Soviet Pact of 1939 as his masterpiece, as it opened the way for the attack on Poland and the Baltic states. He was responsible for the Tripartite Pact (1940) between Germany, Italy, and Japan. He was condemned at the Nuremberg Trials* and executed.

Ridgway, Matthew Bunker (1895–), Chief of Staff of the US army 1953–5. Born in Virginia, he graduated at West Point in 1917, and served in the US army between the wars. He commanded US troops landing in Sicily in 1943, then in campaigns in western Europe and in the Philippines. In 1950, when the situation in Korea deteriorated into war, he was appointed to command the US 8th Army, and organized a UN counter-offensive, before being appointed Supreme Allied Commander in the Far East on the dismissal of Douglas MacArthur*. In 1952 he succeeded General Eisenhower as Supreme Commander of Allied forces in Europe and then became Chief of Staff of the US army.

Robertson, Sir William Robert (1860–1933), British Field Marshal. Joining the army as a private in 1877, he was commissioned and sent to India in 1888. After service there and in South Africa, he became commandant of the Staff College,

Camberley, Surrey, in 1910. He was Chief of the Imperial General Staff from 1915 until criticisms by Lloyd George* of British strategy led to his transfer to the Eastern Command in 1918. He subsequently commanded British troops on the Rhine 1919–20.

Rockefeller family, a wealthy American family. John Davison Rockfeller (1839–1937) was born in New York, but moved to Cleveland, Ohio, when he became a partner in an agriculture-produce business. In 1862 his company joined with the petroleum refiners Clark and Andrews. By 1872 he and his associates, including his brother William, controlled all the refineries of Pennsylvania, and ten years later their Standard Oil Trust controlled 95 per cent of the nation's oil-refining capacity, under a Board of Trustees with Rockefeller as president; this being the first inter-state trust, against which antitrust legislation* against monopolies was to follow. A pious Baptist, he became personally more and more interested in educational philanthropy. In 1892 he founded the University of Chicago, which was followed by Rockefeller University, New York (1901), and the Rockefeller Foundation (1913). Altogether he gave away some $600m. in his lifetime. His son John Davison, Jr. (1874–1960), continued his father's philanthropic interests, building up the Rockefeller Foundation and funding the Rockefeller Centre in Manhattan. He donated the sites for the UN Building and the Lincoln Centre in New York City, together with the restoration of Williamsburg. His son Nelson Aldrich (1908–79) was attracted to public life and was elected for four consecutive terms as Governor of New York (1959–73), and also served as Vice-President to Gerald Ford 1974–7. In turn two of his sons were elected governors of Arkansas and West Virginia.

Rokossovsky, Konstantin Konstantinovich (1887–1968), marshal of the Soviet Union. Born in Warsaw, son of a railwayman, he was conscripted into the Russian Imperial Army in 1914 but in 1917 deserted to join the Bolsheviks and to serve in the Red Army as a junior officer. As a professional soldier he had reached the rank of lieutenant-general by 1938, when he narrowly avoided execution in the Great Purge*, being imprisoned until 1941. He commanded the defence of Moscow, and in the siege of Stalingrad* surrounded the 6th German Army of von Paulus. He took part in the Battle of Kursk* and in August 1944 he won a major battle at Bobruisk in the Ukraine, utterly defeating the German 9th Army. He then directed operations against East Prussia and in Poland, where he ignored appeals for help from the Warsaw Rising*. He returned to Poland in 1945 and was Defence Minister there 1945–56 and deputy Prime Minister 1952–6 under President Bierut*. He led troops in a bloody suppression, in Poznań in June 1956, of Polish workers who were demonstrating for 'bread and freedom'. He was recalled to Moscow and Gomulka* was re-established, able to claim some independence from the Soviet Union. In Moscow he became deputy Minister of Defence 1956–62.

Roman Catholic Church, the largest branch of the Christian Church, with about 750 million members, under the jurisdiction of the Pope, the Bishop of Rome. The First Vatican Council (1869–70) under Pius IX had declared the infallibility of the Pope. At the same time Catholic missionaries, notably from the Society of Jesus (Jesuits), carried the faith through Latin America and into Africa and Asia. During World War II Pope Pius XII* retained the strict neutrality of the Church, and after it the Second Vatican Council* met under John XXIII and Pius VI with intent to modernize teaching, organization, and liturgy. The global mission of the papacy has been emphasized by Pope John Paul II*, with close contacts with the Eastern Orthodox* and Protestant churches. In 1984 it concluded a revision of the Lateran Treaty of 1929 with the Italian government, formalizing separation of Church and state, and in 1989, after fifty years of persecution within the Soviet Union, Pope John Paul met with President Gorbachev. By the year 2000 a majority of Roman Catholics will be living in Latin America, where 'Liberation Theology'*, the identification of the priesthood with the poor, has become a source of controversy within the Church.

Romania. Part of the Ottoman Empire from the 15th century, its independence was recognized in 1878, when Prince Carol Hohenzollern-Sigmaringen was crowned King Carol I (1881–1914). Carol's pro-German policy led him to join the Triple Alliance of Germany, Austria, and Italy. However, under Ferdinand I in World War I it remained neutral until 1916, when it declared war on Austria-Hungary. Romanian troops occupied Transylvania and in August 1919 occupied Budapest, driving out Béla Kun*. Romania was duly rewarded at the Paris Peace Conference*, with the doubling of its territories, mostly from Hungary. The liberal leader Brătianu* was in office 1923–7, and he introduced a new constitution and the secret ballot. When Carol II* returned in 1930 he imposed a right-wing autocratic regime, but rejected the extreme Fascism of the Iron Guard*. Carol was forced to cede much territory to Hungary (Transylvania) and to the Soviet Union (north Bessarabia). He abdicated, and under Antonescu* Romania allied with Hitler in 1941, Romanian troops co-operating with the German invasion of the Soviet Union. After Stalingrad, however, the Red Army began its great advance west, and Romania lost territory to the USSR and Bulgaria; but it then allied with the Soviet Union and was able to regain Transylvania. During the next forty years it was a Soviet satellite. However, it retained a degree of independence, which increased when Nicolae Ceauşescu* became President (1967–89). It benefited from industrialization, assisted by its own oil deposits. In 1971 it became the only Soviet-bloc country to join the International Monetary Fund. Despite this, stringent economic measures had to be enforced in 1987. During 1989, as the movement towards democracy was gathering pace in Eastern Europe, Romania seemed to stand alone. On 16 December hundreds of demonstrators in the eastern city of Timisoara were gunned down by tanks and helicopters. The killings sparked off strikes and demonstrations and a National Salvation Front was hastily formed in the national television studio. Ceauşescu and his wife were captured, tried before a military tribunal, and shot on Christmas Day. After three days of civil strife, peace was restored and the National Salvation Front, whose leader was Ion Iliescu, formed a provisional government, with Iliescu provisional President. The death penalty was abolished and elections promised. These were held in May 1990, when the NSF won. Iliescu, however, and many of his colleagues had been prominent Communists, and there were soon violent demonstrations in Bucharest against alleged corruption and the presence in government of ex-Communists.

Rome, Treaties of (25 March 1957), two international agreements signed in Rome between Belgium, France, Italy, Luxembourg, The Netherlands, and the Federal Republic of Germany. They established the European Economic Community* and the European Atomic Energy Community*. New members would be required to adhere to the terms of the treaties, which were scheduled for revision in 1991.

Rommel, Erwin (1891–1944), German field marshal. Educated at Tübingen, he joined the army in 1910 and rose through the ranks, being awarded the highest German military decoration for bravery in World War I, having been commissioned officer. He remained a regular soldier, and attracted the attention of Nazi Party leaders for his efficiency. By 1940 he held the rank of general and commanded a Panzer division in a brilliant assault through the Ardennes to the Channel. In 1941 he was sent to command the Afrika Korps, an élite tank formation which bore the brunt of the fighting in Libya, where he earned the respect of his opponent Montgomery* and the nickname 'the Desert Fox'. In 1942 he advanced to El Alamein, but British resistance and lack of supplies impeded him and eventually he had to retreat from North Africa. In 1944 he was entrusted with defence of the Channel coast in northern France against possible invasion. Wounded in the Normandy campaign, he was recalled to Germany. The Gestapo believed he was connected with the July Plot*. No accusations were made publicly against him, but he was obliged to commit suicide by taking poison. His son became a notable Mayor of Stuttgart.

Romulo, Carlos Peña (1899–1985), Philippine politician. During the 1930s he was publisher of a chain of Philippine newspapers and in 1941 he joined General MacArthur*, in charge of his propaganda broadcasts. He accompanied him to Australia and in 1944 was given the rank of general in the US army, returning to the Philippines in 1945. He represented the Philippines in the UN, where he was president of the General Assembly 1949–50 and of the Security Council 1956. He took an active part in the Bandung Conference* of 1955. Under President Marcos* he held various political offices, being Secretary and then Minister for Foreign Affairs 1968–78 and 1978–84. His early enthusiasm for democracy waned, and he supported the imposition of martial law in 1972.

Roosevelt, Eleanor (1884–1962), UN diplomat. From a wealthy New York family, and niece of President Theodore Roosevelt, she was educated in England, and in 1905 married her cousin Franklin D. Roosevelt. When the latter developed poliomyelitis in 1921, she helped him to overcome the disease and took on many public duties, as he campaigned first for the New York governorship and then as President. A strong supporter of New Deal* policies, she helped to democratize the White House by her Press conferences and her journalism. Actively involved in numerous social projects, from equal rights for women and minority groups, to child welfare and slum clearance, her strong personality aroused some criticism. A delegate to the UN, she was appointed chairman of the UN Commission on Human Rights (1946–51). As such, she helped to draft the Universal Declaration of Human Rights* of 1948, and travelled the world tirelessly as a UN representative.

Roosevelt, Franklin Delano (1882–1945), President of the USA 1933–45. From a wealthy New York family, he studied law at Harvard and Columbia universities before entering politics, being elected to the New York Senate in 1910. In 1905 he had married his cousin Eleanor*, and in 1913 he became Assistant Secretary for the Navy under President Wilson (1913–20). In 1921 he was stricken with poliomyelitis, and henceforth was confined to a wheelchair. A reforming Democrat, he was elected Governor of New York in 1928, and began his long presidency in 1933, having beaten the Republican incumbent Herbert Hoover. His New Deal* programme tackled with confidence the crisis of the Great Depression*, while in foreign affairs he recognized the Soviet Union, with whom he established diplomatic relations. In the 1936 presidential election he won a crushing victory, gaining every state except Maine and Vermont, thanks partly to his innovation of the regular intimate 'fireside chat' on radio. In his second term (1937–41), inherent weaknesses of the New Deal became more obvious, and hostility towards him from business and other sections of the community grew. His attempts to 'pack' the US Supreme Court*, which had rejected some of the New Deal legislation, caused considerable dismay. However, he carefully steered the USA away from policies favoured by isolationists, who had succeeded in passing a series of

Neutrality Acts* through Congress. After the fall of France in 1940 he made the USA a powerful supporter of Britain's war effort, while remaining a non-belligerent. In August 1940, by his Destroyer Transfer Agreement* with Winston Churchill, he exchanged fifty pre-1914 US destroyers for naval bases in the West Indies, Newfoundland, and British Guiana, thus providing Britain with much-needed convoy escorts. While the Battle of Britain* was being fought he ran for an unprecedented third term against the Republican Wendell Wilkie. He met much opposition from Southern Democrats and his victory was by no means automatic; prosperity resulting from expansion of the munitions industry may well have been the key to his success. In March 1941 he obtained from Congress the Lend-Lease Act*, and in August 1941 met Winston Churchill at sea to issue the Atlantic Charter*. A firm supporter of China in its war with Japan, he denied the latter war supplies, a policy which helped to precipitate Japan's action on Pearl Harbor* in December 1941. The USA now found itself in alliance not only with Britain, but also with the Soviet Union. He was to meet Stalin both at Tehran in November 1943 and at Yalta* in February 1945; he was later to be much criticized for allegedly 'taking Stalin at his word'. Other conferences, at Casablanca (January 1943), Quebec (August 1943 and September 1944), and Cairo (November 1943), when Far East plans were discussed with Jiang Jiehi* (Chiang Kai-shek), all concerned themselves with the Allied war strategy. On this he clashed with Churchill, who wanted to concentrate the assault on Germany from the Mediterranean, while Roosevelt and his advisers pressed for a cross-Channel attack. But in spite of these differences he and Churchill remained close friends. He is the only US President to stand for a third term and then a fourth in 1944, since when this has been made unconstitutional. By 1945 he was a sick man, and he died two months after Yalta (12 April 1945), being succeeded by his Vice-President Truman*. Although capable of provoking critics into accusations of 'shallow incompetence' and 'dictatorial ambition', his administrations are broadly accepted to have been remarkable for their integrity and efficiency.

Root, Elihu (1845–1937), US Secretary of State 1905–9. A Republican and leading corporate lawyer, he became Secretary of War 1899–1903, and made drastic reforms in army organization. As Secretary of State for Theodore Roosevelt, his policy was for US commercial interests to expand in China. From 1909 until 1915 he was Senator for New York, opposing President Wilson's policy of neutrality in 1914 and openly supporting the Allied cause. After the war he was one of the few Republican statesmen to support the League of Nations*, serving on the League's Commission of Jurists, which established the Court of International Justice. As a US representative to the Washington Conference* 1921–2, he worked consistently for international peace.

Rosenberg, Alfred (1893–1946), German journalist and Nazi administrator. He was the son of a cobbler and grew up in Reval, Estonia. He studied architecture in Moscow, and then went to Munich, where he met Hitler, Röhm, and Hess in 1920. He became an enthusiastic recruit to the young National Socialist Party. He became its chief ideologist, strongly anti-Semitic and drawing on Nordic and Teutonic mythology. In 1941 he was appointed Minister of Occupied Countries, responsible for the looting of works of art. He was tried at Nuremberg and hanged.

Rosenberg Case (USA). In March 1951 Julius Rosenberg, an electrical engineer from New York, and his wife Ethel were convicted of obtaining information concerning atomic weapons in 1944–5 and passing it on to agents of the Soviet Union, which was then of course an ally of the USA. They became the first American civilians to be sentenced to death for espionage by a US court. The only seriously incriminating evidence had come from a confessed spy, and the lack of clemency shown to them was an example of the intense anti-Communist feeling that gripped the USA during the Red Scare* of the 1950s.

Round Table Conferences, meetings held in London 1930–2 to discuss Indian constitutional developments. The procedure was suggested by the Viceroy Lord Irwin (later Lord Halifax*) in 1929. The Indian National Congress* boycotted the first session (November 1930–January 1931), but, following the Gandhi*–Irwin Pact (March 1931), Gandhi attended the second session (September–December 1931). With the renewal of a non-co-operation campaign, Gandhi was imprisoned and Congress took no part in the final session (November–December 1932). The constitutional discussions formed the basis of the 1935 Government of India Act*, with its plan for a federal organization involving the Indian princely states.

Rowlatt Act (1919), legislation in British India. A committee under Mr Justice Rowlatt issued a report in 1919 which recommended the continuation of special wartime powers in India for use against 'revolutionary conspiracy' and terrorist activity. The report was implemented, but the Act aroused strong opposition from Indian nationalists, and this was channelled by the Mahatma Gandhi* into nation-wide satyagraha* or civil disobedience. This 'Rowlatt agitation' culminated in the Amritsar Massacre* of 13 April 1919.

Rowntree, Benjamin Seebohm (1871–1954), British industrialist and philanthropist. Born in York, where the family chocolate firm of Rowntrees was based, as a young man he made a study of poverty within the city, which shocked public opinion: *Poverty: A Study of Town Life* (1901). As a director of the firm he made a number of managerial innovations for which he was criticized by more traditional companies, for example: a five-day week, employee pension and profit-sharing schemes, a works council, joint management/worker consultations and disclosure of all management information, payment of full-time shop stewards. He believed strongly that the industry from which people earn their living should provide the means to a life worth living. Many of his innovations remain still to be adopted within much of British industry.

Roxas, Manuel (1892–1948), President of the Philippines Republic 1946–8. From a prosperous Philippines family, he trained as a lawyer before entering local politics in 1917, going on to be Speaker of the House of Representatives. In 1932 he led a Philippines Independence Mission to Washington, DC, which helped to achieve the 1935 Constitution. He became Secretary of Finance in the government of Manuel Quezón* and personal friend of General MacArthur*. He served in the pro-Japanese wartime government, but avoided trial as a collaborator. After Quezón's death in 1944 he became the leading candidate for President, being backed by MacArthur. He was inaugurated 4 July 1946, but his brief presidency was marred by corruption and the alienation of the Hukbalahaps*.

Ruhr, the, the valley of the River Ruhr which flows south into the Rhine. It was the industrial heart of Bismarck's united Germany after 1871 and includes the vast Krupps* factories of Essen. After World War I France feared that it might again become an armaments centre, and in 1923 it was occupied by French and Belgian troops when Germany defaulted on reparations payments. The loss of resources, production, and confidence resulted in soaring inflation in Germany that year. Two years later the French accepted the Dawes Plan* and withdrew. After 1933 war industries were re-established as Germany rearmed. During World War II it was the target for heavy bombing, and afterwards its recovery was monitored by an international control commission*. Control passed to the European Coal and Steel Community* in 1952 and then to the Federal Republic in 1954.

Rundstedt, Karl Rudolf Gerd von (1875–1953), German field marshal. Born into a wealthy *Junker* family in Prussia, he was commissioned into the German Imperial Army and served in World War I as Chief of Staff of an army corps and also on the Turkish General Staff. He remained in the army as a non-political figure until retirement in 1938 as senior field commander. He was called from retirement in 1939 to command an army corps in the Polish and French campaigns. In 1941 he commanded the Army Group South in the invasion of the Soviet Union, but was dismissed after he had withdrawn from Rostov against Hitler's orders, in order to improve his chances of resisting the Soviet counter-offensive. From 1942 to 1945 he commanded the forces occupying France, and in December 1944 launched the counter-offensive of the Ardennes campaign*. Again relieved of his command in March 1945, he was captured by US troops in May, but soon released. He had refused to take part in any of the various plots against Hitler and was described by his opponent Eisenhower as 'the ablest of the German generals'.

Rusk, David Dean (1909–), US Secretary of State 1961–9. Born in Georgia, he won a Rhodes Scholarship to Oxford in 1931, and then taught political science at Mills College, California, until the outbreak of war. He enlisted in the army and served under General Stilwell* in China 1944–5. In 1946 he joined the US diplomatic service, and in 1950 became an Assistant Secretary of State concerned with the Far East, strongly backing the case for war in Korea. In 1961 President Kennedy appointed him Secretary of State, an office he held for eight years, during all of which he consistently advocated a policy of US involvement in the Vietnam War* and opposition to Communist China. After leaving office he became a Professor of International Law, being strongly criticized retrospectively for his intransigence and the lack of flexibility in his policies.

Russell, Bertrand Arthur William, 3rd Earl Russell (1872–1970), mathematician and philosopher. Grandson of a Prime Minister, he grew up and was educated at Cambridge, where he became a Fellow of Trinity College, publishing what many regard as his greatest work, *The Principles of Mathematics*, in 1903. He met and taught Wittgenstein*, and wrote *The Problems of Philosophy* in 1912. A member of the Bloomsbury Group* he left his first wife for the group's patron Lady Ottoline Morrell, who became his mistress. A pacifist, he was imprisoned in 1918 for the vehemence of his views and lost his Fellowship. While in prison he wrote *Analysis of Mind*, which sees mind and matter as different 'structurings' of the same neutral elements. After living and teaching in the USA, he had his Trinity Fellowship restored in 1945, and taught again at Cambridge. The hydrogen bomb* tests in 1954 on Bikini prompted him to found the CND movement and also Pugwash*. He was imprisoned again in 1961 for demonstrating against nuclear weapons, having become terrified of their threat to humanity. In the last years of his long life he vehemently attacked US policies in Vietnam, founding an International War Crimes Tribunal. His work as a mathematician probably represents his greatest contribution to human thought, but his clarity of mind and skills as a teacher are reflected in his book *The History of Western Philosophy*, which enabled millions to reach an understanding of their intellectual heritage which would have otherwise been denied them.

Russia. By early in the 20th century the Russian Empire ruled by the Romanov dynasty stretched deep into Asia to the Pacific, and south into the Caucasus. The Trans-Siberian railway, financed largely by French capital, was completed in 1904, but communications remained inadequate for governing so vast an area. There were increasing political tensions culminating in the first Russian Revolution of 1905. Russian support for Slavonic nationalism within the Austro-Hungarian Empire was a powerful factor for war in the years 1912, 1913, and 1914. Although heavily defeated at Tannenberg* the Imperial Army performed well, and in 1916 Brusilov* defeated the Austrians. Logistical problems of supplies, however, during the winter 1916–17, led to a loss of army morale and to the Revolutions of 1917 and the end of the Romanov dynasty. When the Soviet Union was created in 1922, by far the largest republic was that of Russia, itself a federation of autonomous republics, provinces, and territories, and containing some 70 per cent of the population. Until 1989 the Communist Party of the Soviet Union maintained a firm control over this federation, although it gained a new constitution in 1978. Pressures developed however for greater independence, and in September 1989 it was agreed that this constitution be revised. When this was introduced in March 1990, with a Congress of People's Deputies and a Supreme Soviet, Boris Yeltsin was elected Chairman of the latter (virtually President of the Republic) on a ticket of Russian nationalism, multi-party democracy, and economic reform. This Russian Soviet Federal Socialist Republic (RSFSR) stretches from the Baltic to the Arctic and the Pacific, sharing its capital Moscow with the Soviet Union.

Russian Civil War (1918–21), sometimes referred to as the War of Allied Intervention. In December 1917 counter-revolutionary forces led by General Kornilov, who had escaped from prison, began to be organized into a so-called White* Army. They clashed with a hastily raised Red Army organized by Trotsky*. In northern Russia a multinational force, made up of British, French, German, US, and Canadian troops, landed at Murmansk and occupied Archangel 1918–20; France occupied Odessa; the British, operating from Iraq, seized oil-rich Baku and tried to establish a bridgehead in the Caucasus; in the Far East a Japanese and US force held Vladivostok until 1922. Nationalist revolts in the Baltic states led to the secession of Estonia and Latvia and, after fierce fighting, of Finland and Lithuania, while the Polish army, with French support, successfully advanced the frontier to the Russian Ukraine, gaining an area not reoccupied by the Soviet Union until 1939. In Siberia, based at Omsk, Admiral Kolchak* acted as Minister of War in an 'All Russian Government' and, with the aid of a Czech legion made up of released prisoners of war, gained control over sectors of the Trans-Siberian railway. He however was betrayed by the Czechs and murdered, the leadership passing to General Denikin*. The latter mounted a major offensive in the Ukraine in 1919 aided by Don Cossacks, only to be driven back to the Caucasus, where he held out until March 1920. In the Crimea the war continued under General Wrangel* until November 1920. A famine in that year caused peasant risings, while a mutiny of sailors at Kronstadt in 1921 was suppressed by the Red Army with heavy loss of life. To win the war Lenin* imposed his ruthless policy of 'war communism'. Lack of co-operation between counter-revolutionary forces, together with an innate Russian patriotism against foreign invaders, led to the final collapse of the White Army and the establishment of the USSR.

Russian Revolutions (1917). In the so-called 'February' Revolution strikes and riots in Petrograd (8 March), supported by imperial troops, led to the abdication of the Tsar and thus to the end of Romanov rule after 300 years. The provisional government* of Lvov was established, which Kerensky* took over in July. Lenin had returned in April, but was exiled again to Finland by Kerensky. He returned in October when Kerensky faced rising opposition from the Petrograd Soviet of Workers' and Soldiers' Deputies. Early in November a nearly bloodless coup by Bolshevik soviets*, orchestrated by Lenin (the 'October' Revolution), took control of the major cities, and a cease-fire was arranged with the Germans. In July 1918 a Soviet constitution was proclaimed and Lenin transferred the government from Petrograd to Moscow. The Russian Civil War* was to continue for three years, to end in the supremacy of the Bolsheviks and the establishment of the Union of Soviet Socialist Republics in 1922.

Russo-Polish War (1919–21). Joseph Pilsudski* was determined to gain a frontier for the new Republic of Poland as far to the east as possible. After intermittent fighting during 1919, he allied with Ukrainian nationalists in April 1920 and overran the Ukraine*, occupying Kiev. A counter-attack by the Red Army swept into Poland in July, reaching the outskirts of Warsaw. The British Foreign Secretary Lord Curzon* had recommended a pacification which would have given the Russians all the Ukraine; but France sent General Weygand* to Warsaw to advise the Poles and organize a new attack. By mid-August this was pushing back the Soviet troops, and an armistice was arranged in October. The Treaty of Riga in March 1921 gave Poland substantial parts of the Ukraine and of Belorussia, which it retained until 1939.

Rwanda, a central African republic. It obtained its present boundaries in the late 19th century under pastoral Tutsi* kings who ruled over agriculturalist Bahutu (Hutu*). In 1890 Germany claimed it as part of German East Africa, and in 1916 Belgian forces occupied it. Belgium then administered it under a League of Nations mandate* as Ruanda-Urundi. Following civil strife (1959) between the Tutsi and Hutu peoples, the Tutsi King Mwami Kigeri V was deposed. Rwanda was declared a republic in 1961 and became independent in 1962, separate from the Tutsi-dominated Burundi*. The now dominant Bahutu forced large numbers of Tutsi into exile, but after the accession to power of President Juvénal Habyarimana in 1973, domestic stability gradually improved. In 1975 Habyarimana's party (MRND) declared itself the single political organization, and he was re-elected in 1978, 1983, and 1988. In October 1990 there was a Tutsi invasion from Uganda. Belgian and French forces helped Habyarimana to end civil strife.

Sa'adist Party, an Egyptian political party, founded in 1937 as an offshoot of the Wafd* Party. Led by Ahmad Maher, its members claimed that the Wafd had betrayed the principles of their founder Sa'd Zaghlul. They were in office under Farouk 1944–6 and 1946–9, but disappeared after the fall of the monarchy.

Saar, the, a river flowing north through France and south-west Germany. In 1792 its valley was occupied by French troops, but in 1815 it was awarded to Prussia. It became a major industrial area after 1871, when Germany acquired Alsace and Lorraine with their iron and coal deposits. As part of the Versailles Treaty, the area was placed under League of Nations administration, with its mines going to France. In 1935 a plebiscite voted for its restoration to Germany, but it was again occupied by French troops in 1945. In 1955 another referendum voted for restoration to Germany, and in 1959 Saarland became the tenth *Land* (state) of the Federal Republic.

Sabah, a state of Malaysia* on the island of Borneo*. As British North Borneo it was united with Labuan in 1946, when it became a Crown Colony. In 1963 it adopted the name Sabah, when it joined the Federation of Malaysia.

Sacco–Vanzetti case (USA). In 1920 two Italian immigrants Nicola Sacco and Bartolomeo Vanzetti were found guilty of murder. There were allegations that their conviction had resulted from prejudice against immigrants, anarchists, and evaders of military service. There were anti-USA demonstrations in Rome, Lisbon, and Montevideo, and one in Paris, where a bomb killed twenty people. For six years efforts were made without success to obtain a retrial, although the judge had been officially criticized for his conduct during the trial. The two men were electrocuted in August 1927. The affair helped to mobilize opinion against the prevailing anti-foreigner isolationism and conservatism of post-war America. Later evidence pointed to the crime having been committed by members of a gang led by Joe Morrelli.

Sadat, Muhammad Anwar (1918–81), President of Egypt 1970–81. He graduated from the Cairo Military Academy in 1938, and in 1942 was imprisoned for plotting to expel the British from Egypt with the help of the Germans. He escaped, and after the war joined the Free Officers' movement which in 1952 deposed Farouk*. He was a close friend and ally of President Nasser*, being Vice-President 1964–6 and 1969–70, and becoming President when Nasser suddenly died. By 1972 he had dismissed the Soviet military mission and confirmed Egypt's alliance with Syria, with whom he launched the Yom Kippur War* in October 1973. In 1974, following UN intervention, he regained for Egypt the Suez Canal area and reopened the Canal in 1975. By now high military expenditure was causing grave problems, and in 1977 he offered to visit Israel to address the Knesset (Parliament). Prime Minister Begin* agreed, and this 'Sadat Initiative' marked the beginning of a peace process. Israeli–Egyptian discussions took place in Leeds Castle, Kent, and then at Camp David with President Carter*, culminating in the peace treaty of March 1979. The 'initiative' was violently attacked by the PLO* and was condemned by Syria, Libya, and Algeria. On 6 October 1981 he was shot by four assassins while reviewing a military parade, but Israel and Egypt remained at peace.

SADCC *see* SOUTHERN AFRICAN DEVELOPMENT CO-ORDINATION CONFERENCE.

Saigon, fall of (30 April 1975). Badly damaged in the 1968 fighting, Saigon had remained the capital of South Vietnam under President Thieu*. Some 1 million refugees are estimated to have come into the city during the Vietnam War. After the Paris Peace Accords* (1973) the bulk of US personnel were withdrawn, but President Thieu determined to carry on the war. In January 1975 Communist VietMinh troops captured Phuoc Binh, 80 miles north of the city. Ban Me Thuot in the central highlands fell next, to the NLF Vietcong*, while further north Hue fell on 26 March and Da Nang, which had been the US naval base, on 29 March. In theory Thieu still had over a million men under arms. In practice they began to melt away, trying desperately to reach any port to escape. In April, after meeting stubborn resistance, 40 miles east of the city, General Giap's* troops moved in. Thieu resigned and fled on 20 April. By 28 April Saigon's port had fallen, whereupon US helicopters carried away the last US personnel and as many Vietnamese refugees as could crowd into them. On 30 April North Vietnamese troops entered the city, now to be known as Ho Chi Minh city.

St Christopher and Nevis. A British colony of two islands (St Kitts–Nevis) from 1816, these formed part of the Leeward Islands Federation 1871–1956. In the West Indies Federation* (1958–62) they were joined by Anguilla*. When the Federation collapsed they were granted 'Associated Statehood' in 1967, although Anguilla departed. In 1983 the Federation of St Christopher and Nevis finally became independent within the Commonwealth. A Labour Party had been formed in 1950, but in 1980 a coalition of two parties, the People's Action Movement, led by Dr Kennedy Simmonds, and the Nevis Reformation Party, won power which it retained in 1984, with Dr Simmonds as Prime Minister.

St-Germain, Treaty of (10 September 1919), peace treaty for Austria. The Austrian Republic had been proclaimed on 12 November 1918 when the Emperor Charles I* abdicated. Hence at the Paris Peace Conference* it was on delegates from the Republic that the treaty was imposed. Although the Republic had voted in March 1919 for Anschluss* with Germany, this was forbidden. The new frontiers were laid down, whereby the Slavonic provinces of Dalmatia, Slovenia, and Bosnia-Hercegovina would go to the new Kingdom of Serbs, Croats, and Slovenes; Galicia would go to Poland, and Bukovina to Romania. South Tyrol would go to Italy, thus enclosing the Republic behind the Alps, with no access to the sea except via the Danube. Reparations were to be duly paid when agreed by the Reparations Commission. The Austrian army was to be limited to 30,000 men.

St Laurent, Louis Stephen (1882–1973), Prime Minister of Canada 1948–57. Born and educated in Quebec, he became a successful corporate lawyer before being invited by Mackenzie King* to join the federal government as Minister of Justice. Elected to Parliament in a by-election, he was always a staunch supporter of King, and as Minister for External Affairs 1946–8 he spoke strongly of the need for a North

Atlantic Treaty. As Prime Minister he led Canada through a period of fast economic growth and prosperity. His government extended old-age pensions to all over 65 and created the Canada Council for Arts and Sciences. In Foreign Affairs it supported NATO and Canadian participation in the Korean War*, while its most significant achievement was the agreement with the USA to build the St Lawrence Seaway*. During his years of office the title dominion was dropped and Newfoundland became the tenth province. He was defeated in the general election of 1957 and retired a year later.

St Lawrence Seaway. Following an agreement in 1954 between the governments of President Eisenhower and Prime Minister St Laurent, a St Lawrence Seaway Authority was established in 1956 to construct and maintain the seaway between Montreal and Lake Erie, in order to carry deep-draught ships to Chicago and the Lake Ports. The enterprise involved widening and deepening a series of 19th-century canals, as well as the construction of dams and joint hydroelectric power-stations. The Seaway was officially opened on 26 June 1959. It is used from 1 April to 18 September as a major route for world trade.

St Lucia, a small Caribbean state of 616 sq. km. and population of some quarter-million. The British took it from the French in 1814 and a French patois remains the language. It formed part of the West Indies Federation* 1958–62 and in 1967 became an 'Associated State'. Two political parties emerged, the St Lucia Labour Party (SLP) and the more conservative United Workers's Party (UWP). When independence within the Commonwealth was gained in 1979, the SLP came to power. John Compton of the UWP has held office as Prime Minister since 1982. Tourism remains a major source of income.

St Vincent and the Grenadines, an island state in the Caribbean. It consists of the island of St Vincent (344 sq. km. and 140,000 pop.) and a group of small islands, the North Grenadines, the largest being Bequia, Mustique, Canouan, Mayreau, and Union, with a combined population of some 5,000. The islands were taken by the British in the 18th century and formed part of the West Indies Federation* 1958–62. They then became a self-governing 'Associated State' in 1969, by which time two political parties had emerged, the St Vincent Labour Party (SVLP) and the People's Political Party (PPP). Independence within the Commonwealth came in 1979, with the SVLP elected to power. Its policies of economic retrenchment, together with corruption scandals, lost its popularity to a new centrist party, the New Democratic Party (NDP), which has held office under its leader James Mitchell since 1984.

Sakharov, Andrey Dmitrievich (1921–89), Soviet scientist and human rights activist. Born in Moscow, he graduated from the university in physics in 1942 and went on to become a research physicist. From 1947 until 1953 he worked on the project which gave the Russians the hydrogen bomb*, and was elected to the USSR Academy of Science, their youngest member. He then worked on peaceful applications of nuclear energy. During the 1960s he became increasingly concerned at the threat of nuclear war, advocating a world-wide ban on nuclear weapons, warning of the mortal danger to the human race that they posed, and criticizing the division of the world into capitalist and socialist blocs. He supported the Prague Spring* and in 1970 was co-author of an open letter to President Brezhnev* calling for democratization of the Soviet Union. He won the Nobel* Peace Prize in 1975 and criticized the Soviet invasion of Afghanistan in 1979. He was exiled to Gorki and put under house arrest. His wife was allowed to travel to the West for health reasons and he was released in 1986, after a hunger-strike. In 1989 he was elected to the Congress of People's Deputies, where he criticized the government.

He died suddenly on 14 December, and 100,000 people, including President Gorbachev, paid their respects at his lying-in-state. The Sakharov Prize for Defence of Human Rights was established by the European Community in 1987.

Salan, Raoul Albin Louis (1899–1984), French general. Born near Nîmes, he graduated from the French Military Academy of St-Cyr in 1919. He rose steadily in rank and joined de Gaulle's Free French* army in 1940. He was sent to West Africa and after the war to Indo-China, where his policies quite failed to halt the success of the VietMinh*. After recall he was appointed Commander-in-Chief in Algeria, where again his uncompromisingly tough attitudes worsened the situation. In February 1959 he became military governor of Paris. When he realized that de Gaulle was seeking a political solution to the Algerian question, he resigned and went to Spain to organize his OAS*. His revolt in Algiers in 1961 failed. He was arrested and imprisoned for six years before being granted a pardon in 1968.

Salazar, Antonio de Oliveira (1889–1970), Prime Minister of Portugal 1932–68. Son of an innkeeper, he began to train as a priest, but then became an academic, being appointed a Professor of Economics in 1919. He entered politics and became a highly successful Finance Minister in 1928. In 1932 he was invited to become Prime Minister and was virtual dictator of Portugal for the next thirty-six years. He introduced a new constitution in 1933, creating the *Estado Novo* (New State) along neo-Fascist lines and outlawing all political parties except his Portuguese National Union. He used his authority to achieve some social and economic reforms in a stubbornly conservative society. During the Spanish Civil War and World War II he was also Minister of War and Foreign Minister, and maintained a policy of neutrality. His determination to defend Portugal's African colonies against mounting nationalism embittered his military leaders, who were obliged to wage difficult and costly campaigns in Africa. In 1968 he suffered a stroke and was succeeded by Caetano*.

Salonika campaign, World War I (October 1915–November 1918). Salonika was annexed by Greece from the Ottoman Empire in 1913. In 1915 French troops landed there to try to help Serbia against Austria-Hungary and in 1916 Venizélos* came there to lead a protest against Greece's pro-German King Constantine I. He proclaimed a new Greek government which the Allies recognized, Constantine abdicating. After the failure of the Gallipoli campaign* the French contingent was reinforced and Salonika became the base for an Allied Macedonian front against Bulgaria. In fact there was little action until late in the war. In September 1918 a combined force of French, Serbs, British, Greeks, and Russians attacked, under the command of General Franchet d'Esperey*. They advanced so rapidly that they had reached the Danube and Constantinople by the time the war ended in November.

SALT *see* STRATEGIC ARMS LIMITATION TREATIES.

Salt March (12 March–6 April 1930), a protest march in British India. The private manufacture of salt violated the salt tax system imposed by the British. In a campaign of non-violent civil disobedience the Mahatma Gandhi* led his followers from his ashram (retreat) at Sabarmati to make salt from the sea at Dandi, a distance of 320 km. The government took no action until the protesters marched on to a government salt depot, when Gandhi was arrested (5 May). His followers continued the protest action and a total of 60,000 were arrested and imprisoned.

Salvation Army, international organization for evangelical and social work. In 1865 William Booth had founded the Christian Revival Association to work in the slums of London. In 1878 it took the name Salvation Army, to be run on quasi-military lines with a fundamentalist approach to religion, and placing emphasis on welfare work and care of the destitute. In 1896 a splinter group Volunteers of America was founded by Ballington Booth, one of the sons of the founder. Another son, William Bramwell, succeeded his father as general, and he extended the Army's work into Austria, Czechoslovakia, China, Burma, and elsewhere. In 1929 Edward Higgins became general, and the work was again extended, particularly into Africa. In 1935 Evangeline Booth, fourth daughter of the founder, was elected general and further advances made in Africa. The Army successfully survived the disruption of World War II and by 1960 was working in eighty-four countries in eighty-one languages, with some 800 day schools and over 1,750 other institutions including hospitals, clinics, and maternity, children's, and eventide homes.

Samoa, a group of Polynesian islands in the central Pacific colonized by Germany and the USA in the 19th century. In 1899 the seven eastern islands were formed into **American Samoa**, to be administered by the US Navy Department 1900–51 and since then by the US Department of the Interior. In 1978 the first elected Governor took office, with a bicameral legislature for home affairs: a House of Representatives with twenty members and a Senate of eighteen members nominated by chiefs.

In August 1914 New Zealand naval forces seized **west Samoa** and from 1920 until 1961 New Zealand administered the area, first under a League of Nations mandate*, and then as a UN Trust Territory. Already in the 1920s there was agitation for autonomy, to which New Zealand was not unsympathetic. In 1960 **Western Samoa** became the first Polynesian state to gain full independence, with a unicameral legislature and retaining a treaty of friendship with New Zealand. It is a member of the Commonwealth of Nations and of the South Pacific Forum*, of which in 1987 its Prime Minister Vakai Kolone was the chairman. He criticized French policies in New Caledonia and also, as a strong supporter of the Forum's non-nuclear policy, nuclear testing in French Polynesia.

Sanctions, an international boycott to coerce a nation to conform to acceptable international behaviour. The concept was devised by the League of Nations* and incorporated in Article XVI of the Covenant. Limited sanctions were unsuccessfully applied against Italy in 1935 over the issue of the Italo-Ethiopian War*. The UN adopted the use of sanctions in its Charter and imposed them against Rhodesia in 1965, again without success, there being too many facilities for 'sanctions-busting', particularly in the supply of oil. Strong pressure for sanctions against South Africa was mounted by the Front-Line States (see SADCC) in Africa during the 1980s; they were resisted by Mrs Thatcher for Britain, but imposed by the US Congress and most European Community states. It was claimed that South Africa's reluctant willingness to begin negotiations with the ANC in 1990 was a direct result. In 1990 the UN Security Council imposed stringent sanctions against Iraq for its seizure of Kuwait.

Sandino, Augusto César (1896–1934). Son of a wealthy plantation-owner in Nicaragua, he worked for his father until he was 25, when, after wounding a man for insulting his Indian mother, he fled to Mexico. He returned to Nicaragua to work in a gold-mine, joining a Liberal insurgent unit against a US puppet Conservative government. In 1927, when the USA sent marines to restore order in the country, he withdrew into mountainous north Nicaragua. From here, with guerrilla operations, he harassed the Nicaraguan National Guard and US forces until the latter were withdrawn by President Roosevelt in 1933. For a short while he became leader of a co-operative farming scheme and signed a peace agreement with President Sacasa's government. He met Anastasio Somoza*, head of the National Guard, who then had him abducted, tortured, and murdered in April 1934. He became a popular hero and the Sandinista Liberation Front considers itself his spiritual heir.

San Francisco Conferences (1945, 1951). There were two conferences in San Francisco at the conclusion of World War II. The first was attended by delegates from fifty nations, when the design and structure of the United Nations Organization was agreed. It was opened on 15 April, and concluded on 26 June with the signing of the UN Charter. The second conference was to devise a peace treaty with Japan, which came into force in April 1952, when Japanese occupation ended and Japanese sovereignty was restored. Japan recognized the independence of Korea and renounced its rights to Taiwan, the Pescadores, the Kuriles, southern Sakhalin, and those Pacific islands mandated to it after World War I. These latter all went as Trust Territories* to the USA. The USA would maintain its forces in Japan until it could shoulder its own defensive responsibilities. The Soviet Union did not sign the peace treaty, but diplomatic relations were restored in 1956. Peace treaties with Asian nations conquered by the Japanese during the war were signed through the 1950s, as reparations* issues were settled.

São Tomé and Príncipe, a small republic in the Gulf of Guinea consisting of an archipelago of islands. It was a Portuguese colony from 1521 until 1974, when it gained independence. There is a single political party, Movimento de Libertaçao de São Tomé e Príncipe, which nominates candidates for the People's Assembly.

Sarajevo, principal city of Bosnia which had been occupied by Austria-Hungary in 1908. Here on 28 June 1914 the heir to the Austrian throne Francis Ferdinand and his wife were shot by a Serbian student, Gavrilo Princip, a member of a terrorist organization Black Hand. Austria-Hungary accused Serbia of complicity, of which it was in fact innocent, and imposed harsh demands in retaliation. Although Serbia met most of these, war followed on 28 July.

Sarawak, a state of Malaysia* on the island of Borneo*. Ceded by the Sultan of Brunei to a wealthy British adventurer, Sir James Brooke, in 1841, it had remained in the Brooke family until the Japanese occupation 1942–5. In 1946 it became a British Crown Colony. After a guerrilla war in 1962–3, during the run up to independence, it joined the Federation of Malaysia in 1963.

Sarekat Islam, an Islamic political organization in the Dutch East Indies*. It had been formed in 1911 as an association of Javanese traders anxious to protect themselves against Chinese competition. By the time of its first party congress in 1913, it had developed into a mass organization dedicated to self-government through constitutional means. Its leader, H. Q. S. Cokroaminoto (1882–1934), had an immense popular appeal; but the movement was weakened from within by the challenge of the emergent PKI* in the early 1920s, as well as by more radical nationalist parties such as the PNI (see INDONESIA). Its influence waned rapidly after Cokroaminoto's death.

Sato Eisaku (1901–75), Prime Minister of Japan 1964–72. As a supporter of Yoshida Shigeru* he advocated co-operation with the USA in the immediate post-war period. Forced from the cabinet over allegations of corruption in 1954, he returned to office four years later and then served as Prime Minister. He overcame student unrest in 1968, oversaw the extension of the revised US Security Treaty (1970), negotiated for the return of Okinawa* and other Ryuku islands (1972), and normalized relations with South Korea. After leaving office he was awarded a Nobel* Peace Prize for his efforts to make Japan a nuclear-free zone.

Satyagraha (Hindi; 'holding to the truth'), a tactic of civil disobedience, passive resistance, and non-co-operation developed by Mohandas Gandhi* in South Africa (1907–14), and widely used by him in India as a weapon there against the British. Campaigns of civil disobedience tend to degenerate into violence, always deplored by Gandhi; but the method had some success against liberal governments reluctant to use force. The technique continued to be used in India and elsewhere after 1947, for example in Goa* in 1955, when the satyagrahis were fired on by Portuguese police, but were eventually successful.

Saudi Arabia, a vast desert kingdom in Arabia. In 1927 the Treaty of Jeddah recognized Abd al-Aziz ibn Saud* as an independent ruler, King of the Hejaz, the Nejd and its peoples, and of the Principality of Asir. In 1932 the country was proclaimed the Kingdom of Saudi Arabia. Until the discovery of oil it was a backward state whose only wealth came from the pilgrimages to the holy cities of Mecca and Medina. An oil concession was awarded to a US firm, Standard Oil of California, in 1933, and oil was first exported in 1938. In 1944 the oil company was re-formed as the Arabia American Oil Company (ARAMCO), and Saudi Arabia was recognized as having the world's largest reserves of oil. Since the death of Ibn Saud (1953) efforts have been made to modernize the administration, by a series of new codes of conduct, to conform with both Islamic tradition and 20th-century developments. The Minister for Petroleum and Natural Resources, Sheikh Ahmad Yemani, ably led OPEC* in controlling oil prices during the 1970s. King Fahd succeeded to the throne on the death (1982) of his half-brother Khalid*. Various Saudi initiatives to bring peace to the Middle East all failed, partly because of Syrian suspicions of the close links between Saudi Arabia and the USA. In 1987 over 400 Shia Muslim pilgrims from Iran were killed in riots in Mecca, and there remained a threat of political instability from Islamic fundamentalism*. In 1989 there were renewed Saudi attempts to resolve the crisis in Lebanon and in 1990 an international force, led by the USA, arrived to protect the country from Iraqi aggression.

Savage, Michael Joseph (1872–1940), Prime Minister of New Zealand 1935–40. Born in Australia, where he became a gold-miner, he emigrated to New Zealand in 1907, becoming a trade union official. In 1916 he joined the Labour Party, which was formed that year, and entered Parliament in 1919. He became deputy leader of his party in 1923, and on the death of its earlier leader Harold Holland succeeded as leader in 1933. With Walter Nash* he put forward an expansionist programme which won the election of 1935, gaining support from the small farmers and farm workers, as well as the urban vote and Ratana*. This first Labour government introduced extensive public works as well as a comprehensive Social Security Act and other social welfare legislation. He was re-elected in 1938, but died in office in March 1940, being succeeded by Peter Fraser*.

Savang Vatthana (1907–81?), King of Laos (Setha Khatya) 1959–75. A devout Buddhist, he remained in the holy city of Luang Prabang throughout the years when his two cousins Souvanna Phouma* and Souphanouvong* were struggling to preserve Laos from chaos during the years of the Vietnam War*. He tried in vain to win international guarantees for Laos and in 1975 was obliged to abdicate. In 1977 he was sent to a 're-education seminary' in north-east Laos and was not heard of again. It is thought he may have died there in 1981.

Savimbi, Jonas Malheiro (1934–), Angolan revolutionary. Son of the first Black postmaster in Portuguese Angola, he was educated at mission school and in Portugal. Here he was imprisoned, but escaped and completed his education in Switzerland. He returned to Angola*, where in 1966 he formed UNITA (see ANGOLA), winning the military backing of China. As a guerrilla fighter he remained in Angola until independence in 1975. His UNITA movement now spent ten years fighting the rival MPLA, backed by the Soviet Union and Cuba. To the fury of many Africans Savimbi negotiated with South Africa for arms and support, and in 1986 obtained military aid from the USA. With South African troops supporting him he continued to conduct a civil war within Angola until August 1988, when representatives of interested powers met in Geneva. Although an accord was signed, Savimbi refused to accept that he had agreed to go into exile. Fighting resumed in June 1989 and efforts at peace-making by President Mobutu of Zaïre failed, Savimbi demanding UNITA participation in any future Angola government. In November 1990 he won $60m. aid from the US Congress.

SCAP (Supreme Command of the Allied Powers). General Douglas MacArthur* was appointed Supreme Commander of the Allied Powers on 15 August 1945 and as such signed the Instrument of Surrender on board the battleship *Missouri* on 2 September. A complex command SCAP was established in Tokyo, with numerous administrative sections, for example Operations, Economic and Scientific, Education and Information, and so on. SCAP was in practice a US command, and its officials worked closely with Japanese government officials. Among its tasks were the repatriation of some 3 million troops and another 3 million civilians to Japan; the demobilization of 5½ million troops; the destruction of all military installations; the organization of trials of so-called war criminals. Its greatest achievement was to persuade the Japanese to accept their new Japanese Constitution*, which took effect 3 May 1947. The occupation of Japan lasted until April 1952.

Schacht, Hjalmar (1877–1970), German financier and statesman. Born in Kiel, he became a highly successful banker in Bremen. A member of the German Democratic Party, he became Commissioner of Currency in 1923, and his rigorous monetary policy stabilized the Mark after its collapse that year. From 1923 until 1930 he was president of the Reichsbank. He took part in reparations negotiations, but rejected the Young Plan*. In 1930 he left the Democrats and joined the National Socialists. Under Hitler he became Minister of Economics (1934–9), responsible for Nazi programmes on unemployment and rearmament. In 1939 he urged Hitler to reduce armaments expenditure. This and his rivalry with Goering resulted in his dismissal. In 1944 he was imprisoned in a concentration camp for his alleged part in the July Plot*. At the Nuremberg Trials* he was acquitted, and in 1949 released from internment after appealing against a de-Nazification sentence. He returned to his earlier career in banking.

Scheer, Reinhard (1863–1928), German admiral. Born in Hanover, he enlisted in the German navy and by 1910 was Chief of Staff of the High Seas fleet. After winning fame as a submarine expert, he was appointed Commander of the High Seas fleet in 1916. His brilliant manœuvring at the Battle of Jutland* saved the fleet, in what was a drawn battle. In October 1918 his fleet mutinied at Kiel, refusing to put out to sea.

The mutiny spread rapidly and by November Germany had accepted defeat. He served as chief of the German Admiralty staff until his retirement.

Schirach, Baldur von (1907–74), founder and leader of the Hitler Youth. He became an enthusiastic Nazi while a student and from 1933 onwards led the Hitler Youth*. In 1940 he was appointed Governor of Vienna, where he organized the shipment of Jews to the concentration camps. He was found guilty at the Nuremberg Trials* and sentenced to twenty years' imprisonment.

Schleicher, Kurt von (1882–1934), Chancellor of the Weimar Republic of Germany December 1932–January 1933. Born in Prussia of an aristocratic family, he joined the Imperial Army and was a staff officer throughout World War I, after which he served in the War Ministry. In 1932 he became Minister of Defence, a determined opponent of the paramilitary activities of the National Socialists. In December 1932, following von Papen's* resignation, President Hindenburg* invited him to be Chancellor. Hindenburg failed, however, to back him, and in January 1933 he resigned. He was murdered by the SS* on 30 June 1934.

Schleswig-Holstein, a state (*Land*) of the Federal Republic of Germany. In 1866 Prussia had annexed both duchies. After World War I there were plebiscites and much of north Schleswig passed to Denmark, as the province of South Jutland. Between the wars the presence of a German minority in this province created considerable tension. After World War II over 3 million refugees from the East crowded into Schleswig-Holstein. A referendum voted for incorporation into Germany and it became a West German state.

Schlieffen Plan, plan devised by Field Marshal Alfred von Schlieffen (1833–1913). The latter believed that Germany could fight a war on two fronts by rapidly descending through Belgium and neutralizing France, and then attacking Russia. It failed, due to British and French resistance and to lack of military manœuvrability. It was abandoned in September 1914, when Germany's leaders decided to withdraw forces from the Western Front to stem continued Russian advances into East Prussia. In 1940 Hitler successfully adopted the principles of the plan in his blitzkrieg* in the west. By then columns were motorized rather than moving on foot, as in 1914.

Schmidt, Helmut (1918–), Chancellor of the Federal Republic of Germany 1974–82. Born in Hamburg, son of a schoolteacher, he was still a student when war broke out. He served in an armoured division on the Eastern Front, and after the war completed his economics studies at Hamburg. He joined the Social Democrat Party and worked in city administration 1949–53. He was elected to the Bundestag in 1953 and served as Minister of Defence 1969–72 and Minister of Finance 1972–4. Elected Chancellor in 1974, following the resignation of his friend Willy Brandt*, he sought to continue the policy of *Ostpolitik** with the Soviet Union and the German Democratic Republic, at a time when relationships became soured by the Soviet invasion of Afghanistan. During his second term he lost the support of the left wing of his party and of the Green Party which was emerging.

School systems, for the provision of primary and secondary education. By the late 20th century primary education had become universal throughout the world, while secondary and technical education were everywhere a high priority. France (1791) and Prussia (1807) were the first two countries to establish systems of secular education. The USA quickly followed, while Scotland had for long had a well-established parish system. England and Wales only established a complete system in 1902, while secondary education did not become available for all until 1944. Systems vary in terms of administration, curriculum, and teaching methods, but they broadly divide between the centralized Prussian and French models and the more decentralized US and British, for example: the Prussian system in 19th-century Russia and Turkey; the French in much of Western Europe, Egypt, the Middle East, French-speaking Africa, and much of South-East Asia; the US system in China, Japan, and much of Latin America; the British in India and throughout the Commonwealth. To a greater or lesser extent curricula have been affected by political ideology and/or religious belief, the latter being particularly so in Islamic Iran and Pakistan. In both Britain and the USA private fee-paying schools retained a significant social standing throughout the 20th century, while in France strong, independent religious foundations, both Protestant and Catholic, retained some influence. Secondary education is selective in much of Europe, but comprehensive systems have been adopted in parts of Germany, in Scotland, Wales, and most of England. A feature of teaching method everywhere during the century was for it to become more child-centred, more relaxed, and more concerned with problem-solving than memorization. The introduction of a centralized national curriculum in England and Wales in the late 1980s was partly a political attempt to reverse this trend.

Schumacher, Kurt (1895–1952), German politician. Born in Prussia, he served in World War I and from 1924 until 1931 was a Deputy in the Württemberg Parliament. In 1930 he was elected as a Social Democrat to the Reichstag of the Weimar Republic. Arrested in 1933, he spent eleven years in a concentration camp. On liberation he set about rebuilding the Social Democratic Party (SDP), being leader of the opposition in the Bundestag of the new Federal Republic against the Christian Democrats. He criticized Chancellor Adenauer's* determination to obtain German rearmament, arguing that it would widen the divisions with the Soviet Union and Eastern Europe and prevent reunification.

Schuman, Robert (1886–1963), Prime Minister of France 1947, 1948. Born in Luxembourg of a wealthy Lorraine family, he studied law at Strasbourg and Bonn and practised as a lawyer. After World War I he entered French politics as a Deputy for Metz and was chairman of the parliamentary finance commission for seventeen years. In September 1940 he was arrested and imprisoned by the Gestapo. He escaped in 1942 and joined the French Resistance, helping to found the underground Mouvement Républicain Populaire, which after the war emerged as the MRP. Elected to the Constituent Assembly in 1945, he became Minister of Finance 1946–7 and briefly Prime Minister 1947, when he faced widespread strikes. Prime Minister again in 1948, he was Foreign Minister 1948–52, a strong advocate of Franco-German amity. He sponsored Jean Monnet's* plan for combining coal and steel production which resulted in the establishment of the ECSC*. He helped in the creation of NATO in 1949. He became first president of the European Parliamentary Assembly 1958–60, forebear of the European Parliament*.

Schuschnigg, Kurt von (1897–1977), Chancellor of Austria 1934–8. Born in Riva del Garda, then part of the Austro-Hungarian Empire, he was educated at Innsbruck and became a lawyer. In 1927 he was elected to the Austrian Parliament as a member of the Christian Social Party. He became Minister of Justice 1932 and Minister of Education 1933, in the Dollfuss* government. When he succeeded Dollfuss as Chancellor, he considered his main aim to be the prevention of German absorption of

Austria. In July 1936 he signed an Austro-German agreement which guaranteed Austrian independence. But, determined to prevent annexation, he was in fact negotiating for the restoration of the Habsburgs. Hitler accused him of breaking the agreement, and in February 1938 obliged him to accept Austrian Nazis into his cabinet. His attempt to hold a plebiscite was prevented and he was forced to resign 11 March. Next day German troops invaded without resistance in the Anschluss*. He was imprisoned throughout the war and after liberation in May 1945 became a Professor of Political Science in the USA 1948–67.

Schweitzer, Albert (1875–1965), German missionary and theologian. From 1902 to 1906 he was Professor of Theology at Strasbourg. He then trained as a doctor and in 1913 went as a Lutheran missionary to Lambaréné in Gabon, specializing in the treatment of leprosy and sleeping sickness. He spent much of the rest of his life there, with occasional visits to Europe, where he would give music recitals, being a distinguished organist. His hospital was run on highly paternalistic lines, but he taught the need for service as an act of atonement to the African peoples.

Scopes case (USA). In July 1925 John T. Scopes, a biology teacher at Dayton High School, Tennessee, went on trial charged with violating state law by teaching Darwin's theory of evolution. The state legislature had enacted in March 1925 that it was unlawful to teach any doctrine which denied the literal truth of the account of the creation as presented in the Authorized, King James Version of the Bible. The judge ruled out any discussion of constitutionality and William J. Bryan* led for the prosecution. Since Scopes had clearly taught Darwin's theory he was convicted and fined $100. On appeal to the state Supreme Court the constitutionality of the state's law was upheld, but Scopes was acquitted on the technicality that he had been 'fined excessively'. The law was repealed in 1967.

Scotland. The Act of Union of 1707 had left Scotland with its own legal system, its own Church, and its own schools system. Intense industrialization allied to religious evangelicalism of Clydeside, and elsewhere in the Lowlands, had resulted in rapid growth of radical and later socialist politics. Scotland had received its own Secretary of State in 1885, but demands for political independence, or at least home rule, developed before World War II, with the formation of the Scottish National Party (SNP) in 1934. The SNP won its first seat in Parliament in 1945 and in 1974 gained eleven seats. The Kilbrandon Report had recommended in 1973 the creation of a Scottish Assembly, and a Bill providing for this was passed in Parliament in July 1978. It required however at least 40 per cent of the total electorate to support it. The referendum of March 1979 failed to gain this, although a majority of those who did vote gave their support. SNP support fell, with only two seats won in 1979 and three in 1987 During the 1980s the Labour Party maintained strong support in Scotland, still committed to some form of devolution. During the decade offshore oil brought brief prosperity to Aberdeen, but the decline of shipbuilding and of heavy industry continued. The city of Glasgow was declared the cultural capital of Europe for the year 1990. In that year some form of Scottish separatism seemed closer than ever.

Scottsboro case (USA). In April 1931 nine black youths were accused by two white girls of multiple rape on a train near Scottsboro, Ala. They were found guilty and sentenced to death or long-term imprisonment. The sensational case highlighted race relations in Alabama and across the USA. After two interventions by the US Supreme Court and a series of retrials by the state of Alabama, verdicts of not proven were reluctantly accepted by the state, which paroled all the boys in 1946, except Haywood Patterson, who was to die in prison.

Scullin, James Henry (1876–1953), Prime Minister of Australia 1929–31. A devout Catholic from Victoria, he had been a gold-miner, shopkeeper, and journalist before becoming an official of the Australian Workers' Union. In 1910 he was elected to the federal Parliament as a staunch Labour supporter. He helped to create the Australian Labor Party* in 1917, and was re-elected to Parliament in 1922, becoming leader of his party in 1928 and Prime Minister in 1929. The crisis of the Great Depression* in 1931 split his party and, when J. A. Lyons* joined a group of Nationalists to form the United Australia Party, Scullin was left to lead a Labor rump in opposition until 1935, when he resigned, retiring from Parliament 1949.

SDI *see* STATEGIC DEFENSE INITIATIVE.

SEAC (South East Asia Command). In May 1943 the Allied Command system in the Far East was reorganized. Vice-Admiral Lord Louis Mounbatten* was appointed Supreme Commander of South-East Asia, with General Stilwell* as his deputy; the latter remained directly responsible to the US Joint Chiefs of Staff for his role in China. The idea was almost certainly that of Winston Churchill, who planned an early recovery of Singapore by means of amphibious operations. Its headquarters were first in New Delhi, but Mountbatten later moved them to Kandy (Sri Lanka). There were considerable tensions within the command, not only between Mountbatten and Stilwell, but also as to whether land-based operations in Burma were to precede the recapture of Singapore. Mountbatten would have preferred the latter, but pressure from Jiang Jiehi* (Chiang Kai-shek) resulted in the decision to recapture Burma and reopen the Burma Road. A campaign 1942–3 in the Arakan had failed, and Mountbatten's first campaign there in 1944 came near to disaster; but after a forty-day battle defending Indian Imphal, General Slim's* 14th Army took the offensive, advancing on Mandalay. Meanwhile, along the Arakan coast successful amphibious operations, with air-drop supplies to jungle troops, moved south in parallel, joining the final assault on Rangoon May 1945. SEAC's next role was to have been the recapture of Malaya and Singapore, but for Japan's surrender.

Seaga, Edward (1930–), Prime Minister of Jamaica 1980–9. Born in Boston, Mass., he was educated in Jamaica and at Harvard University, where his postgraduate studies were in child development, for which he did his fieldwork in Jamaica. He joined the Jamaica Labour Party and became its secretary in 1962, when he was first elected to the Jamaica Parliament. He held office as Minister of Development and Social Welfare 1962–7, and was leader of the opposition 1974–80. As Prime Minister he faced increasing economic problems, with stringent IMF* demands for austerity. In foreign affairs he supported US policies, severing relations with Cuba and sending a contingent to Grenada* in 1983.

SEATO *see* SOUTH EAST ASIA TREATY ORGANIZATION.

Second Front, World War II. During 1942 the Soviet Union pressed for an early opening of a second front in west Europe, as a means of relieving heavy German pressure in the east. Churchill insisted that there was insufficient expertise or shipping. The disaster of the Dieppe Raid* in August confirmed this, although the Soviet and US governments continued to criticize British hesitancy through 1943. When the Normandy landings eventually opened a second front in June 1944, it was clear that this immense enterprise could easily have failed if undertaken too hastily.

Secularization, a term used to describe a decline of the prestige and power of religious belief/practice. During the earlier 20th century it appeared to be an unstoppable process, for example in Christian Western society, where churchgoing dramatically dropped, in Islam, where the Republic of Turkey* was to be a secular state, and in Hinduism, where the new state of India was also to be secular. The second half of the century, however, was to see a remarkable recovery, with fundamentalist movements in Christianity, Islam, and Judaism claiming increasing support, with a world-wide appeal of the Catholic papacy under John Paul II*, and with Hinduism continuing to retain the support of millions of devout Indians.

Security Council, UN. Charged with the responsibility of keeping world peace, it was established in 1945 with five permanent members (Britain, the United States, the Soviet Union, China, and France) and ten members elected to two-year terms by the General Assembly. It may investigate any international dispute, and its recommendations, which might involve the imposition of trade sanctions, or a request to members to provide military forces, are to be accepted by all member countries. In order to act the Council requires the votes of nine members, but each of the five permanent members has a right of veto. During the Cold War 1946–89 this veto in many instances prevented UN action. By 1990 the Council's composition, based on the 1945 status quo, was being criticized as no longer reflecting the political and economic realities of the world.

Seeckt, Hans von (1866–1936), German general. Born of a Prussian military family, he was commissioned in 1885. He gained his experience of warfare on the east European front and in the Balkans, becoming Chief of Staff for the Turkish army 1917–18. After the Kapp *putsch** he became Commander-in-Chief of the German army in the Weimar Republic*. Although this was limited by the Treaty of Versailles to 100,000 men, he trained his soldiers as an efficient nucleus for a much larger army. The secret agreement concluded after the Treaty of Rapallo* (1922), permitting German soldiers to train in the Soviet Union, enabled him to circumvent the peace treaty. President Hindenburg dismissed him in 1926, but his work enabled Hitler, whom he much admired, to expand the army rapidly after 1933. In 1934 Seeckt went to China as military adviser to Jiang Jiehi* (Chiang Kai-shek).

Seipel, Ignaz (1876–1932), Chancellor of Austria 1922–4, 1926–9. Born in Pernitz, he was ordained a Roman Catholic priest in 1899 and taught philosophy at Salzburg and then Vienna. After World War I he entered Austrian politics, determined to prevent the Christian Social Party from splitting monarchist/republican. As Chancellor he negotiated a League of Nations loan for reconstruction, but he was wounded in an assassination attempt and resigned in 1924. In his second term he moved to the right, linking with neo-Fascists. In 1927 he deployed a paramilitary Fascist unit, the *Heimwehr*, against his socialist opponents. He became increasingly authoritarian, but was admired by Engelbert Dollfuss*, whose own views moved towards a clerical-Fascist dictatorship.

Senanayake, Don Stephen (1884–1952), Prime Minister of Ceylon (Sri Lanka) 1947–52. A devout Buddhist, he worked on his father's rubber plantation before entering politics. He was elected to the Ceylon Legislative Council in 1922 and became Vice-President of the State Council in 1936. In 1931 he was appointed Minister of Agriculture and Land, an office he held until 1947, during which time he encouraged the development of co-operative farms and of hydroelectric schemes. As leader of the United National Party he negotiated full independence from Britain in 1947, and the new Constitution for Ceylon, of which he became Prime Minister. During his years in office he achieved a high degree of racial harmony between Sinhalese, Tamil, and European. He was killed in a riding accident and succeeded by his son Dudley Shelton, who was Prime Minister 1952–4 and again 1965–70.

Sendero Luminoso ('Shining Path'), a revolutionary guerrilla movement in Peru. It was founded in 1970 by Abimael Guzmán, a long-time Communist and philosophy teacher at Ayachuco University. It is described as 'Maoist' in that it accepts Mao Zedong's* vision of a peasant-based revolution and dislike of bourgeois urban society. It first emerged in the early 1980s, with its main following in the Ayachuco region and among the underprivileged in Lima itself. Rural poverty and depopulation had long been a feature of Peru, whose cities contain many thousands of bitter, unemployed young men who left their Andean villages in search of prosperity. Terrorist bomb attacks and assassinations have been the main weapons. In September President García established a Peace Commission, but it resigned a year later. In 1987 the president of a state-owned food corporation was murdered and in May 1989 a leading member of APRA*. In the same month over seventy peasants were ambushed and killed, allegedly for 'collaboration'. The movement seemed unlikely to decline until the extent of rural poverty was reduced if not eliminated. In May 1990 it was claimed they would abandon violence for the 'ideological battle'.

Senegal, a West African republic. Founded by France in the 17th century, the colony sent its first Deputy to the French Assemblée in 1871 and became part of French West Africa in 1895. During World War II the Governor-General Pierre Boisson remained loyal to the Vichy government and Dakar was bombarded by the British in 1940. Boisson changed allegiance to the Free French* in 1942. In 1958 it was made an autonomous republic within the French Community* and then part of the short-lived Mali Federation (1959–60). Under the leadership of Léopold Sédar Senghor* it became an independent republic in 1960. During the years 1966–74 Senegal was a one-party state, but since then a stable pluralist political system has flourished. In 1982 it federated with the Gambia* as Senegambia. The confederation shared certain joint institutions and the integration of defence and security, but was dissolved in 1989. Ethnic disturbances against Moorish Mauritanians resulted in a virtual border war between the two countries during 1990.

Senghor, Léopold Sédar (1906–), President of Senegal 1960–80 and intellectual. During the 1930s he was active in Paris as a writer and poet. Together with the writers Aimé Césaire and Léon Damas he formulated the concept of *négritude*, which he defined as 'the sum total of cultural values of the Negro-African world'. In 1946 he was elected to the French Assemblée Nationale as a Socialist Deputy. In 1959 he sought unsuccessfully to achieve federation among what had been French West Africa, the short-lived Mali Federation. This failed and in 1960, when Senegal became a separate independent republic, he was elected President. A vigorous spokesman for African democracy, he retained close links with France, while also supporting ECOWAS* and the Organization of African Unity*. He was succeeded on his retirement by President Abdou Diouf. In 1984 he became the first Black member of the French Academy.

Serbia, a constituent republic of Yugoslavia, formed from the former Kingdom of Serbia. In 1903 Peter Karageorgević succeeded to the throne and introduced parliamentary government. Austrian fears of Serbian expansion into neighbouring Bosnia-Hercegovina led it to annex the latter in 1908 and to attempt to control Serbia. These policies led to the assassination of the Archduke at Sarajevo and to World War I. Serbia was invaded by Austrian troops, and by the end of 1915 its army had collapsed and its leaders retired to Corfu. Some Serbian troops later joined the Allied force at Salonika*. By December 1918 Serbia had absorbed Bosnia-Hercegovina and the new Kingdom of Serbs, Croats, and Slovenes was established under King Alexander I*, with Nikola Pašić* as Prime Minister. Croatia had been a reluctant partner; it would have preferred a confederation. During World War II the Axis occupying powers established a separate puppet Croatia* and bitter tension developed between Croats and Serbian partisans. Although the two nations were reunited into the Federal Republic of Yugoslavia by Marshal Tito*, rivalry between them continued, reasserting itself in 1990, when civil war threatened. In December 1990 the hard-line Slobodan Milosevic was re-elected President and the League of Communists of Yugoslavia (LCY) retained control. He strongly supported the Yugoslav army, a majority of whose members were Serbs.

Sèvres, Treaty of (1920), signed between the Allies and the Ottoman Empire. Adrianople (Edirne) and most of the hinterland to Istanbul was to pass to Greece; the Bosporus and the Dardanelles were to be internationalized and demilitarized; a short-lived independent Republic of Armenia was created; Smyrna (İzmir) was to become part of Greece. Syria became a French mandate*, while Britain gained the mandates of Iraq, Palestine, and Transjordan, hoping thereby to link Egypt with India. The treaty was rejected by Mustafa Kemal* and the nationalists, who provoked war with Greece in 1922. A revision was gained at the Treaty of Lausanne* in 1923.

Seychelles, a republic of ninety-two islands in the Indian Ocean. Administered by Britain during the 19th century from Mauritius, they became a separate Crown Colony in 1903. The islands gained universal suffrage in 1970, autonomy in 1975, and became an independent republic in 1976. In 1977 there was a coup, the Prime Minister France-Albert René proclaiming himself President, with the sole legal party being the Seychelles People's Progressive Front.

Seyss-Inquart, Arthur (1892–1946), Austrian Nazi leader. Born in German Sudetenland, he qualified and practised as a lawyer in Vienna. In 1926 he joined the Austrian Nazi Party (National Socialists), and in February 1938 was imposed on Chancellor Schuschnigg* as Minister of the Interior. As such he organized the Austrian Anschluss* of March and was then made Governor of Austria by Hitler. In 1939 he was made Governor-General of Poland and then Nazi Commissioner in The Netherlands. Here he was responsible for thousands of deportations to concentration camps and of summary executions. He was sentenced to death at the Nuremberg Trials*.

Shaba, a province of Zaïre. After World War II the Belgian government granted mineral rights, particularly for copper and uranium, until 1990 to a conglomerate of mining companies, the Union Minière. The province was then called Katanga. The sudden decision in 1960 to grant independence brought vigorous reaction. Union Minière recruited a strong force of White mercenaries to support Moïse Tshombe* in a bid for independence in the Congo Crisis* (1960–3). The province was renamed Shaba in 1971, when its capital Élisabethville became Lubumbashi. Fighting was renewed 1977–8, with a force of mercenaries calling themselves the Congolese National Liberation Front invading from Angola. French and Belgian troops went to the support of the Zaïrian government, Union Minière having been nationalized in 1967.

Shamir, Yitzhak (1915–), Prime Minister of Israel 1983–4, 1986– . Born in Poland, he completed his education at the Hebrew University of Jerusalem, when he joined the Irgun* and was later a founder of the anti-British Stern Gang* 1940–1. Arrested and imprisoned by the British, he was exiled to Eritrea, but twice escaped, being granted asylum in France. He returned to Israel in 1948. For twenty years he was a secret service agent and then a civil servant. He returned to politics in 1970 and was elected to the Knesset in 1973. A supporter of the Likud coalition*, he succeeded Menachem Begin* as its leader. He has held various senior government posts, including that of Prime Minister. He has been a staunch opponent of any compromise with the PLO*.

Shanghai Massacre (1927), a mass atrocity in China. In April 1927 units of the victorious National Revolutionary Army* under Jiang Jiehi (Chiang Kai-shek) arrested and imprisoned many hundreds of Communists and trade union leaders in Shanghai. This resulted in mass demonstrations in Nanjing, Nanchang, and other cities controlled by the Guomindang*. In August Jiang ordered mass arrests, thousands of Communists being tortured and killed. In Nanchang Zhou Enlai* and Zhu De* raised an army of supporters, but they were defeated and fled.

Sharecropping system, a system of farm tenancy. It developed in the Southern states of the USA after the abolition of slavery, involving both Black and White tenant farmers who lacked resources to provide their own equipment or stock. In return for their labour, the farmer and his family were entitled to a half-share of the crop. Generally, however, this was diminished in value by the need to obtain further credit from the landlord for family needs. As late as 1940 there were still some 750,000 sharecroppers in the USA, but after World War II their numbers declined as a result of farm mechanization and a reduction in land devoted to cotton.

Sharpeville Massacre. On 21 March 1960 the police opened fire on a demonstration organized by the Pan-Africanist Congress* in the South African township of Sharpeville. Sixty-seven Africans, many of them children, were killed and 186 wounded. There was widespread international condemnation, and a state of emergency declared in South Africa. One thousand seven hundred persons were detained and the political parties, the Pan-Africanist Congress and the African National Congress*, were banned. As international pressure mounted, South Africa became a republic and withdrew from the Commonwealth (1961).

Shastri, Lal Bahadur (1904–66), Prime Minister of India 1964–6. Son of a minor civil servant, he joined the Indian National Congress* in 1920 as a young follower of Mohandas Gandhi* and was first imprisoned in 1921, after which he attended the Hindu University of Kasi Vidyapitha. After taking part in various civil disobedience campaigns resulting in prison sentences, he served in the Legislative Council of the United Provinces (Uttar Pradesh) 1937–46 and then in the government of the new state 1947–52. He then joined the government of Prime Minister Nehru*, serving as Union Minister of Transport and later as Minister of Home Affairs. On Nehru's death he succeeded as Prime Minister, facing the renewed problem of Kashmir* in the second Indo-Pakistan War*. In January 1966 he travelled to Tashkent, where the Soviet Prime Minister Kosygin negotiated a cease-fire. He died there suddenly of a heart attack.

Shaw, George Bernard (1856–1950), Irish playwright and critic. Born and educated in Dublin, he went to London in 1876, working as an art and music critic. He became a socialist and helped to form the Fabian Society* in 1884. From 1904 until he died he successfully staged a series of plays in London. These were both didactic and intellectual, but also for the most part highly acclaimed, for example *Major Barbara* (1905), *The Doctor's Dilemma* (1906), *Pygmalion* (1913), *Heartbreak House* (1919), *St Joan* (1924). He consistently advocated basic social justice against progressive materialism, always asserting the case for the freedom of the human spirit against political, religious, or conformist orthodoxy. In his later plays, for example *The Simpleton of the Unexpected Isles* (1935), he was moving towards symbolic farce and theatre of the absurd, which was to develop after his death. Always the radical, as an essayist and lecturer, as well as playwright, his didacticism influenced generations of students and theatregoers for over half a century.

Shinwell, Emanuel, Baron (1884–1986), British politician. Born in London's East End, of Polish immigrants, he moved to Glasgow when 11, where he was apprenticed to a tailor. He joined a trade union and by World War I was working on Clydeside, where he was imprisoned in 1919, for allegedly inciting seamen to riot. He sat in Parliament 1922–4 and 1928–31 as a Labour MP and held minor office. Re-elected in 1935, he continued in the House of Commons until 1970, when he became a life-peer. He supported the coalition government of 1940–5 but did not join it; he was Minister for Fuel and Power 1945–7 and Minister of Defence 1950–1. 'Manny' was a much-loved figure in Parliament and a man of strong principles. He opposed Britain's entry into the European Community and the drift of the Labour Party to the left in the early 1980s.

Shona. They are a people found in both Mozambique and Zimbabwe, where they form 70 per cent of the population and where they have for long been settled as agriculturalists, cultivating millet, beans, and in more recent centuries maize. Their ancestors achieved high levels of urban civilization, as revealed in the vast granite ruins of Zimbabwe dating back to the 8th century. The invasion of the Ngoni Ndebele* people from Zululand following the *Mfecane* in the early 19th century, and their establishment of Matabeleland, led to tensions which continued to operate into the 1980s.

Showa period. The term Showa, meaning 'bright peace', was given to the period in Japan 1926–45. Artistically it refers to the transition in Japanese art from traditional to Western-influenced styles. Politically the term was used by military extremists in the 1930s seeking a militarist dictatorship supporting a divine emperor.

Sierra Leone. In 1772 Britain had declared that any escaped slave who came to Britain would automatically become free. British philanthropists organized their transport to Cape Sierra Leone, where in 1788 Freetown was established, becoming the first British Crown Colony in Africa in 1806. The hinterland was gradually explored and in 1896 it became a British protectorate, remaining separate from the colony until 1951. During the 1930s an anti-colonial movement, led by Wallace-Johnson*, gained much popular support, and after the war there were moves towards home rule. Sierra Leone gained its independence in 1961 under Prime Minister Sir Milton Margai (1895–1964), but after his death electoral difficulties produced two military coups before some stability was restored by the establishment of a one-party state under Dr Siaka Stevens. Food shortages, corruption, and tribal tensions produced serious violence in the early 1980s. In 1985 Stevens retired in favour of Major-General Dr Joseph Momoh, who retained a civilian cabinet with the All People's Congress (APC) the sole legal party.

Sihanouk, Norodom (1922–), King of Cambodia 1941–55. Placed on the throne in 1941 by the French Governor-General on the death of his grandfather, he supported the Japanese in the hope of winning independence from the French. After the war he successfully negotiated full independence in 1953 and sent representatives to the Geneva Conference in 1954. He abdicated in favour of his father in 1955, becoming Prime Minister. On the death of his father in 1960 he was proclaimed head of state as Prince Sihanouk and leader of the Popular Socialist Community. In 1963 he renounced US aid, trying to be neutral in the growing Vietnam struggle, while at the same time (1967) suppressing the Khmer Rouge*. On 18 March 1969 the first US bombing (Operation Breakfast) of Khmer Rouge positions began and the USA restored relations with Sihanouk; but next year, while he was overseas, he was deposed by a military coup led by Lon Nol*, with CIA* support. It was alleged that Sihanouk had been spending too much time on his interests as a film-maker. He was offered asylum by Zhou Enlai (Chou En-lai) in China, where he founded the National United Front of Cambodia. In 1975 US forces withdrew from Cambodia and the Khmer Rouge took over. He was invited to return as head of state, but within a short while he was under house arrest, which continued for two years. He was released in 1978, when he spoke out in favour of the Pol Pot regime. When, however, this was overthrown by Vietnam in 1979, he once again went into exile, in China and North Korea, from where he sought to overthrow the Vietnam-backed regime of Heng Samrin, in collaboration with nationalist forces led by Son Sann. In 1982 he became President of an exiled coalition government, in alliance with the Khmer Rouge. This was recognized by the UN, since when he has begun negotiations with the Heng Samrin regime.

Sikhism, an eastern religion. It emerged from a community gathered around the guru Nanak in the 15th century. Nanak accepted the mission of the prophet Muhammad, as an agent of the Hindu supreme being Brahman, the path to whom lies through many incarnations. Nanak's successor guru Arjun Mal built the holy city of Amritsar. A later guru Govind repudiated the Hindu caste system, but also fiercely opposed Muslim domination, teaching Sikhs to 'live by the sword' in 'humility and sincerity'. He created an independent Sikh state, which was conquered by the British in 1849, but in 1947, when British India was partitioned, the frontier ran through Sikh territory. Millions moved east into East Punjab, and the Sikh East Punjab States Union was created. Agitation for independence developed in the 1970s. Indira Gandhi* ordered the occupation of the Golden Temple in Amritsar in 1984 and she was assassinated in October. Sikh worship takes place in the *gurdwara* and it has no priests, the *gurdwara* being a room kept for worship and containing a copy of the sacred scriptures the *Adi Granth*. Sikhs keep Hindu festivals, but also celebrate the birthdays of the nine gurus.

Sikkim, a small state in the Republic of India in the eastern Himalayas. Until 1975 it was ruled feudally by *chogyals* (kings) of the Namgyal dynasty. In the past Sikkim suffered continual invasions from its Himalayan neighbours, especially Bhutan* and Nepal*. In 1861 it had become a protectorate of British India. In spite of criticism of his feudal rule, the *Chogyal* hoped to retain internal autonomy when Britain left India, but a referendum of 1975 demanded transfer to the Indian Union, which then followed.

Sikorski, Wladislaw (1881–1943), Polish general and statesman. Born in Galicia, he studied engineering at the universities of Lvov and Cracow before World War I. In 1914 he joined Pilsudski*, who allied himself with Austria-Hungary against the Russians. In July 1917 he and Pilsudski were both interned by Germany. In the Russo-Polish War* 1919–20 he commanded troops against the Bolsheviks at Vilna and then Warsaw. He became Chief of the Polish General Staff and then Prime Minister in Poland, but Pilsudski distrusted him as too liberal and forced him to retire in 1926. In 1939 he fled to France and organized a Polish army in exile, which was to fight throughout the war. He himself headed a government-in-exile in London which succeeded in maintaining tolerable relations with Moscow until news of the Katyn Massacre*. During his ascendancy Polish prisoners of war in the Soviet Union were recruited to form the Polish army in Russia under General Wladyslaw Anders. Sikorski was killed in an air crash.

Silesia, a region of Eastern Europe. In 1815 Prussia was confirmed as having control of Upper and Lower Silesia, while Austria retained a province of Austrian Silesia. In 1918, after a series of plebiscites, Upper Silesia (the coal and steel-producing area) went to Poland and most of Austrian Silesia to Czechoslovakia. Germany retained Lower Silesia. During the inter-war years there was heavy investment by France in Polish Silesia, but also civil unrest as the interests of Germany, Poland, and Czechoslovakia were disputed. In September 1939 German troops occupied Polish Silesia, but with the defeat of the Third Reich the Potsdam Conference* of 1945 decreed that the whole area of Silesia should pass to Poland. German nationals became refugees in the Federal Republic of Germany.

Simon, John Allsebrook, 1st Viscount Simon of Stackpole Elidor (1873–1951), British statesman and Lord Chancellor 1940–5. Born in Bath and educated in Scotland and at Oxford, he became a Fellow of All Souls College and a distinguished lawyer. He entered the House of Commons in 1906 as a Liberal, and in 1910 was appointed Solicitor-General and then Attorney-General. Doubtful about Britain's entry into World War I, he nevertheless served as Home Secretary 1915–16 before resigning over the issue of conscription. He later served in the Royal Flying Corps. He supported Asquith against Lloyd George in the Coupon Election* and lost his seat, but returned to Parliament in 1922. As chairman of a Commission on India 1927–30 he issued the **Simon Report**, which advocated greater Indian participation in government. He took office in the National government under MacDonald* in 1931. As Foreign Secretary 1931–5 he favoured disarmament and was a notable 'appeaser'. The Liberal Party had split in 1932 over the Ottawa Agreements*. Simon led the National Liberals, while Sir Herbert Samuel led the Liberal Nationals, who left the government. Simon was Home Secretary under Baldwin 1935–7, Chancellor of the Exchequer 1937–40, and Lord Chancellor through the war period 1940–5.

Simon Commission, a commission of seven established in 1927 by the government of Stanley Baldwin* to report on the political condition of British India. It was under the joint chairmanship of the Liberal Sir John Simon* and the Labour Party's Clement Attlee*. It was boycotted in India because it had no Indian members. Its report (1930) proposed autonomy for the Indian provinces but rejected parliamentary government for India as a whole. It influenced the drafting of the Government of India Act* (1935), although this was more influenced by the Round Table Conferences*.

Sinai campaign (October–November 1956). During 1955 terrorist (fedayeen) raids were launched into Israel from the Egyptian-held Gaza Strip*, where the ex-Mufti of Jerusalem Haj Amin al-Husseini* had settled. Israel responded by invading and killing thirty-six Egyptians, whereupon Egypt closed the gulf of Aqabah and set up a combined military command with Jordan and Syria. Meanwhile its nationalization of the Suez Canal* had provoked strong reaction from Britain and France. After secret talks with both the latter, Israel attacked on 29 October 1956. Brilliant use of tanks and aircraft by General Dayan* resulted in the whole Sinai peninsula being occupied before 5 November, when the United Nations imposed a cease-fire, ending the Suez Crisis*.

Singapore, an island republic of the Malay peninsula. By 1914 it was a colony within the Straits Settlements of Malaya*, with a large Chinese population, and was the largest British naval base east of Suez. Its capture in February 1942 was seen as the greatest disaster of the British Empire. In 1946, after liberation, the island became a separate colony and in 1959 achieved internal self-government under Lee Kuan Yew*. It was expelled from the Federation of Malaysia* in 1965 for fear that its predominantly Chinese population would gain too much influence; but it has maintained strong commercial links with the Federation as well as being a supporter of the Commonwealth of Nations and a member of ASEAN*.

Singapore, fall of (8–15 February 1942). Although Singapore had strong coastal defences, no fortifications had been built against land attack, apart from provision for the mile-long causeway across the Strait of Johore to be blown up. After swiftly overrunning Malaya, Japanese forces under General Yamashita* massed opposite the island of Singapore at the beginning of February 1942. During the night of 7/8 February armoured landing craft crossed the Strait of Johore, surprising the garrison of Australian troops opposite. Many Japanese troops followed by swimming across the water. The defenders blew up the causeway and retreated, the island's garrison of British, Australian, and Indian troops under General A. E. Percival numbering some 80,000. Incessant air attack destroyed oil-tanks and supplies, and weakened morale. Having repaired the causeway, more Japanese moved on to the island and Percival continued to retreat south-east towards the residential area. On 15 February attempts were made to evacuate key personnel by boat, but few survived and Percival surrendered. The defeat was a significant milestone in the ending of British imperial interests in South-East Asia.

Sinn Fein (Gaelic; 'we ourselves'). Originally founded in Dublin in 1902 by Arthur Griffith* as a cultural revival movement, it became politically active in 1912 and supported the Easter Rising* of 1916. Having won the overwhelming majority of Irish seats in the 1918 general election, Sinn Fein MPs, instead of going to Westminster, met in Dublin and proclaimed Irish independence in 1919. An independent Parliament (Dáil Éireann) was set up, even though many of its members were in prison or on the run. Guerrilla warfare against British troops and police followed. The establishment of the Irish Free State and the partition of Ireland in December 1921 were bitterly resented by Sinn Fein, and the party abstained from the Dáil, supporting civil war 1921–4. When de Valera* formed the Fíanna Fáil* Party in 1926 Sinn Fein became an outlawed organization for many years. Today it openly exists in the Irish Republic but does not contest elections. At the same time Sinn Fein has been the political wing of the Provisional IRA; its leader Gerry Adams supports the acts of violence in Northern Ireland, which he regards as acts of war.

Sino-Japanese War (1931–45). China had been a target for Japanese expansionism since the late 19th century, and after the Mukden Incident (see MANCHUKUO) of 1931 the huge province of Manchuria was occupied. Marauding Japanese troops clashed with nationalist forces in the Marco Polo Bridge Incident* in 1937, following the Xi'an* agreement of Nationalists and Communists to restore the

United Front*. Full hostilities rapidly followed. Shanghai was captured in the autumn of 1937, when Japanese forces penetrated inland along railways and up the Chang Jiang (Yangtze) to take Nanjing in December. Hankou followed, and a rapid coastal advance gained Guangzhou (Canton) in October 1938. In the Nanjing Massacre* perhaps as many as 200,000 civilians were killed. The invaders were opposed by both the National Revolutionary Army* under Generalissimo Jiang Jiehi* (Chiang Kai-shek), and Communist armies under Zhu De*. From 1940 Nationalist troops were being supplied by Britain and the USA along the Burma Road, until this was cut in 1942. By then the conflict had been absorbed into World War II, but war had reached a state of near stalemate; Japanese military and aerial superiority was insufficient to overcome tenacious Chinese resistance and the problems posed by massive distances and poor communications which were constantly the target of guerrillas. The Chinese kept over 2 million Japanese troops tied down for the entire war, inflicting a heavy defeat upon them at Jiangxi in 1942, and successfully survived a final series of offensives in 1944 and 1945. In August 1945 the Soviet Union declared war against Japan and the Red Army invaded Manchuria. The Japanese formally surrendered to Jiang Jiehi on 9 September 1945, leaving him to contest the control of China with Mao Zedong* and the Communist forces.

Sino-Soviet frontier dispute (1969). The exact position of the border between north-east China and the Soviet Union had long been a matter of dispute. The disagreement turned into a military confrontation because of worsening ideological disputes between the two countries after 1960, and the militant nationalism which was part of the Cultural Revolution*. In March 1969 two battles were fought for possession of the small island of Zhen Bao (Damansky) in the Ussuri River. The Chinese ultimately retained control, and talks in September 1969 brought the crisis to an end. In 1977 there was a limited agreement on rules of navigation on the river, but tension continued along the frontier until 1989.

Sino-Vietnam border war. Following a deterioration in relations between China and Vietnam in 1978, a border dispute at You yi guan (Friendship Pass) took place on 25 August 1978, when the Chinese claimed that 200 troops of the Vietnamese People's Army had crossed the border. In February 1979 there was a brief if massive Chinese invasion into Vietnam, followed by inconclusive negotiations. China had by now offered asylum to the Khmer Rouge* from Cambodia (Kampuchea), and was opposed to the Vietnamese intervention. Sporadic fighting took place along the border through early 1980, and relations between the two countries remained strained.

Six-Day War (5–10 June 1967), also known as the June War. The immediate causes of the war were Egyptian pressure on the UN Emergency Force in Sinai to withdraw from the Israeli frontier, and a build-up of Egyptian forces in Sinai during May. At the same time a new naval blockade in the Gulf of Aqabah was blockading the Israeli port of Elat. Brilliant use of armour and paratroops brought Israeli troops to the Suez Canal within two days. They crossed the Canal and advanced south into Egypt as far as Luxor and Ras Banas on the Red Sea. At the same time the Egyptian Sinai army was encircled. Old Jerusalem was seized and the whole of the West Bank was cleared of Jordanians. In the north (9–10 June) Israeli tanks occupied the Golan Heights, captured Kuneitra, and advanced 30 miles into Syria. In pre-emptive air strikes the Egyptian air force was destroyed on the ground, while other airfields in Syria, Iraq, and Jordan were also bombed. Jordan accepted a UN cease-fire on 7 June, Egypt on 8 June, Syria on 9 June, and Israel on 10 June. The war was over. Israel now occupied the whole of Sinai, including the Gaza Strip, and Israeli troops remained established on the west bank of the Suez Canal, which was to be closed until 1975. Israel would now have to administer over half a million Arabs in occupied areas, although many fled as refugees to Jordan, Lebanon, and Syria.

Slansky Trial (November 1952). In November 1951 Rudolf Slansky, Vice-Premier of Czechoslovakia and a Jew, was arrested by security forces in Prague. Thirteen other Communist officers were also arrested, all Jews, including the Foreign Minister Vladimir Clementis. They were all charged with Titoism as 'Trotskyist bourgeois traitors'. Evidence was clearly fabricated and all but three were hanged on 2 December. It is thought that Stalin was directly responsible; his anti-Semitism was becoming obsessive just before he died.

Slim, William Joseph, 1st Viscount (1891–1970), British field marshal. Born in Bristol, he became a teacher there until World War I, when he enlisted. He served at Gallipoli and was commissioned. He then joined the Indian army, where he served until World War II. He commanded an Indian division in the conquest of the Vichy French territory of Syria* in 1941. In early 1942 he was sent to Burma, where Japanese forces were approaching the Indian frontier. In 1943 he took command of the 14th Army. After victory at Kohima he pushed down the Irrawaddy River to recapture Rangoon and most of Burma during 1944–5. After the war he became Chief of the Imperial General Staff (1948–52) and Governor-General of Australia (1953–60).

Slovenia. The Slovenes are a Slavonic people living in the upper valley of the River Sava and its surrounding mountains, their lands centred on Ljubljana. They were ruled by the Habsburgs from Vienna from the 14th century until 1918. After World War I the majority of the Slovene people were incorporated into the new Kingdom of Serbs, Croats, and Slovenes, later Yugoslavia. In 1941 their lands were divided between Italy, Hungary, and the Third Reich. In 1945 Slovenes from the Istria peninsula were incorporated from Italy into the Republic of Slovenia, within the Federal Republic of Yugoslavia. Slovenes are Catholic in religion and remain suspicious of the more assertive Serbs, with their differing traditions and Orthodox religion. Following elections, a non-Communist government was formed in May 1990, and in December a referendum declared for secession from the Federal Republic. By now anti-Serbian partisans were reported to be increasingly active.

Smigly-Ridz, Edward (1886–1939), Polish statesman. Born in Galicia, he fought with Pilsudski* in World War I on the Austrian side against the Russians. He remained a close aide of Pilsudski in the Russo-Polish War that followed and in the coup of 1926. In 1935 as marshal and Inspector-General of Forces he succeeded Pilsudski as dictator of Poland. He disappeared in 1939; it is thought he escaped to Romania.

Smith, Alfred Emanuel (1873–1944), Governor of New York 1918–20, 1924–8, always known as 'Al'. An Irish-American Roman Catholic, as a boy he worked in the New York fish-market and then in the headquarters of the New York City Democratic Party in Tammany Hall. Here, when he was only 22, he was given a job as an investigator, the first step towards a political career. He served in the New York State Assembly 1903–15, where he gained a reputation for determination and integrity. In 1918 he was elected state Governor, backed by Roman Catholic, new immigrant, and Irish-American votes. He fought hard for improved housing, child care, factory legislation, and leisure facilities. He failed to win presidential nomination in 1924, but was Democratic candidate in 1928 as champion of 'urban America'. He was defeated by the conservative Republican Herbert Hoover. Smith resented the successful rise of his gubernatorial successor Franklin D. Roosevelt* and in 1936 and 1940 gave his support to the Republican candidates who opposed Roosevelt.

Smith, Ian Douglas (1919–), Prime Minister of Rhodesia 1964–80. Born in Rhodesia, he served in the RAF 1941–6 and then entered politics, being a member of the Southern Rhodesia Legislature 1948–53 and the federal Parliament for Southern Rhodesia 1953–62. In 1962 he was a founder member of the Rhodesia Front Party and in 1964 became Prime Minister, following Northern Rhodesia's independence as Zambia. In 1965 he announced a Unilateral Declaration of Independence (UDI) from Britain, and after unsuccessful talks with the British Prime Minister Harold Wilson in HMS *Tiger* in 1966 and HMS *Fearless* in 1968 he introduced apartheid* legislation into Rhodesia, which was declared a republic in 1970. Following abortive discussions in Geneva in 1976, he agreed to joint Black/White rule under Bishop Muzorewa* in 1978. When this failed he attended the Lancaster House Conference of 1980, which led to independent Zimbabwe. He has served in the Zimbabwe Parliament since 1980, but was suspended 1987–8.

Smuts, Jan Christian (1870–1950), Prime Minister of the Union of South Africa 1919–24, 1939–48. In 1899, as a young Johannesburg lawyer, he contributed to a propaganda pamphlet *A Century of Wrong*, explaining the Boer case against Britain. During the Boer War which followed he was a guerrilla leader of exceptional talent. He was a leading negotiator at the Treaty of Vereeniging, believing that the future lay in co-operation with Britain. He held a succession of cabinet posts under President Botha*, but in 1914 rejoined the army and served in South Africa's campaign against German East Africa. In 1917 he joined the imperial war cabinet in London, and helped to establish the RAF. He was an advocate of the League of Nations* at the Paris Peace Conference, returning to South Africa in 1919 to become Prime Minister. He led the opposition 1924–33 and was deputy Prime Minister to Hertzog* 1933–9. As Prime Minister 1939–48 he was a close friend and adviser to Winston Churchill, who created him field marshal. Among his many achievements in these years was the drafting of the UN Covenant (1945). Defeated in 1948 by Dr Malan* and the National Party*, he became Chancellor of Cambridge University. His struggle against extreme nationalism found expression in his philosophic study *Holism and Evolution* (1926).

Snowden, Philip, 1st Viscount (1864–1937), British politician. Educated at a York elementary school, he became a civil servant in 1886. Permanently crippled in a bicycle accident, he joined the Independent Labour Party*, for whom he next worked as a journalist. He was elected to Parliament for Blackburn in 1906, opposing British entry into World War I and supporting Indian demands for self-government. He became Chancellor of the Exchequer in 1924 and again 1929–31, remaining so for a short while in the 1931 National government*. He did not support the General Strike* of 1926, and his cautious approach to welfare spending alienated many Labour supporters. His budget in 1931 reducing unemployment benefits because of the alarming international financial crisis further antagonized them. Lord Privy Seal in 1932, he resigned from the government on the abandonment of free trade at Ottawa*.

Social Credit parties. A theory of Social Credit was advanced by the economist Clifford Douglas during and after World War I, to eliminate the concentration of economic power. He argued that since in every productive establishment the total cash issued in wages, salaries, and dividends is less than the market price of the product, deficiencies of purchasing power should be remedied by the payment of subsidies. His ideas became popular in Canada and New Zealand, particularly among hard-pressed farmers and small businessmen at the time of the Great Depression*. In Canada a Social Credit Party, led by William Aberhart, won an overwhelming victory in Alberta in 1935 and remained in power until 1971, without however implementing many of Douglas's ideas. In 1952 it won the election in British Columbia, where it has remained in office, apart from 1972–5. The party won thirty federal seats in Ottawa in 1962 but largely disappeared after 1980. A New Zealand Social Credit Party was formed in 1953 and has held one to three seats since in the New Zealand Parliament. Social Credit parties have been more a gesture of rural protest than of radical capitalism.

Social Darwinism, the application of Darwin's evolutionary theories to human societies. As with species these are seen as subject to the laws of natural selection, whereby continual progress is achieved even at some human cost. Herbert Spencer and other late 19th-century thinkers were attracted to such an ideology. In the 20th century it has been used to justify racist ideologies and power politics, as well as to explain the operation of 'the free market economy'.

Social Democratic parties. The term Social Democrat was first adopted by Wilhelm Liebknecht and August Bebel in Germany, when they founded the German Social Democratic Labour Party in 1869, based on the tenets of Karl Marx, but advocating reform by democratic and constitutional means. In 1875 this became the Social Democratic Party of Germany. Others followed, for example Denmark (1878); Britain (1883)—a short-lived Social Democratic Federation; Norway (1887); Austria (1889); the USA (1897)—later under Eugene Debs* becoming a Socialist Party; and Russia (1898), where a split followed in 1903 between Bolsheviks and Mensheviks. In other countries, the term Socialist Party was more often adopted by the left wing, for example in France, Italy, and Spain. The German SDP was the largest party in the Weimar Republic*, governing the country until 1933, when it was banned. It was re-formed (SDP) in West Germany after World War II, with a new constitution in 1959 which ended all Marxist connections, its chairman being Willy Brandt* 1964–87. In East Germany a revived SDP returned to campaigning for office after the collapse of the Communist regime in 1989. In Sweden the SDP has been the dominant party since the 1930s, although out of office 1976–82. In Britain a short-lived Social Democratic Party (UK)* was formed in 1981 by Roy Jenkins* and three disillusioned Labour Party colleagues.

Social Democratic Party (UK). Founded in London in 1981 as a break-away party from Labour, by four leading Labour politicians, Roy Jenkins*, David Owen, Bill Rodgers, and Shirley Williams, it formed an Alliance with the Liberal Party led by David Steel. As such it made some political headway during the early 1980s, especially in local politics. After the 1987 general election, however, the Alliance split. David Steel retired and David Owen rejected the idea of a merger with the Liberals. He remained leader of a tiny rump of three SDP MP's while the remainder, the official SDP, merged in March 1988 to form the Social and Liberal Democrats.

Socialism. By the late 19th century an ideology of socialism had emerged, and was being adopted by Social Democratic, Labour, and Socialist parties throughout Europe and beyond. Its tenets were that the powers of government, both local and central, must be strengthened in order to eliminate poverty and to provide health care and social amenities, including an education service. At the same time, in order to achieve a more just society, there must be a redistribution of wealth through taxation, an end to class privilege and the hereditary principle, and a removal from private capitalists of basic industries, including agriculture, which should be handed to 'the people of the nation' by a process of nationalization. 20th-century Communist regimes all to a greater or lesser extent adopted socialist principles, while in Western societies many aspects of socialism, including wide welfare provision, were implemented. By the end of the century, however, socialism had ceased to be a coherent ideology and had become little more than a rhetorical term of decreasing attraction, especially in what had been the Communist regimes of Eastern Europe.

Socialist realism, an aesthetic ideology first proclaimed by Maxim Gorky in the 19th century. It was adopted by socialist and Communist activists early in the 20th century. They demanded that artists should not just 'reflect reality in its revolutionary development', but seek to achieve social change through their art in a both 'optimistic and intelligible manner' and in line with party doctrine. Although a few artists, for example Brecht*, were able to work successfully within the confines of the ideology, for a majority it was anathema. In the Soviet Union it stunted and distorted many of the arts (with perhaps the exception of the ballet) throughout the Stalinist era, producing no more than sentimental cliché.

Social realism. In general this term can be applied to any work of art which seeks to make a social statement in realistic or naturalist terms, for example a Dickens novel. In a more specific sense it was used by American artists of the 'Ashcan School' in the early 20th century, who sought to depict the commonplace and unglamorous realities of urban life, and to make social statements on poverty, unemployment, and injustice. It differed from socialist realism in not being required by party ideology to be optimistic. In fact it was usually the reverse.

Solh, Takieddine al- (1909–88), Lebanese Sunni politician. The Solh family was for long active in the struggle of the Arab Peoples, first against the Ottoman Turks and then against the French and British, who gained the mandates in 1920. As a *lycée* professor 1935–43, Takieddine agitated against French control, but also joined the Anti-Fascist League in Beirut, always inspired by the democratic ideals of the French Revolution. A moderate, who sought conciliation between Christian and Muslim communities in Lebanon, he was a Lebanese representative to the Arab League and was first elected to the Lebanon National Assembly in 1957. Between 1964 and 1974 he held various government offices, including Prime Minister 1973–4, but was unable to form a government in 1980.

Solidarity (Polish; 'Solidarnosc'), a trade union movement in Poland. It emerged out of a wave of strikes at Gdansk* in 1980, when demands included the right for trade unions to be independent of the Communist Party. Its leader was Lech Walesa*. Membership rose rapidly, as Poles began to demand political as well as economic concessions. In 1981, following further unrest aggravated by bad harvests and poor distribution, General Jaruzelski was appointed Prime Minister. He proclaimed martial law, and arrested and imprisoned Solidarity leaders including Walesa. Solidarity was outlawed in 1982, but continued as an underground movement. It was restored to legality in January 1989 and invited to take part in round-table discussions in February, when Poland was in deep economic crisis. In August it joined the moderate coalition government of Tadeusz Mazowiecki. Itself a coalition of anti-Communist views, from strong Catholic on the right to secular radicalism on the left, it began to suffer internal tensions, as the tough policies of the new government began to lower an already low standard of living within Poland.

Solomon Islands, an archipelago of islands in the Pacific, most of whose people are Melanesian. In 1914 the German New Guinea Company had control of the islands at the north-west end, including Bougainville and Buka, adjoining the Bismarck Archipelago. These passed to the mandate of Australia in 1920 and now form part of Papua New Guinea*. The remaining islands on the south-west of the archipelago had been made a British protectorate in 1893. Occupied by Japan in 1942, they were the scene of very fierce fighting in World War II, particularly on Guadalcanal. Britain steadily introduced a parliamentary system of government from 1960 onwards, establishing a unicameral legislature under a Governor-General. The islands became independent as The Solomon Islands within the Commonwealth of Nations in July 1978, joining the South Pacific Forum*.

Solzhenitsyn, Alexander (1919–), Russian writer. Born in the Caucasus, he was educated at Rostov and Moscow before World War II, in which he served in the Red Army as an artillery officer. He began work as a writer and journalist and was imprisoned 1945–53 and then exiled to Siberia 1953–6. His novel *A Day in the Life of Ivan Denisovich* was acceptable to Khrushchev* on publication in 1962, as exposing the horrors of prison-camp life which he himself had been criticizing. In 1971 he published *August 1914* and *The Gulag Archipelago*. By now Khrushchev had been replaced by Brezhnev*, and his exposure of the corruption of the Soviet system led to his expulsion in 1974. He settled in the USA, where he became an ever-more embittered critic of the Soviet Union. In 1990 he was restored to favour and invited to return home, an invitation which he rejected.

Somali Democratic Republic. The area of the Horn of Africa was divided between British, Italian, and French spheres in the later 19th century. French Somalia became the Republic of Jibouti* in 1977. Somalia Italia was occupied by British troops 1941–9 and restored to Italian responsibility in 1950 as a UN Trust Territory. In 1960 the Somali Republic was created by the unification of the British Somali protectorate and the Italian Trusteeship Territory. Since independence there have been border disputes with Kenya and Ethiopia, intermittent war with the latter lasting over the years 1977–88, concerning grazing rights in the Ogaden desert. In 1969 President Shermarke was assassinated in a left-wing coup, the Marxist Somali Revolutionary Socialist Party taking power, led by General Mohammed Siyad Barre, who renamed the country the Somali Democratic Republic. Constitutional government was restored in 1979, President Barre being elected in 1980 and again in 1988. By November 1990 the country was in a state of civil war between Siyad Barre's supporters and a rebel Somali National Movement.

Somali-Ethiopian War (1977–8). In 1977 Somali troops invaded the Ogaden desert over the grazing rights of its Somali inhabitants. Fierce fighting lasted through the year, in an unsuccessful attempt to annex the Ogaden from Ethiopia, which was supported by the Soviet Union. Somalia had also had Soviet backing since 1969, but now expelled 3,000 Soviet personnel. Faced by famine and refugee problems, it now turned to the USA for backing and aid. Sporadic fighting continued through the 1980s, especially in 1987. In 1988, however, Presidents Barre and Mengistu signed a peace treaty, re-established diplomatic relations, and began repatriation of prisoners of war from the 1977–8 war.

Somme, Battle of the, World War I (1 July–18 November 1916). This bloody battle was planned by Joffre* and Haig* to relieve pressure on Verdun*, whose defence had nearly destroyed the French army. Fighting extended over a 20-mile front, with the French 6th Army on the right, but the brunt of the offensive fell on the British. A preliminary eight days' bombardment poured 52,000 tonnes of ammunition on the German positions. On 1 July the British advanced from their trenches, almost shoulder to shoulder, a perfect target for German machine-gunners, 20,000 being killed on that day. A series of offensives over 4½ months forced the Germans to retreat a few kilometres. They fell back on the Hindenburg Line defences (a barrier of concrete pillboxes armed with machine guns), while for the loss of some 600,000 men the Allies had gained a sea of mud. Tanks were used for the first time in this battle on 15 September, but they became bogged down in the mud.

Somoza family, a Nicaraguan family which held power 1933–79. Anastasio Somoza (1896–1956) was the son of a wealthy coffee planter. He was educated in the USA, where he married and always had support. He returned to Nicaragua in 1926, when he joined the Liberal revolt. As Minister of War and Foreign Relations, he established good relations with US occupiers, who were engaged in creating a Nicaraguan National Guard (Guardia Nacional). In 1933 he became head of the guard and began to eliminate his rivals as well as having Augusto Sandino* murdered. By 1936 he was able to overthrow the elected President Juan Sacasa and have himself proclaimed President. Although his reforms to diversify Nicaragua's economy were sound, he set about amassing a huge personal fortune, using corruption and violence when necessary and turning the National Guard into a personal bodyguard of his family. His regime enthusiastically backed US policies and welcomed US bases. He remained virtual dictator until September 1956, when he was shot by a young poet. His older son Luis, educated at US universities, and president of the Nicaraguan Chamber of Deputies at the time, immediately succeeded him. His brother Anastasio Debayle Somoza, a graduate of West Point and head of the National Guard, imprisoned all likely opponents. Many of the reforms of Luis—in housing, education, social security—were welcomed, and he became an enthusiastic supporter of the Alliance for Progress*. But the regime was still corrupt and all elections rigged. From 1962 a guerrilla group, the FSLN (see NICARAGUA) began to operate. Luis died in 1967, when Anastasio Debayle won a rigged election as President, backed by his National Guard. His rule was a disaster, concerned only with expanding the family fortune, even plundering an international earthquake relief fund in 1972. From now on opposition groups gathered momentum, apart from the open guerrilla opposition of the FSLN, with the Catholic Church denouncing the regime for flagrant disregard of human rights. In 1978 a Broad Opposition Group was formed, and only US support kept Somoza in power. In July 1979 FSLN troops occupied Managua and Somoza fled into exile, where he was assassinated.

Song Qingling (Soong Ch'ing-ling) (1892–1981), Chinese leader. Sister of T. V. Soong* and educated in Connecticut, USA, she returned to China to become secretary to Sun Yixian*. When the latter's Guomindang was banned in 1913, she accompanied him to Japan, where they were married. After Sun's death she became a leader of the left wing of the Guomindang and in 1927 strongly condemned the rift with the Communists. In 1931 she founded the League for the Protection of Human Rights and in 1938 in Hong Kong the China Defence League, which raised money for medical relief. In 1941 she was active in Communist-controlled north China setting up military hospitals. With the establishment of the People's Republic she became a government official, holding various posts. She was criticized by the Red Guards in the Cultural Revolution*, but remained in office.

Soong, T. V. (born Sung Tzu-wen) (1894–1971), Chinese financier. Son of a wealthy Chinese industrialist in Shanghai, he was educated at Harvard before returning to China, when he gave his support to Sun Yixian*, his brother-in-law, establishing the Central Bank of China in Guangzhou (Canton). He served as Finance Minister in the National government* of the Guomindang 1928–31, resigning after a disagreement with Jiang Jiehi (Chiang Kai-shek), who had married one of his sisters Soong Mei-ling. During the 1930s he was active in Chinese banking, trying to standardize the currency, centralize banking, and encourage investment. He also amassed one of the world's largest fortunes. During the war years 1941–5 he was China's Foreign Minister and chaired the San Francisco* Conference setting up the UN in 1945. He served for a short while as Premier of the National government 1945–7, but failed either to curb inflation or to negotiate successfully with the Communists. In 1949 he went to live in the USA, where he was active in banking.

Souphanouvong, Tiao, Prince (1902–   ), President of the Laos People's Democratic Republic 1975–86. One of the nephews of King Sisavang Vong (1946–59), he studied engineering in Paris and then worked as an engineer in Laos 1938–45. He joined those opposed to French efforts at regaining control and helped to found the Pathet Lao* in 1950, fighting French forces until the Geneva Agreements* of 1954. He joined a coalition government in 1962, but in the following year was active in opposition to the US-backed government of his half-brother Prince Souvanna Phouma*. He was imprisoned, but escaped and led the Pathet Lao forces in the mountains in the north. In 1973, following the Paris Peace Accords*, he became President of a coalition National Political Consultative Council, with Souvanna as Prime Minister. In 1975 the Pathet Lao abolished the monarchy in Vientiane and established the Lao People's Democratic Republic with Kaysone Phomvihane*, party leader and Prime Minister, and Souphanouvong as head of state and Chairman of the Supreme Council.

South Africa, a federal republic consisting of four provinces—Cape Province, Natal, Orange Free State, and Transvaal—together with ten African 'Bantustan homelands'*. Of these four are considered by the South African government to be independent—Transkei, Bophuthatswana, Venda, and Ciskei. The Union of South Africa was formed in 1910 as a self-governing dominion of the British Crown and consisting of the former colonies of Cape Colony and Natal and the Boer republics of Orange Free State and Transvaal, recently defeated in the Boer War 1899–1902. Politically dominated by its small White minority, it supported Britain in the two world wars, its troops fighting on a number of fronts. From 1948 onwards the right-wing Afrikaner-dominated National Party* has been in power. It instituted a strict system of apartheid*, intensifying discrimination against the disenfranchised non-White majority. South Africa left the Commonwealth and, following a referendum, became a republic in May 1961. Its economic strength enabled it to dominate the southern half of the continent. Its moves to establish independent Bantu homelands were denounced by the International Court of Justice at The Hague. The rise of Black nationalism both at home and in surrounding countries, including both Namibia* and Mozambique*, produced increasing violence and emphasized South Africa's isolation in the diplomatic world. In 1985 the regime of President P. W. Botha* made some attempts to liberalize apartheid. They failed, however, to satisfy either the increasingly militant non-White population or the extremist right-wing groups within the White élite. In 1986 a state of emergency was proclaimed and several thousands imprisoned without trial. The domestic and international aspects of the problem remained inseparable, with South African troops fighting against SWAPO* guerrillas in Namibia and in Angola, while support by surrounding Front-Line States (see SADCC) for the African National Congress* produced a number of cross-border incidents. In 1988 the US Congress voted to support the Front-Line States in their demand for international sanctions. President Botha retired in 1989 and his successor President de Klerk* made a number of positive moves towards racial reconciliation.

South-East Asia Treaty Organization (SEATO) (1954–77). A treaty was signed in Manila in September 1954 between Australia, New Zealand, the USA, Pakistan, Thailand, the Philippines, Britain, and France. It provided for mutual collective action if any signatory be attacked or be subject to internal subversion. A defence organization was established, based in Bangkok, with the USA seeking to use it as part of its policy of containment of Communism. Dispute arose as to whether the treaty should apply to Vietnam, Cambodia, and Laos, all of whom experienced civil war during the years 1954–75. Many prominent Asian countries refused to join, for example India, Indonesia, Sri Lanka, while Pakistan withdrew in 1973 and France in 1974. In September 1975 it was agreed 'to phase out' the organization due 'to changed conditions'. It was formally dissolved in 1977.

Southern African Development Co-ordination Conference (SADCC). Founded in 1980, its members consist of Angola, Botswana, Lesotho, Malawi, Mozambique, Swaziland, Tanzania, Zambia, and Zimbabwe, the so-called 'Front-Line States'. Its aim is to finance and co-ordinate a number of projects in agriculture, animal husbandry, mining, transport, energy, telecommunications, trade, and disease control, and to reduce dependence upon South Africa, particularly regarding access to sea-ports. There is an annual conference of heads of states and a permanent Council of Ministers.

South Pacific Commission. It was formed in 1947 at Canberra by the five nations Australia, New Zealand, Britain, France, and the USA, to co-ordinate programmes of economic development and social welfare of the South Pacific. Since then Western Samoa, Nauru, and Fiji have become members. It consists of two Commissioners, with technical advisers, from the member states; it now uses the disused US headquarters on Noumea and sponsors an annual South Pacific Conference. The first of these was held in 1950, when leaders from Polynesia*, Melanesia*, and Micronesia* were all brought together for the first time. It conducts research in a variety of areas: economic, ecological, medical, and educational.

South Pacific Forum. This was formed in 1971, largely on New Zealand's initiative. It consists of irregular meetings of heads of governments of the following: Australia, Cook Islands, Fiji, Kiribati, the Republic of the Marshall Islands, the Federated States of Micronesia, Nauru, New Zealand, Niue, Papua New Guinea, Solomon Islands, Tonga, Tuvalu, Vanuatu, Western Samoa. In 1985 it adopted a treaty for a nuclear-free zone in the Pacific, which became operative in December 1986. The USSR and China endorsed this Treaty of Raratonga, but the USA, France, and Britain have refused to do so.

South West Africa People's Organization (SWAPO). Formed in 1959 in response to attempts by the South African government to extend formal authority in South West Africa, in 1966 it began a guerrilla campaign under the presidency of Sam Nujoma*, operating largely from neighbouring Angola. Efforts at mediation by the UN for long failed to find an acceptable formula for Namibian independence, and the guerrilla war continued until 1988. SWAPO gained a majority in the election of November 1989, which preceded the independence of Namibia*.

Souvanna Phouma, Prince (1901–84), Prime Minister of Laos 1951–4, 1956–8, 1960, 1962–73. Nephew of King Sisavang Vong (1946–59), he studied as an engineer in Paris and Grenoble before returning to Indo-China in 1931. After the collapse of Japan in 1945 he served briefly in the provisional government, but, when his uncle welcomed the return of the French in 1946, he joined a Free Laos (Lao Issara) movement, though he opposed the Pathet Lao's use of guerrilla tactics. He became Prime Minister in 1951 during the French Indo-China War*, and welcomed the Geneva Agreements* of 1954. He was Prime Minister again 1956–8 in coalition with the Pathet Lao. When this collapsed, civil war lasted until 1962, when he again tried to form a Pathet Lao coalition. This collapsed in 1963 and, as Prime Minister until 1973, he relied increasingly on US aid. After the Paris Peace Accords* he allied himself again with his half-brother Prince Souphanouvang* and the Pathet Lao, and became an adviser to the new people's Democratic Republic of Laos in 1975.

Soviet (Russian; 'Council'), an elected governing council. In 1905 a soviet of workers' deputies was formed in St Petersburg to co-ordinate strikes and other anti-government activities in factories during the Revolution of that year, each factory sending delegates. Other Russian cities were also for a time dominated by soviets. Both Bolsheviks* and Mensheviks realized the potential of soviets and duly appointed delegates. In 1917 a soviet modelled on that of 1905, but now including deserting soldiers, was formed in Petrograd (previously St Petersburg), sufficiently powerful to dictate industrial action. It did not at first try to overthrow the provisional government*, but grew increasingly powerful as representing opposition to continuation of the war. It consisted of some 3,000 members; power was exercised by the executive committee, which the Bolsheviks now sought to control. Soviets were established in other cities and the provinces and in June 1917 the first All Russia Congress of Soviets met. In October the Bolsheviks seized control and the Russian Civil War* followed, during which village soviets controlled local affairs. The national soviet is called the Supreme Soviet of the Union, comprising delegates from all the soviet republics, each of which in turn has its own Supreme Soviet. During 1989–90 profound constitutional changes were taking place in the Soviet Union, with multi-party politics emerging, together with debate as to the relationship of the Supreme Soviets of the Republics with that of the Union.

Soviet-Japanese frontier disputes. Having occupied Manchuria and established the puppet regime of Manchukuo* (1933), Japan found itself with thousands of miles of frontier with the Soviet Union. The latter signed a non-aggression pact with China, following the establishment of the United Front*, and during 1938 and 1939 there were numerous border disputes, which twice developed into full-scale battle. Following the Nazi–Soviet Pact of August 1939, however, these subsided, and in April 1941 Japan negotiated a neutrality pact with the Soviet Union, as plans were already developing for war against the Dutch East Indies. In 1945 the Soviet Union invaded and Japan accepted defeat.

Soviet Union. As of 1990 it consists of a Union of fifteen socialist republics: Armenia, Azerbaijan, Belorussia, Estonia, Georgia, Kazakhstan, Kirghizia, Latvia, Lithuania, Moldavia, Russia (RSFSR), Tadzhikistan, Turkmenistan, Ukraine, Uzbekistan. Of these by far the largest is Russia*, which under its constitution of 1978 has the full title of Russian Soviet Federal Socialist Republic (RSFSR), consisting of six territories, forty-nine provinces, sixteen autonomous republics, and five autonomous regions. By 1990 ten of these fifteen republics had requested a constitutional revision whereby they would gain greater independence from the Supreme Soviet of the Union. The original Union had consisted of four republics, the RSFSR, the Ukraine, Belorussia, and Transcaucasia, which in 1922 agreed on a constitution under Lenin. Ultimate legislative power was to be in the hands of the Supreme Soviet of the Union, itself largely controlled by the Politburo* of the party. Under Stalin a series of five-year plans was initiated in 1928, the first involving collectivization* of agriculture, when rebellious kulak* villages were obliterated and unwilling kulaks transported to unpopulated regions of Siberia. Heavy industry was developed, especially in the Urals and beyond. The second plan 1933–7 allowed more consumer goods, but towards its end a series of political purges took place to ensure Stalin's position. The third plan 1938–42 involved heavy rearmament, as well as massive hydroelectric schemes. It is estimated that some 9 million or more people died as a direct result of the vast social revolution of these years 1928–41. The Union expanded during the same period, with a number of constitutional changes, especially in 1924 and 1936. In 1939 the Soviet Union signed the Nazi–Soviet Pact, annexing eastern Poland to the Curzon Line* and going on to fight the Finnish-Russian War*. In its struggle against German armies which invaded in June 1941 it lost some 20 million as war casualties. It took part in the conferences of Tehran, Yalta, and Potsdam, by means of which it was able to establish a wall of satellite nations on its western frontier, which in 1955 formed the Warsaw Pact*. Soviet troops were sent to Hungary in 1956 and to Czechoslovakia in 1968 to prevent attempts in those countries to gain greater independence. Ideological differences developed with China in the late 1950s, and lasted until the late 1980s. In the developing world the Soviet Union gave aid to pro-Soviet regimes, while troops were engaged in Afghanistan from 1979 until 1989. A pervasive element of Soviet society from early days until the late 1980s remained a high degree of police surveillance and state control of people's lives. The appointment of Mikhail Gorbachev* as general secretary of the Communist Party in 1985 represented a fundamental change in style of Soviet leadership, committed to modernization of technology, elimination of bureaucratic corruption, liberalization of the economy, and international arms control. In 1990 Gorbachev became Executive President as part of profound constitutional changes being initiated. Multi-party politics were permitted and a debate began as to the future of the Union, against a background of near economic collapse. By December 1990, with the Cold War ended, democratic regimes established among its western neighbours and German reunification achieved, the future of the Soviet Union was far from clear.

Soweto, an amalgamation of several Black townships south-west of Johannesburg in South Africa. In January 1976 Black schoolchildren demonstrated against legislation making Afrikaans the compulsory language of instruction. Police broke up the demonstrations using tear gas and guns. It triggered off a wave of violence. On 16 June there were further riots with police killing two children and four adults. By the end of June some 236 non-Whites had been killed and over 1,190 injured. By the end of the year over 500 Blacks and Coloureds had been killed, many of them children. The plans for compulsory use of Afrikaans were dropped. Since then each anniversary of the demonstration has led to further riots and violence.

Soyinka, Oluwole (1934–), African playwright. Born in Yorubaland, Nigeria, he studied in Ibadan and then at Leeds University. He became a director of the Royal Court Theatre, London, producing his play *The Lion and the Jewel* in 1959. In 1960 he returned to Nigeria to write *The Dance of the Forests* for Independence Day.

Interned in 1967, allegedly for sympathy for Biafra, he spent twenty-seven months, mostly in solitary confinement, without trial, his account of the experience being written in *The Man Died.* Since then he has resumed his post as a lecturer at Ibadan University, continuing to write poetry and plays, but concerned more for universals—hope, hatred, defeat, sacrifice—and less with the particulars of the corruption and violence of Nigerian politics.

Spaak, Paul-Henri (1899–1972), Prime Minister of Belgium 1938–9, 1947–9. Born at Schaerbeek, he qualified as a lawyer in 1922 and practised until entering Parliament as a Socialist in 1932. He became Belgium's first Socialist Prime Minister in 1938 and after the German invasion moved to London, where he was Foreign Minister of the government-in-exile. Here he played a major role in the establishment of the UN, and became the first president of the General Assembly. He was one of the architects of the Benelux Customs Union* of January 1948 and became president of the consultative assembly of the Council of Europe* 1949–51. He was Belgian Foreign Minister 1954–7 and again 1961–6, in between which offices he was the secretary-general of NATO 1957–61. Spaak was a firm supporter of the concept of a united Europe, and his early post-war proposals for an economic association based on free trade and movement of labour, as well as joint social and financial policies, were to form the basis of the European Community.

Space exploration. In 1903 the Russian physicist Konstantin Tsiolkovsky was developing ideas for space rockets fuelled by liquid gas. In 1923 the German scientist Hermann Oberth published *The Rocket and Interplanetary Space*, and by 1926 the American Robert Goddard had successfully designed the first liquid-fuelled rocket. In 1942 Wernher von Braun successfully designed the V2 rocket missile, and after the war he went to work in the USA. In 1957 the Soviet Union launched the first artificial satellite *Sputnik I.* This was followed by the US *Explorer I* a year later. Yuri Gagarin was the first man in space in 1961 and in 1969 the US spacecraft *Apollo 11* landed Neil Armstrong and Edwin 'Buzz' Aldrin on the moon. There were five more moon landings, the last by Apollo 17 in 1972. Meanwhile space probes to investigate Venus, Mars, Jupiter, Saturn, Uranus, and Neptune followed. Space stations and spacelabs have enabled astronomers to gain much new knowledge of the universe, but probably the main effect of space technology has been the placing of communications satellites, which have revolutionized business communication and the world's television diet. They have also been used to predict weather, detect military installations, and search for mineral resources.

Spain. Spanish history through the 19th century had been one of almost continuous tension between absolutist monarchists and liberals, and this tension continued into the reign of Alfonso XIII*. The virtual dictatorship of Primo de Rivera* (1924–30) was followed by a republican interlude, during which demands for autonomy from both the Basques and Catalonia caused renewed tension, as did anti-clerical acts of violence and Communist-inspired strikes, as in the Asturias in 1934. It was to restore what it regarded as reasonable law and order that the army launched the Civil War in 1936. Nationalist victory in 1939 resulted in the dictatorship of Francisco Franco*, Caudillo of the Realm and Chief of State 1939–75. His gradual liberalization of government during the late 1960s was continued under the restored monarchy, Juan Carlos I being successful in establishing a liberal, democratic, constitutional monarchy. The first general election since 1936, in 1979, brought the moderate Democratic Centre Party to power, under Adolfo Suárez, Prime Minister 1977–81. Suárez envisaged autonomous regional parliaments for Catalonia, Andalusia, and the Basques. He and the King survived two attempted military coups, 1978 and 1981, by Colonel Molina, with a number of senior military officers being implicated. In 1982 a Socialist government led by Felipe González was elected, with González remaining in office after the 1986 elections and again, by a narrow majority, in 1989. Of its remaining colonies, Spain granted independence to Spanish Sahara in 1976, which was then divided between Morocco and Mauritania. Spain joined the European Community in 1986.

Spanish Civil War (1936–9). After the fall of Primo de Rivera and the abdication of Alfonso XIII* in 1931, Spain was split. On the one hand were the landowners, strong supporters of the Church and the Falange Party*, and other privileged groups such as the military; on the other hand were the Republicans, Catalan and Basque separatists, anti-clerical socialists, Communists, and anarchists. The elections of February 1936 established a left-wing Popular Front government under President Azaña*; riots and military plots followed. In July 1936 generals José Sanjurjo and Francisco Franco in Spanish Morocco led a coup against the Republic and civil war followed. Most of Castille declared for the rebel Nationalists, Burgos becoming their headquarters. Valencia and much of Andalusia, Barcelona and Catalonia, and Basque Bilbao remained loyal to the Republic. Madrid was internally divided in loyalty. The government of the Republic moved to Valencia and then in 1937 to Barcelona, and Madrid became the main target of Nationalist attack throughout the war, being besieged for 2½ years. In 1937 Franco's Nationalist troops overran the Basque region and captured Bilbao. His forces held the important town of Teruel, which enabled him, with German and Italian assistance, to divide the Republican forces by conquering the territory between Valencia and Barcelona. The Republicans, weakened by internal intrigues between rival factions and by the withdrawal of Soviet support, attempted a desperate counter-attack in the Battle of the Ebro, August–November 1938. It failed, and Barcelona fell on 26 January 1939, quickly followed by Madrid. Franco became head of state and the Falange the sole legal party. The war had inspired international support on both sides: the Soviet Union sent advisers and supplies to the Republicans, while some 50,000 Italian soldiers fought with Franco. Germany supplied some 10,000 men in the Condor Legion*, mostly in the aviation and tank services. Bombing of civilians by the latter and the destruction of the Basque town of Guernica* became a symbol of Fascist ruthlessness and inspired Picasso's masterpiece. Left-wing and Communist volunteers from many countries fought for the Republicans in the international brigades*. The war cost some 700,000 lives in battle, 30,000 executed or assassinated, and 15,000 killed in air raids.

Spartakist movement, a group of German left-wing socialists. It was formed in 1915, led by Rosa Luxemburg* and Karl Liebknecht*, in order to overthrow the German imperial government and obtain international peace. The name was coined by Liebknecht, who called on the modern 'wage slave' to revolt like the Roman gladiator Spartacus, against international capitalism. In December 1918 the Spartakists formed the German Communist Party in Berlin. In January the Defence Minister of the new Weimar Republic Gustav Noske ordered the suppression of all uprisings throughout Germany. There was a week of street violence in Berlin which was suppressed by the paramilitary Freikorps, who captured and shot its leaders Luxemburg and Liebknecht. There were further Spartakist riots in the Ruhr in 1920.

Speer, Albert (1905–81), Nazi architect and politician. Born in Mannheim, he trained as an architect and practised there. He joined the National Socialist Party and was appointed its official architect, designing the grandiose stadium at Nuremberg in 1934. An efficient organizer, in 1942 he became Minister for Armaments and was mainly responsible for the planning of Germany's war economy, marshalling conscripted and slave labour in his *Organization Todt* to build strategic roads and defence lines. By 1944 he was responsible for all industrial output. He was imprisoned for twenty years at the Nuremberg Trials*.

Spence, William Guthrie (1846–1926), Australian trade unionist, from Queensland. He had helped to found a number of trade unions at the end of the 19th century, including the Shearers' Union (1886) and the Australian Workers' Union (1894), Australia's largest union, of which he was president 1898–1917. He was a member of the federal House of Representatives for Darling 1901–19.

Sri Lanka (formerly Ceylon). By 1914 the island of Ceylon had been a British colony for a century, during which time land had been steadily transferred to companies which had developed tea, coffee, and rubber plantations. Many of the workers on these were Tamils brought from southern India. By 1914 approximately three-quarters of the population were Sinhalese and followers of Buddhism, and one-quarter Hindu Tamils, themselves divided between Indian and Ceylonese (later Sri Lankan) Tamils. A nominated Legislative Council had been established in 1833 and in 1920 it was agreed that some members of this be elected, in response to pressure from a small, English-educated middle class which by now had emerged. Further pressure resulted in the Donoughmore Commission of 1928, whose recommendations were implemented in 1931. The Legislative Council would now be elected on universal adult suffrage, but the British Governor-General would retain control over law, finance, and the Civil Service, while defence remained a British responsibility. After World War II pressure mounted for full independence, which was granted by the Ceylon Independence Act of 1947, Ceylon becoming a dominion within the Commonwealth. The United National Party (UNP) of Don Stephen Senanayake* created a regime, secular and liberal, which successfully united the ethnic groups, European, Sinhalese, and Tamil. In 1951 Solomon West Ridgeway Bandaranaike* formed a new political party, the Sri Lanka Freedom Party, a mildly socialist and strongly nationalist party which was in power 1956–65 and introduced a number of reforms. British military bases were closed, Ceylon adopting a neutralist position, and the Buddhist cultural tradition was encouraged. Sinhalese replaced English as the official language, the latter resulting in Tamil riots in 1958. Although the Tamil language was to be given recognition under the government of Donald Senanayake (1965–70), from now on ethnic tensions were steadily to increase. In May 1970 a Marxist United Left Front (ULF) government came to power. Ceylon became a republic. The ULF was replaced in 1977 by the UNP under Junius Jayawardene (President 1978–89), whose years in office saw a struggle between Tamil and Sinhalese deteriorate into civil war. The government of India, under strong pressure from the state of Tamil Nadu, sent in troops to try to bring about a cease-fire in 1987, but a tense situation remained, with the banned Sinhalese JVP* bringing fear throughout the island. Indian troops withdrew in 1989, but Tamil violence continued. During 1989–90 the UNP, under President Ranasinghe Premadasa, initiated all-party talks in an effort to end civil strife.

SS (the *Schutzstaffel* or 'protective echelon'). Founded in 1925 as Hitler's personal bodyguard, the SS was schooled in absolute loyalty and obedience, and in total ruthlessness towards opponents. From 1929 onwards it was headed by Heinrich Himmler*, who divided it into two groups: the *Allgemeine SS* (General SS), and the

Waffen-SS (Armed SS). Initially subordinated to the SA (Brownshirts*) the SS assisted Hitler in the 'Night of the Long Knives'* which eliminated its rivals. By 1936 Himmler, with the help of Reinhard Heydrich*, had gained control of all police forces. Subdivisions of the SS included the Gestapo* and the *Sicherheitsdienst* (SD), in charge of foreign and domestic intelligence. The *Waffen-SS* served as an élite combat troop alongside, but independent of, the armed forces. Together with the Gestapo it also administered the concentration camps.

Stalin, Josef Vissarionovich (1879–1953), Soviet dictator. Born in Georgia, son of a shoemaker, he attended a seminary for priests, from which he was expelled for his revolutionary views in 1898. In the same year he joined the Russian Social Democratic Party and sided with the Bolsheviks in 1903. He worked underground in Transcaucasia and in 1913 was co-opted by Lenin* and Zinoviev on to the Bolshevik Central Committee. He was banished to Siberia on six occasions but each time escaped, the last time being after the 'February' Revolution in 1917, when he returned to Petrograd. He became editor of *Pravda* and when Lenin returned accepted his plans for the seizure of power by the Bolsheviks. He was now Lenin's right-hand man and in 1923 seized control of the party machine at the Twelfth Party Congress. Lenin died next year, and by 1925 he had won his struggle with Trotsky* for the leadership, proclaiming his doctrine of 'Socialism in one country'. In 1928 he initiated his five-year plans, which turned the Soviet Union into one of the most powerful nations on earth. His utter ruthlessness is seen in the purge technique adopted in the later 1930s against all potential rivals. The show trials and executions of the *Yezhovshchina** not only removed those he considered as personal rivals, but also placed millions of Soviet citizens in the prison camps of the Gulag Archipelago. In May 1941 he made himself Prime Minister, as well as Commissar of Defence and marshal of the Soviet Union. In 1942 he signed the Anglo-Soviet Treaty, and from 1941 onwards was in receipt of Lend-Lease* materials from the USA. This was ended abruptly in September 1945, and he rejected its alternative as provided through the Marshall Plan*. Although he won major concessions at the wartime conferences on Yalta* and Potsdam* for Soviet dominance in Eastern Europe, he rejected requests from Communists in Greece and other Western states for direct Soviet intervention. In 1948 he broke with Tito*, whose party policies in Yugoslavia had rejected his control. Increasingly the victim of his own paranoia, he ordered the execution of colleagues and of satellite allies, such as in the Slansky* affair, anti-Semitism becoming an obsession in his later years. At the Twentieth Party Congress* of 1956 Khrushchev attacked the cult of Stalinism, accusing him of tyranny and terror. In the later 1980s the policy of *glasnost* enabled Soviet scholars to begin to unravel the enormity of many of his actions.

Stalingrad, Battle of, World War II (September 1942–January 1943). During 1942 the German 6th Army under von Paulus* and 4th Panzer Army under Ewald von Kleist occupied Kursk, Kharkov, all the Crimea, and the Maikop oilfields, reaching the key city of Stalingrad (now Volgograd) on the Volga in September. Soviet resistance successfully prevented the river being crossed, with grim and prolonged house-to-house fighting. Meanwhile a Soviet counter-attack was organized, and in November Stalin launched a winter offensive of six Soviet armies under Marshals Zhukov*, Koniev*, and Rokossovsky* advancing from north and south and annihilating Romanian forces. By January von Paulus with some 91,000 troops was surrounded, but he fought on to enable von Kleist to escape. On 31 January he surrendered, total German and Romanian casualties killed and captured being some 300,000. The Russians now advanced to recapture Kursk*, and this defeat marked the beginning of the end of German success on the Eastern Front.

Stamboliisky, Alexander (1879–1923), Prime Minister of Bulgaria 1919–23. Son of a wealthy peasant family in Bulgaria, he studied agriculture in Germany and returned to Bulgaria in 1908 as a socialist agitator, demanding agrarian reform. He was imprisoned in 1915, but in 1918 emerged to establish a brief republic. In October 1918 Boris III* was established as King by the army and Stamboliisky agreed to act as Prime Minister. He instituted a virtual agrarian dictatorship, imposing high urban taxation but exempting the peasantry. Although opposed to Communism, he became the target of right-wing nationalists, one of whom assassinated him in 1923.

Stavisky scandal (1934). During 1933 Serge Stavisky, a naturalized Frenchman from Russia, floated a huge sum on bogus bonds on the Paris Bourse. He committed suicide 3 January 1934, but enquiries revealed that he had been backed and protected by a number of Deputies and Ministers of the Third Republic*. An official of the Public Prosecutor's Office was mysteriously murdered. Thousands had lost money and there were riots and demonstrations by right-wing neo-Fascists in Paris 6–9 February. They claimed that the scandal revealed the corruption of French democratic government, a view later used by the Vichy government to justify its autocratic regime.

Stepinac, Aloysius (1898–1960), Archbishop of Zagreb and Primate of Croatia. Born in Krasić, he served in the Austrian army during World War I, after which he decided to become a priest, studying in Rome. He had been sympathetic to the idea of a southern Slav republic, but became disenchanted by Alexander I* and Serbian dominance. He became Archbishop of Zagreb in 1937 and increasingly resented the Serbian Orthodox Church. In 1941 he welcomed the establishment of an 'independent' Croatia, but was later shocked by Pavelić's* regime. In 1945 he met Tito* to try to reach an understanding on Church–state relationships, but a year later he was arrested and imprisoned for collaboration with the Germans. He was released in 1951 and made a cardinal in 1952, although he could not go to Rome for his 'Red Hat', as Tito would not grant a visa. His successor Archbishop Seper achieved a reconciliation between the Catholic Church and Belgrade.

Stern Gang, the name given by the British to a Zionist terrorist group, Lohamei Herut Israel Lehi ('Fighters for the Freedom of Israel'). It was founded in 1940 by Abraham Stern, following a split in the underground movement Irgun* Zvai Leumi, dedicated to the expulsion of the British and the creation of a state of Israel. Stern was killed by British counter-insurgency forces in 1942 and the gang never numbered more than a few hundred. They operated in small groups, attacking British service personnel and organizing the assassination of government officials, as well as destroying installations. Their victims included Lord Moyne, the British Minister for the Middle East in Cairo (1944), and Count Bernadotte*, the UN mediator in Palestine (1948). They were suppressed after 1949, although some members joined the Israeli defence forces.

Stevenson, Adlai Ewing (1900–65), US statesman. Born in Los Angeles, he graduated from Howard Law School in 1926 and practised law in Chicago 1927–41. During World War II he held various government appointments and in 1948 he was elected Governor of Illinois, with the largest majority in the state's history. His energetic administration attacked gambling and corruption and imposed greater efficiency on the bureaucracy. Chosen as the Democrat candidate for the presidency in the elections of

1952 and 1956, on both occasions he was badly beaten by Dwight D. Eisenhower*. A liberal reformer and internationalist, his presidential campaigns were marked by brilliant and witty speeches. President Kennedy appointed him US ambassador to the UN (1961–5), with cabinet rank.

Stilwell, Joseph Warren (1883–1946), US general. Born in Florida, he graduated at West Point in 1904 and spent much of his time between the wars in China, being military attaché for the USA in Beijing 1935–9. In 1941 Jiang Jiehi* (Chiang Kai-shek) asked him to take command of the Chinese 5th and 6th armies on the Burmese border. He was defeated by the Japanese and arrived in India after a 140-mile escape through the jungle. In March 1942 he was appointed commanding general of all US forces in China, Burma, and India, and Chief of Staff under Jiang Jiehi of all Allied armies in the Chinese theatre of operations. Known as 'Vinegar Joe' for his tactlessness, he frequently clashed with Louis Mountbatten*, Supreme Allied Commander of South-East Asia, pressing hard for a Burmese land campaign as against Mountbatten's preferred amphibious operations. His view prevailed. In 1944 the Burma Road into China was reopened and in January 1945 the Ledo Road, a major US-built highway linking it with north India, carried its first trucks back into China; it was renamed the Stilwell Road. There were differences with Jiang Jiehi, and in October 1944 Stilwell had been recalled to Washington and appointed commander of the US 10th Army in the Pacific, receiving the surrender of the Japanese in the Rikuyu Islands. He remained in active command until his death.

Stimson, Henry Lewis (1867–1950), US Secretary of State 1929–33, Secretary of War 1911–13, 1940–5. Born in New York, he was a graduate of Harvard and admitted to the New York Bar in 1891. He worked for President Theodore Roosevelt and in 1911 was first appointed Secretary of War. He served in World War I, after which he returned to his law practice until appointed Governor-General of the Philippines 1927–9, where he pursued a policy of conciliation. In 1929 President Hoover appointed him Secretary of State. As such he formulated the 'Stimson Doctrine' in response to the Japanese invasion of Manchuria in 1931: (1) a refusal to grant diplomatic recognition to actions which threatened the territorial integrity of the Republic of China; (2) a refusal to recognize any territory or agreement obtained as a result of aggression, in violation of the Kellogg–Briand Pact*. He again returned to his law practice in 1933, but though a life-long Republican he supported F. D. Roosevelt's* foreign policies, in 1939 recommending aid for Britain and a resumption of conscription. Appointed Secretary of War in July 1940, he served until September 1945, his chief role being to guide the recruitment and training of the US armed services for war. He supported the recommendation in 1945 to drop the atomic bomb, arguing that its use would save more lives than it cost.

Stock Market Crash (October 1929). The post-war US economy was on a narrow base with fundamental flaws. Older basic industries such as mining and textiles were weak; agriculture was depressed; unemployment at 4 million was unacceptably high; international loans were often poorly secured. A new rich class enjoyed the flamboyant life-style of the Jazz Age, but too many over-stretched their credit and depended on stock-market speculation. During the first half of 1929 there was an unprecedented boom on the New York Stock Market. Prices began to fall, however, in September and panic selling set in through October; stock value fell by 40 per cent and the fall continued over the next three years. Real estate values collapsed; factories closed; banks began to call in loans. The world-wide Great Depression* followed.

Stormont, suburb of Belfast and seat of the former Parliament of Northern Ireland. Created by the Government of Ireland Act of 1920, as a subordinate body to Westminster, the Stormont Parliament, through a skilful manipulation of its constituencies, had a built-in Unionist majority. Following the breakdown of law and order in the late 1960s, it was suspended in March 1972. Direct rule from Westminster was imposed, to be administered by civil servants of the Northern Ireland Office based in Stormont Castle.

Strategic Arms Limitation Treaties (SALT), two treaties between the USA and the Soviet Union aimed at limiting the production and deployment of nuclear weapons. A first round of meetings November 1969 to May 1972 produced the SALT I Treaty, which prevented the construction of comprehensive anti-ballistic missile (ABM) systems and placed limits on the construction of strategic, that is, intercontinental, ballistic missiles (ICBMs) for an initial period of five years. Talks began again in 1974 and a SALT II Treaty was agreed between Carter* and Brezhnev* in June 1979, which sought to set limits on the numbers and testing of new types of intercontinental missiles. This treaty was never ratified by the US Senate, but arms-control talks were again resumed in 1982 (START). They continued intermittently through the 1980s, but were hampered by the US insistence on its SDI* programme.

Strategic Defense Initiative (SDI), US defence system. Extensive development took place under President Reagan during the 1980s. It is intended to protect the USA from ICBMs (Inter Continental Ballistic Missiles) by destroying them in space before they reach their target. US critics argued that this massively expensive system could never be infallible, while Soviet critics claimed that development of the programme breached the SALT* agreements. Insistence on it by Reagan for long prevented progress on arms control. Funding for the programme was much reduced after President Bush came to office.

Strauss, Franz Josef (1915–88), Minister- President of Bavaria 1980–8. Born in Munich, he distinguished himself as a student there by graduating in 1939 with highest honours. Commissioned into the artillery, he served through World War II until captured by US forces. On release he immediately entered local politics in Munich and Bavaria. His politics were right wing, but he was a robust democrat. He founded the Christian Social Union in Bavaria and allied with Adenauer's* Christian Democrats, becoming a member of the federal Bundestag 1949–78. He held the offices of Minister of Atomic Affairs 1955 and Minister of Defence 1956–62 at a time of crucial importance, when German rearmament, for which he had staunchly campaigned, had begun. He resigned in 1962 over a newspaper article which alleged corruption, but was federal Finance Minister 1966–9. He resigned in 1978 to become the forceful Minister-President of Bavaria. As a realist he was a cautious supporter of *Ostpolitik**, as well as of his neighbouring socialist state of Austria. He failed in a bid for the federal Chancellorship in 1980.

Streicher, Julius (1885–1946), Nazi leader and propagandist. Born in Fleinhausen, he trained as a schoolmaster, and after World War I he taught in Nuremberg, where Hitler made him party gauleiter in 1925. He founded a weekly newspaper *Der Stürmer*, in which he expounded his violently anti-Semitic views. In 1933 Hitler appointed him *Aktionsführer* (riot leader) against Jewish shops, and later made him Governor of Franconia. He continued to function as a party propagandist and was sentenced to death at the Nuremberg Trials*.

Stresa Conference (April 1935), a conference held on Lake Maggiore at Stresa between Britain, France, and Italy. It issued a strong protest at Hitler's* action in openly rearming in defiance of the Treaty of Versailles. Together these countries proposed to form the 'Stresa Front' against future German aggression; but the Front soon disintegrated. In June 1935 Britain negotiated unilaterally a naval agreement with Germany; in November Mussolini invaded Ethiopia; and in November 1936 he proclaimed the Rome–Berlin Axis* with Hitler.

Stresemann, Gustav (1878–1929), German statesman. Born in Berlin, of a wealthy brewer's family, he was educated at Leipzig and at 24 founded the Union of Saxon Industrialists. He first entered the Reichstag as a National Liberal in 1906. He was an ardent supporter of German involvement in World War I but afterwards became an enthusiast for the Weimar Republic*. He was briefly its Chancellor in 1923 and then Foreign Minister 1923–9. He ended passive resistance to the French and Belgian occupation of the Ruhr and negotiated their withdrawal. He readily accepted both the Dawes* and Young* Plans on reparations. Personal friendship with Briand* and Austen Chamberlain* helped him to play a leading part in negotiations at Locarno which led to the admission of Germany to the League of Nations in 1926. In 1928 he signed the Kellogg–Briand Pact*. At the same time he believed that Germany's eastern frontier had to be revised, advocating that Danzig (Gdansk*), the Polish Corridor*, and Upper Silesia be returned from Poland.

Strijdom, Johannes Gerhardus (1893–1958), Prime Minister of the Union of South Africa 1954–8. An Afrikaner born in Cape Colony, he was educated at Stellenbosch and Pretoria, graduating in law. In 1929 he was elected to Parliament as a loyal supporter of Prime Minister Hertzog*. In 1934 he helped Daniel Malan* to create the Purified National Party, when Hertzog joined Smuts in a coalition. He was totally committed to White supremacy and served in Malan's government 1948–54, helping to implement apartheid*. When he succeeded as Prime Minister he was responsible for ending multi-racial traditions in the universities, for disenfranchising Coloured voters, and for the Treason Trial, which was to last from 1956 until 1961. A racial fanatic, ill health forced him to retire in December 1957.

Stroessner, Alfredo (1912–   ), President of Paraguay 1954–89. Son of a German immigrant, he joined the army, attended military college, and was commissioned in 1932. He fought in the Chaco War* and rose steadily in rank to become a general and Commander-in-Chief of the armed forces 1951–4. He was responsible for the overthrow of President Frederico Chávez (1949–54) and then, as sole candidate, was elected President. Strongly supportive of the large landowners and international commercial interests, he used foreign aid to develop schools, hospitals, highways, and hydroelectric power. His regime remained strongly backed by the army, and was both authoritarian and corrupt. While allowing some political criticism, it was guilty of harsh and repressive methods, which by February 1989 had generated sufficient opposition for him to be deposed and sent into exile.

Structuralism, a 20th-century term in anthropology, linguistics, and psychology. It originated in the anthropological theories of Claude Lévi-Strauss (b. 1908), who stressed the universality of kinship structures, with their basic sexual relationships, in all cultures. Structural Linguistics has been concerned with the relational structure of language. Its tools have been morphology (word structure) and phonology (sound structure) and its concern that of universality. In Psychology the term was developed by Edward Titchener (1867–1927) to describe the anatomy of the adult mind, while being in no way concerned with how or why that mind functions.

Student revolts. There is a long tradition of student political involvement, going back at least to the German university student movement opposing Metternich, and in Russia to student opponents of tsarism being exiled or executed. In the period between the two world wars the universities in Germany and Japan had movements supporting mainly right-wing militarist agitation. After 1945 universities in the developing countries often fostered strong nationalist and Marxist movements, while in the 1960s left-wing movements were also predominant in many European, US, and Japanese universities and colleges. Protests at the University of California's Berkeley campus in 1964, and the Kent State shootings*, with a nation-wide strike at some 200 US campuses, in 1970, successfully challenged US policy over Vietnam. In Paris in 1968 French students joined workers in the movement to challenge the de Gaulle regime, while in Japan students acted militantly against the Westernization of Japanese society and the continued US military presence. Students in Hungary, Poland, and Czechoslovakia were often in the forefront of protests against their countries' authoritarian regimes, while demonstrations by South Korean university students in 1987 led to constitutional amendments and the release of political prisoners. A student hunger-strike triggered the tragedy of Tiananmen Square*, while students were everywhere active in the East European revolutions of 1989. In South Africa students as young as 10 have been actively involved in the protests against apartheid and compulsory Afrikaans, as at Soweto* in 1976.

Sudan, the. In 1898 Kitchener had avenged the death of Gordon (1885) at the Battle of Omdurman and the huge area of the Upper Nile became linked with Egypt, which Britain already controlled, as the Anglo-Egyptian Sudan. This 'condominium' was to be administered by a British Governor-General and British and Egyptian administrative officers. Under condominium rule the Sudanese economy expanded and mission schools, especially in the south, flourished. The University of Khartoum (originally the Gordon Memorial College, founded 1902, and the Kitchener School of Medicine) began to produce a generation of students anxious for political independence. During the 1920s, Civil Service posts were increasingly allocated to Sudanese, generally from northern Sudan, where Islam was dominant. By 1945 two groups were demanding rapid constitutional development, the Ummah Party, headed by the grandson of the Mahdi, which stood for immediate independence, and the Ashiqqa (Brothers), allied to the Khatmiyyah sect which favoured constitutional links with Egypt. The Muslim Brotherhood* grew rapidly at this time, but entirely in the north. The south Sudan, which accounted for possibly a third of the population, was animist or Christian and largely politically unconscious. In 1948 Britain accepted the principle of Sudanese autonomy, but in 1951 Farouk* proclaimed himself King of the Sudan. After his fall, Egypt accepted the Sudan's right to self-determination. Self-government was granted in 1953, despite objections from the south, and full independence was declared unilaterally in 1955, with Ismail al-Azhari* as first Sudanese Prime Minister. In 1958 there was a military coup, and the parliamentary system was replaced by a military dictatorship under General Ibrahim Abboud*, who concluded a Nile Waters Agreement in 1959. Abboud's policies of imposing Arabic on the south and the expulsion of missionaries led to civil strife. Abboud resigned in 1964 but disorder continued, and in 1969 a new coup brought Colonel Jaafar al-Nemieri* to power. He declared the Sudan a one-party state, the Sudan Socialist Union. In 1972, following negotiations in Addis Ababa, unrest in the south eased, but the Sudanese Civil War* resumed in the 1980s, bringing great suffering to millions. Nemieri was deposed in April 1985, and a transitional constitution restored civilian rule in 1986, under Prime Minister Sadiq al-Mahdi, leader of the Ummah Party. In July 1989 continued civil strife in the south, a failure of the economy, and new allegations of corruption resulted in yet another military coup, led by General Omar Hassan Ahmad al-Bashir. The Sudan under his rule was reported to be a 'human rights disaster', while famine and civil war continued to rack the south.

Sudanese Civil War. The policy of the Sudanese army which seized power in 1958 was, in the name of national unity, to encourage the spread of Arabic and Islam at the expense of English and Christianity. In October 1962 there was a widespread strike in southern Sudanese schools, and this was followed in 1963 by the emergence of a guerrilla movement and of armed rebellion. Conflict continued after the resignation of General Ibrahim Abboud*, with increasing tension between the Sudanese Communist Party and the traditionalist Islamic party, the Ummah. Colonel Nemieri* negotiated the Addis Ababa agreement 1971 with southern rebels, granting wide regional autonomy. In the north, however, continued influence of the Muslim Brotherhood* resulted in a further strengthening of Islamic law. This produced renewed conflict in the 1980s, southern guerrilla forces claiming that Nemieri had reneged. The Sudan People's Liberation Army militarized much of the south, which resulted in dislocation of the economy; this, together with drought, precipitated famine. After Nemieri was overthrown in 1985 negotiations between the new Prime Minister Sadiq al-Mahdi and the SPLA broke down over the basic issue of the imposition of Islamic law (shariah) on the south. The Sudan's economy continued to deteriorate, there were international efforts to restore peace and provide aid, and in 1989 a new military coup.

Sudetenland, the north-west frontier region of Czechoslovakia. It had attracted German settlers for centuries and by 1914 had become highly industrialized. In 1919 it was transferred from the Austro-Hungarian Empire to the new state of Czechoslovakia. It seems that most of the population was content with this arrangement, but in 1935 a Sudeten German Party was formed, led by the Nazi Konrad Henlein (1898–1945) and financed by Berlin. Henlein whipped up opinion, especially after the Austrian Anschluss*, and in September 1938 the Munich Agreement* agreed that there should be 'an orderly taking over of the Sudetenland by Germany'. This immediately followed in October, when it was annexed into the Third Reich. In 1945 Czechoslovakia regained the territory and by the Potsdam Agreement* was authorized to expel some 3.3 million German-speaking inhabitants.

Suez Canal. The 106-mile long canal had been opened in 1869, the largest share-holding (40 per cent) being acquired by Britain in 1875. British troops had been stationed along it in 1882, and the Constantinople Convention of 1888 provided that the canal be 'free and open' in 'time of war as in time of peace'. In spite of this Britain closed it to enemy ships in 1914 and again in 1939. In 1948 Egyptian authorities refused access to Israeli shipping. British troops left the Canal Zone in June 1956, whereupon Egypt nationalized the Suez Canal Company, precipitating the Suez Crisis*, during which ships were sunk, closing the canal until April 1957. It was again closed in 1967, when Israeli tank units occupied the western bank, remaining there until 1974. It was again cleared and reopened in June 1975; by that time many giant oil-tankers were too large to use it.

Suez Crisis (July–November 1956). The Suez Canal Company was nationalized by Egypt on 26 July 1956, its shares having been largely held by British and

French investors. President Gamal Nasser's* justification was that he needed to finance the Aswan Dam Project* from canal dues. Britain and France immediately began secret military discussions to invade Egypt and depose the President. Having failed, however, to win US support for this venture, Britain abandoned the idea. Instead, it agreed to a secret plan between France and Israel, whereby the latter would invade Egypt and a combined Anglo-French force be launched to 'restore order'. Claiming Egyptian provocation, the Israeli invasion began on 29 October, and French and British forces from Malta and Cyprus landed at Port Said and Port Faud on 5 and 6 November, to occupy the Canal Zone. A collapse of sterling and strong opposition from the USA, however, resulted in the operation being halted the following day. The UN, with US and Soviet support, threatened intervention, and the Anglo-French forces withdrew in December. A UN peace-keeping force was sent to the area and Israeli troops were evacuated in March 1957. The crisis resulted in increased prestige for Egypt, with President Nasser relying more on aid from the Soviet Union. British and French influence in the Middle East suffered severely, as did the reputation of the British Prime Minister Anthony Eden*.

Suharto, Raden (1921–   ), President of Indonesia 1968–   . Having trained as a bank clerk, he joined the Dutch colonial army, but in 1942 switched to a Java Defence Force organized by the Japanese, by whom he was commissioned. He distinguished himself in guerrilla fighting against the Dutch 1946–9. Strongly anti-Communist, he had reached the rank of major-general and head of strategic command in 1965, when the army staged a counter-coup against the PKI*. In power from 1966, he became officially President in 1968, since when he has been re-elected four times. He ended Sukarno's Konfrontasi* against Malaysia, and his policies have been to work closely with the UN and the USA in order to establish Indonesia as a leading power in South-East Asia. Increasingly dictatorial, he has faced considerable domestic opposition, as well as problems arising from the Islamic fundamentalist* movement.

Sukarno, Achmad (1901–70), first President of Indonesia 1949–68. Son of poor schoolteachers, he was to be known as Bung Karno ('The Leader') in later years. Brought up a Muslim, as a student he showed great linguistic ability—Arabic, Dutch, and the languages of Javanese, Sundanese, and Balinese, which were to form the basis of modern Indonesian. He studied engineering in Bandung, where he became involved in the foundation of the PNI (Partai Nasionalis Indonesia) in 1927. He was imprisoned 1929–31 and then exiled to Sumatra in 1933, from where he welcomed the Japanese in 1942, becoming an adviser. He became increasingly disillusioned by Japanese rule and on 17 August 1945 declared Indonesian independence. Four years of struggle with the Dutch followed, before he returned to Jakarta in December 1949 as President, residing in the splendid palace of the Governor-General, and seeking to preside over a process that he described as 'finding a national identity'. The Bandung Conference* of 1955 was perhaps his greatest achievement. Economic problems followed this and he became increasingly autocratic; as inflation rose there were riots in both 1957 and 1958. In 1963 he embarked on the Konfrontasi* against what he called an 'imperialist plot of encirclement'. In September 1965 a group of Communist conspirators kidnapped and murdered six army generals. General Suharto* used this as a pretext to move against the Partai Kommunis Indonesia* (PKI), perhaps 300,000 being slaughtered. A power-struggle developed, with Sukarno suspected of implication in the September coup. He lost, and effective power had passed to Suharto by March 1967, although Sukarno retained the title of president until 1968.

Summit conferences. During the 1950s there were repeated calls for a resumption of the wartime practice of meetings of heads of government of the four victorious powers, the USA, the Soviet Union, Britain, and France, to resolve international problems. There was one such conference in Geneva in 1955, which made some progress towards reducing Cold War tensions, and another in Paris in 1960, attended by Eisenhower, Khrushchev, Macmillan, and de Gaulle, which did not. After this there were numerous bilateral meetings between heads of states from both Eastern and Western blocs. Kennedy had a disastrous meeting in Vienna in June 1961. There were further presidential 'summit' meetings between the USA and USSR in 1972 and 1979; since 1987 these have become increasingly frequent.

Sun Yixian (Sun Yat-sen) (1866–1925), founder of the Guomindang*. Son of a peasant in south China near Macao, he was educated by missionaries and was for several years a US resident, training as a doctor in Hong Kong. In 1895 he organized an unsuccessful rising against the Qing dynasty in China and fled the country. Briefly imprisoned in 1896 in the Chinese legation in London, his release was negotiated by the British government, and he went to Tokyo, where he formed a revolutionary society, the Tongmenghui ('United League'), based on the three principles of nationalism, democracy, and socialism, which became the nucleus of the Guomindang in 1912. When the Chinese Revolution broke out (1911) he returned from the USA, and an assembly in Nanjing (Nanking) elected him President of 'the United Provinces of China'. After a few months he resigned in favour of Yuan Shikai* in the forlorn hope that the latter could unite China as a republic. When Yuan suppressed the Guomindang in 1913, he fled to Japan and later set up a secessionist government in Guangzhou (Canton). In 1923 he modified his teachings in order to accommodate Communists, and agreed to accept Soviet help in reorganizing the Guomindang, thus inaugurating a period of uneasy co-operation with the Chinese Communist Party. He died of cancer in Beijing, trying to negotiate a unified Chinese government. Both Chinese Nationalists of Taiwan and Chinese Communists regard him as the founder of modern China.

Supply-side economics, a fashionable economic theory in the 1960s and 1970s in Western capitalism. It repudiated the theory of J. M. Keynes* that it is demand which 'is the lever that controls the economy'. Instead it stresses the need to emphasize production of goods and services (supply) at given prices, to provide incentives to encourage savings and investment, and to discourage enforced leisure, that is, unemployment. This can all best be done by reduction of tax rates, which will stimulate purchases and sales, increase output and employment, and also reduce inflation. Monetarists* argued against the last proposition. In 1961 the Kennedy administration reduced taxation and there was an expansion of real economic activity. It was claimed that this confirmed the supply-side theory. When the Reagan and Thatcher administrations both introduced income-tax cuts in the early 1980s, however, there was little evidence that these produced all that they were supposed to.

Supreme Court (USA). This is established by Article III of the US Constitution, its members being appointed by the President, with advice and consent of the Senate. Early in its history it established its right to judge whether Bills passed by Congress or by state legislatures conform to the provisions of the Constitution, with power to declare them unconstitutional if they do not; but it can only do so when specific cases are referred to it. During the early 19th century it established itself as also the highest Court of Appeal. The decisions of the Court have played a central role in the history of the USA, not only balancing the relationships between executive and legislature, and between states and the federal government, but also contributing to the evolution of social, economic, and legal policies. The Commerce clause of the Constitution has enabled it powerfully to influence the economy by invalidation of any state legislation deemed likely 'unduly to burden interstate commerce'; while its interpretation of the Fourteenth Amendment of the Constitution has enabled racial discrimination steadily to be eliminated. Judges hold office 'during good behaviour', but since 1937 have been entitled to retire at 70. In that year President Roosevelt sought to reorganize and expand the Court from nine to fifteen. His additional younger judges would have counter-balanced many sitting conservatives, who were over 70 and declined to retire. He was accused of attempting to 'pack' the Court and was defeated. In 1990 President Bush's appointment of Judge David Sowter, on the retirement of the liberal Judge William Brennan, was expected to shift the balance of the Court to the conservative right.

Surinam, a republic on the northern coast of South America, known until 1948 as Dutch Guiana. The territory alternated between British and Dutch control until 1815, when it was assigned to The Netherlands. The emancipation of slaves led to plantation labour being recruited from Java and from India, the resulting ethnic diversity causing increasing racial tension in the early 20th century. Although the sugar plantations declined, the discovery of bauxite deposits boosted the economy after World War I. With The Netherlands occupied by Germany 1940–5 there were increasing demands for an end to colonial status by the end of World War II. Universal suffrage was granted in 1948 and the Nationale Partij Suriname (NPS) emerged. In 1954 the country became 'an equal partner in the Kingdom of the Netherlands' and a second party emerged, the Verenigde Hindoostaanse Partij (VHP). A coalition between the NPS and VHP provided a period of stability through the 1960s. In the early 1970s there was a rise in unemployment to 35 per cent, coupled with demands for independence. Forty thousand of the wealthier Creoles, out of a population of some half-million, emigrated to The Netherlands, which granted independence in November 1975, together with a package of 3,500m. guilders of aid. A confused period of strikes, corruption, and political power-struggles culminated in a revolt by non-commissioned officers of the army in January 1980, led by Sergeant-Major Desire Bouterse (later lieutenant-colonel). Bouterse spent a decade trying to maintain some sort of balance in a very confused situation. Dutch aid ended, subject to a restoration of democracy; the principal bauxite mines and smelting plants were closed; and a guerrilla movement demanding a restoration of democracy staged a series of hit and run attacks. A 'Front for Democracy and Development' was formed, and in September 1987 a draft constitution was approved by a 93 per cent referendum vote. There were elections in November 1987 and the country returned to democracy in January 1988, when President Ramsewak Shankar was installed. In July 1989 he reached an agreement with the guerrilla Jungle Commando, many of whose members operated from French Guyane. But a confused and violent situation remained.

Surrealism, an artistic term which began as a literary movement in France. During the early 1920s its progenitor André Breton, editor of *Littérature*, proclaimed that 'psychic automatism', which facilitated 'automatic' or 'stream of consciousness' writing, together with the magical, the irrational, and the hallucinatory, were the true means by which to explore the human psyche. Painters such as Arp, Miró, Magritte, and above all Dali adopted the credo of Surrealism. Their vision in turn was powerfully to influence the theatre and cinema, but also the whole apparatus of late 20th-century advertising techniques.

Suzuki, Kantaro (1867–1948), Prime Minister of Japan 1945. Graduating from the Japanese Naval Academy, he distinguished himself in the Sino-Japanese War 1894–95, the Russo-Japanese War 1904–5, and in World War I, reaching the rank of admiral in 1923. Appointed a Privy Councillor, he was seriously wounded in the February 1936 military rising. In April 1945 he was called upon by Emperor Hirohito to form the government which negotiated Japan's surrender, after which he resigned.

Suzuki Zenko (1911–), Prime Minister of Japan 1980–2. Son of a fisherman, he created a successful fishing business, becoming director of the Japan Fisheries Club. In 1947 he entered politics as a socialist, but joined the Liberal Democratic Party in 1959, where he displayed great skill as a mediator. On the sudden death of Ohira Masayoshi he became president of his party and Prime Minister, dedicated to the 'politics of harmony'. In 1982, however, he relinquished office while remaining a member of the Diet.

Svoboda, Ludovik (1895–1979), President of Czechoslovakia 1968–75. Born near Bratislava, he fought in World War I and remained in the new Czechoslovak army in 1919. By 1939 he was a general and escaped to the Soviet Union, where he commanded a Czech Army Group which in 1945 liberated Prague. Stalin feared his prestige, and he was imprisoned and then worked on a collective farm. He was reinstated by Khrushchev* and proclaimed a Hero of the Czechoslovak Republic. In March 1968 he replaced Novotný* as President. He endeavoured to support Dubček's* reforms; when these were prevented by Moscow, he tried to persuade the latter to adopt a conciliatory line. He remained a much-respected figure until his retirement in 1975.

Swahili, a language and culture found through much of Tanzania, Kenya, and along the east coast of Africa as far as Mozambique. The peoples of these lands interbred with Arabic Muslim traders navigating the coast over the centuries. The Bantu language Kiswahili which emerged, with its many Arabic words, is spoken throughout East Africa and into Zaïre.

SWAPO *see* SOUTH WEST AFRICA PEOPLE'S ORGANIZATION.

Swaraj Party ('self-ruling'), a political party in British India. It was formed in 1922 by C. R. Das, Motilal Nehru*, and Vallabhbhai Patel* to win control of the legislative councils for the Indian National Congress*, and then paralyse their working by obstructionist tactics. In 1923 the Swarajists won a majority in the central legislature and in several provinces. However, the policy of obstruction was not generally successful, and the party was weakened by the resignation of members who felt that Congress was not sufficiently safeguarding Hindu interests. The party came to an end in 1929, but was briefly revived for the 1937 elections.

Swaziland, a small land-locked monarchical state in southern Africa. It takes its name from the Swazis who occupied it in the mid-18th century. It came under British rule in 1902 as a High Commission Territory, following the end of the Boer War, retaining its monarchy. In 1968 it became a fully independent kingdom under Sobhuza II, King of the Swazi 1921–82. Revisions of the Constitution in 1973 and again in 1976 have given the monarchy very wide powers, King Mswati III being crowned in April 1986. The country's economy relies on co-operation with South Africa. Nevertheless, it has associated itself with demands for the ending of apartheid and is a member of SADCC*.

Sweden. Sweden's constitutional monarchy is descended from a Napoleonic general, Jean-Baptiste Bernadotte, who became Crown Prince in 1810. Until 1905 it also ruled Norway. The policy of neutrality began during the 19th century and was maintained through both world wars, during which it hosted a number of abortive peace missions. Its governments in the 20th century have been from either the Liberals or the Social Democrats, the latter being in power almost continuously since 1932. A coalition government maintained strict neutrality during World War I, universal suffrage and an eight hour day being introduced in 1917. A series of social reforms was introduced by the Social Democrats in 1924, the Liberals at that time being split by a prohibition debate. During the 1930s a National Defence Commission recommended the expansion of the navy and the air force, in order to maintain neutrality. This did not however prevent Sweden from having to give Germany transit facilities for men and materials 1940–3. After World War II the already extensive social welfare provision was further extended, particularly under Tage Erlander*.

Sweden has taken its membership of the League of Nations and the UN conscientiously, accepting the former's decision over the Åland Islands*. After 1945 it did much to provide immediate relief and reconstruction aid to war-ravaged Europe. It has provided troops and observers for a number of UN peace-keeping missions, for example Egypt, Cyprus, the Congo, Namibia; and it provided the UN with perhaps its most charismatic secretary-general in Dag Hammarskjöld*. In 1952 Sweden helped to form the Nordic Council* and in 1959 hosted the conference from which EFTA emerged. Politically the long Social Democrat hegemony came under challenge during the 1970s. There was a Conservative government 1976–82, before the Social Democrats regained office under Olof Palme*. The continued failure, after three separate parliamentary commissions, to identify the latter's assassin became an increasing political scandal.

Switzerland. Originally a federation of twenty-two cantons, its Constitution basically goes back to 1848. Despite linguistic ties to both Germany and France, it managed to preserve its neutrality through both world wars and it has achieved one of the highest standards of living of any European state. The canton of Jura was added to the Federation in 1979 and three other cantons split, so that it is sometimes now referred to as consisting of twenty-six members. The cantons are all self-governing, but elect members to a federal Assembly, composed of a National Council and a Council of States. The Assembly is responsible for defence, the Civil Service, and the law, and it elects a Supreme Court; Councils have tended to be coalitions of Christian Democrats and Radical Democrats. Women were not allowed to vote on a federal basis until 1971, and suffrage remains restricted in some cantons. Because of its deep tradition of neutrality the International Red Cross was housed in Geneva, as was the Palace of the Nations for the League of Nations; the city has continued to host many UN activities and international conferences.

Sykes–Picot Agreement (May 1916), a secret agreement negotiated between two Middle East diplomats, Sir Mark Sykes and François Georges-Picot. When the Ottoman Empire was defeated France would be pre-eminent in Syria (including Lebanon), southern Anatolia, and northern Mesopotamia (Mosul). Britain would establish protectorates in southern Mesopotamia (Baghdad and Basra), the Persian Gulf, Arabia and the Hejaz, Palestine, and the Jordan Valley. Thus Egypt would be linked with the British Indian Empire. Russia was to have a free hand in Armenia and northern Kurdistan*. A copy of the agreement was published by the Bolsheviks after the Russian Revolution, causing international dismay and Arab anger. It was to form the basis of the League of Nations settlement for the Middle East in 1920, in spite of earlier undertakings made by the British High Commissioner Henry McMahon to Sherif

Hussein ibn Ali* at the time of the Arab Revolt. In the event Britain was to gain control of the oil-rich region of Mosul.

Syndicalism, a movement among industrial workers to abolish capitalism and replace it with groups of workers, or syndicates, which would control 'units of production'. It was influential in the early 20th century, putting forward its basic claims in the Charter of Amiens in 1906 and being powerfully influenced by the writings of Georges Sorel, whose *Réflexions sur la violence* followed in 1908. Syndicalists differed from socialists in believing essentially in anarchism. Advocating direct action through the general strike, they were active in France, Spain, Italy, Russia, and the USA. The growing complexity of industrial organization and the attraction of Communism after World War I reduced the influence of syndicalism after 1918.

Syria. This ancient land had been a province of the Ottoman Empire 1516–1918, although French interest had begun during the 19th century. Turkish troops were defeated in September 1918 at the Battle of Megiddo, whereupon its capital Damascus was entered on 1 October by Allied troops under General Allenby* and Arab troops under Emir Faisal. The latter was proclaimed King of Syria. France claimed the mandate for Syria and this was accepted by the League of Nations. French troops drove out Faisal in July 1920, and he became King Faisal I* of Iraq instead. French rule was never popular, and there were several uprisings, including that of the Syrian Druzes* 1925–7. In July 1941 British and Free French forces moved into Syria against Vichy France, anxious to prevent German air bases being constructed there. Syrian independence was proclaimed 1 January 1944, but French troops were reluctant to leave, only doing so in April 1946. Syrian troops took part in the Arab–Jewish fighting of 1948, but a boundary with Israel was agreed at the armistice of 1949. In that year there were three military coups, and Colonel Shishkali established himself, surviving until 1954, when a presidential republic was restored. In 1956 relations with the Soviet Union were improved, and Syria took part in the Suez War, having signed a pact for a joint military command with Jordan and Egypt. In 1958 the short-lived UAR* was formed with Egypt, by which time the Ba'ath* Party was in the ascendant. The link with Egypt was ended in 1961 and two years later the Ba'ath Socialist Party came to power, as the Arab Socialist Renaissance Party. In 1966 its radical wing established Hafiz al-Asad* in power, and the party's moderate founder Michel Aflaq* left the country. In 1971 Asad was named President. In 1973 Syria adopted a new constitution, as the Syrian Arab Republic, 'a democratic, popular, socialist state'. Syrian troops were twice defeated by Israel, 1967 and 1973, and have become more and more deeply involved in Lebanon, having occupied much of the country in 1976. Syria opposed the Camp David Accord* (1978) and has remained closely allied with the Soviet Union. In 1990 Syrian troops imposed the Taif Accord* in Lebanon.

t

Taft–Hartley Act (USA). Passed by Congress in June 1947, over the veto of President Truman, this act banned the closed shop in US labour unions and the secondary boycott, allowed employers to sue unions for breach of contract and for damages inflicted on them through strikes, empowered the President to order a sixty-day 'cooling-off period' before strike action, and required union leaders to take oaths stating that they were not Communists. Despite protests from the unions, it has remained relatively unchanged.

Tagore, Rabindranath (1861–1941), Indian writer. Born in Calcutta, son of Devendranath Tagore, he has been described as 'the greatest Bengali of all time'. Poet, playwright, philosopher, musician, storyteller, he championed the Bengali language, its art and music and way of life, and had a profound effect on 20th-century Indian nationalism. After publication of the English translation of his Bengali poems *Gitanjali Song Offerings* (1913), Tagore became briefly fashionable in Europe. However, although he is hardly read any longer in the West, his writings continued to be widely read in his native Bengal.

Taif Accord. When President Amin Jumayyil* ended his term of office in Lebanon in September 1988, the country was without a government. The Maronite Commander-in-Chief of the Lebanese army, General Michel Aoun, established himself in the presidential palace and launched a strong attack against Syrian forces in Beirut. In October 1989 a peace plan was put forward by a 'troika' of the Arab League, King Fahd of Saudi Arabia, King Hassan of Morocco, and President Chadli of Algeria. The seventy surviving MPs from the 1972 election assembled in Taif in Saudi Arabia, where they reached an agreement by which the Maronite Christian domination in government would be reduced. The President would become subject to a cabinet equally divided between Christian and Muslim, and to a National Assembly also equally divided. The Accord was criticized by both the Druzes* and the Shia Amal Party and rejected by General Aoun. The MPs duly elected a Maronite, René Mouawad, as President.

Taisho era, the period in Japan covering the reign of the Emperor Yoshihito 1912–26. The Emperor took no part in politics, but his reign saw significant developments. At home it was a period of modest political liberalism, with a Reform Act of 1919 doubling the electorate and universal male suffrage being granted in 1925. The startling advance of Japan's economy continued in spite of some industrial unrest. In foreign affairs Japan's role in World War I greatly extended its influence in China, while its participation in the Russian War of Intervention established an antagonism with the Soviet Union. In Korea* a protest revolt against occupation was ruthlessly suppressed. The term is derived from the Emperor's full name Taishō Tennō Yoshihito.

Taiwan (Republic of China), previously known by its Portuguese name of Formosa. It was occupied by Japan as a result of the Treaty of Shimonoseki in 1895 and remained under Japanese control until 1945, when it was occupied by Nationalist forces of Jiang Jiehi* (Chiang Kai-shek) in September. Resentment at the administration of Governor Chen Yi produced a revolt, which had to be suppressed by force of arms. When the Chinese Civil War began to turn against the Nationalists in 1948, arrangements were made to transfer the National government to Taiwan, and by 1950 almost 2 million refugees from the mainland had arrived. Supported militarily by the USA, Taiwan maintained its independence from Communist China, as the Republic of China, and, until expelled in 1971, sat as the sole representative of the Chinese people in the UN. Jiang remained President until his death in 1975, when he was succeeded by Yan Jiangan* and then by Chiang Ching-kuo* (1978–88). Since 1950, with the support of US capital, it has undergone remarkable industrialization. In March 1990 President Lee Teng-hui was re-elected.

Tamils, a Hindu people originating in southern India who speak a Dravidian language. They are concentrated in the state of Tamil Nadu (Madras), but Tamil culture is found in other Indian states and in Sri Lanka*, where the so-called Indian Tamils have lived for centuries. Until recently they regarded the so-called Sri Lankan Tamils, descended from labourers imported in the 19th century to work British tea plantations, as low-caste foreigners; but with the development of tension and civil disorder the two Tamil groups have co-operated against the Sinhalese, demanding autonomy for the Tamil northern and eastern provinces of the island. The presence of Indian peace-keeping forces 1987–9 did not prevent continued acts of terrorism.

Tanaka Kakuei (1918–), Prime Minister of Japan 1972–4. A graduate of technical school in Tokyo, he became a successful designer and founded the Tanaka Construction Company, before entering politics in 1947. His career was briefly interrupted by a bribery scandal soon after, but he remained a member of the Diet and in 1957 became Minister of Communications. He went on to be Minister of Finance in three different cabinets and then Minister of International Trade. In 1972 he became Japan's youngest Prime Minister, but had to resign in December 1974 and in 1976 to face accusations for responsibility for the Lockheed scandal*. Despite lengthy legal proceedings he remained a powerful force within the ruling Liberal Democratic Party* until disabled by a severe stroke in 1985.

Tangier. By 1914 France and Spain had succeeded in establishing protectorates in Morocco*, the port of Tangier having been declared an international zone by the Treaty of Algeçiras 1906. In 1923, at a time of considerable unrest in Morocco, it was declared a 'free port'. It was occupied by Spanish troops 1940–5, and united to Morocco on independence in 1956. In 1962 it was again declared a 'free port', while remaining an integral part of the Kingdom of Morocco.

Tannenberg, Battle of, World War I (26–30 August 1914). When war broke out in 1914 the German Schlieffen Plan* was to defeat France before Russia had time fully to mobilize. In fact the Russians moved more quickly than expected, sending two armies under Generals Rennenkampf and Samsonov into East Prussia. The German commander Prittwitz proved ineffective, and Hindenburg* and Ludendorff* were sent east, the German offensive in the west losing impetus. Co-operation between the Russian armies was poor, they were separated by the Masurian Lakes, and the Germans succeeded in intercepting radio messages. Under a plan devised by Colonel Max Hoffmann, Samsonov's army was surrounded on 26 August near Uzdowo; he lost some

100,000 prisoners and shot himself. In September Rennenkampf was defeated by Hindenburg. In the spring of 1915 Ludendorff resumed the offensive and, taking advantage of frozen swamps, captured another four Russian divisions in the region of the Masurian Lakes. These victories effectively halted Russian aims to conquer East Prussia.

Tan-Zam railway. Its construction was formally agreed between Zambia, Tanzania, and China in August 1970. Its main purpose was to free Zambia from dependence on South Africa for rail transport of its vital copper and other exports. China provided an interest-free loan of £169m. together with technical aid. It is 1,860 km. (1,162 miles) long and was completed in October 1975, one year ahead of schedule. It links Dar es Salaam with Lusaka in Zambia.

Tanzania. A German colony from the late 19th century, Tanganyika became a British mandate after World War I and a UN Trust Territory administered by Britain after World War II. It became independent in 1961, followed by Zanzibar* in 1963. The two countries united in 1964 to become the United Republic of Tanzania, under its first President Julius Nyerere*. In the Arusha Declaration* of 1967 Nyerere stated his policy of equality and independence for his country. Recurring economic problems have caused some political tension, as did Tanzania's involvement in war (1979) in Uganda. Under Nyerere Tanzania gave strong support to political exiles from neighbouring states struggling for independence, such as Zimbabwe, Angola, and Namibia. On his retirement in 1985 President Nyerere was succeeded by President Ndugu Ali Hassan Mwinyi; Nyerere however remained chairman of TANU (Tanganyika African National Union) in this one-party state.

Tariff reform (UK). Joseph Chamberlain (1835–1914) believed that 19th-century free trade had to end and that tariffs would strengthen links within the Empire by a policy of imperial preference (the application of lower rates of duty between member countries). He was strengthened in this view by the Boer War. His campaign, however, divided the Conservatives, defeated in Parliament in 1905, and was rejected by the Liberals. Tariff reform was rejected again in 1923, when Stanley Baldwin and the Conservatives failed to secure a majority in Parliament in an election primarily on that issue. However, the shock caused by the international financial crisis of 1929–31, and the intensification of nationalist political and economic rivalries, made Britain's free trade policy seem even more of an anachronism. Protectionism was adopted by MacDonald's National government by the Ottawa Agreement* of 1932. After World War II GATT* was to be established by the UN to try to reduce the levels of international protectionism, but it had only limited effect. Imperial preference was ended when Britain entered the European Common Market in 1973.

Taruc, Luis (1913–   ), leader of the Philippines Hukbalahap* 1942–54. Son of poor peasants, he was educated at Manila University and in the 1930s took up the cause of landless peasants at the time of the Great Depression*, which adversely affected the economy of the islands. In 1935 he joined the Socialist Party and the Philippines Anti-Fascist Front. In 1942 he formed the Hukbalahap on Luzon, becoming its Commander-in-Chief. By 1945 his forces had almost total control of the rural areas of the island, but in 1946, although elected to the liberated House of Representatives, he was banned from his seat by US pressure. After failure in negotiations with President Manuel Roxas*, he went underground and led the Huk movement until its collapse in 1954. Imprisoned, he was released in 1968.

Tata family, an Indian Parsee* commercial and industrial family. It is one of the two (with the Birla family*) most important merchant families in modern India. It began in Far East trade, but diversified operations under Jahangir Ratanji Dadabhai Tata (1904–), creating an airline which became Air India. The Tatas have spent extensively on education as sponsors of research, founding the Tata Institute of Fundamental Research in Bombay.

Tawney, Richard Henry (1880–1962), British economic historian and social critic. Born in India, he was educated at Rugby School and Balliol College, Oxford. He worked at Toynbee Hall* and then for the Worker's Education Association, whose president he was 1928–44. In 1913 he began teaching at the London School of Economics, where he remained all his professional life, becoming Professor of Economic History in 1931. His academic work was mostly in the field of 16th- and 17th-century economics, for example *Religion and the Rise of Capitalism* (1926). Politically he twice failed to win election to Parliament, but he had an immense influence on generations of liberally minded students, especially urging expansion of educational provision. In *The Acquisitive Society* (1920) he argued that material acquisitiveness was morally wrong, corrupting both rich and poor. His book *Equality* (1929) had a powerful influence on social thought thereafter, particularly in, but not confined to, the Labour Party.

Teamsters' Union, the largest labour union in the USA. Formed in 1903, its full title is the International Brotherhood of Teamsters, Chauffeurs, Warehousemen and Helpers in America. In the late 1950s its president David Beck was indicted for having links with criminals, and the union was expelled from the AFL-CIO. Congress then passed the Landrum–Griffin Act, giving the Secretary of Labor considerable powers over the finances of unions. Beck's successor James R. Hoffa was found guilty of trying to influence a federal jury, while on trial in 1964 for misusing union funds, whereupon the Teamsters sought to recover their reputation by co-operating closely with such responsible union leaders as Walter Reuther* of the United Automobile Workers. The sheer size and wealth of the union, with its ability to paralyse interstate commerce, has given it great bargaining power within the USA.

Teapot Dome scandal (USA). When Warren G. Harding, who had been a Senator for Ohio, was elected President of the USA in 1920, he brought into his administration (1921–3) many of his self-seeking friends, who became known as the 'Ohio gang'. Harding transferred the management of the US navy's oil reserves at Teapot Dome, Wy., from the Secretary of the Navy to the Secretary of the Interior, his old friend Albert B. Fall, who proceeded to siphon oil reserves at Wyoming into the Mammoth Oil Company, and other reserves at Elk Hills, Calif., into the Pan-American Petroleum and Transportation Company. Harding, who was in no way implicated, died before the full extent of scandal and the involvement of Fall was exposed by Senator Thomas J. Walsh during the years 1922–4. Fall was found guilty of accepting a $100,000 bribe and imprisoned 1929–32.

Tedder, Arthur William, 1st Baron Tedder of Glenguin (1890–1967), marshal of the RAF. Born in Scotland, he was educated at Whitgift School, Croydon, and at Cambridge. He joined the colonial service and then the army. In 1916 he transferred to the Royal Flying Corps, which in 1918 became the RAF, with which he served all his professional life. By 1936 he was RAF Commander in the Far East. He was then transferred to the Middle East, and worked closely with Allied troops in the Sicily and Italy landings. In 1944 he became deputy to General Eisenhower* and his bombing tactics at the Normandy landings contributed greatly to their success. During the war he attended summit meetings with Winston Churchill, and on 8 May 1945 he was in Berlin where, with Marshal Zhukov*, he signed the German Instrument of Surrender. He was Chief of the Air Staff 1946–9 and served with NATO until 1951. He was described as both diplomatic and unobtrusive, with 'an unrivalled power to manage contemporaries, both military and civil'. He became Chancellor of the University of Cambridge and held various offices of distinction in retirement.

Tehran Conference (28 November–1 December 1943), a meeting between Churchill, Roosevelt, and Stalin in the Iranian capital. Here Stalin, invited for the first time to an inter-Allied conference, was told of the plans for a second front in France in the summer of 1944, to coincide with a resumed Soviet offensive against Germany. The three leaders discussed the establishment of the UN after the war, and Stalin pressed for a future Soviet sphere of influence in the Baltic states and Eastern Europe, while guaranteeing the independence of Iran.

Telecommunications, a term that became fashionable during the 1970s to describe the increasingly sophisticated communications technology. The key invention which had shrunk the world had been the electric telegraph in 1837, which fundamentally altered business and banking, as well as the conduct of international affairs. Wireless telegraphy followed in 1897, with the first transatlantic radio signal in 1901. Commercial radio began in 1920, and the first teleprinter was invented in 1924. A drum-scanner facsimile telegraph was invented in 1934, commercial television began in 1936, and videotaping was used in television in 1958. In 1962 the US Telstar satellite made instantaneous world-wide television possible. During the 1980s electronic facsimile machines (FAX) became an office commonplace, using satellite-linked telephones. The latter, using laser technology, were developing optical fibre transmission, while the cordless telephone became a commonplace in home and motor car.

Teleki, Pál, Count (1879–1941), Prime Minister of Hungary 1939–41. Born in Budapest, he had a distinguished career as a geographer at the university before being elected to the Hungarian Parliament in 1909. He attended the Paris Peace Conference* but then returned to academic life. In February 1939 he was persuaded to re-enter politics and become Prime Minister under Admiral Horthy*. He managed to persuade the latter to disband more extreme Fascist groups, but allowed anti-Semitic laws to stand. He believed that the only way to regain territory lost by Hungary at the Treaty of Trianon* was to negotiate through Hitler. In this he was successful, via the two Vienna Awards*. He had negotiated a treaty of friendship with Yugoslavia, and when German troops invaded the latter in April 1941 Hitler demanded Hungary's support. Teleki committed suicide.

Temple, William (1881–1944), Archbishop of Canterbury. Son of a former Archbishop of Canterbury, he taught at Oxford before becoming ordained a priest in 1909. He was a headmaster, and then successively Bishop of Manchester and Archbishop of York, before going to Canterbury 1942–4. The achievement of greater equality of opportunity in the educational system, springing in part from his early activities with the Workers' Educational Association, was an important objective for him. He worked closely with R. A. Butler* on his Education Bill, which became law in 1944. He also sought a greater sense of common purpose between different religious denominations, and helped to found the World Council of Churches. He was sympathetic to the Labour Party and a strong supporter of the Beveridge Report*.

Templer, Sir Gerald (1898–1979), British field marshal. From an Irish family, he was educated at Wellington College and Sandhurst, and joined the Irish Fusiliers in 1916, serving in France. After wide service overseas between the wars, he commanded the 1st and 56th Divisions in North Africa and then the 6th Armoured Division in Italy during World War II. After the war he was Vice-Chief of the Imperial General Staff before being appointed Commander-in-Chief and High Commissioner in Malaya 1952–4. Through a combination of military efficiency, adaptability to local circumstances, and the fostering of good relations with village populations by 'winning hearts and minds', he turned the tide of war decisively against the Communist guerrillas. In 1955 he became Chief of the Imperial General Staff until he retired in 1958. He played a significant role in the Suez Crisis* of 1956.

Te Puea Herangi (1883–1952), a member of the Young Maori* movement of New Zealand; she was described as 'possibly the most influential woman' of New Zealand's political history. Born into a paramount family of the Waikato Maori, she first gained prominence by leading a campaign against conscription of Maori in World War I. Her home settlement ('pa') of Turangawaewae became something of a national centre in the 1920s, and with her ally Ngata* she taught her people, the Waikato, to restore and preserve their traditional culture while subsisting within their own territory. At the same time she in no way rejected European (pakeha) institutions (schools, hospitals), nor participation in the political process, but encouraged her people to register and to vote.

Terrorism. During the 19th and 20th centuries acts by terrorists (also called guerrillas and freedom fighters) have included indiscriminate bombings, kidnapping, hijacking, lynching, and assassination, in order to produce fear among opponents and the general public. During the 19th century anarchist terrorists had used bombings in tsarist Russia, while the Ku Klux Klan had used lynchings to intimidate the Black population. Twentieth-century dictators, including Hitler and Mussolini, have come to power through the use of terror tactics. The period after World War II witnessed the growth of nationalist and liberation groups which used terrorism as part of their struggle against an occupying power: for example Cyprus, Palestine, and many countries in Africa, Asia, and the Middle East. A further development was the appearance of terrorist groups struggling against their countries' social and political structure, for example the Baader–Meinhof Gang* and the Red Army Faction in West Germany, the Red Brigade and its offshoots in Italy, Action Directe in France, ETA in Spain, the IRA* in Northern Ireland, and the Tupamaros* in Uruguay. These and other groups, not least those in Palestine, presented an international problem, as they began to co-ordinate attacks across frontiers.

Teschen dispute. Teschen is an ancient Silesian city in what was the Duchy of Teschen, ruled by the Habsburgs from the 18th century until World War I. It became highly industrialized in the later 19th century. During 1919 there was fighting between Polish and Czech troops, and in 1920 the League of Nations* arbitrated, dividing the area along the River Olza. The northern half, including the city, went to Poland, and the southern half, with rich coalfields, went to Czechoslovakia. The issue soured the relations between Poland and Czechoslovakia for twenty years. The Munich Agreement* allowed Poland to seize it all. In 1945 the Soviet Union decreed that the 1920–38 border be restored.

Tet Offensive (29 January–25 February 1968). Launched in the Vietnam War* by Vietcong guerrillas and North Vietnamese army units, it was timed to coincide with the first day of the Tet holiday (the Lunar New Year). Under the command of General Giap* attacks were mounted against Saigon, Hue, some ninety towns, and hundreds of villages. There were heavy casualties on both sides and the main attacks were successfully repulsed by US troops. But the extent of the offensive shocked US public opinion and convinced the administration of President Johnson* of the need to end US involvement. The war was to grind on for another five years, but talks were begun in Paris which were to lead to the Paris Peace Accords* of 1973.

Thailand. Known in the West until 1939 as Siam, in 1914 it was an absolute monarchy which, under King Chulalongkorn (1868–1910) had seen extensive Europeanization, close links being established with France, to whom had been ceded the vassal states of Laos* and Cambodia*. The middle class produced by the modernization process became intolerant of absolute royal rule, and an economic crisis in 1932 produced a bloodless coup, which established a Parliament under a constitutional monarchy in what came popularly to be called 'The Land of the Free' (Muang Thai). By 1938, however, a *de facto* dictatorship was emerging under Pibul Songgram*, who in 1940 took advantage of the fall of France to try to re-establish control over Laos and Cambodia. Hostilities were ended after Japanese mediation in return for which Songgram declared war against Britain and the USA in January 1942. Bangkok suffered from US air raids, but in rural areas anti-Japanese guerrilla resistance was strong. Peace was made in 1946, Indo-Chinese territories being restored to France and the country joining the UN. On 9 November 1948 Songgram staged a military coup and returned to power on a wave of anti-Communist hysteria. He remained there until 1957, a strong supporter of SEATO* and US policies in the area. The 1932 Constitution had been theoretically restored in 1944, but in practice civilian government has alternated with periods of often corrupt military rule. During the Vietnam War US bombing raids were extensively carried out from bases within the country, the turmoil in neighbouring Laos and Cambodia being used as justification for military rule. Riots in 1973 resulted in a more democratic constitution in 1974, but the struggle between Khmer Rouge* and Vietnam forces in Cambodia produced new tensions, as did the arrival of thousands of refugees, many of them boat people*. In 1981 a brief military uprising was forcibly suppressed and a delicate balance between civilian and military power has survived, with King Bhumibol* universally respected.

Thant, U (1909–74), secretary-general of the UN 1961–71. On graduation from Rangoon University he became a schoolteacher (1928), and a headmaster 1931–47. In 1948 he joined the Burmese Civil Service, at first as Director of Broadcasting. He was a Burmese delegate to the UN in 1952 and became permanent representative in 1957 until he succeeded Dag Hammarskjöld* as secretary-general in 1961, filling the post with distinction. His achievements included assistance in the resolution of the Cuban Missile Crisis* in 1962, and of the Congo Crisis* in 1964, the formation of a UN peace-keeping force in Cyprus 1964; the negotiation of an armistice to end the Six-Day War* (June 1967); and work towards the admission of Communist China to full UN membership and to the Security Council in 1972. In addition he worked tirelessly to try to end the war in Vietnam. His Burmese origins gave him a particular interest in underdeveloped countries and in the need for genuinely non-aligned policies.

Thatcher, Margaret Hilda (1925–), British Prime Minister 1979–90. Born Margaret Roberts in Grantham, Lincs., where she attended school, she graduated in chemistry at Somerville College, Oxford. She later qualified as a barrister, having married a wealthy businessman, Dennis Thatcher. She entered Parliament in 1959 as a

Conservative and served as Minister of Education 1970–4. A year later she successfully ousted the Conservative leader Edward Heath*, adopting many right-wing, populist policies. In 1979 she became the first European woman Prime Minister. During the first year of office the Rhodesian crisis was resolved and Zimbabwe* established, but the main thrust of her policy was to reduce British inflation and industrial inefficiency. A severe monetary policy was adopted which reduced overmanning, but also resulted in many bankruptcies and a shrinking of Britain's manufacturing base. A determination to curb public spending led to increasing friction between central and local government and to the curbing of trade union power. Unemployment rose to over 3 million, levels not seen since the Great Depression*. Her determination not to compromise over the issue of 'sovereignty' of the Falkland Islands was popular and the resulting Falklands War* produced a mood of national pride that helped to secure a landslide victory for the Conservatives in the 1983 election. During the second term (1983–7), Nigel Lawson as Chancellor of the Exchequer adopted a less rigid economic policy. Unemployment fell, as did inflation, helped by lower commodity prices, while a lower pound helped manufacturing industry. Major trade union legislation was challenged by the National Union of Mineworkers, which conducted an unsuccessful and often violent strike (1984–5). A wide-ranging programme of public asset sales, including council housing, together with profits from North Sea oil, helped to reduce public borrowing. Throughout her years as Prime Minister she steadily opposed numerous policies of the European Community and until October 1990 refused, in spite of pressure from financial institutions, to allow Britain to enter the European Monetary System. In spite of IRA* activity, including an attempt to blow up the cabinet at the 1984 Conservative Party conference, the Hillsborough Agreement was signed in 1985, with the Irish Republic. In 1987 the Conservatives were returned to power for a third term with a majority of 101, making her the first modern British party leader to face three consecutive new Parliaments as Prime Minister. Reforms in education, the legal profession, and the National Health Service followed. All met strong professional opposition, while the introduction of a Poll Tax or Community Charge provoked widespread unrest early in 1990. Her autocratic style in cabinet, together with firm rejection of many aspects of European Community policy, led in November 1990 to a leadership challenge by Michael Heseltine—a challenge which developed into a three-way struggle to be won by John Major, when she resigned as Prime Minister.

Thatcherism, a term current in Britain in the later 1980s, being used mostly by critics of the Thatcher* administrations. In spite of her rhetoric, a characteristic of Mrs Thatcher was in fact the absence of any deeply felt ideological beliefs, apart perhaps from a strong sense of patriotism, together with a loose attraction to *laissez-faire*. This enabled her quickly to respond to those shifts in popular prejudice which gained her office and kept her there. Thus Thatcherism as a coherent ideology did not really exist. Policies of Thatcher governments, however, were to strengthen the powers of central government and reduce those of local government, to give maximum support to multinational corporations, while privatizing state corporations, to cut public spending in the fields of health, social welfare, and education, while maintaining that on defence, and to show extreme caution towards any move which could further integrate Britain within the European Community. To her credit Mrs Thatcher never sought to impose her personal preferences for tougher racial policies and the restoration of capital punishment, even though both of these would have had much popular support.

Theatre. This has been the centre of a continuous dialectic in fashion throughout the 20th century, which opened with the powerful naturalistic writing of such social realists* as Ibsen and Strindberg. 'Theatre in the Round', contrasted with the 'picture-frame' setting of naturalistic tradition, began in Moscow in 1933. Surrealist* theatre aimed to go back to the primitive ritual, but led also to the 'theatre of cruelty', the 'living theatre', 'documentary theatre', and the 'theatre of the absurd', whose masters were Beckett, Ionesco, and Pinter. While pessimists continued to predict the demise of live theatre in the age of video/television, its stubborn survival to the end of the century reflected the basic human need for live group participation.

Thieu, Nguyen Van (1923–), President of South Vietnam 1967–75. Of modest family origin, he joined the VietMinh* in 1945, but left after the return of the French to Indo-China in 1946. He enlisted in the military academy in Saigon and then served in the French colonial army 1949–54, fighting his old VietMinh comrades. From 1954 he was commander of the Vietnam National Military Academy in Saigon. He reluctantly took part in the military coup which overthrew President Ngo Dinh Diem* in 1963, becoming head of state of the military government which followed and then President of South Vietnam from 1967 onwards. Despite massive US military and economic aid, his corrupt administration failed to win the fight against the Vietcong and North Vietnamese forces. He vainly tried to continue the war after the Paris Peace Accords* of 1973 and was highly critical of the USA for what he considered desertion. As Saigon was surrounded in April 1975 he fled to Taiwan and then to Britain.

Third French Republic. This had been established in 1871 on the collapse of the Second French Empire. It had been deeply divided by the Dreyfus Affair (1894–1906), anti-Semitism being a continuous issue in France. The dominance of Clemenceau's* personality helped to preserve it through World War I, when by 1917 troops on the Western Front were in a state of mutiny. Between the wars it experienced forty-four governments and twenty prime ministers from a range of political parties covering the whole political spectrum. Coalition governments tended to be either conservative, from the 'bloc national', or left wing, from the 'bloc des gauches', which itself moved further left in the 1930s to form the 'front populaire', a grouping of Socialists, Radicals, and Communists. Extremists from both ends of the spectrum, Fascist and Communist, regularly staged strikes, riots, and demonstrations, as for example over the Stavisky scandal* of 1934. Technically the Republic lasted until 1946, but in reality it collapsed in 1940 when the neo-Fascist Vichy* regime was established, opposed by a Resistance, where the Communists often dominated.

Third Reich *see* GERMAN THIRD EMPIRE.

Third World, a term in common usage soon after World War II. It was applied to those non-aligned countries, mostly in Asia and Africa, whose economies were neither capitalist (First World) nor Communist (Second World). In the 1970s the term was widened to include all countries with relatively low GNP, including many in Latin America as well as Greece, Turkey, and Yugoslavia. A later term to describe countries with lowest GNP was Least Developed Countries (LDCs).

Tho, Nguyen Huu (1910–), Vice-President of the Council of State of Vietnam, 1981–4, 1986– . Son of a rubber plantation manager, he studied law in Paris and began a law practice in Saigon. In 1949 he led an anti-French demonstration there and in 1950 was imprisoned for criticism of US involvement in the area, when he went on hunger-strike. After the Geneva Agreements* of 1954 he tried to co-operate with Ngo Dinh Diem*, but quarrelled over the issue of nation-wide elections, which Diem refused to allow. He was in prison from 1958 until 1961, when he escaped. A non-Communist, he joined other anti-Diem refugees and became chairman of the National Liberation Front* (1962), which had been formed two years earlier. As the military campaign developed and the North Vietnamese became more involved, his influence temporarily declined. In 1969 he became chairman of the Advisory Council of the Provisional Revolutionary Government of South Vietnam. Since 1976 he has held various offices, including that of Vice-President of Vietnam 1976–80 and of Acting President 1980–1.

Thorez, Maurice (1900–64), French politician. Son of a coal-miner from the Pas de Calais, he also became a miner, and joined the French Communist Party in 1920 on its foundation. He was imprisoned a number of times and in 1930 became general secretary of the party. He was elected a Deputy in 1932, and his party supported, but never joined, the Popular Front* coalition government of 1936 which enacted important social and labour reforms. Conscripted in 1939, he deserted and went to Moscow, where he allegedly 'organized European Resistance movements'. Sentenced *in absentia* to six years' imprisonment, he was pardoned in 1944, returned to France, and was elected to the Constituent Assembly. He continued to sit as a Deputy throughout the Fourth Republic*. He led the Communist Party in the elections of 1945 and 1946, and became Vice-Premier 1946–7. Although his health was ailing, he succeeded in 1956 in putting through his party congress a resolution supporting de-Stalinization. He was briefly a member of the Chamber of Deputies for the Fifth Republic, elected in 1958.

Three Mile Island, a nuclear power-station on an island in the River Susquehanna, Pennsylvania, USA. In March 1979 a series of human errors and instrumental failures resulted in the overloading of the reactor core of the plant, releasing radioactive gases. The accident resulted in the temporary closure of all nuclear power-stations of a similar design, a moratorium on the licensing of new reactors, and a new public awareness of the potential dangers of nuclear power, confirmed by the much more serious disaster of Chernobyl* in April 1986. The clean-up operation cost $1000m. and the US nuclear industry only slowly regained any momentum.

Three Principles of the People. First formulated by the republicans as the goals of the Chinese Revolution (1911), they were popularized by Sun Yixian* in the 1920s, when he re-established the Guomindang. He spent his last years trying to sharpen and refine them: nationalism, or the right of self-determination by the Chinese people, first opposed to the Qing dynasty and then to foreign imperialists; democracy, or the right of the Chinese people to control their own government; social reform, or the establishment of a just system of taxation and equitable land ownership for the people's livelihood. The Nationalists claimed allegiance to the Principles as did the Communists.

Tiananmen Square, Massacre of, a mass atrocity in China. During the early months of 1989 there was mounting student criticism in China, with demands for greater democracy and an end to 'corruption'. In April, following the death of the former General Secretary Hu Yaobang*, there were demonstrations in Tiananmen Square before the Great Hall of the People in Beijing. When demands failed to be met, many students began a hunger-strike, which precipitated a vast popular movement, uniting professionals and intellectuals with workers, shopkeepers, clerks, peasants, pedlars, and even the Beijing thieves. Martial law was proclaimed on 20 May and a clear power-struggle developed within the government, between the 'hard-liners' led by Premier Li Peng and the conciliators led by the then party secretary Zhao Ziyang. The latter had the support of some of the military, a group of army generals writing that 'the People's Army belongs to the People'. Early attempts by local military to clear the

Square failed, and on 3 June units of the 27th Army were moved in. Changan Avenue, leading to the Square, was totally blocked by 100,000 citizens and, when tear gas failed to move them, the killing began on the orders of Deng Xiaoping*. Students and demonstrators were shot and crushed alive by tanks under the eyes of the world's television. Some 300,000 troops occupied the city and the Square was closed for twelve days while helicopters flew out dead bodies to be burned. The numbers killed ran into thousands and the massacre was followed by mass arrests and executions of alleged 'counter-revolutionaries'.

Tibet (Xizang), an autonomous region of the People's Republic of China. Fears of Russian influence led to a British invasion in 1904 and the negotiation of an Anglo-Tibetan treaty. When the Chinese empire collapsed in 1911 Tibet became autonomous under British control, and remained so until 1950, when Chinese troops returned, completely occupying the country a year later. After a rebellion in 1959 the Dalai Lama* and thousands of his subjects fled. Tibet was administered as a Chinese province until 1965, when it was reconstituted an autonomous region within the People's Republic. A further revolt was suppressed in 1987 and the situation remained tense and unhappy.

Tilak Bal Gangadhar (1856–1920), Indian Hindu scholar and politician. Known as Lokamanya ('revered by the people'), he had been imprisoned 1897–9 for alleged sedition in his weekly Marathi-language newspaper *Kesari* ('Lion'). In the Indian National Congress* he had strongly supported the radical wing. In 1914 with Annie Besant* he formed the Indian Home Rule League, and subsequently advocated moderate policies of co-operation with the Muslim League*. In the Lucknow Pact of 1916 Congress recognized the need for separate electorates for Muslim minorities.

Tillett, Benjamin (1860–1943), British trade union leader and politician. Born in Bristol, son of a railwayman, he became a bricklayer and then joined the Royal Navy. Later he worked as a docker, and his powers as a speaker made him influential in the London Dockers' Strike of 1889. He concentrated on union activities, organizing a national dock strike in 1911, and working tirelessly for the Trades Union Congress, of which he became president in 1929. He was a Labour MP 1917–24 and 1929–31, but never held office.

Timor. By 1914 the eastern half of this East Indies island was a Portuguese colony and the western half part of the Dutch East Indies*. The island remained divided between the two powers until after World War II (except during the Japanese occupation). Dutch Timor became part of Indonesia* in 1949, but Portuguese Timor remained an Overseas Province of Portugal until 1975, when the left-wing Fretilin movement proclaimed independence after a brief civil war. Shortly afterwards Indonesian troops invaded, occupying it after heavy fighting. This eastern part of Timor was formally integrated into Indonesia in 1976, when thousands of Fretilin supporters were interned. Guerrilla resistance continued, however, and there was a massive purge by the Indonesian army in 1981, after which tension on the island was somewhat reduced, although Fretilin continued its underground activities.

Tiritakene, Sir Eruera Tihema (1895–1967), the first Ratana* candidate to win a seat in the New Zealand Parliament. He had become an enthusiastic follower of Tahupotiki Ratana* during the 1920s and in 1932 stood as a candidate in a by-election for the Southern Maori electorate, which he won. When he was sworn in, his two sponsors were the Labour whips, thus symbolizing the alliance which was to help win the 1935 election. He pressed for an end to long-standing Maori* grievances, many of which were to be righted after the 1935 victory, including equal pay for relief work. Before then Maori rates were half those of Europeans. He was chairman of the Maori War Effort Organization 1939–45 and Associate Minister for Maori Affairs 1957–60. His daughter succeeded him as a distinguished Maori politician.

Tirpitz, Alfred von (1849–1930), German grand-admiral. Born in Prussia, he joined the Prussian navy in 1865 and quickly advanced in rank. In 1897 Kaiser William II appointed him Secretary of State for the Navy, and his first Navy Bill 1898 began the expansion which led to a naval race with Britain. In 1907 he began a large programme of construction of *Dreadnought*-class battleships for the High Seas fleet. During World War I he made full use of submarines, but, following the sinking of the *Lusitania** (1915), unrestricted submarine warfare was abandoned by order of the Kaiser. Tirpitz disagreed with this and resigned in 1916, spending the rest of the war in retirement. He sat in the Reichstag of the Weimar Republic 1924–8 as a Nationalist, a strong admirer of Adolf Hitler*.

Tiso, Josef (1887–1947), Catholic priest and President of the Slovak Republic 1939–45. Born in Bratislava, he was ordained a priest in 1910. After World War I he became interested in politics, helping to build the Slovak People's Party within the new Czechoslovakia and demanding autonomy. He was briefly Minister of Health in a coalition government 1927–9. He became leader of his party in 1938, and in March 1939 proclaimed Slovakia an independent republic, with himself as President. The new Republic was a protectorate of the Third Reich* and declared war on the Soviet Union in June 1941. Tiso was arrested in hiding in May 1945, sentenced, and hanged, although a Monsignor.

Tito, Josip 'Broz' (1892–1980), Prime Minister of Yugoslavia 1945–53; President 1953–80. Born on the border between Slovenia and Croatia, when these were Austrian provinces, he served in the Imperial Army during World War I and was captured by the Russians in 1915. He escaped and went to Petrograd, where he served with Trotsky's* new Red Army. In 1920 he returned to Zagreb, where he worked in a metal foundry and helped to organize a Communist Party. He was arrested and imprisoned in 1928. After his release in 1934 he went to Moscow, where he became general secretary of the Yugoslav Communist Party. When German troops invaded in 1941 he went to the mountains in south Serbia and from Drvav organized the partisans into a National Liberation Front. This was so successful that by 1943 he could convene the Jajce Congress* in Hercegovina, when, in spite of chetnik* opposition, he was proclaimed marshal of Yugoslavia and leader of a federal republic; this was immediately recognized after the war ended. He rejected Stalin's demand that he control the Communist-governed states of Eastern Europe. As a result Yugoslavia was expelled from the Cominform* and Tito became a leading exponent of non-alignment* in the Cold War. Normal relations with the Soviet Union were resumed in 1955, although he criticized Soviet actions in Hungary in 1956 and in Czechoslovakia in 1968. His determined independence enabled him to experiment with different styles of Communist economic organization, including worker participation in the management of factories. Following his death on 4 May 1980, his funeral was attended by kings and presidents from all over the world. The office of President was to be replaced by collective government by the leaders of the six republics, with a rotational presidency.

Tobruk, Siege of, World War II (April 1941–June 1942). When General Wavell's* army captured Tobruk in January 1941 some 25,000 Italian prisoners were taken. The Afrika Korps of General Rommel* arrived in Libya in March 1941, and the balance of forces in the desert radically changed, as British reinforcements were sent to Greece. The British withdrew east, and in April a largely Australian garrison was left to defend Tobruk, which was subjected to an eight-month siege and bombardment. In November 1941, after being reinforced by sea, the garrison broke out, capturing Rezegh and linking up with the 8th Army troops of General Auchinleck*. In June 1942, however, Rommel counter-attacked and after heavy defeats the British again withdrew, leaving a garrison of two divisions, mostly South African and Australian, in Tobruk. They were subjected to a massive onslaught by German and Italian troops and on 20 June capitulated. Twenty-three thousand prisoners and vast quantities of stores were lost. Tobruk was recaptured on 13 November by the troops of General Montgomery*.

Togliatti, Palmiro (1893–1964), Italian politician. Born in Genoa of a middle-class family, he was educated at Turin, and was commissioned into the Italian army in World War I. He was wounded, and became a socialist journalist. In 1921 he and a group of followers split from the Socialists to form the Italian Communist Party; he became a member of the central committee and then general secretary 1926–64. In 1926 Mussolini banned all Communists, and he lived mainly in Moscow 1926–44. During the Spanish Civil War he was chief of the Comintern* in Spain. In 1944 he returned to Italy, and was Vice-Premier 1944–5 and Minister of Justice 1945–6; he survived an assassination attempt in 1948. The Italian Communist Party became the largest in Western Europe, and won 135 seats in the 1948 elections, as well as control of many municipal councils, especially in the south, which it retained. Togliatti was pragmatic and undogmatic in his Communism. He recognized Roman Catholicism as the dominant religion of his country and propounded the idea of 'polycentrism', that is, allowing for national differences in the implementation of Communism, for example a belief in 'an Italian road to Socialism'. He inspired Eurocommunists in other countries and helped a general trend towards liberalization, to be realized in 1989–90.

Togo, a West African republic. Annexed by Germany in 1884, Togoland was mandated between France and Britain after World War I. After a referendum the western section joined Ghana as the Volta region on the latter's independence in 1957. The remainder of the area under French UN mandate* achieved independence as the Republic of Togo in 1960. After two civilian regimes were overthrown in 1963 and 1967, Togo achieved some political stability under President Gnassinbe Eyadema. In 1979 it staged its first election in sixteen years, with a single legal political party Rassemblement du Peuple Togolais. President Eyadema was elected, and re-elected in 1986.

Tojo Hideki (1884–1948), general and Prime Minister of Japan 1941–4. From a military family, he graduated from the Army Staff College in 1915 and then held various staff appointments, until being posted to the Guandong Army*, where he built a power-base of like-minded officers. From 1931 he was virtual leader of the faction prepared to support all but the most fanatical militarists in Tokyo. He was Chief-of-Staff of the Guandong Army 1937–8, before becoming Vice-Minister of War 1938–9 and then War Minister 1940–4. As such he secured Vichy France agreement to Japanese bases in Indo-China, from which the assault of December 1941 was to be launched. In October he engineered the fall of Prime Minister Konoe Fuminaro's* government, and, as Prime Minister, gave orders for Pearl Harbor*. During 1942 he gradually took increased powers, creating a virtual military dictatorship. When US forces took the Mariana Islands, from which they could mount their bombing offensive against Japanese cities, Emperor Hirohito sought Tojo's resignation (July 1944). He was found guilty at the Tokyo Trials* of permitting 'barbarous treatment' of Japan's enemies and hanged.

Tokyo Trials. Between May 1946 and November 1948 twenty-seven Japanese leaders appeared before an international tribunal, charged with crimes ranging from murder and atrocities to responsibility for causing the war. Seven, including the former Prime Minister Tojo Hideki*, were sentenced to death and sixteen to life imprisonment (two others receiving shorter terms). General MacArthur* refused to allow the Emperor Hirohito to be tried for fear of undermining the post-war Japanese state. Of the many hundreds of atrocities committed by Japanese military forces during the war few resulted in prosecution, although a military tribunal in Manila did prosecute some of the most notorious perpetrators, including General Yamashita*.

Tonga, a Pacific islands kingdom consisting of some 180 small Polynesian island, the largest being Tongatapu. Named the Friendly Islands by James Cook, they were under British protection from 1900 until 1970, when they became an independent kingdom within the Commonwealth of Nations. The 1970 Constitution confirmed that of 1875, providing a Privy Council appointed by the sovereign and a Legislative Assembly consisting of sovereign, Privy Council, nine chiefs chosen by their peers, and nine representatives elected on universal suffrage. Queen Salote Tupou III (1918–65), always a popular figure when visiting Britain, was succeeded by the present sovereign Taufa'ahau Tupou IV.

Tonkin Gulf Resolution (7 August 1964), a resolution of the US Congress approving US 'retaliatory' air raids on North Vietnamese targets and giving President Johnson authority to 'take all necessary steps including the use of armed forces' to help members of SEATO* to 'defend their freedom'. It was passed in response to an alleged attack on 2 August by North Vietnamese patrol boats against the US destroyer *Maddox* in the Gulf of Tonkin. Subsequent investigation revealed that the intelligence information on which it was based was inaccurate. US military involvement in the Vietnam War* followed. After the war a War Powers Act was passed in 1973, which restricts the time a President can commit US forces without congressional approval to sixty days.

Totalitarianism, a 20th-century term to describe the ideology of those political regimes which permit no institutions to be autonomous and free from control of the state. It is a loose term, always used pejoratively, for example of Nazi Germany or the Stalinist Soviet Union. It does not necessarily resort to violence and is not synonymous with absolutism or autocracy. As state control of school curricula and of the universities increased, and the media were manipulated, while the influence of such autonomous institutions as the churches decreased, it can be argued that Britain became increasingly totalitarian during the later 20th century.

Touré, Ahmed Sékou (1922–84), President of Guinea 1958–84. From a poor Muslim family, he attended a French technical school in Conakry and then became a trade union organizer. He helped to found the Federation of Workers' Unions in Guinea, in 1948 became secretary-general of the CGT (Confédération Générale de Travail) for Africa, and later became a vice-president of the World Federation of Trade Unions. In 1946, together with Houphouët-Boigny* and other African leaders, he was a founder of the Rassemblement Démocratique Africain (RDA). In 1955 he was elected

Mayor of Conakry and took his seat in the French Assemblée Nationale in 1956. In 1957 he became Vice-President within the Guinea cabinet. He was elected President when Guinea became independent in 1958 and broke all links with the French Community*. A convinced Marxist, he received aid for Guinea from the Soviet bloc when France withdrew all support. Although his domestic policies were harsh, he was consistently re-elected. He re-established links with France, was a strong supporter of the OAU*, and in 1982 led an unsuccessful bid to end the Iran–Iraq War*. Following his death in 1984 the armed forces staged a coup and established a military regime.

Townsend, Francis Everett (1867–1960), US physician and reformer. After a lifetime in the medical profession, mostly in California, he became involved in politics at the time of the Great Depression* in the early 1930s. He devised the Townsend Plan, or Old Age Revolving Pension Scheme, which called for payments of $200 a month to all aged 60 or over and to the disabled, to be funded by a federal tax on commercial transactions. He secured over 10 million signatures to a national petition, which may have persuaded President Roosevelt* to adopt more far-reaching social policies including old-age pensions in his Social Security Act of 1935. In 1936 he helped to found the short-lived Union for Social Justice, and remained an active spokesman for the retired until his death at 93.

Toynbee Hall, a settlement in the East End of London. Founded in 1884 by Samuel and Henrietta Barnett, aimed at creating a settlement to improve the lives of the urban deprived, it was named after a young Oxford philosopher, Arnold Toynbee, and has always had close links with that university. It has attracted a wide range of reformers, among them William Beveridge*, J. M. Keynes*, Clement Attlee*, and R. H. Tawney*. Its programmes have influenced such far-reaching social reforms as old-age pensions, labour exchanges, changed attitudes to young offenders and child care, the development of a National Health Service, and improvements in race relations. Since World War II, when it was bombed, it has been a centre for a wide range of adult education and activities, and also for pensioners. It has always provided a base for those training to work in the social services.

Trade/labor unions. These are referred to as 'labor unions' in the USA. They had succeeded in gaining recognition among skilled craftsmen, together with some legal protection, throughout the industrialized world—Europe, the USA, and Australasia—by the end of the 19th century. With the development of mass-production methods large numbers of semi-skilled and unskilled workers were recruited into industry from the 1880s onwards, and attempts were made to organize these into unions such as the Transport and General Workers' Union in Britain and the Teamsters* in the USA. These attempts were more successful in Britain, Europe, and Australasia than in the USA, where immigrant labour was cheap and easily available. As industrialization has proceeded in other countries so have trade unions developed, although in South Africa union activity among Black workers was illegal until 1980. In the Soviet Union and Eastern Europe 90 per cent have belonged to government-controlled unions, which until 1990 concerned themselves with training, economic planning, and the administration of social insurance, but not with wage negotiation. In October 1990 the General Confederation of USSR Trade Unions declared itself independent 'of political and public structures', no longer being prepared to work 'under the guidance' of the CPSU. The Polish trade union movement Solidarity* was illegal 1980–9. In Britain trade union legislation during the 1980s introduced compulsory ballots, but also outlawed secondary picketing. At the same time the Thatcher administrations ceased to involve trade unions in economic planning to the extent that they had been since World War II.

Trades Union Congress (TUC), founded in 1868 as an annual assembly. In 1871 it established a Parliamentary Committee whose aim was to advance the interests of unions with Members of Parliament. This was replaced in 1921 by a General Council, elected by trade union members and with a wide range of responsibilities. The Congress could urge support from other unions, when a union could not reach a settlement with an employer in an industrial dispute, but it had no powers of direction or arbitration, and could not recommend secondary picketing by sympathetic unions. After the General Strike of 1926 relations between the Congress and the government of whatever party were cautiously conciliatory. It was closely involved in industrial planning and management in Britain during World War II, and to some degree under successive Labour and Conservative governments until 1979. Its prestige suffered during repeated strikes in the 1960s and early 1970s. Since then it has tended to be on the defensive, particularly following legislation by the Thatcher governments designed to weaken trade union power in industrial disputes, while its relations with the Labour Party became somewhat more distant.

Transkei, a 'homeland' of the Ngoni* Xhosa nation, consisting of three separate areas. It was the first Bantustan* to be created as an autonomous 'state' in 1963. In 1976 it became 'independent' with its own National Assembly under Paramount Chief Ndamase, when the 1½ million Transkeians in the Republic lost their South African citizenship to become citizens of the new 'state'. In 1988 there was a bloodless military coup by Major-General Bantu Holomisa, Chief of the Defence Force. He declared martial law and suspended the Constitution, but in February 1990 promised a referendum on integration with South Africa.

Transylvania, a fertile plateau area of the Carpathian mountains, rich with mineral deposits. It was an integral part of Hungary in the Austro-Hungarian Empire, but with a majority of Romanian-speaking people. In 1918 these proclaimed adherence to Romania and this was confirmed by the Treaty of Trianon* 1920. In 1940 Carol II of Romania* was obliged by the Vienna Award* to cede some three-quarters of the region to Hungary, including the mining areas. The Red Army entered it in March 1944 and in 1947 Hungary was obliged to restore it all to Romania. Redistribution of land and enforced cultural assimilation in turn by Romanians and Hungarians have remained the cause of friction between the two countries; there were riots in Bucharest by disaffected ethnic Hungarians in March 1990.

Trenchard, Hugh Montague, 1st Viscount (1873–1956), creator of the British RAF. From an ancient West Country family, he failed to gain entry into the navy, but was commissioned into the army in 1893 and served in India and in the Boer War 1899–1902. In 1912 he obtained an air pilot's certificate and in 1913 joined the Royal Flying Corps, a branch of the army. In August 1915, with the rank of major, he became RFC commander in France, where he organized fighter battles against the superior German Fokker monoplane. In addition he began to develop the use of bombers aimed at military targets in Germany and occupied France. In April 1918 he won his fight for the RFC to become independent of the army as the RAF. Chief of Air Staff 1919–29, he steadily built this up, continually resisting inter-service rivalry from the army and navy. In 1927 he was created the first air marshal. As Commissioner of Police (1932–5) he reorganized the Metropolitan Police Force, establishing a short-lived Police College and a Forensic Laboratory at Hendon.

Trianon, Treaty of (4 June 1920), peace treaty imposed on Hungary after World War I. In 1914 the Kingdom of Hungary as part of the Austro-Hungarian Empire had a population of 21 million. By this treaty the new Republic of Hungary under the regency of Admiral Horthy* had to accept a reduction to some 8 million, losing two-thirds of pre-war territory. Slovakia and Ruthenia would go to the new Czechoslovakia; Croatia to the new Kingdom of Serbs, Croats, and Slovenes; Romania would vastly expand by absorbing Transylvania*. The arguments used at the Paris Peace Conference* were that the new Hungary reflected ethnic realities, but the treaty was deeply resented. The Vienna Awards* of 1938 and 1940 restored the Felvidék and two-thirds of Transylvania in return for Hungary's support for the Axis, but these were lost again by the Treaty of Paris* in 1947.

Trieste, a port at the head of the Adriatic. It flourished as the main port of the Austro-Hungarian Empire and after World War I was awarded to Italy. It was occupied by German troops during World War II, after which an attempt was made to create a 'Free Territory of Trieste', acceptable to both Italy and Yugoslavia under the protection of the UN. There was deadlock over the rival Italian and Yugoslav claims until 1954, when negotiations in London led to settlement. Italy received the city and port of Trieste and Yugoslavia the adjoining coastal strip. In 1975 an Italo-Yugoslav treaty of friendship settled residual claims in the region.

Trinidad and Tobago, an island republic in the Caribbean. The small island of Tobago was ceded to Britain from France in 1763 and that of Trinidad from Spain in 1802. The two became administratively united as a British colony in 1888. By that time indentured Indians, Chinese, and Madeirans were being brought in to supplement emancipated Africans working the sugar plantations. A resulting ethnic mix of 41 per cent African and 40 per cent Indian descent was to lead to racial tensions in the early 20th century. The islands were fortunate however to have deposits of asphalt in a great pitch lake (visited by Walter Raleigh in 1585!) in southern Trinidad, which from 1888 began to be exploited for early road surfacing and which boosted the economy. The first oil well had been drilled on the lake in 1857, and in 1910 the Trinidad Oilfields Company began major exploitation. Since then its reserves of petroleum and natural gas have made it the richest and most sophisticated West Indian state. In 1923 elected members of the Legislative Council, on a limited franchise, replaced nominations, and universal suffrage was introduced in 1945. Eric Williams*, who founded the People's National Movement (PNM), gave his support to the idea of a West Indies Federation* as a step towards political independence, and the Federation established its headquarters in Port of Spain. When Jamaica voted to withdraw, however, the Federation collapsed. In 1976 Trinidad and Tobago became a republic within the Commonwealth, with its first President Ellis Clark, the country gaining economic buoyancy in the 1970s following oil price rises. After Williams died his successor as leader of the PNM was George Chambers, but during the 1980s the PNM's political grip was challenged by an alliance of opposition groups, forming the 'National Alliance for Reconstruction' (NAR) with strong Indian support. In December 1986 this won a general election, breaking the long rule of the PNM, in which drug-trafficking scandals emerged. Noor Hassanali became President with Arthur N. R. Robinson as Prime Minister. The latter was held hostage by a Muslim sect 27 July–1 August 1990.

Trotsky, Leon (1879–1940). Russian revolutionary leader. Born of Jewish parents in the Ukraine, when he was 17 he joined the Russian Social Democrats. He was arrested and sent to Siberia. He escaped and joined Lenin* in London, working with the

Iskra organization. Here he sided with the Mensheviks, and in 1905 he returned to St Petersburg to help set up the soviet. He was again arrested and again sent to Siberia. When he escaped this time he went to Paris, where he tried to organize international Communism; he was expelled and went to the USA, and then in May 1917 returned to Petrograd to take charge of the soviet which was to spearhead the 'October' Revolution. By now he had joined with the Bolsheviks, and he worked closely with Lenin. They faced two dangers: war with Germany and internal civil war. In the first Soviet government he was Commissar for Foreign Affairs and negotiated the Peace of Brest-Litovsk*, by which Russia withdrew from the war. As Commissar for War 1918–24 his greatest achievement was the creation of the Red Army, which he not only recruited but turned into a force which defeated the White Russian armies and saved the Bolshevik revolution. On the death of Lenin in 1924, he seemed the obvious successor; but he lacked Lenin's prestige within the party and had not the guile of the new general secretary Stalin*. He was an internationalist, dedicated to permanent world revolution, and strongly disagreed with Stalin's more cautious policy of 'socialism in one country'. Steadily losing influence, he was expelled from the party in 1927 and exiled. He went to Turkey, and then lived for some years in Paris. In 1938 he founded the Fourth International, and soon afterwards moved to Mexico to support the Revolution there. He was murdered by an agent of Stalin in August 1940.

Trotskyism, a loose term associated with the doctrines and policies of Leon Trotsky*. Within the Soviet Union it was adopted by opponents of Stalin, who claimed more respect for democratic procedures and human rights and opposed the centralized bureaucracy of the regime. Elsewhere Trotsky's belief in the need for an international socialist revolution has appealed, but too often the term has simply been one of abuse between left-wing politicians.

Trucial States, a group of seven emirates in the lower Persian Gulf which during the early 19th century made a perpetual 'truce' with Britain to abstain from piracy. By 1914 they were a British protectorate, Britain providing defence and controlling all external affairs. In 1968 abortive discussions were held for a federation with Bahrain and Qatar. Britain signed a Treaty of Friendship (1971) with the seven emirates, which proceeded to negotiate (December 1971–February 1972) an independent federation, the United Arab Emirates*.

Trudeau, Pierre Elliott (1919–), Prime Minister of Canada 1968–72, 1972–4, 1974–9, 1980–4. From a wealthy Montreal family, he practised as a barrister before getting involved in Quebec politics by becoming an active political journalist; he founded the magazine *Cité libre*, opposed to the tight clerical control of the Union Nationale*. From 1960 he was a strong supporter of Jean Lesage and his 'Quiet Revolution'*. In 1965 he entered federal politics and became Minister of Justice in 1967. Next year he succeeded Lester Pearson* as leader of the Liberals and won the general election. While supporting a policy of bilingualism, he took a tough line against Quebec separatism. From 1972 to 1974 his government was dependent on the New Democratic Party in Parliament, and it gained popularity by increased government social spending. Following another general election in 1974, however, it found itself battling with inflation, and imposed unpopular wage and price controls. Trudeau lost the election of 1979, but returned to office for the last time in 1980, when he concentrated on the issues of Quebec separatism and the 'patriation' of the Canadian Constitution, which was achieved in 1982, together with a 'Charter of Freedoms'. A man of immense intellectual energy and political skills, he retired in February 1984, being succeeded as Liberal leader by John Turner.

Trujillo (Molina), Rafael (Leónidas) (1891–1961), Dictator ('Benefactor of his Country') of the Dominican Republic 1930–61. He joined the Dominican Republic army in 1918 and was trained by US marines. Between 1919 and 1925 he was attached to the national police, reaching the rank of colonel. In 1927 he became a general in the army and in 1930 seized power against President Vázquez. Officially President 1930–8, 1943–52, and ruling through his brother Hector 1952–60, when not President he remained all-powerful as Commander of the armed forces. His regime dominated all aspects of Dominican life, including the economy, employing authoritarian means to accomplish some material progress and terrorist methods to suppress opposition. In 1937, fearing Haitian infiltration, his troops crossed the border and massacred many thousands of itinerant Haitian labourers. With high profits during World War II for Dominican goods, he was able to liquidate the country's debt and embark on a programme of industrialization and public works. Using blackmail, bribery, torture, and murder, he and his family amassed an enormous fortune, becoming the country's largest landowner and owning one-third of the sugar industry. After alienating all Latin America as well as the USA, he was assassinated in 1961 by discontented military.

Truman, Harry S. (1884–1972), President of the USA 1945–53. From a farming community in Missouri; his parents could never decide what the S. was to stand for and it remained just an initial. From working on the farm, he enlisted for World War I in the US army, and then opened a haberdashery store. He studied law at night school in Kansas City (1923–5), and quickly built a law practice and entered local politics as a Democrat. He became a presiding judge at Jackson County Court and then Senator for Missouri in 1935, backed by a notoriously corrupt party machine. In the Senate he quickly gained a reputation for scrupulous integrity, and was made chairman of a Special Commission Investigating National Defense which uncovered considerable graft, waste, and inefficiency in the federal administration of President Roosevelt. In 1944 he was invited to run as Roosevelt's Vice-President. Having met the President only twice, he himself became President after eighty-two days in office and with little experience of government. At home he aimed to develop Roosevelt's New Deal* policies, but he was to meet opposition from Southern Democrats allied with Republicans in Congress. In foreign affairs he was immediately faced with problems. In July he found himself attending the Potsdam Conference*, and in August he authorized the use of the atom bomb against Japan. His abrupt termination of Lend-Lease* in September 1945 was damaging to East–West relations, which were not improved when the Soviet Union rejected its alternative, the Marshall Plan*. In 1947 the Truman Doctrine* was adopted in response to a perceived threat of Soviet expansion during the early Cold War. In 1948 he won the presidential election against the Republican Thomas E. Dewey, contrary to the prediction of the polls. In his State of the Union message in January 1949 he put forward his Point Four* and Fair Deal* programmes. Although Congress allowed little of the latter to pass into law, he did manage to achieve his 1949 Housing Act, providing for low-cost housing. By his executive authority he had already ended, in July 1948, racial segregation in the armed forces and in schools financed by the federal government. He took the USA into its first peacetime military pact, NATO*, tried to give technical aid to less-developed nations through his Point Four scheme, and in the Korean War* ensured that Western intervention would be under UN rather than US auspices. In 1951 he dismissed General MacArthur* for publicly advocating all-out war with Communist China. He did not run for re-election in 1953, although he remained active in politics long after his retirement, regarded in retrospect as one of the strongest of US presidents.

Truman Doctrine. This was enunciated by President Truman* in a speech to the US Congress on 14 March 1947, when he specifically called for aid to Greece and Turkey. He pledged that the USA would 'support free peoples who are resisting attempted subjugation by armed minorities or by outside pressures'. Congress voted large sums for military and economic aid to countries whose stability was perceived to be threatened by communism.

Trust Territories of the United Nations. In 1945 there were twelve mandates* of the League of Nations which still had not become self-governing. These were placed under the Trusteeship Council of the UN, except for South West Africa, for which South Africa refused to accept the Trusteeship system, although ultimately it was to become independent Namibia in 1990. In the Pacific the Japanese mandated islands became the Trust Territory of the Pacific Islands under the USA. By 1980 all Trust Territories had gained independence except for the Republic of Belau.

Tshombe, Moïse Kapenda (1920–69), African politician in the Belgian Congo. Educated at an American Methodist mission, he was the son of wealthy parents, inheriting hotels and stores, and having close links with Belgian and US interests. In the later 1950s, backed by Belgian funds, he founded the Conakat Party, which advocated an independent but loosely federated Congo. He took part in talks which led to Congo independence in 1960, but almost at once declared the province of Katanga (Shaba*) to be independent of the rest of the country. He maintained his position as self-styled President of Katanga (1960–3) with the help of White mercenaries and the support of the Belgian mining concern Union Minière. On 14 January 1963 Katanga was technically reunited with the Congo Republic and Tshombe fled to Europe. He returned a year later with US backing, and was briefly Prime Minister of the Congo Republic (1964–5). In May 1965 he was accused of the murder of Patrice Lumumba*, of corruption, and of rigging the 1965 elections. He again fled the country, when General Mobutu* seized power. In 1967 he was kidnapped and taken to Algeria, where he died in prison.

Tswana, a people living in Botswana and the Republic of South Africa, speaking a Bantu language. Their traditional life- style was based on animal husbandry and farming, with a social structure of extended families under chiefs subjected to general councils of all adult males. They were the target of Christian missionaries from Cape Colony from the early 19th century; throughout the 20th century many have moved from their grassland villages to work in South Africa in mines and industry. Bophuthatswana, a Bantustan homeland*, was created in 1977.

Tubman, William Vacanarat Shadrach (1895–1971), President of Liberia (1944–71). From a poor Americo-Liberian family, he became a Methodist lay-pastor and then a successful lawyer. He was elected to the Liberian Senate in 1930 and became President in 1944. He encouraged economic development to remove Liberia's financial dependence on the USA, and successfully integrated the inhabitants of the country's interior into an administration which had hitherto extended little beyond the coastline. He died in London after twenty-seven years in office.

TUC *see* TRADES UNION CONGRESS.

Tudeh, the Communist Party of Iran, founded in 1920. Crushed in 1930, it was refounded in 1941, when the USSR entered World War II and occupied north Iran.

In 1945 it helped to create two new, short-lived republics, Azerbaijan* and Kurdistan*. These were suppressed in December 1946, after which the party was again banned. It supported Musaddiq*, but after his fall (1953) was again driven underground by the Shah. Its members mostly became exiles in Eastern Europe.

Tunisia, a North African country. Nominally under Ottoman rule since the 16th century; the Bey of Tunis had become virtually independent by the time Tunisia became a French protectorate in 1881. Opposition, especially among urban Tunisians educated in France, began to develop after World War I, when the Destour Party was formed (1920). The French Resident-General suppressed this by force, but nationalist agitation grew stronger in the 1930s, when Habib ibn Ali Bourguiba* founded the Neo-Destour Party (Parti Socialiste Destourien) in 1934. There were serious disturbances 1938–9, when the Neo-Destour leaders were deported to France. Released by the Vichy government, they were sent to Rome, from where they were allowed back to Tunisia. Occupied by German troops 1942, Tunisia was liberated by Free French and Anglo-US forces in 1943, when French colonists sought to reimpose control. A decade of unrest followed before France offered autonomy (1954) and a customs union, followed by independence (1956). A constitution gave considerable powers to the Bey of Tunis, advised by a Prime Minister. In 1957 the National Assembly deposed the Bey and proclaimed Tunisia a republic, with Bourguiba as President. During the Algerian War* (1954–62) France operated from military and naval bases retained in Tunisia, where many Algerians sought refuge. Some 1,000 Tunisians were killed in incidents during the war, at the end of which the last French base Bizerta was handed over. The nationalization of many French-owned estates resulted in all French aid being suspended 1964–6, after which trade links were restored with France and extended to the EEC in 1969. In spite of efforts to diversify the essentially agricultural economy, poverty and unemployment have caused unrest, especially in the south, where Libyan intervention in 1980 nearly precipitated civil war. In 1981 Prime Minister Mohammed Mzali persuaded President Bourguiba to grant greater political freedom, and multi-party elections took place. There was further unrest, however, 1984–5, after which Bourguiba, who in 1975 had been proclaimed President for life, was deposed (1987). He was succeeded by President Zine el-Abidine Ben Ali. The latter introduced a multi-party system in April 1988 and Tunisia was a founder of the Maghreb Union* in 1989. During 1990 a growing Islamic fundamentalist* movement Nahdah was suppressed.

Tupamaros, members of the Movimento de Liberación Nacional in Uruguay. An urban guerrilla organization, it was founded in Montevideo in 1963 and led by Rául Sendic. It began as a 'Robin Hood' affair, robbing banks and giving to the poor, but from 1968 it became more overtly political, seeking violently to overthrow the government and establish a socialist state. Its robberies, bombings, kidnappings, and assassinations of officials continued until the early 1970s, when it was severely weakened by police and military repression, with some 3,000 members imprisoned. The survivors, including Sendic, were released in 1985, when they agreed to form a legitimate political party. The Tupamaros derived their name from the 18th-century Inca revolutionary against Spanish rule, Tupac Amarú.

Turkey. The Republic of Turkey evolved from the defeated Ottoman Empire* The Treaty of Sèvres* (1920) triggered off fierce national resistance led by Mustafa Kemal*. A Greek army marched inland from Smyrna (Izmir), but was defeated. The city was captured, Armenia was occupied, and the new Treaty of Lausanne* negotiated, following which the Republic was formally proclaimed in October 1923, with Kemal its

President. The one-party rule of his Republican People's Party continued under his lieutenant Ismet Inönü* until 1950, when in the first open elections the free enterprise opposition Democratic Party entered a decade of power, ending with an army coup and the judicial murder of the Prime Minister Menderes*. Civilian rule was resumed in 1961, but there was a further period of military rule 1971–3. Atatürk's neutralist policy had been abandoned in 1952, when Turkey joined NATO*. Relations with allies, however, were strained by the problem of Cyprus*, which was invaded in 1974. A US trade embargo resulting from this lasted until 1978. Tension between left-wing and right-wing factions, hostility to Westernization by the minority shiites, who seek to enforce Islamic puritanism, and fighting between Turks, Kurds, and Armenians, have all continued to trouble the country. Under Presidents Kenan Evren (1982–7) and Turqut Özal (1987–) some political stability has developed, with rather more concern for human rights. Martial law was lifted in 1987 and the state of emergency ended in 1988, political parties having been legalized. Nevertheless treatment of Kurdish and Armenian minorities has remained repressive. Many expatriate Turkish workers have moved to West Germany and their status there has been one of the factors prompting Turkey to apply for membership of the European Community.

Tutsi, an ethnic group in East/Central Africa, probably of Nilotic stock. Their ancestors came south in the 15th century to what was to become the German colony of Ruanda-Urundi, seeking rangelands for their cattle. Their superior technology enabled them to dominate the indigenous Hutu* peasantry, whose living by work with the hoe the aristocratic Tutsi disdained. In 1959 the Tutsi monarchy was overthrown in Rwanda* and in 1966 the rule of the Tutsi *mwamis* was also ended in Burundi*. Many Tutsi were expelled from Rwanda, but in Burundi an unsuccessful Hutu rebellion resulted in over 100,000 deaths. Burundi remained under Tutsi dominance, the Tutsi people having adopted the Bantu languages of Ruanda and Rundi.

Tutu, Rt. Revd Desmond Mpilo (1931–), Archbishop of Cape Town 1986– . Born in the Transvaal of mixed Xhosa and Tswana descent, he was educated at mission schools headed by his father and then trained as a teacher. He was ordained Anglican priest in 1961, became a lecturer at a theological college, and then moved to King's College, London, for further study. He became Associate Director of the World Council of Churches 1972–5 and then Dean of Johannesburg in 1976, and general secretary of the South African Council of Churches and Associate Bishop of Johannesburg in 1978. While demanding an end to apartheid, he always forcefully repudiated anti-White action. His passport was withdrawn for a while in 1981 and he was awarded the Nobel* Peace Prize in 1984. In 1986 he had talks with President Botha* on 'the deteriorating situation', and a few weeks later became Archbishop of Cape Town. Since then he has fearlessly continued to be involved in efforts to achieve peace in South Africa.

Tuvalu, a Pacific islands state of some nine Polynesian islands, formerly called the Ellice Islands. Made a British protectorate in 1892, it was incorporated into the British Crown Colony of the Gilbert and Ellice Islands in 1916. This was never popular. The islanders developed island councils before being occupied by the Japanese in 1942. After liberation, a movement for secession and independence developed, culminating in a referendum in 1974. The islands seceded from the Micronesian Gilbertese (*see* KIRIBATI) in 1976 and gained independence within the Commonwealth of Nations in 1978, and as a member of the South Pacific Forum*.

Twentieth Congress (February 1956). This Congress of the Communist Party of the Soviet Union was noted for Khrushchev's* denunciation of Stalin. In the first open session Khrushchev made three doctrinal points: that peaceful coexistence between East and West was possible; that war between them was not inevitable; and that there were 'different roads to Socialism', not only the Soviet route. This was followed by a secret session, when Khrushchev denounced the Stalinist 'cult of personality' and the acts of terror of Stalin's regime. The speech was carefully constructed to emphasize Stalin's bad treatment of the party rather than of the country at large. A fervour of de-Stalinization swept through Eastern Europe, as well as the Soviet Union. It directly prompted abortive risings in Poland and in Hungary.

Twenty-One Demands (January 1915), a Japanese ultimatum to China. After entering World War I on the Allied side, Japan occupied the German base of Qingdao and in January 1915 issued twenty-one demands upon China, then diplomatically isolated and torn by civil unrest following the Revolution of 1911. These would have imposed virtual protectorate status on China. Although a key demand, imposing Japanese advisers throughout Chinese government, was not enforced, the threat of war left President Yuan Shikai* little choice but to concede to others. These included extension of Japanese leases in Manchuria, take-over of the German concessions in Jiaozhou, substantial interests in Chinese mining concerns, and an embargo on future territorial concessions to any other power. The Demands greatly extended Japanese power in China, although provoking deep resentment within the country. They aroused US fears of Japanese expansionism.

Tyrol, the Alpine region of the Brenner Pass. In 1810 Napoleon had awarded Italian-speaking south Tyrol to Italy, but in 1815 it was restored to the Habsburgs. Italian nationalists continued to claim it, and after World War I it was handed to Italy. An agreement between Hitler and Mussolini in 1938 provided for extensive forced migration of its German-speaking minority to Germany. After World War II it became an autonomous region of Italy, but tension developed with Austria. In 1971 a treaty concerning the Trentino-Alto Adige region was ratified between the two countries, which agreed to refer any future dispute to the International Court of Justice.

U

U-2 Incident. On 1 May 1960 a US high-altitude Lockheed U-2 spy plane was shot down by Soviet forces and its pilot Gary Powers taken prisoner. Such flights had been going on for several years and at the Summit Conference* in Paris later in the month President Eisenhower refused to apologise. Khrushchev* then used the incident to break up the conference. Later in the year, during the presidential elections, the US Democrats were to argue the incompetence of CIA intelligence activity. There were no more flights, and Powers was exchanged for a Soviet spy in February 1962.

UAR *see* UNITED ARAB REPUBLIC.

Uganda, an East African republic. During the 19th century the Kingdom of Buganda on Lake Victoria became the dominant power in the area, under its *kabaka.* In 1890 there was an Anglo-German agreement that the area be administered by the British, and the newly formed East Africa Company placed Buganda and the western kingdoms of Ankole and Toro under its protection. In 1896 the British government took over this protectorate, adding the kingdoms of Bunyoro and Busoga, placing four of the regions under direct British rule, but allowing Buganda to retain much self-government. After World War II nationalist agitation for independence developed, with Kabaka Mutesa II* being deported for allegedly refusing to co-operate with the British administration of the protectorate. In 1962 full self-government was granted, with Uganda becoming a federation of the five kingdoms. In September 1962 the Prime Minister Milton Obote* renounced this Constitution and declared Uganda a republic, with an elected President. Mutesa II was elected first President, but in 1966 he was deposed by Obote, who became President himself, only to be deposed in turn by General Idi Amin* in 1971. Amin's rule was tyrannical, and in 1980, after an invasion by Tanzanian forces, he fled the country. Obote returned in 1980 and was re-elected President. But his failure to restore order led to a coup in 1985, the resulting military regime lasting only six months before being overthrown by the National Resistance Army led by Yoweri Museveni, who became President in 1986. Since then the country has been trying to recover from the devastating and disastrous years 1971–80, which ruined the economy and cost hundreds of thousands of lives.

Ujamaa. Following the Arusha Declaration*, the Tanzanian government strongly supported the policy of Ujamaa Uijijini, establishing co-operative villages involving collective production and equality of opportunity, with village families sharing decision-making, costs, and benefits. In 1977 the Ujamaa Villages Act made the village in Tanzania the main administrative and development unit; by 1990 it had manifestly failed to achieve its aims.

Ukraine, a rich steppe region of Eastern Europe. The area was ruled by Russia after the partition of Poland in the 18th century, but Ukrainian nationalism remained strong in spite of repression. In 1918 independence was proclaimed, but by 1922 the region had been reconquered by the Red Army. It became one of the first four constituent republics of the Soviet Union in 1922, as the Ukrainian Socialist Soviet Republic. Stalin imposed collectivization* on the region, which suffered grievously from his purges. It was devastated by fighting during the years 1941–4, although many nationalists welcomed the Germans. Territorial gains from Romania, Poland, and Czechoslovakia completed the union of all Ukrainian lands into one republic by 1945, the Crimea being added in 1954. The Ukrainian Soviet Republic is separately represented at the UN, and in 1990 was demanding independence from the Soviet Union.

Ulbricht, Walter (1893–1973), East German statesman. Born in Leipzig, he was an active trade unionist during World War I, became a Communist, and was elected to the Reichstag in the Weimar Republic 1928–33. He then emigrated to the Soviet Union, when Hitler came to power. After World War II he returned, to become a leading member of the Communist-dominated Socialist Unity Party in the Soviet zone of Germany, which subsequently became the German Democratic Republic*. He became party secretary in 1950, and remained so until 1971. In 1960 he became chairman of the Council of State, which he remained until his death. Although in later years the regime became somewhat more relaxed, his Stalinist policies had been stern and uncompromising, as was revealed in 1953 when there were serious riots, only suppressed by Soviet troops.

Ulster Peace Movement (1976). Following an IRA incident in August 1976, in which three children were accidentally killed, a group of both Protestant and Catholic mothers in Belfast organized meetings with prayers for an end to violence in the province. The movement spread to Londonderry, Dublin, Liverpool, Glasgow, and overseas to the USA. Two leaders, Mairead Corrigan and Betty Williams, were awarded the 1976 Nobel* Peace Prize, but the movement later faded.

Ulster Unionist Party. In 1886 Lord Hartington and Joseph Chamberlain formed the Liberal Unionists and allied with the Conservatives. They pledged to maintain the Union of Ireland with the rest of the United Kingdom. The party supported the formation of the Ulster Volunteers* in 1912, but could not prevent the passage of the Government of Ireland Bill in 1920. After the division of Ireland in 1921 this Unionist wing of the Conservative Party, now calling itself the Ulster Unionists, became the majority party in Northern Ireland. Its leader James Craig was Prime Minister for twenty years 1921–40. In 1969 Ian Paisley* led a break-away group, the Protestant Unionist Party, but the Official Unionist Party continued to rule until the imposition of direct rule in 1972. Under its leader James Molyneux it bitterly opposed the Anglo-Irish Agreement of 1985, but in 1987, together with Ian Paisley, it ended a nineteen-month boycott and entered discussions with the British Secretary of State Tom King. In 1990 it very reluctantly agreed to round-table discussions with the Irish Republic.

Ulster Volunteers. This was formed as an Irish paramilitary organization in 1912, in order to exclude Ulster from the Home Rule Bill then about to go through Parliament. Its supporters pledged themselves 'to use all means' to resist this. They were given every encouragement by Sir Edward Carson* and several prominent Conservatives. They were openly drilled and armed, thousands of rifles being smuggled into Ireland for their use. A clash between these volunteers and the nationalist **Irish Volunteers** formed in Dublin in 1913 became probable, but was averted by the start of World War I. After the war they co-operated with the Black and Tans*, until the Anglo-Irish Treaty of December 1921, while many of the Irish Volunteers joined the IRA*.

Union Nationale, a Quebec political party founded in 1935 and largely the creation of Maurice Duplessis (1890–1959). It held power 1936–9 and 1944–60 with Duplessis as Premier. Strongly opposed to any extension of federal power within the province, the party drew its support from conservative rural areas and from the Catholic hierarchy. It suppressed religious dissent and was attacked for its highly repressive legislation during the 1950s, being fanatically anti-Communist. It began to disintegrate after Duplessis's death, although it came to power again 1966–70 under Daniel Johnson, since when it has largely disappeared.

United Arab Emirates, a federation of Gulf states. In 1968 Britain sought to create a federation of Arab emirates consisting of Bahrain, Qatar, and the seven Trucial States*. When this failed the six emirates of Abu Dhabi, Dubai, Sharjah, Ajman, Umm al-Qaiwain, and Fujairah negotiated in 1971 to form the UAE, Ras al-Khaimah joining in 1972. Movements towards centralization under a single federal government were at first resisted, but after Sheikh Rashid bin Said al-Maktoum, ruler of Dubai, became Prime Minister in 1979, advances were made towards a single federal state with Sheikh Zayed bin Sultan al-Nahayan, ruler of Abu Dhabi, as President.

United Arab Republic (UAR), a short-lived political union between Egypt and Syria. It was proclaimed on 1 February 1958, when Nasser's* reputation was at its height after the Suez Crisis*. There were plebiscites in both countries and Nasser was elected President. Other Arab states were invited to join, but only the Yemen* Arab Republic did so, in a loose association 1962–7, at the time of its civil war. Syria in fact seceded in September 1961, but Egypt retained the name and flag until September 1971, when President Sadat* adopted the title Arab Republic of Egypt.

United Front, a term used in China to describe the co-operation between the Chinese Communist Party and the Guomindang or Nationalist Party. The first United Front was established in 1924 by Sun Yixian*. After his death it quickly disintegrated, and on 10 April 1927 Jiang Jiehi* (Chiang Kai-shek) began his systematic campaign against Communists, the CCP formally ending the alliance in July. The second United Front was created in 1937 following the Xi'an Incident*. In the Sino-Japanese War Japanese forces proved superior to Guomindang troops, and it was CCP guerrillas behind the lines who achieved most. The United Front, in many ways a skilful Communist tactic, ended in civil war in 1946.

United Fruit Company, a major US food company which in 1970 was merged into the United Brands Company. Begun by Minor C. Keith of Boston, Mass., in 1872, its first operations were in Costa Rica, developing banana plantations. By 1890 it had moved into sugar, especially in Cuba, and it began to influence political affairs in Central America and the Caribbean. In return for tax exemption the Company would move into large tracts of forest-land which it cleared and planted, building port facilities and its own Great White Fleet. It opened up plantations in many South American states and extended its interests into cocoa (having a monopoly in Costa Rica, Ecuador, and Panama), hemp (in Guatemala), tropical woods, quinine, natural oils, and other tropical crops; as well as banana plantations in Honduras, Costa Rica, and Guatemala.

Popularly known as *el pulpo* (the octopus), it was a target throughout the earlier 20th century of left-wing criticism, accused of exploitation and bribery. One of its more notorious political acts was the promotion of a revolt to unseat President Arbenz Guzmán* of Guatemala in 1954, while its estates were one of the early targets of Castro's Cuban Revolution in 1959. Elsewhere since then it has sought to survive by programmes of land transfer, whereby producers gained rights over the land which they worked, while the Company provided technological advice and acted as marketing agent.

United Kingdom. Between 1801 and 1921 the British Islands were termed the United Kingdom of Great Britain and Ireland. Since the Irish Treaty of 1921 the United Kingdom has consisted of England, Scotland, Wales, and Northern Ireland. The growth of nationalist feeling since World War II has somewhat weakened its appeal.

United Malays National Organization (UMNO), Malaysian political party. Formed by Dato Onn bin Jaafar, then Prime Minister of Johore, in 1946 in response to British attempts to form the Union of Malaya*, its aim was to fight for national independence, while protecting the interests of the indigenous Malay population. Since independence in 1957, UMNO has been the dominant party in Malaysia, forming the cornerstone of successive electoral alliances, notably the Alliance Party of the 1960s and more recently the National Front.

United Nations Conference on Trade and Development (UNCTAD), a permanent organ of the UN General Assembly, with headquarters in Geneva. It was established in 1964 to promote trade between countries in different stages of development and to negotiate trade agreements. The Conference meets every four years, but there is a permanent Secretariat and Trade and Development Board. It has sought positive discrimination in favour of the developing countries, and in 1968 proposed that developed countries give 1 per cent of their GNP in aid. The gap between rich and poor countries continued to widen, as noted in the two Brandt Reports*, but at a conference in September 1990 of representatives from forty-one Least Developed Countries (LDCs), the developed countries undertook to assist development policies, 'provided these were linked to political reform and respect for human rights'.

United Nations Educational, Scientific, and Cultural Organization (UNESCO), a specialized agency of the UN. It was formed in November 1946, following an international conference in London a year previously. Its aim has been extension and improvement of education, within the concept of the world community. It has established two centres, one in Mexico and one in Egypt, to train teachers for fundamental education among the least-developed nations. It also has centres, in Montevideo, Cairo, New Delhi, Nairobi, and Jakarta, for scientific research into improved living conditions of mankind, research emphasis being focused on peace, human rights, and youth needs. In 1984 its directorate under Amadou Mahtar M'Bow was accused by the USA of an overtly political, that is left-wing, approach to its educational and cultural activities, and the governments of Ronald Reagan and Margaret Thatcher withdrew support and funding (30 per cent). In 1989 it had 161 members and a new director-general, but neither Britain nor the USA had rejoined.

United Nations Organization. A conference in San Francisco April–June 1945 drafted a United Nations Charter, which was signed by fifty nations on 26 June and ratified at the first meeting of the General Assembly in London on 24 October. Since then over 100 other nations have joined, the exceptions being Switzerland and

North and South Korea. The General Assembly* would meet annually, with all members sending five delegates and electing a president each session, as well as a secretary-general every five years. The Security Council* would have powers to execute UN policies. An Economic and Social Council would co-ordinate and establish commissions on specific issues, with a number of specialized agencies to deal with social, economic, health, and financial issues. An International Court of Justice, based at The Hague, would deal with legal disputes, and an International Monetary Fund* would promote monetary co-operation. In many ways it was to be the work of the specialized agencies, particularly the World Health Organization*, the Food and Agriculture Organization*, and the International Labour Organization*, which achieved the UN's greatest successes. Politically it successfully mediated in Palestine (1947), Kashmir (1948), Indonesia (1962), Cyprus (1964), the Middle East (1956, 1967, 1973), and the Congo (1960–4). The Korean War* (1950–3) was fought in its name. It failed to prevent Soviet intervention in Hungary (1956) or Czechoslovakia (1968); it failed to impose its wishes on Rhodesia (1965–80) or on South Africa; it failed to prevent the Falklands War* (1982), but contributed to the cease-fire of the Iran–Iraq War in 1988. With a shift in membership during the early 1960s, the General Assembly aligned itself into voting blocks, including NATO and Warsaw Pact nations, the Arab nations, and the Afro-Asian nations. Resolutions passed by the Assembly have however had little effect on world politics, while during the Cold War* the Security Council was seldom able effectively to take positive action. As the Cold War subsided in the later 1980s the Council was able to act more effectively, numerous world trouble spots, particularly in Africa, responding to the new situation. The decision of the Council to face Saddam Hussein* over Kuwait in 1990 was generally accepted as the most critical challenge it has had to face.

United Party (South Africa). Officially the United South African National Party, it was established in 1934 as a coalition between the followers of Hertzog's* National Party and Smuts's* South African Party. Although it had Afrikaner and British backing, it split when Prime Minister Hertzog attempted to declare South Africa neutral when war broke out. From 1948 it was the principal opposition party in South Africa until the mid-1970s. With the rise of the Progressive Party (later known as the Progressive Federal Party), support fell away and it was dissolved in 1977.

United States of America, a federal republic consisting of fifty states and the District of Columbia, in which stands the capital Washington: Alabama (1819), Alaska (1959), Arizona (1912), Arkansas (1836), California (1850), Colorado (1876), Connecticut (1776), Delaware (1776), Florida (1845), Georgia (1776), Hawaii (1959), Idaho (1890), Illinois (1818), Indiana (1816), Iowa (1846), Kansas (1861), Kentucky (1792), Louisiana (1812), Maine (1820), Maryland (1776), Massachusetts (1776), Michigan (1837), Minnesota (1858), Mississippi (1817), Missouri (1821), Montana (1889), Nebraska (1867), Nevada (1864), New Hampshire (1776), New Jersey (1776), New Mexico (1912), New York (1776), North Carolina (1776), North Dakota (1889), Ohio (1803), Oklahoma (1907), Oregon (1859), Pennsylvania (1776), Rhode Island (1776), South Carolina (1776), South Dakota (1889), Tennessee (1796), Texas (1845), Utah (1896), Vermont (1791), Virginia (1776), Washington (1889), West Virginia (1863), Wisconsin (1848), Wyoming (1890). In addition there are the following non-self-governing 'Outlying Territories' Balau (1947), the Commonwealths of American Samoa (1900), Guam (1898), Puerto Rico (1898), and the Virgin Islands (1917); the Northern Mariana Islands (1947), the Republic of the Marshall Islands (1947), the Federated States of Micronesia (1947), Johnston and Sand Islands (1858),

Midway Island (1867), Wake Island (1898). The structure of government was set out in the Constitution of 1787, which established a federal system, dividing power between central government and constituent states, with an executive President, a Legislature (the Congress) made up of two houses, Senate and House of Representatives, and an independent judiciary headed by a Supreme Court*. There have been twenty-six amendments to the original Constitution, the most recent being in 1971, extending the vote to 18-year-olds. Each state is responsible for a range of fiscal and social policies and provision, all (bar Nebraska) having a bicameral Legislature (Senate and House of Representatives), all possessing a state Governor and state Supreme Court. The acquisition of Spanish overseas territories, Puerto Rico and the Philippines, after the Spanish-American war of 1898, established the USA as a major world power. From the early 19th century the USA has maintained close links with Latin America through the Monroe Doctrine* and later policies of 'dollar diplomacy' and occasional military intervention, and has sometimes supported a number of unsavoury and right-wing regimes, especially in Cuba, Nicaragua, and Chile. At the same time its strict enforcement of the Monroe Doctrine of non-intervention in the Western hemisphere by European powers has been challenged by the Soviet Union in Cuba and elsewhere. Between 1945 and 1990 the USA was seen globally as a superpower pitted against its rival, the Soviet Union. As the Cold War's global division subsided, so the international role of the United States would clearly need reassessment. Its domestic politics has generally been dominated by competition between at first Federalists and Republicans; then Whigs and Democrats; and since the 1850s Republicans* and Democrats*.

Universities. It is estimated that some 160 universities existed in 1800 and that by 1980 there were some 1,320, including over 400 in the USA, together with several thousand colleges, polytechnics, and institutes of higher education, 500 of these being founded in the years 1945–80, particularly in developing countries. The Open University system, launched in Britain in 1971, has since been successfully introduced in many other countries. A characteristic feature of 20th-century intellectual history has been the proliferation of university academic disciplines. In 1914 the range was still relatively narrow: classics, theology, mathematics, natural science, physical science, medicine, engineering, history, languages, and so on. Not only did many of these divide and subdivide, but there was a massive growth of a new range of social sciences including economics, sociology, and anthropology. It is estimated that university research resulted in a doubling, in every decade of the century, of the totality of human knowledge, but not, alas, of human wisdom!

Urbanization. Technically this is defined as the increase in the proportion of a population living in urban as opposed to rural areas. The term is also used loosely to describe the rapid transformation of certain rural areas, such as the coastlines of Florida or southern Spain, usually for the benefit of relatively wealthy, often retired people. In the more technical sense it is in the developing world that urbanization has taken place most rapidly since 1950—for example in Indonesia, North and South Korea, Iraq, and Latin America. By 1980 65 per cent of Latin American peoples were urbanized, often in appalling shanty towns, with Mexico City the world's largest city, having a population of 20 million.

Uruguay, a South American republic. It became independent of both Brazil and Argentina in 1828. During the later 19th century two political parties emerged, the Colorados and the Blancos, the former having an eighty-six-year run in office from 1872. During the first three decades of the 20th century José Batlle y Ordónez*, while in and out of office, helped to mould Uruguay into South America's first Welfare State under a National Council of Administration. Numerous measures for promoting governmental social services and a state-dominated economy were enacted. In 1958 the elections were won by the more conservative Blancos. Economic and political unrest plagued the nation throughout the 1960s, at the end of which the Marxist Tupamaros* emerged. In 1966 a national plebiscite voted to return to presidential government. The military took over in 1973 and a return to civilian rule took place in 1985, when Julio Sanguinetta became President. He immediately faced a long struggle over his desire to grant an amnesty to political prisoners held for violation of human rights during the period of military rule, an Amnesty Law not being finally approved by referendum until April 1989. Uruguay has the highest level of GNP per capita in Latin America, the highest levels of literacy, and the lowest levels of population growth, with an 84 per cent urban population. Nevertheless high inflation and a fall in world commodity prices have caused problems. In elections in March 1990 Luis Alberto Lacalle Herrera, leader of the Blanco Party, was elected President, forming a coalition government with the Colorado Party.

US–Canada relations. After World War I the USA gradually replaced Britain as Canada's main trading partner, a reciprocal trading agreement being made in 1935. Since 1945 Canadian governments have steadily welcomed US investment, so that by the mid-1980s over half Canada's manufacturing industry and most of its mining industry were owned and controlled by US corporations. Since 1965 there has been continental free trade in increasing areas of the economy, with strong US pressures for access to Canada's water. The policies of Brian Mulroney* are for further extension of free trade. In matters of foreign policy, Canada has taken an increasingly separate stance from that of the USA, most notably over Vietnam. At the same time, however, it accepted, if reluctantly, NORAD* and the installation of the DEW line and other early warning systems against nuclear attack from the Arctic.

Uštaše, Croatian independence movement. Originally a terrorist organization against Austria-Hungary, it was revived in the 1920s in opposition to Serbian dominance of Yugoslavia, conniving at the murder of King Alexander I* in 1934. Its leader Ante Pavelić* collaborated with the Axis powers and in 1941 set up the Independent Croatian State. Appalling atrocities were committed against Communist partisans and others, and in 1945 Pavelić fled. The movement survived, however, being responsible for terrorist acts in the 1960s and 1970s, and for renewed agitation for separation from Serbia in 1990.

Vandenberg, Arthur Hendrick (1884–1951), US Senator. Born in Michigan, he worked as a journalist in his home town of Grand Rapids until elected to the US Senate in 1928. He supported much of President F. D. Roosevelt's* domestic legislation, although himself a Republican, but in foreign affairs at that time he was an avowed isolationist. In 1945, however, responding to changing world conditions, he worked for a bipartisan foreign policy and supported US membership of the UN. Thereafter he was instrumental in gaining Senate approval for the Truman Doctrine*, the Marshall Plan*, and the North Atlantic Treaty Organization*.

Vanuatu, a Pacific islands state consisting of some eighty islands in Melanesia, formerly called the New Hebrides, with Vila on the island of Efate its capital. It is rich in mineral deposits, and during the 19th century thousands of islanders had been forcibly abducted to work on Queensland sugar plantations, while at the same time Asian workers were imported by French companies to work in the mines. In 1906 an Anglo-French 'condominium' was established over the islands, which was confirmed in 1922. They served as an Allied base during World War II, after which pressure developed for self-government, with all political groupings concerned with land tenure. A general election took place in 1975 for a new Legislative Assembly, which was to elect a President. Vanuatu gained full independence in 1980 as a member of the Commonwealth of Nations, and of the South Pacific Forum*. Both the French and English educational systems remain on the islands. In January 1989 George Ati Sokomann, the country's first President, was dismissed from office, charged with incitement to mutiny.

Vargas, Getúlio (Dornelles) (1883–1954), Dictator and President of Brazil 1930–45, 1950–4. Born in Rio Grande do Sul of a prominent family, he studied and practised law until entering state politics in 1908. In 1922 he was elected to the Federal Congress and became a Finance Minister. In 1928 he became Governor of his home state. In October 1930, with a group of friends and the backing of the army, he led a coup which overthrew the old Republic of Brazil. The fraudulence of elections earlier in 1930, political corruption, and the growing impact of the Great Depression* on Brazil's agricultural economy, had all precipitated the coup, when he proclaimed himself head of a provisional government. In 1934 he was elected President, but in 1937 he announced a state of emergency, the dissolution of Congress, and a ban on all political organizations. Claiming a desire to rise above factional strife, he announced a new constitution which would create a nationalist, corporate, unified 'New State' (*Estado Novo*). For this he was backed by the military. The economic strategy of the *Estado Novo* concentrated on the diversification of agricultural production, improvements in transport and communication, the promotion of technical education, the implementation of a new labour code, the national ownership of mineral resources and key industries, and the promotion of industrial expansion. World War II stimulated economic growth, and diplomatic and commercial co-operation with the USA led to Brazil joining the war in 1942. With the defeat of the Axis there was pressure on Vargas to relax the authoritarianism of his regime, the *Estado Novo* being criticized as neo-Fascist. His reluctance to do so led to a military coup in 1945. However his continued popularity, especially on the left, enabled him to win the presidential election of 1950. A growing economic crisis from 1952, together with accusations of corruption, led him to commit suicide in 1954, but he is still remembered as 'the Father of the Poor'.

Vatican City. In 1870 the former Papal States became incorporated into a unified Italy, and the temporal power of the Pope was suspended. In 1929 the Lateran Treaty* (Concordat) between Mussolini and Pope Pius XI recognized the full and independent sovereignty of the Holy See in the City of the Vatican. This covers an area of 44 hectares (109 acres), with its own police force, postal service, coinage, radio, and diplomatic corps.

Vatican Council 1962–5, council of the hierarchy of the Catholic Church. The first Vatican Council had been held 1869–70; in October 1962 Pope John XXII* opened the second Council, which was attended by more than 8,000 bishops of the Roman Catholic Church, together with observers from the Anglican and Orthodox churches. It had been summoned to 'discuss renewal of the faith and ways of promoting Christian unity'. The Council's first sessions lasted a year, producing sixteen decrees which represented a more tolerant approach to other churches and encouraged less formality in church services (for example, the Mass to be celebrated facing west and in the vernacular). Pope John died in 1963, but his successor Pope Paul VI* reconvened the Council, which sat for two more years and whose post-conciliar commissions further developed liturgical reform.

Venezuela, a large South American state along the Caribbean coast. It was in its capital Caracas that the Colombian Independence movement began in 1806, resulting in the creation of Gran Colombia by Simón Bolívar. By the 20th century this had broken up, and the present Republic was being ruled by a series of 'caudillo' dictators such as Cipriano Castro (1899–1908) and Juan Vicente Gómez* (1909–35). Oil was discovered before World War I, and by 1920 Venezuela was the world's leading exporter of oil. Military juntas continued to dominate until Rómulo Betancourt* completed a full term as a civilian President (1959–64), to be peacefully succeeded by Dr Raúl Leoni (1964–9). Since then two democratic parties, Acción Democrática and Partido Social-Cristiano, have alternated in power, even though extremists of left and right have harassed them with terrorism. A post-war boom brought considerable prosperity, with minimum personal taxation and high standards of health and welfare. A founding member of OPEC*, Venezuela nationalized its oil industry in 1976, though without undue confrontation with the oil companies concerned. During the 1980s the government under Dr Jaime Lusinchi of the Acción Democrática (1983–8) faced problems of rising population, high inflation, a fall in oil prices, and drug-trafficking scandals. His successor, former President Carlos Andrés Pérez (1989–   ), found that none of these problems had gone away, and there were serious riots in February 1989 against austerity measures. In addition there was a border dispute with Colombia. Nevertheless Venezuela's rich mineral deposits and continuing reserves of petroleum and natural gas, on which the USA more and more depends, remained a huge asset.

Venizélos, Eleuthérios (1864–1936), Prime Minister of Greece 1910–20, 1924, 1928–32, 1933. Born in Crete, he entered politics there and negotiated the union of Crete with Greece in 1905. He moved to mainland Greece in 1909, where he became Prime Minister the following year, modernizing Greek institutions and joining the Balkan League war against Turkey in 1912–13. In 1914 his wish to join the Allies in war against the Central powers was thwarted by King Constantine I, who sympathized with Germany; but when he abdicated Venizélos got his way and Greek troops took part in campaigns on the Macedonian front. In the Treaty of Sèvres (1920) he won promises from Lloyd George* of considerable territorial gains for Greece, but Atatürk's* army attacked and he resigned in 1920. Greece now alternated between monarchy and republic, and his years as Premier alternated with periods in exile. He died in Paris following the establishment of the dictatorship of Metaxas*.

Verdun, Battle of (21 February–16 December 1916), the longest and bloodiest battle of World War I. It was based on a plan of General von Falkenhayn, Chief of the German General Staff, to concentrate the whole weight of his resources against the French fortified city of Verdun. After the heaviest artillery bombardment to date, and with exceptionally heavy casualties, he captured the forts of Douaumont and Vaux, and morale in the French army fell, with widespread mutiny. Yet under Generals Nivelle and Pétain* the city never fell. By calling up all available resources, the French launched a counter-attack in October and regained the lost forts and the wasteland devastated by a summer's fighting. The Battle of the Somme* to the west had, at great cost, succeeded in easing the pressure, and in December the fighting subsided. French casualties were some 400,000, while the Germans lost almost as many, including some of their finest regiments.

Versailles, Treaty of (28 June 1919), the treaty between the Allied powers (except for the USA, whose Senate refused to ratify it) and Germany, whose representatives were required to sign it without negotiation. The armistice of 1918 had been concluded on the Fourteen Points* of President Wilson, but a new 'war-guilt' clause required Germany to accept responsibility for provoking war. Various German-speaking territories were to be surrendered, including Alsace-Lorraine to France; parts of east Silesia to Czechoslovakia; Moresnet and Eupen-Malmédy* to Belgium; and Memel* (now Klaipeda) to Lithuania. In the east, Poland was resurrected and given parts of Upper Silesia and the Polish Corridor* to the Baltic Sea, with Danzig (Gdansk) declared a free city. There would be a plebiscite in Schleswig-Holstein* to settle the Danish frontier. The Saar* valley would be occupied by the French and placed under international control for fifteen years, as would the Rhineland, which, together with Heligoland, was to be demilitarized. Union (anschluss) with the new Republic of Austria was forbidden. The German army was to be limited to 100,000 men, with no conscription, no tanks, no heavy artillery, and there were to be no submarines or military aircraft. Overseas colonies in Africa and the Far East were to be mandated to Britain, France, Japan, Belgium, South Africa, and Australia. Reparations* were to be fixed by a Reparations Commission. The treaty was regarded by the French and many in Britain as too lenient, while its unpopularity in Germany created a political and economic climate that enabled Hitler to come to power. It also established the League of Nations* and the International Labour Organization*.

Verwoerd, Hendrik Frensch (1901–66), Prime Minister of South Africa 1958–66. His parents emigrated from Holland to South Africa when he was a boy. He was educated there and at the University of Berlin. From 1927 until 1937 he taught at the University of Stellenbosch, and then became a newspaper editor before entering politics as a member of the National Party*. He was Minister for Native Affairs 1950–8 and was responsible for enforcing the policy of apartheid*. He became leader of his party in 1958 and Prime Minister, surviving a first attempt at assassination in 1960. During his government, in the aftermath of Sharpeville*, South Africa became a republic and left the Commonwealth. Harsh measures were taken to silence Black opposition, including the banning of the African National Congress*. He was assassinated in Parliament by a dissatisfied East African Portuguese from Mozambique.

Vichy government (France). After the Germans had occupied Paris in June 1940, it was established in the spa town of Vichy by remnants of the National Assembly, 'to administer France and the colonies'. Having unconstitutionally dissolved the Third Republic, it issued a new constitution establishing an autocratic state with Marshal Pétain* 'Head of the French State'. It was never recognized by the Allies. It was dominated first by Laval* (1940–1) as Pétain's deputy and then by Darlan* (1941–2). In 1942 German troops occupied all of France and Hitler obliged Pétain to reinstate Laval, with French Fascist groups gaining more and more influence. After the Allied landings in Normandy and on the south coast of France, the Vichy government established itself under Pétain at Sigmaringen in Germany, where it collapsed in 1945.

Victor Emanuel III (1869–1947), King of Italy 1900–46. Succeeding Humbert I, he retained good relations with France and Britain, although a member of the Triple Alliance with Germany and Austria-Hungary. He maintained neutrality in World War I until joining the Allies in 1915, many Italians having ambitions on Austro-Hungarian lands. With the breakdown of parliamentary government 1921–2, he refused to suppress Fascist riots, but asked Mussolini* to form a government, fearing the alternative to be civil war and Communism. He was created Emperor of Ethiopia in 1936 and King of Albania in 1939. In 1943 he dismissed Mussolini, replacing him with Badoglio* and concluding an armistice with the Allies. In October 1943 he declared war on Germany. He abdicated in 1946, dying in exile in Egypt.

Vienna Awards (1938, 1940), two territorial awards in favour of Hungary. Soon after the Munich Agreement* in 1938, Hungary successfully gained from Czechoslovakia the region of Felvidék in southern Slovakia, which had a Magyar-speaking population. It had previously been part of the Austro-Hungarian Empire and in 1945 was to become part of the Ukrainian Soviet Socialist Republic. By the Second Award of August 1940, Hungary gained some two-thirds of the long-disputed Transylvania* from Romania. This it was obliged to return to Romania in 1947, a decision challenged in 1990, when the two countries returned to democratic rule.

Vietcong, a term used for the military arm of the National Liberation Front* (NLF), or National Front for the Liberation of South Vietnam, founded in 1960.

VietMinh (League for the Independence of Vietnam). Formed in China in 1941 by exiled Ho Chi Minh* and Nguyen vo Giap*, it came into Vietnam in 1943, in collaboration with the USA and the Allies, engaging in guerrilla activities against the Japanese. It liberated much of North Vietnam and entered Hanoi in August 1945. It began as a broad-based movement (nationalists, socialists, Catholics, etc.), but from 1951 onwards the Dalongor Vietnamese Workers' Party (Communist) had increasing control. After the Geneva Conference (1954), VietMinh support continued to grow in South Vietnam. In the early 1960s it linked up with the military wing of the National Liberation Front* (the Vietcong), and the Vietnam War* followed.

Vietnam, Socialist Republic of. The three provinces of Vietnam—Tongking, Annam, and Cochin-China—had been reunited in 1802 by Gia Long, who pronounced himself Emperor. French occupation in the 19th century had allowed emperors still to reign in Hue, but Vietnam was no longer united, being administered as three provinces within French Indo-China. Determination to achieve a united, independent country developed in the early 20th century, stimulated by the writings of Phan Boi Chau*. However, it was only after the French Indo-China War* (1946–54) and civil war in South Vietnam, leading to the Vietnam War* (1964–75), that unification was finally achieved. Although Ho Chi Minh* had successfully kept the support of both China and the Soviet Union, his successors were less adroit. In June 1976 the Socialist Republic of Vietnam was proclaimed in Hanoi, becoming a firm ally of the Soviet Union and joining Comecon* in June 1978. Vietnam and China engaged in a short but costly war February–March 1979, at the time when Vietnam acted against the Khmer Rouge* in Cambodia*. A massive programme of reconstruction and reforestation was begun in the south, the Mekong delta having suffered particularly from the US policy of 'defoliation'. International opinion was shocked by the flood of refugees, mostly Chinese from Saigon, many of whom fled as 'boat people'*. From 1988 onwards strong efforts were being made to disengage from Cambodia and to improve relations with China.

Vietnam, North. The Geneva Agreements* of 1954, which ended the French Indo-China War*, accepted the existence of the 'Democratic Republic of Vietnam' under Ho Chi Minh. Technically this never claimed to be confined only to the north. Since its proclamation in August 1945 it had regarded itself as representing all three provinces of Tongking, Annam, and Cochin-China. In practice, under the Agreements Vietnam was partitioned along the 17th parallel, south of which emerged in 1955 the regime of Ngo Dinh Diem*, calling itself the Republic of Vietnam, so that effectively North Vietnam consisted of Tongking and a small portion of Annam. Here, in the decade 1955–65, very rapid industrialization took place, with aid coming from both China and the Soviet Union. At the same time, under the control of General Giap*, the North Vietnamese army was expanded, rearmed, and equipped. Against some opposition, agriculture was collectivized and productivity greatly increased. Geneva had promised free elections in 1956. When these were refused by Diem in the south, VietMinh* support for unification grew there, especially in the Mekong delta area. From 1960 onwards arms and supplies began to pass along the Ho Chi Minh Trail for Vietcong guerrillas. From 1965 North Vietnam was consistently bombed, while more and more VietMinh troops under General Giap penetrated the south to fight US and South Vietnamese forces. Following the capture of Saigon in April 1975, North and South Vietnam were united as the Socialist Republic of Vietnam*.

Vietnam, South. The French colony of Cochin-China was somewhat less antagonistic to French rule after 1914 than were the protectorates of Annam and Tongking further north. Nevertheless the Communist Party of Indo-China steadily gained recruits there in the 1930s, when the French made rigorous efforts to suppress them. The VietMinh* were active in the delta against the Japanese 1943–5. French rule was re-established in Saigon in September 1945, followed by violent fighting between Communist guerrillas and British, French, and even Japanese troops. In February 1946 General Leclerc* claimed that 'the conquest of the south had been completed', and certainly during the French Indo-China War* the military campaigns were concentrated in the north. In 1949 Bao Dai* returned to Saigon and was proclaimed head of state of Vietnam; but his was a decadent rule and support for his regime steadily eroded, so that by the time of the Geneva Agreements* it could lay no effective claim over North

Vietnam, where the Democratic Republic of Ho Chi Minh* was recognized *de facto*. In 1955 Bao Dai was deposed by Ngo Dinh Diem*, who proclaimed himself President of Vietnam, by which in practice he meant Cochin-China and most of Annam, including the imperial city of Hue, south of the 17th parallel. The promised elections for 1956 were refused. In 1963 Diem fell to a military coup. By then a state of civil war existed between the regime and the National Liberation Front*, which had been formed in 1960 and was aided by the north. There were nine changes of government between 1963 and 1965, when Nguyen van Thieu* and Air Vice-Marshal Nguyen Cao Ky* between them broke the influence of militant Buddhists and imprisoned all political opponents without trial. By now civil war had widened into the Vietnam War*, with a massive US military intervention. In 1967 President Thieu proclaimed a 'Second Republic of Vietnam', but this gained no support from the north. In 1973, when US forces were withdrawn, South Vietnam remained at war with North Vietnam, whose forces captured Saigon in April 1975. After this South Vietnam was absorbed into the new Socialist Republic of Vietnam*.

Vietnam War (1964–75). Civil war, sometimes called the Second Indo-China War, had begun in South Vietnam* in 1960, and in 1961 President Ngo Dinh Diem* declared a state of emergency. Already large numbers of US 'advisers' were in Saigon, heavily committed to the so-called 'domino theory'—that if one South-East Asian area fell to Communism the whole area would. After an alleged attack on a US warship, the US Congress passed the Tonkin Gulf Resolution* (August 1964), giving President Johnson* powers to take military action. US marines landed at Da Nang on 7 March 1965, and by the summer of that year there were some 125,000 US troops serving in South Vietnam, based all along the coast from Da Nang in the north to Cai Ngai in the Mekong delta. Within two years the figure was over half a million, serving with a South Vietnamese army the same size. Contingents from South Korea, Australia, New Zealand, and Thailand fought alongside the US troops. Attacks by Vietcong* guerrillas on US bases at Pleiku and Qui Nhon in February 1965 prompted the first bombing raids of North Vietnam*. By 1966 these were averaging at 164 missions a day, aimed not only at industrial targets, but at the civilian population as well. As the war advanced and Vietcong bases in the Mekong delta were established, napalm bombing and defoliation missions using 'agent orange' were employed. In spite of massive US bombing the Tet Offensive* was successfully launched in January 1968. Although only partially successful, the offensive shook US official belief in the possibility of victory, and attempts began to find a formula for peace. Bombing raids were temporarily curtailed in March 1968, and from January 1969 President Nixon began to withdraw US troops, concentrating on trying to build up Vietnamese units. By now talks had begun in Paris, which ultimately were to lead to a cease-fire and the Paris Accords* of 27 January 1973. Before then however there had been further heavy ground fighting in the spring of both 1971 and 1972, while air raids on Khmer Rouge* bases in Cambodia and Pathet Lao* in Laos lasted from 1968 until 1973. By February 1973 all US combat troops had been withdrawn, although advisers remained in both Saigon and Phnom Penh until April 1975. Talks between North and South Vietnam continued after the cease-fire, but these collapsed in April 1974, when fighting resumed. In March 1975 a new North Vietnamese offensive defeated the South Vietnamese army and on 30 April captured Saigon. Elections to a National Assembly followed, and in June 1976 this approved the unification of Vietnam as the Socialist Republic of Vietnam*.

Villa, Francisco 'Pancho' (1877–1923), a leader of the Mexican Revolution*. Son of a field labourer, he spent his youth as a fugitive in the north Mexican mountains and as a bandit. In 1909 he joined Francisco Madero in his attempted revolution against Porfirio Díaz, and helped to provide the military leadership which resulted in Madero's defeat of Díaz in 1911. After Madero's assassination in 1913, Villa joined Carranza's* 'constitutional' opposition against the usurper Victoriano Huerta*. He and Zapata* however broke with Carranza, and he returned north, where he held sway in the province of Chihuahua with his cavalry, *los dorados*, expropriating the hacienda holdings of large landowners and using their revenues to equip his revolutionary forces. In 1915, however, he was badly defeated by Carranza and retreated into the mountains. In 1916, for reasons still controversial, he ordered an attack on the US town of Columbus in New Mexico, killing seventeen Americans. This provoked retaliation from a punitive expedition under General Pershing*, but he escaped. He continued to harass the regime of Carranza until the latter's death. He then lived in retirement on a farm until he was assassinated in 1923.

Vimy Ridge, Battle of, World War I (9–14 April 1917). The ridge, lying north-east of Arras in France, had been occupied by the Germans in 1914 and had long resisted French attacks. On 9 April Canadian troops under Lieutenant-General Julian Byng launched a massive artillery barrage, and then attacked with four divisions. They took Hill 145 and then within fifteen minutes gained the long ridge itself, capturing 4,000 German prisoners. They successfully held it through the next few days, although with casualties of some 11,297. It became a key defensive position against the German offensive on Arras and Amiens in March 1918.

Virgin Islands, a group of Caribbean islands at the extremity of the Lesser Antilles*. They were discovered by Columbus and settled by Danish and British sugar planters. In 1917 Denmark sold its islands to the USA for their strategic value. The British Virgin Island Group, with Tortola the largest, formed part of the colony of the Leeward Islands 1871–1956, and since then have been administered separately by appointed governors, under whom self-government has been gradually extended. The US islands are an 'Incorporated Territory' of the USA. Their inhabitants are US citizens and since 1954 have elected their own Governor. Demands for greater autonomy have not met electoral support. The islands are all very popular as tourist resorts.

Vittorio Veneto, Battle of, World War I (24 October–4 November 1918). The town of Vittorio Veneto in north-east Italy is named after Victor Emanuel II, in whose reign Venetia had been gained from Austria. In the summer of 1918 the Austro-Hungarian army had been forced to retreat from its gains following Caporetto*, and in October Italian forces under General Diaz, supported by a British force under Lord Cavan, avenged that disaster. By separating the Austrians in the mountains from those in the plains they broke the enemy line, and the result was an Austrian request for an armistice.

Vojvodina, a province of Yugoslavia. It formed part of the Austro-Hungarian Empire, when it was the centre of strong Serbian nationalism. It was united to Serbia in 1918 but occupied by Hungary in 1941. In 1945 it was returned to Yugoslavia.

Vorster, Balthazar Johannes (1915–83), Prime Minister of South Africa 1966–78; President 1978–9. Born in Cape Province, son of a wealthy sheep farmer, he studied law at Stellenbosch University, where he became involved in student politics before becoming a practising lawyer. In World War II he founded a neo-Fascist movement Ossewa Brandwag ('Ox-wagon Guard'), and became a 'general' in their extremist wing the *Stormjagers*, being interned for fourteen months 1942–3. After the war he tried to enter politics, but was at first rejected by the National Party* as too extreme. He was accepted, however, in 1953, and entered Parliament on the right wing of the party, helping Verwoerd* to become Prime Minister in 1958. As his Minister of Justice he harassed and persecuted all Black politicians, especially after the Sharpeville* Massacre. He succeeded Verwoerd in 1966. Although firmly enforcing apartheid, his regime did in fact make some concessions to international opinion. He established relations with neighbouring Black leaders and helped to persuade Ian Smith* to change course in 1979. However, his decision to send South African troops into Angola* was bitterly resented. He won a landslide victory in 1977, but resigned as Prime Minister a year later for reasons of health, becoming President. He resigned from this in 1979 when criticized over the so-called 'Muldergate scandal', in which public funds had been misused by the Department of Information.

Vyshinsky, Andrey Yanuarievich (1883–1955), Soviet Foreign Minister 1949–53. Born in Odessa of Polish descent, he studied law at Moscow University, where he joined the Russian Social Democrats in 1902. He supported the Mensheviks and took part in the 1905 Revolution. He served under Trotsky in the Red Army 1918–21, having by now joined the Bolsheviks*. He became Professor of Law at Moscow University, and in 1935 State Prosecutor of the USSR. He became notorious during the series of trials 1928–38, including the Great Purge (Yezhovshchina*) trials. At the university he taught that a prisoner's guilt is absolute once he or she has confessed. In 1940 he became deputy Foreign Minister, and then Foreign Minister in 1949. During these years he consistently put forward the policies of Stalin and the Politburo.

Wafd (al-Wafd al-Misri), an Egyptian political party. In 1918 a delegation of Egyptian nationals led by Zaghlul Pasha (1860–1927) petitioned Britain for the end of the protectorate. They were rejected, and the term Wafd ('Delegation') was adopted by the nationalist party which then emerged. In 1919 Zaghlul was banished to Malta, but after widespread anti-British disorder he was released and travelled to London. Allowed to return to Egypt, he was again arrested and exiled, but released in 1923 to take part in Egypt's first general election, easily won by the Wafd. During 1924 Zaghlul was Egypt's Prime Minister, but he was an old man and the Wafd leadership passed to Mustafa al-Nahhas*. The party dominated Egyptian politics until 1952, when it was criticized as corrupt by Nasser's Free Officers' Association. It was dissolved in 1953.

Wahabism. This was a doctrine of Islamic reform which had developed during the 18th century in Nejd in central Arabia. It demanded a rigorous, puritanical interpretation of Sunni teaching. During the 19th century Wahabis were persecuted by the Ottoman Turks, but the cult was revived by Ibn Saud (Abd al-Aziz ibn Saud*), who exploited their fanaticism in his successful conquests. In 1929, however, with British help from Iraq, he crushed them in the Battle of Sibilla as too extremist. Since 1945 Sunni Wahabism has revived with the growth of Islamic fundamentalism*. Wahabi Mujaheddin were fierce participants in the Afghan Civil War 1979–89.

Waitangi Tribunal, a New Zealand tribunal established in 1975, at first consisting of the Chief Justice of the Maori Court and two appointees, with powers of recommendation. Among other things it concerned itself with Maori fishing rights, land, and the preservation and extension of the Maori language. The Tribunal failed to assuage a sense of dissatisfaction among younger Maori politicians, who in 1979 formed a new political party, Mana Motuhake. In 1982 there were scenes of violence on Waitangi Day (6 February) by Maori groups resentful of the traditional form of celebration of the signing of the Waitangi Treaty in 1840. To meet growing Maori protest the Tribunal was revised in 1985, with six appointees. In 1986 Parliament ordered that Waitangi Day be celebrated as an expression of the establishment of a bicultural society.

Waldheim, Kurt (1918–), President of Austria 1986– , secretary-general of the UN 1971–82. Born near Vienna, he attended the university there before and after World War II, in which he served as a young staff officer on the Eastern Front. He was involved in military activities resulting in the deportation of Jews and the shooting of Yugoslav partisans. He entered the Austrian foreign service from university and was Foreign Minister 1968–70, before being elected successor to U Thant* at the UN. His period in office was at a time when the Security Council was regularly deadlocked, so that his efforts at international conciliation seldom succeeded, for example over the Iran Hostage Crisis*, over Namibia, and in the Middle East. He was elected President of Austria at a time when his wartime record was being severely criticized by the World Jewish Congress. Their accusations actually increased his popularity in Austria and unleashed an unfortunate new wave of anti-Semitism there.

Wales. The Principality had been a stronghold of Nonconformity in religion since the late 18th century, especially Baptist and Calvinistic Methodist, and early in the 20th century the position of the Anglican Church was a dominant question in Welsh politics. Disestablishment Bills were passed by the Commons in 1912 and 1914. After considerable bitterness the Church was finally disestablished by Lloyd George* in 1920, released funds going towards the establishment of the University of Wales and other institutions. The Industrial Revolution had brought prosperity to south Wales, which became the world's biggest coal- exporting area. After the end of World War I this prosperity waned, with serious industrial unrest a result. During the 1920s there was a steady exodus from the mining valleys, to England and much further afield, to the Commonwealth and even to Patagonia. After 1945, however, there was something of an industrial revival. In 1949 a National Council for Wales was established, and in 1951 a Minister for Welsh Affairs. But demands for greater autonomy were growing, and during the 1960s and 1970s there were acts of violence against reservoirs feeding English cities and against English 'second homes'. A Secretary of State for Wales was established in 1964. In 1925 a nationalist party, Plaid Genedlaethol Cymru, had been founded, dedicated to cultural and linguistic revival. From 1960 this became overtly political, demanding a separate representative assembly for Wales. The Welsh language movement also became very active. In July 1966 Plaid's chairman Gwynfor Evans won a by-election in Carmarthen, and three Plaid Cymru MPs were elected in 1974. A Bill to establish a Welsh Assembly passed Parliament in 1978, subject to a referendum giving support from 40 per cent of the electorate. In fact only 11.9 per cent gave support. The major towns and cities continued to favour established political provision, supporting mostly Labour, while the rural areas of north, mid-, and west Wales, where the Welsh language is strongest, favoured devolution. The collapse of the older industries and closure of depleted coalfields in the 1980s was partially alleviated by the introduction of light and service industries. Welsh language revival, stimulated by an annual National Eisteddfod, continues to receive political prominence. A Welsh Television Channel (SC4) was a notable concession to its advocates.

Walesa, Lech (1943–), Polish politician; President of Poland 1990– . Born in Popowo, son of a carpenter, he entered the Gdansk shipyard in 1967, having trained as an electrician. In 1970 he witnessed the frustration of the Polish people when twenty were shot in the street for demonstrating; he became politically orientated, working first with his official trade union. In 1976 he drew up a list of workers' grievances in the yard and presented it to the management. All were rejected and he was dismissed. He set about creating an independent trade union movement. Now unemployed, in August 1980 he climbed over the wall of the shipyard and addressed 17,000 workers, calling on them to strike. Plants were seized and a total industrial stoppage occurred along the Baltic seaboard until the government conceded the right of workers to organize themselves independently. On 16 December 200,000 people, including dignitaries from Church and state, heard him speak, when he pleaded for national reconciliation. As chairman of the new Solidarity* Union he won important concessions, but one year hence, in December 1981, a state of emergency was declared, Solidarity was outlawed, and he spent a year in prison. In 1983 he was awarded the Nobel* Peace Prize, which he was to donate to social welfare. When Solidarity was legalized in January 1989 he took part in round-table discussions, while refusing to be a candidate for Prime Minister. In April 1990 he was re-elected chairman of Solidarity and demanded the withdrawal of Soviet troops from Poland. Increasingly on the right wing of the movement, in November 1990 he defeated Tadeuz Mazowiecki in the presidential election.

Wallace, George Corley (1919–), Governor of Alabama 1963–7, 1971–9, 1983–7. A farmer's son, he worked his way through the University of Alabama (1942) and, after military service, became a state attorney. In 1947 he was elected an Alabama state Congressman and then a District Judge (1953–8). When first elected Governor of Alabama in 1962 he resisted the desegregation of state schools and universities. He stood in the presidential campaign of 1968 as leader of the newly established Dixiecrat* American Independent Party. His main support was in the South, and he polled over 10 million votes. In 1972 he sought the Democratic Party's presidential nomination, but his campaign ended when an assassination attempt left him paralysed. He stood again unsuccessfully in 1976, by which time he was becoming reconciled to the issue of civil rights. For the 1982 election as Governor he publicly recanted his opposition to desegregation, polled a substantial number of Black votes and was re-elected. He retired from politics in 1987 through ill health.

Wallace, Henry Agard (1888–1965), US Secretary of Agriculture 1933–41; Vice-President of the USA 1941–5. Born in Iowa, he became an agricultural expert, developing successful varieties of corn which were adopted throughout the USA. In 1932 he entered politics and helped to swing the state of Iowa to the Democratic Party. He became Roosevelt's Secretary of Agriculture, formulating the New Deal* agriculture policy through the controversial Agricultural Adjustment Acts (1933 and 1938) to stabilize prices. In 1940 he was elected Vice-President and in 1945 became Secretary of Commerce in Roosevelt's last administration, continuing under Truman*. A visionary liberal, Wallace soon fell out with Truman's Cold War policies, and he resigned in 1946. He moved considerably to the left, exposing himself to charges of 'fellow-travelling' with the Communists. In 1948 he formed his own 'Progressive Party' and ran against Truman for the presidency. He won only 1.2 million votes and carried no state. He then retired to continue his agricultural research.

Wallace-Johnson, Isaac T. A. (1895–1965), journalist and politician of mixed Creole*/African descent from Sierra Leone. He left mission school early to support his family, and by 1914 was already a trade union organizer in Freetown. He served in the British army in World War I, and then returned to trade union activity in Freetown. In 1931 he founded the African Workers' Union of Nigeria and wrote for the *African Morning Post*. Later he was invited to Moscow and spent some years in Europe, where his detractors claimed he was recruited by the KGB. He returned to West Africa in 1935 and was arrested and tried as an agitator. His appeal to the Privy Council failed, and after release he spent some months in London. In 1938 he returned to Sierra Leone and formed a radical party, the Sierra Leone Youth League 'for the oppressed inhabitants of the country'. It held mass rallies and won much popular support. He was interned by the British in 1939 as an 'unscrupulous agitator'. Released in 1944, he was a key organizer of the fifth Pan-African* Conference in Manchester in 1945. He joined the National Council of Sierra Leone in 1950, and his Radical Democratic Party, formed in 1958, was surprisingly conservative. He was killed in a motor-car crash.

Wang Jingwei (Wang Ching-wei) (1883–1944), head of Japanese puppet regime in China 1940–45. His father was a junior civil servant and he was the youngest of ten children in Guangzhou (Canton). In 1903 he won a scholarship to study in Japan, where he became an associate of Sun Yixian*, whose Chinese patriotism inspired him. Imprisoned in 1910 for an assassination plot, he became a popular hero in 1911 when the Republican Revolution began. He rejoined Sun in 1917 and was his close assistant until 1925. On Sun's death he became a rival with Jiang Jiehi* (Chiang Kai-shek) for leadership of the Guomindang. The two men were reconciled in 1932, and Wang became administrative head of the Nationalist* government for the next four years. When war began with Japan, however, he collaborated, and in March 1940 was recognized by Japan as head of a Chinese puppet government based in Nanjing. While having to accept continued military and economic dominance by Japan, his government was not unattractive to many Chinese who held Pan-Asian, anti-Western views. He died in Japan under medical treatment.

Ward, Sir Joseph George (1856–1930), Prime Minister of New Zealand 1906–12, 1928–30. Born in Melbourne, he had emigrated to New Zealand in 1877, becoming a successful grain merchant. He first entered Parliament as a Liberal in 1887 and held office 1899–1906, when he gained a reputation for financial skill. As Liberal Prime Minister 1906–12 he represented New Zealand at the London Imperial Conferences 1907, 1908, and 1911, being a strong imperialist. In 1915 he joined William Massey in a coalition government, but lost the 1919 election. He returned to Parliament in 1925, and with G. W. Forbes* sought to widen the appeal of the Liberal Party, which changed its name to the 'United Party'. In spite of his financial abilities, his government of 1928 failed to solve the growing economic problems facing New Zealand. He died in office in May 1930 and was succeeded by Forbes.

Warlords, regional military rulers in China. Following the death of Yuan Shikai* in 1916, much of China came under the control of the personal armies of local rulers. These warlords might have been senior officers of the imperial or republican armies, bandits, or local officials who had managed to recruit an army. They depended on taxation of the towns and rural areas which they controlled. The largest of the many wars between rival cliques of warlords witnessed the mobilization of hundreds of thousands of soldiers. Jiang Jiehi* (Chiang Kai-shek) succeeded in establishing central authority over many of these warlords in the decade 1928–37, but military rulers persisted in the far west of China until the late 1940s.

Warren, Earl (1891–1974), Chief Justice of the US Supreme Court 1953–69. Son of a railroad worker, he graduated from the University of California and was then appointed a district attorney and state Attorney-General before being elected Governor of California 1943–52. In 1948 he was the unsuccessful Republican vice-presidential candidate. In 1953 President Eisenhower appointed him Chief Justice. In 1954 he wrote the Supreme Court ruling that racial segregation in public schools was unconstitutional. In 1964 he was appointed chairman of the Commission to investigate the assassination of President Kennedy*. By establishing the sole responsibility of Lee Harvey Oswald, the Commission allayed the nation's fears of either Communist or extreme right-wing conspiracies. His ruling of 1966 in *Miranda* v. *State of Arizona** marks a landmark in US criminal law.

Warsaw Pact (Warsaw Treaty of Friendship, Co-operation, and Mutual Assistance). It was signed in 1955 by Albania, Bulgaria, Czechoslovakia, the German Democratic Republic, Hungary, Poland, Romania, and the Soviet Union, after the Paris agreement between the Western powers admitting the Federal Republic of Germany to NATO*. Albania formally withdrew in 1968. The Pact provided for a unified military command, the maintenance of Soviet army units in member states, and mutual assistance. The latter provision was used by the Soviet Union to launch a multinational invasion of Czechoslovakia against the Dubček regime in 1968. In 1989–90 the Pact disintegrated as the East European Communist regimes broke up.

Warsaw Rising (August–October 1944). As the Red Army advanced into Poland in the summer of 1944, Soviet contacts in Warsaw encouraged the underground Home Army, supported by the exiled Polish government in London, to stage an uprising. Resistance troops led by General Tadeusz Komorowski gained control of the city against a weak German garrison. Heavy German air raids, lasting sixty-three days, preceded a strong German counter-attack. The Soviet army under Rokossovsky* reached a suburb of the city, but failed to give help to the insurgents, or to allow the Western Allies to use Soviet air bases to airlift supplies to the hard-pressed Poles. Supplies ran out and, as famine threatened, the Poles surrendered on 2 October. The Germans then systematically deported Warsaw's population and destroyed the city itself. The main body of Poles who had supported the London government-in-exile was thus destroyed. As the Red Army resumed its advance, the Soviet-sponsored Polish Committee of National Liberation was able to impose a Communist provisional government on Poland (1 January 1945) without resistance.

Washington Conference (November 1921–February 1922), a conference summoned on US initiative to discuss both political stability in the Far East and naval disarmament. It was attended by Belgium, Britain, China, France, Italy, Japan, The Netherlands, Portugal, and the USA, and resulted in a series of treaties. These included a Nine-Power Treaty guaranteeing China's independence and territorial integrity; a Japanese undertaking to return the region around Qingdao to China; an Anglo-French–Japanese–US agreement to guarantee each other's existing Pacific territories. Naval discussions resulted in a ten-year moratorium on capital-ship construction. The Conference successfully placed restraints on both the naval arms race and Japanese expansionism, but by the 1930s both problems had broken out afresh.

Watergate scandal (USA). In 1972 five employees of a Republican Party organization were arrested for breaking into the headquarters of the Democratic Party's National Committee. This had been meeting in the Watergate complex in Washington, DC, and they aimed to wire-tap the proceedings. It was soon discovered that their actions formed part of a campaign to help President Nixon* to win the 1972 election. At first the White House denied all knowledge of the incident, but after intensive investigations, initially led by journalists of the *Washington Post*, it became apparent that several of the President's staff had been involved in illegal activities and in an attempt to cover up the whole operation. Several White House officials and aides were prosecuted and convicted on criminal charges. Attention then focused on President Nixon, who was ordered by the Supreme Court to release tapes of White House conversations or face impeachment. As extracts from the conversations were released, it became clear that he too had been involved. This he eventually admitted, and on 9 August 1974 he resigned. He was pardoned for any federal offences he might have committed by the new President Gerald Ford*, but the pardon did not apply to members of his staff, some of whom were later tried and imprisoned.

Watts Riots (USA) (August 1965). The Watts district of south-western Los Angeles had become a notorious centre of slum housing for Blacks and Mexicans. On 11 August 1965 fierce rioting broke out, as a protest against long-standing social injustice aggravated by resentment against draft conscription for the Vietnam War.

Some thirty-four people died and over 1,000 were injured during a week of riot, which was quelled by the National Guard*. The riots helped to create a White backlash against gains that were being made by Blacks as a result of the civil rights movement*.

Wavell, Archibald Percival, 1st Earl Wavell of Cyrenaica (1883–1950), British field marshal and Viceroy of India 1943–7. Born in Colchester and educated at Winchester and Sandhurst, he was commissioned into the Black Watch in 1901 and fought in the Boer War. He served in France in World War I, where he lost an eye, and then in Palestine under Allenby*. In 1937 he returned to command British forces in Palestine, and in July 1939 became Commander-in-Chief in the Middle East. When Italy entered World War II in June 1940, his troops won some substantial victories, including the Battle of Sidi Barrani. In 1941, however, forced to divert troops to Greece, and facing new German formations, he had to retreat in North Africa and was dismissed by Churchill. He then served in India, first as Commander-in-Chief from 1941, and then as Viceroy, where he made it his main task to prepare the country for independence; but he was dismissed in favour of Mountbatten, perhaps unfairly, in 1947.

Webb, Sidney James, 1st Baron Passfield (1859–1947), British social reformer and statesman. Initially a civil servant, he qualified as a lawyer and was called to the Bar in 1885. He became a socialist and with his wife Beatrice (née Potter) a founding member of the Fabian Society*, of which Lady Passfield was later to become president. As members of the Royal Commission on the Poor Law (1905–9), they were the moving spirit behind its minority report that poverty should be dealt with by setting up government organizations to concentrate on specific causes. This later became the basic approach to the problem by successive governments. He was instrumental in the foundation of the Imperial College of Science and of the London School of Economics, where he was Professor of Public Administration 1912–27. He helped to found the Labour Party and largely wrote its constitution in 1918. He served as a Labour Member of Parliament 1922–9, was President of the Board of Trade in 1924, and Dominion and Colonial Secretary 1929–31, when Palestine gave him much difficulty. His writings, many of them joint studies with his wife (v. *Our Partnership* (1948)), on trade unionism, labour history, and local government, exerted considerable influence on political theory and practice. He had a particular influence on the development of secondary and technical education, and on educational administration. However, a naïve account of the Soviet Union in the 1930s somewhat marred his reputation.

Weimar Republic. On 9 November 1918 a republic was proclaimed in Berlin under the moderate socialist Friedrich Ebert*. An elected National Assembly met in Weimar in January 1919 and agreed on a constitution, with Ebert first President. It had to meet the fierce terms of the Versailles Treaty*, terms which were so unpopular as to provoke the brief right-wing Kapp *putsch**. With rising inflation it was unable to meet reparation costs, whereupon France and Belgium occupied the Ruhr*. The Mark collapsed, and in Bavaria right-wing extremists tried to restore the monarchy. Gustav Stresemann* was briefly Chancellor and then Foreign Minister November 1923. He succeeded in persuading the USA to act as mediator; the Dawes Plan* adjusted reparations payments and France withdrew from the Ruhr. It was followed by the Young Plan* of 1929. For a time the Republic appeared more stable, but as the effects of the Great Depression* began to be felt, with rising unemployment, powerful financial and industrial groups turned to the National Socialists as their best hope against Communism. In the presidential election of 1932 Hitler* gained 13 million votes, although Hindenburg* was re-elected. On the night of 27 February 1933 the building of the Reichstag was burnt down, the arsonist being a half-crazed Dutch Communist, van der Lubbe. The fire was used by the new Chancellor Hitler as an excuse to issue a decree banning all civil liberties and declaring a state of emergency on 28 February. This was to last until 1945. On the death of President Hindenburg in August 1934 Hitler declared himself President of the Third Reich. The Weimar Republic was at an end.

Weizmann, Chaim Azriel (1874–1952), President of Israel 1948–52. Born in Poland, he came to England to work in biochemistry at the University of Manchester, becoming a British subject in 1910. During World War I his scientific work as director of the Admiralty Laboratories brought him to the notice of Lloyd George. His contacts helped to obtain the Balfour Declaration* for the Zionist cause in 1917. In 1918 he headed a Zionist* Commission to Palestine, and was chief Zionist spokesman at the Paris Peace Conference* in 1919. He was president of the World Zionist Organization 1920–31, and played a major role in shaping the policy of the British Palestine mandate*. In 1929 he succeeded in obtaining from Britain recognition of the Jewish Agency*, in which he played a major part. He gave evidence to the Peel Commission*, whose recommendation of partition he supported. He returned to Britain as scientific adviser to the Ministry of Supply in 1939. When the state of Israel came into being he became its first President.

Welensky, Sir Roy (1907–), Prime Minister of the Central African Federation* 1955–63. Born in Rhodesia of Jewish Lithuanian refugees, he worked as a railway official, and in 1938 helped to form the Labour Party of Northern Rhodesia. He served on the Northern Rhodesian Legislative Council 1938–53. He founded the Federal Party in 1953, dedicated to racial partnership, and gave his full support to the Central African Federation, created largely as a result of his own negotiations with Godfrey Huggins*, whom he succeeded as Prime Minister. When the Federation dissolved in 1963, he lost the support of White Southern Rhodesians, who gave their allegiance to the Rhodesian Front and to Ian Smith*. A strong supporter of the Commonwealth, his New Rhodesia Party failed (1964) to gain any following in Rhodesia. He retired to independent Zambia and then to southern England.

Welfare State/Society. The term Welfare State is believed to have been coined by Archbishop William Temple* in 1941, to describe a country with a comprehensive system of social welfare funded by both taxation and schemes of national insurance. The emergence of the strong secular state during the 19th century was characterized by the development of state involvement in an increasing number of areas of social activity, for example education, health, and housing. A scheme of social insurance against unemployment, sickness, and old age was pioneered by Germany in the 1880s, and other European states followed. In Britain a variety of social welfare measures was introduced by Liberal governments 1906–14. Between the wars significant developments towards its establishment took place in New Zealand, while in the USA the New Deal* created a series of federal social welfare agencies. In 1942 a report by William Beveridge* proposed that the British system of national social insurance be extended to provide for the entire population 'from the cradle to the grave'. His proposals were implemented by the Attlee* ministries, which added other reforms, such as the National Health Service. In the Soviet Union and the East European states, state welfare provision became an official part of the fabric of society, while Sweden developed the most generous and efficient of all schemes. In the USA and elsewhere in the Western world, the concept of social welfare support remains selective; while social-security old-age pensions are universal, other benefits are highly selective, to be given only to 'those in need'. In Britain the heavy public expenditure required to distribute social benefits irrespective of means was increasingly challenged from the mid-1970s. Right-wing critics talked of 'the nanny state' and Thatcher* governments steadily reduced rates of welfare support.

West Bank, an area of Palestine west of the River Jordan, mandated to Britain 1920–47, but allocated by a UN Partition Plan of 1947 as a separate Arab state. In 1948 it was occupied by Abdullah ibn Hussein*, King of Hashemite Jordan, and, following an armistice of 1949, formally annexed by him, although only Britain and Pakistan recognized this. Some tension existed between Palestinian Arabs of the area and the Hashemite government, which they regarded as autocratic, but in the Arab–Israeli Six-Day War of 1967 the whole area was occupied by Israel. PLO* forces, operating from Jordan until 1971 and later from Lebanon, mounted continuous guerrilla attacks. At the same time Israeli occupation became more aggressive, increasing numbers of Jewish settlers expropriating Arab land. Thousands of Arab refugees fled to Lebanon and Jordan, and some were forcibly evicted. From December 1987 onwards Israeli troops used ever harsher methods to suppress the Palestinian protest of *Intifadah**. This began on the Gaza Strip* but quickly spread to the West Bank, stone-throwing Arab youths being beaten and shot in increasing numbers.

Western European Union (WEU). In March 1948 the Dunkirk Treaty* was enlarged when Britain, France, and the Benelux countries signed the Treaty of Brussels. In 1949 a parallel organization NATO* was established by the USA at the time of the Berlin Airlift*. The USA wanted West Germany within NATO, but France at first resisted, proposing instead a European Defence Community*. When this scheme collapsed in 1954, Germany and Italy were admitted to the Brussels Pact, now renamed the Western European Union, and Germany was admitted to NATO, the primary function of the WEU being at first to supervise the rearmament of West Germany as it joined NATO. Social and cultural activities were also envisaged, but these were transferred to the Council of Europe in 1960, leaving the WEU with the rather vague task of improving defence co-operation among countries of Western Europe. It was reactivated in 1984 and joined by Spain and Portugal in 1989. Enthusiasts for the European Community* regarded it as potentially a military arm of the Community.

Western Front, World War I, a line stretching from the Vosges mountain through Verdun and Amiens to Ostend in Belgium. In September 1914 German forces were checked in the first Battle of the Marne* and the attempt to reach the Channel ports was defeated in the first Battle of Ypres*. Thereafter from November 1914 both sides settled down to trench warfare: miles of parallel trenches, linked by intricate systems of communication trenches, protected by barbed wire, and with dug-out accommodation for those not on duty. To break the stalemate various new weapons were introduced: hand-grenades, poison gas, trench mortars, tanks, and massive artillery barrages. Highly lethal machine guns were developed, and casualties hitherto undreamed of followed every mass infantry attack. 1915 saw inconclusive battles, with heavy casualties: Neuve Chapelle (March), the second Battle of Ypres* (April/May), and Loos* (September). 1916 was dominated by the massacres of Verdun* and the Somme*. In 1917 the Germans withdrew to a new set of prepared trenches, the Hindenburg, or Siegfried, Line. In August 1917 the British launched yet another major offensive, the Battle of Passchendaele*, at the cost of 300,000 men, to be followed by the Battle of Cambrai*. The USA had entered the war in April 1917, and their troops began to land under General Pershing* in June. In March 1918 the German final offensive was launched. It failed to take Amiens, but further east reached the Marne, being stemmed at Château-Thierry by the Americans. In the July counter-offensive the British broke through at St-Quentin and the Americans in the Argonne region. By now superior Allied tanks had enabled a four-year stalemate to be broken, and by October the Germans were seeking an armistice.

West Indies, three groups of islands in the West Atlantic and the Caribbean Sea: the Bahamas*, the Greater Antilles*; the Lesser Antilles*. After the abolition of slavery, the introduction of free trade in the mid-19th century resulted in economic hardship and growing racial tension between White minorities and the poor, largely illiterate, Black majority. The development of sugar-beet in Europe reduced the demands for West Indian cane-sugar and resulted in further widespread unrest in the early 20th century. After World War II, emigration, the growth of the tourist industry, improved education, and attempts to diversify economies all helped to ease the situation. However, the combined effects of oil price rises, sharp falls in commodity prices, US protectionism, and world recession were responsible for economic decline in most islands and serious debt problems, notably in Jamaica*.

West Indies Federation (1958–62). There had been two attempts in the later 19th century at federating Britain's smaller West Indian colonies—the Leeward Islands Federation based on Antigua, and the Windward Islands Federation based on Grenada. Neither was very popular. After World War I the British came to accept the need for some element of self-government, but were convinced that no single island was large enough for separate political existence. Following a period of racial tension in the 1930s, a Colonial Development and Welfare Act was passed in 1940, while the University of the West Indies was founded in Jamaica, under the guidance of the University of London. Jamaica was granted universal suffrage in 1944, the first time that a Black majority would have the vote, the need for self-government having been accepted. In 1947 a Conference on the Federation of the British West Indies took place in Montego Bay, and a second Conference in London in 1953 finalized recommendations. A majority in Jamaica at that time supported it, as did Eric Williams* of Trinidad. There would be a federal capital (Port of Spain in Trinidad) and a Governor-General. Lord Hailes was appointed. The Federation lasted three years. In 1961 Alexander Bustamente* in Jamaica organized a referendum which voted against federation but for separate independence, which was granted. Trinidad and Tobago at once withdrew, whereupon the smaller islands were duly granted 'Associated Status', leading mostly to independence. Considerable debate followed as to the reasons for failure, which certainly went beyond the ambitions of Bustamente. Federation was a British-imposed import from the London Colonial Office and lacked any spirit of West Indianism to give it energy and drive. Below the Governor-General there was a pyramid of governors, commissioners, and ministers spread around the islands, with little but a complex bureaucratic machine (mostly of British officers) to hold the system together. West Indian nationalists were to remain strongly attached to Crown and Commonwealth, but they were in a hurry for independence, which few in London in 1947 ever envisaged within their lifetime.

Westminster, Statute of (1931). At the 1926 and 1930 Imperial Conferences* pressure was exerted by the dominions, Canada, New Zealand, the Commonwealth of Australia, the Union of South Africa, Eire, and Newfoundland for full and complete autonomy within the British Commonwealth. The result was the Statute of Westminster, accepted by each dominion Parliament. This recognized the right of each dominion to control its own foreign as well as domestic affairs, to establish a diplomatic corps, and to be represented at the League of Nations. It still left unresolved certain legal and constitutional questions, not least the status of the British Crown. The Consequential Provisions Act of 1949 allowed republics such as India to remain members of the Commonwealth.

Weygand, Maxime (1867–1965), French general. Born in Brussels, he was commissioned into the French army in 1888. He was Chief of Staff to Foch in World War I, and in 1920 was sent by the French government to aid the Poles in their ultimately successful defence against the advancing Red Army. After service in Syria he became Chief of the French General Staff 1930–5. In the military crisis of 1940 he was called from retirement to command the French armies trying to stem the German advance. Advising capitulation, he was sent by the Vichy* government to command in North Africa. He was dismissed at the request of the Germans, arrested by the Gestapo, and interned 1942–5. A sentence of infamy was passed on him for collaboration with the Vichy government, but this was quashed in 1948. He became a critic of all policies of decolonization in Africa.

White Russians, those who fought as counter-revolutionaries in the Russian Civil War*. The term was derived from the royalist opponents of the French Revolution, known as the Whites because they adopted the white flag of the Bourbon dynasty. The White Army, though smaller than the Red, was better equipped and had an abundance of tsarist officers, some of whom offered to serve as ordinary soldiers. Its two main bases were in the south, where it was led successively by Kornilov, Denikin*, and Wrangel*, and in Siberia, where Kolchak* was nominally head of a provisional government at Omsk. The Whites were ultimately defeated by their own internal quarrels, by their refusal to grant land reform in those areas under their control, and by the organizing genius of Trotsky*, who raised the Red Army and who gained control of the railways.

Whitlam, Gough (1916–), Prime Minister of Australia 1972–5. Born in Melbourne, he served in the Royal Australian Air Force in World War II, before becoming a successful lawyer. He entered the federal Parliament in 1952 as a Labor member, became leader of his party in 1967, and won the election of 1972. He immediately ended conscription, Australia having recently withdrawn from the Vietnam War. Although he had a majority in the House of Representatives, he did not in the Senate, and his government fell during the so-called 'Whitlam Crisis'. Members of his government had been seeking overseas loans without reference to himself or the cabinet, and he dismissed his deputy Prime Minister and Energy Minister. The Senate demanded a dissolution. Whitlam refused, whereupon the Senate held up Appropriation Bills (unlike in Britain these have to pass through both Houses). The Governor-General Sir John Kerr* now dismissed Whitlam, appointed Malcolm Fraser* as caretaker Prime Minister, and called a general election for December 1975. Whitlam lost this and a second election in 1977, after which he resigned the Labor leadership.

Wilhelmina (1880–1962), Queen of The Netherlands 1890–1948. The daughter of King William III, she probably had a large share in maintaining the neutrality of The Netherlands during World War I. When her country was invaded in 1940, she and her ministers maintained a government-in-exile in London. Through frequent radio talks she became a symbol of resistance to the Dutch people. She returned in 1945, but abdicated in favour of her daughter Juliana.

Wilkins, Roy (1901–81), US civil rights leader. The grandson of a slave, he graduated from the University of Minnesota and then worked as a journalist on the *Kansas City Call*, a newspaper for the Black community. In 1931 he joined the National Association for the Advancement of Colored People* (NAACP), editing its journal *Crisis* 1934–49. Appointed chief executive of NAACP in 1955, he consistently advocated the policy of legal redress in order to gain civil rights. Deeply committed to non-violence, in August 1963 he was one of the organizers of the March on Washington with Martin Luther King* and A. Philip Randolph*. In 1968 he served on the US delegation to the International Conference on Human Rights. In his later years at the NAACP he came under increasing pressure from more militant Blacks to diversify his non-violent policies. He retired in 1977, remaining a director emeritus of NAACP.

William II (1859–1941), German Kaiser 1888–1918. After dismissing Bismarck in 1890 he embarked on a policy of personal government, enthusiastically supporting Tirpitz* in his programme for naval expansion. He was genuinely keen to maintain friendship with Britain, but from 1908 took a less active part in government. He had little personal responsibility for war in 1914, although agreeing that Germany should honour its alliance with Austria-Hungary. From 1916 onwards he was a somewhat isolated figure, and fled to Holland on 8 November 1918. Queen Wilhelmina* steadfastly refused all requests that he be handed over for trial, and he died at Doorn after twenty years in exile.

Williams, Eric Eustace (1911–81), Premier of Trinidad and Tobago 1961–2, Prime Minister 1962–81. Educated at Queen's Royal College in Trinidad and at Oxford, in 1939 he went from there to teach at Howard University, USA, where he became associated with a Commission discussing the future of the Caribbean. He returned to Trinidad in 1955 to found the People's National Movement (PNM), which remained unsuccessful during the period of the West Indies Federation* (1958–62), towards which Williams was never more than luke-warm. In 1961, however, the PNM won a landslide electoral victory and he became Premier. In 1962 his country left the Federation and became independent, Williams becoming Prime Minister. An 'empirical' socialist, stressing the need for education and social welfare, he attracted foreign capital through tax incentives, making Trinidad and Tobago the wealthiest Commonwealth nation in the West Indies. He faced increasing opposition from militant Black Power groups before his death in 1981.

Wilson, James Harold, Baron Wilson of Rievaulx (1916–), British Prime Minister 1964–70, 1974–6. Born in Yorkshire, he was educated there and at Oxford, where he taught economics before World War II, when he became a civil servant. Elected to Parliament in 1945, he served as President of the Board of Trade from 1947 until February 1951, when he resigned over the imposition of health charges by Hugh Gaitskell*. He succeeded the latter as leader of the Labour Party in 1963, and narrowly won the general election of October 1964, increasing his majority to almost 100 in 1966. His policies were noted for their non-doctrinal approach, pragmatic according to his supporters, unprincipled according to detractors. Early economic problems took the form of a balance-of-payments deficit and a sterling crisis, the latter leading to a devaluation of the Pound in 1967. Experiments to create a prices and incomes policy collapsed, and the White Paper *In Place of Strife* (1969), which sought to end unofficial strikes and to establish a permanent Industrial Relations Commission, was withdrawn. In February 1968 his government passed an Immigration Act* to prevent mass entry into Britain from Uganda and Kenya of Asians who held British passports, and to restrict entry by all Commonwealth citizens. A Race Relations Bill was introduced two months later. Important regional development and social reforms were achieved by his governments 1964–70, including the introduction of comprehensive education, the expansion of higher education, changes in the law on sexual relations, divorce, and abortion, the end of the death penalty, and the reduction of the age of adulthood to 18. In 1968 the decision was made to withdraw British military forces from east of Suez. His government failed, however, to solve the problem of Rhodesian UDI (see ZIMBABWE); two series of talks with Ian Smith*, in HMS *Tiger* in 1966 and HMS *Fearless* in 1968, both collapsed. In 1974 he inherited a balance-of-payments deficit from Edward Heath five times as serious as that of 1964. During the next two years the major problem was inflation, which his government unsuccessfully tried to cure by abating wage demands through a 'Social Compact' with the trade unions. In 1975 his administration somewhat reluctantly confirmed British membership of the European Community after a referendum. Wilson's health was failing and on his 60th birthday he unexpectedly announced his retirement, handing over to James Callaghan*, who inherited a domestic inflation rate of over 20 per cent.

Wilson, Thomas Woodrow (1856–1924), President of the USA 1913–21. Son of a stern Presbyterian minister, he enrolled at Princeton University in 1875 and returned in 1890 as Professor of Jurisprudence. In 1902 he was appointed president of the university, and was responsible for a number of major changes in its educational and social organization. In 1910 he resigned to run as Governor of New Jersey, and was elected. He became a successful reform Governor, and earned a reputation which helped him to gain the Democratic nomination for the President in 1912. Once in office, Wilson determined to effect a programme known as the 'New Freedom', designed to stimulate competition by reducing the power of the Trusts, to promote equal opportunity, and to check corruption. Faced with the outbreak of World War I in 1914, he at first concentrated on conserving US neutrality. Gradually, however, he came to the view that the USA should enter the war as a 'Co-belligerent' ally of France and Britain. The German policy of resumed unrestricted submarine warfare from February 1917 led to the declaration of war in April. From then on he worked to realize his vision, proposed in the Fourteen Points*, of a peaceful post-war world. His Presbyterian background and respect for legal traditions made him favour an international peace-keeping forum, but he fell foul of American isolationism*, which saw his proposed League of Nations* as a tool of British and French imperialism. The isolationists in the Senate defeated him on the issue of US participation in the League, while his exertions in negotiating the Versailles Peace Settlement* and trying to win its acceptance by the Senate brought on a severe stroke in September 1919. He never fully recovered, and for the last year of his presidency Mrs Wilson, a lady of powerful personality, largely directed such business as could not be avoided or postponed.

'Wind of Change', a phrase used by the British Prime Minister Harold Macmillan*. In a speech to both Houses of Parliament in South Africa on 3 February 1960, he drew attention to the growth of national consciousness, which was sweeping like a 'wind of change' across the African continent, and warned that South Africa should take account of it.

Windsor, House of, official name of the British royal family. During World War I anti-German feeling was sufficiently strong for George V to feel in 1917 that it would be appropriate to remove all references to the German titles of Saxe-Coburg in his family name, originally derived from the marriage of Queen Victoria to Prince

Albert of Saxe-Coburg-Gotha. The King took the name Windsor, as Windsor Castle, Berks., has long been one of the homes of the royal family.

Wingate, Orde Charles (1903–44), British major-general. Born in India, he was educated at Charterhouse and the Royal Military Academy, being commissioned in 1922 and then serving in the Sudan. During the 1930s he helped to establish and train Jewish irregular forces operating in Palestine against Arabs, and became a brilliant exponent of guerrilla warfare. In 1941 he organized Somali and Ethiopian irregulars to fight Italian occupiers and restore Emperor Haile Selassie* to the throne. In 1942 he was sent to India. Here he created and led the Chindits, a Burmese guerrilla group that operated behind the Japanese lines. He died in an air crash in 1944, at the outset of his second greatly enlarged Chindit operation.

Wittgenstein, Ludwig Josef Johann (1889–1951), philosopher. Born and educated in Vienna, son of a rich industrialist, he came to England in 1911 to study aeronautical engineering, but then, at Cambridge, he developed a philosophical interest in mathematics. He was taught by Bertrand Russell*, although the latter soon admitted he was learning as much from Wittgenstein as he was teaching him. He served in the Austrian army in World War I and was captured in Italy in 1918. As a prisoner he wrote his most famous pamphlet, the *Tractatus Logico-Philosophicus*, and also read Leo Tolstoy. The latter's influence led him to give away his fortune and to live ascetically as a gardener and elementary schoolteacher for ten years. In 1929 he was persuaded to return to philosophy and to Cambridge, where he was appointed professor in 1939. The *Tractatus* and his later writings are all concerned with language: how is it, and the complex thinking for which it is used, possible? Equally, what lies beyond the limits of language, the unthinkable? since 'unsayable things do indeed exist'. His work influenced logical positivism* in Vienna and later 20th-century linguistic philosophy in Britain and the USA. At the same time it also influenced both educational theory and literary criticism.

Women's liberation, a radical feminist movement. Its analysis of the social relationship between the sexes goes deeper than earlier demands for education and the suffrage. During the 1960s and 1970s it was especially vocal in the USA, where the National Organization for Women (NOW) was formed and has remained active. Its demands were taken up in other industrialized countries, notably Britain and Australia. In Britain the Sex Discrimination Act and the creation of the Equal Opportunities Commission in 1975 gave legal effect to some demands, although many employment practices and financial rewards remained tilted in favour of men. Women's liberation movements in Islamic countries suffered a set-back with the revival of Islamic fundamentalism* in the 1970s, which re-established the traditional segregation and restriction of women.

Women's suffrage, the right of women to take part in political life and to vote in an election. The state of Wyoming, USA, first introduced it in 1869, and it was first attained at a national level in New Zealand in 1893. In 1920 all women over 21 were given the vote in the USA. The first European nation to grant female suffrage was Finland in 1906, with Norway following in 1913 and Germany in 1919. In Britain, as a result of suffragette agitation before 1914, the vote was granted in 1918 to those over 30, and in 1928 to the 'flappers' (women over 21). In 1918 Constance Markievicz became the first woman to be elected to the British House of Commons, though, as a member of Sinn Fein*, she never took her seat, becoming instead the world's first woman Minister of Labour in de Valera's Dáil Éireann. The Roman Catholic Church was reluctant to support women's suffrage and in many Catholic countries it was not gained until after World War II. In France it was granted in 1944, in Belgium in 1948, while in Switzerland not until 1971. It was won in Russia in 1917 after the Revolution, and in the new European republics after World War I. Following World War II it came to the rest of Eastern Europe. In Third World countries it was usually obtained with independence, and women have also been granted suffrage in most Muslim countries. One result of the suffrage has been the emergence of some outstanding women politicians, for example Golda Meir*, Indira Gandhi*, and Margaret Thatcher*, although the proportion of women taking an active part in politics has remained low.

Woodsworth, James Shaver (1874–1942), Canadian politician. He was ordained a Methodist minister in 1896 and devoted his ministry to work in Winnipeg among immigrants and Indians. In 1918 he resigned from his Church, being a strong pacifist, and was briefly imprisoned in 1919 because of his support for a general strike in Winnipeg. In 1921 he entered politics and was elected to the federal House of Commons, where in 1926 he was largely responsible for the Old Age Pensions Act. In that year he became leader of the Co-operative Commonwealth Federation*, but broke with the party in 1939, when once again he opposed Canada's participation in war against Germany.

Woomera Rocket Range. Some 300 miles from Adelaide in central Australia, it began in 1946 as the Australian Weapons Research Establishment. In 1949 it started to be used for guided missiles developed in the United Kingdom, which could be fired into the Australian desert, or even into the Indian Ocean 3,000 miles away. It was also used for testing early atomic bombs at Emu and Maralinga. This was done with such minimum precautions against radiation that a high cancer incidence among staff resulted. The tests ended in 1963, but it remains a site for testing, launching, and tracking space satellites for the European Space Vehicle Launcher Development Organization (ELDO). A Royal Commission in 1985 confirmed that both servicemen and Aborigines had died from radiation exposure, and Britain was ordered to 'clean up' both the Maralinga area and the island of Monte Bello.

World Bank (International Bank for Reconstruction and Development), a specialized agency of the UN. Proposed by the Bretton Woods Conference*, it was constituted in 1945. It has over 130 members, whose economic interests it is pledged to advance. It receives funds from member countries and from borrowing on the world money markets. After a series of reconstructional loans following the end of World War II, since 1949 it has concentrated on development loans, particularly to less-developed countries. Many early projects were grandiose and very expensive, and, even though loan interest rates are below the market rate, heavy debts were incurred. Sensitive to the problems of the North–South divide as highlighted in the Brandt Report* in 1977, it established a Special Fund to help Least Developed Countries with debt-service relief. At the same time it has moved towards supporting projects which involve simpler 'intermediate technology'. It has concentrated on agricultural and rural development, education, health, and public hygiene, while also helping to plan realistic strategies for industrialization.

World debt burden. During the 1980s the mounting debt burden of the Third World countries resulted in the transfer of $15 billion to the developed countries through servicing on a $1.3 trillion debt. This resulted in a sharp decline in expenditure on health and education: for example, in Nigeria the latter fell from 10 per cent of the GNP to under 2 per cent. At the same time twenty-five developing countries spent more on arms imports and military than on health and education, with an overall 7.5 annual percentage military increase 1960–86.

World Health Organization (WHO), a specialized agency of the UN, founded in 1948, with its headquarters in Geneva. Its aim is to promote 'the highest possible level of health' of all peoples. It undertakes the establishment of health services in less-developed countries, organizes campaigns against epidemic diseases, implements international quarantine and sanitation rules, funds research programmes and the collection and collation of statistics, and helps to sponsor the training of medical specialists. A notable success has been the eradication of smallpox throughout the world.

World War I (1914–18), war fought between the Allied powers (Britain, France, Russia, Japan, and Serbia, who were joined in the course of the war by Italy (1915), Portugal and Romania (1916), the USA and Greece (1917)) against the Central powers (Germany, Austria-Hungary, the Ottoman Empire, and Bulgaria (from 1915)). Its two principal causes were the fear of Germany's colonial ambitions and European tensions arising from shifting diplomatic divisions and nationalist agitation, especially in the Balkans. On the Western Front* both sides believed that superiority in numbers must ultimately prevail, despite the greater power of mechanized defence. A four-year stalemate of trench warfare was the result. Aerial warfare, still in its early stages, involved mainly military aircraft in combat, although German Zeppelin airships bombed Paris in March 1915 and London in May, before defence arrangements could be made. On the Eastern Front the initial Russian advance was defeated at Tannenberg* (1914). The Dardanelles* expedition was planned to provide relief, but it failed. Temporary Russian success against Austria-Hungary was followed by military disaster and the Russian revolutions. Serbia was occupied by Austrian, Hungarian, and German forces, and a Macedonian front against Bulgaria was opened with French, Serbian, and British troops based in Salonika*. The Mesopotamian campaign* was prompted by the need to protect oil installations and to conquer outlying parts of the Ottoman Empire. A British advance against the Turks in Palestine, aided by the Arab Revolt*, was successful. In north-east Italy a long and disastrous campaign, after Italy joined the Allies, was waged against Austria-Hungary, with success only coming late in 1918. Campaigns against Germany's colonial possessions in Africa and the Pacific were less demanding. At sea there was only one major encounter, the inconclusive Battle of Jutland* (1916), although the German U-boat offensive almost brought Britain to defeat in 1917. A conservative estimate of casualties gives 10 million killed and 20 million wounded. Its major result was the disintegration of the two empires of Austria-Hungary and the Ottomans, together with the resurrection of Poland. Armistices* were signed in November 1918 and peace terms imposed at the Paris Peace Conference*.

World War II (1939–45), a war fought between the Axis powers*, together with Thailand, and the Allied powers, which included Britain, France, the USA, Brazil, Mexico, and the Soviet Union. Having secretly rearmed Germany and signed the Nazi–Soviet Pact (August 1939), Hitler felt free to invade the Polish Corridor* and to divide Poland between Germany and the Soviet Union. Britain and France, which until 1939 had adopted a policy of diplomatic appeasement, declared war on 3 September. The Soviet Union attacked Finland and occupied the Baltic states. In 1940 Norway, Denmark, Belgium, The Netherlands, and three-fifths of France fell to Germany in rapid succession, while the rest of France was established as a neutral state under the Vichy* government. A bombing offensive was launched against Britain, but the planned invasion was postponed after the Battle of Britain*. Pro-Nazi governments in Hungary, Bulgaria, Romania, and Slovakia supported the Axis powers. Albania had been occupied by Italy, and in March–April 1941 Yugoslavia and Greece were overrun. In June 1941, breaking his pact with Stalin, Hitler invaded the Soviet Union, reaching the outskirts of Leningrad and Moscow. Without declaring war, Japan attacked the US fleet at Pearl Harbor* in December 1941, provoking the USA to enter the war. Burma, Malaya, the Philippines, and the Dutch East Indies all fell to Japanese forces. In 1942 the first Allied counter-offensives began: the USA gradually asserted naval superiority in the Pacific, while in North Africa Rommel* was defeated and US troops landed. In 1943 Allied armies began the invasion and conquest of Italy, resulting in the overthrow of Mussolini's government. On the Eastern Front decisive battles at Stalingrad and Kursk broke the German hold. In June 1944 the Allied invasion of Europe was launched, and Germany surrendered in May 1945, after Hitler's suicide in Berlin. The Pacific campaigns* had eliminated the Japanese navy, and the heavy strategic bombing of Japan by the USA, culminating with atomic bombs, induced Japan to surrender on 2 September 1945. The dead have been estimated at 15 million military, of which 2 million were Soviet prisoners of war. An estimated 35 million civilians died, with between 4 and 5 million Jews perishing in concentration camps, and perhaps 2 million in mass murders in Eastern Europe. Refugees from the Soviet Union, East Germany and the rest of Eastern Europe numbered many millions. The war resulted in the forty-year division of Germany, the restoration to the Soviet Union of lands lost in 1919–20, together with the creation of Communist buffer-states along the Soviet frontier. Britain had accumulated a $20 billion debt. In the Far East nationalist Resistance forces were to ensure the decolonization* of South-East Asian countries. The war hastened the independence of India, Pakistan, and Ceylon, and stimulated a demand for independence in Africa. The USA and the Soviet Union emerged as the two dominant global powers. Their wartime alliance collapsed within three years, and each embarked on a programme of rearmament, with nuclear capability, as the Cold War* developed.

Wrangel, Piotr Nikolayevich (1878–1928), Russian general. Born in St Petersburg, he was commissioned into the tsarist Imperial Army and served as a professional soldier through the Russo-Japanese War 1904–5 and in World War I with the Cossacks. In 1917 he moved to Siberia to join the counter-revolutionary forces there. He served first under Kolchak* and then Denikin*. When the latter withdrew to the Caucasus in March 1920, he became Commander-in-Chief of all White armies operating from the Crimea, where he established a provisional government. In November 1920 the Red Army broke through his defences and he fled to Turkey and later to Belgium, where he died. His defeat ended White Russian* resistance to the Bolshevik revolution.

Wuchang Uprising (10 October 1911), the revolt which began the Chinese Revolution*. An accidental explosion in the city of Wuchang forced republican revolutionaries to begin a planned uprising earlier than intended; on the next day army units won over to the rebel cause took over the city. The Qing government failed to respond swiftly to the uprising, and further provincial risings followed, leading to the formation of a Provisional Republican Government on 1 January 1912, and the abdication of the Qing Emperor Pu Yi*.

Xhosa, a Ngoni people of southern Africa speaking a Bantu language. They were traditionally herdsmen (sheep and cattle) and during the 19th century were victims of European expansion in Cape Colony during the so-called 'Kaffir Wars' (1811–79), being forced to move to the east. They became concentrated in what became the two Bantustan* homelands of Ciskei and Transkei. They are long-time rivals of the Zulu* people.

Xi'an Incident (December 1936). While visiting disaffected Manchurian troops in Xi'an, President Jiang Jiehi* (Chiang Kai-shek) was kidnapped by a group of officers under orders from Chang Hsueh-liang*. They attempted to force him to give up his campaigns against the Communists and lead a national war against the Japanese, who had occupied Manchuria in 1931. After Jiang had refused to accede to their demands, the Communist leader Zhou Enlai* became involved in the negotiations. After being held for thirteen days Jiang was released, having promised to take a more active role against the Japanese and to allow local autonomy to the Communists. The incident led to the limited co-operation between the Communists and the Guomindang of the United Front* during the years 1937–45.

Yahya Khan, Agha Mohammed (1917–80), President of Pakistan 1969–71. Educated at Punjab University and the Indian Military Academy, he served in the Indian army on the North-West Frontier, in the Middle East, and in Italy. In 1947 he created the Pakistan Staff College and in 1966 became Commander-in-Chief of the Pakistan army. As such he was chief administrator of martial law. In 1969 he was invited by Ayub Kahn* to succeed him as President. In 1971 the brutality with which his army tried to suppress the Awami League* induced India to invade East Pakistan, precipitating the third Indo-Pakistan War*. With the loss of Bangladesh he resigned, from office and was succeeded by Zulfikar Bhutto*, who placed him under house arrest, soon after which he suffered from a stroke.

Yalta Conference (4–11 February 1945), a meeting between the Allied leaders Stalin, Roosevelt, and Churchill at Yalta in the Soviet Union. The final stages of strategy for World War II were discussed, as well as the proposed occupation of Germany. Stalin obtained agreement for the Ukraine and Outer Mongolia to be admitted as full members of the UN, whose founding conference was planned for San Francisco two months later. Stalin undertook to enter the war against Japan after the surrender of Germany, and was promised the Kurile Islands and an occupation zone in Korea. The conference was to be followed five months later by the Potsdam Conference*, by which time Truman would have replaced Roosevelt as US President.

Yamagata Aritomo (1838–1922), first Japanese Prime Minister under the Meiji Constitution 1889–91 and again 1898–1900. A member of a samurai family, he had been an early opponent of the Westernization of Japan but later became a strong advocate of modernization. He was the prime architect of the modern Japanese army, based on conscription and an unswerving loyalty to the Emperor. He exercised great influence and power behind the scenes in the years before and during World War I, but in 1921 he was publicly censured for meddling in the Crown Prince's marriage, dying in disgrace.

Yamamoto Isoruku (1884–1943), Japanese admiral. Educated at the Japanese Naval College, he was wounded in the Russo-Japanese War (1904–5), after which he went on to naval staff college. He served as a naval attaché in Washington, and in the 1930s was responsible for developing the Japanese Navy Air Corps. As Vice-Minister for the Navy he spoke against the Axis agreement with Germany and Italy (1936), but was appointed Commander-in-Chief of a combined squadron in 1939. He still opposed Japan's participation in the war, but on a decision being taken he devised the successful Pearl Harbor* attack and directed the Battle of the Java Sea. He suffered defeat, however, in the Battles of the Coral Sea* and Midway Island*. In April 1943 US intelligence identified his movements and his plane was destroyed over the Solomon Islands.

Yamashita Tomoyuki (1888–1946), Japanese general. Born near Tokyo, as a young officer he served in the Russo-Japanese War (1904–5) and in World War I. He attended staff officers' college (1926) and then held a series of staff appointments, by 1937 serving in Manchuria, with the rank of lieutenant-general. In 1940 he headed a military mission to Germany and on return was given command of the 25th Army, which invaded Malaya in December 1941, capturing Singapore in February 1942 and going on to campaign in Burma. In 1944 he was given command in the Philippines, organizing fierce opposition to US landings at Leyte and on Luzon, where he continued to fight on until the surrender of 2 September. He was tried and executed in Manila for atrocities committed by his troops during the campaign.

Yan'an (Yen-an), a town and district in the north Shanxi province of China, which became the headquarters of the Chinese Communist Party armies after the Long March*. It remained their headquarters throughout the Japanese war, until captured by the Nationalists in 1947. It became a symbol of the pioneer spirit of the early Communist Revolution in China, when agrarian reforms were first introduced. It was recaptured by Peng Dehuai* in 1948.

Yan Jiangan (Yen Chia'kan) (1905–), President of the Republic of China (Taiwan) 1975–8. A banking expert, he served in the Chinese National government and after World War II moved to Taiwan. Here he was chairman of the Bank of Taiwan and held various government financial posts, showing great skill in managing the Taiwanese economy. He served as Prime Minister 1963–72 and Vice-President 1966–75 before succeeding Jiang Jiehi (Chiang Kai-shek) as President. He was succeeded in 1978 by Chiang Ching-kuo*, the latter's son, and retired into private life.

Yemen, Republic of. At the base of the Red Sea on the Arabian peninsula, the Yemen was under Turkish rule 1517–1918. It was the scene of fighting in World War I between Turkish troops and the British garrison of Aden*. In 1918, with British support, it was proclaimed a kingdom under Imam Yahya, its borders with both Aden and Saudi Arabia being matters for dispute. Yahya was assassinated in 1948, and his son Ahmad ruled until 1962. On his death the army under General Abdullah al-Sallal proclaimed the Yemen Arab Republic, backed by both Syria and Egypt. Saudi Arabia supported those tribes who gave their loyalty to Ahmad's son Imam Muhammad al-Badr. Civil war lasted until 1967, when Nasser withdrew Egyptian troops after the defeat of the Six-Day War*. Sallal resigned and a more moderate government was formed. In April 1970 there was a general pacification, but in 1979 a month-long war broke out with the neighbouring People's Democratic Republic of Yemen (see ADEN). Intermittent talks to unify with the latter followed, with a draft constitution agreed in December 1989. The unified state was proclaimed in May 1990, its political capital being San'aa and commercial capital Aden. A five-member Council was headed by President Ali Abdullah Saleh. The new Republic was welcomed to the UN and found itself a member of the Security Council at the time of the Gulf Crisis*.

Yezhovshchina or Great Purge. Nikolay Yezhov, as Commissar for Internal Affairs 1936–8, was responsible for the bloodiest period of the Stalinist regime. There had been spectacular state trials before 1934, but the murder of Kirov* (probably itself on Stalin's orders) unleashed a terror in which millions died, including all those senior members of the Bolshevik Party thought to rival Stalin. Six members of the Politburo, more than half the generals of the Red Army including Tukachevsky, Chief of the General Staff, and even the head of the NKVD itself, Yakoda, were all victims. Most trials were brief. After arrest and torture to obtain a 'confession', they were summarily sentenced to death or long-term imprisonment by a three-man NKVD committee. There were in addition some 'show trials', such as that of Bukharin* in March 1938, perhaps staged for propaganda purposes. When Beria* replaced Yezhov the pace slackened, but German minorities suffered badly after war broke out in 1941. Perhaps as many as 10 million were arrested in these years, of whom at least 3 million were executed, while at least as many died in prison camps.

Yom Kippur War (October 1973), known by Arabs as the October War. It began on 6 October, the Day of Atonement (Yom Kippur), the holiest day of prayer and fasting in the Jewish year. In their surprise attack Egypt and Syria won some initial success—Egyptian troops advancing over the Suez Canal and the Syrians advancing along the Golan Heights, which had been occupied by Israel in 1967. Although caught unawares Israel forcefully counter-attacked within two days. On 8 October Israeli forces recrossed the Suez Canal further east, encircling part of the Egyptian army and advancing on Cairo; at the same time other Israelis recovered the Golan Heights, with tanks advancing to within 35 miles of Damascus. Alarmed by Israeli successes, oil-rich Saudi Arabia put pressure on its main customer the USA to persuade Israel to halt its advances and accept UN mediation. A cease-fire was arranged on 24 October, followed by disengagement along the Suez Canal and the establishment of a UN peace-force on the Golan Heights. Prime Minister Golda Meir* was criticized for accepting peace, but her critic and successor Menachem Begin was to go on to agree to the Camp David Accord*(1978).

Yoruba, a Nigerian people who are also found in Benin and Togo and whose traditional life-style involved agriculture, especially cocoa farming. During the 19th century a sophisticated urban society developed whose main cities were Oyo, Ife, and Ibadan. A loose confederation of kingdoms had extensive trade links and high achievements in sculpture and ceramics, with a single spiritual leader, the Oni of Ife. In 1888 the British imposed 'protection' over all Yoruba kingdoms, which were then incorporated into Southern Nigeria. Distinctive Yoruba culture partly disintegrated, although the elaborate traditional religion survived well against both Christianity and Islam. At the same time Yoruba people quickly adapted to Western mores and Ibadan became a leading African city. The Western Region of Nigeria, formed in 1954, was dominated by Yoruba Ibadan.

Yoshida Shigeru (1878–1967), Prime Minister of Japan 1946–7, 1949–54. A graduate of Tokyo University, he entered the Japanese diplomatic corps and held various posts. In 1930 he was appointed ambassador to Italy. A Liberal-conservative, his appointment as Foreign Minister was blocked by militarists in 1936, and instead he became ambassador to Britain until the war. In the closing stages of war he was imprisoned for advocating surrender, but he emerged afterwards as leader of the Liberal Party. As Prime Minister he was a major architect of post-war Japan and its socio-economic recovery. He worked closely with MacArthur*, successfully negotiating the peace treaty of 1951 and espousing Western policies. His popularity peaked with the ending of Allied occupation in 1952 and thereafter rapidly declined. When the Liberal Democratic Party was formed in 1955 he retired into private life.

Young Maori Party. In 1914 there still survived a small group of followers of Rua Kenana who demanded Maori independence. These were crushed by police action in 1916. A group from the next generation of Maori, many educated at Te Aute College at the turn of the century, believed in the need to 'emulate the pakeha'. They became urbanized on the east coast and successfully trained in the professions—teaching, law, and medicine. They formed themselves into a political party, the Young Maori Party, in alliance with the Liberal Party, entering Parliament from 1905 onwards to represent the four Maori electoral districts which had been created in 1867. One outstanding member of the group was Apirana Ngata*, who became Minister for Native Affairs in 1928. Much was achieved, but by the mid-1930s the party was losing support, with the rise of the Ratana* movement.

Young Plan (February 1929), proposals for settlement of German reparations debt. The plan was embodied in recommendations from a committee which met in Paris under the chairmanship of the US financier Owen D. Young, to revise the Dawes Plan*. The total sum due from Germany was reduced by 75 per cent to 121 billion Reichsmarks, to be paid in fifty-nine instalments. Foreign investment in the German economy was to be made easier. Germany accepted the plan and made a first payment in May 1930. The collapse of the world economy followed and in 1932 the Lausanne Pact accepted that further payments would have to be suspended. The following year Hitler repudiated all reparations obligations.

Young Turks, European name for the reformers of the Ottoman Empire who carried out a revolution in 1908. They should be distinguished from the Ottoman reformers of an earlier generation who were active 1865–76, and from the military reformers under Atatürk* who seized power after the collapse of the Ottoman Empire in 1918. Their Committee of Union and Progress seized power in 1913, and under a triumvirate of Enver Pasha*, Jamal Pasha, and Talaat Bey ruled the Empire until its collapse.

Ypres, Battles of, World War I. There were four so-called Battles of Ypres. The first, following the retreat of the British Expeditionary Force* from Mons, raged 12 October–11 November 1914, the Germans capturing the Messines ridge, but failing to take Ypres or to reach the Channel ports. In the second (22 April–24 May 1915), a German assault, using poison gas, failed to break the British line. The third, which began on 7 June 1917 with a British, Canadian, and Australian mining operation of the Messines ridge, continued into the Passchendaele* offensive August–November 1917. The fourth was part of the German offensive of March–April 1918, when the Germans were held at the River Lys and once again failed to capture the devastated city of Ypres. It is estimated that over 500,000 British and Commonwealth troops died fighting on the Ypres salient during World War I.

Yuan Shikai (Yuan Shih-k'ai) (1859–1916), Chinese soldier and statesman. He had served in Korea before returning to China in 1898 to undertake a programme of army reform. Dismissed after the death of the Empress Dowager Cixi in 1908, he retired to northern China. He was recalled by the court when the Chinese Revolution* began, but he temporarily sided with the republicans and advised the Emperor to abdicate. In 1912 he became President of the Republic in place of Sun Yixian*, having some success in restoring central control. But his suppression of the Guomindang* in 1913, dismissal of Parliament, and submission to Japan's Twenty-One Demands* provoked a second revolution in the Chang Jiang (Yangtze) region. In 1916 he had himself proclaimed Emperor, provoking severe opposition, but he died shortly afterwards, leaving China divided between rival warlords*.

Yugoslavia. After World War I the Kingdom of Serbs, Croats, and Slovenes was formed from the Austro-Hungarian provinces of Slovenia, Croatia, and Bosnia-Hercegovina, together with Serbia, Montenegro, and Macedonian lands ceded from Bulgaria. The King of Serbia Peter I was to rule, and he was succeeded in 1921 by his son Alexander I*. After the death of Nikola Pašić* in 1927 political turmoil caused the King to establish a royal dictatorship, renaming the country Yugoslavia (January 1929). His autocratic, pro-Serbian policies ended with his assassination in 1934. Alexander was succeeded by his infant son Peter II*, with his uncle Prince Paul as Regent. The latter maintained the right-wing autocratic policies of his brother and in 1941 (25 March) signed the Anti-Comintern Pact* with Hitler. Peter II claimed his right to rule, and a new government under General Simovic was established. At the beginning of April German troops invaded, and Belgrade fell on 13 April, the German forces having been aided by Italians from Albania, Bulgarians, and Hungarians. The King fled to London and dismemberment of the country followed. The puppet 'Kingdom of Croatia' was created by Ante Pavelić* and tension followed between his supporters, the royalist chetnik* supporters, and the Communist partisans of Tito*. Peter lost the support of the Allies, who were favouring Tito. The latter held the Congress of Jajce* in November 1943, and by 1944 was in virtual control as the Red Army advanced and the Germans withdrew. Subasic, Premier of the exiled government in London, returned to work for a while with Tito, but he resigned in November 1945, allowing Tito to proclaim the Socialist Federal Republic of Yugoslavia, supported by the Soviet Union. It soon became clear, however, that he would not accept Soviet domination, and in June 1948 the party was expelled by Stalin from 'the family of fraternal Communist Parties'. Yugoslavia became the leader of the non-aligned states and champion of 'positive neutrality'. Improved relations with the West followed. In 1953 Tito became President under a new constitution and, following Stalin's death that year, diplomatic and economic ties with the Soviet Union were renewed. After he died in 1980 a rotating presidency was established. However, ethnic and cultural differences within the Federal Republic caused it increasingly to disintegrate during the 1980s. Albanians in the district of Kosovo staged rebellion and a state of emergency was proclaimed February–July 1989; there was a fierce leadership struggle within the plenum of the League of Communists of Yugoslavia. The hard-line Serbian leader Slobodan Milosevic pressed for greater Serbian powers over Kosovo. There were demonstrations in Montenegro and Croatia and a rift between Serbia and Slovenia. In 1990 the situation worsened. Macedonian demands for independence revived, and in May non-Communist governments were formed in Slovenia and Croatia, as they were in Bosnia-Hercegovina and Macedonia in December. In June the Serbian Borisav Jovic assumed the office of collective President, warning against a possible disintegration of the Federal Republic. By December in both Slovenia and Croatia anti-Serbian partisans were active, prepared to fight the Yugoslav army.

Yukon Territory. A territory of 482,515 sq. km. in the north-west of Canada with a population in 1989 of 26,166. Following the Klondike Gold Rush of 1896 it had been separated from the North West Territories and given its own system of administration, which today consists of a Commissioner, responsible to the federal Minister for Indian Affairs and Northern Development, an elected Legislative Assembly of sixteen members, with an Executive Council of five members, and the right to elect one member to the federal Parliament. Its mineral resources of gold, silver, asbestos, lead, zinc, copper, tungsten, and coal have been steadily exploited since World War II; the Territory has become increasingly popular for tourists, many of whom come for fishing or big game hunting.

Zaharoff, Sir Basil (1850–1936), international financier and munitions manufacturer. Born Zacharias Basileios and originating from Anatolia, he became known as 'the mystery man of Europe'. Before World War I he built up profitable connections with British, German, and Swedish armaments firms and amassed great wealth. He was accused of fomenting warfare and of political intrigue. During the war his sympathies lay with the Allies, who rewarded him with civil decorations for supplying them with arms and intelligence information. After the war he endowed the universities of Paris, Oxford, and London from his vast fortune.

Zaibatsu, Japanese business conglomerates or 'financial cliques'. These large concerns, with ownership concentrated in the hands of a single family, were responsible for the rapid industrialization of Japan in the late 19th century. They had their origins in the activities of the *seisho*, or 'political merchants', who made their fortunes by exploiting business links with government. As government played a less direct role in economic activity, *zaibatsu* like the Mitsui and Mitsubishi expanded to fill the gap, through the ownership of interrelated mining, transport, industrial, commercial, and financial concerns. They dominated the business sector in a fashion which had no near equivalent elsewhere in the industrialized world. Despite efforts to break up their power after World War II, they continued in a modified form to provide the characteristic pattern of Japanese industry into the 1980s. They are now more usually known in Japan as *keiretsu*.

Zaïre (formerly Belgian Congo). H. M. Stanley's voyage down the River Zaire (Congo) in 1871 prompted King Leopold II of Belgium to found the Congo Free State, to open up the vast mineral resources of the area. Maladministration by Leopold's agents obliged him in 1906 to hand the state over to the Belgian Parliament as the Belgian Congo. In the next fifty years virtually nothing was done to prepare the country for self-government, there being only a few Catholic mission schools. An outbreak of unrest in 1959 led to the hasty granting of independence in the following year. The regime of Patrice Lumumba* was however undermined by the Congo Crisis*, and disorder in the newly named Congo Republic remained endemic until the coup of General Mobutu* in 1965. In 1967 the Union Minière, the largest copper-mining concern, was nationalized and Mobutu achieved some measure of economic recovery. In 1971 the name of the country was changed to Zaïre. Falling copper prices and centralized policies undermined foreign business confidence. Two revolts, involving mercenaries and probably backed by outside interests, in the province of Shaba* (formerly Katanga) in 1977 and 1978 were put down with French military assistance. The Constitution of 1980 recognized the sole political party, Mouvement Populaire de la

Révolution (MPR), and President Mobutu was re-elected in 1977 and 1984. Multi-party elections were promised for May 1991.

Zambia. Formerly Northern Rhodesia, it had been settled in the early 19th century by Ngoni* people in flight from the Zulu*. In 1890 agents from Cecil Rhodes entered the country (known at this time as Barotseland). Rhodes's British South Africa Company had been granted responsibility for it in its charter of 1889, and it began to open up the rich copper deposits of Broken Hill from 1902. The country was named Northern Rhodesia in 1911. It became a British protectorate in 1924, and between 1953 and 1963 it was federated with Southern Rhodesia and Nyasaland as part of the Central African Federation*. It became the independent Republic of Zambia in 1964 under President Kaunda*. Dependent on its large copper-mining industry, Zambia has experienced persistent economic difficulties due to its lack of a coastline and port facilities and to low copper prices. It suffered from the economic sanctions against Rhodesia 1965–80, but was assisted by the construction of the Tan-Zam railway*. It has given refuge to political exiles from its neighbours Rhodesia (Zimbabwe), Angola, Namibia, and Mozambique, and played a prominent part in the formation and organization of SADCC*. In May 1990 Kaunda yielded to pressure to hold a referendum on the introduction of a multi-party system.

Zamindar, a holder or occupier of land in India. In Bihar and West Bengal the term acquired a special meaning after the British had created a class of hereditary tax collectors in the 18th century. They became large and wealthy landowners, whose estates were not broken up until 1947.

Zanzibar, an island off the East African coast. British and German trading interests were developing there in the later 19th century, but in 1890 Germany conceded British autonomy in exchange for the North Sea island of Heligoland. It became a British protectorate under its traditional sultans. In December 1963 it became an independent member of the Commonwealth, but in January 1964 the last Sultan was deposed and a republic proclaimed. Union with Tanganyika, to form the United Republic of Tanzania*, followed in April. It retained its own internal administration through a Revolution Council and a certain degree of autonomy. After the assassination of Sheikh Karume in 1972, Aboud Jumbe and the ruling Afro-Shirazi Party ruthlessly suppressed all forms of political opposition until growing resentment forced Jumbe to resign in 1984, Ali Hassan Mwinyi succeeding him. When the latter became President of Tanzania he was succeeded by Idris Abdul Wakil. There has been growing opposition on the island from a banned Islamic fundamentalist* movement Bismillahi.

Zapata, Emiliano (1879–1919), a leader of the Mexican Revolution*. Son of a mestizo peasant, in 1909 he forcefully occupied estates of the great haciendas and distributed the land among his peasant band. He joined Madero's revolution, and his tiny band of guerrillas succeeded in toppling the Díaz regime. On the latter's overthrow he sought to have all land returned to the *ejidos* (the former Indian communal system of ownership). When he realized that Madero was not prepared to embark upon major programmes of agrarian reform, he declared against him. After Madero's death, he occupied Mexico City with the so-called Liberation Army of the South and was joined by Pancho Villa* from the north. The two men quarrelled with Carranza*, however, and were forced to leave the city, Carranza's forces defeating Villa in 1915. In 1917 Zapata captured Puebla and his troops widened the southern zone under his control. Here he opened a Rural Loan Bank and implemented his land reforms. In 1917 Carranza again defeated Villa's guerrillas, isolating Zapata's southern zone. The latter however continued successfully to control this, until in 1919 he was ambushed and shot by Carranza's troops at Chinameco. For eight years he had led his peasant guerrilla armies against the haciendas and successive heads of state. His creed, *zapatismo*, became one with *agrarismo*, calling for the return of the land to the Indians. With the fundamental tenet of *indianismo*, it was the cultural, nationalist movement of the Mexican Indian.

Zeebrugge Raid, World War I (23 April 1918). During the night of 22–3 April 1918 a daring raid was carried out by a British unit under Admiral Keyes against the German U-boat base at Zeebrugge in Belgium. A submarine loaded with explosives was blown up below the viaduct and three blockships were sunk in the channel, nearly but not completely closing it.

Zeppelins, German airships first designed and built by Count Ferdinand Zeppelin (1838–1917). His first such aircraft LZ-1 was built in 1900, taking off from a floating hangar on Lake Constance. It had two external cars, each with a 16 hp engine, and reached a speed of 20 m.p.h. After a twenty-four-hour flight in 1906, Zeppelin was commissioned to build a fleet of 100 ships, and he opened a passenger service in 1910. Zeppelin bombing raids took place on various French and English towns, including Paris and London, during World War I, after which the *Graf Zeppelin* was built in 1928. This successfully crossed the Atlantic 144 times, to be followed by the *Hindenburg* in 1936. This travelled at 84 m.p.h. and completed ten round trips between Germany and the USA before exploding in May 1937. Allied bombing destroyed the Zeppelin factory in World War II and it was never rebuilt.

Zhdanov, Andrey (1896–1948), Soviet politician. He joined the Bolsheviks* when a young man in 1915, and took an active part in the 1917 revolutions. In 1924 he became responsible for the Communist Party in Novgorod, moving to Leningrad in a similar role in 1934, and becoming secretary of the Central Committee. He survived the Great Purge* and became a member of the Politburo in 1939. He was responsible for the ideological aspects of Stalinism, enforcing socialist realism* in the arts and a Bolshevik historiography, while campaigning against Western cultural 'decadence'. In 1947 he played an important part in the formation of the Cominform*.

Zhivkov, Todor (1911–), President of Bulgaria 1971–89. Son of poor peasants, he drifted to Sofia when a boy, where he joined the outlawed youth section of the Communist Party. By 1937 he had reached senior rank within the party and he helped to organize the Resistance People's Liberation Insurgent Army through World War II, welcoming the Red Army in September 1944. During 1945, as commander of the militia, he arrested thousands of political opponents, who disappeared. In 1946 the Bulgarian monarchy was abolished and the new Republic was ruled by the Communist Party until 1989. Zhivkov became a member of the Politburo in 1951 and, as a protégé of Khrushchev, first secretary of the party. He became Premier of Bulgaria 1961–71, surviving an attempted coup in 1968, and was then President from 1971 until ousted from office in November 1989.

Zhou Enlai (Chou En-lai) (1898–1976), Premier of the People's Republic of China 1949–76. From a relatively affluent family in the Jiangsu province, he attended a number of schools before going to further his studies in Japan. He returned to China with the May Fourth movement*, when he first met Mao Zedong* in a study-group at Beijing University and became editor of a student magazine. He was imprisoned for anti-Japanese rioting and was then sponsored to go to Paris. He was in Europe for four years, and while there became a Communist and founded a branch of the Chinese

Communist Party in Paris. On his return to China he was appointed secretary of the Guangdong (Kwangtung) provincial committee of the party and political director of the Whampoa Military Academy. In 1926 he went in secret to Shanghai to organize Communist-controlled unions, and here he was arrested in April 1927 when Jiang Jiehi* (Chiang Kai-shek) struck against the Communists. Unlike many he escaped. He spent the next four years moving around China, being involved in unsuccessful Communist risings in Nanchang and Guangzhou (Canton) and also going to Moscow. In 1931 he joined Mao Zedong in the Jiangxi Soviet*, being a member of the Politburo. He played a key role in planning the first phase of the Long March* and, in December 1936, used his brilliant negotiating skills at Xi'an to persuade Jiang Jiehi to re-establish the United Front* against the Japanese. When the National government* established a People's Political Council in 1938 Zhou headed the Communist delegation, and he spent much of the war-years in Chongqing (Chungking) as a liaison officer. Throughout the war he was the leading spokesman for the Communist cause, taking part in the US attempt to mediate in the civil war in 1946. Appointed Premier on the establishment of the People's Republic, he was also Foreign Minister 1949–58. He played a major role in negotiations leading to the Panmunjom Armistice in Korea and in the Geneva Conference ending the French Indo-China War. At the Bandung Conference* (1955) he won wide support for his moderation and became an influential figure in post-imperial Asia and Africa. He was one of the earliest proponents (1963) of the Four Modernizations* policy, which later became associated with Deng Xiaoping*. During the Cultural Revolution* he actively restrained extremists and helped to restore order. He was the architect behind the reconciliation with the USA, and was instrumental in bringing about the 1972 Nixon visit, perhaps one of his most notable achievements. He died of cancer, a much-loved and revered figure.

Zhu De (Chy Teh) (1886–1976), Commander-in-Chief of the Chinese Communist army. Born in Sichuan province of a wealthy landowning family, he was commissioned into the Imperial Army, but in 1911 gave his support to Sun Yixian* in the Revolution. He then spent ten years serving with various warlords in north China. After World War I he travelled in Europe. Here he became interested in politics, became a Communist, and was expelled from Germany. On his return to China he joined the Chinese Communist Party, to which he gave all his inherited wealth. He became an officer in the National Revolutionary Army*, but when the Guomindang* split in 1927 he sided against the Nationalists, taking part in an unsuccessful uprising in Nanchang, after which he led his surviving 5,000 comrades to join Mao in the establishment of the Jianxi Soviet*. Here he built an army which successfully defied Jiang Jiehi's (Chiang Kai-shek) troops for three years, using characteristic guerrilla tactics. In 1934 he commanded the army on the Long March* into the Shanxi province, where he set about rebuilding an army. From 1937 until 1945 he commanded this 8th Route Army in north China against the Japanese, and on the resumption of civil war was Supreme Commander of the newly named People's Liberation Army*, when ranks were introduced into the PLA, he was appointed marshal. From 1954 he was chairman of the standing Committee of the National People's Congress*. At the beginning of the Cultural Revolution* he was 'purged', but was restored to favour in 1967 and lived out his life in honoured retirement. Zhu perfected large-scale guerrilla warfare, destroying his enemy by attrition rather than by pitched battle. He would win the support of the rural populace by the high discipline and courage of his troops.

Zhukov, Georgi Konstantinovich (1896–1974), marshal of the Soviet Union. Of peasant origin, he joined the Bolsheviks when a young man and enlisted in Trotsky's

Red Army in 1918. He became a junior officer and then a professional soldier. He trained as a tank specialist under German instructors at Kazan 1921–2, and went on during the 1930s to gain a reputation for defeating Japanese infiltrators into Mongolia. By now a senior general who had survived the Great Purge*, he was responsible for much of the planning of the Soviet Union's campaigns after the German invasion of 1941. He defeated the Germans at Stalingrad and went on to lift the siege of Leningrad. In 1945 he led the final assault on Germany, capturing Berlin and becoming commander of the Soviet zone of occupation. In 1947 he was demoted by Stalin, to become commander of the Odessa military district; but after the latter's death he rose to become Defence Minister in 1955. He supported Khrushchev* against his political enemies, but was nevertheless dismissed in 1957, only to be reinstated in 1964 after Khrushchev was deposed. He was restored in rank and allowed to live in honoured retirement. He is generally accepted to have been one of the outstanding commanders of World War II.

Zia ul-Haq, General Mohammed (1924–88), President of Pakistan 1978–88. After serving in the Indian army during World War II he was commissioned in 1945 and rose through the Pakistan army, until in 1977 he led the military coup deposing Zulfikar Bhutto*. When proclaimed President in 1978 he outlawed all political parties and industrial strikes and enforced Press censorship. A zealous Muslim, he introduced a full Islamic code of laws (*Shariah*) and an Islamic welfare system, while insisting on greater Islamic control of the school system. After the Soviet invasion of Afghanistan in December 1979, millions of refugees flooded Pakistan, and Zia received increasing amounts of economic and military aid from the USA. From 1982 he was under some pressure to allow wider participation in government and at the time of his death was in serious dispute with his provincial governments. It is not clear who was responsible for his assassination on 17 August 1988.

Zimbabwe. The Shona* people of the country had been settled for centuries when, in the early 19th century, the Ndebele*, under their leader Mzilikazi, invaded from the south. He created the Kingdom of Matabeleland, which for the next fifty years was in a state of tension with the Shona of Mashonaland. In 1890 Cecil Rhodes sent a Pioneer Column into Mashonaland and, following a bloody Matabele War in 1893, the British South Africa Company declared Mashonaland and Matabeleland united. Rebellion erupted in 1893 but it was ruthlessly suppressed. The country became the Crown Colony of Southern Rhodesia in 1911 and a self-governing colony in 1923, after rejecting union with South Africa as a fifth province. Rapid economic development followed, with British immigrants establishing both coffee and tobacco plantations. In World War II it strongly supported Britain, being extensively used for RAF training. After the collapse of the Central African Federation* in 1963, a right-wing Rhodesia Front (formed in 1962) demanded immediate independence. It refused British demands for Black participation in government and, under Prime Minister Ian Smith*, issued a Unilateral Declaration of Independence (UDI) in 1965. It renounced colonial status and declared Rhodesia to be independent. British-sponsored attempts to negotiate a political compromise failed, and Joshua Nkomo's* ZAPU (founded in 1961) began increasingly successful guerrilla operations. In 1979 Smith conceded the principle of Black majority rule, but the regime of the moderate Bishop Muzorewa* could not come to terms with the guerrilla leaders of the Patriotic Front*, Robert Mugabe* and Joshua Nkomo. Following a conference at Lancaster House in London (1979), Mugabe was elected Prime Minister and Rhodesia became the Republic of Zimbabwe in 1980. The decade of the 1980s saw a revival of tension between Shona and Ndebele, personified by the friction between Mugabe and Nkomo, who was out of the government from 1982 until

1987. In that year a new constitution created the office of executive President (Mugabe) and ended racial representation. Zimbabwe achieved some success in easing domestic internal tensions and played a leading role in the politics of southern Africa. Its five-year development plan 1986–90 aimed to expand the public-sector involvement in the economy, whose major exports are minerals, tobacco, and coffee. In July 1990 the state of emergency of 1965 was finally ended.

Zimmermann Note (19 January 1917), a secret telegram containing a coded message from the German Foreign Secretary Alfred Zimmermann, to the German minister in Mexico City. It instructed the minister to propose an alliance with Mexico, if war broke out between the USA and Germany, Mexico being offered the territories lost in 1848 to the USA. The British intercepted and decoded the message, and passed it to the US State Department. It was released on 1 March 1917 as German–US relationships were deteriorating over unrestricted submarine warfare. With the possibility of a German-supported attack by Mexico, the isolationists lost ground, and on 6 April the US Congress voted to enter the war against Germany.

Zinoviev, Grigori Yevseyevich (1883–1936), Soviet politician. Of Jewish origin, he joined the Russian Social Democratic Party in 1901 and supported the Bolshevik* faction in 1903, becoming a member of the party's Central Committee. He emigrated after the 1905 Revolution and returned to Petrograd with Lenin* in April 1917. He at first opposed Lenin's plan to seize power, but then supported him. From 1919 until 1926 he was chairman of the External Committee of the Comintern*. As such he was supposed to have sent the famous **Zinoviev Letter** of 1924 in the British election campaign, urging revolutionary activity within the British army and in Ireland. This allegedly helped to bring about the defeat of the Labour Party. Defeated in the power-struggle which followed Lenin's death in 1924 by Stalin and Bukharin, he was allowed to emigrate, but returned to Russia in 1935 and was imprisoned for 'moral complicity' in the death of Kirov*. In August 1936 he was retried in a show trial of the Great Purge* and executed.

Zionism, a movement advocating the return of Jews to their homeland of Palestine. Following the Russian pogroms* it assumed a political character, notably through Theodore Herzl's book *Der Judenstaat* (1896) and the establishment of the World Zionist organization (1897). The Balfour Declaration* in 1917 and the grant of a mandate* for Palestine to Britain gave a crucial impetus to the movement. During the mandate (1920–48) the World Zionist Organization under Chaim Weizmann* played a major part in the development of the Jewish community in Palestine by facilitating immigration, by investment (especially in land), and through the Jewish Agency*. Zionist activities in the USA helped to gain the support of Congress and the presidency in the period 1946–8 and for the creation of the state of Israel. It has continued to be a major factor in US politics.

Zog (1895–1961), King of Albania 1928–39. Born Ahmed Bey Zogu, a wealthy Albanian landowner, he supported Austria in World War I, after which he served as Premier of the Albanian Republic 1922–4, President 1925–8, and then announced himself King. He championed modernization of the country, instituting language reforms, educational development, and religious independence. He relied on Italy for financial help, and the Treaty of Tirana of 1926 provided extensive Italian loans in return for Albanian concessions. By 1939 Italy controlled the entire economy of Albania and Mussolini sent in his army to take over the country, forcing Zog into exile.

Zulu, a Ngoni people of southern Africa living mostly in Natal, whose farming lands were steadily taken over by Europeans during the 19th century. Their King Shaka (1816–28) had developed a force of warriors, the impi, who had driven out other Ngoni in the so-called *Mfecane* (Time of Troubles). It was these highly skilled warriors who fought the classic Zulu wars against the British 1878–9, 1888, 1906. Although in the 20th century deracinated and forced to live and work on White farms or in cities, the Zulu people retained a strong pride in their history and culture, while also being enthusiasts for Christianity, their Chief Luthuli* being an outstanding Christian. The Bantustan* homeland of KwaZulu has been governed since 1972 by Chief Buthelezi*. His Inkatha movement, founded to preserve cultural heritage, has become a political party in strong rivalry with the ANC*.